Early Indian Metallurgy

The Production of Lead, Silver and Zinc through Three Millennia in North West India

Dedicated to H.V. Paliwal

who instigated and supported this project

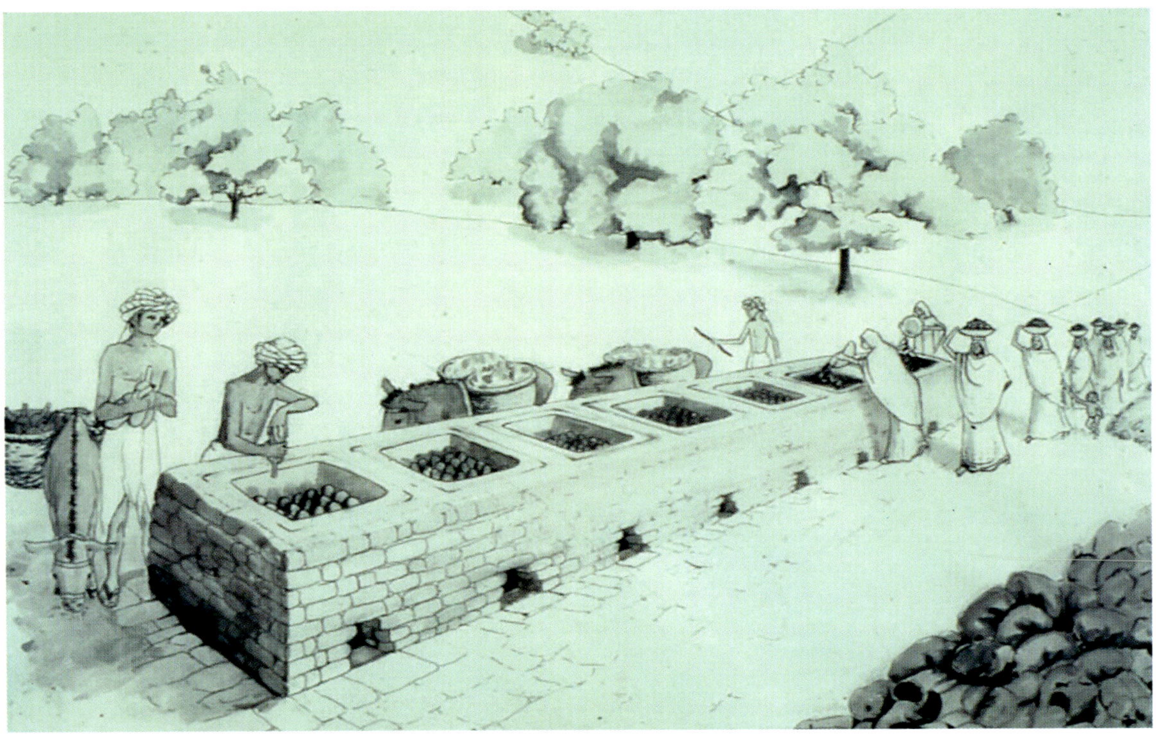

Reconstruction of how the Site 30 furnace block may have appeared while being loaded with retorts ready for smelting. The block was probably protected from weather by some form of light shelter.

Early Indian Metallurgy

The Production of Lead, Silver and Zinc through Three Millennia in North West India

P.T. Craddock

with K.T.M. Hegde, L.K. Gurjar and L. Willies

Archetype Publications

First published 2017 by Archetype Publications Ltd

Archetype Publications Ltd
c/o International Academic Projects
1 Birdcage Walk
London SW1H 9JJ
www.archetype.co.uk

ISBN: 978-1-909492-51-6

British Library Cataloguing in Publication Data
A catalogue record for this book is available from the British Library.

Printed on acid-free paper

Designed by Marcus Nichols at PDQ Digital Media Solutions Ltd
Typeset by PDQ Digital Media Solutions Ltd, Bungay
Printed and bound in Great Britain by Latimer Trend & Co Ltd, Plymouth

Contents

Chapter 1 Introduction

P.T. Craddock

This publication describes the survey, excavation and scientific study of three major mines and metal production sites in the Aravalli Hills of Rajasthan (formerly Rajputana) in north west India (Fig. 1.1).[1] All three sites have very extensive remains of mining and smelting activity that occurred through three millennia. The major minerals exploited at these mines are mixed argentiferous lead/zinc sulphides. Zinc is currently the principal metal extracted together with minor amounts of lead and silver. In antiquity the driving force behind the establishment by the Mauryans of the major mines at Agucha and Dariba was the production of silver but at Zawar it is more likely to have been zinc oxide. In the medieval and post-medieval periods some lead and copper were produced at Dariba, mining had ceased altogether at Agucha but at Zawar zinc metal was produced by advanced processes of high temperature distillation, almost certainly the first production of zinc on an industrial scale anywhere in the world. The production of metallic zinc by distillation is described in several iatrochemical texts from the first millennium AD, and the translation of these laboratory methods into a viable commercial chemical process in the medieval period was a remarkable achievement. The project was initiated to investigate the remains of this process, but rapidly broadened to include the silver production at Dariba and Agucha. The scale and sophistication of these enterprises and their demise necessitated broader considerations.

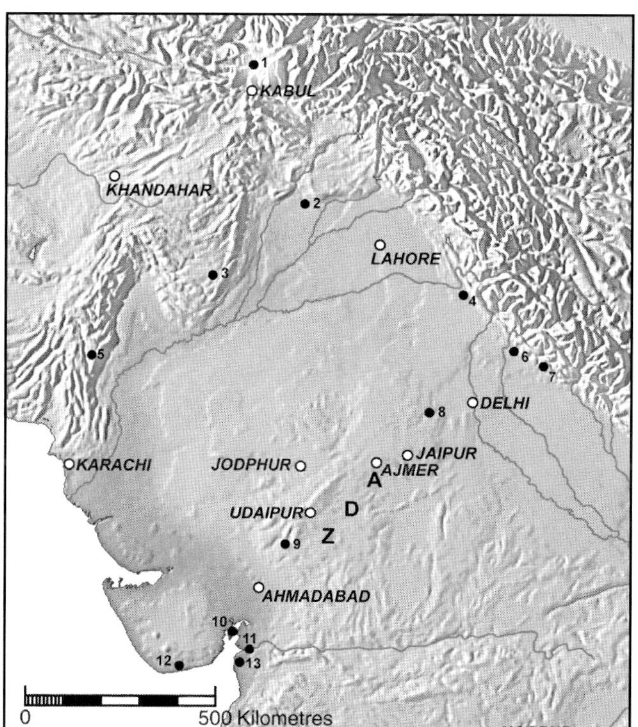

Fig. 1.1 Location of Zawar, Dariba and Agucha (Z, D and A) together with some sites in north western India and surroundings mentioned in the text (the Rajasthan sites are covered in more detail in Fig. 2.1). 1 Panjhīr & Jārbāya; 2 Taxila; 3 Mehgarh; 4 Kwanu; 5 Nal; 6 Chamoli District; 7 Kumaon District; 8 Khetri-Singhana copper belt; 9 Ambaji; 10 Khambhat (ancient Cambay); 11 Bharuch (ancient Barugaza or Broach); 12 Diu; 13 Surat (map: A. Simpson).

These were major and complex industrial enterprises taking place well outside the Eurocentric Middle Eastern areas that are the source of most studies of early technology. The scale and technical sophistication of the processes represented were quite unexpected. However, by the post-medieval period these mines had been abandoned or were in serious decline. The decline and extinction of indigenous technologies when exposed to more advanced western European technologies in the 18th and 19th centuries was a common occurrence across the world. The reasons behind this decline and ultimate demise, well exemplified and documented at Zawar, led to a study of the impact of European production on mining and smelting throughout South Asia. This, in turn initiated a reconsideration of the usual Eurocentric perceptions of indigenous technologies generally.

Origins and participants

All three mines were operated by Hindustan Zinc Ltd. (HZL) and all three sites exhibited abundant evidence of earlier working both underground and on the surface. These remains not unnaturally had led to speculation as to the age and function of the earlier activity. Such interest is common to mines around the world where remains of earlier working are extant. These, especially underground, can sometimes be quite intact, making it very easy to identify with the earlier workers who, after all, had similar objectives and faced similar difficulties. The interest of the staff at HZL went far beyond a general curiosity and even before the present project a number of serious historical studies had been made.[2] In 1980 through the good offices of Mr Kenneth Bush of Zinc International, R.P. Kapur of HZL and P.T. Craddock of the British Museum met in London to discuss the feasibility of a project. Following this P.T. Craddock and H.V. Paliwal of HZL, together with K.T.M. Hegde of the Maharaja Sayajirao University of Baroda at Vadodara, Gujarat, and L. Willies of the Peak District Mining Museum, Matlock Bath, Derbyshire, initiated a joint project of research into the history of zinc production at Zawar.

The British Museum Research Laboratory (latterly the Department of Conservation and Science) was already engaged in research into the development of copper alloys, especially the copper-zinc alloy brass (Craddock 1978). In the course of these studies there were occasional references to the extensive and intriguing remains of early zinc production at Zawar in Rajasthan. The site had never been archaeologically investigated but the early occurrence of brass in India, the fairly detailed descriptions of the production of zinc by distillation processes in the early Indian iatrochemical literature (Rây 1956), and the introduction of metallic zinc into Europe from the East as part of the cargoes of the European trading vessels from the 16th century onwards, all suggested that the remains at Zawar could be of considerable significance.

The Department of Archaeology at the MS University of Baroda had a long-term interest in all aspects of scientific archaeology, especially in the technology of early metallurgy in India (Hegde 1991). In the course of this work it had already carried out surveys of early mining activity in the Aravallis (Hegde and Ericson 1985). None of the organisations referred to so far had expertise in the underground exploration and recording of early mine workings so the team of mining archaeologists based at the Peak District Mining Museum was invited to join. In addition to the exploration of the old mines of the Peak District this team had also undertaken exploration work in the ancient copper mines in the Wadi Timna in Israel (Willies 1991a) and the great Phoenician and Roman copper and silver mines at Rio Tinto in the south of Spain (Willies 1997).

Therefore the research interests of all four participants were aroused by the potential of Zawar and a joint project to study the area was set up in 1982. This project was completed in seven years with four quite short seasons in the field and much longer sessions in laboratories and libraries both in Britain and India. In December 1982 a very short reconnaissance trip was made by P.T. Craddock, L. Willies and K.T.M. Hegde to Zawar with H.V. Paliwal. In December 1983 detailed surveys of Zawar both above and below ground were carried out, followed by excavation of the small retort and large retort zinc smelting units (Sites 30 and 37 respectively). In December 1985 there was further work underground and the excavation of a selection of the slag heaps at Zawar. Following this the team moved to Dariba where extensive surveys were made of the mining systems as well as surveys and excavation of a selection of the slag heaps. The team moved next to Agucha to carry out a rapid inspection and survey of the surface remains and excavation of some interesting features exposed in the side of a well. In 1987 the mines survey at Zawar was completed and the exposed remains of an intermediate retort furnace (Site 33) investigated. Following this, the team moved on to Dariba to undertake the principal task of recording the then recently exposed timber revetment in the East Lode opencast. Finally, at Agucha the team recorded the remains of ancient mining then being exposed as the modern mine was being developed.

Survey, sampling and scientific methods adopted

Underground survey

Survey methods have changed considerably since the 1980s surveys were made. Excavation inside the mines can prove useful, especially in some adits and levels which may have remained in use for access and drainage for extensive periods, allowing for a considerable build-up of stratified layers. However, in the limited time available for this project, work concentrated on recording what was easily visible with the very minimum of excavation, which was usually undertaken to obtain charcoal for radiocarbon dating. Some 1500 colour and black and white photographs have been taken, almost a third underground. Digital records of these are held by HZL, the British Museum and the authors of this work. Survey, when this was possible, used the methods available in the 1980s, namely fixed Fibron tapes with elevations and directions taken using a hand compass and clinometer, with corrections sometimes possible using fixed points provided by HZL surveyors. Cross-tapes at right angles were used to fix the plan, and from these floor and ceilings were fixed on sections again using tapes. The accuracy, therefore, is not always of a high standard since access and conditions were often difficult, sometimes requiring estimation. The objective of presenting an adequate view of the remains has, however, probably been achieved by this technique in conjunction with the photographs and field sketches. Significantly higher survey standards would have required many months if not years to complete with the recording techniques then available and the loss meanwhile of much of what was temporarily revealed.

At Agucha, where an opencast mine was being developed at the time of the 1987 survey, the benches at nominal 5-metre intervals were used as base levels. Tapes were stretched horizontally along these to plot temporary exposures resulting from the modern open pit operation. Details were then sketched into the sections, which were also photographed. Following these surveys Mr K.S. Choudhary, the General Manager (Mining) at Agucha, carried out a watching brief as the modern mine progressed and produced a report, *Resource Identification at Agucha and Assessment through Interpretation of Ancient Mining*, in 1999 (see Chapter 5). His schematic plans and elevations of the ancient mining operations at Agucha, some of which are included here, are based on the HZL mining surveys.

Ore sampling

It might be thought that to determine which metals were being sought in the past could be easily ascertained by sampling the ore remaining in the walls of the contemporary mine workings. This seemingly simple task is actually very difficult as exemplified by Zawar in antiquity: the potential ores of zinc, silver and lead were all mined, but which were actually sought at particular phases in the mine's history is still not known for certain. First, there is the obvious problem that the ore that remains today was the ore which the previous miners did not want. This is well exemplified by Agucha where the argentiferous minerals were removed in antiquity but the zinc-rich deposits were left. The remaining mineral in the deposits that were worked can give some indication of the ore grade below which it was considered unprofitable to continue. In major mine systems with large worked-out stopes, where much of the original mined ore would have been at some distance from the nearest surviving mineral, it is clearly impossible to gauge the grade of the original ore or even in some instances its mineralogy. At Zawar the silver content of the sphalerite and galena minerals left in the huge ancient galleries is low, but it might have been much higher in the ore that was removed, which once occupied space now several metres from the present walls.

The surviving mineral in the walls of ancient mines is susceptible to oxidation therefore an original sulphidic mineral can now be an oxide, or more usually a hydroxy-carbonate, at least at the surface (although the minerals exposed in the Aravalli mines studied here do not seem to have been unduly affected). The near-surface mineral deposits are often made up of carbonates and hydroxy-carbonates brought about by long-term oxidation over the geological life of the deposit. No evidence of such deposits was found at the mines studied, however, possibly because they had all been mined away or had never existed. At both Dariba and Agucha there are extensive gossan deposits where the original near-surface iron sulphide minerals has oxidised principally to hematite and goethite, and the remaining host rock has metamorphosed to calc silicates from the original dolomite. It is possible that the original lead and zinc minerals associated with these deposits may also have been oxidised rather than the primary sulphides which constitute the remaining ore. At Zawar there is little evidence of intense weathering, there are no gossan deposits and the host rock is predominantly unchanged dolomite, suggesting that only the primary sulphidic minerals were present. Sometimes fragments of unchanged ore minerals survive in the slags, presumably having been charged to the smelting unit just before the slag was tapped, as evidenced by some of the slags from Site 5 at Zawar, providing direct evidence of the ore processed.

Mineral samples should be taken from a variety of locations in the main ore body. Picking up random pieces that look interesting is of little relevance in establishing what was actually mined deliberately. The mineralogist, Bob Ixer, has warned against what he termed 'the magpie school of ore collecting'. In antiquity the smelter required a regular supply of ore of consistent grade.

When exploring old galleries, especially in ancient copper mines, it is not uncommon to encounter brightly coloured pieces of apparently good ore. However, when they are broken up these usually turn out to be just lumps of sterile rock that have become coated with a thin layer of copper minerals, deposited by the mine waters since the workings were abandoned. Their weight is a more certain indicator.

Recording smelting sites

The methods of rapid investigation of major mining and metal production sites evolved out of investigation of other major mining and smelting sites, notably at Timna in Israel and at Rio Tinto in Spain (Craddock *et al.* 1985b; Craddock 1989; 1995: 8–22; Willies 1997). Early smelting sites present an especially daunting prospect to the conventional archaeologist. In general, very little in the way of furnaces or other technical structures can be expected to survive *in situ* (the zinc smelting furnaces at Zawar are exceptional in this respect). The principal surviving process evidence is the featureless heaps of production debris sometimes extending over several square kilometres to a depth of several metres at major mines (as at Dariba). The heaps consist mainly of slags and are largely devoid of recognisable artefacts that could offer information on the process or dating evidence. In the past, archaeologists charged with the investigation of the sites of early metal production tended to avoid the slag heaps – which was unfortunate as that is where the process information can be found – to concentrate more on ancillary buildings, fortifications, road systems and burials, all of whose association with the actual process was secondary but could at least be dealt with in the usual archaeological manner.

The information on the smelting process lies in the debris itself. The slags and the associated refractory fragments contain a direct and permanent record of the smelting processes from the study of which the principal operating parameters can be established. The key to this information lies in their scientific examination as explained in Chapter 7 and below. A major problem arises with the association and dating of the layers. Compared to most archaeological deposition situations, slag heaps tend to form very quickly and the processes that created them are carried on in areas away from ordinary domestic activities. There may be very little standard dating material such as coins or even pottery. In this project we had to rely almost exclusively on radiocarbon dating. Fortunately there are usually abundant quantities of charcoal, although in heaps of large loosely stacked slags most if not all of the charcoal can eventually get washed down to the bottom (as exemplified by Site 7 at Dariba).

A detailed description of the layers and composition of the trial sections and trenches should be made, together with section drawing and photography in order to record the stratigraphical context of the samples; although stratigaphically important, the swiftly deposited layers have little overall chronological significance and their deposition can be misleading.[3] There is not much to be gained from planning the surfaces of the layers.

Sampling smelting sites

Clearly the archaeological objective is to obtain a representative sample of associated debris from around the site together with dating evidence. This usually involves an initial careful survey of the area, followed by a number of trial cuts into the heaps to expose sections. Very often more recent disturbances can provide some exposures readymade, as illustrated by the modern road cutting through the retort heaps at Old Zawar (see Chapter 6, Plate 6.26) and the sections exposed in the sides of wells sunk at both Dariba and Agucha (see Chapter 6, Plate 6.49). Having carefully cleaned and recorded these by both section drawing and photography they can be sampled for representative production debris, usually the slags and the refractory fragments together with any unusual or distinctive minerals (the weight is a good field indicator that a piece is likely to prove interesting). It is essential that the layer from which samples are taken is carefully noted so that subsequently they can be studied together as part of one process. Slag itself is the principal component of most heaps and a selection of three typical pieces from every layer, each about 8 cm^3, should suffice for scientific study. Examination of larger pieces can reveal whether the slags are likely to have set within the smelting unit as 'furnace slags' or were run or bailed out as a liquid tap slag. The refractory fragments are also collected, quantified and a selection made for scientific study. In this investigation the majority of the fragments were photographed and drawn immediately – only a small number were taken to Britain for scientific study. The remains of these after sectioning and sampling are preserved in the Department of Conservation and Science sample collection at the British Museum. Where mentioned in the text their sample number is given. It is also important to obtain a collection of the local contemporary domestic coarseware pottery for petrological and analytical comparison. Describing the refractory properties of the metallurgical ceramics only makes sense when compared to the contemporary local pottery; very often they are the same (Freestone and Tite 1986). Fragments of furnace wall or linings should include those with well-preserved profiles from the surface exposed to the process through to the rear. Pieces that preserve some of their undistorted original shape or form are also important. These can be especially useful for estimating the shape and dimensions of the smelting units, whether standing furnaces or hearths. As already noted, it is usually unlikely that actual units will survive *in situ*. Fragments that are so badly eroded or vitrified as to be unidentifiable are of little use. Conversely, fragments with recognisable features such as holes for the bellows nozzle (tuyeres) or rims are very useful as they give an indication of where in the smelting unit the fragment originated. Evidence of most of the various refractory types, smelting unit walls, tuyeres, crucibles etc. will be present, although frequently damaged and rather difficult to recognise, and the contemporary layers from more than one heap will often have to be sampled before a representative collection is obtained from which the form of the process units can be reconstructed. This is important: scientific study can reveal the chemical and physical parameters of the process but careful sampling is still required in order to establish the shape of the components of the process units.

The original aim of the second season at Zawar was to sample the selection of retort and slag heaps made in the first survey but the discovery of the almost intact furnaces still bearing their last load of retorts led to their excavation in a more conventional archaeological fashion. Following their excavation and recording they were rigorously sampled to obtain refractory pieces from precise locations in and around the furnace structures to establish the heat regimes in their different parts.

Scientific investigations

A range of analytical methods was used including optical microscopy (OM), thin-section petrography, scanning electron microscopy with energy dispersive X-ray (SEM-EDX) analysis, X-ray diffraction (XRD) and atomic absorption spectroscopy (AAS) at the laboratories of HZL and the British Museum, as well as inductively coupled plasma atomic emission spectroscopy (ICP-AES).[4] Details are given below.

Optical and electron microscopy (with elemental analysis)

All artefacts were examined using a binocular microscope and selected samples were mounted on glass and prepared either as conventional (covered) or as (uncovered) polished thin sections, ground to a thickness of 30 µm. In the first instance, the thin sections were examined using a petrographic microscope to gain information on texture and mineralogy. For ceramic materials, these observations provided details of the nature of the clay pastes used, the general level of firing, and the nature of any glazed or vitrified surfaces.

A range of samples was selected for further examination using an SEM equipped with an EDX analyser to perform more detailed observations of the fine-scale textures and also to analyse chemically the individual phases. By analysing larger areas of the samples, estimates of bulk composition were obtained (see below for more details).

The degree of vitrification of ceramic materials visible in the SEM offered some estimate of the operating parameters of the metallurgical processes. When a clay body is exposed to successively higher temperatures, progressive changes occur to its mineralogy and texture. Tite and others (Maniatis and Tite 1981; Tite *et al.* 1982a, b) have shown that the textural changes, in particular the degree of vitrification and bloating, can be related to the maximum temperature reached by the ceramic material (see Chapter 7, Table 7.9). The temperatures at which the various changes occur depend upon the composition of the ceramic (e.g. calcareous/non-calcareous, high/low refractory) and also on the firing atmosphere (i.e. oxidising or reducing). The textures developed in the excavated ceramics can be compared using the SEM with those of sherds of similar composition which have been refired in a laboratory furnace. This comparison allows some estimate to be made of the temperatures to which the ceramics used for metallurgical processes had been exposed. Accordingly, a limited programme of refiring of the metallurgical ceramics was carried out. Samples were heated in a laboratory furnace for 1–2 hours in a tube furnace to a maximum temperature of 1100°C in a neutral to reducing atmosphere, generated by passing nitrogen through the furnace and placing a tray of flour close to the samples. These refired materials were mounted in resin (Araldite MY753 resin and HY951 hardener) and polished to a 1 µm finish using diamond pastes before being coated with carbon for examination in the SEM.

Information on the duration of high temperature processes may be obtained by investigating the changes that occur in the degree of vitrification through the thickness of the ceramic from the 'hot' to the 'cold' face (see Chapter 7, Table 7.9). These changes reflect the thermal gradient that was established during use; in general, the more rapidly the degree of vitrification diminishes away from the hot face, the steeper the thermal gradient and the shorter the duration of the process (Tite *et al.* 1982a: 68–70; also Middleton 1991: 30–31 and further discussion below).

Some of the ceramic samples were examined under a biological optical microscope in order to identify the organic tempering materials used. Fragments of plant material and diagnostically shaped voids where organic temper had burnt out could often be seen in the hand specimens. Examination and identification of this material was carried out initially on fractured surfaces of the samples using reflected light at magnifications ranging from ×10 to ×100. Subsequently, examination of selected thin sections of the samples took place using transmitted light at magnifications ranging from ×50 to ×150. A range of plant materials was identified and the results incorporated into the fabric descriptions.

X-ray diffraction

XRD was used to determine the crystalline phases present in some of the samples. Small samples were removed and mounted in Debye-Scherrer powder diffraction cameras. The resulting diffraction patterns were matched against the standard ICDD database (ICDD n.d.) in order to identify the phases present.

Elemental analysis

Selected samples were analysed using air-path energy dispersive XRF. In addition to the semi-quantitative analysis performed on the sample of zinc metal from Zawar Site 30, surface analysis of the refractories was carried out without any sample preparation and is only qualitative in nature. Nevertheless, they were found to be useful, for example to obtain conveniently an indication of the nature of vitrified layers and traces of metal and corrosion products present on some of the ceramic samples. In particular the presence of tin, as well as copper and lead, on the contact surfaces of most of the refractories from Site 10 at Dariba (see Chapters 6 and 7) enabled that site to be identified as a metal-working area rather than one of primary extractive metallurgy.

The composition of the ores and slags can provide information on the materials charged into the furnace (or retort) while the amount of metal left in the debris gives an indication of the overall efficiency of the process. Samples of slags and other production debris from each of the three sites were quantitatively analysed using a combination of ICP-AES and AAS. The samples were crushed and homogenised using equipment at the Natural History Museum, London. Most samples were prepared for analysis using a lithium tetraborate fusion: 25 mg samples of powder were weighed into high purity graphite crucibles (Johnson Matthey C007220/Uf-4S) followed by 100 mg of specpure 100 lithium tetraborate. Batches of crucibles were heated in a furnace at 1050°C for 15 minutes to form a glass bead. During fusion some of the more volatile elements may have been lost, notably arsenic (As), cadmium (Cd), mercury (Hg), sulphur (S) and antimony (Sb). The entire crucibles were then plunged into 4% nitric acid and the resulting solutions stirred to aid dissolution of the glass bead, made up to 50 ml and filtered into polythene bottles. Some silica may have been lost during dissolution using lithium tetraborate due to the nitric acid strength being too high; silica has therefore been quoted by difference. The solutions were analysed for 24 elements by ICP-AES using a specially prepared analytical procedure on the ARL 2410 instrument at the Natural History Museum. The resulting count data were transformed into percentages (usually expressed as the oxide) using a quadratic curve fitting program. In addition to the metallurgical debris, samples of four commercial standard materials were also analysed. Duplicate synthetic standards were run after every eight unknowns, acting as drift monitors. Further samples of the debris were dissolved using nitric acid, hydrofluoric acid and aqua regia, and made up to volume (20 ml) using 40% hydrochloric acid. These second solutions were employed for the analysis of gold and silver using AAS. A duplicate sample of galena was dissolved in hydrochloric acid and analysed with AAS as a check on the ICP analyses.

The tabulated results (see Chapter 7, Tables 7.1, 7.2, 7.10–7.12, 7.15, 7.17–7.19) have precisions of approximately ± 2% for the major elements (i.e. present at levels above 30%), ± 5% for elements present between 10 and 30%, ± 10% between 1 and 10%, deteriorating to ± 50% at the respective detection limit.

Objectives

The primary objective, as stated above, was to establish the technology and dates of the extractive metallurgy employed at the three sites. These mines, operating in the seemingly remote Aravalli Hills through three millennia, produced four metals by very different processes on a massive scale. Most studies of ancient technology have hitherto been based on the civilisations of the Mediterranean and Middle East, especially those of Egypt, Greece and Rome.[5] This project provided an opportunity to study the long-term development of major industry well away from the Graeco-Roman world.

Although large-scale industrial enterprises were in progress at these mines contemporary with the major mines of the Classical world, they have previously received scant recognition, even in antiquity while they were in operation.[6] The postulated production of zinc oxide at Zawar by methods similar to those described by contemporary classical authors exemplifies this.

Moving forward – but still with the theme of expanding the study of industrial technical innovation and the establishment of large-scale production beyond the usual Eurocentric view – the development of the zinc distillation process at Zawar is important. It seems most likely that during the first millennium AD, methods of producing and condensing metallic zinc were established in India. The surviving early accounts clearly describe laboratory-scale processes, and it is possible that the metal being produced was intended for burning to produce pure zinc oxide for pharmaceutical use, free from lead contamination. Meanwhile at the Zawar mines, production had recommenced, probably with the majority of the zinc oxide still intended for brass production by the cementation process. It is possible that around a thousand years ago serious attempts were being made to develop systems to collect the zinc vapour as a metal on an industrial scale, based on the pre-existing laboratory methods. The real innovation was to develop these into viable industrial processes as described in Chapter 11. The mines at Zawar would always have been under the control of the ruling authority. Clearly here it was encouraging and supporting real innovation, not so much of theoretical science, but rather of efficient, workable high-tech processes. This was to be achieved by standardised industrial-scale production of the necessary components of the process. At Zawar there must have been hundreds if not thousands of the furnace blocks, fitted with components such as the perorated plates with identical dimensions. Clearly tens of thousands, if not hundreds of thousands of these components must have been made in central workshops to maximise efficiency and also to ensure maintenance of overall control.

This spirit of innovative experimental development is not usually associated with medieval India, and was certainly not the impression gained by the European geologists and mining experts on their surveys in the 19th century (Chapters 2 and 13). On first visiting the underground mines at Zawar in the 1980s, the present team was impressed by the scale and obvious systematic organisation of the early operation, so much at variance with the European descriptions of the last indigenous mining operations still taking place in 19th-century India (Chapter 2), and it must be confessed, of our expectations. This surprise was even greater because it had been assumed from the excellent state of preservation that these must be late workings contemporary with the last recorded operations in the 18th and early 19th century at Zawar. Perhaps there should have been less surprise when the radiocarbon dates showed the underground workings to be ancient – previous millennia operations had been conducted in a more superior fashion to those employed in the 19th century. Clearly there had been a regression. What Europeans had observed in the 19th century was not the unchanging primitive technology of a system incapable of improvement,[7] but rather a relatively recent response to collapsed authority. It therefore became part of the overall brief of this project to examine these conflicting paradigms of stasis and innovation as exemplified by the three mines throughout their long histories.

It must be stressed again that these are enormous sites and our visits in total only amounted to a few short weeks spread over five years so we concentrated very specifically on the mining and extractive metallurgy technology, well aware that important issues such as the usual arrangements of occupation and infrastructure were not being addressed, even at Zawar with its excellent surviving medieval temples, fortress and post-medieval house ruins. We did not have the time, resources or the necessary expertise to attempt these objectives.[8] We hope that our surveys and excavations, based in large part on the previous observations of geologists and mining engineers at HZL over the years, have at least exposed a representative sample of the mining and metal production activities carried on at the three sites over the millennia and will form a basis for further more targeted work on specific areas.[9]

Acknowledgements

During the project all logistical requirements within India were organised and undertaken by HZL, together with much of the subsequent scientific work. In addition, the American Friends of the British Museum generously donated £1,000 which enabled the plans, sections and other drawings to be produced to publication standard. The work has also benefited from the careful reading of Professor M. Wayman (University of Alberta), Dr A. Mongiatti (British Museum) and Philip Andrews (Wessex Archaeology), although, of course, such errors and infelicities that remain are entirely the responsibility of the authors. The lead isotope work was carried out by Noel Gale of the Isotrace Laboratory, University of Oxford. The team members included:

- P.T. Craddock, B.R. Craddock, A. Kurlis, I.C. Freestone and A.P. Middleton (British Museum)
- L.K. Gurjar and S.M. Gandhi (HZL)
- K.T.M. Hegde with members of the Department of Archaeology and research students (Maharaja Sayajirao University of Baroda).
- L. Willies, S. Willies, C. Willies, T. Worthington and P. Andrews (Peak District Mining Museum).

Unless otherwise credited all the section plans and drawings are by B.R. Craddock, photographs of the surface sites are by P.T. Craddock using a 30 cm scale, and the mining photographs are by L. Willies and his team, using a metre pole or 30 cm rule.

Notes

1. There have been many and continuing changes in individual states and place names since India gained independence. It was decided to retain the old city names as given in *The Times Atlas* for 1987 to maintain continuity with the old, mainly pre-independence sources referred to in the text and also for general readers outside India who may not be familiar with all of the new names.
2. This interest led to a number of radiocarbon determinations being carried out on timbers encountered in the mines (see Chapter 5, Tables 5.1 and 5.2), and to a number of publications including: Straczek and Srikantan 1967; Friedlaender and Choo 1970; Gandhi 1983, 2000; Paliwal *et al.* 1989; Gandhi *et al.* 1984; Gangopadhayay *et al.* 1984; Kumar and Biswas 1986; Aery and Tiagi 1988.
3. For example in one important section in the ancient slag heaps at the Rio Tinto mines there was a metre-thick layer of sterile clay which was initially interpreted as representing a major chronological break in the deposition. On cutting back the face, the layer petered out completely after a few centimetres. Far from documenting a long-term disruption over the whole site, it was probably just a large puddle.
4. All analyses are expressed in weight percentage or parts per million.
5. This emphasis on a Graeco-Roman perspective where it is inappropriate is well exemplified by the many publications, including those penned by Indian authors, in which the Indian zinc distillation and condensation process is usually described as *per descensum*, rather than 'distillation by descending' in English or the original *tiryakpatana yantram* in Sanskrit, which would have been understood by those who established the process. Few readers now read Latin and even fewer read the language of the original operatives.
6. Strabo in his *Geography* (Jones 1930: 51–4) made dismissive remarks on Indian metal production (Book 15, chapter 1, section 30), quoted in Chapter 2, p. 6. This was a common European misconception through the ages but the mine systems examined in this project suggest otherwise.
7. John Percy (1870: 293) described Indian smelting techniques as 'representing the infancy of the art' (Chapter 12, p. 234 and Chapter 2, p. 17).
8. Subsequently there have been investigations at Zawar which addressed some of these areas, notably the study of the temples (Kumar forthcoming). In general, the settlements at Mauryan Zawar and Dariba are as yet unlocated, and at Agucha there is evidence of Mauryan and Gupta occupation beneath the present settlement (Tiwari and Kavdia 1984) (see Chapter 3, Plates 3.21 and 3.22). Similarly, no evidence of medieval settlement at Zawar was identified and no settlement evidence at all was found at Dariba.
9. During the years since the investigations began many interim reports have been published, among the more significant: Craddock 1987, 2009a, b, c, 2013; Craddock *et al.* 1983, 1985a, 1998a, b, 1989, 2013, 2015; Craddock and Gurjar 1991; Craddock and Eckstein 2003; Eckstein *et al.* 1996; Falk 1991; Freestone 1989; Freestone *et al.* 1985a, b, 1991; Freestone and Tite 1986; Gurjar *et al.* 2001; Hegde 1986, 1989, 1991; Hegde *et al.* 1986; Paliwal *et al.* 1989; Tite *et al.* 1985; Willies 1983, 1987, 1988, 1989; Willies *et al.* 1984.

Chapter 2 Early metallurgy in the Aravalli Hills and Rajasthan

P.T. Craddock

This chapter[1] provides the background to aspects of early metal production in the north west of India contemporary to the ancient activity at the Aravalli mines studied, followed by an overview of some of the principal evidence for primitive mining and extractive metallurgy as it existed, particularly in the 19th century in the Aravalli Hills of Rajasthan. The latter is to a large degree based on the descriptions of European geologists, soldiers and administrators[2] but they are also of great value to the understanding of the processes carried on at Zawar and Dariba in the post-medieval period. Many of the accounts make clear that the primitive organisation and technology encountered were believed to be unchanged practices that had survived from antiquity. In fact, as discussed more thoroughly in Chapter 12, they are likely to have been the relatively recent local responses to the adverse political and economic conditions over much of north India that followed the collapse of the Mughal Empire at the beginning of the 18th century. This had brought about the cessation of more sophisticated and organised production; simple lead smelting may have continued at Ajmer, but at nearby Zawar high-tech zinc distillation had ceased.

The Aravallis, a treasure house of minerals of all kinds (Fig. 2.1), are considered in more detail in Chapter 4. Evidence of extensive old workings abound but by the 19th century the mines were in serious decline or, as at Zawar, already moribund. Ball (1881: 259) summed up the state of mining in Rajasthan as follows:

> Mining has been practised on a large scale, but at present the trade of the miner is becoming extinct, and the operations, which are only carried on in a few localities, are of a very petty nature.

Prehistory

Metallurgy probably had a long history in Rajasthan but so far little evidence has been found for the earliest phases commensurate with sites such as Mehrgarh (Jarrige 1985; Kenoyer and Miller 1999; Moulherat *et al.* 2002) and Lal (Fairservis 1971), now in Pakistan (see Chapter 1, Fig. 1.1), dating back to the fifth millennium BC.

The long debate concerning the source of metals used by the Harappan civilisation in the third and second millennium BC still continues. The Aravallis lie within the Harappan area, and in particular the Khetri mines are frequently claimed as a source of their copper (Reedy 1997: 104; Agrawal and Seshadri 1998). Without detailed survey and dating programmes, especially among the smaller mines, this can only be regarded as speculation.[3] Mughal (1980) claimed that copper was smelted at Harappan-period sites in the Cholstan region, which lies close to Khetri, and Chakrabarti (1985/6), and others have taken this as evidence that they were being worked during this remote period. However, the published evidence for smelting here, as at other Harappan sites, is tenuous in the extreme (Kenoyer and Miller 1999).

There is little evidence for pre-Harappan or Harappan activity at Khetri or indeed at any of the other big mines in the Aravallis that have been investigated. In particular the apparent absence of the distinctive stone mining hammers, which are found on Copper and Bronze Age mining sites around the world, is significant (Craddock 1995: 37–46). However, it is likely that the contemporary copper- and bronze-using cultures such as the Ganeshwar-Jodhpura in the northern Aravallis together with the early Ahar cultures, exemplified by Balathal, 42 km east of Udaipur in the central Aravallis (Allchin and Allchin 1982; Agrawal and Kharakwal 2003) and the Banas culture in the south, were exploiting their local copper resources, but as yet their mines have not been located. Majumdar and Rajaguna (1964) studied the evidence for early copper mining in the vicinity of Udaipur.

Silver

Silver/lead deposits were worked from the third millennium BC (Fairservis 1971: 159–65) and later (Siddiqui and Sharp 1993) near the ancient site of Nal in the Karlat region of Baluchistan in Pakistan (see Chapter 1, Fig. 1.1) (Heron 1954; Bender and Raza 1995: 261) but otherwise there is little surviving evidence for the production of silver in prehistoric South Asia. Silver was used quite extensively by the Harappans (Kenoyer and Miller 1999: 119–20) although little is known of the sources or smelting technology, but it is of potential importance given the exploitation of silver

Table 2.1 Analyses of Harappan silver items.

	Mohenjodaro		Lothal		Rojdi	
%	DK 5774	DK 11337	5034	4176	15114	
Silver	94.52	95.52*	54.65	86.53	71.20	90.2
Lead	0.42	1.40	1.64	–	–	2.02
Copper	3.68	3.08	2.67	7.87	4.13	2.84
Iron			3.29		Trace	Trace
Insolubles	0.85	–	3.06	2.32	16.29	0.95
Total	99.47	100	65.31	96.72	91.62	94.93

All analyses performed using wet chemical methods. The high iron content of Lothal sample 5034 is almost certainly a contamination from the burial environment in this heavily corroded sample.
*Silver by difference.

at Dariba and Agucha in later antiquity. Other sources to the north in the Hindu Kush and to the west in Baluchistan, Iran and Afghanistan have also been mentioned as possible sources, as well as to the east in the Aravallis (Allchin and Allchin 1982). Recent lead isotope studies have suggested that the most likely sources of Harappan silver lay in Baluchistan (Law and Burton 2008).

Elemental analyses of two silver artefacts from Mohenjodaro (Hamid and Sanallah in Mackay 1938: 480, 599), three from Lothal (Lal 1985: 656, 658) and a bangle from Rojdi in Gujarat (Bhowmik 1964) are given in Table 2.1. The metal has only minor traces of gold and zinc, but usually contains a quite appreciable lead content of between 0.42% and 2%. This suggests that the silver is likely to have been smelted from argentiferous lead, followed by cupellation (Chapter 10) rather than having been smelted from a silver mineral or recovered from gold refining (Craddock 1995: 212–13). Cupellation has a very long history in the Middle East, dating back to the fourth millennium BC (Hess *et al.* 1999; Pernicka *et al.* 1998). All the silver contains small amounts of copper, almost certainly a deliberate addition.

Silver from the Panjhīr sources in Afghanistan (see Chapter 1, Fig.1.1) usually contains substantial traces of bismuth which seem not to be present here (Cowell and Lowick 1988).

Overall, although presently based on a limited number of analyses, there is little evidence for the exploitation of metals by the Harappans in the Aravallis and more specifically no evidence for the exploitation of lead or silver during the Harappan period at Zawar, Dariba or Agucha.

Despite some early radiocarbon dates on a number of unprovenanced large timbers from within the mines, the exploitation of silver on a massive scale only really seems to have begun in the mid-first millennium BC for reasons not difficult to find. In common with the Mediterranean and the Middle East, the introduction of silver coinage after the mid-first millennium BC created an unprecedented demand for the metal. The Indian punch-marked silver coins date back to the 5th century BC, more or less coincident with the earliest dates for the major workings at the Aravalli mines (Gupta 1969; Cribb 1985, 2003). The source of the silver used in these and succeeding coin series has been one of the major problems of Indian numismatics for many years, given that it was believed that India had never produced silver in quantity (Bhardwaj 1979: 139–40; Mani 1998; Deyell 1999). It is now clear that the Aravallis must have been a significant source.[4]

Foreign reports on Indian metals

Following Alexander's expeditions in the 4th century BC, commercial contacts between India and the West were established. Several classical geographers apparently described India in some detail, notably Megasthenes and Nearchus, although these are now largely lost. The surviving fragments were quoted by later writers in the Roman period such as Strabo, Pliny and Arrian (Giumlia Mair *et al.* 2009; Warmington 1974). Megasthenes (McCrindle 1877) as an ambassador from Alexander's successor, Seleucus, visited the Mauryan Emperor Chandragupta several times at his capital around 300 BC and enough survives of his account to make it clear that metals were being produced on a considerable scale:

> India has underground numerous veins of all sorts of metals, for it contains much gold and silver and copper and iron in no small quantities, and even tin and other metals which are employed in making articles of use and ornament as well as implements and accoutrements of war (Megasthenes from McCrindle 1877: 31).

Strabo in his *Geography* (15.1.30) (Jones 1930: 51–4) refers to the country of the Sopeithes, which lay beyond the Acesines and the Hyarotis rivers (i.e. to their east), identified as the rivers presently known as the Chenab and the Jhelum, both tributaries of the Indus, that course through modern Punjab in Pakistan. He noted specifically that:

> And gold and silver mines are reported in other mountains, not far away, excellent mines, as has been plainly shown by Gorgus the mining expert from Alexandria. But since the Indians are inexperienced in mining and smelting, they also do not know what their resources are, and handle the business in a rather simple manner.

Pejorative remarks on Indian metallurgy by western commentators apparently have a long history!

Pliny in his *Natural History* (6.73–5) (Rackham 1942: 393–4), also using Megasthenes as his principal source, noted gold and silver mines which could possibly be located in the Aravallis:

> Below these deserts (to be identified as the Thar Desert) are the Maltaecorae, Singae, Moroae, Rarungae, and Moruni. These people are the inhabitants of the mountains that stretch in a continuous range on the coast of

the ocean; they are a free people having no kings, and they occupy the mountain slopes with a number of cities. Next come the Nareae, who are shut in by the Capitalia range, the highest of the mountains of India. The inhabitants of the other side of this mountain work a wide range of gold and silver mines.

The Capitalia range could well be the Aravallis: Ball (1881: 177, 231) went so far as to identify Capitalia with Mt. Abu, but believed the silver mines referred to were likely to be in southern India, as in the 19th century no argentiferous deposits were believed to exist in the Aravallis.

Another, very different source, was the trade manual known as the *Periplus of the Erythraean Sea*, which was probably written in the 1st or 2nd century AD (Huntingford 1980; Casson 1989). This contains very specific information on trading contacts and lists commodities required or for purchase at ports all around the Arabian or Erythraean Sea.

The *Periplus* describes the trade of the major port of Barugaza, now known as Bharuch or Broach, in the Bharuch District at the mouth of the River Narbada on the Gujarat coast in some detail (see Chapter 1, Fig. 1.1), stating that tin, copper and lead were imported together with gold and silver coins and fine silverware, but with no mention of any metals or minerals among the exports (Casson 1989: 27; Huntingford 1980: 44–8). This suggested to later scholars that the Indians did not have these metals available locally. This understandably puzzled Casson, who pointed out that there are abundant ore deposits of copper and lead close to the major trading ports, and that Strabo (*Geography* 15.1.69) (Jones 1930: 122–3) had specifically mentioned items made of 'Indian copper'. This is true, but Strabo, together with Pliny, based his account on that of Megasthenes compiled in the late 4th century BC when the major mines in the Aravallis were flourishing. As will be shown (see Chapter 5, Tables 5.1 and 5.2), these mines had probably ceased production by the first centuries AD, creating shortages that could only be met by imports.

Coin evidence

Gold and silver were imported into South Asia principally in the form of coinage from the Mediterranean world as recorded in the *Periplus* (Casson 1989: 209). Roman coins are well known in South Asia, especially from the south. The majority of the coins are of early Imperial date with some earlier coins leading Turner (1989) to believe that these hoards were deposited in the two centuries around the beginning of the Christian era. Critical reviews (MacDowall 1990) have pointed out that some of the hoards also contain a few later coins. The consensus of opinion is that the majority of the hoards were deposited no earlier than the 2nd century AD; however it would still seem tenable that the bulk of the coins actually came to South Asia somewhat earlier, particularly to Gujarat, where large numbers of *denarii* of the time of Emperor Augustus have been found. There was a marked scarcity of locally minted silver coins at this time, especially in north India, suggesting that the enormous quantities of imported coins, initially imported as bullion, were themselves used as currency, replacing the local coins of previous centuries. This has a direct bearing on the silver mines of the Aravallis, where the dates cluster around the last centuries BC with little activity in the succeeding centuries, at just the period when the Roman coins were most prevalent, suggesting that they were filling a serious shortfall in local silver production.[5] Thereafter there is little to suggest silver production in northern India (Chapter 13) until the post-medieval period when there is some evidence for production at Zawar (Chapters 6 and 7).

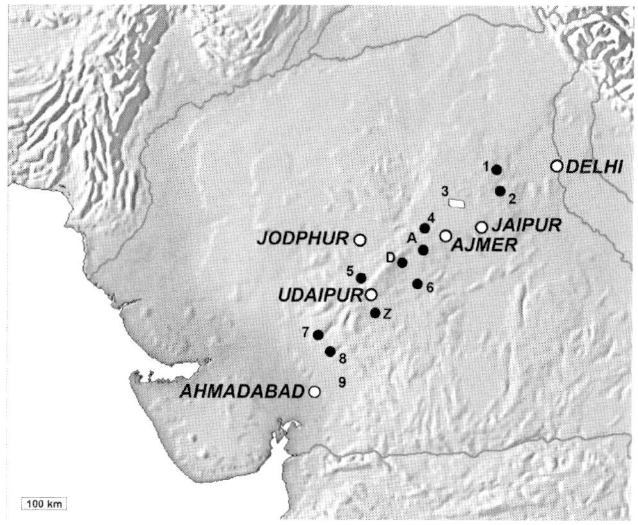

Fig. 2.1 Location map of some early Aravalli mines and other places mentioned in the text: Z Zawar; D Dariba; A Agucha. 1 Singhana; 2 Khetri; 3 Lake Sambhar; 4 Taragarh Mine; 5 Samoli; 6 Chittaurgarh (Chittor); 7 Banaskantha Dist.; 8 Ambamata and Kumbaria mines; 9 Ambaji Mine (map: A. Simpson).

Records of 19th-century copper production

Although the main deposits of copper lie towards the north end of the Aravallis, old copper workings are found throughout the range from Gujarat in the south (Shekar 1983). In the Banaskantha District, in the north of Gujarat (Fig. 2.1), these include Kumbaria. Charcoal from the old mine workings there has given a radiocarbon date of 670 ± 110 BP (PRL-549) and from the slag heaps behind the Jain temple dates of 905 ± 85 BP (TF 1222) and 535 ± 90 BP (TF 1221). Timbers from the underground workings at Ambamata have radiocarbon dates of 2110 ± 200 BP (PRL 53) and 850 ± 100 BP (PRL 66) (Agrawal *et al.* 1976; Agrawal and Seshadri 1998).

The most famous and productive copper mines lie at the other end of the Aravalli Hills around Khetri and Singhana in Rajasthan and have been much published (Boileau 1831; Lockett 1831; Brooke 1864; Roy 1958: 120–41; Straczek and Srikantan 1966: 47–53; Kakoti 1971; Willies 1992a). The mines were visited in the early 19th century by Lockett and Boileau and both kept journals, but whereas parts of Boileau's account were published (1831), Lockett's only survives as a manuscript in Delhi. The copper belt extends for about 80 km in total with evidence of old but undated small-scale workings all the way along. The major historic and modern mines are confined to the last northerly 12 km between Khetri and Singhana. Old workings, often on a considerable scale, are almost continuous throughout this region, culminating in the colossal workings and slag heaps at Singhana (Plates 2.1 and 2.2). Clearly copper was produced here on a very considerable scale, and the mines were one of the main sources of Indian copper in the historic past. In the 16th century Abū `LFaẓl Allāmī, for example, recorded that the mine paid the enormous sum of 11,881,620 rupees in revenue, specified to be paid in coin (Phillott 1927, II: 193, 205).

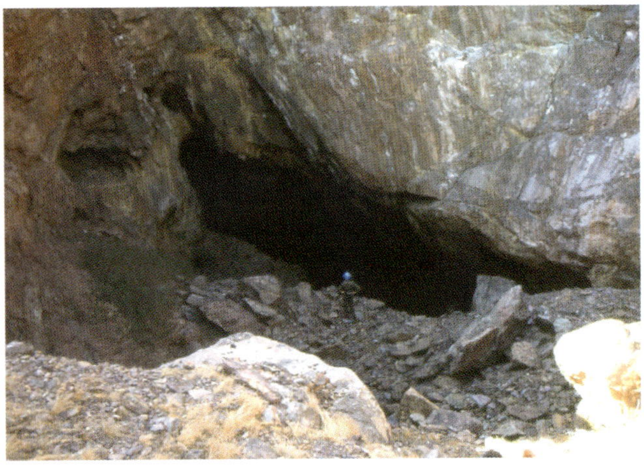

Plate 2.1 Entrance to the huge stope at the Madhan copper mine above Singhana in the Aravallis.

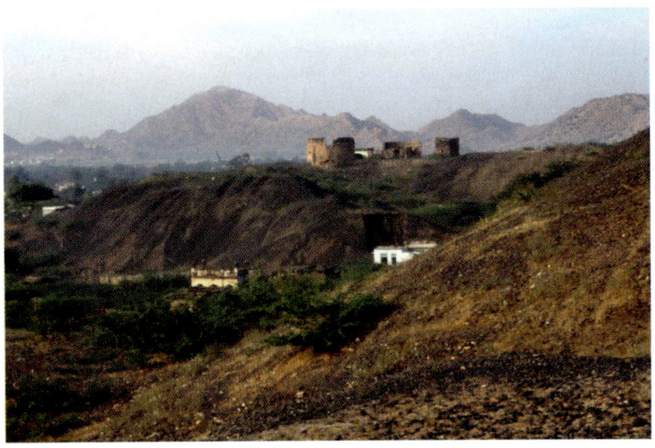

Plate 2.2 One of the bastions or small forts on the enormous heaps of accumulated smelting debris at Singhana, as mentioned in the 19th-century descriptions.

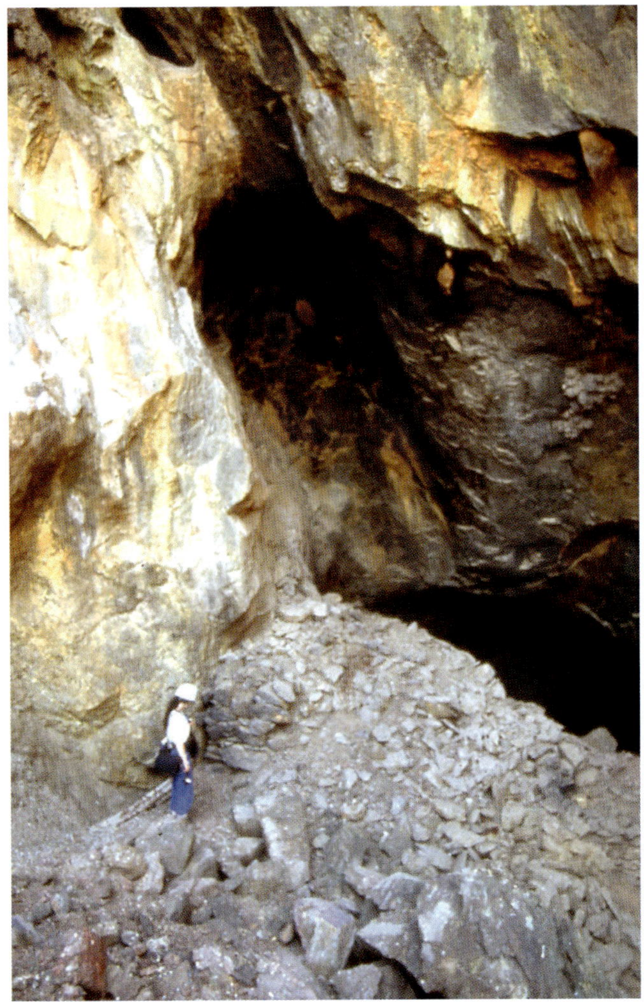

Plate 2.3 Platform built of stone and brushwood to position the fires against the rocks for precision firesetting at the Madhan mine.

Coins continued to be minted there until the 19th century (Ball 1881: 260–62) (see below). This provided considerable wealth to the area as attested by the fine buildings in Khetri itself, which date mainly to between the mid-18th and the mid-19th century (Cooper 2008) after Bhopal Singh established Khetri as his capital in 1755. At Singhana such buildings of architectural merit as survive tend to be rather earlier; by the early 19th century the town was described as being in decay (Boileau 1831). It is perhaps significant that there is no contemporary record or memory of some of the very largest workings being in operation (Chakrabarti 1985/6), notably the vast chamber of the Madhan mine above Singhana (Willies 1992a) (Plate 2.1).

This chamber is perhaps to be identified with that described by geologist C.A. Hacket of the Geological Survey of India (GSI) in the 1870s as a large descending gallery through which other workings were reached, although already abandoned through lack of ore (Hacket 1880, quoted in Ball 1881: 261). The mines at Singhana were worked almost exclusively by firesetting. The early 19th-century descriptions state that the resistant rocks were split by large fires left to burn against them for up to three days (Boileau 1831, 1845). In 1831 Boileau recorded that:

> At the head of the mine, *kan*, a great quantity of firewood is collected and stacked in a compact manner to the amount of 150–200 maunds (400–550 kg), which being set on fire, the miners make their escape as fast as they can to avoid the chance of their being smothered in the gallery, for some fatal accidents are said to have happened in this manner. On the third day, the workmen descend into the mine and find the interior of the mine rent and split into large masses by the violence of the heat, and they then proceed, if the mine be sufficiently cooled, to carry on the drift.

The great platforms, built of drystone bonded with acaia twigs, that were constructed to position the fires against the sides and roof of the chambers survive in quantity (Plate 2.3). Radiocarbon determination on the twigs collected in our surveys at the mine gave a calibrated date of AD 1680–1740 or 1805–1940 (BM 2796), that is, the late 17th–early 18th century or 19th–20th century, although the latter date can presumably be discounted by Hacket's comments. It must be borne in mind that these relate to the very last mining operations and the origins of the working are at present totally unknown. The Madhan gallery and other workings between Khetri and Singhana dwarf even the largest chambers at Zawar or Dariba, and must be among the largest preindustrial underground workings anywhere. A calibrated date of AD 1420–1535 or 1550–1635 (BM 2794) was obtained from a small timber in a shaft (no. 3) in the nearby Kudhan mine, attesting that the mine was operational in the late/post-medieval period (Willies 1992a).

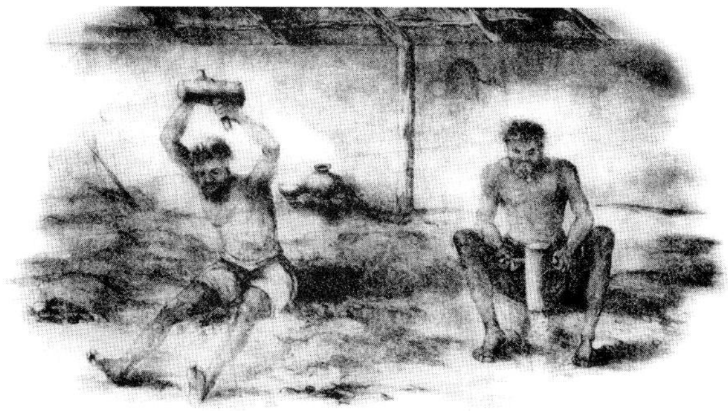

Fig. 2.2 Pounding the copper ore at Khetri with heavy two-handled hammers (from Brooke 1864).

It is important to keep the scale of these early workings in mind when reading the pejorative comments of the 19th-century European writers on the contemporary mining operations that they were then observing (discussed further in Chapter 12). Boileau's 1831 report provided further information on the mining operations:

> Each labourer is provided with a lamp, *derja*; a hammer, *hatori*; a mining chisel or gad, *tanki*; and a small wicker basket, *tokri*. The lamp is placed on the workman's head, and not only affords light for the execution of his work, but enables him also by watching the glistening particles of metal, to detach only such portions of the ore as may promise to give him the best remuneration for his labour. While at work he seats himself upon his heels, with the lamp upon his head as mentioned above, the hammer is in his right hand, the gad in his left, and the little basket upon his knees, in which he receives all the fragments of ore that are struck off by the chisel.

At the contemporary mines in Sikkim the gads were described as being held in handles of bamboo (Ball 1881: 276) (see Chapter 8, Fig. 8.1).

Boileau also recorded that the mines were worked for about eight months of the year up to the onset of the monsoons, and that the miners preferred to work during the night because of the excessive heat. The difference between night and day time temperatures cannot have been that marked underground but the air flow through the mines was probably better at night.

Hacket also described mines in Khetri itself with one gallery that was said by the miners to be 'upwards of two miles in length!' (his exclamation). The Kulhari or Kolihan mine was described as a tortuous warren, enlarged only where pockets of ore were discovered. Captain J.C. Brooke, an officer in the Mewar Bhil Corps, described a human chain passing pitchers of water to the surface to drain the mines when work recommenced in October after the rains: in one place 27 miners were employed at an average spacing of 8 feet (Brooke 1864). Similarly the workings in the Babai mine, which lay some 12 km south of Khetri, were described as 'quite irregular commencing as narrow inclines, widening out into large chambers, contracting to narrow burrows and ramifying as the ore led the miners' (Heron 1917, quoted in Roy 1958).

The smelting process at Singhana was described by Boileau (1831) and Lockett (1831), accounts of which are given in both Ball (1881: 260–61) and Percy (1861: 391–2); another account was prepared latterly by Brooke (1864). The more detailed description of the operations recorded by Percy is given here, with significant additional detail from Ball's account added in { }, original comments in [ ] and present comments in ( ).

The first stage was to crush and beneficiate the ore:

> The ore was copper-pyrites with a matrix of quartz. {The pounding or crushing was effected on a stone anvil with a hammer weighing eight or ten seers (7–9 kg)}.

Brooke describes this stage in considerably more detail:

> The Borah (the Muslim factor or businessman who controlled the smelting side of the business) having concluded his purchase from the miners (as described below), employs a man with a small hammer, who receives Rs. 3 per month, to separate the ore from the schistose rock and quartz [with which it is intermingled in about equal quantities], and to break it into small pieces. ...The ore now has to be finely powdered. ...This is done by men using 'Glums' or heavy hammers, weighing from 32 to 34 pounds (14.5 to 15.4 kg) each. The hammer is lifted with both hands, one on either side of the hammer head and brought down with great force on a small heap of the ore, raked into place with the toes, that never failing substitute for the hands among natives: as this is the most laborious operation in the whole process, only the strongest men are employed (Fig. 2.2). The ore has to undergo the hammering three times before it is fine enough for the roasting process. A Glum man on coming to his work very early in the cool of the morning has five maunds (185 kg) of ore weighed out to him, this is his proper quantity for a day's work, and is as much as can be supplied daily by the coarse breaker. Preparing this properly, gives the Glum laborer six hours of unremitting work, and his wages are proportionately high, viz, five rupees a month.

The smelting process was basically a dead roasting, suitable for treating the sulphidic ores obtained from the mines, followed by reduction and refining.

Boileau's account continued:

> It was reduced to powder, mixed with cow dung, and kneaded by hand into sausage-shaped pieces 5 inches (13 cm) long and 1¼ (3 cm) in diameter. These were sun-dried and roasted in circular heaps 4 feet (120 cm) and 1½ (45 cm) high. The fire was lighted in the evening, and on the following morning the roasted ore, which had a red colour (Plate 2.4), was smelted with charcoal in a small blast-furnace of the following construction (Figs 2.3–2.5). A quantity of common sand was spread upon the floor of a circular hut, and in the centre a small hollow was made from 12 to 15 inches (30 to 40 cm) in diameter and from 2 to 3 (5 to 8 cm) deep. In this was laid a stratum of fine yellow sand and then another of ashes. A sand bottom, or hearth, was thus formed, which, by the action of heat and the alkali in the ashes would become firmly consolidated. Two clay nozzles were placed on opposite sides of the hearth, and a third one midway between them, the fourth side being left for the escape of the melted slag. The nozzles were connected together with moist clay, so as to form a little circular wall a few inches high to serve as a basement for the upper part of the furnace, which consisted of three

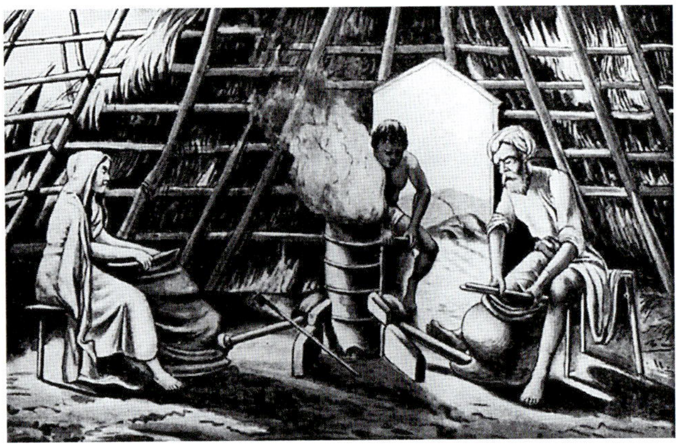

Fig. 2.3 Copper smelting at Singhana in the 19th century using three simple bag bellows set at right angles, with a large poker in the foreground. Note the apparent ruinous state of the hut. In fact the thatch was probably partially removed during smelting to allow the fumes to escape (from Ball 1881)

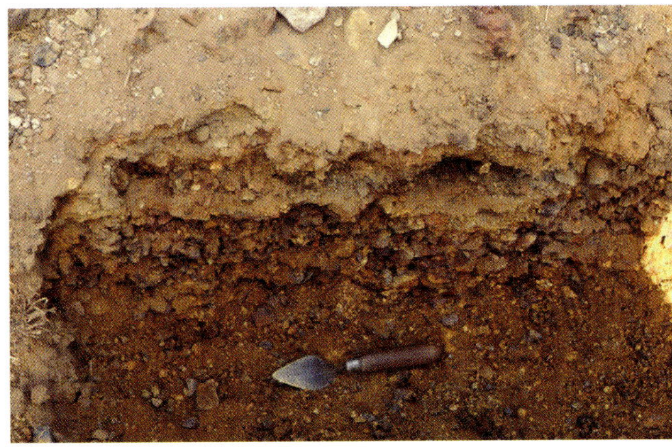

Plate 2.4 Layers of burnt debris from the roasting of the sulphidic ores, as described in the 19th century, in the great heaps of smelting debris at Singhana (cf. Plate 2.2).

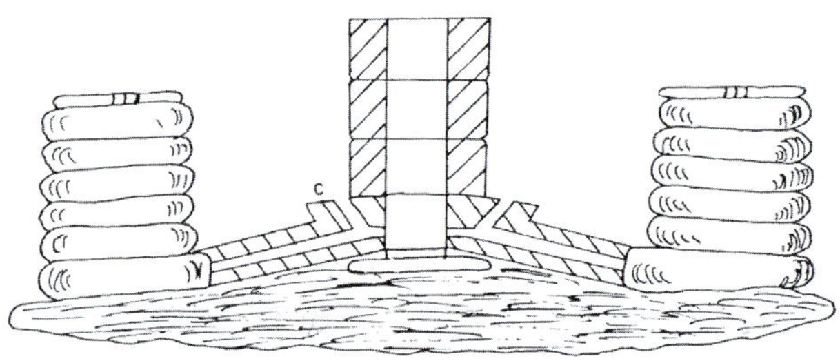

Fig. 2.4 Section through the Singhana furnace showing blowing arrangements but using concertina-type bellows. Note the separate inspection hole (c) in the tuyeres allowing, when unplugged, the interior of the furnace to be inspected, and any obstructions such as setting slag to be cleared with the poker depicted in Fig. 2.3. Separate holes were necessary as the tuyeres were angled (from Subbarayappa 1971).

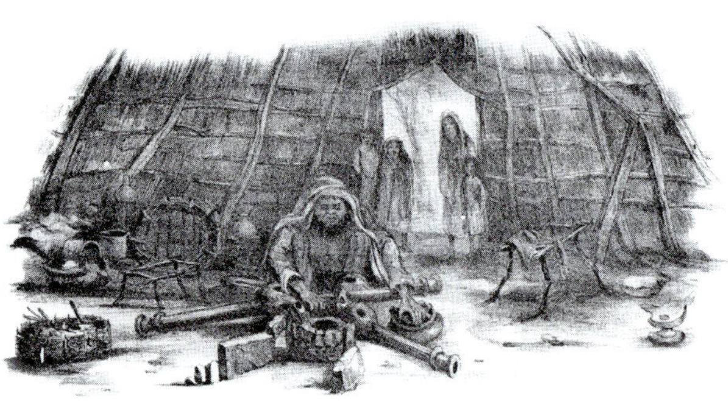

Fig. 2.5 Constructing the copper smelting furnace. Note the long tuyeres (from Brooke 1864).

> annular vessels of fireclay placed one above another. Each of these vessels was about 15 inches (40 cm) in external diameter, 9 or 10 inches (23–26 cm) high, and about 3 inches (8 cm) thick. They were used over and over again, but the bottom of the furnace required to be reconstructed daily in the manner above described. Holes were made round the basement, through which a poker might be introduced into the furnace, and there was one such hole obliquely directed through each nozzle near its junction with the furnace, so that a clear passage for the blast might always be maintained. These holes were stopped with clay, which could easily be removed when necessary. The blast was produced by three ordinary goat-skin bellows, of which one was attached to each nozzle. They were worked by men, women and even children, all Mussulmans. Before the furnace was charged, a quantity of charcoal was burned in the hearth in order to dry it. It is reported that in the day [9 to 10 hours] a single furnace would consume 3 maunds of charcoal [one mun or maund = 80 lb avoirdupois (36 kg)], during which time were added 2½ maunds (90.2 kg) of the sausage-shaped pieces of ore and cow-dung and 2 or 3 maunds (80 to 100 kg) of iron-slag, which was required as a flux, and was brought from a distance {the slag was drawn off and the smelted copper which had accumulated at the bottom of the furnace was removed the following day}. Four persons were employed at each furnace – perhaps a man with his wife and two children – who received for their united services 10 rupees a month. The head man prepared the furnace and occasionally relieved one of the party working the bellows, of which all three were kept constantly in action. On the morning after the first melting, the mass of copper which had collected in the hearth was taken out and sent to the refining furnace (Fig. 2.6). This is described as a small vessel which received the blast of a single bellows. {It was then re-melted and refined in an open furnace under a strong blast from bellows, and cast into small bars or ingots, which were subsequently removed to the Mint and cut up and fashioned into coins.} The refined copper was cast into shallow, clay moulds, each about 1 foot in length. The ingots thus obtained weighed 2 or 3 seers each [1 seer = 2 lbs avoirdupois](more accurately about 2.4 lb or 1.1 kg). The copper was lilac-coloured and brittle. {The quality of the metal was said to be inferior, this being attributed to the use of iron as a flux.} Copper smelting must have been carried on in this locality during a very long period, as the slags had accumulated to such an extent as to form a line of small hills several hundred feet in length and from 30 to 60 feet in height. There were four isolated stone bastions built on one of these artificial mounds (Plate 2.2).

Brooke's description of the smelting process was made some 30 years later, but is essentially similar:

> The ore having been reduced to a proper state, is next mixed with cow-dung, and made into rolls about four

inches (10 cm) long, which are first dried in the sun, and then roasted in the open air, in a fire of cow-dung cakes. This is an inexpensive process, costing only a few annas for cakes to roast five maunds (185 kg) of ore.

The ore is now ready for the smelting furnace. For this, Koomhars or potters are employed. The potter builds and works his own furnace, and supplies the bellows; in fact extracts the metal. Four people, one of whom may be a child 12 or 14 years old, are required for each furnace. They receive collectively Rs.11 a month. The furnace is about 3½ feet (105 cm) high and 12 inches (30 cm) in diameter, built of pieces of slag cemented with clay in a most primitive manner (Figs 2.4 and 2.5); and the nozzles of the bellows are built up in it. The nozzles are earthen tubes which are thickest at the furnace end, and at the top of the thick part is a small air hole, usually closed with a piece of wet rag, but opened now and then to clear the tubes (Fig. 2.4). The other end of the tube is fixed to the bellows bag. The bellows valve is formed by two sticks at the mouth, which are opened when the bag is raised for the admission of air, and closed when the bellows are pressed down with force by the bellows men, who use both hands for the purpose. The upper part of the furnace is formed with rings of fire clay, about 10 inches (25 cm) deep. The bellows are worked on three sides, while on the fourth is the opening to the furnace, in which a plate of fire clay is placed, at the lower part of which is a hole for stirring the molten metal and allowing it to flow out (Fig. 2.3).

The furnace is prepared daily, each smelting occupying about 12 to 14 hours. After the furnace has been lit and well heated, the roasted ore is gradually introduced, alternately with charcoal and the flux which is called 'Reet'. This is the refuse from the old iron furnaces, of which hills of debris still remain, the iron having been worked for ages before the copper ore was discovered (*keet* is the current Nepali word for slag). At each operation, five maunds (185 kg) of roasted ore is gradually introduced into the furnace; this requires an equal amount of the 'Reet', and 4 maunds (150 kg) of charcoal to smelt it.

The statement that the lower parts of the furnaces were made of slag is surprising and has not been encountered elsewhere.

Brooke's description of the copper production process concluded with the refining procedure for treating the black copper:

After the ore has been smelted, the metal has to be refined, and the sulphur driven off. This is done by passing a very strong current of heated air over the liquid mass, and constantly skimming it (Fig. 2.6). To obtain the blast a single bellows is used, which is worked by one man opening and drawing it up, and two others pressing it forcibly down with their feet, placing their whole weight on the bellows, and maintaining their balance by means of ropes fastened to the roof of the building (a somewhat similar arrangement was used by the bellows operatives to smelt iron in the Nonkreem Valley in the Khāsi Hills, see Chapter 9, Fig. 9.2).

About one maund (37 kg) is refined at a time, which produces about 20 to 25 seers (22 to 28 kg) of good copper. The refining is contracted for at 8 annas the maund. The process requires about 3 hours, and the men are paid 1½ annas per diem each. When the pot in which the refining has been conducted is ready, the ore (?metal) is poured into small earthen troughs prepared on the ground for the purpose, and is then taken to the mint for weighment and duty.

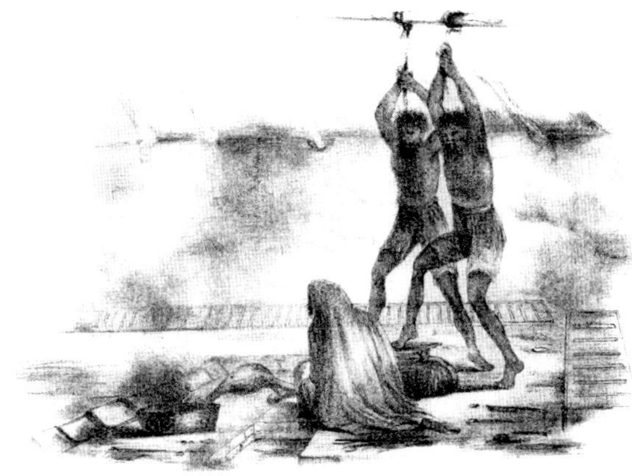

Fig. 2.6 Refining the black copper to burn out sulphur. The unusual mode of operating the bellows was also recorded at the iron smelting furnaces in the Nonkreem Valley in the Khāsi Hills (see Chapter 9, Fig. 9.2) (from Brooke 1864).

In the illustration it is not clear how the copper was kept molten. Possibly the pot and the charcoal were inside the low cylindrical structure as depicted in Fig. 2.6, or perhaps that was the pot and the charcoal and metal were inside. This would have been fixed and the metal released into the tile-covered channel depicted to the left and into the moulds.

This is basically a three-stage process. In the first stage the sulphidic ore was oxidised, and then in the second, smelted to reduce the copper oxides to metal. The iron slags added as a flux would have contained quantities of wustite (FeO). The function of this would have been partly to remove sulphur from the copper, forming iron sulphide, and also to form an iron silicate slag with the quartz in the ore (see Chapters 7 and 9). The resulting metal would have been iron-rich and seems to have been refined by remelting in a large open vessel to oxidise the iron, resulting in a very considerable loss in weight, but even so the refined metal was still brittle. Lockett (1831: 95) noted that imported English sheet copper was used to make the better quality *pucka pice*, but that the local imperfect copper was employed to produce a *pice* of lower value (*pice* were low denomination coins, 4 to an *anna*, 64 to a *rupee*).

The smelting and refining processes described here are the sort of operation it is envisaged took place at Dariba at Sites 13, 14 and 15 (see Chapter 6). However, although the furnace dimensions are similar, the internal diameter of the Singhana furnace being 23 cm (compared to our estimate of 20–25 cm for those at Dariba), the actual construction was apparently different. In particular at Dariba there is no evidence for the reusable 25-cm-tall clay rings: the furnace walls were built of coils of clay approximately 4 cm in diameter. The clay nozzles as described are clearly set in some sort of curved collar such that three of them can be brought together to form three-quarters of a circle. These sound very like the nozzle and collars found by Hegde and Ericson (1985) at traditional copper smelting sites in the Aravalli Hills.

As Boileau's account stated, the scale of the copper smelting in previous eras at Singhana itself is attested by the huge mounds of slags, calcined ores and other smelting debris which still occupy about 2 km^2 in the centre of the town. The heaps had clearly built up over a considerable period, and little can be said at present about the origins of copper working here. In addition to the bastions mentioned above, a small fortified residence, which was in existence in 1820, sits on top of one of the major heaps, showing that deposition had finished there by that date. Charcoal from a small trial trench excavated by the side of this building (Plate 2.4) gave a calibrated radiocarbon date of AD 1670–1755 or 1795–1945 (BM 2793) (Willies 1992a). The latter date can be rejected because of the nearby early 19th-century fort so this date would suggest that activity ceased in the vicinity of this part of the mound in the 17th or 18th century AD, similar to the date of the last firesetting operations in the Madhan mine.

After briefly describing the mining of cobalt minerals for use as an enamel colorant at Jaipur and elsewhere, Brooke concluded his description of the copper mining and smelting:

> The above is a short description of the processes employed in smelting the rich copper ore found in Khetree. The miners are wretchedly poor and ignorant: the mines are choked with rubbish, and worked without system, so that the more valuable ores are not reached: the ore only passes through the furnace once. The metal then separated is the only part kept, but the layer next above the regulus (likely to have been largely of iron sulphides together with some copper sulphide) which is also rich in metal, is thrown away: of this, vast heaps or rather hills exist, both at Khetree and Singhana, and the present furnaces are on these mounds, from which a little enterprise and knowledge would extract a large produce.
>
> There is no means of knowing what the produce of the copper mines would be, if worked on European principles, or whether the fuel, which suffices for the insignificant native works would not soon be exhausted. The fuel is charcoal from the 'phog' plant (*Calligonum polygonaides*). It grows freely all over the neighbouring sandy deserts. The 'phog', on which camels feed freely, is a low succulent plant about a foot and half (45 cm) high, but the roots of which swell out to a large size and make excellent fuel for all purposes.

In fact, the bulk of the heaps comprises slag and calcine debris, and does not contain sufficient copper to be worth reprocessing. In the early 20th century copper smelting 'on European principles' was established and Hindustan Copper Ltd. continues to mine and smelt the copper there.

It seems likely that in earlier times the mines would have been state controlled as inferred from both contemporary Mauryan and Mughal accounts (see Chapter 8). However, by the 19th century the miners had obtained a degree of control, although this was not necessarily to their advantage and certainly not conducive to the overall development and maintenance of the mine. The mines at Khetri were already falling into disrepair as chronicled by Lockett (1831). The following is a transcription of the entry made at Singhana on 27 April 1831:

> The copper mines in this neighbourhood are very numerous. Some of the miners assured me, there were not less than 305, but of these only 30 are now worked in the Hills about 2 miles south west of Singhana. The *Neele-yhotha*, or blue-vitriol so common here, is prepared in the mines of Khetree. The alum is prepared at the Copper Mines near Singhana. The most productive copper mine is situated between Khetree and Bubaee, and is called Ak Walee. A *Kaith* named Runjeet Rae, has the superintendence of all these mines. The labour in these mines is very great, as well as the danger. I understand some are killed by the falling down of stone and on two occasions almost the whole of the workmen were destroyed – 360 having been crushed to death by the falling in of the mine. About two years ago about 60 men they say perished in the same manner. The government share of the profits is said to be one sixth, – viz., 9,000 rupees – the total amount derived from the mines and alum of Singhana, being 54,000 rupees per annum (about £2,700).
>
> The receipts were formerly very considerable, but have been decreasing yearly. The mines are constantly getting out of order, and some falling in, and there is no one to repair them. The Khetree Govt., and the other two Panahs connected with it, will not incur the expense, and the poor miners and ryots have not the means.[6]

Brooke described the organisation of the mines some 30 years later:

> The copper mines are owned by the miners themselves, whose ancestors discovered them in former times. The larger are managed by a punchayet, on behalf of the mining community, who are descended from the ancient discoverers. Some of the smaller mines are owned by traders who have bought them up, either from the original discoverers, or else gradually from their descendants, as these have become involved in difficulties, and have pawned or made over their shares to the traders; who pander, for their own interests, to the unthriftiness for which all such men are noted. The larger mines do not seem to have shared this fate.
>
> Each year, after the rainy season the various branches of each mine are put up to 'auction' by the punchayet. The *Koolhán* mine has six or seven branches. The miners themselves are the bidders. Each branch of the Koolhan mine sells for from Rs. 50 to 100 a year, and the whole mine fetches from Rs. 400 to Rs. 600, which is a small sum considering the treasure that is within it (a rupee was worth approximately one shilling sterling and so the mines were valued at from £20 to £30 per annum).
>
> Each branch of the mine is jealously watched by the miner who purchases it, and who hires other miners as labourers on two annas per diem. Were the mine not guarded, these miners might purloin the ore and sell it.
>
> The miners work in gangs, and a party of eight, starting in the morning at about 8 o'clock, will bring back about from 2½ to 3 maunds (93 to 110 kg) of ore by the evening. The ore is brought in small baskets, weighing about 6 lbs. (2.2 kg) each, and is then put up to auction, in the same manner as is done with the ore obtained from the mines still in the hands of the original proprietors, or the traders. The auction takes place at the town of Khetree itself, and furnishes a scene of much excitement.

In this system of working it was not the responsibility of the miners to maintain the workings, much less to cut drainage adits or shafts. This would have made mining progressively more difficult, and Brooke commented that some of the richest old workings were flooded and no one had the means to drain them. The ultimate cause for the decline and unwillingness to invest in the mines is not difficult to find. At the nearby town of Churawah, in 1831, Lockett observed:

> There is a mint for *pice* in the town which I visited in the evening, while the workmen were employed. There they make the *pukka pice* from English sheet copper from Buwanee. Sixteen or seventeen of these *pice* make a *rupee*. The *kutha pice* (there are two kinds of *kucha pice*, one is called *kur*, from 40 to 45 in the *rupee*; the other *lowa*, from 25 to 27 in the *rupee*) are made in Khetree, of the imperfect copper, brought from the mines near Bubee and Singhana.

Therefore even though the mines lay hundreds of kilometres from the coast, imported copper was still able to take the cream of the market, leaving the local mines only able to survive at all by producing an inferior product at low prices, but with no money for investment, they soon went into total decline. By the mid-19th century the main activity seems to have been the recovery of copper and copperas from the waters running through the old workings and leaching the old tips for copper and alum. There was some mining for the cobalt ore, jaipurite (*saita*) at the Babai mine with only a little mining for copper (Brooke 1864).

Other major early copper mines in Rajasthan can be found in the Alwar District at Daribo, Khoh Dariba, Prat Abgarth and at Bairai, where the modern town sits athwart the huge slag heaps. Chakrabarti (1985/6) believed that copper production was important in the Alwar area from the medieval period.

The smelting operations at Daribo were described by Major Cadell, and incorporated into the 1878 *State Gazetteer*, compiled by P.W. Powlett (p. 82) and quoted in Ball (1881: 259) and Chakrabarti and Lahiri (1996: 160):

> The manufacture is carried out in thatched sheds, which are generally in a very dilapidated state. The ore is chipped out of the solid rock with hammer and chisel; and, having been beaten with hammers into powder, is mixed with double its weight of powdered iron slag. This mixture is then made into small cakes with an equal amount of cow-dung; and, after, being roasted in a fire made of grass (surely brushwood?) and cow-dung, is placed, like the iron ore, in the smelting furnace, in alternate layers with charcoal. When the ore is melted, the furnace cylinder is broken down, and the mass of copper which has collected at the bottom, after being allowed to cool, is lifted out. It is then taken to another shed, and is placed in an open charcoal fire, where it is melted a second time with the aid of the bellows, which is worked by two men standing, and which is opened and closed at the proper moment by the man who also attends to the fire. It is then poured into a mould in bars, and out of these bars the copper currency of the state is coined.

The ore required roasting to remove some of the sulphur before the reduction stage in the furnace. The addition of iron slags as a flux is well attested in many of the accounts of traditional non-ferrous metal smelting in South Asia, including at Khetri-Singhana as noted above. As there, the iron oxide in the iron smelting slag would have removed sulphur from the copper and also acted as a slag-former with the quartz gangue.

Cadell's comment on the thatched huts in a dilapidated state is interesting and possibly significant, beyond the usual European pejorative comment on the state of native manufactories. Huts containing furnaces are depicted in several illustrations of smelting operations and they too have thatched roofs with much of the thatch removed (Fig. 2.3). A light roof such as those depicted would keep the furnace dry when not in use, but could be easily removed during smelting to allow the noxious fumes to escape. This was especially relevant here as the ore is described elsewhere (Ball 1881: 259) as being of pyrites mixed with arsenical iron in a vein of quartzite. In addition, Ball noted that an average furnace load of 30 lb (13.6 kg) of ore required four times that weight of charcoal and produced 5.5 lb (2.5 kg) of copper. This account was accompanied by a breakdown of the costs of the various stages of the operation. To produce one smelt of copper, the costs were as follows:

- Rupees
- 30 lb ore .31
- 120 lb charcoal .38
- Breaking up ore .06
- Making ore into cakes with iron ore and dung .12
- One skilled workman .19
- Bellow's men .13
- Refining .06

At the time that Cadell produced his report copper smelting was the principal occupation for 32 families, numbering some 88 persons, but in 1881 Ball noted that during the previous 12 years the average annual output had been only 3 tons and 8 hundredweight (3.5 metric tons), and was diminishing because of the influx of European copper. The copper so produced would still have contained some iron or sulphur that was removed by the second, oxidising smelt.

Lead and zinc

The ores of lead and zinc as well as silver are normally associated, often making it difficult to be certain which metals were being exploited at any particular period in the past in any old abandoned mine or deposit without conducting detailed investigations. Accordingly some of the mines listed in the 'Lead' section below may also have produced some silver, although by the time of the first European geological surveys silver production seems to have ceased. Lead/zinc ores are quite common in the subcontinent (Coggin Brown and Dey 1955; Raghunandan *et al.* 1981; Krishnasswamy 1988; Bhatnagar and Mathur 1989), although commercially exploitable deposits are concentrated in the Aravallis including, of course, Agucha, Dariba and Zawar, which are now all major producers of lead and zinc and to a lesser extent, silver.

The only other major lead/zinc deposit worked on a substantial scale in antiquity was located at the southern end of the Aravallis at Ambaji, in Gujarat (Fig. 2.1). As with the copper deposits at Singhana-Khetri, these silver/lead mines have been assumed to be of great age but previously without any firm evidence.

Lead

The mines at Taragarh, near Ajmer (Fig. 2.1) were described in detail by Captain C.G. Dixon (1831) after his visit in 1830, and later reproduced, with comments, by Percy (1870: 293–4). As at Dariba the principal ore was galena, although apparently it was not argentiferous. The deposit was large and had been worked for many centuries. The overall mining strategy having located a vein at surface was to follow it into the hill. Where a number of veins were encountered the larger was worked out first before returning to take out the more minor deposits. This is believed to have been the strategy adopted at Zawar in Mauryan times although on a much larger scale. Dixon's account is followed closely by Percy. His additional comments in the text are in { } and the present text comments in ( ).

> The excavation in following a vein of ore is made only large enough to admit a man moving on his hands. Shafts, galleries, and adits, if not quite unknown, are altogether neglected. The ore is found in a continuous vein averaging from 3" to 6" (7.5 to 15 cm) in thickness, the bearing being generally N.E., or, rather, between N.E. and E. It is easily traced from the innumerable shining particles of which it is composed. Sometimes it is black, porous or spongy; and at others deep red; but the prevailing colour is yellow or ochreous. This variation in colour is ascribed to the nature of the soil in which the ore is imbedded. The author mentions also the occurrence of ore in beautiful crystals imbedded in quartz, called by the natives *surma*, and sold, as taken from the mines, as antimony; and, it is added, 'lumps of this ore, weighing from 2 to 3 lbs. (0.8 to 1.3 kg) have been used as antimony, for laboratory purposes, in the Ajmer magazine, in their natural state'. The author was evidently no mineralogist; but it is clear that the substance to which he here refers is galena, though it may be very puzzling to explain how galena could have served as a substitute for sulphide of antimony in the Ajmer magazine. The strata traversed by the veins are described as 'sandstone, free and slaty; chalky {?} (Percy later suggested that potstone, which is a form of impure steatite was meant) and quartz rock. Immediately afterwards the author defines chalk as 'a greasy saponaceous earth, sometimes beautifully streaked, at others perfectly white, and of little density.'
>
> There is no blasting with gunpowder, and the hard quartz ore rock is excavated by the old method of *fire-setting*, which consists of lighting a fire of wood against the working face, whereby the rock is cracked to a certain depth and fragments of it are split off. This method is resorted to, on leaving off work for the day, about noon; when, on their return the following morning, the object has been effected, and the smoke dissipated.
>
> The mining tools are few and simple in construction. They consist of two or three hammers of different weights, of iron chisels or wedges edged with steel, and of a few one-handed picks (see Chapter 8, Fig. 8.1). The miners begin work about sunrise, and enter the mine stark naked, each man carrying a small lamp in one hand. The ore is broken as it is gotten into small pieces, for convenience of carriage, and put into small bamboo baskets (see Chapter 8, Plate 8.6). As the gallery is not high enough to allow the miners to stand upright, they sit on their haunches in a row and pass the baskets on from one to another. When the whole has reached the foremost man, they move upwards and again seat themselves in a row; and this is repeated until the mouth of the mine is reached.
>
> Mining is chiefly carried on by persons to whom it has descended hereditarily from time immemorial. The miners being very limited in their means were formerly obliged to resort to money-lenders in order to get the funds needful to enable them to carry on their operations; but the interest was most exorbitant, and absorbed most of their profits. But, when Captain Dixon visited the mines, advances were made without interest by the government, and 'the miners were wholly emancipated from the rapacious grasp of the money-lenders;' and 'the result' says the author, 'has fully justified the expedient'. The government as lord of the manor received a share of the metal and purchased the balance 'at an assigned and rather favourable price'. The annual production is about 850 cwt (45 tons) of lead.

It is interesting to compare these descriptions of the ores with those given in the *Arthaśāstra* some 2000 years previously (see Chapter 8). Ball (1881: 299) stated that the crystals of pure galena called *surma* were sold as an eye cosmetic. Galena was a common substitute for stibnite, antimony sulphide, both in 19th-century India and worldwide to the present day (Craddock *et al.* 1997), and it seems more likely that this was what was being sold to the Ajmer magazine, presumably to make bullets.

The original miners seem simply to have followed the ore from where it outcropped, as was the practice in most early mines (Craddock 1995: 31) (Chapter 8). In the admittedly brief description of firesetting there is no mention of dowsing the heated rock to bring about splitting. Captain Dixon's report continued, describing the smelting, which once again was reported by Percy (1870: 294–6):

> The ore is spread out on the hill, levelled for a few square yards at the entrance of the mine and there left, if wet, until it becomes quite dry. It is then beaten with cylindrical bars of wood from 3" to 4" in diameter (8 to 10 cm), and about 2½' long (75 cm), one end being thinner and curved for convenience of grasping. Holding the handle with both hands, the workmen continue to beat the ore until the whole of the metallic particles have been separated from the matrix, which is collected from time to time and thrown down the face of the hill. After complete pulverization, which is effected much sooner than would be imagined from the rude manner of execution, and after separation of the matrix, the ore is put into baskets and subjected to levigation in water. In the absence of running streams pits from 10' to 12' square (3 to 3.6 m), and 3' to 4' deep (90 to 120 cm) are made, and filled with water by bullock power as is usual in irrigating land. Each ore-washer has a wooden tub shaped like the fustrum of a cone, 3' in diameter at the top (90 cm) and 1' at the bottom (30 cm), and 1' 6" deep (45 cm). The ore-washer places himself on the edge of the pit and floats his tub containing pounded ore and water. The contents

are 'violently stirred up with the hands', while the tub, owing to its shape, is kept floating all the time. After this stirring, the tub is gradually drawn backwards and forwards on the surface of the water, when the heavier or metalliferous particles subside quickly to the bottom, leaving the specifically lighter sand and other impurities at the top, which are easily skimmed off by hand. The operation is repeated 20 or 30 times until the ore is completely cleansed. The sediment, which accumulates in the pits, is from time to time removed, and the water in the pits is renewed about once a week, or oftener when evaporation is considerable. The washed ore is delivered to a separate set of people, who mix it intimately with its own weight of fresh cow-dung and roll the mixture into balls of about the size of a pigeon's egg, which are left exposed to the sun until perfectly dry.

The furnace is nearly cylindrical, 11" (28 cm) in diameter at the top and 10" (25.5 cm) at the bottom, and 3' (90 cm) high. Three somewhat angular clay pipes or nozzles are luted to the circumference {*i.e.* I presume, at or near the bottom on the same level, and one at each third of the circumference}, and serve to convey the blast from as many bellows, made of half-dressed goat-skin, each bellows requiring the labour of one man. In each nozzle a hole is left for cleansing the end or stirring up the ignited ore, as also for enabling the smelters to observe the interior of the furnace. At the bottom is a tap-hole, and on the opposite side 5" (13 cm) higher is a hole for the outflow of slag. The author states that the three bellows are worked simultaneously, and 'the blast is so disposed as to fall in the centre of the base'; but the bellows were doubtless worked in alternation, so as to produce an unremitting blast (this important point is discussed below). Charcoal is the fuel, of which a half to two thirds of the yield of lead is required. After first lighting the furnace, which is fed alternately with balls and charcoal, tapping takes place in about 3 or 4 hours, the slag being drawn off before the lead, which runs into moulds. The subsequent tappings when the furnace is in full working condition recur about every 12 hours. The quantity of lead smelted in such a furnace in one day of about 9 hours, averages from 2 to 3 cwt. (100 to 150 kg). The lowest yield is about 31% of lead, the average 40%, and the highest from 50% to 58%.

Captain Dixon's seemingly very detailed account did not find favour with Percy who, somewhat unfairly one might think, attacked its accuracy, lack of detail and Dixon's competence. In the section above, Dixon's description of the blowing operations and Percy's sceptical comments do require some discussion as they have important implications for the physical chemistry of the smelting process, and are discussed in more detail in Chapter 9. Suffice it here to note that the three blowers very probably did operate simultaneously to deliver a pulsed blast (Craddock 1995: 186; Merkel 1990: 102–3).

Production at Ajmer in 1818 was reported at around 350 tons of lead per annum, but by the time of Dixon's 1830 visit this had apparently fallen to only about 43 tons. In this instance there is no mystery concerning the fate of the mines. Ball (1881: 299) states that by the 1840s the lead they produced, although considered to be purer than English lead, was one-eighth more expensive, even at Agra, and could not compete. The mines finally closed at the time of the Mutiny in the 1850s, with 'an anxiety on the part of authorities to make lead for bullets scarce'.

Percy used Dixon's account quite specifically as 'representing the infancy of the art' of lead production, yet the mines were clearly much less technically and organisationally advanced than those at Agucha, for example, lying only 50 km to the south but worked two millennia earlier. In fact, the mining technology, if such it could be called at this stage in the Taragarh mine's long life, was typical of that of a mine in its terminal phase: poorly organised, low technology, requiring minimal investment, mining for the pockets of ore left by the earlier and larger scale operations (Craddock 1997) (see Chapter 12).

There are two important lead-zinc deposits in the Chittaurgarh District, east of Udaipur (Raghunandan *et al.* 1981) – at Karjiakhera and Rewara – with evidence of early mining and lead smelting.

Zinc

Zawar, together with nearby Kaya, is the only site of early zinc exploitation that has so far been positively identified in India, although other sources must have existed (see Chapter 13). Major deposits of zinc are not common in South Asia compared to the lead deposits with which they are always associated (Ball 1881: 312–13; Chakrabarti and Lahiri 1996: 30–31; Agrawal 2000: 204–21). Zinc was mined on a very considerable scale at the Bawdwin Mines in the Northern Shan State of Burma in the first half of the 20th century in deposits that had previously been worked for silver/lead only (Coggin Brown and Dey 1955: 156–61; Bender 1983: 177–82).

There were reports of zinc production near Jodphur (mentioned in the *Imperial Gazette* for 1900) but with little other evidence for exploitation. Medieval zinc *jittal* coins from the Kangra kingdoms of present-day Punjab–Uttarakhand with lead isotope values that are incompatible with any source in the Aravallis demonstrate that other sources must have been worked in the past (Cribb 2014: Craddock *et al.* 2015) (see Chapter 13).

The only deposits in India itself that have been commercially worked in the recent past for zinc are found in the Aravallis with minor production at the lead mines at Sargipalli (Orissa) and Agnigundala (Andhra Pradesh). In another multi-ore deposit near a village called Mamandur, west of Turukoilur in the Kallakurichchi taluk of South Arcot District in Tamil Nadu, there is a large opencast working of unknown date which is now predominantly of zinc, but lead and copper are also present and may have been the principal metals mined in the past (Krishnaswamy 1988: 457–8). Certainly there has been no report of zinc smelting retorts there and brass usage was relatively late in south India. In any case, the majority of the sources of zinc found in South Asia would not have been usable by early metalsmiths because the zinc minerals are intimately mixed with lead and thus not separable by hand picking.

Major zinc deposits are reported from Afghanistan, including deposits of smithsonite in the vicinity of Kandahar with evidence of early working (Shareq *et al.* 1977: 134–46); Wertime (1968) reported deposits in the east of Persia (now Iran).

Small deposits of zinc ore have been reported at Darabi and Riasi Tahsil in the Udhampur District of Jammu and Kashmir.

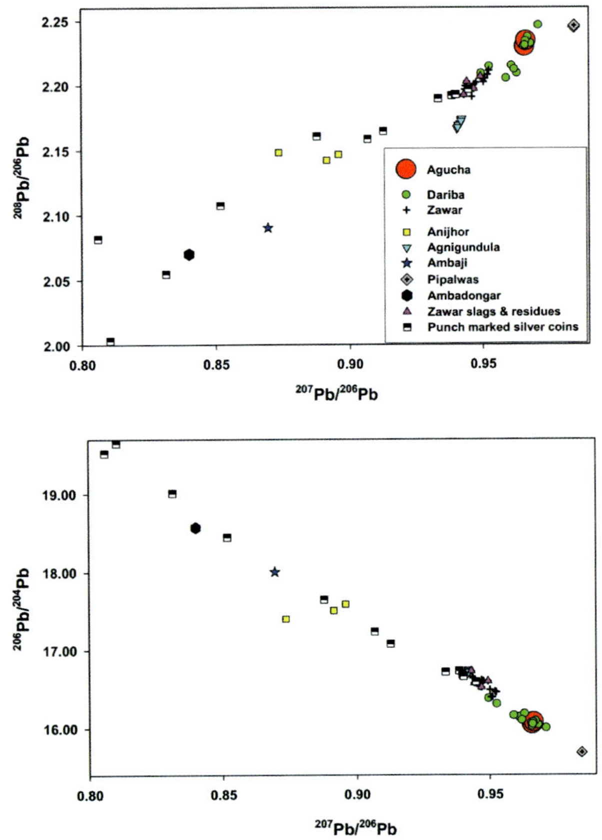

Fig. 2.7 Lead isotope values for a variety of potential silver sources in northern India and a small section of punch-marked silver coins of Mauryan date.

Abū 'LFaẓl Allāmī (Phillott 1927: 317) in his description of the Lahore Subah (present-day north Pakistan) mentions 'washing the soil for gold, silver copper, *rui* (a lead and copper alloy), zinc, brass and lead'. It seems inherently unlikely that zinc and copper would have originated from alluvial deposits, and even less likely that alloys such as *rui* or brass were being 'washed', and the number of metals does come to seven, always a propitious number in early metallurgy. However, there is at least a suggestion that zinc was being produced in the area.

Notes

1. Principal sources: Ball 1881; Percy 1861, 1870; Dharampal 1971; Chakrabarti and Lahiri 1996; Grover 2005.
2. The various dimensions and weights given in the originals have been converted as far as possible into metric units, following the table given in Appendix G in Ball (1881: 640) (there is also the problem that imperial and metric units for apparently the same thing can differ – an imperial ton is 2,2240 lb, but a metric tonne is 1000 kg, i.e. 2,204.6 lb.) There were many local units: at one point for example, Brooke (1864) refers to the Khetree maund as a unit of weight when describing the operations at that town, which was different from the Shahjehanee maund used in the Khetri mint. Also the local units themselves were not necessarily consistent. Thus, for example, in the two accounts of silver assay where the authors actually define the Indian units, O'Shaughnessy (1847) and Ball both define the *tola* as being 180 grains troy, but Viccajee (1908) defines it as 172.5 grains (Chapter 10, p. 208). A further problem is that it is not always made clear whether the troy or avoirdupois system of weights is being followed.
3. Agrawal (1984), Agrawala (1984), Hegde (1984) and many others have stated that the pattern of trace elements found in the copper ore, particularly at the Khetri mines but also at other sources in the northern Aravallis, matches that found in the Harappan copper. However the trace elements quoted are common to most copper ores. None of the Aravalli copper ores analysed to date have the very high nickel contents found in some Harappan copper-based artefacts but which do occur in ores and metal from the Oman and other Gulf sources and in the early flat axe from Nal (see Chapter 1, Fig.1.1), in southern Pakistan with 4.9% of nickel (Piggott 1950: 49).
4. To test this hypothesis lead isotope determinations were carried out on a selection of punch-marked coins and some galena ores from the Aravallis. Twelve coins were selected including eight from the Vatsa hoard, two Kosmati coins and two other Mauryan coins, which should all date between the 4th and 2nd centuries BC. The lead isotope compositions of the silver coins, plotted together with the isotopic data for the ores in Fig. 2.7, are much more spread out than the Aravalli sources and do not correlate to any of them with the exception of one coin, Vatsa 3, which matches the ores from Zawar. Alhough three other Vatsa coins fall near (but not quite in) the Zawar field, as a whole the Vatsa coins exhibit quite a spread of isotopic compositions which would most simply be interpreted in terms of this group of coins having been made of silver from a range of as yet unknown silver ore sources. Similarly, the two other Mauryan coins have very different lead isotope compositions, so it seems likely that a wide range of ore deposits produced the silver for these coins. There is an additional problem: the punch-marked coins contain substantial quantities of copper and the lead together with this copper is likely to confuse attempts to provenance the silver. Analytical studies on punch-marked coins have reported copper contents of between 15 and 25% (Rao *et al.* 1990; Agrawal *et al.* 1990). Furthermore the *Arthaśāstra* (2.12.24; Kangle 1965–72, II: 108–9) stated that the punch-marked silver coins should have one part of copper to three parts of silver. Six of the coins in this project were analysed by EDXRF on the uncleaned surfaces. They were not corroded but even so some surface depletion of the copper is likely. The coins contained between 5 and 23% of copper and between 0.7 and 2.3% of lead therefore the lead isotope values obtained were from an irresolvable mixture of lead derived from the silver and copper.
5. Hegde (1967) believed that the silver of some silver coins of the Ksatrapa period of about the 2nd to the 4th century AD, found at Devnimora in Gujarat, did not come from the Aravalli mines but this was based only on the comparison of the qualitative spectrochemical trace element analysis of the coins with a sample of ore from Zawar. Hegde (1968) also analysed two lead coins of the Ksatrapa period, c. AD 100–400 from Nagara in the Kaira District of Gujarat, approximately 290 km from Zawar, and concluded that they were likely to be from Zawar because they contained 1.18% and 1.26% of zinc. This association is not valid but does suggest that the ores were smelted in a primitive manner that retained the volatile zinc. The lead isotope ratios of one of these coins were also determined by Brill and Shields (1972) and compared to those from a single sample of galena from Zawar. They found that the Pb^{208}/Pb^{206} was 2.1318, the Pb^{207}/Pb^{206} was 0.8833 and the Pb^{204}/Pb^{206} figure was 0.05632. Subsequent lead isotope analysis showed that the ratios equate much better with the Ambaji mines in Gujarat, considerably closer to Nagara than Zawar, suggesting a site in the southern Aravallis if not Ambaji itself.
6. With thanks to Ilay Cooper for the transcription.

Chapter 3 The history and monuments of Zawar, Rajpura-Dariba and Rampura-Agucha and the foundation of Hindustan Zinc Limited[1]

L.K. Gurjar and P.T. Craddock

The time of maximum activity at Zawar, Dariba and Agucha in antiquity coincided with the Mauryan Empire from the 4th to the 2nd century BC. At Dariba and Agucha it is likely that the Mauryan state initiated and ran the mines directly as state-run enterprises of the kind outlined in the *Arthaśāstra* (see Chapter 8). It is possible to envisage the mines and all the necessary infrastructure as having been created as state enterprises – pioneer settlements planted in the wilderness as Rajasthan in the Iron Age is likely to have been very backward (Possehl and Gullapalli 1999). The situation at Zawar is less certain. The other two mines were engaged primarily in silver production, mainly of coins, in which the state had a very direct interest. At Zawar at this time it seems that the principal product was zinc oxide, which was used for medicinal purposes, for the production of brass and possibly as a pigment, in which activities the state may not have been so specifically concerned.

Zawar

In the following account for convenience the deserted and ruined settlement is referred to as Old Zawar (with the Ramanath temple) and the present traditional settlement (with the Zawarmata temple) as New Zawar, following Kumar (forthcoming), although both are of considerable antiquity (see Chapter 6, Fig. 6.2).

One of the mysteries of Zawar is the absence of early remains of occupation contemporary with the main period of working of the deep mines and the few small associated smelting sites. However, field surveys undertaken by J.S. Kharakwal (2011: 189) between 2002 and 2004 located spreads of Iron Age pottery in the vicinity. The sheer scale of the highly organised mining activities at Zawar make it likely that the Mauryan state was involved in their organisation. The operations must have employed hundreds of workers as a very minimum, all of whom would have required accommodation and supplies together with elements of administration, ritual and security, evidence for which has not so far been located. There are no mounds of occupation deposits, architectural or sculptural fragments, reports of stray finds, pottery sherds or indeed anything above ground earlier than about the 12th century AD (Plate 3.1) except for some of the relatively small slag heaps.

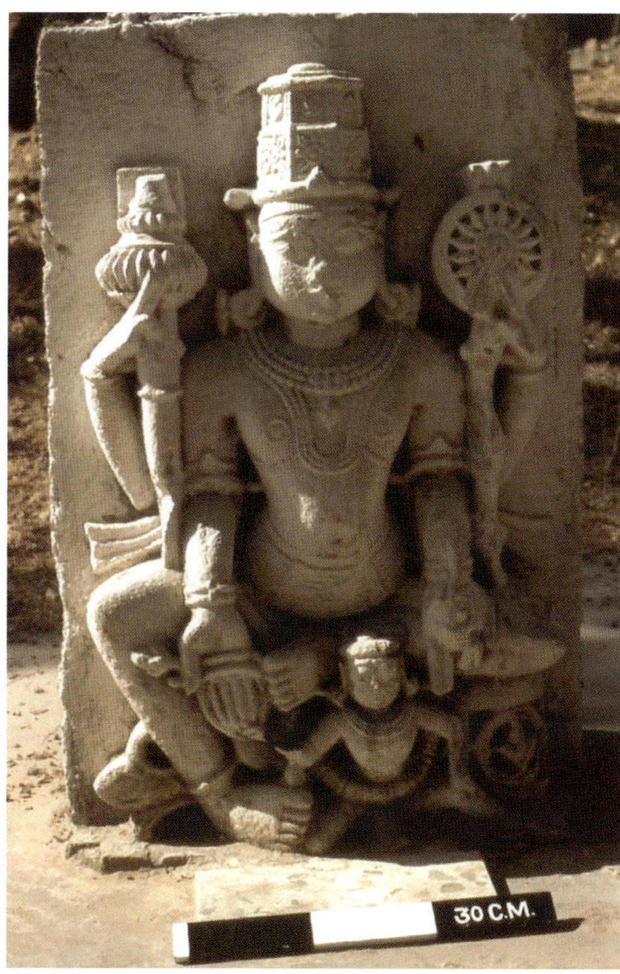

Plate 3.1 Limestone figure of Vishnu with eagle in anthropomorphic form beneath, 12th century AD. Found at Zawar and kept in front of the company guest house in 1980s.

1 ओं नमः ।। पुनातुदिनकृमरीचिविच्छुरितपादपद्मपत्रच्छविदु-
रितमाशुश्चण्डिकापादद्व
2 यं ।। हरेशिखिशिखा भकेसरस्थितमपास्तरजनूपुराभयाः च्छुरित-
दिविभाव स
3 दाः असुरोरस्थलशूलविनिर्भिनमुद्गिररुधिरनिवहं । मवालोक्य
केसरि वहृति ति
4 रश्च चापलममप्येव भयमुद्विजनिवः ।। जयति विजयी रिपूनां
देवद्विजगुरु
5 जणानन्दीः श्रीशीलादित्यो नरपति स्वकुलावर चन्द्रमा पृथ्वीः।।
जयति वट
6 नगर विनर्गतमहाजनं जेकप्रमुखं । येनास्य लोकजीवनं आगरमु
7 प्तादि[त्पादि]तंमारण्यकूपगिरौः । नानादिदेशमागत अष्टा-
दशवैतालिलेक[लिलोक]विख्यातं
8 धनधान्यहृष्टपुष्टविष्टजन नित्यसंबाधं ।। एभिर्गुणैयुतं तत्र(जेन्त)
9 कमहत्तर श्री अरण्यवासिण्या देवकुलं चक्रे महाजनादिष्ट ।।
देवी च प्रति
10 ष्टाप्यंमनुपालयतु चिरंः स च जेन्तकमहतरः आस·······
11 वस्वत[वैवस्वत]दूता समवेक्ष । देबुवक सिधायत (ने)...........
12 लनं प्रविष्ट ।। 703 ।। कति (क) f... ....

Fig. 3.1 The Samoli inscription (from Halder 1929–30).

According to tradition, Zawar was founded by the Jains in the 4th century AD and was known as Shryamangiri (literally the place of Shryman, a Jain saint) (Sharma1966: 316; Kharakwal 2011: 172). This could refer to the re-establishment of the mines after their closure at the end of the Mauryan period, as evidenced archaeologically by the small workings inside the existing galleries in Zawar Mala Magra (Magra is a hill or ridge) dated to the 7th century AD and the corresponding smelting site at Site 7, also dated to the 7th century AD, beneath the north end of the Zawar Mala ridge, adjacent to the present-day small village of New Zawar (see Chapter 6, Fig. 6.2), which is the earliest surviving settlement in the Zawar area.

Sometime towards the end of the first millennium AD, Nemisena, a follower of the Ramasenacharya religious group, is said to have spent some time at Zawar meditating in the huge system of caves (Kasaliwal 1982: 79, 146). This must be an early reference to the mine system and from their description as being caves rather than mines suggests they were not fully operational at this time. Perhaps the best known and potentially the most important inscription believed to refer to the Zawar mines is the dedicatory stone inscription to the goddess Chandika found at Samoli (Halder 1929–30; Sharma 1966). Samoli is now a tiny settlement about 50 km west of Udaipur and approximately 60 km from Zawar as the crow flies (see Chapter 2, Fig. 2.1), although there has never been a direct route of communication across the mountains. The quite small stone (about 24 × 32 cm) was inscribed in the reign of King Shiladitya with a *samvat* date of 703 (AD 646), and refers to a mining community. Because this has sometimes been assumed to be Zawar it must be discussed here in some detail (Fig. 3.1). The first four lines praise the goddess who is depicted seated on a throne. Following this is an eulogy to King Shiladitya and praise for Jek (Jaintak?), the chief of the Mahajan Sangh, the merchant group. He is credited with providing a livelihood through mining operations for the local population at a settlement referred to as Koopgiri, the town of wells or shafts (*koop*), situated in Aranya Pradesh, 'the land of forest', and for building a temple to Aranyavasini there. It should be pointed out that Aranyavasini, which literally means the goddess who lives in the forest, is an alternative name for Chandika, and there is a temple to Chandika in New Zawar. The inscription continues with remarks on *vaitalilokvikhyalatam* (those with knowledge of underground things), presumably mining engineers, who lived at Koopgiri together with healthy and prosperous labourers.

The inscription refers to a sizeable mining establishment in forested lands where the ground was already covered with shafts. It must be stressed that there is no direct association with Zawar in the inscription and there is no other reference to Zawar as Koopgiri. The presence of the temple to Chandika at New Zawar, adjacent to a ridge of hills that was already riddled with mine workings, is suggestive. However, there are other possibilities, for example there are remains of extensive early copper mines at Basantgarh, which are only about 20 km to the north west of Samoli with which it had direct communications and where there is also an old temple dedicated to Chandika.

There is little evidence for activity at Zawar between the 7th and the 14th century although it is possible that the development of commercially viable methods for the production of zinc metal from the existing production of zinc oxide took place during this time. The Polo brothers on their return journey from the Far East in 1293 passed through Cambay, now Khambhat in Gujarat (see Chapter 1, Fig. 1.1), and Marco Polo specifically mentioned *tutty* as one of the materials being imported (Latham 1972: 267). The interpretation of *tutty* is often problematic, but the only other occasion the term was used by Marco Polo was as the name of the zinc oxide produced at Kerman in Iran (Craddock *et al.* 1998c). The interpretation must be that this rather unusual commodity had previously been available in the region but by that time local production had been disrupted.

The first historic references specifically to Zawar occur in the reign of Maharana Lakha Singh (1382–1421) (Hooja 2006: 331; Tod 1978, I: 221–3). The territory in which Zawar lies is described as having been captured from the Bhils (the indigenous nomadic tribal people) and that silver and tin deposits had been discovered. Furthermore, mines had been opened and silver, tin, copper, lead and antimony were being produced (Tod 1978, I: 117, 223ff.). The 'tin' (likely to have been zinc) and copper was probably produced in the general vicinity of Zawar, but there is no evidence of antimony mineralisation so possibly lead was meant. Ojha (1928, 1931) stated that in Lakha Singh's time, silver and lead mines were discovered and substantial amounts of these metals produced, thereby increasing the state's income. Tod (1978, I: 222) stated that 'the riches of the mines of Jawura were expended to rebuild the temples and palaces levelled by Alla-o-din (the Tartar, Alauddin Khilji, Sultan of Delhi)'. It is possible that Lakha Singh established the production of zinc on a large scale and developed a major chemical industry, a truly remarkable achievement.

As a result of this industrial activity in the 14th century Zawar became a prosperous town and a number of temples were constructed (Ojha 1931, II: 260). From their dedicatory inscriptions it is clear that they were supported by a number of professional groups in the town representing a wide range of trades. For the next century or so revenue from the mines is frequently recorded in documents preserved in the Mewar State Archives[2] at Udaipur, evidence that the mines were obviously flourishing. In addition Zawar was also a sacred place

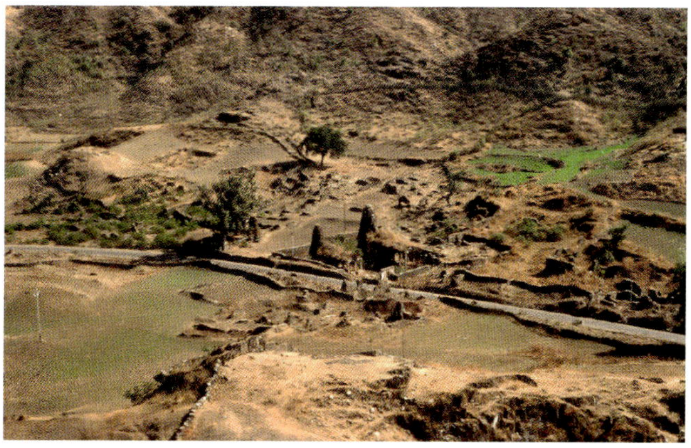

Plate 3.2 Old Zawar: temples and ruins from the watch tower (see Chapter 6, Fig. 6.3).

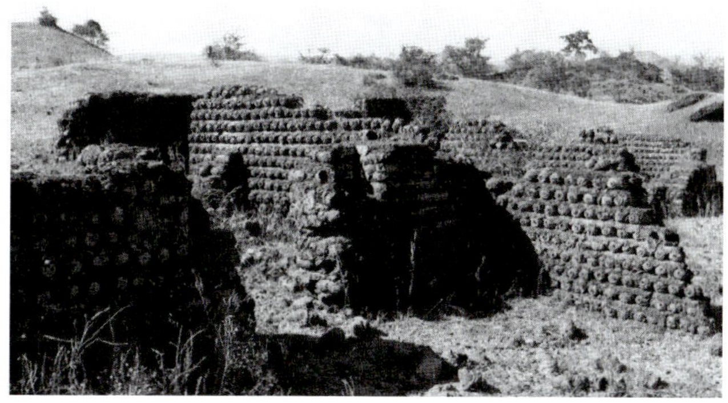

Plate 3.4 Old Zawar: ruins of buildings built of reused retorts used as hollow bricks.

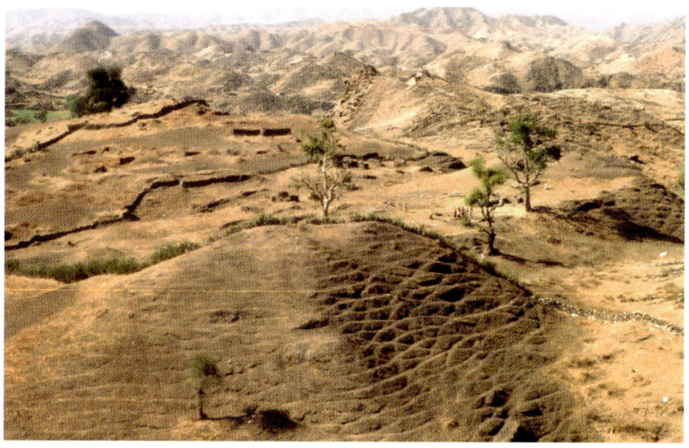

Plate 3.3 Old Zawar: major spoil heaps from the watch tower (see Chapter 6, Fig. 6.3). Site 37 is in the centre behind the major mound.

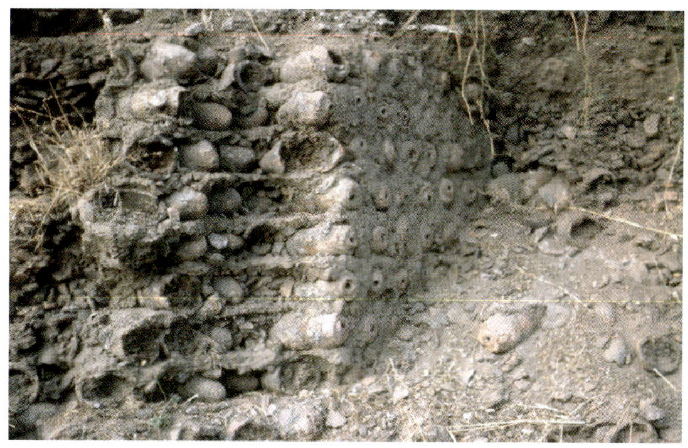

Plate 3.5 Wall showing how the retorts were laid in a double thickness with the pointed ends inwards on thick layers of mortar made from the retort debris.

of Jain pilgrimage to the various Jain *Munis* or *Yogini* who took up residence there. At this period Zawar was referred to as Yoginipattan or Yoginipur in Jain documents and this appears to be the original name of the settlement (Chakravarti 1934/5). The name Yoginipur is also cited in an alchemy text written in Zawar in 1474 (Bhatnagar 1979–80: 129). There is a mid-15th-century inscription at Kumbhalgarh dated 1460 and an inscribed figure of Ganesa from Zawar, now in the Udaipur Museum, dated 1523, both of which refer to Zawar as Yoginipur (Agrawala 1958: 221–2). The inscription on the Ramanath temple at Zawar refers to it as Yoginipattan (Sharma 1972: 156).

This historically recorded activity is matched archaeologically by the settlement ruins and the huge heaps of metallurgical debris in the vicinity of Old Zawar (Plates 3.2 and 3.3), and the opencast mines running for kilometres along the hill ridges (see Chapter 5, Plates 5.39 and 5.46, and Chapter 6, Fig. 6.2). The settlement up to this time had probably been located mainly at New Zawar on the north bank of the River Tiri (also known as the Tidi), centred on the temple there. This was joined by the now-abandoned settlement at Old Zawar, with the familiar walls built of retorts used to distil the zinc reused as a form of hollow brick (Plate 3.4). The retorts were normally laid in double thickness with their flatter ends outwards and the pointed ends toward the centre embedded in a thick mortar made of material from the ore roasting heaps (Freestone *et al.* 1985a) (Plate 3.5).

The settlement's most prosperous period was probably the late 14th to the early 16th century judging by the dated inscriptions on the temples which serve it and from the industrial activity excavated as part of this project on the edge of the settlement. As Kumar (forthcoming) points out, it is possible that some of the existing structures incorporate earlier structures destroyed or damaged in the Tartar raids. Some major structures were constructed at this time including the great dam across the Tiri (Plates 3.6 and 3.7), which must have provided a more assured water supply for what was very likely an increased population. Additionally, the silts deposited in the ponded water behind the dam could have served as a source of the clay needed in substantial quantities for the refractories. A major tank was constructed adjacent to the Ramanath temple consecrated in 1497.

Increased activity and wealth inevitably brought problems of security, especially in this area of Mewar, and a great fort was constructed on the Harn Magra ridge above Old Zawar (Kharakwal 2011: 85) (Plate 3.8). This major structure could have been built after the troubled times of Maharana Mokal (the son of Maharana Lakha), when the surrounding states of Bundi, Sirohi and Nagaur attacked Mewar and briefly occupied the area. Mokal defeated Sultan Feroz of Nagaur near Zawar in 1421 (Hooja 2006: 335–6), but was assassinated in 1433 and in the ensuing confusion the Maharawal of Dungerpur Gaipa (Gopinath) temporarily annexed the southern part of Mewar

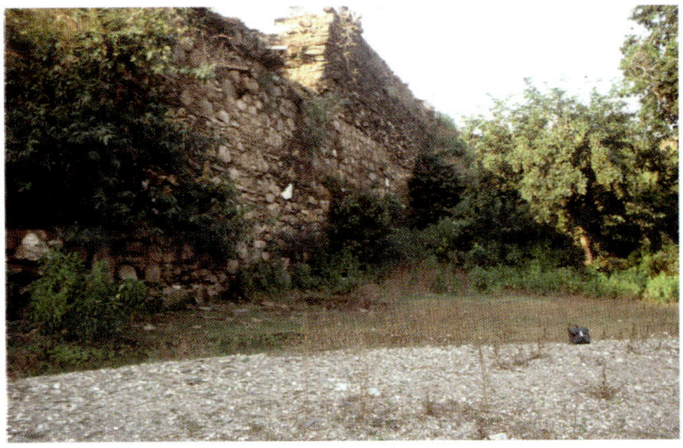

Plate 3.6 Downstream wall of the great dam across the River Tiri (see Chapter 6, Fig. 6.2).

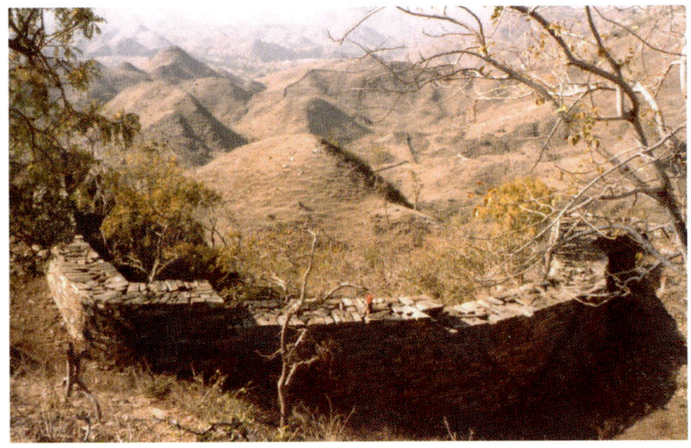

Plate 3.8 Harn Fort: massive wall across a ravine. Note the figure in red acting as a scale.

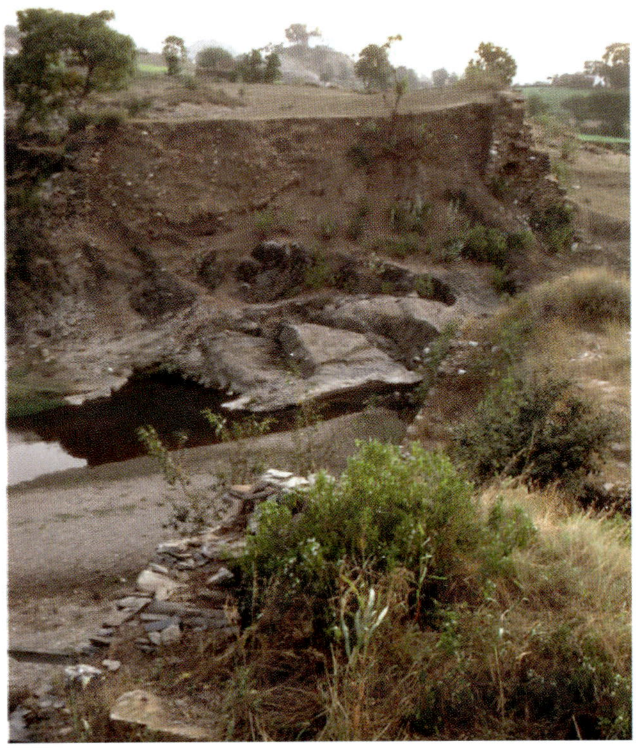

Plate 3.7 After the River Tiri breached the dam it created a section through the great thickness of the earth fill between the stone walls to the front and rear. The very small retorts were found in this earth fill of the dam and thus predate it.

including Zawar. Mokal was succeeded by his son, Kumbha (1433–1468) who recaptured the southern portion of Mewar, including Zawar, in 1460.

Kumar (forthcoming) estimates that altogether there are about 20 temples in the Zawar area, and in the Tiri Valley alone, between the village of Tiri and Old Zawar, there are ten temples, all of which seem to substantially date from this period to judge from their dedicatory inscriptions (Plate 3.9). These were studied by the Chief Superintendent of Archaeology and Museums at Jaipur and summarised in Straczek and Srikantan (1967: 48) as follows:

> Some of the ruined Temples, both Jain and Brahmanical, contain inscriptions which are mostly engraved on the lintels over the door frames of sanctums and record the names of the donors and their family members who constructed them, together with the names of their religious preceptors under whose guidance the construction took place. There are in all about a dozen such inscriptions which also bear the inaugural dates with the relevant details. These dates range between the 15th and 17th centuries of the Vikrama era [57 years ahead of the Christian Calendar] showing that this site was in full flourish during these centuries when a number of temples, though almost plain and inornate were constructed by the Jain and Brahmanical residents of the place.

One might take issue with the description of the temples as being 'almost plain and inornate', but the date range is correct. The earliest inscription is on the Ambamata temple recording a donation by Maharana Lakha in 1404. At the Parsnath temple there is an inscription of *samvat* 1478 (AD 1421) stating that Dhanpal, a leader of the Mahajana Sangh (the Merchant Group), constructed the temple after he had been on a pilgrimage to holy sites around western India, and the consecration was performed by Guru Samodeva, a follower of Acharya Devsundersuri of the Tapagachcha sect of the Jains (Sharma 1966). The Suparasvanath temple has an inscription dating from *samvat* 1489 (AD 1432) recording that the temple was built by Shah Padam Oswal and that the consecration ceremony was performed by Shri Jin Sagar Suri of the Khartargachcha sect of Jainism (Plate 3.9).

Finally, on the most important temple of all, the Ramanath temple (Plate 3.10), dedicated to Lord Vishnu, and the accompanying tank, an inscription records that the consecration ceremony took place in *samvat* 1554, *Chaitra Sudi 7 Ravia* (Sunday 12 March 1497). The temple and tank were constructed by Ramabai, daughter of Khumba and sister of Rana Raimal, ruler of Mewar. She had been married to Mandlik of Girmar, at Saurashtra in Gujarat, but after differences between them she had retired to Zawar, granted to her by Rana Raimal where, according to the inscription, she lived for many years (Agrawala 1958). This is of some importance, suggesting that Zawar was not just a thriving mining town, but also a place where the sister of a monarch could consider making her permanent abode. This was probably in the Harn fort, which as well as being a major military structure was also of some considerable architectural and artistic pretension.

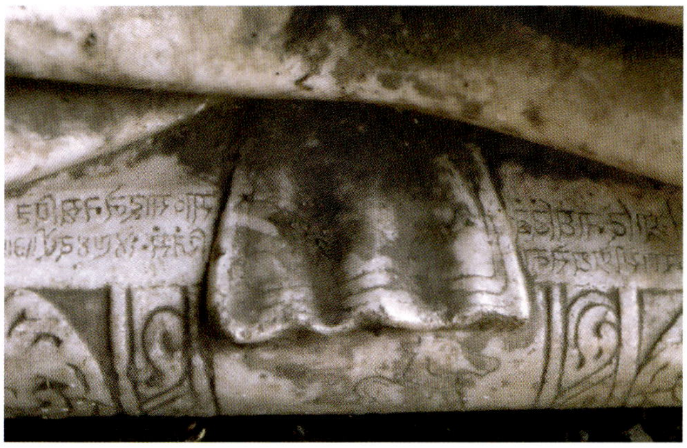

Plate 3.9 Dedicatory inscription in Jain script beneath two statues at the Suparshwanath temple between Old Zawar and New Zawar. The inscriptions read that the images of the deities were the meritious donations of the *sadhu* Sādula, the beloved son of Lalatā Devī, on the 5th day of the bright fortnight of the *phālguna* month of the *samvat* year 1494 (AD 1437).

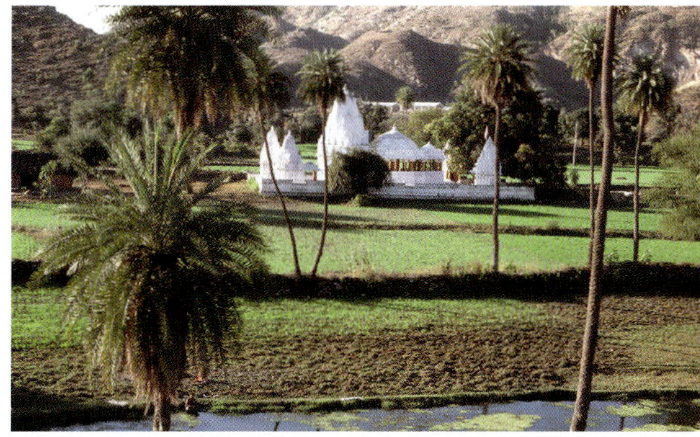

Plate 3.10 The Ramanath temple from Site 5.

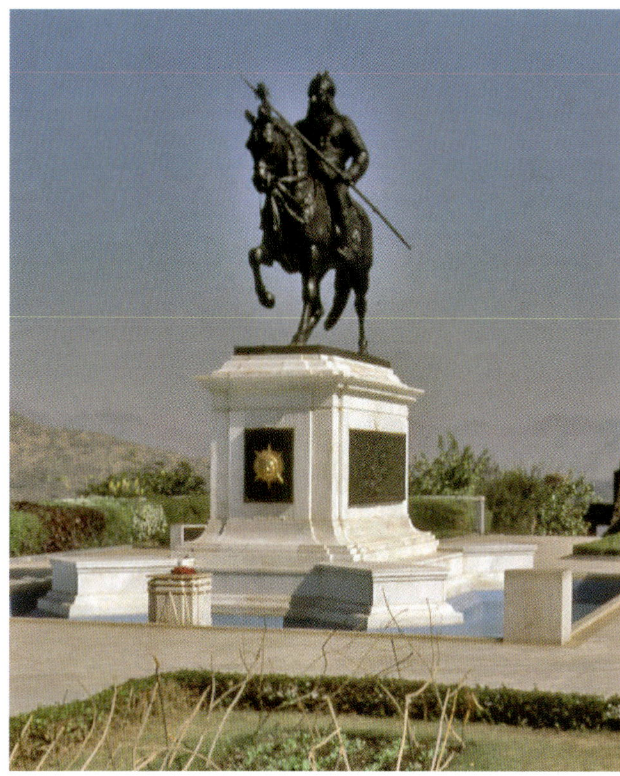

Plate 3.11 Bronze equestrian statue of Pratap, Rajasthan's King Alfred, Udaipur.

Things had been a little disturbed in southern Mewar in the 15th century, but were set to become very disrupted indeed for the whole of Rajputana in the later 16th century with the rise of the Mughals. It is not appropriate to chart the progressive domination of the Hindu Rajput states by the Mughals throughout the 16th century here except to record some incidents in the final stages when the subterranean galleries of the mines of Zawar Mala were the scene of some of the most brave and romantic episodes of Rajput history, as the secret base for the continued resistance of Pratap, the Maharana of Mewar, and the great hero of Rajasthan (Plate 3.11).

Rana Raimal's successor, Rana Singh, popularly known as Maharana Sanga of Mewar, was defeated by the Mughal Emperor Babar in 1527, and thereafter, while the rest of the Hindu states submitted, Mewar, under the successive maharanas, Udaisingh (1540–1572) and Pratap (1572–1597), continued to defy the burgeoning Mughal Empire. Finally, in 1576, Emperor Akbar, determined to end Mewar's independence, attacked with overwhelming strength and rapidly occupied the whole country. Rana Pratap, however, refused to submit and guerrilla warfare continued for many years with some actions taking place at Zawar itself (Hooja 2006: 472).

According to tradition, Pratap retreated to the hills with his queen, young family and a few loyal supporters, and quite literally went underground into the mines of Zawar Mala to bide his time and attempt to rally support while being hunted by the Mughal forces (Tod 1978, I: 272). The stories of their privations have been recounted many times, together with the legends of the hideout in Zawar Mala. Burgess, in an anonymous article in the *Indian Antiquary*, visited this place and provided a romantic description (Anon. 1872: 63, quoted below).

Today, the villagers on Zawar Mala still refer to one of the ancient mines as Pratap Khan and another opposite as Rani Khan. The former was almost certainly Pratap's headquarters during his long sojourn as a fugitive (see Chapter 5, Fig. 5.2 and Plate 5.7). Burgess and others, notably the previously mentioned Captain J.C. Brooke (1850) (see Chapter 2), seem to have assumed that the mine workings themselves were contemporary with, and the result of, Pratap's visit, even suggesting that he might have indulged in some mining on the side himself. Our investigations have shown that in fact he was occupying long-abandoned ancient workings, with no evidence of later reworking for ore. At Pratap Khan there is a relatively broad flight of steps of carefully cut stones, otherwise unparalleled in the mines, which descend to what is still referred to as the Audience Chamber (see Chapter 5, Plate 5.7). This is in an ancient worked-out stope that forms a great circular domed chamber and in this a substantial stone platform had been constructed which could not have served a mining function. This could have been the hideout for some of the hostile Bhil nomads, but it seems too grand a conception for this and it is more likely that it is indeed the base of Pratap's last resistance.

The coin hoard found in 1984 buried in an abandoned large retort furnace that was part of the group that also contained our Site 37 at Old Zawar is evidence of the troubled times (Plates 3.12 and 3.13). The hoard consisted of 29 freshly struck silver

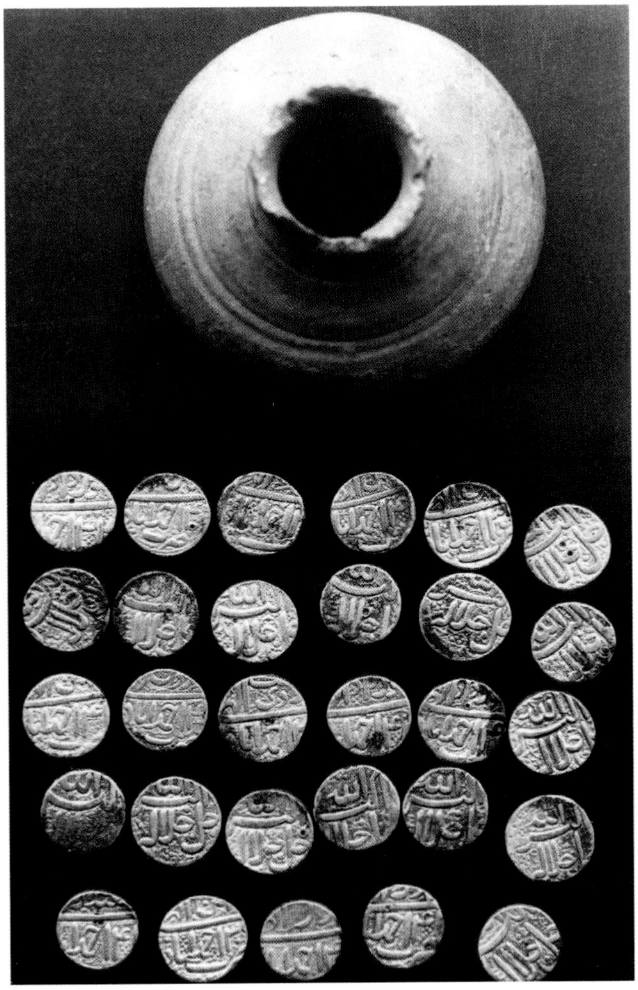

Plate 3.12 Hoard of silver coins and the pot in which they were contained, found hidden in a large retort furnace near to Site 37 (photo: HZL).

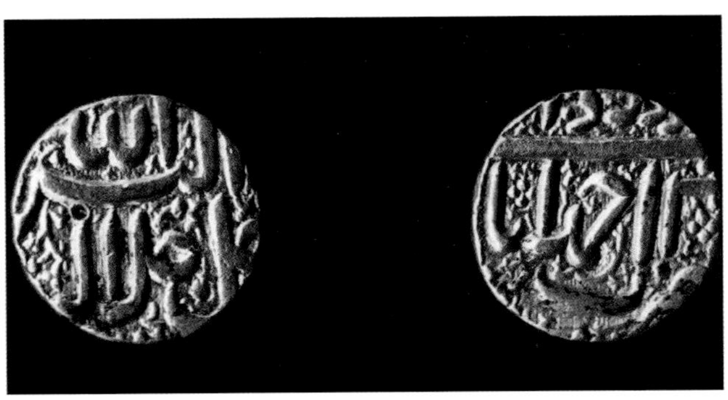

Plate 3.13 Two of the coins shown in Plate 3.12, minted at Ahmadabad in AD 1593 (photo: HZL).

coins of Akbar, minted at Ahmadabad in Gujarat between the 40th and 43rd years of his reign (1596–98). It is possible that these coins could well have been hidden shortly afterwards during the unrest caused by the hit-and-run skirmishes between Pratap's guerrillas and the Mughal army at about the time of Pratap's death in 1597.

Despite his efforts, Mewar finally succumbed to the Mughals and Pratap's son, Amar Singh, signed a treaty with Jahangir in 1615, which brought peace and tranquillity to the region but effectively ended Mewar's independence for a century or more. According to the terms of the treaty the Mughals respected Mewar's sovereignty, but Mughal coins became the legal tender in the state, Mewar ceasing to mint its own coins, and Mewar was forbidden to repair or reinforce the damaged walls of the great fort at Chittor (Ojha 1931). During the struggles with the Mughals, trade and industry were severely disrupted so production at Zawar was probably at the very least severely curtailed – certainly during this period importation of zinc from China commenced, and once established, remained the principal source of zinc in India (Chapter 13).

Abū 'LFaẓl Allāmī makes two references to Zawar in his monumental survey of the Mughal Empire, the *Ā-īn-i-Akbarī* (Phillott 1927) completed in 1596. In the first (I: 41–2) in a section on the various metals he states that:

> *Jast* which according to the opinion of some is *Rūḥ-i-tūtiyā* (that is, made from or originating from *tūtiyā*, zinc oxide), and, resembles lead, is nowhere found in philosophical books, but there is a mine of it in Hindustan, in the territory of Jālor, which is a dependency of the Sūba of Ājmīr.

This is of some importance as the first specific mention of mining for zinc at Zawar, showing that some production must have continued during the difficult years. It may seem strange that the mine should be described as being subject to Ajmer, but at the time that Abū L-Faẓl was writing much of Mewar was ruled by the Mughals from Ajmer.

The second reference (II: 273), dealing with the Sūba of Ājmīr states: 'In the village of Jāwar, one of the dependencies of Chainpur is a zinc mine.' This seems to be the first use of the current name for the site (Jawar continues to be used locally in both the written and spoken form alongside Zawar to this day). The report of the Chief Superintendent of Archaeology and Museums, Jaipur, on the history of Zawar (quoted in Straczek and Srikantan 1967: 48) considered that:

> The name Jawar was evidently derived from Japura found in a few earlier inscriptions. In an inscription dated *samvat* 1492 (AD 1435) it was called Japurapura where an extra pura is superfluously added at the end of the Prakit form of Japura apparently in ignorance of this original derivation. In later inscriptions, however, the name Yoginipura is regularly found although it is doubtful whether the old name applied to altogether a new habitation. At any rate, the name Yoginipura was put out of use in course of time, while the original name still clings to the place.

Alternatively the two different traditional settlements, New Zawar and Old Zawar, could be being referred to separately.

Munhot Nais, the first minister (*deevan*) through much of the second half of the 17th century of Jaswant Singh, Maharaja of Marwar, the state lying to the west of Mewar, with its capital at Jodhpur (see Chapter 1, Fig. 1.1), compiled a history of his own times, the *Khayat*. This included substantial comments on the neighbouring state of Mewar (*Munhot Nainse-Ri-Khyat*, Pt. 1 and II, MS. Nagri Pracharmi Sabba, 1925, *prachya vidhya sansthan*, City Palace, Udaipur). He stated that at Zawar there was a silver mine, the daily income from which was between 400 and 500 rupees, and that both silver and zinc were produced (Ojha 1931, II: 1123). It is probable that at this time the mines at Baroi Magra (see Chapter 5) were opened principally for the exploitation of argentiferous lead, as exemplified by the smelting sites at Site 5 (see Chapter 6). It is also recorded that from the 1640s Maharana Jagat Singh was weighed annually and had his weight in silver distributed to the poor, suggesting that silver was not in short supply (Hooja 2006: 615).

These references can be supplemented with some accounts preserved in the State Archives at Udaipur where, for example, it is recorded that in *samvat* 1713 for the *Bhadva* month (August and September 1656) an income of 170,967 rupees was made (Bakhikhana Bahi 92, Bhadva) (Bhati 1995 a, b). Clearly, despite the political vicissitudes of the late 16th and early 17th century, Zawar seems to have been able to continue some production of both zinc and silver, albeit on a reduced scale, as also evidenced by the physical remains of metal production (Kharakwal 2011: 210). There are also some fine architectural monuments of this period, such as the elegant pavilion in New Zawar, which suggest some return to prosperity (Plate 3.14).

Following the death of Aurangazeb in 1707 the Mughal Empire went into rapid decline, culminating in the sack of Delhi by Nadir Shah in 1739, and Mewar was able to regain its independence. This included the right to issue its own coins in 1713, and the production of lead, zinc and silver continued at Zawar. It is probably to this period that the small fort near New Zawar belongs (Plate 3.15) (see also Chapter 6, Fig. 6.2).

The next substantial historical account is to be found in James Tod's monumental *The Annals and History of Rajputana*. After many years spent surveying Mewar and studying the history and antiquities, in 1817 Tod was appointed political agent to the Western Rajput States, specifically to the Maharana of Mewar, following the Dosti London (Friendship with London) treaty. He was based in Udaipur, and for the next five years carried out an intensive survey of the surrounding country including Zawar (Head 1985; Tillotson 2007). In 1822 he retired to England to compile *The Annals*, the first volume appearing in 1829, the second in 1832 (republished 1978), shortly before his death in 1835. Some of the historical and architectural dating is rather suspect, but the strength of the work is in the local knowledge based on his fieldwork, and he was also able describe at first hand the recovery of Mewar after the decades of Maratha depredations (described below) culminating in the catastrophic drought and famine of 1812.

The late 18th century saw increasing confusion in western India with the collapse of Mughal authority, the rise of the Marathas and the rivalry between the European powers, notably Britain and France. Rajputana managed to avoid the European powers at this stage but was devastated by the Marathas. These bands of warriors, originally from central India, attacked the Rajput states with some success from 1735 on, exacting huge tributes known as *chouth*. For example, in the short reign of Maharana Raj Singh II (1754–61) almost 11 million rupees were paid out. For over half a century their ever-increasing depredations brought about the cessation of law and order and of commerce, and for long periods government itself ceased. Into this general lawlessness another group of raiding horsemen, known as the Pindaris, appeared in the early 19th century and actually occupied Udaipur in 1816 (Hooja 2006: 774). As late as 1759 records available to Tod showed that the annual revenue from Zawar was 222,000 rupees, less than it had been during the 17th century, but still substantial (Tod 1978, I: 399; Ojha 1931: 11; Erskine 1908: 53; *Imperial Gazette* 1908: 52). After the defeat of Maharana Bheem Singh in 1788, organised industry including the mines went into rapid decline (Tod 1978, I: 352). Tod quoted these figures of 60 years previously to contrast with the abandoned mines which he visited, already filled with water. Similarly, Hamilton (1820, I: 548), describing the few mineral resources available in the Udaipur District of Mewar, does not even mention Zawar. The sufferings of Mewar were graphically recorded by Tod at first hand (1978, I: 363–4) in the first decade of the 19th century, culminating in a mass evacuation after the famine of 1812 (Hamilton 1820, I: 547–52; *Imperial Gazette* 1908; Erskine 1908).

Burgess (Anon. 1872) suggested that the rulers of Mewar may have deliberately ceased production at Zawar so as not to attract the attention of the Maratha marauders, but the real reason was the inability to protect the vulnerable operations from attack. The sophisticated technology and products of the mines required peace and stability both to continue the training and support of the necessary technical staff and to maintain the markets for the zinc. Once these had been interrupted for any length of time it would have been very difficult to re-establish them. Such events had probably taken place before, during the Mughal occupation, but then some activity was resumed, albeit substantially reduced.

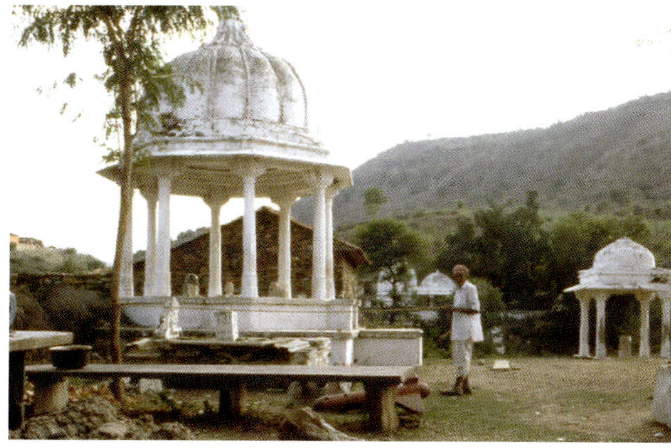

Plate 3.14 Limestone pavilion of 16th or 17th century date in the village of New Zawar.

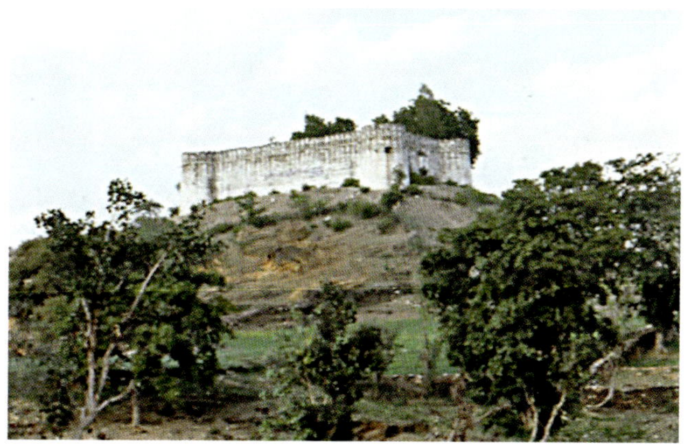

Plate 3.15 Small fort near New Zawar, probably of 18th century date (1983).

The 19th century: abandonment and first attempts at re-establishment of the mines

The British defeated both the Pindaris and the Marathas in 1817–18 and thereafter largely controlled the Mewar administration although the old spirit of Mewar independence continued to prevail: the locally minted coins never bore the name of the reigning British monarch; instead they bore the legend *Dosti Landon* (friend of London). Despite Tod's confidence that peace and prosperity would flourish after the intervention of the British, parts of Mewar were to remain effectively beyond the control of the government of the Maharana through much of the 19th century (Brooke 1859; Pinhey 1909). In particular the Bhils, who were active in the hilly country in which Zawar lies, were a constant source of trouble. The Mewar Bhil Corps was founded in 1840 drawn from the Bhils of southern Mewar, both to provide employment for some Bhils and hopefully to keep the others in check (Hooja 2006: 785). However as late as the 1880s, Bhils were involved in active armed revolt in Paduna just to the south of Zawar (Pinhey 1909: 66). Kumar (forthcoming) claims that the small fort above New Zawar was constructed in 1879 specifically to protect the area, although architecturally this seems unlikely. Given these circumstances it was not possible to reopen the mines, even though there were apparently still people who knew of the process surviving into the 1840s (Brooke 1850) (see below). The problem was compounded, as Brooke himself noted, because the market for zinc had been irretrievably lost, initially to imports from China and latterly from Europe.

Tod described the mines at Zawar as quite deserted at the time of his residence: 'the caste of miners is extinct' (1978, I: 10) and 'They (the mines) have long been abandoned, the miners are extinct, and the protecting deities of mines are unable even to get a flower placed on their shrines' (1978, I: 222). Thereafter the nomadic Bhils comprised the main population of the area. Brooke (1850: 213) stated the same in a little more detail:

> They became closed during the great famine, which devastated western India in A.D. 1812–13, during which the miners, dependent on the surrounding country for food, were obliged to leave a locality situated in the heart of the Bheel country, whose starving population seized all the grain intended for the city of Jáwar. The government of U´dypura (Udaipur), too weak to defend itself, and at the time oppressed by the Maharattas and other freebooters, failed to forward the requisite assistance, and the town suffered the same fate as that of many other places.

Local legend at Zawar attributes the end of the settlement to a dust cloud that poisoned all the land. Given the very high heavy metal content, notably cadmium, in some of the heaps of debris (Aery and Tiagi 1988), it is possible that dust clouds from these, borne by the hot dry winds during the drought, could have wrought yet more havoc to the parched farmland nearby. Despite these troubles, through the 19th and first half of the 20th century there were sporadic surveys and suggestions that perhaps the mines could be reopened (Straczek and Srikantan 1967: 53–4). As early as 1829 James Hardie, an army surgeon, reported that:

> There is a lead mine at the village of Jawar, twelve miles south of Udaipur which occurs in a series of alternations of quartz rock and argillaceous schist and it is said that a considerable proportion of silver was found combined with this metal which was worked at one time to considerable advantage. Were this country properly explored, I have no doubt but that many rich mines might be discovered.

In the 1840s Brooke visited the abandoned mines and gave the following description:

> Jáwar lies half way between Kherwárá (Khairwara) and U´dypura (Udaipur), or some 25 miles due south of the latter place. It is situated in an irregular valley surrounded by hills rising to a height of 1000 to 1500 feet, clothed in rich verdure to their summits, and overlooking an irregular-shaped plain covered over for a space of 5 or 6 square miles with monuments of former wealth and importance. Many of these ruins consist of ancient buildings and temples on hills rising in several instances to the height of upwards of a hundred feet, and composed entirely of ashes (the Old Zawar retort heaps), which alone fully attest the distant period from which these mines must have been worked. The small river Thirí flows through the plain. It has been in one place bunded up with a masonry bund now in ruins (Plates 3.6 and 3.7), the excellence of which is attested by the age of the temples built hundreds of years ago on the alluvium of the lake that must have formerly formed an extensive and lovely sheet of water. The chief hill which was used for mining has been worked into a mass of excavations, from which myriad of bats and a stray bear now and then serve to startle the incautious intruder (Brooke 1850).

The temples are not actually built on the slag heaps but some have been partially buried by retort debris, that is, the slag heaps postdate the temples not the other way round. Similarly, none of the temples at Zawar upstream from the bund or dam are actually built on the alluvium of its lake. Fortunately at the time of our explorations the bears had long since gone, but the bats were still very much in evidence in the old workings!

Brooke attempted to interest Maharana Sarup Singh (1842–61) in reopening the mines. He was no doubt aware of the lead mines that were still operating at this time near Ajmer and described by Captain C.G. Dixon (see Chapter 2). Indeed Brooke contacted Dixon who asked for ore samples to assess the quality of the deposit, but there were problems:

> It is some 6 years since I first visited Jáwar, and it then seemed to me to induce work-people to re-open the mines, but I was unable to procure specimens of the ores from the jealousy of the then Minister of U´dypura, as to my intention in bringing these hidden treasures to light. On a subsequent occasion about 2 years ago, I proposed to the present Máhá Ráná of U´dypura, the expediency of opening the mines; he appeared very anxious to do so, and authorized me to obtain miners from Ajmere. Having addressed Lieut.-Col. Dixon on the subject, that officer with great kindness entered fully into my views and pointed out to me the little use of commencing mining operations at all, unless perfectly sure of the value and richness of the ores to be worked. He mentioned the

> various descriptions of ore which were most likely to be found, and gave me several useful hints on the subject.

Mindful of Dixon's caution, Brooke obtained his samples from the last miner, who also claimed to know the secrets of the zinc distillation process:

> The Ráná gave me to understand, that the previous year he had directed the old man to smelt a little ore which he had done, and brought to U´dypura (Udaipur), but afraid of losing his influence, should he divulge the secret of preparing the ore for the furnace, he had refused to give information as to the manner of working it, – the Ráná thereon tried to burn a little himself, but all his crucibles broke.
>
> A few days afterwards on my visiting Jáwar, the old man came and at my request brought a basket of ore in its rough state. He described it as being found in veins three or four inches (8 to 10 cm) thick and sometimes in bunches, in quartz rock and mixed with other stone. He broke some pieces with a hammer, showing me the good ore (later identified as being sphalerite) and the simple way it was freed from the quartz rock, with which it was mixed. The pure ore being very friable is then pounded and freed from quartz and placed in crucibles some 8 or 9 inches (20 to 23 cm) high and 3 inches (8 cm) in diameter; with necks 6 inches (15 cm) long and half an inch (1.5 cm) in diameter. The mouth being fastened up, the crucibles were inverted and placed in rows on a charcoal furnace when the ore was fused in about 3 or 4 hours. If pieces of the quartz are allowed to remain with the ore the crucibles of course break, and hence the cause of the Ráná's failure. From each crucible the quantity of metal does not vary very much, a specimen of that extracted for the Ráná from one crucible is sent, as are also remains of ancient crucibles found among the ruins.
>
> I could not discover whether any flux was used in the smelting, or whether the crucibles are entirely or only partially filled previous to the operation.

The description is devoid of detail, for example there was no mention of the essential preliminary roasting, and the smelt was not very successful. Brooke sent specimens of ore and the product of the smelt to James Prinseps of the Asiatic Society of Bengal, who noted that 'the specimens are very small, and apparently contain very little metal'. The statement that the inclusion of quartz in the charge would cause the crucibles to rupture is hard to understand. The description that the mouth of the crucibles was fastened up presumably refers to the attachment of the condensing necks to the crucibles. If the crucible mouth really was sealed then this would have caused a rupture as soon as it was heated. The only interesting point is that the operation lasted for three to four hours.

This failure before the Maharana cannot have improved the chances of the venture being supported. Brooke concluded his article rather forlornly but accurately that:

> It is very probable that with the immense import of all kinds of metals from England, the working of these ancient mines may not now be profitable, still the subject is itself interesting, as bringing to notice another part of India in which mineral riches abound.

Burgess (Anon. 1872) emphasised the sense of abandonment and decay in the rather high flown account he gave for the *The Indian Antiquary*. First he described the temples before moving on to a description of the mines:

> At the south corner of the valley stands a temple dedicated to a Devi, a goddess [so her votaries say] of wonderful power; but some are found to own that they have their doubts as to her being present in this particular shrine in these days of degeneracy and dissipation. The high priest is there, than whom it would be difficult to find one of more clerical cut. But he, with doleful countenance, because of glory departed, will assure you, in trembling accents, that the goddess has withdrawn her patronage from the caste of miners, hence they have forgotten their business. Be this true or not, the miners no longer exist – probably enough because they fled when the war-cry of the Maratha invader re-echoed through the land.

After describing some of the temples and noting the houses built of old retorts, he continued:

> But let us to the mines, which may yet be again laid under contribution by the rulers of Mewar. Giant heaps of refuse point the road to the principal of them. The path leads through the thickest and most beautiful jungle, over rivulets, through defiles, through thickets which electrify the botanist, and dense jungle which afford the huntsman true earnest that royal game is well within reach of his rifle. The mines themselves are mere clefts in the solid rock – apparently natural, yet doubtless superficial. In some places, attempts at regular work appear to have been made; but as a rule, the miners were content with boring only far enough into the hill to obtain ore without trouble, never venturing where natural solidity of the stone would not suffice for the support of the rocks above. Veins of almost pure lead ramify through the primitive rock; whilst beautifully coloured ores of the mineral sparkle overhead. Silver is obtained in small quantities, whilst gold has been found, it is said, on several occasions. A very short time sufficed to find numerous valuable specimens; no doubt more extended research would have shown that these mines could be worked with profit. A hundred years ago, in 1760, Jowara (Zawar) alone poured two lakhs (200,000) of rupees into the Maharána's treasury. About the beginning of this century an attempt was made to re-work these mines; but it failed – it may have been from an imperfect knowledge of the science of mining, or from the fear that increase of wealth would attract the spoiler from the plains (i.e. the Marathas). Such a feeling still exists. When these mines were visited with a view to obtaining some little material for this letter, they had not been visited by any native for five or six years, although the hill is almost entirely perforated. There the tiger had its lair, and the panther its habitation.

From the description of the open clefts it would seem that these were the opencasts on Mochia, Balaria and Baroi Magras that are being described. The reference to a small quantity of silver being worked is interesting and there are persistent and continuing rumours of gold in the mines. One of the hills is called Sonaria, literally hill of gold, but there is no record of gold ever having been produced in the Zawar mines. The description continued:

> A steep and rugged path winds up a great hill, overlooking the Temple of Kali, crosses a wall of great thickness (the Harn Magra fort, Plate 3.8), and finally is lost in a plain which is surrounded by a girdle of other hills, wherein are dug the mines (Baroi and Zawar Mala).

A description of Pratap's hiding place on Zawar Mala follows:

> In these caves there were preserved, for generations, the rings and bolts to which the cradles of Pratáp Singh's children had been attached; and even now the inhabitants of the village on this elevated plain speak of their being there still. Mines and shafts are innumerable. There are traces of steps, walls and outlets for water which shows that more attention had been paid to the working – probably because the ore was not so abundant. Perhaps Pratáp himself opened the mines when in exile here. However that may be, the extensive ruins show that he must have had a large host with him.

As already noted, our investigation of the large chamber, still known as Pratáp's Chamber, suggests that it was adapted from existing ancient workings and that no new workings were undertaken at that time.

Maharana Shambhu Singh (1861–74) was interested in the mines and employed Professor Bushell in 1873 to investigate the possibility of reopening them. He actually built an access road to the mines and cleared some of the galleries, but soon realised that the ore had all been removed from the upper levels and he lacked the means to drain the lower levels where he quite correctly predicted untouched orebodies could be located. To dewater the mines would have been very expensive, and the analysis of ore samples found that the silver content was about 10 oz to the ton (about 300 ppm), which was judged to be low. As it was the potential silver content of the ores that was the main stimulus, rather than the zinc, the project was abandoned after 10 months and the expenditure of 15,000 rupees (as reported in the *Rajputana Administration Report*, Mewar Agency, 1873–4: 137).

Very probably as a result of Bushell's activities, in the annual statistical report issued by the Maharana's office to the political agent for 1875, a somewhat negative assessment was made of the mines' prospects (MS Udaipur State Archives). Given the unsettled conditions prevailing in the area at this time, this is perhaps not surprising. It was stated that in the valley of Zawar where zinc, lead, copper, iron and sulphur had been formerly produced there were between 5 and 10 mines, but all had been defunct for some time, and there was no advantage to be gained in reopening them. They were described as being full of water. In the hills there were plenty of mines and one could travel for up to a mile in them. The report continued that in olden times the metal was smelted in crucibles and that there were big heaps of them. At Zawar there was also green-coloured building stone and at Tiri a marble quarry. At Samat (now Hameta) Magra there was an ochre mine.

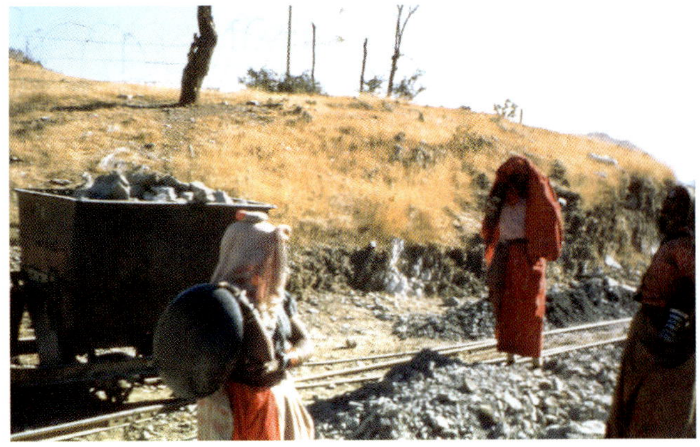

Plate 3.16 Women sorting galena ore at Mochia Magra in 1954 (photo: H.W.A. Sommerlatte).

During the reign of Sajjan Singh (1874–84) the previously mentioned geologist, C.A. Hacket, who had also investigated the copper mines at Khetri, examined the deposit and was reported to have found that 'the ore worked had not been in continuous veins, but undetached pockets or hollows near the surface in the contortions of limestone or other rocks'. This suggested that the deposits in the ancient mines were likely to have been thoroughly worked out and that exploration should concentrate on locating previously undiscovered near-surface pockets of ore in the adjacent hills. Hacket's 1880 report went on to recommend offering a substantial reward to the Bhils for the discovery of a workable deposit (*Rajputana Administration Report*, Mewar Residency, 1881–2: 153–4). However, nothing seems to have happened apart from the occasional survey such as that of Heron (Anon. 1930) of Mewar's mineral resources in the 1920s.

The 20th century: war and the re-establishment of the mines

In 1932 a mining department was established at Udaipur specifically to develop the exploitation of Mewar's mineral resources. In 1938 the Mewar Mineral Corporation of Mohatas of Udaipur, led by R.K. Mehta, was formed and in 1940 a prospecting licence was issued. They carried out some exploration at Baroi Magra and attempted to open some of the old mines but without success (File no. 312, Basta No. 13, Rajasthan State Archives, Bikaner).

The Bawdwin mines in Burma had been the principal source of zinc in the Indian Empire (Coggin Brown and Dey 1955: 156–61) but the capture of Burma by the Japanese in 1942 changed everything, creating an urgent need for new sources of zinc ore. The GSI recommended a fresh exploration of the Zawar mines, and in 1942 acquired the lease from the Mewar Mineral Company. Between 1942 and 1945, with the help of mining engineers and miners evacuated from the Mawchi mine in Burma, and Ghurkas recruited from Nepal, seven adits were driven into Mochia Magra, together with drives, cross-cuts and winzes, totalling some 3100 m, but viable deposits of zinc were encountered in only one of them (Crookshank 1947; C.V. Thornton 1945, *Final Report on Mining at Mochia Magra*, Rajasthan State Archive, Bikaner). Following the end of the war, there was no longer an urgent need for zinc and the mines were

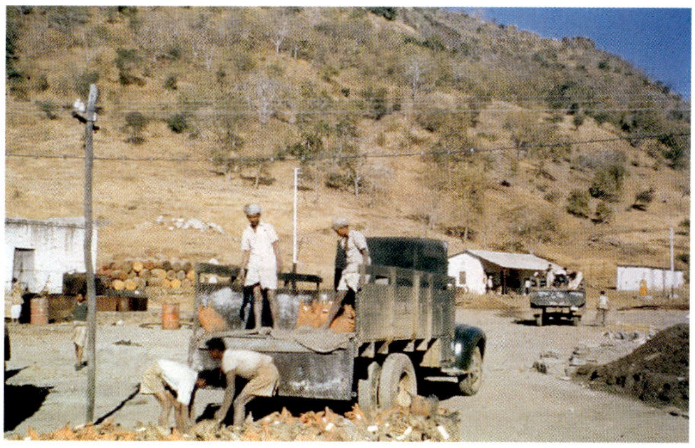

Plate 3.17 Loading gunny sacks of lead ore onto a lorry for transportation to the smelter in Bihar in 1954 (photo: H.W.A. Sommerlatte).

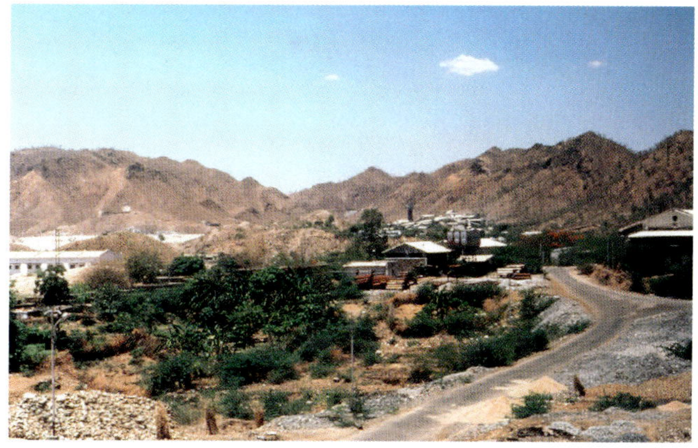

Plate 3.18 The mines at Zawar in 1954: note the headstock for the Mochia mine in the background. Compare with the view (Chapter 5, Plate 5.32) taken 30 years later (photo: H.W.A. Sommerlatte).

handed back to the Mewar government (Crookshank 1947; G.N. Dutta, *Progress Report 1944–5*, Rajasthan State Archives, Bikaner).

In 1945 the area was taken over by the Metal Corporation of India Ltd. (MCI) that had itself been established in 1942 and which operated a lead smelter at Tundoo near Dhanbad in Bihar (now Jharkhand) (Fig. 3.2). This initially worked exclusively recycling scrap lead, but was later converted into a primary smelter to treat the lead concentrate from Zawar, many hundreds of kilometres distant, with a licensed capacity of 3600 tonnes per annum. At this time there was no plant in India to smelt the zinc concentrate which instead was sent to Japan for toll smelting.

Initially MCI was content to stope out the lead ore that the previous exploratory workings in Mochia Magra had encountered. Only about 1 tonne per day was output from the current mining operations – the rest was from the accumulated stocks (Crookshank 1947). This was hand-sorted by local women (Plate 3.16) and about 3.5 tons per day sent out in gunny sacks by lorry to the nearest railhead at Udaipur (Plate 3.17). The present metre gauge railway that links Ahmenabad in Gujarat with Udaipur and actually passes through Zawar was only constructed in the 1960s and was never envisaged as a means of transporting ore. Similarly the other HZL mines in the Aravallis have never utilised the local rail system.

Crookshank also reported that a beneficiation plant with a target output of 175 tonnes per day was being constructed and would shortly become operational. Although the mine had so far been worked by imported labour, it was his opinion that the local labour 'is above average intelligence and hard working'. He did, however, foresee that local labour would be insufficient to work the mines once expansion began. MCI invested about 30 *lakh* of rupees (3,000,000) between 1945 and 1947 with a further 50 *lakh* in 1952 (Gurjar *et al.* 2001). As a result, a proper mine with a permanent settlement began to be developed by the company (Plate 3.18). The capacity was subsequently enhanced to 500 tonnes per day by 1958, which was the capacity in 1966 when HZL was formed. Further exploration continued under the direction of Kerr-Cross (1949 and 1952), and work on the beneficiation of the ore was carried out by Fleming *et al.* (1949).[3]

In the early 1960s MCI embarked on an ambitious scheme to expand the mining and milling capacity at Zawar to 2000 tonnes of lead-zinc ore per day. At the same time as the mine expansion programme, the company in its efforts to create indigenous zinc smelting capacity began to construct an electrolytic smelter at Debari, a village just to the east of Udaipur. This was planned to have a production capacity of 18,000 tonnes zinc per annum. However, due to limited resources MCI ran into acute financial difficulties and by 1964 all work came to a halt.

The creation of Hindustan Zinc Limited (HZL)

To focus attention on the production and development of non-ferrous metals a new Department of Mines and Metals was established within the Ministry of Steel and Mines of the Government of India in 1963 with N.C. Shrivastava as the first Secretary. At this time the scenario was quite depressing; following independence there was a desire to make India self-sufficient in as many materials as possible, but instead almost the entire requirement of non-ferrous metals was met through imports while domestic ore resources remained largely undeveloped. MCI, a private company, clearly did not have the resources to carry out the smooth operation of the mines and complete the planned expansion work therefore it was considered appropriate that its operations should be nationalised. Accordingly on 22 October 1965 MCI was nationalised and HZL came into existence as a public sector company (Shrivastava 1991).

Straczek and Srikantan (1967) provide a good overview of the operations at Zawar just prior to the formation of HZL, and Anon. (1983c) outlines the first 15 years of HZL's activities at Zawar. In 1970 HZL began development work at Rajpura-Dariba and in 1979 exploration commenced at Rampura-Agucha. A brief description of the early history of these two sites is apposite before continuing with the story of HZL.

The Rajpura-Dariba mines

There seems to have been very little archaeological or historical research within the vicinity of these mines despite the obvious scale of the early mining operations in the first millennium BC (Jain and Rakesh 1984). Neither of the present villages of Dariba or Rajpura, in the Rajsamand District of Rajasthan, is of any great age. The shrine to the Incarnation of Shiva at the north end of South Lode is reputed to be of 12th century date, and

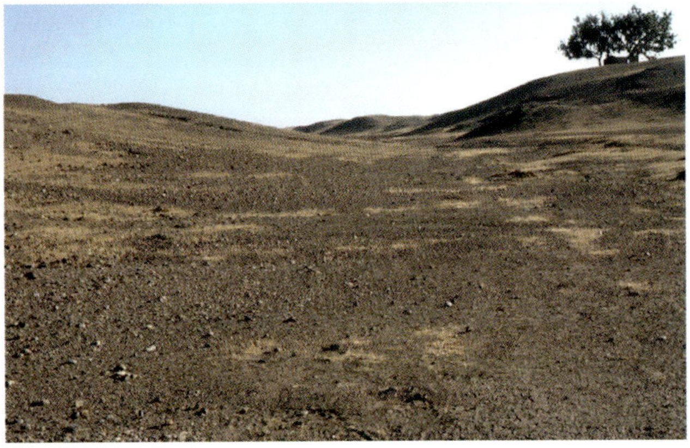

Plate 3.19 The severely eroded opencast workings at Agucha in 1985 just before the modern development of the mine.

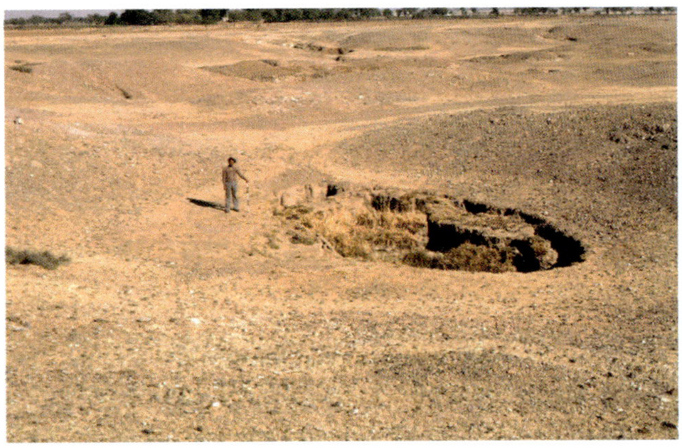

Plate 3.20 Ancient eroded slag tips and heaps of mine waste at Agucha in1985 with a typical cave-in caused by the collapse of mine workings beneath.

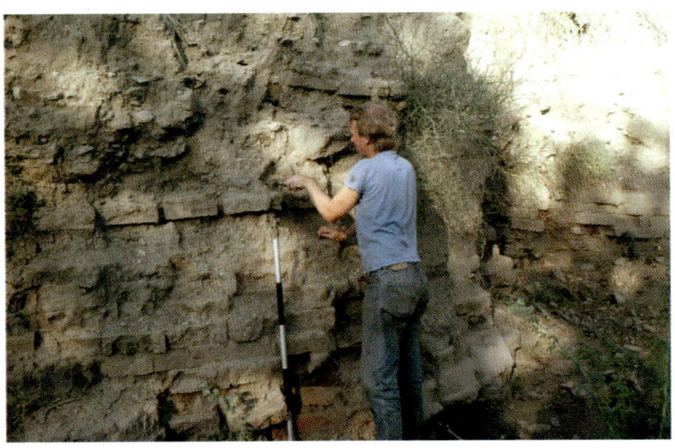

Plate 3.21 Exposed section in Agucha village revealing walls of large early bricks.

some distinctive large rectangular bricks of probable Mauryan age or of the succeeding Kushan or Gupta empires of the first millennium AD have been incorporated in its walls. The mounds that form Site 8 in the ancient metal smelting areas (see Chapter 6, Plate 6.53) contain walls built of mud bricks. As discussed above, it is very likely that at their period of maximum activity in the late first millennium BC, the mines here and at Agucha (and probably at Zawar as well) were worked directly by the Mauryan state.

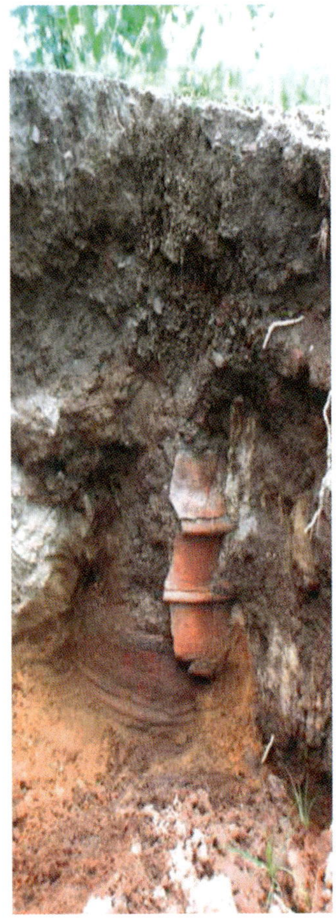

Plate 3.22 Large ceramic soakaway drains (from unpublished report, *Resource Identification at Agucha and Assessment through Interpretation of Ancient Mining*, by K.S. Choudhary).

There is abundant evidence of post-medieval smelting activity as evidenced by Sites 13, 14 and 15, where apparently both lead and copper were smelted. There is a brief mention of 'Dureba' in Tod (1978, I: 618) to the effect that the mines were still producing copper and 'tin' at the time of his visit in 1821. There is no tin mineralisation and no evidence that zinc was extracted before the present. In an unpublished statistical report compiled by the Maharana's office to the political agent of Mewar for *samvat* 1932 (AD 1875), it states that the lead and copper mines at Dariba had been closed for some time. Note that none of the 19th-century accounts mention the extraction of silver at Dariba.

The deposits at Dariba were described for the first time in a GSI report (Gupta 1934). Test drilling by the GSI in the 1960s confirmed the size and importance of the lead-zinc deposit and in 1970 HZL acquired the site and commenced development. The present mine was commissioned and started production in 1984 (see below).

The Rampura-Agucha mines

In contrast to both Zawar and Dariba, there is no evidence of metallurgical activity in the recent past although the highly eroded remains of ancient activity abound (Plates 3.19 and 3.20) (Gandhi 1983; Gandhi *et al.* 1984; Tiwari and Kavdia 1984). The present village of Agucha itself is clearly of great age (Nagarch and Trivedi 1981). The settlement and its fort sit on a mound and there are two others nearby that could well represent the accumulated debris of previous occupation as digging in the town has revealed walls and other occupation debris down to a considerable depth. The local inhabitants reported that large distinctive bricks had been recovered from a depth of 20 feet in the village. The Agucha mound itself is the largest at about 1200 × 800 m rising to a maximum height of 15 m. On the north east side of the village is the Gardar Ka-Bargh mound (about 400 × 200 m with a maximum height of 7 m), while the Rampura mound, which lies just to the east of the present Rampura village, is around 600 × 150 m with a maximum height of 3 m. In these mounds brick walls are also exposed (Plate 3.21). The bricks – the large, smooth, rectangular type typical of the Mauryan, Kushan and Gupta periods – are between 30 and 55 cm long by 25–30 cm wide, and between 6 and 9 cm thick. In addition large quantities of pottery and many square and round punch-marked coins and later issues have been found over the years.

Most of the temples and mosques in Agucha are recent, but to the south near the River Mansi there are many ancient sculpture fragments, one dated *samvat* 1000 (AD 1057), and more early sculpture can be found near to the Raneswar Mahadev on the banks of the Mansi, about 1.5 km west of Agucha.

Metallurgical and occupation material is frequently found at depths of up to 10 m where exposed in the sides of well shafts all over the area around Agucha, especially to the east, very similar to the situation at Dariba, suggesting that the original ancient occupation is likely to have been extensive. During construction of the present mine, several soakaway drains were found made from sections of ceramic cylinders about 40–50 cm in diameter placed one on top of another in the ground to a depth of about a metre, suggestive of a substantial and organised settlement (Plate 3.22). The archaeological investigations of Nagarch and Trivedi (1981) in the Agucha mound identified three main periods. Period I was pre-Mauryan and Mauryan of about the 3rd century BC, with plain red, black and red, and black slipped wares, as well as Painted Grey and Northern Black Polished ceramics in a wide range of shapes and types. Two punch-marked silver coins and various terracotta pieces were recovered including beads, a mother goddess, a male figure and animals. Slag pieces of similar composition to those in the mines area were also found, suggesting that the settlement was involved in the working of the mine and the production of silver/lead (Tiwari and Kavdia 1984: table 1). Period II was assigned to the Sunga and Kushan periods (approximately from the 2nd century BC to the 3rd century AD) and was characterised by a brick structure measuring 5.3 m wide with walls surviving to a height of 1.6 m. The finds included various terracotta beads, figures (both human and animal) and a stone plaque of Vishnu. The third period was the Gupta phase (from the 3rd to the 6th century AD), and a brick floor was uncovered with more terracotta beads, figures and a head of a Saiva ascetic.

Agucha continued to be occupied throughout the medieval period up to the present, as revealed by the limited excavations of Nagarch and Trivedi (1978/79), even though mining had long ceased. The fort and the tank are clearly post-medieval in their present forms, but both are likely to be much earlier in origin. Agucha was once the most important settlement in what was latterly the north eastern part of Mewar on the border with the neighbouring Rajput state of Shapura. In the past, control of Agucha changed hands many times therefore a strong fort would be essential for whoever held it. The fort provided local sanctuary during the Maratha wars in the late 18th and early 19th century. It is about 110 m long by 85 m wide and was once surrounded by a wet moat around 10 m wide. The present buildings in the fort comprise a small multi-chambered residence, a well, stables, a mosque and a temple dedicated to Lakshminarayana. The platform before the fort is made of huge slabs of granite, which are also incorporated into the walls of the tank. The tank was rebuilt and partly realigned in the early 20th century, but is clearly of ancient origin if for no other reason than to stop monsoon rains from flooding the mines lying immediately to the south east.

The present settlement at Rampura was only established in the mid-20th century as an offshoot of Agucha. In fact, up until the recent past, Agucha seems to have been the only settlement for some considerable distance around. The current mining activity is very recent but the high grade of the ore (1.9% lead and 13.54% zinc) has ensured the rapid development of the deposit (Gandhi 1983). The deposit was only rediscovered in 1977 during an inspection of a nearby garnet pit. The visiting geologist noted the gossan outcrops and slag spreads, and recognised that the shallow valley was the remains of major ancient mining activity (Plate 3.19). Drilling began almost immediately, the deposit explored and the mine developed during the 1980s. The mine was commissioned and began production in 1991 (Agrawal 1999, and see below).

The development of HZL

This is a story of rapid development and expansion, not only in Rajasthan but throughout India.

Mining and milling

Soon after incorporation the company continued the task of completing the remaining construction work for the expansion of the mining and milling capacity at Zawar, and the international mining companies Krabs and de Penarroya, which had been working with MCI, continued their involvement (Bansal and Ahluwalia 1990; Ravindranath 1991). By 1972–73 the

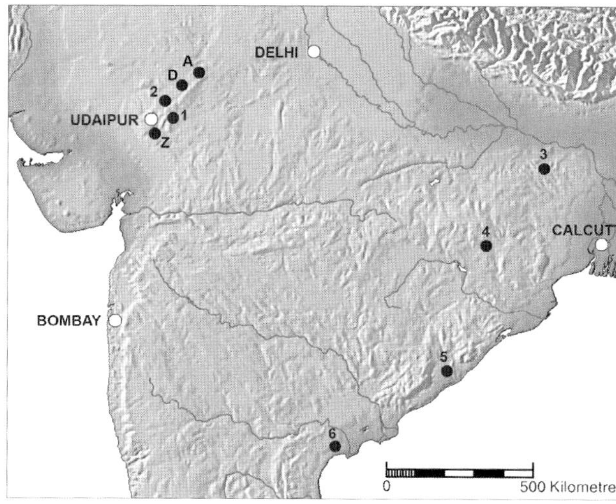

Fig. 3.2 Places linked with the development of HZL. 1 Chanderiya (Chittaurgarh); 2 Debari (Udaipur); 3 Tundoo; 4 Sargipali; 5 Visakhapatnam; 6 Agnigundala.

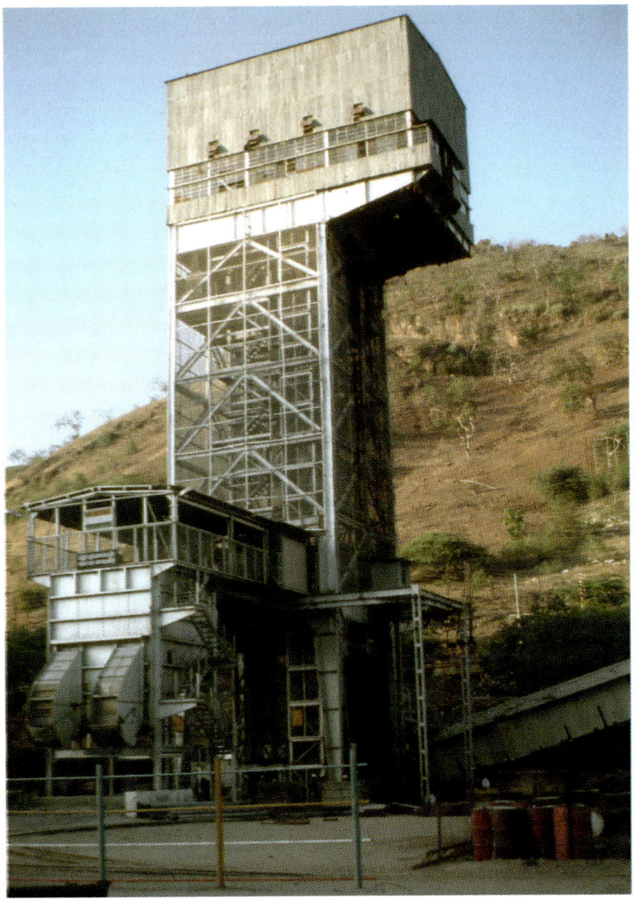

Plate 3.23 Mochia headstock constructed in 1966.

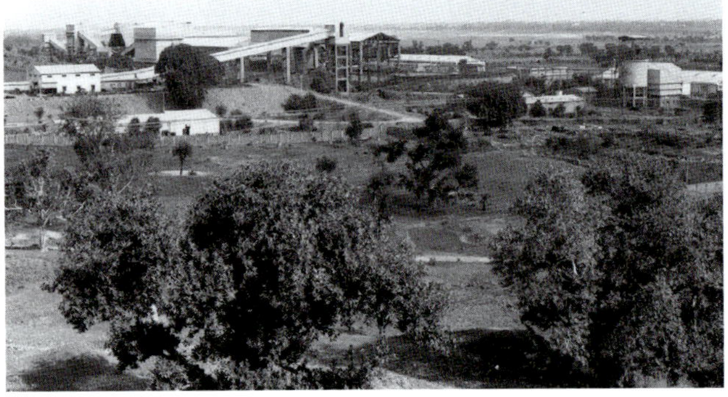

Plate 3.24 Main shaft and beneficiation plant at South Lode, Dariba, in1985.

production capacity of Mochia mine had increased to 2000 tonnes of ore per day with matching beneficiation facilities (Plate 3.23). The mining and milling capacities of the company were further increased to 4000 tonnes of ore per day in 1978 when ore production at the newly developed Balaria mine, with a capacity of 2000 tonnes of ore per day, was commissioned (Anon. 1989; Jakhu 1990). In the same year the Agnigundala lead mine, situated in the Guntur District of Andhra Pradesh (Fig. 3.2), with a capacity of 120 tonnes of ore per day, was acquired from Hindustan Copper Limited. Its capacity was subsequently raised to 240 tonnes of ore per day.

The mining and milling capacity of the company saw a quantum jump in 1984 when the newly developed mines at Dariba, with an ore production of 3000 tonnes of per day capacity (Plate 3.24), and at Sargipali, in the Sundergarh District of Orissa (Fig. 3.2), with an ore production of 500 tonnes per day capacity, were commissioned. These measures increased the mining and milling capacity from 500 tonnes per day in 1966 to 7740 tonnes per day in 1984.

Exploration and development of the lead-zinc deposit at Agucha began in 1979 (Gandhi 1983) enabling the company to plan for an integrated project of mines and smelters which has gone a long way to meet the national demand. The Agucha deposit was explored and developed, and in the year 1990–91 a 3000 tonnes of ore per day capacity mine with matching beneficiation capacity was commissioned, thereby enhancing the company's overall mining and milling capacity by 39% to 10,740 tonnes per day. The production capacity of the Agucha mine was further increased to 4500 tonnes of ore per day in 1997–98 (Agrawal 1999).

Smelting capacity

As already stated, at the time of the incorporation of HZL a lead smelter was in operation at Tundoo in Bihar (Fig. 3.2), while the zinc concentrate was toll smelted in Japan (Parvathisem 1990; Mukherjee *et al.* 1991). This was clearly unsatisfactory and expansion of the Tundoo smelter was completed in two phases – to 6000 tonnes and finally to 8000 tonnes of lead metal per annum – together with modernisation of the existing plant. In the early 1980s another lead smelter was commissioned at Visakhapatnam in Andhra Pradesh (Fig. 3.2), with a capacity of 22,000 tonnes, thereby increasing the company's total capacity to 30,000 tonnes of lead metal per annum.

The electrolytic zinc smelting plant at Debari was commissioned in 1968 (Fig. 3.2). Capacity was expanded from 18,000 tonnes to 45,000 tonnes per annum in the late 1970s to treat the increased supply of zinc concentrate from Zawar resulting from the development of the Balaria mine (Parvathisem 1990). In the early 1970s the company started construction of another electrolytic zinc smelter of 30,000 tonnes capacity at Visakhapatnam, Andhra Pradesh (Fig. 3.2), based on imported zinc concentrate. It was commissioned in 1977 thereby increasing the company's zinc smelting capacity to 75,000 tonnes per annum, a fourfold increase since 1968. In 1984 the assimilation of a leach residue plant further enhanced the capacity of the Debari plant to 49,000 tonnes per annum from 45,000 tonnes, thereby expanding the company's total capacity from 75,000 tonnes to 79,000 tonnes per annum (Parvathisem 1990).

To treat the concentrate produced at Agucha, construction of an integrated lead-zinc smelter based on the Imperial Smelting Process technology (Morgan 1985, 1998) was built at Chanderiya in the Chittaurgarh District of Rajasthan (Fig. 3.2). The plant was commissioned in 1991 (Mukherjee *et al.* 1991) to produce 70,000 tonnes of zinc and 35,000 tonnes of lead metal per annum, and development work continued through the 1990s (Gupta 2000). The zinc smelting capacity was further enhanced to 169,000 tonnes per annum in 2001–2002 by the expansion of the production capacity of the Debari and Visakhapatnam smelters (Jakhu *et al.* 2001).

Currently HZL is India's leading producer of lead, zinc and silver. In India, galvanising accounts for about 75% of zinc usage, followed by about 10% in die castings, 10% in dry batteries with the remainder in alloys, mainly brass, and chemicals, some pharmaceutical, including the traditional calamine lotion. HZL has come a long way and grown into a multi-unit, multi-product company with a mining capacity of 4.2 million tonnes of ore and a smelting capacity in the region of 500,000 tonnes of zinc and 60,000 tonnes of lead per annum. These latest developments in zinc smelting technology now operate close to the vicinity where the whole concept of high temperature distillation and condensation technique was invented and developed many centuries ago at Zawar.

Notes

1. In the preparation of this account the authors have used the works of Tod (1978), Sharma (1966), Hooja (2006) for Rajasthan in general and Straczek and Srikantan (1967) for Zawar. The authors are grateful to Arvind Kumar for allowing us to consult his *History of Zawar* before publication.
2. Mewar was the traditional state in Rajputana within which Zawar was situated and survived until independence in 1947.
3. The zinc and lead minerals were separated by flotation but the silver content of the ores was then considered too low to be economically recovered; subsequently it was recovered from the lead.

Chapter 4 Geography and geology

L.K. Gurjar and P.T. Craddock

Introduction

The mines discussed in this section lie in the Aravalli Hills of north west India. These hills run as a grid approximately north east–south west for about 700 km from Palanpur in the north east of Gujarat to Delhi, through most of Rajasthan, forming the most prominent topographical feature of that state (see Chapter 5, Plates 5.1 and 5.42). Due to repeated tectonic folding, the hills are arranged in a series of parallel ridges from Palanpur to Udaipur, often in the form of a grid, and in places the ridges branch off in multiple disorderly ranges, as at Zawar (see Chapter 6, Figs 6.1 and 6.2). The range attains a maximum width of over 100 km in this region. In historic times these ranges formed a considerable barrier to communications as well as providing a refuge from attack, factors that have had a significant effect on the history of Mewar, the historic state in southern Rajasthan through which the hills run. As the hills continue to the north beyond Udaipur they become steadily less prominent in the landscape, fading away into the dry lands of north eastern Rajasthan but still occasionally breaking through the surrounding plains as prominent ridges, almost to Delhi. This general statement is well exemplified by the three mines that are the subject of this report. At Zawar the hills form a series of more or less continuous ridges rising hundreds of metres above the valley bottoms. At Dariba, lying somewhat to the east of the main Aravalli range, only a few isolated hills are prominent, such as Dariba Hill, above South Lode, which rises to over 100 m above the surrounding plain (see Chapter 6, Plate 6.48), although the ridge can still be followed as a line of mined mineralisation almost at the level of the surrounding undulating countryside (see Chapter 5, Plate 5.47). At Agucha, the northernmost of the three mines investigated, located to the east of main Aravalli range, the rocks do not rise above the surface at all, but the ancient opencast, following the strike of the mineralisation, still runs along the same north east–south west orientation (see Chapter 3, Plate 3.19). The copper-bearing hills in the Singhana-Khetri region are the north-eastern outliers of the chain and are quite prominent (see Chapter 2, Plate 2.2).

Geography

The hills provide a striking contrast to the semi-desert lands to the north and west. Although the climate is quite dry, broken only by the Arabian Gulf monsoon in the summer months, agriculture can be practised by the judicious use of tanks and wells to conserve and supply water. The farming is typically mixed arable, growing grains, oil plants and vegetables on quite small plots (Plate 4.1). Cattle and buffalo are kept in the valleys and the slopes are grazed by sheep and goats. In addition to the permanent farms there is a considerable seasonal population of the semi-nomadic Bhils with large herds of sheep and goats (Plate 4.2).

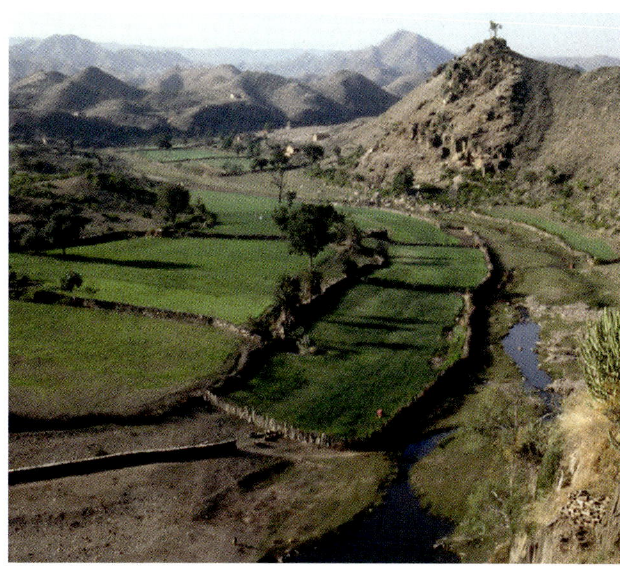

Plate 4.1 The River Tiri flowing beneath Old Zawar surrounded by small arable plots (1983).

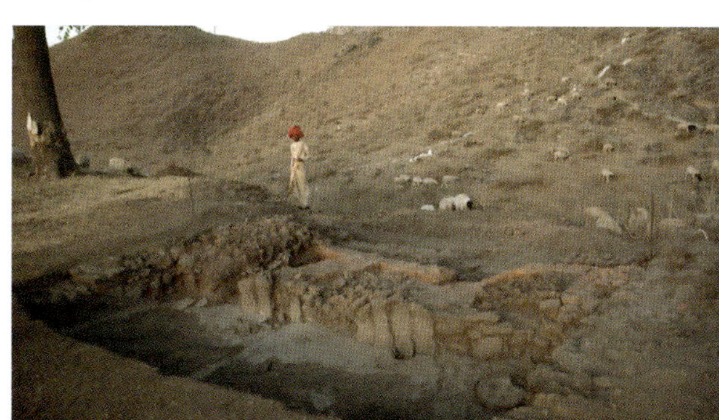

Plate 4.2 A Bhil shepherd leading some of his flock through the retort heaps at Site 37, Old Zawar (1983).

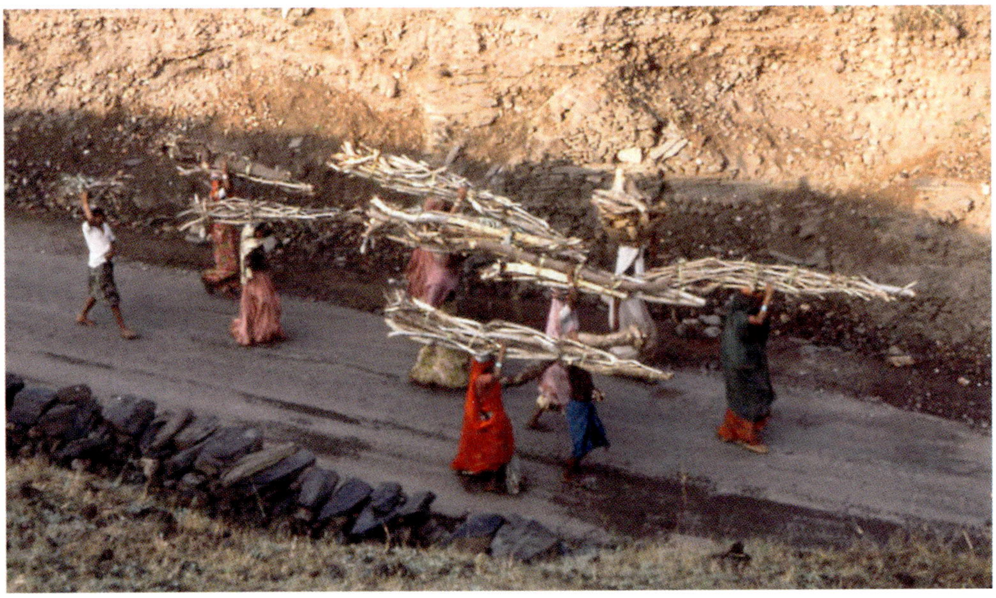

Plate 4.3 Women carrying timber cut for firewood from the hill slopes at Zawar (winter 1985).

The hills are generally extensively wooded, although these have been progressively depleted in the vicinity of settlements by felling for firewood and charcoal. In the past, the mining and smelting activities must have used very large amounts of timber and charcoal thus it is likely that managed woodlands existed on the slopes (see Chapter 8). The mines at Zawar were abandoned in the early 19th century and by the middle of the century jungle had become established. In the description of the mines that appeared in *The Indian Antiquary* in 1872 (quoted in Chapter 3), the area is described rather picturesquely as being covered in jungle, the haunt of tigers, the Bhils and their cattle. The area was still heavily wooded in the 1950s when mining recommenced, but the hill slopes have been steadily denuded over the years, mainly for domestic firewood and charcoal (Plate 4.3). More remote areas, such as Kaya some kilometres north of Zawar (see Chapter 6) are still heavily wooded although unofficial felling for lumber, firewood and charcoal in the reserved forest are taking their toll (see Chapter 9, Plate 9.1).

Regional geology

The geological framework of the Aravalli range is based on the pioneering work of Heron (1953), who identified three major orogenic cycles now represented by the rocks of the Bhilwara, Aravalli and Delhi Supergroups of Archaen to Proterozoic age (Bhatnagar and Mathur 1989) (Figs 4.1–4.3). The majority of commercially viable deposits of lead, zinc and copper occur in the Proterozoic mobile belts of Rajasthan and Gujarat (Fig. 4.2), which comprise the rocks of the Bhilwara, Aravalli and Delhi Supergroups.

The rocks of the Bhilwara Supergroup (the banded gneissic complex of Heron 1953, hereafter referred to as the BGC) is a major unit in the region and is predominantly gneissic, with a few enclaves of less metamorphosed sedimentary rocks; it also contains a few granitic rocks considered to be equivalent to pre-Aravalli, post-Aravalli and post-Delhi in age. The BGC contains basaltic amphibolites and granitic/tonalitic gneisses that are apparently of Archaen age (~ 3500 Ma) (Macdougall *et al.* 1983) therefore it undoubtedly constitutes in part a basement to the Aravalli sequence.

A predominantly argillaceous sequence with some carbonates, quartzites and basic volcanic, which lies unconformably over the BGC, is classified as the Aravalli Supergroup of rocks. A definite unconformity has been recognised at certain locations around Udaipur and further south east near Jhamarkotada (Roy *et al.* 1980; Roy and Paliwal 1981) between the rocks of the BGC and the Aravalli Supergroup. The basal quartzites and conglomerates are overlain by a predominantly carbonate unit containing phosphate-bearing stromalites which are observed around the Jhamarkotada phosphorite deposit to the south east of Udaipur. Carbonaceous phyllites containing low-grade uranium mineralisation are associated with the carbonates. These are followed upward by a thick sequence of greywacke and phyllite, the western fringe of which is marked by a linear zone of serpentinites and amphibolites, known as the Rakhbdev lineament. West of this lineament is a monotonous sequence of low-grade metapelites with slices of quartzites and serpentinites which comprise the Jharol group. The Aravalli rocks thin rapidly towards the north and their exposure gives way to those of carbonate facies of rocks that are distributed within an oval area around Lake Rajsamand to the west of Jaipur. The latter rocks were considered to overlie the Aravalli succession and were termed Raialo by Heron (1953) (Gupta *et al.* 1980, 1997). However, the stratigraphic status of the Raialo rocks remains controversial.

The rocks of the Delhi Supergroup form the rib of the Aravalli mountain range and are divided into two groups (Heron 1953): the Alwar group of quartzites, mica schists, basal grits and conglomerates constitute the lower part while the rocks of the overlying Ajabgarh group, which are predominantly calcareous and pelatic in nature, form the upper part. Amphibolites locally retaining the pillow structures are found extensively within the Ajabgarh group, indicating contemporaneous volcanism (Deb *et al.* 1989).

The granites occurring along the core of the Arivalli chain and to the west, in the Mt. Abu area, are considered as intrusive into the Delhi Supergroup of rocks, and are referred as Erinpura

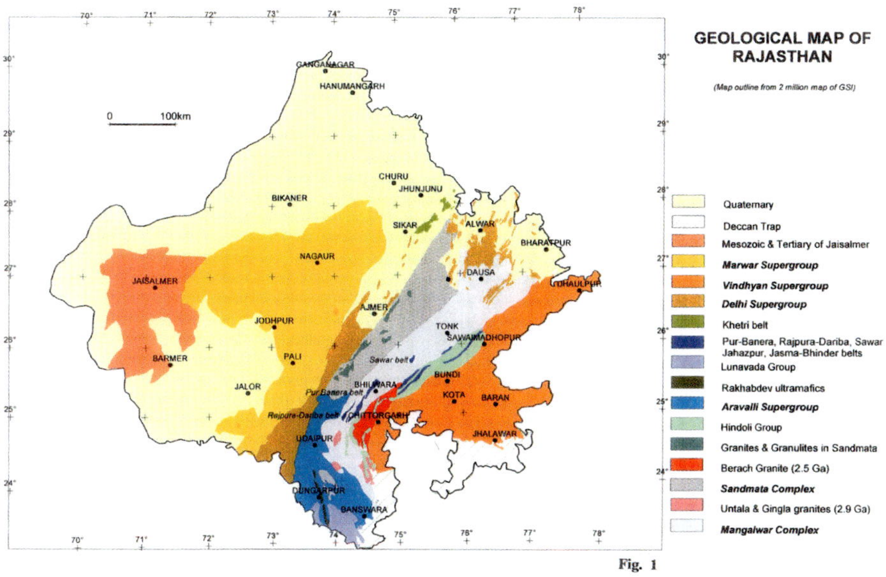

Fig. 4.1 Surface geology of Rajasthan (*HindZinc Tech* 2000: 7, fig. 4).

granite. The rhyolite flows and minor granites occurring in western Rajasthan are considered younger than the Erinpura granite and are termed Malani rhyolites.

The host rocks of the metalliferous ores are sandy limestones and dolomites. At Zawar they are largely unaltered but at Dariba and Agucha they have metamorphosed to calc-silicates. Extensive gossanisation of the deposits has also occurred at the latter sites. The ores there are located in both the calc-silicates and in the immediately adjacent complex graphite schists.

Mineral resources

The Aravalli Hills have traditionally been one of main sources of metals in South Asia (Roy 1958; Gandhi 2000) (Fig. 4.2). More than 80% of the known base metal (lead, zinc and copper) deposits and 95% of the known lead-zinc resources of India are located in the Aravallis. The prominent deposits include: Rampura-Agucha, Rajpura-Dariba and several others in the Pur-Banera belt (Bhilwara Supergroup), Zawar (Aravalli Supergroup), the Deri-Ambaji belt, the Ghugara-Kayar belt and the Khetri belt (Delhi Supergroup). The northern parts of the Aravalli Hills tend to be rich in copper with minor or negligible zinc and lead contents. The Khetri-Singhana mines have been one of the providers of the main sources of copper in South Asia for many centuries (Chapter 2). In contrast, the central parts of the Aravallis are rich in lead and zinc but poor in copper (Zawar, Dariba, Agucha, etc.). The southern parts are rich in lead, zinc and copper (Ambaji, Deri, Basantgarh, etc.). Copper, silver and lead seem to have been mined and smelted on a considerable scale at Ambaji for at least the last thousand years. In addition, good quality iron ores are abundant throughout including the hematite schists in the central Aravallis. There are also smaller occurrences of rich magnetite ores near Amargarh in the Jahajpur Hills which run parallel to the main Aravalli chain. These ores once supplied a specialist iron smelting industry at the nearby town of Bigod.

The mining of mica was important throughout much of the 20th century from the pegmatites in the central Aravallis, particularly in the Ajmer-Bhilwara, Jaipur and Udaipur districts

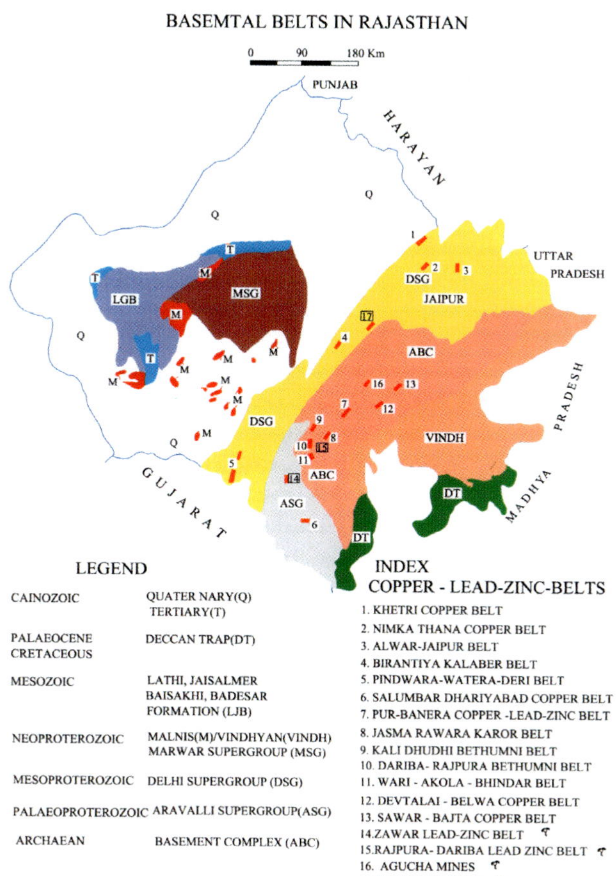

Fig. 4.2 Base metal belts in Rajasthan (*HindZinc Tech* 1989: 18, fig. 5).

(Mull Rajgarhia 1951: 6–8; Coggin Brown and Dey 1955: 543–5; Krishnaswamy 1988: 363), although production has declined latterly. Apatite and rock phosphates occur in economically viable quantities in Rajasthan, both in the Jaisalmer area and around Udaipur. Rock phosphate was mined by HZL at Maton, 3–4 km east of Kanpur, near to Udaipur (Krishnaswamy 1988: 394–400; Roy *et al.* 1980).

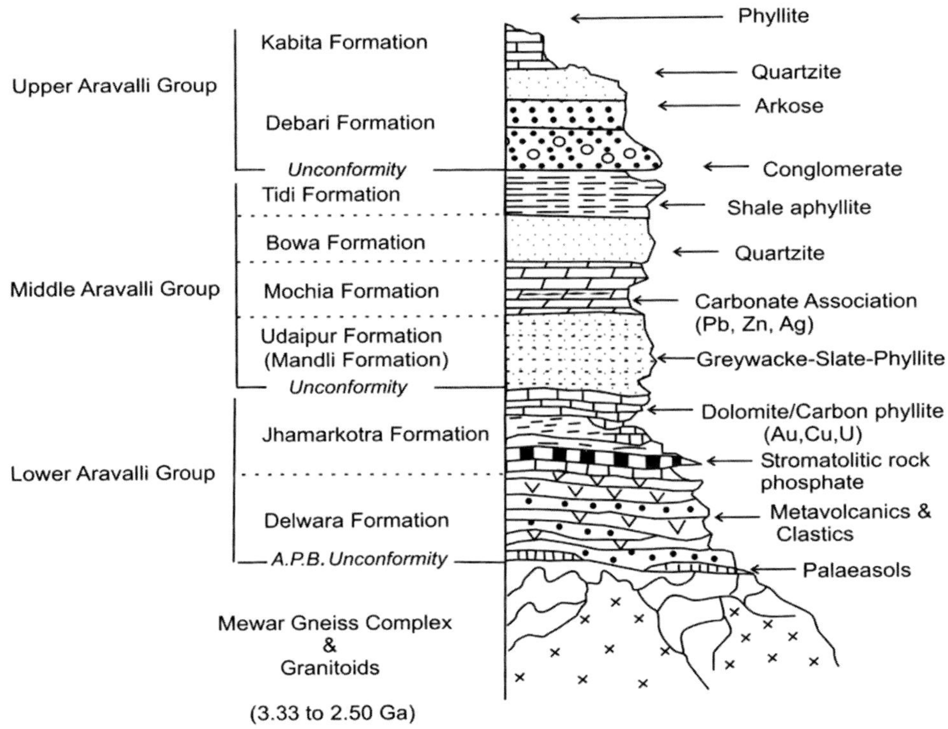

Fig. 4.3 Generalised stratigraphic column of the rocks of Aravalli Hills as represented at Zawar (from Roy 1995).

Salt has been extracted from the salt lakes at Sambhar in Jaipur District (see Chapter 2, Fig. 2.1) and Didwana in Jodphur District for many centuries (Aggarwal 1956: 173–232). The lakes continue to be an important source of common salt as well as sodium carbonate, bicarbonate and sulphate for the traditional glass industry (Sen Mamata Chaudhuri 1985: 135). The cobalt ore, jaipurite, which is a form of cobaltite, the sulpharsenide of cobalt (CoAsS), was mined in the Aravallis to the south of Khetri at the Babai copper mines such as Akulali, where it occurs with copper and iron sulphides as irregular stringers, layers and lenticles through the host rock of black slates (Coggin Brown and Dey 1955: 227–8). It was used as a blue colorant in the enamels for which Jaipur was famous (Untracht 1997: 357–62).

The Aravallis are now an important source of precious and semi-precious stones, especially in the Ajmer-Jaipur area. The exploitation of gemstones in the Aravallis apparently does not have a long history, although the production of fine jewellery, incorporating precious stones, has been carried on in Rajasthan for many centuries at centres such as Jaipur (Untracht 1997: 352–6). Emeralds and garnets are now mined in the Ajmer and Udaipur districts (Krishnaswamy 1988: 462–6). It was during prospecting for garnets in the 1970s that the ancient workings at Agucha were discovered.

The geography and geology of the Zawar area

Geography

The Zawar lead-zinc mines lie about 42 km south of Udaipur (see Chapter 6, Fig. 6.1). The climate is semi-temperate with an annual temperature range between 5 and 40°C. The mean annual rainfall lies between 60 and 65 cm, with most falling in the summer monsoon between July and September. Zawar is set in the midst of the typical Aravalli grid of ridges (see Chapter 5, Plate 5.1), drained by the River Tiri (Plate 4.1), a tributary of the River Mahi. The ridges rise to as much as 200 m above the valley floors, which are typically between about 350 to 400 m above sea level.

Regional geology

The Zawar area is occupied by a thick pile of calcareous, arenaceous and argillaceous metasediments belonging to the Aravalli Supergroup of Protoerzoic age (Figs 4.3–4.5). The principal rock types are conglomerate, quartzite, greywacke, phyllite, slate and dolomite, showing a wide variation in grain size and mineral constituents.

Local geology

The Zawar mines were described by Crookshank in 1947, but the first major paper on the geology of the mines was produced by Mookherjee in 1964. A detailed geological succession of the area has been given by Straczeck and Srikantan (1967) (Table 4.1) and latterly by Roy (1995) (Fig. 4.3). The present structural disposition of the area is the manifestation of two distinct major cycles of tectonic activity, each of which was characterised by intense folding and faulting. The first tectonic cycle was post-Aravalli and has folded the rock into a north–south trending, northerly plunging and tight isoclinal fold, represented by the Baroi and Zawa Mala ridges (Figs 4.3 and 4.4). The second tectonic cycle was the post-Delhi orogeny which refolded the structures of the first tectonic cycle. The post-Delhi orogeny induced a totally different trend giving rise to west plunging, east–west trending, isoclinal folds, resulting in a cross-fold structure represented by the Mochia and Balaria ridges. The area is affected by a number of low-angle folded thrust faults of the first, and high-angle longitudinal faults of the second, tectonic cycle.

Mineralisation

Base metal sulphide deposits occur throughout Zawar solely within the dolomitic horizon of the Baroi Magra formation (Raghunandan *et al.* 1981). Most of these occurrences are manifested at the surface by ancient workings (see Chapter 5). The ancient mine systems offer considerable information to the mining geologist concerning the shape, size and attitude of the exploited or explored ore lenses and of the deposits below the surface deposits, but they provide very little data on the grade of the ore mined in the past.

The distribution of the ore through the deposit is controlled by the shear fractures and tensional fractures related to the fold patterns, as well as by fault contacts. The ores form lenticular bodies confined to the dolomite. They are generally parallel with the host formations except in Balaria, where the lenses in the south zone are very askew. In this area the lenses trend in a north west–south east direction, which gradually becomes east–west along the strike further west towards Mochia. The average strike length of the lenses varies between 100 and 300 m and the width ranges between 2 and 45 m.

The zinc occurs as sphalerite, zinc blende or as marmatite in which the zinc sulphide is in solid solution with iron sulphide, with the latter over 10%. The varying iron content of the sphalerite causes it to exhibit myriad colours from light yellow to brownish black. The sphalerite is medium- to coarse-grained and occurs variously as veins, veinlets, brecciated zones, disseminations and stringers. It is also present to a very limited extent in secondary minerals such as hydrozincite in the oxidation zone near the surface, although some of this secondary mineralisation may have occurred since the mining took place some 2000 years ago. It is significant that Zawar lacks the major gossanised deposits that are so prominent at Dariba and Agucha. Lead, which occurs as galena, is more locally concentrated as big crystalline patches in the sphalerite and pyrite, as well as in veins and stringers together with the sphalerite. The veins are continuous along the strike while the crystalline galena is restricted to the open fractures, voids and along the quartz veins. The dimensions of the patches of galena vary from a few centimetres to tens of metres, and it is possible that some near the surface may have been oxidised to the hydroxy sulphate, anglesite, or to cerussite, the hydroxy carbonate, possibly with some enhancement of silver content, although none is now visible. Pyrite is also very common and occurs as thick veins and stringers with or without sphalerite and galena.

The galena is argentiferous while cadmium is associated with the sphalerite. The average ratios Zn:Pb, Zn:Cd and Pb:Ag are 3:1, 1: 40–45 and 1:40 (ppm) respectively. All the deposits in the Zawar area are zinc-rich except Baroi, which is rich in lead. Native silver occurs as thin films and flakes along fractures and joints in the galena-rich areas of all the deposits. In most of the areas the galena and sphalerite are quite distinct macroscopically and after crushing can easily be separated by hand picking. This is rather unusual, in South Asia at least, and as the separation of the zinc and lead minerals is essential in the production of zinc, this seems to have been one of the reasons why the industry developed at Zawar.

Crude mineral zoning is apparent in some of the lenses, with richer mineralisation occurring in the central portion of the lens. This is of some significance when trying to estimate the original mineral content of the orebodies mined in antiquity from the ore that remains at the edges of the worked-out lenses. The pyrite usually increases towards the tips of lenses and is indicative of the pinching or narrowing of the ore mineralisation.

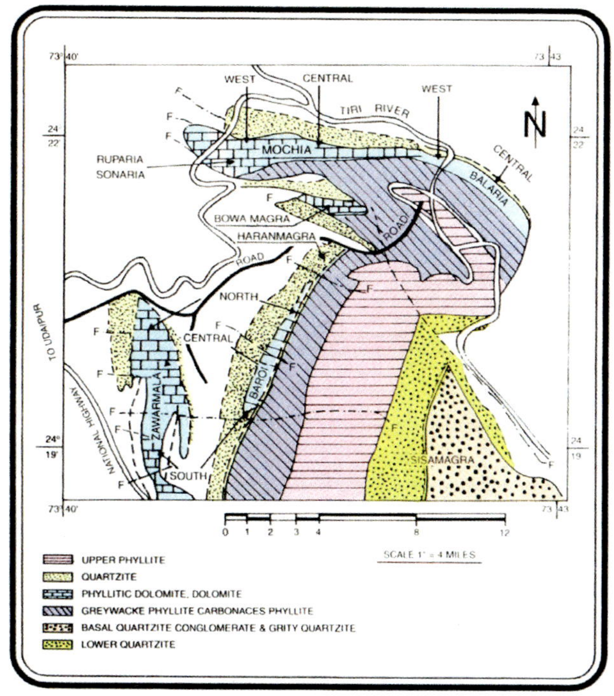

Fig. 4.4 Geology of the Zawar area (*HindZinc Tech* 2000: 28, fig. 2).

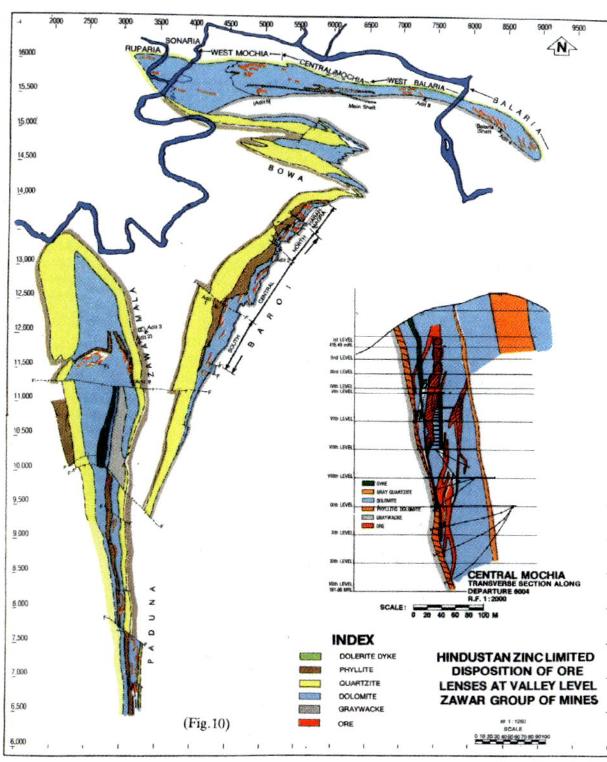

Fig. 4.5 Disposition of the ore lenses at valley floor level, together with a section through the central Mochia deposit (*HindZinc Tech* 1989: 23, fig. 10).

Brief description of individual mines

Mochia and Balaria mines (Fig. 4.5)

The rocks of the Mochia mine area are part of the Aravalli Supergroup, forming part of the second, post-Delhi tectonic cycle. The general geological stratigraphy is as follows (see also Table 4.1):

- Dolerite dyke
- Upper phyllite
- Quartzite
- Upper dolomite
- Arkosic dolomite
- Lower dolomite (host rock)
- Lower phyllite
- Greywackes

The principal rock types are greywacke, phyllite, dolomite and quartzite with dolerite dykes occasionally intruding into these. Dolomite, the main host rock for ore mineralisation, is further subdivided into lower dolomite, middle arkosic dolomite and upper dolomite. Generally the lithological contacts are sharp except for the contacts between the dolomite and siliceous/ arkosic dolomite, where the boundary contact is gradational. The maximum concentration of the mineralisation is in the lower dolomite where current mining activity is concentrated.

Mineralisation

The principal sulphide minerals in order of abundance are sphalerite, pyrite and galena with very minor amounts of pyrrhotite (ferrous sulphide (FeS)) and chalcopyrite (copper iron sulphide ($CuFeS_2$)). Cadmium and silver are associated with the sphalerite and galena respectively. Native silver occurs as thin films, wires, foils and flakes along the fractures and joint planes in the crystalline galena and in the galena-rich areas. The concentration of native silver is higher in the upper levels of the mine and seems to decrease with depth. Dolomite, quartz and sericite are the main gangue minerals. The lead and zinc minerals in the ores are macroscopically distinct and can easily be separated by hand picking after crushing.

The sphalerite is fine- to medium-grained and occurs as veins and veinlets, stringers and brecciated zones, and occasionally as disseminations with or without galena. The galena occurs as fine-grained veins along with the sphalerite and as coarse-grained crystalline patches occurring alone or with the sphalerite. The veins are continuous along the strike of the lode, while the crystalline galena is restricted to the open fractures, voids and along quartz veins. The dimensions of the patchy galena vary from a few centimetres to tens of metres.

Pyrite occurs as veins along the sphalerite and galena lenses, varying in thickness from a few millimetres to 1.5 m. It is concentrated towards the tips of these lenses. Massive pyrite bands have developed especially at the eastern ends of the lenses indicative of pinching (narrowing) of the mineralisation.

Zawa Mala mine (Fig. 4.5)

The local geology at Zawa Mala is similar to that at the Mochia and Balaria mines. The general geological stratigraphy is given below (see also Fig. 4.3 and Table 4.1):

- *Upper phyllite*: grey phyllites and slates
- *Quartzite*: fine- to medium-grained quartzite
- *Carbonaceous phyllite and dolomite*: carbonaceous phyllites with interbedded marly slates and dolomite bands
- *Dolomite*: variants of dolomite i.e. siliceous, arkosic and sericitic dolomite with intercalations of quartzite (host rock)
- *Lower phyllite*: gritty, locally conglomeratic greywacke, with intercalations of phyllites

The rocks of the Zawa Mala mine are in the antiformal fold of the first post-Aravalli tectonic cycle. The ancient workings are mainly confined to the eastern limb and in the central hinge zone of the fold, while in the western limb the ancient workings are small and widely separated.

Mineralisation

Sulphide mineralisation occurs throughout the dolomite horizon but is confined to structural openings, that is in the shears and tensional fractures. The principal sulphide minerals, in order of abundance, are pyrite, sphalerite and galena with minor amounts of pyrrhotite and traces of chalcopyrite. Exploratory mining has revealed the presence of an apparent mineral zonation from high sphalerite content in the eastern part to galena-sphalerite in the central part and sphalerite-pyrite in the western part, corresponding to the eastern limb, hinge zone and western limb of the antiformal fold described above.

The pyrite occurs as thick veins both with and without sphalerite. The sphalerite is honey yellow to dark brown in colour, medium-grained and occurs as streaks, veins and stringers with pyrite. The galena occurs as coarse crystalline pockets of varying dimensions within the sphalerite-pyrite mass and occasionally as massive veins along faults or shears. The zinc to lead ratio is roughly 3:1. Native silver is found in the form of thin films, flakes, foils and wires, and is concentrated along the fractures near to the patches or pockets of crystalline galena.

Baroi mine (Fig. 4.5 and Table 4.1)

The rocks of the Baroi mine are part of the Mandli, Baroi Magra and Zawar formation of the Tiri Series. The general geological stratigraphy is given in Fig. 4.3.

Mineralisation

Baroi is the one mine in the Zawar group where lead is more abundant than zinc. The principal sulphide minerals in order of abundance are galena and sphalerite with minor amounts of pyrite, pyrrhotite, and traces of chalcopyrite. Dolomite, quartz and sericite are the main gangue minerals. Silver occurs in solid solution in the galena and cadmium is associated with the sphalerite. Native silver occurs as thin films, foils, and flakes and occasionally as wires and strings. It is generally concentrated in the fracture zones and shears near the galena-rich patches or pockets. Flakes of native silver also occur with the stringers of pyrrhotite in the areas devoid of lead/ zinc mineralisation.

Galena occurs as fine to medium grains disseminated along foliations and as patches and irregular fracture fillings. The grain sizes range from 2 to 105 μm. The sphalerite occurs as massive independent veins, finely disseminated in association with galena, and as inclusions in the recrystallised galena veins and veinlets. The grain sizes range from 2 to 86 μm. The average zinc to lead ratio is 1:4 in marked contrast to the other mines where zinc is predominant. Pyrite and pyrrhotite occur together sometimes disseminated along with the other sulphides.

Table 4.1 Stratigraphic succession at Zawar (after Straczeck and Srikantan 1967).

Recent to sub-recent		Alluvial soil and river deposits
Pleistocene		Terrace boulder deposit
Eocene or late Cretaceous		Olivine dolerite dykes
	Unconformity	
Intrusives		Meta dolerite dykes and stocks. Quartz plagioclase rock. Apparently altered graphite dykes and sills, quartz and quartz microcline veins
	Unconformity	
Zawar formation		Phyllites and slates, interbedded dolomite, quartzite, quartzite-dolomite breccias
Baroi Magra formation	Haran Member	Quartzite
	Mochia Member	Mainly dolomite, gritty arkosic dolomite, dolomitic arkose, subordinate phyllites and quartzite
Mandli formation		Gritty and conglomeratic greywackes phyllite intercalated with greywackes
Kathalia formation		Carbonaceous phyllite, phyllite and interbedded dolomite
Kanpur formation		Quartzite, dolomite
	Unconformity	
Dantali formation		Dantali quartzite
	Unconformity	
Sisa Magra formation		Gritty and conglomeratic quartzite

The geography and geology of the Rajpura-Dariba mines

Geography

The Rajpura-Dariba lead-zinc mines lie about 76 km north east of Udaipur. The climate is semi-arid with an annual precipitation of around 50–55 cm per annum, most of which falls in the summer monsoon from June to September. The main source of water is from wells. The annual temperatures range between about 5 and 42°C. The topography of the region is a gentle peneplained flat terrain, lying at about 500 m above sea level, broken only by the long ridges containing the mineralisation. The prominent hills include Dariba Hill above South Lode (see Chapter 6, Plate 6.48), Rajpura Hill and the Sindesar Ridge.

Regional geology

The area lies in a linear belt of high-grade metasedimentary rock of the Bhilwara Supergroup of Precambrian age.

Local geology

The deposit is part of a 17-km-long linear metasedimentary sequence that trends north–south, dipping steeply (Fig. 4.6). It extends from Dariba in the south to Bethumni in the north and hosts sizeable lead-zinc deposits in various blocks (Fig. 4.7). Outcrops of gossan as well as extensive early mine workings, mine waste dumps and slag heaps occur along the ridges and conveniently mark on the surface the extent of the economic mineralisation (Raja Rao *et al.* 1972; Raghunandan *et al.* 1981; Gupta *et al.* 1997). The lithounits consist of calc-silicate-bearing dolomite, calc-biotite schist, actinolite schist, garnetiferous mica schist, graphitic mica schist, and staurolite/kaynite-bearing graphitic mica schist, quartzite, ferruginous chert/braccia and banded chert. The zinc-lead sulphide bands are associated with the calc-silicate-bearing dolomite chert and graphite mica schist. They are stratiform and synsedimentary in nature. The calc-silicate dolomite and

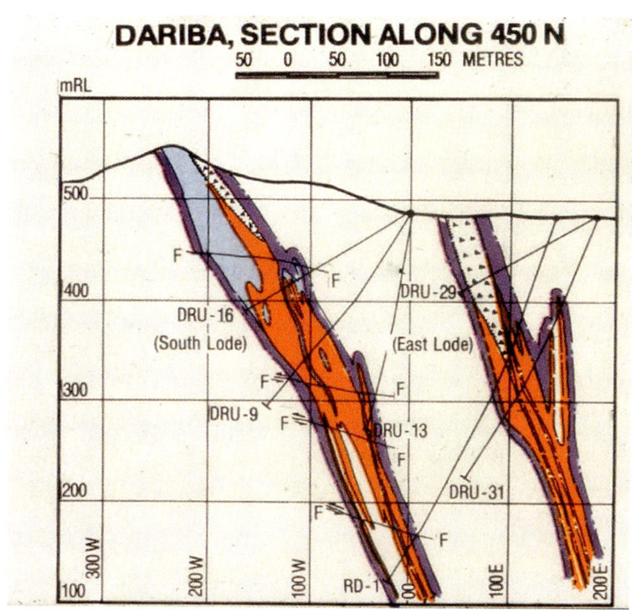

Fig. 4.6 Geological section showing Dariba Main Lode and East Lode at 450 N (*HindZinc Tech* 1989: 19, fig. 6 inset).

cherty quartzite outcrops form the ridges in the plain that is itself formed from the great pile of easily weathered graphitic and micaceous schists.

The general stratigraphic sequence of the metasedimentary sequence of the belt suggested by the GSI (Gandhi *et al.* 1984) is as follows:

- *Upper metapelite*: quartz sericite schist, phyllite with thin bands of chert and quartzite-arid metamarl
- *Pyroclastic sequence*: carbonaceous phyllite/graphitic schist, calc-silicate rocks, with thin clastic bands of arenaceous and meta-argillites

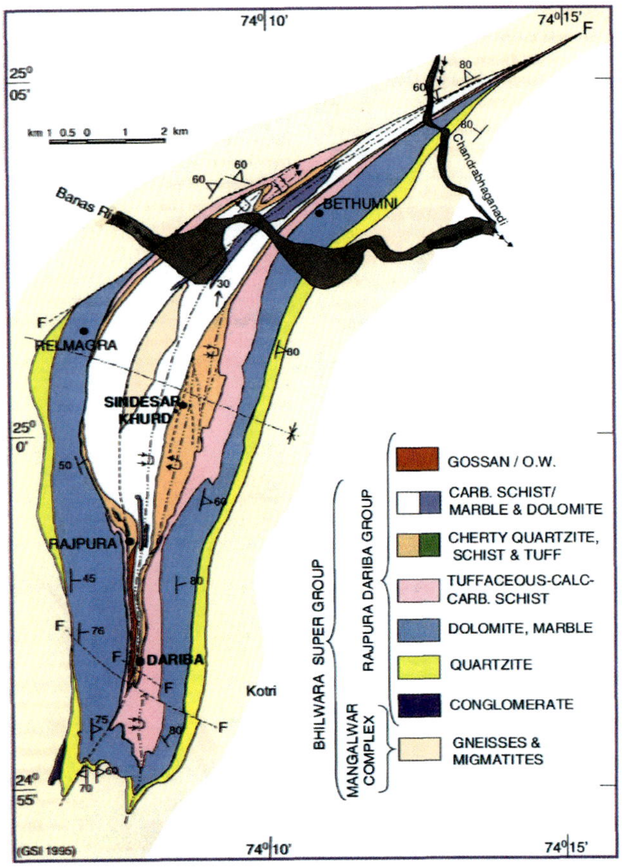

Fig. 4.7 Geological map of the Dariba-Bethumeni belt (*HindZinc Tech* 2001: 5, fig. 3).

- *Lower metapelite*: quartz mica-schist with garnet, staurolite and kyanite with bands of quartzite

Mineralisation

Surface evidence

Intense chemical weathering and oxidation of the sulphidic lode and its host rock have left a pronounced *in situ* gossan which stands out as a prominent quartzitic ridge between Dariba and Rajpura, known as Dariba Hill (see Chapter 5, Fig. 5.19 and Chapter 6, Plate 6.48). The gossan displays a variety of colours – dark brown, bright reddish yellow, vermilion, brick red, red, purple, pale yellow, grass green, bluish green, azure blue, black, grey and white – indicating the presence of various minerals such as limonite, goethite, jarosite, jasper, malachite, azurite, hematite, manganese oxides, clay minerals etc. Cubic, cellular and honeycomb boxworks, together with diamond and spear-shaped, globular and rounded voids and dense limonite are all found in the gossan, indicating the presence of sphalerite, galena, pyrite, pyrrhotite and chalcopyrite. The gossan outcrops at Dariba are unique in India and have been declared a national geological monument.

Orebody disposition

The mineralisation is confined to the upper layers of the calc-silicates and extends to the overlying graphite mica schist in the southern part of the deposit. The mica schist contains a variety of minerals such as graphite, quartz, mica, sillimanite and potassium feldspar as well as the metalliferous minerals. The average depth of oxidation in the Dariba block is 30 m but is over 400 m in the Rajpura B and C blocks to the north. Surface and subsurface exploration has revealed the existence of two sub-parallel orebodies, namely the Dariba Main Lode and Dariba East Lode, separated by 250–300 m of barren rock (Fig. 4.6).

The Dariba Main Lode itself splits into two well-defined lodes – the South and North Lodes, separated by 300 m of barren or lean grade ground along the strike (see Chapter 5, Figs 5.19 and 5.20). The Dariba South Lode corresponds to the gossan zone exposed on Dariba Hill and extends over a strike length of 550 m from the southern end of the hill. The width of the deposit varies between 2 and 68 m with a mean width of 18 m. The North Lode has a strike extension of 1000 m and the width of the lode varies from 3 to 30 m with an average of 13 m.

Dariba East Lode is situated below the large ancient opencast (see Chapter 5) east of Dariba Hill, and extends over a strike length of 800 m. It ranges in width from 2 to 35 m and has the shape of an elongated lens. The average depth of oxidation is 50 m from the surface. Drilling has indicated that the lode was mined extensively down to a depth of 263 m from the surface.

Mineral/metal distribution

South Lode

Two major types of ore, zinc-lead-copper and zinc-lead, occupy characteristic positions with respect to the enclosing rocks. The zinc-lead-copper zone occurs in the footwall in the calc-silicate host with an extent limited to about 250 m long and 75 m deep, whereas the zinc-lead zone occupies the central and hanging wall side with wide and consistent metal distribution. At the contact of calc-silicate and graphite schist in the hanging wall a coarsely crystalline geocronite-rich lead-zinc ore along with diopside patches occurs. This has a higher concentration of antimony and silver. Although argentiferous antimony and lead, freibergite and galena minerals are widespread and uniformly distributed throughout the orebody, they appear to be most concentrated at the hanging wall contact. It seems that the copper was only mined from the upper oxidised portion of the deposit on Dariba Hill; certainly the only extensive deposits of copper slags are to be found on its slopes (Sites 13 and 14, see Chapter 6).

East Lode

The entire ancient underground mining activity was confined to a quartz vein 5–8 m thick, which contained coarse crystals of galena with sphalerite. The patches of crystalline galena occurring in the centre of the lode have significant silver values.

The geography and geology of the Rampura-Agucha mines

Geography

Rampura-Agucha is located 245 km north east of Udaipur and 210 km from Jaipur. The climate is sub-arid, subtropical, with a very wide annual temperature range of between 3 and 46°C. The main source of water is from the tank, but wells and the River Mansi are also important. The annual rainfall is 36 cm, the majority of which falls between July and September in the south west monsoon. The topography is of a generally flat peneplain which slopes gently down to the River Mansi about 2 km to the south or to the River Khari around 7 km to the north east. The height of the plain at Agucha is about 390 m above sea level.

Regional geology

The stratigraphic position of the Rampura-Agucha orebody is uncertain, primarily because of the difficulty in distinguishing between older members of the BGC and the upper amphibolte to granulite facies metasedimentary rocks of the Bhilwara belt, a problem further aggravated by extensive migmatisation in the area. According to Heron's (1953) map, the deposit lies well within the BGC of Archaean age. However, recent work by the GSI (Gupta *et al.* 1980, 1997; Roy *et al.* 2004) indicates that the deposit lies at the contact between the high grade gneisses, migmatites and granulites of the Sandmata complex and metasedimentary rocks of the Mangalwar complex of the Bhilwara Supergroup (that is, the northern block of the BGC and Bhilwara metasedimentary belt) (Deb *et al.* 1989).

Local geology

The deposit occupies a northerly plunging synformal structure within an oval patch of graphite mica-sillimanite schist, similar to that at Dariba (Gandhi 1983; Gandhi *et al.* 1984). A major shear zone occurs on the regional scale in the footwall of the orebody, which is evidenced by a conspicuous zone of mylonites. The rock sequence from the hanging wall to the footwall (from east to west) can be broadly grouped as follows:

- Garnet-biotite-sillimanite gneiss
- Graphite-mica-sillimanite gneiss (the host for lead-zinc mineralisation)
- Garnet-biotite gneiss with intermittent bands of calc-silicate, quartzo-feldspathic, amphibolite, pegmatite and aplite
- Granite gneiss
- Mylonite rock

The general strike of the rock formation is north east–south west, dipping 62° due south east with a steepening dip near the surface (Fig. 4.8).

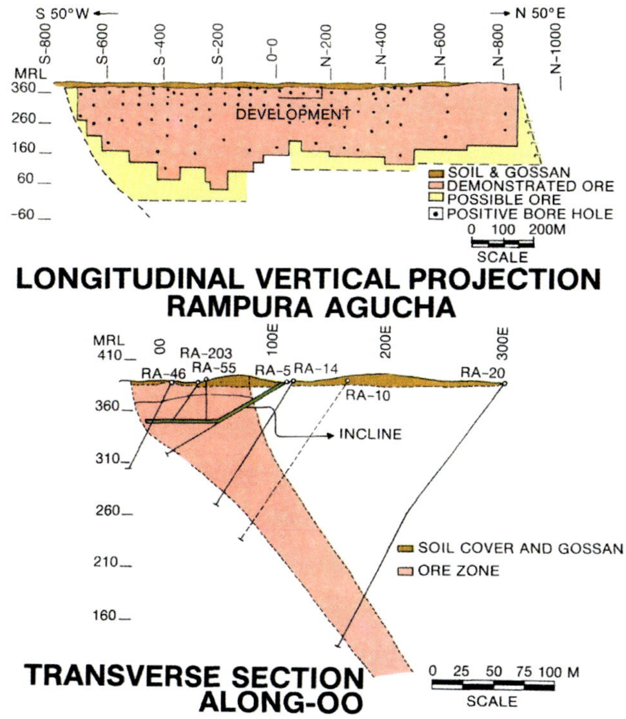

Fig. 4.8 Geological section across the Main Lode at Agucha (lower) with detail of projected mining development (above) (*HindZinc Tech* 1989: 22, fig. 9).

Mineralisation

The principal ore minerals are sphalerite, pyrrhotite, pyrite and galena with minor or trace amounts of arsenopyrite, marcasite, chalcopyrite, boulangerite, digenite, tetrahedrite-tennantite, freibergite, löllingite and ilmenite. The galena and freibergite are argentiferous and were worked in antiquity. Sphalerite is by far the predominant sulphide mineral in the deposit, occurring as dark brown to light brown varieties in the form of veinlets and mineral bands, associated with pyrite, pyrrhotite or laths of mica. Galena generally occurs as fine- to medium-grained disseminations and veinlets. Fine-grained galena forms a matrix cementing subrounded to rounded gangue minerals of various sizes. Occasionally coarse crystalline galena patches occur in the hanging wall associated with lenses of calc-silicate rocks. Minor amounts of freibergite, tetrahedrite-tennantite, are closely associated with the galena in the hanging wall calc-silicate patches. Pyrite, pyrrhotite and minor amounts of arsenopyrite are also found in close association with the sphalerite and galena.

Surface evidence

The surface indications of the presence of economically viable sulphide mineralisation beneath which were the gossanised cap rocks with malachite encrustations, together with mined debris and the remains of ancient workings. These take the form of a long shallow depression extending for 1600 m in a north east–south west direction (see Chapter 3, Plate 3.19). Gossan outcrops occur on either side of the linear shallow depression along the strike and also as discontinuous *in situ* exposures within the depression. The gossan showed various shades of yellow, brown, red, grey, maroon and purple colours along the 1500 m strike of the orebody. At places cellular and spongy boxwork developments were observed. The depth of the gossan zone varied between 15 and 22 m from the surface. The weathered/gossanised zone varies in thickness from 10 to 25 m above the unweathered sulphides. Subsurface exploration by diamond drilling indicated that the lead-zinc mineralisation is confined to the 1500 m-long linear oval-shaped outcrop of graphite-mica-sillimanite gneiss/schist between the S-725 to N-825 grid (see Chapter 5, Fig. 5. 31). The ore zone has a sharp contact with the hanging and footwall rocks. The average width of the orebody is 60 m, ranging from 2 to 120 m (Fig. 4.8).

Orebody disposition

A few thin barren patches of host rock with low-grade metal contents occur at places within the orebody. The orebody is massive and plunging due north east with varying degrees of plunge. Crude mineral zoning is noticeable within the orebody. Although galena is well distributed across the orebody it tends to concentrate along the hanging wall. Along the strike of the orebody the galena concentration increases towards the north east.

Chapter 5 Investigation of the mining operations at Zawar, Dariba and Agucha

L. Willies, L.K. Gurjar and P.T. Craddock

Introduction

This chapter is based on the preliminary published accounts in Willies (1983, 1987, 1988, 1989), Willies *et al.* (1984) and Craddock *et al.* (1985a, 1989), augmented by further unpublished fieldwork undertaken in December 1990, together with various unpublished accounts by the staff of HZL, particularly that of K.S. Choudhary at Agucha. The underground investigations were necessarily confined to mining remains located by modern underground or open pit operations, supplemented by those which were exposed at the surface and the results of the drilling programme. These have been recorded photographically and as far as feasible by surveying and very limited excavation.

There was a consciousness, shared by HZL, that the present mining operations were necessarily and rapidly exposing ancient remains and then obliterating them, particularly at Agucha. The investigations concentrated on recording workings that were due to be lost as a result of modern operations. There is very considerable scope for further archaeological work, both at the surface and underground, on a long-term and rescue basis.

Zawar

The early mines at Zawar include both opencast and underground workings. The host rock is a sheared dolomite, the result of metamorphism of sedimentary dolomites, which were laid down in Precambrian times, with a much greater thickness of shales and greywackes (Chapter 4). Today the sheared dolomites are found as highly inclined isoclinal folds forming conspicuous ridges at an altitude of about 600 m, flanked by rounded hills of softer phyllite and slate rocks (Plate 5.1) (see also Chapter 4, Fig. 4.4). The ore seems to have had a submarine–exhalative origin and was deposited syngenetically as disseminated lenses within the dolomite beds. The zinc content can be as high as 20% (the average mined in the 1980s was 8%, currently the lead and zinc combined are about 5.5%) of the whole rock over a width of up to some tens of metres, with a total strike length of some 18 km. The ore can contain up to 2% lead (except at Baroi Magra), and there is some native silver on the joints, but overall this constitutes only a very low percentage. This is in contrast to the ore at Dariba and Agucha that is argentiferous, and this should be borne in mind when considering which metals were being exploited at Zawar in the early phases of the mine. Some remobilisation has taken place during metamorphism, especially of the lead, which is found mainly in shears and joints.

A rough and conservative calculation suggests that perhaps 2500 tonnes of zinc concentrates were extracted from some 2.5

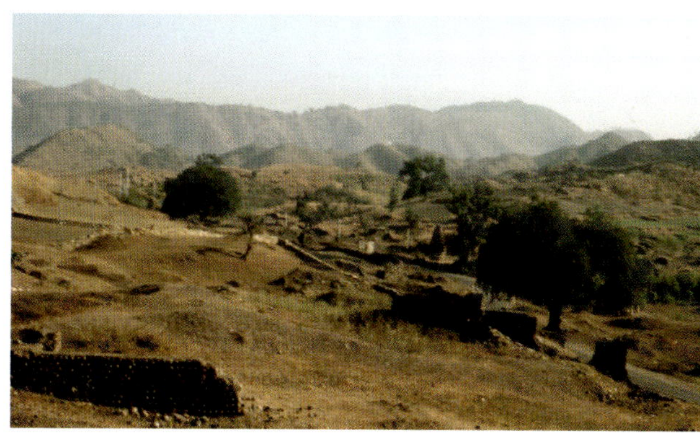

Plate 5.1 Zawar Mala Magra seen from Old Zawar, some 3 km to the east.

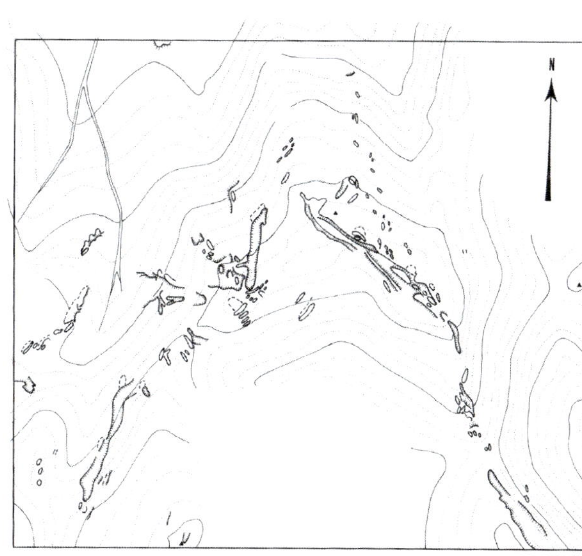

Fig. 5.1 Zawar Mala: plan of the ancient workings on the ridges surrounding the enclosed valley on Zawar Mala Magra.

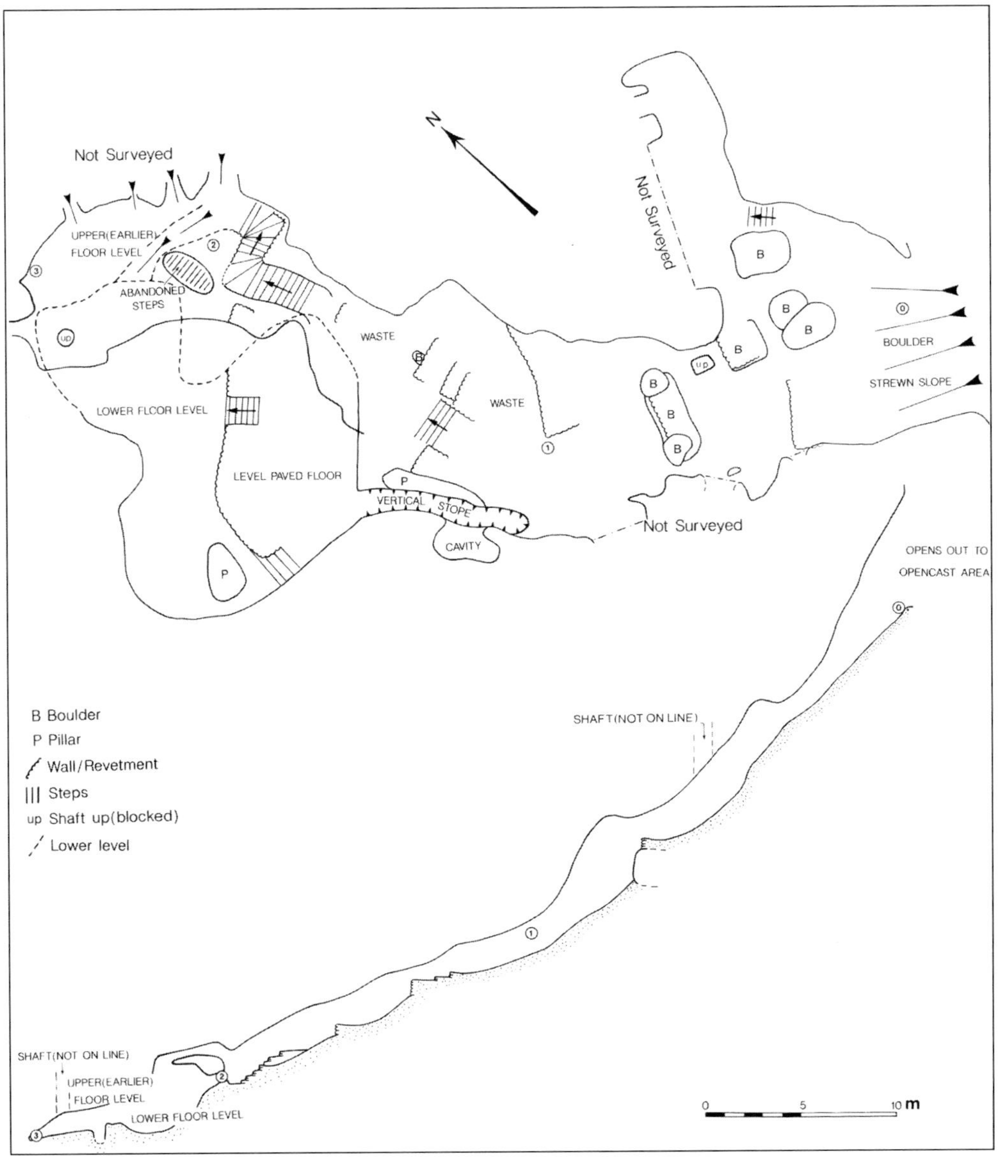

Fig. 5.2 Zawar Mala, Pratap Khan: plan and profile of the main mine.

million tonnes of ore mined in the area before modern mining commenced. Having calculated that the concentrate in the early zinc smelting process was similar to that in the modern concentrate with about 50% zinc (see Chapter 7), this would suggest that about 100,000 tonnes of zinc were produced. However much of the mine debris belongs to the early mines when it is believed that the main product was zinc oxide and the recovery was likely to have been much lower, and in the post-medieval period at least, it is likely that argentiferous lead was also being mined.

Zinc is predominant at Zawar Mala, Zawar Mochia and Balaria, and is being mined today. Lead is the main metal at Baroi Magra, but despite an exploration programme, it is not currently being mined. These mines are described individually below.

Zawar Mala Magra

Ancient workings are found at outcrops on the ridge of Zawar Mala (Plate 5.1) (see also Chapter 6, Fig. 6.2), penetrating as deep as 120 m from the top of the ridges. At the surface mine openings occur at intervals of 50 m or so on either side of a narrow valley formed by denudation of an anticline at the top of the ridge (Fig. 5.1); most are very small, opening directly into the host beds, inclined here at up to 40°. Only a few shafts are now visible at the surface. Immediately adjacent to the mines are scatters of mine debris of all sizes, but in the valley below the mines and above the smelter at Site 30 (see Chapter 6) are large alluvially altered heaps of more regularly sized beneficiation debris, clearly derived from the workings on either side.

Pratap Khan mine

One mine of more commodious proportions that could be entered from the surface was examined in some detail (Fig. 5.2). This is known as Pratap Khan (Pratap's mine), after Maharana Pratap Singh who, as previously mentioned (Chapter 3), is believed to have taken refuge there with his immediate family in the late 16th century.

Entry to the mine is from a small quarry, evidently the result of opencast working. It is possible that this later reworking and quarrying are medieval, and could have supplied the smelters at

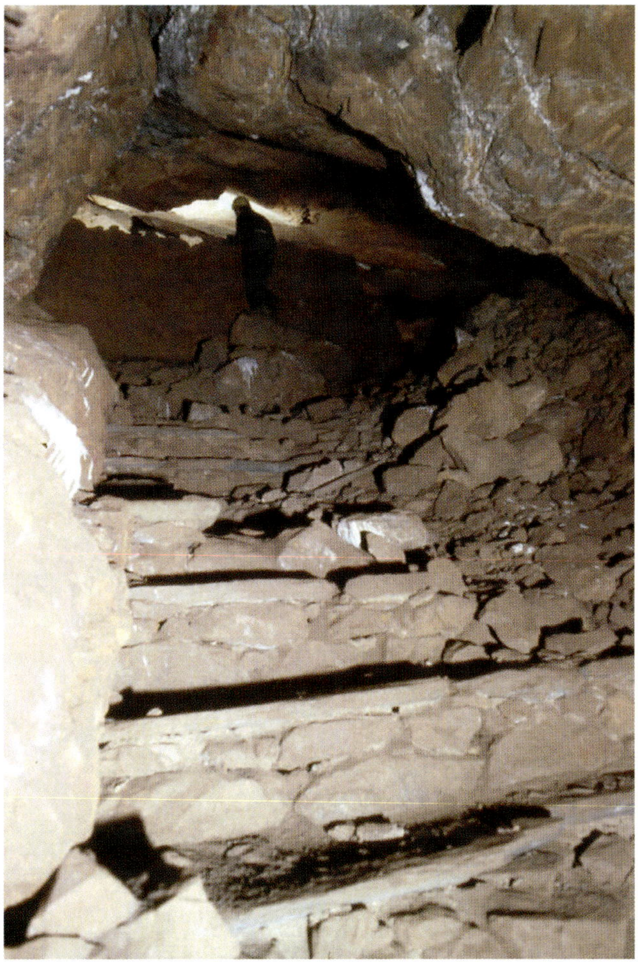

Plate 5.2 Zawar Mala, Pratap Khan: broad stone stairway leading down from the surface to the audience chamber. Note the stone slabs on the threads.

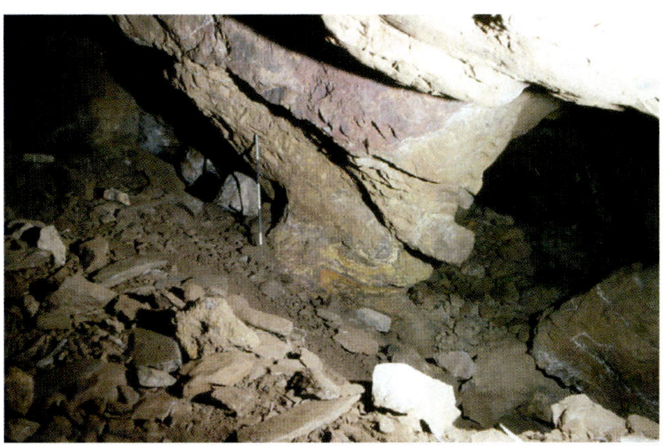

Plate 5.3 Zawar Mala, Pratap Khan: pillar supporting the roof (scale: 1 m). Note the steep inclination of the beds, typical of the Zawar Mala deposits, and also the pick marks on the left-hand pillar.

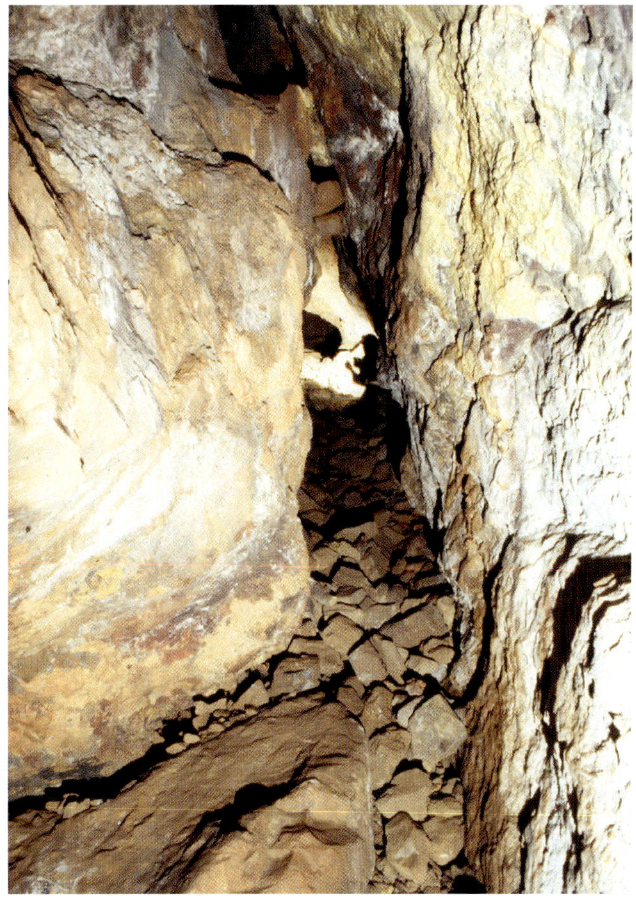

Plate 5.4 Zawar Mala, Pratap Khan: stope in vertical mineralised joint.

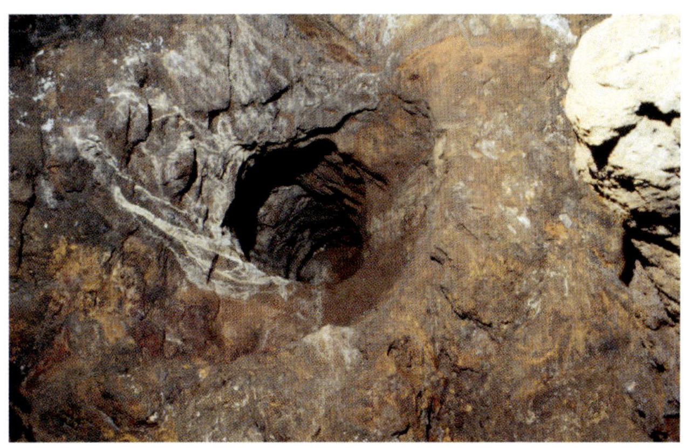

Plate 5.5 Zawar Mala, Pratap Khan: shaft in the roof now blocked by a large boulder, looking up from the bottom chamber.

Site 30 which lies just below in the narrow valley. Part of the cliff face above the mine entrance has a joint which has been opened out in the past using gads (pointed chisels), and traces of gad marks are common in the quarry (see Chapter 8, Plate 8.4) and in the mine itself. There is a well-marked path into the mine, with rock-cut steps, and the most notable feature of the mine itself is the development of pathways and steps. Since one flight of steps has been mined away and a new set built, this suggests that the former were contemporaneous with the early mining and that the more accomplished steps shown here (Plate 5.2), with the treads paved with stone slabs, may be associated with the later use as a refuge, for which there is other evidence (see below).

It is evident that there were at least two stages of underground mining since the main section with the pathway, which is 10–20 m in width and generally at least 2 m high, seems to cut through older, much smaller remains of low tunnels, small chambers and small climbing shafts, infilled with alluvium, which are visible at the sides (Plates 5.3 and 5.4). Most of the work seems to have been done with hammer and

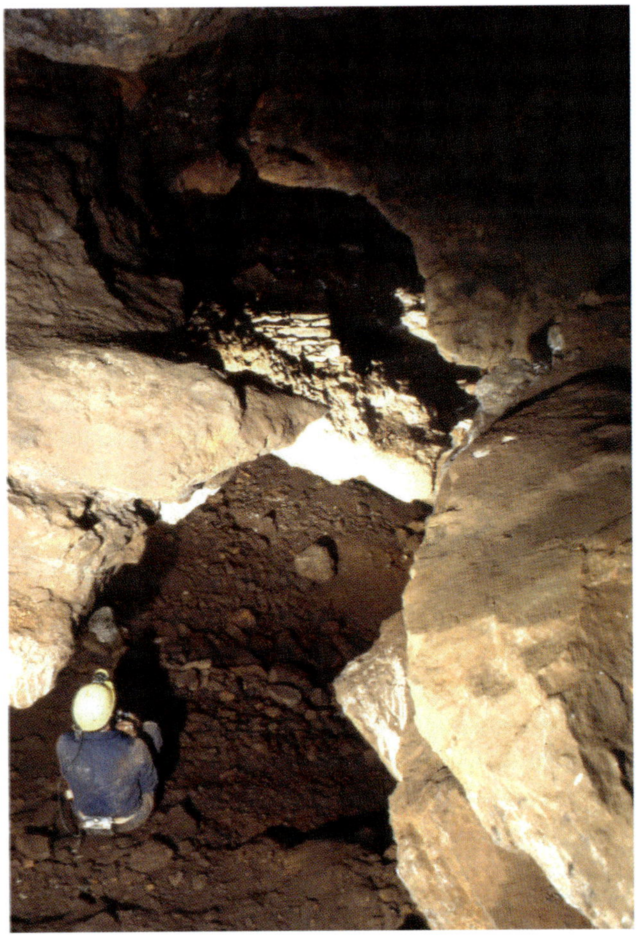

Plate 5.6 Zawar Mala, Pratap Khan: upper and lower parts of the lower chamber.

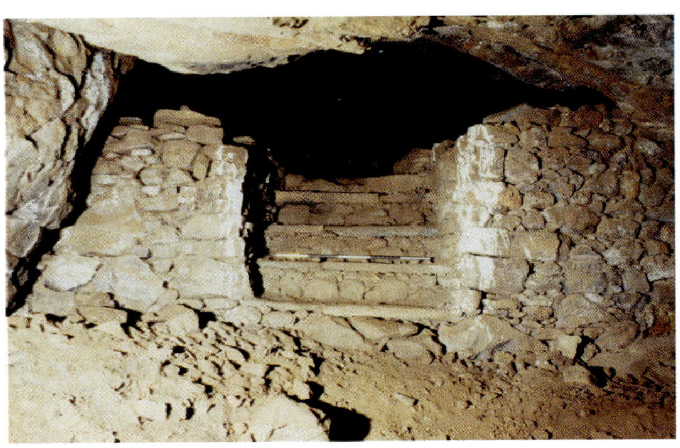

Plate 5.7 Zawar Mala, Pratap Khan: carefully constructed flight of steps ascending the retaining wall of the large platform in the so-called Audience Chamber, built into an ancient stope (cf. Fig. 5.2).

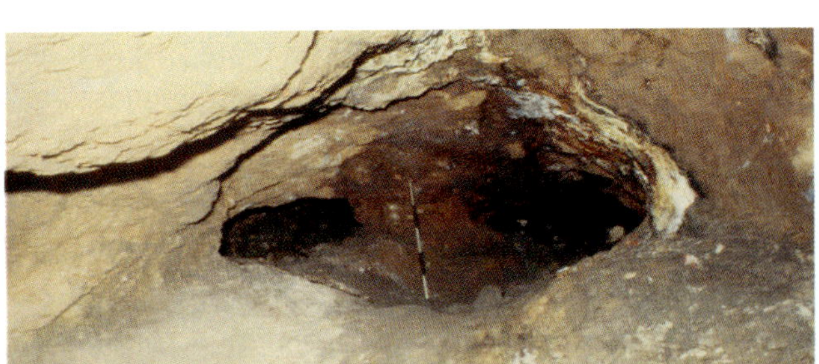

Plate 5.8 Zawar Mala mine: small upper workings.

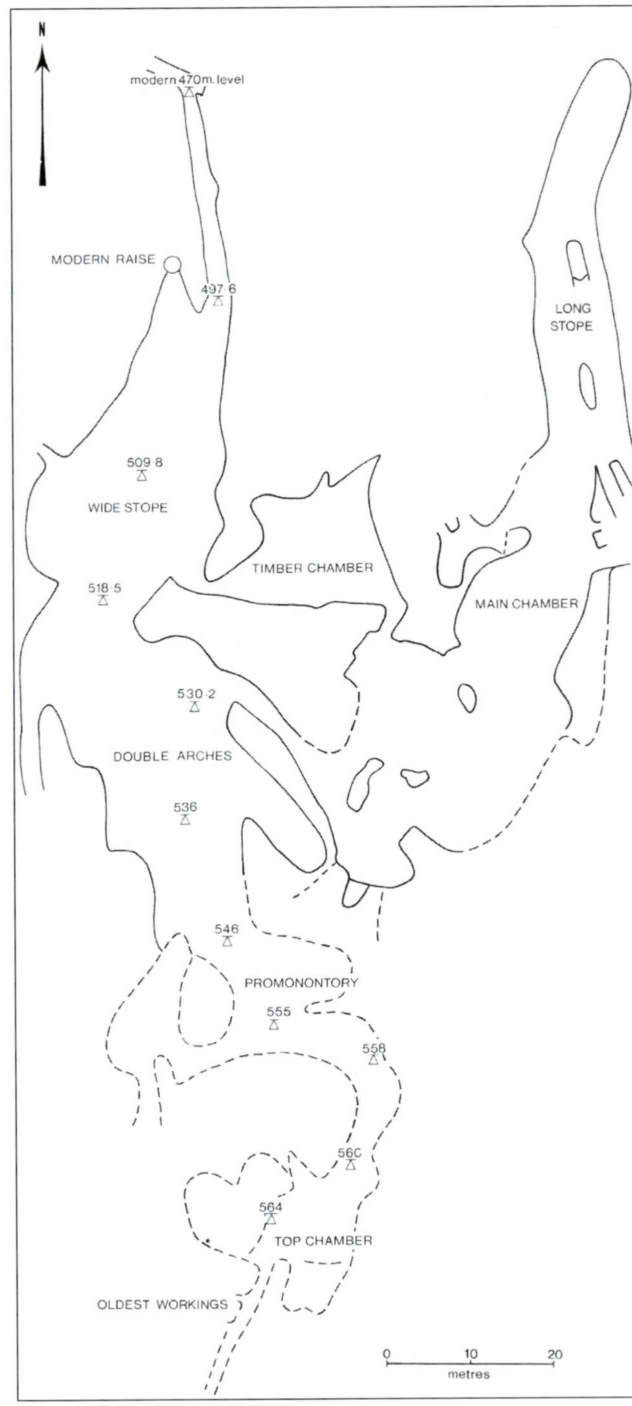

Fig. 5.3 Zawar Mala mine: overall plan of the main areas surveyed.

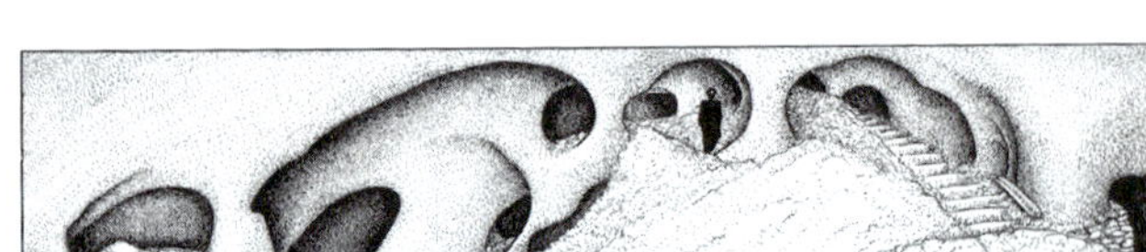

Fig. 5.4 Zawar Mala mine, arches area: 270° profile of these small workings that are believed to belong to the earliest phase of mining, typical of fireset workings.

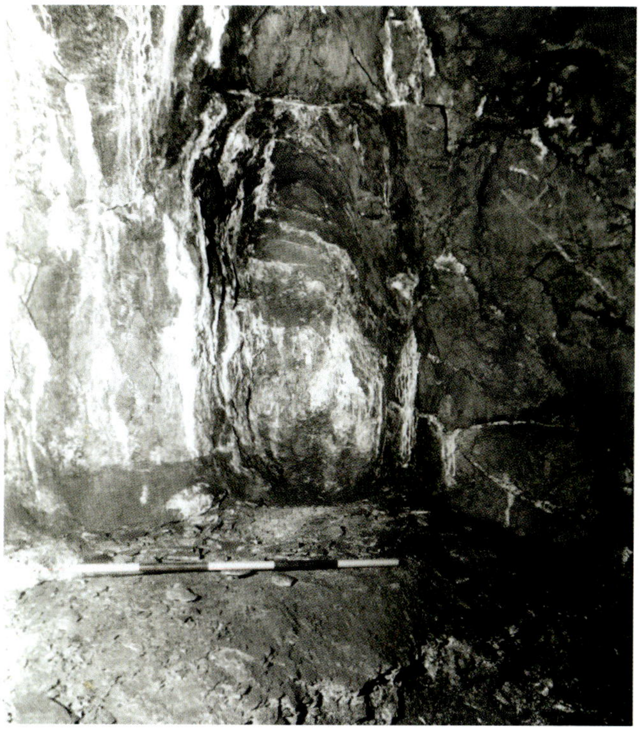

Plate 5.9 Zawar Mala mine, top chamber: firehole in the wall.

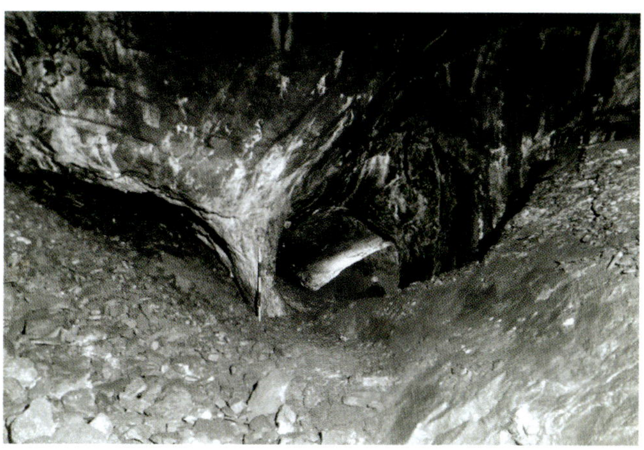

Plate 5.10 Zawar Mala mine, top chamber: lower end with fireset arches and pillar (scale: 1 m).

gad, exploiting limonitic boxworks in the partially weathered sheared dolomite. There is evidence of some firesetting near the entrance in hollows in the roof and at the bottom of the extant workings. Near the bottom, in a wide chamber, is a shaft to the surface although it is now blocked by a boulder (Plate 5.5). The surface above has a stone-lined shaft in an appropriate position, also blocked, which is likely to be the upper section. The bottom chamber has been worked in two levels (Plate 5.6 and Fig. 5.2), the bottom cut removing the steps to the upper floor. Waste debris, of which there is a great deal, is held back by carefully built stone walls, bonded with clay. One of these walls, just above the bottom, has been used to form a flat platform or living area originally floored with phyllite, slabs of some of which still remain, approached by a short but carefully made and imposing flight of steps (Plate 5.7). Since this is a very unusual mining feature, it is inferred that it is likely to have

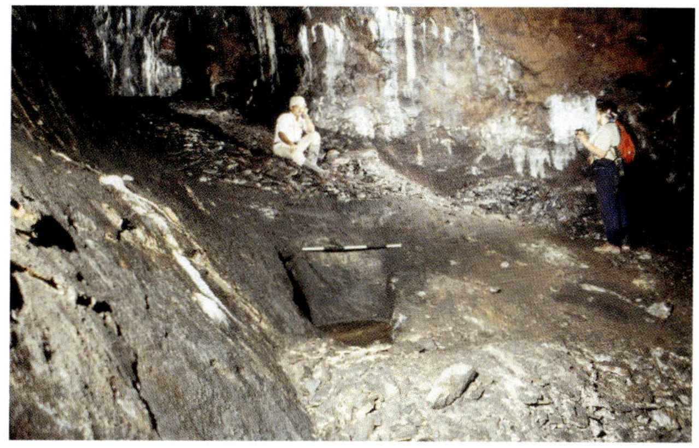

Plate 5.11 Zawar Mala mine, top chamber: note the small firehole in the floor with waste carefully stacked behind on which a team member sits.

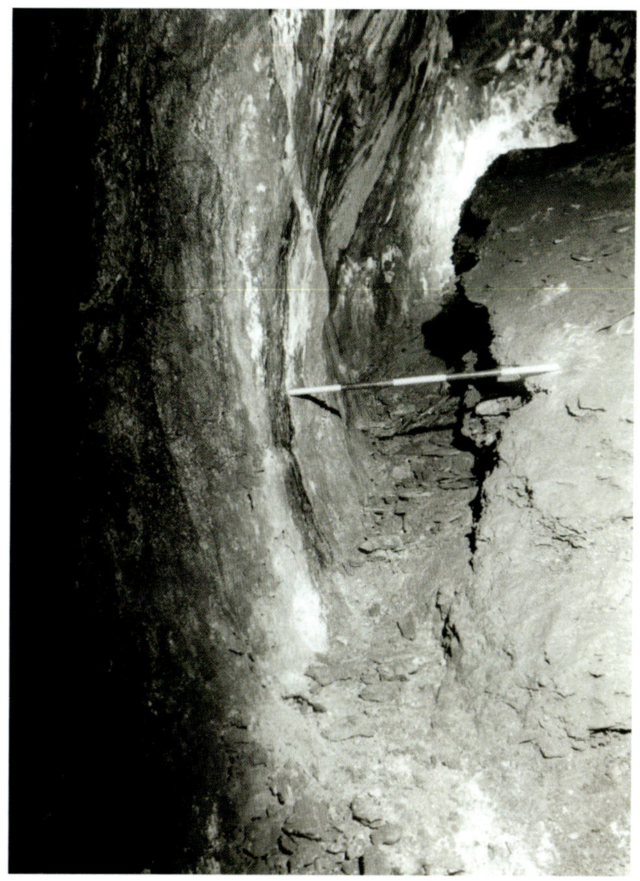

Plate 5.12 Zawar Mala mine, top chamber: the debris from earlier operations has been removed to expose the wall for further extraction.

been constructed as a refuge, probably for Pratap Singh and his family and supporters, or may at least have given rise to the legend. Charcoal from the earlier firesetting operations, near to the platform, gave a calibrated radiocarbon date of 350–310 BC or 210 BC–AD 120 (BM-2639).

Zawar Mala mine

Extensive underground workings were intersected by the modern adit driven in from the hillside (Fig. 5.3). These can be followed up-dip to very close to the surface (about 125 m above), as indicated by washed-in animal dung and wood debris, but despite the presence of a very strong draught no connection with the

ZAWAR MINES

Zawar Mala Mine

	Material	Lab. number	Radiocarbon Result BP	Possible Calibrated Age Ranges Calendar Years 68% confidence	95% confidence	Ref
ZM/LW/85/13 from small chamber in main gallery	charcoal	BM-2481	modern			1989, **31**(1),24
ZM/LW/85/14 from ladderway	wood	BM-2482	2150 ± 110	370 to 90 BC	410 BC to 80 AD	1989, **31**(1),24
ZM/LW/85/8	wood	BM-2483	2180 ± 35	360 to 290 BC or 250 to 190 BC	380 to 160 BC or 140 to 120 BC	1989, **31**(1),24
LW/82/2 from scaffold in escape stope	wood	BM-2148R	2350 ± 120	800 to 250 BC	800 to 150 BC	1985, **26**(1),67
LW/82/1 from launder in escape stope	wood	BM-2149R	2140 ± 110	370 to 80 BC	400BC to 80 AD	1985, **26**(1),67
470 MRL, RL 456 MRL	charcoal	PRL-932	2410 ± 100	760 to 680 BC or 660 to 630 BC or 600 to 580 BC or 560 to 390 BC	850 to 250 BC	
RL 566 MRL	charcoal	PRL-933	1990 ± 140	200 BC to 180 AD	400 BC to 350 AD	
ZW/LW/22 Pratap Khan mine, from fire near "Audience Chamber"	charcoal	BM-2639	2040 ± 70	170 BC to 30 AD	350 to 310 BC or 210 BC to 120 AD	1994, **36**(1), 108
ZW/LW/87/26 from firehole in top chamber	charcoal	BM-2634	1340 ± 100	590 to 800 AD	450 to 900 AD or 920 to 950 AD	1994, **36**(1), 108
Mochia Mine						
Pit prop from gallery	wood *Terminalia* sp	BM-2381	2360 ± 50	750 to 740 BC or 520 to 390 BC	770 to 680 BC or 660 to 630 BC or 600 to 360 BC	1987, **29**(2). 189
ZM/LW/87/32, from loose heap of debris in opencast mine	charcoal	BM-2666	390 ± 50	1440 to 1520 AD or 1590 to 1630 AD	1430 to 1640 AD	1994, **36**(1), 108
Balaria Mine						
10 West stope 415 MRL	wood *Terminalia* sp	BM-2338	170 ± 50	1660 to 1700 AD or 1720 to 1820 AD or 1850 to 1870 AD or 1920 AD to modern	1650 to 1890 AD or 1910 AD to modern	1987, **29**(2), 189

Table 5.1 Radiocarbon dates for the Zawar mines. Calibrated using OxCal v2. 18 (Bronk Ramsey 1995) and the calibration curve of Pearson and Stuiver (1986). References in the far right column are to the original publication of the figures in *Radiocarbon*.

surface was found. The highest and possibly oldest workings are formed of small arched chambers, some 3 m across, and were the waste backfill to be removed, probably also as high (Plate 5.8 and Fig. 5.4). Access is very constricted, at times limited to a low, flat-out crawl until blocked entirely at the upper end. The method of excavation seems to have combined firesetting and hammer-gad work with possibly the use of a small pick to produce 'woodpecker work' in narrow places. This form of mining may once have penetrated to much greater depths, and it is suspected that some of the later workings on a much larger scale lower down may have removed the older work.

In addition to the radiocarbon dates given below (Table 5.1), charcoal samples from unspecified locations within Zawar Mala, where the modern mine cut into the lower part of the ancient workings, gave calibrated dates of 850–250 BC, and a sample from near the surface of the Zawar Mala mine (if that is where the sample was located) gave a date of 400 BC–AD 350 (PRL 932 and PRL 933).

Currently access to the highest workings is via a large chamber, designated the Top Chamber (Plates 5.9–5.12 and Fig. 5.3), which has broken into the older workings from below. The chamber is some 30 m long and 5–8 m high and wide. There

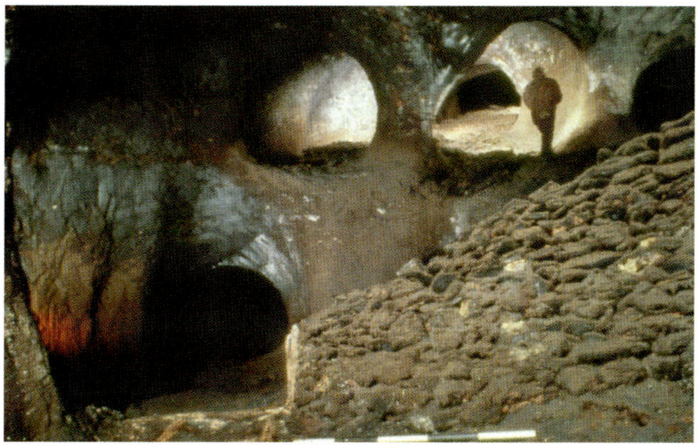

Plate 5.13 Zawar Mala mine, Double Arches: the lower arches are later.

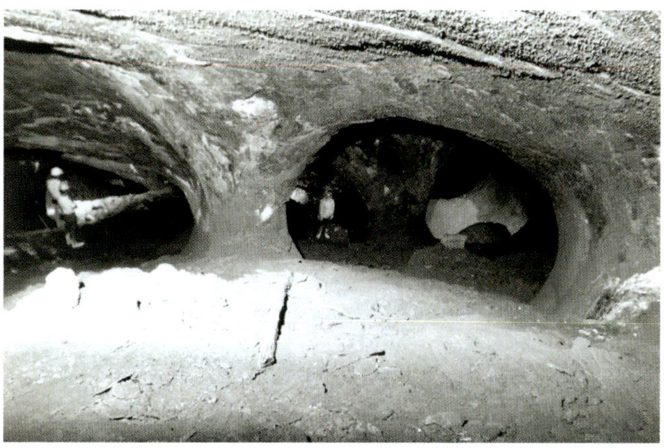

Plate 5.14 Zawar Mala mine: view down-dip from above the Double Arches. Note the encrusted soot from firesetting on the roof.

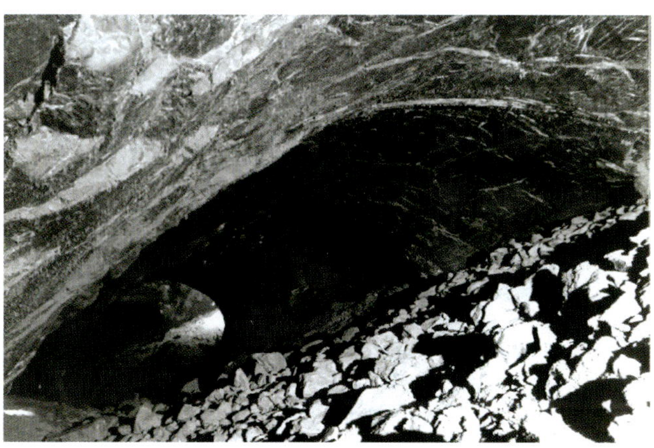

Plate 5.15 Zawar Mala mine: mine waste tipped into abandoned stope near the Double Arches. Note the tip does not fill the stope, suggesting that the height may have been deliberate to control the air flow.

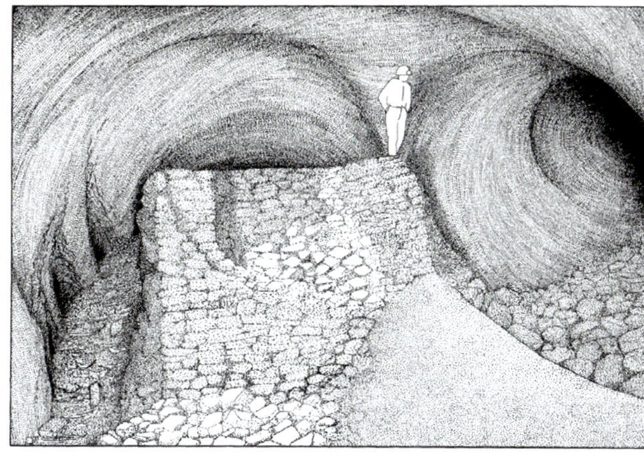

Fig. 5.5 Zawar Mala mine, Top Chamber: platform built up of deads and brushwood to enable the roof to be fireset.

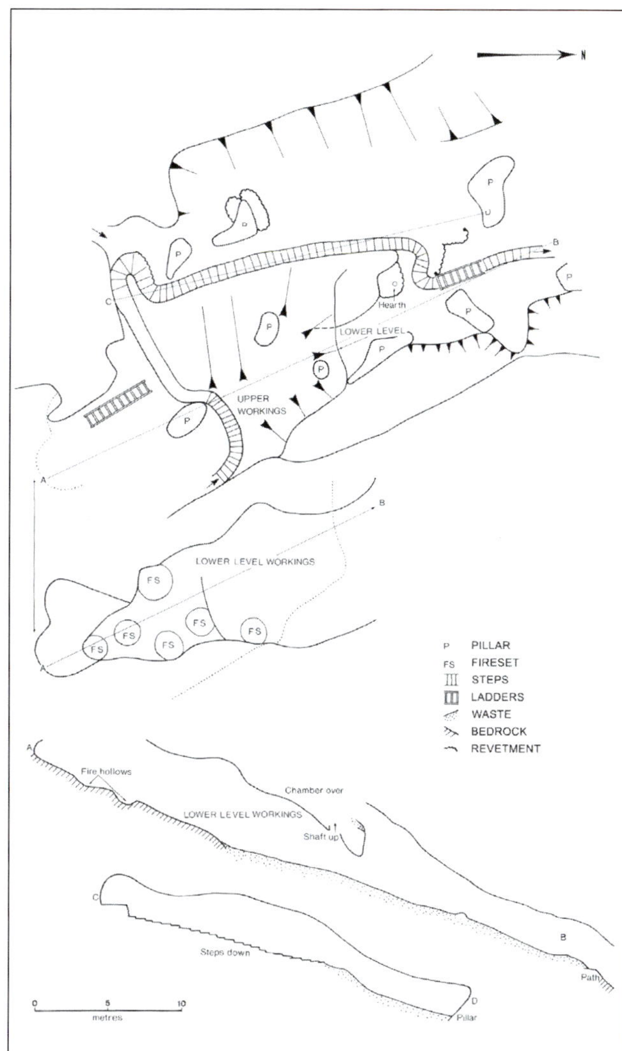

Fig. 5.6 Zawar Mala mine: plan and profile of the area below the Top Chamber showing the winding path/ladderway, various fireholes in the floor and the putative hearth (cf. Plates 5.8–5.24).

are characteristic fire-hollows in the walls of the chamber (Plate 5.9), and the rounded roof and pillars are typical of firesetting (Plate 5.10). The sloping floor was worked up-dip by means of benching, using fireholes on three benches, following each other. Debris from the last burn is still neatly placed at the back of one of the holes (Plate 5.11). Platforms of stone debris interlaced with brushwood were built to enable the roof to be fireset (Fig. 5.5). A radiocarbon date on charcoal from this fireset debris gave a calibrated date range between AD 450 and 900 or AD 920–950 (BM-2634). Older debris has been stacked at one side of the lower end of the chamber, and later – possibly much later – has had a trench dug through it, presumably to allow renewed attack on the walls (Plate 5.12).

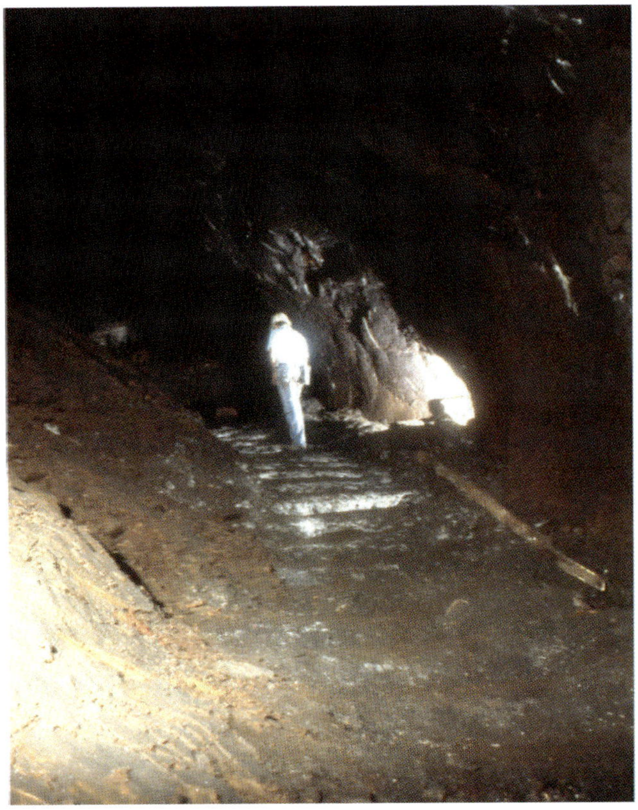

Plate 5.16 Zawar Mala mine: stairway near the Double Arches. Note the great width.

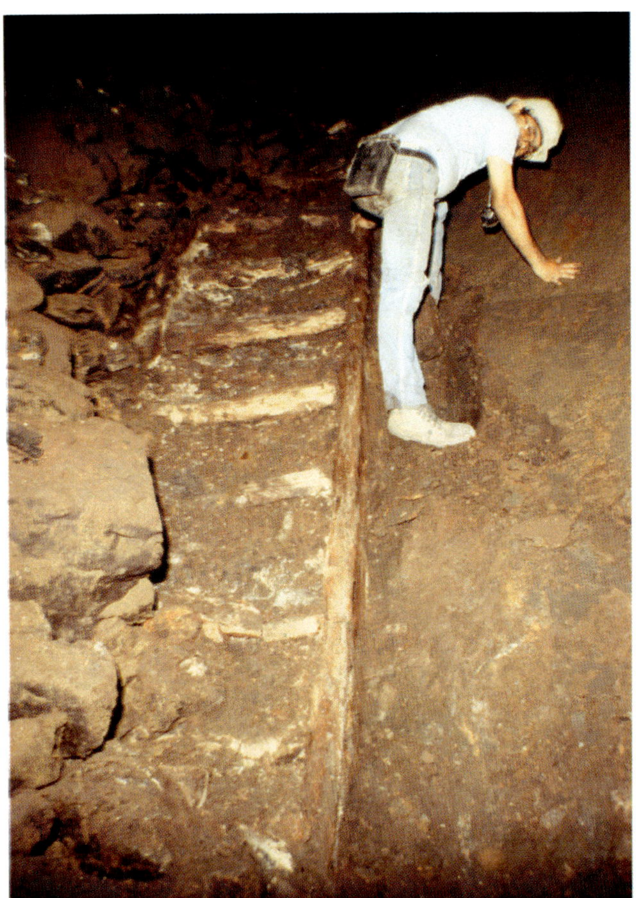

Plate 5.17 Zawar Mala mine: ladderway where the path becomes steep in order to gain access to the stone promontory.

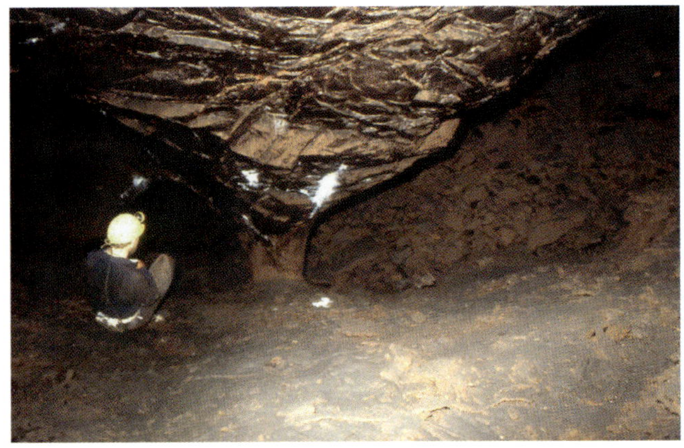

Plate 5.18 Zawar Mala mine: pillar worked by picking after firesetting.

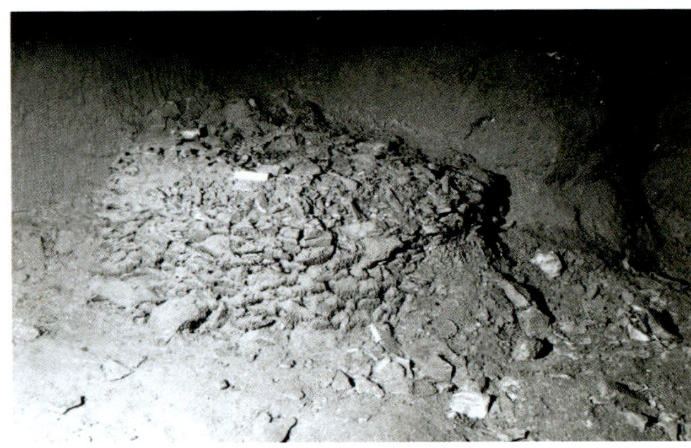

Plate 5.19 Zawar Mala mine: possible hearth near the Double Arches.

Lower still in the workings, the mine has been worked almost entirely by firesetting with very little gadwork evident. The scale is grand, with pillars 5–10 m apart and up to 5 m high, although the spaces between are often partially and sometimes wholly infilled with dumped waste (Plates 5.13–5.18). Through the area a clear and well-made path has been formed, with stone and wood steps where it is steep (Plates 5.16 and 5.17 and Fig. 5.6). A sample of timber from a set of steps in this area was dated to 410 BC–AD 80 (BM-2482). At the top of the area, a probable collapse has obliterated the likely exit route of the original mine. Nearby, a promontory built of stone rubble and approached from below by a steep ladderway (Plate 5.17) juts out from the waste heaps, possibly where the roof was to be reworked (Fig. 5.7). A small chamber off this has an inflow of water which originally ran into a large earthenware container the size and shape of a baby bath. Fragments of this pot remain in the shaped hollow in which the vessel must originally have sat. It was obviously used to collect water, possibly for cooling the fireset rock. The stacked deads piled on the steep slopes appear to have been deliberately mortared together with fireset debris, the fines of which made an effective cement.

The path through the area leads down to a spectacular area of double arching, one set above the other (Plate 5.13). Evidently the upper arches were formed in the original advance down-dip, the lower by reworking later towards the up-dip. Fireholes in the floor suggest the period of work was similar to that in the top chamber, which is part of the final phase. The debris from these operations was neatly stacked near the fireholes

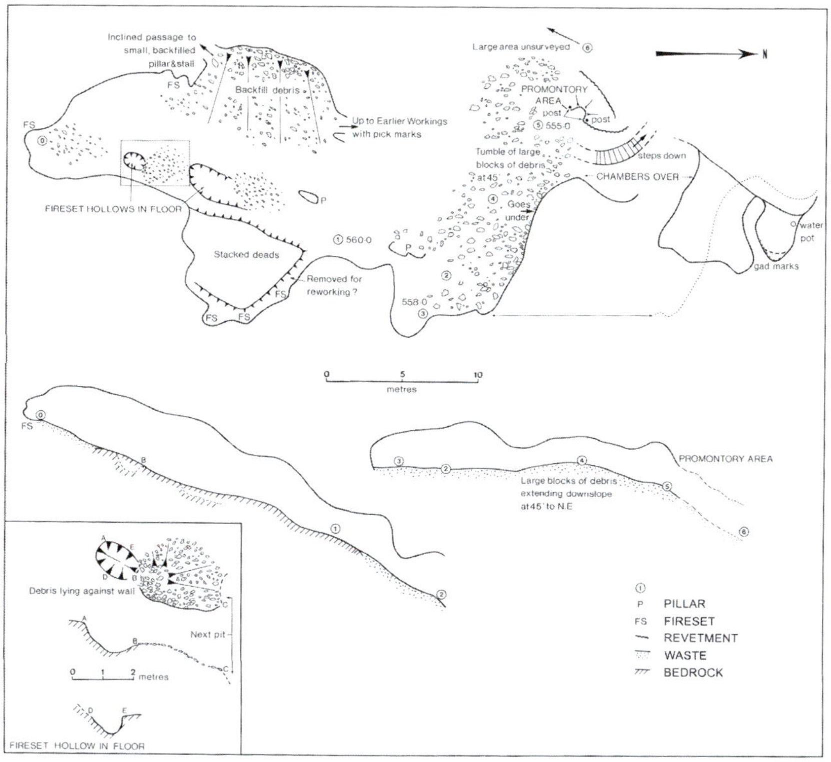

Fig. 5.7 Zawar Mala mine: plan and profile of the promontory area with deads apparently built to allow the roof to be fireset. Inset: plan and section of one of the fireholes in the floor of the Double Arches area.

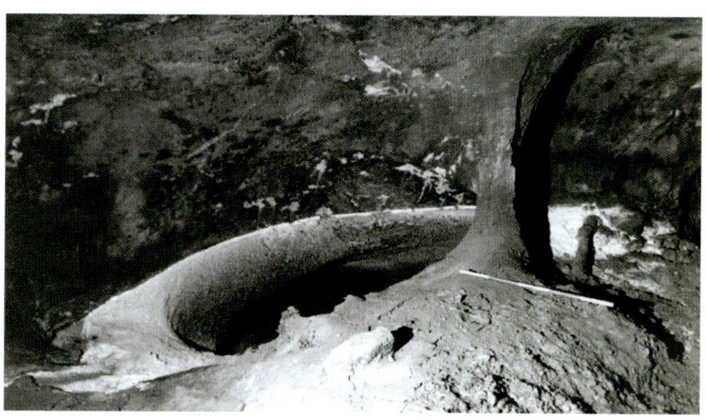

Plate 5.20 Zawar Mala mine, wide stope: large fireset hole in the floor.

and the floor has otherwise been kept clear (Plate 5.11 and Fig. 5.7 inset), presumably to allow the miners a better view of the underlying minerals in order to assess whether they were worth exploiting. Other debris from this working seems to have been dumped in adjacent areas that may have been worked out (Plate 5.15). Some pillars had been worked by picking after firesetting (Plate 5.18). A sub-circular stacked heap of rock debris against a pillar seems to have functioned as a raised hearth (Plate 5.19) and contained the ashes of a small fire and fragments of an earthenware pot.

From here the path leads both down-dip and eastwards across the strike of the beds. Down-dip the area opens out into a wide stope some 20–30 m across, about 2 m high on average, with only occasional pillars. Fireset holes are found at the sides, both in the walls and in the floor (Plate 5.20), and also in the roof. Timber props have been used both to hold up roof flakes (Plate 5.21) and to stabilise the waste stacked on the steep slope

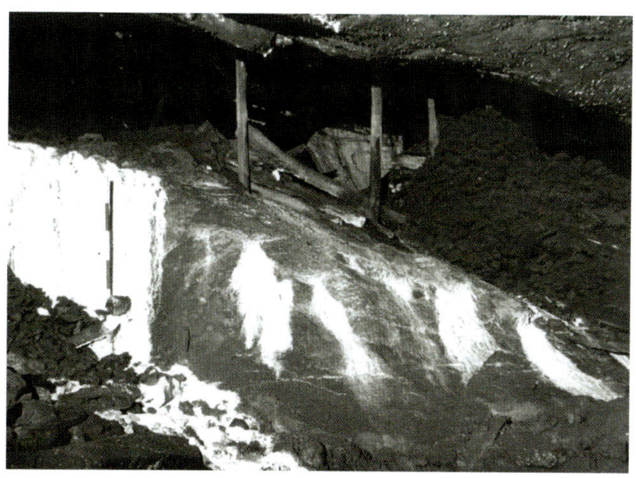

Plate 5.21 Zawar Mala mine, wide stope: wooden posts supporting a roof flake. Note the post-mining white calcite flows in the foreground.

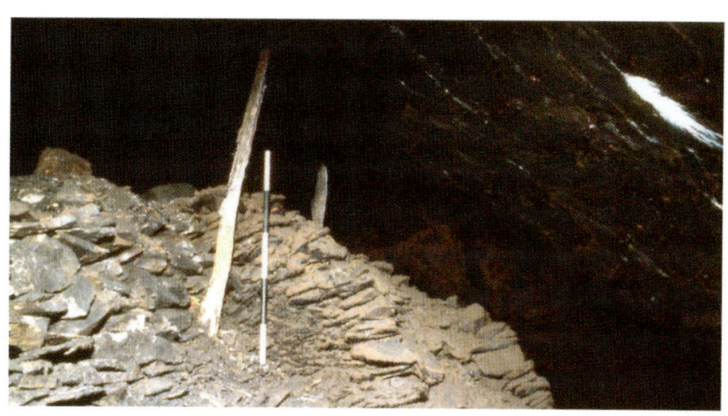

Plate 5.22 Zawar Mala mine: timbers holding the waste material stacked on a slope.

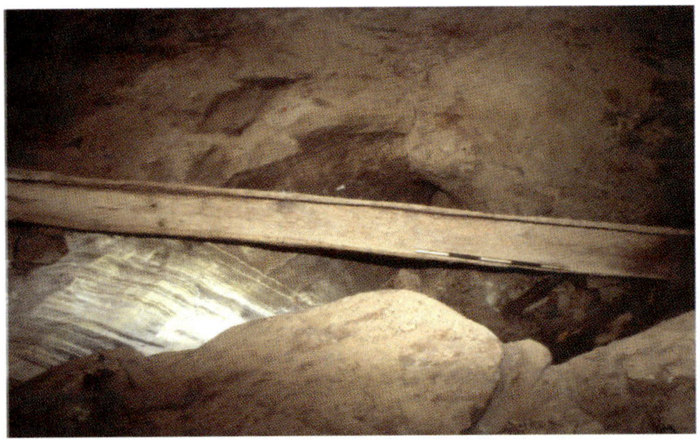

Plate 5.23 Zawar Mala mine, wide stope: wooden launder.

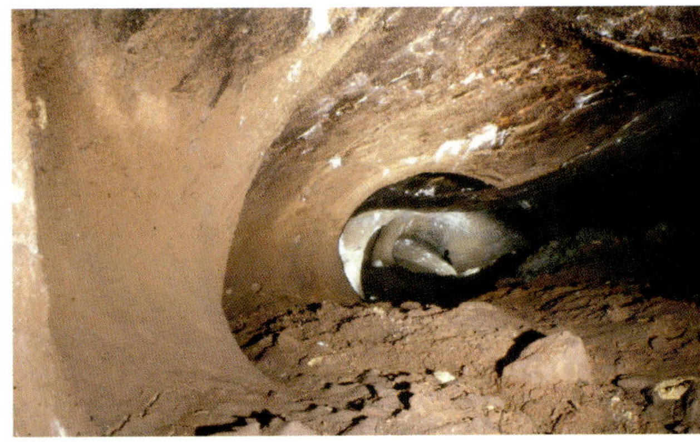

Plate 5.26 Zawar Mala mine: long stope.

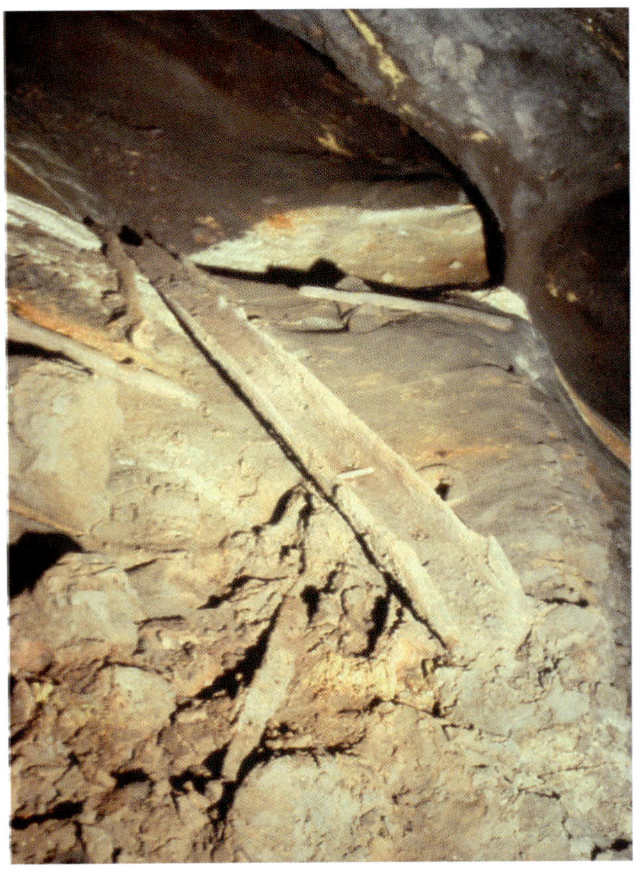

Plate 5.24 Zawar Mala Mine, wide stope: wooden launder.

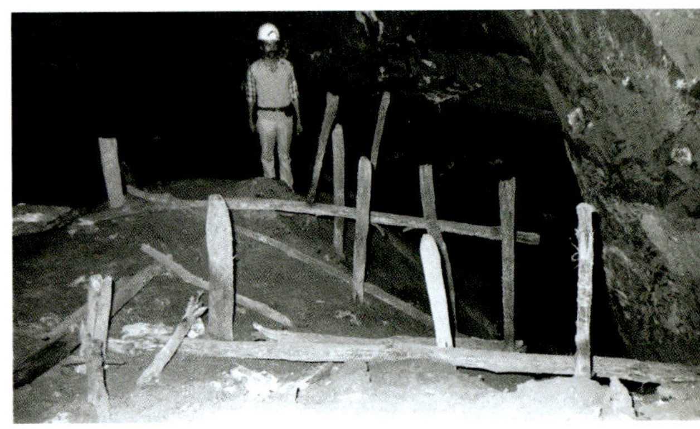

Plate 5.27 Zawar Mala mine: Timber Chamber.

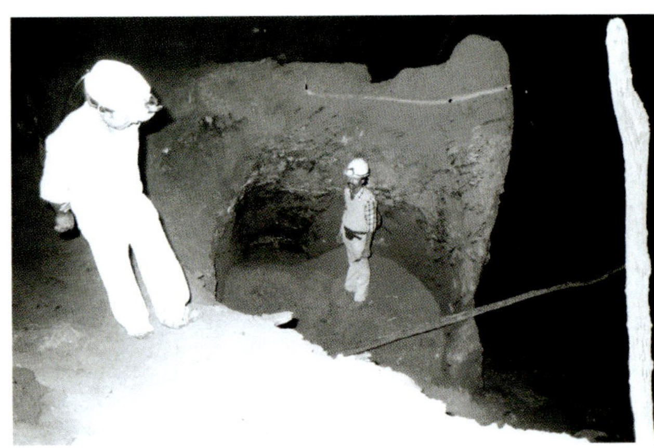

Plate 5.28 Zawar Mala mine, Timber Chamber: remains of a pillar with fireset depression inside.

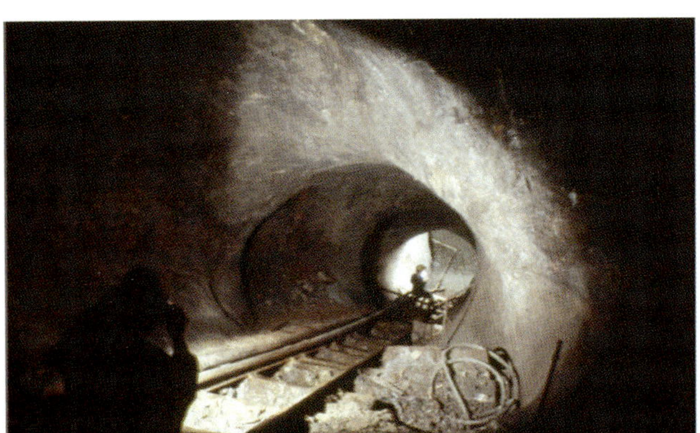

Plate 5.25 Zawar Mala mine: finger stope used as an escape route etc. in the modern mine.

(Plate 5.22). Shallow trenches had been dug to convey water inflows to the path, and several wooden launders were used for a similar purpose. They were several metres in length with a U-section about 30 × 30 cm (Plates 5.23 and 5.24). Some had been displaced following flooding in the mine. One of the launders has a radiocarbon date of 400 BC–AD 80 (BM- 2149R).

At the bottom of the 'wide stope', a working referred to here as a 'finger stope' was developed (Plate 5.25) and another stope, designated the 'long stope' (Plate 5.26), led from the Main Chamber. These are fireset inclines of characteristic rounded section, about 3 m in diameter, but sometimes 5 m or more, driven down-dip, which is hereabouts 30–40°. They seem to have followed the richer shoots of ore down, and presumably

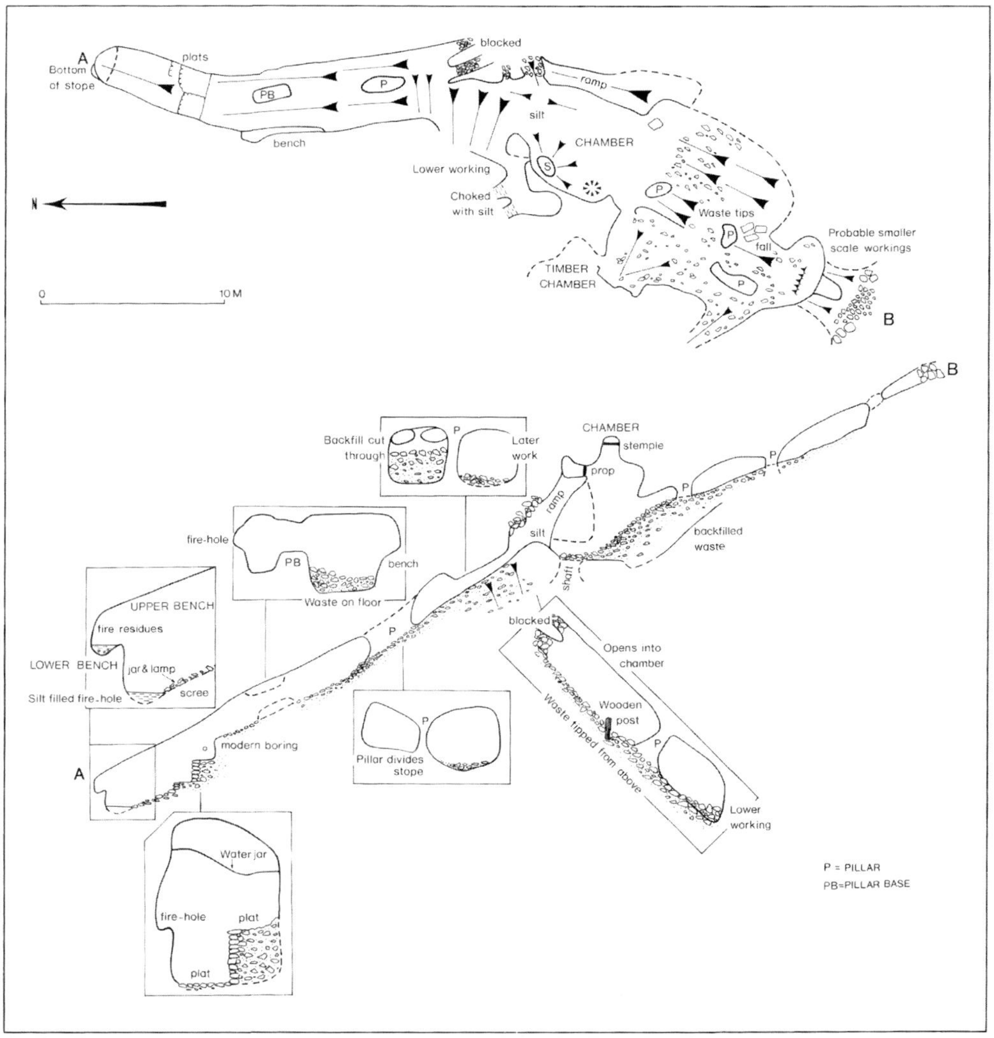

Fig. 5.8 Zawar Mala mine: plan and profile of the long finger stope.

allowed control of water inflow by their isolation from other workings that run roughly parallel. They had all been worked by firesetting. The substantial height in these inclines would presumably have allowed a flow of cold air near the floor to run down to the fires and hot gases to escape at roof level. No form of brattice to separate air flows was found. The finger stope has fireholes at the base and partly up the face of the forefield, that is, it was worked on two benches and with one side slightly in front of the other (Fig. 5.8). This stope had been cut by the modern adit and the bottom had already been entirely removed before adequate photography was possible. One, possibly two other finger stopes, now infilled with rubble and alluvial debris, appeared to penetrate down-dip from the chamber. A further large finger stope, referred to here as the long stope (Figs 5.3 and 5.8), extends slightly eastwards across the dip downwards for over 80 m. This runs parallel to another stope and a short distance down the two unite. The remains of pillars between them suggest that they were used to create separate airways before their removal. Near the bottom is a 10-m high step with adjacent platforms built up of stone rubble, which would have allowed short ladders to surmount it.

The step can only have been made by working up-dip from below, on the retreat. That is, after the first stage of mining had worked out the richest ore, removing more by working back up the slope of the stope, resulting in debris being thrown back to the presumably abandoned down-dip forefield. This forefield had also been worked in two benches. The depth, close to or a little lower than the modern adit level, was presumably limited by the inflow of water. A radiocarbon date on a sample of burnt wood gave a calibrated date for the retreat mining of between 380 and 160 BC or 140–120 BC (BM-2483), and the previous sample (PRL 932) with a date of 850–250 BC is likely to have come from this area. It is probable that work continued using firesetting to widen the stope after the limiting depth was reached, no doubt coinciding with the removal of pillars between the two finger stopes that together constitute the long stope.

The floor in the long stope, as elsewhere in the mine except where firesetting in the floor was still in progress when mining ceased, was covered in debris. This was composed of mechanically broken material, possibly from the use of hammers and gads, and maybe a battering ram together with considerable firesetting debris. The latter was made up of thin flakes of material, much of it small but with some pieces up to 30 cm in one or more directions and 5 cm or even thicker. There are large amounts of burnt wood fragments ranging in size from sticks to small logs 50 cm or more long and 5–6 cm in diameter, including bamboo and acacia. Fragments of baskets made of reed and/or split bamboo for carrying the ore were found. Unfortunately, none of the baskets recovered at Zawar Mala

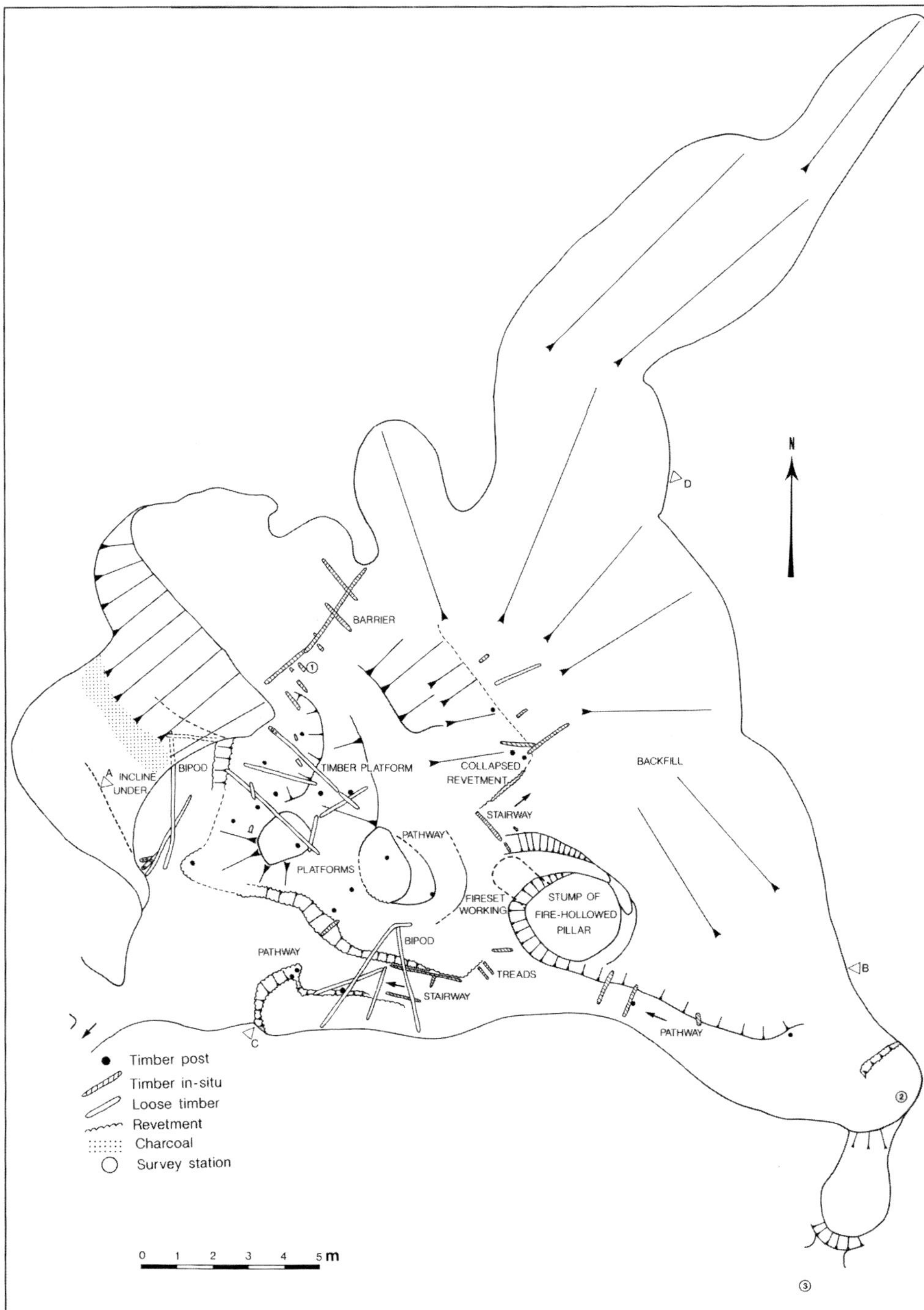

Fig. 5.9 Zawar Mala mine, Timber Chamber: plan.

was in good enough condition to photograph, but undated examples were discovered at Balaria (see Chapter 8, Plate 8.6), Dariba, as well as radiocarbon-dated examples from Agucha (Plate 5.69).

Eastward along the strike from the Double Arches area a path leads to a large chamber of some importance. This was named the 'Timber Chamber' because of the many vertical timbers revetting large heaps of small firesetting debris stacked to form a platform (Plate 5.27 and Figs 5.9 and 5.10). Possibly used as a hauling position from the finger stopes down-dip, it also contained the two timber bipods described below. These bipods were part of the operation to remove a pillar of an earlier working which left a small 'pendule' in the roof (Fig. 5.10) (see Chapter 8, Plate 8.1 for a very similar 'pendule' in the Khetri copper mines). Directly beneath this at the platform level was a small depression in the debris, rather like a shaft or chimney stump hollowed inside by firesetting, and clearly partly broken down mechanically (Plate 5.28). A large pottery sherd found in the 'shaft' may have belonged to a vessel that held water used to cool the heated rock. The two wooden bipods, each with one leg morticed into the other (Plates 5.29 and 5.30), are conjectured to have been part of the support for a battering ram that would have been swung against the pillar, weakened by firesetting, although no other parts were found (Fig. 5.11).

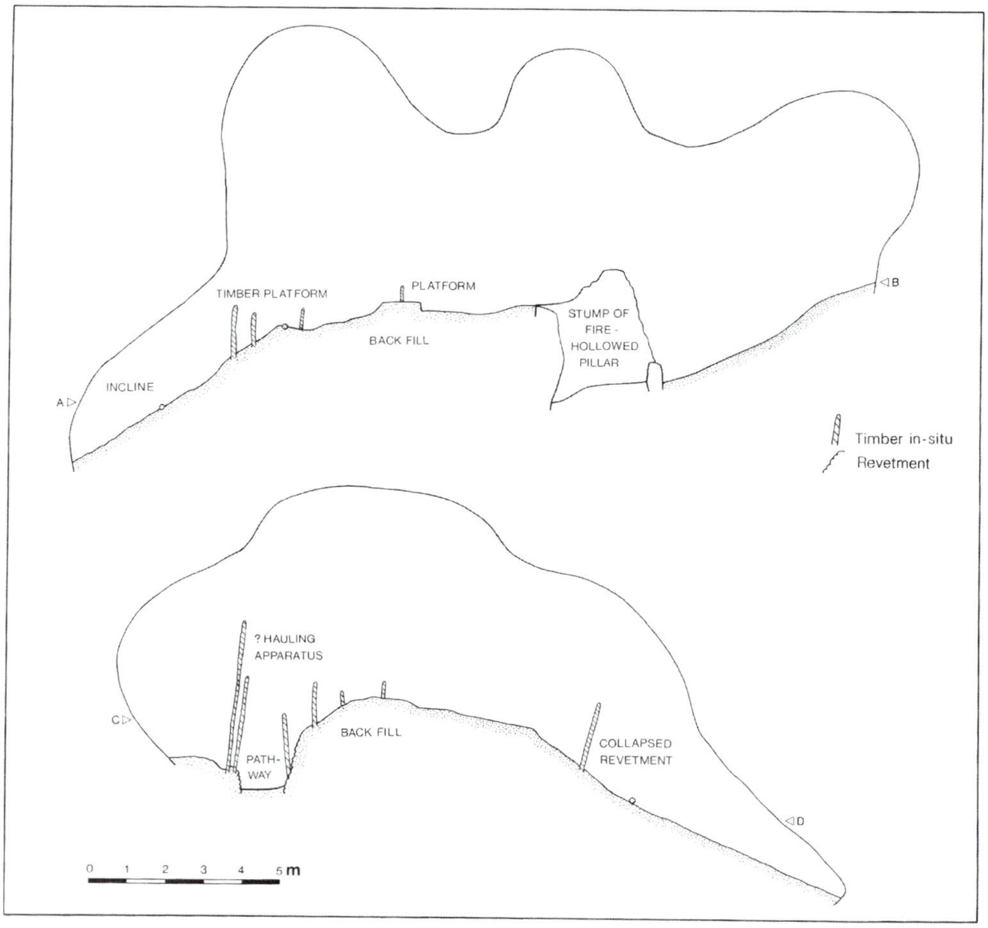

Fig. 5.10 Zawar Mala mine, Timber Chamber: profiles.

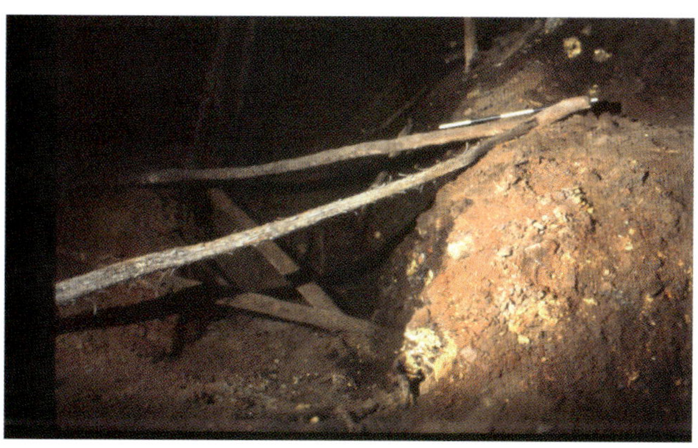

Plate 5.29 Zawar Mala mine, Timber Chamber: bipod lying over pathway.

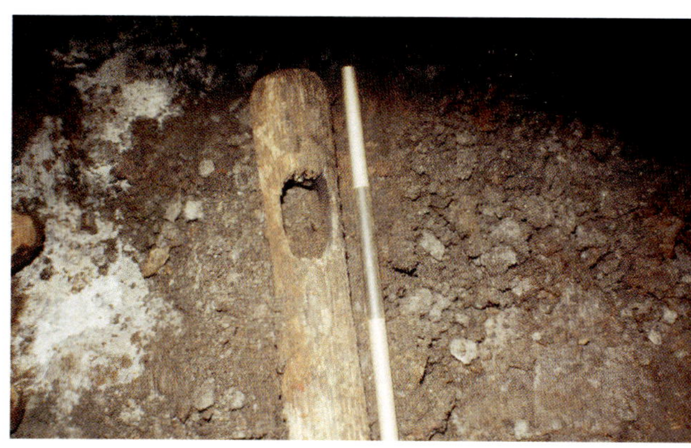

Plate 5.30 Zawar Mala mine, Timber Chamber: detail of mortice joint from one of the bipods.

The path as it entered the chamber was diverted around the revetted platform and protected by a rubble wall (Figs 5.3 and 5.6). As it steepened, wooden steps were used, and above these the steep ramp in rather loose material was stabilised by both vertical and horizontal timbers as already noted (Plate 5.27). The purpose of two very tall timbers, either side of the path, was not determined, but may have been part of a scaffold erected in order to reach the earlier roof above the pillar stump. A sample of wood from the scaffold gave a calibrated date of between 800 and 150 BC (BM-2148R).

From the Timber Chamber, the path led via a metre-high step and circular firehole opening to a further large chamber, for convenience called the Main Chamber. This was of the order of 30 m across and perhaps 20 m high, developed on the dip (Figs 5.3 and 5.6). Removal of a former pillar was obvious from a conspicuous 'pendule' in the roof.

Ceramics were relatively common underground, with many complete or near complete vessels (Plate 5.31 and Fig. 5.12). Some of the pots were identified as type Ahichchtra 10A, datable to the last quarter of the first millennium BC (Hegde 1991: 63).

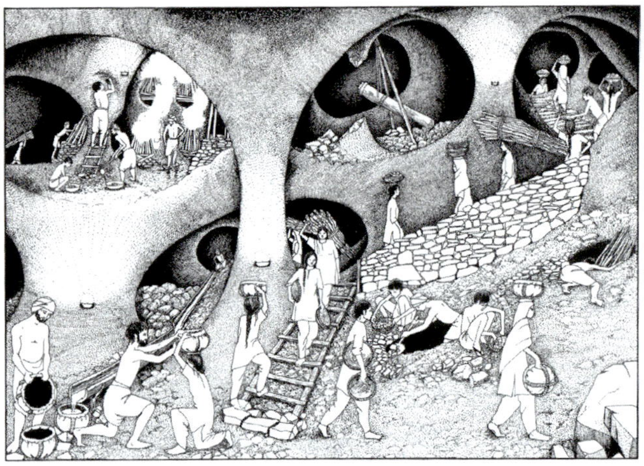

Fig. 5.11 Zawar Mala mine: how the mines might have looked, showing some of the mining operations, although it is unlikely that the firesetting would have taken place while the miners were still in the mine. Note the large number of operatives, evidenced by the wide stairways and lit galleries.

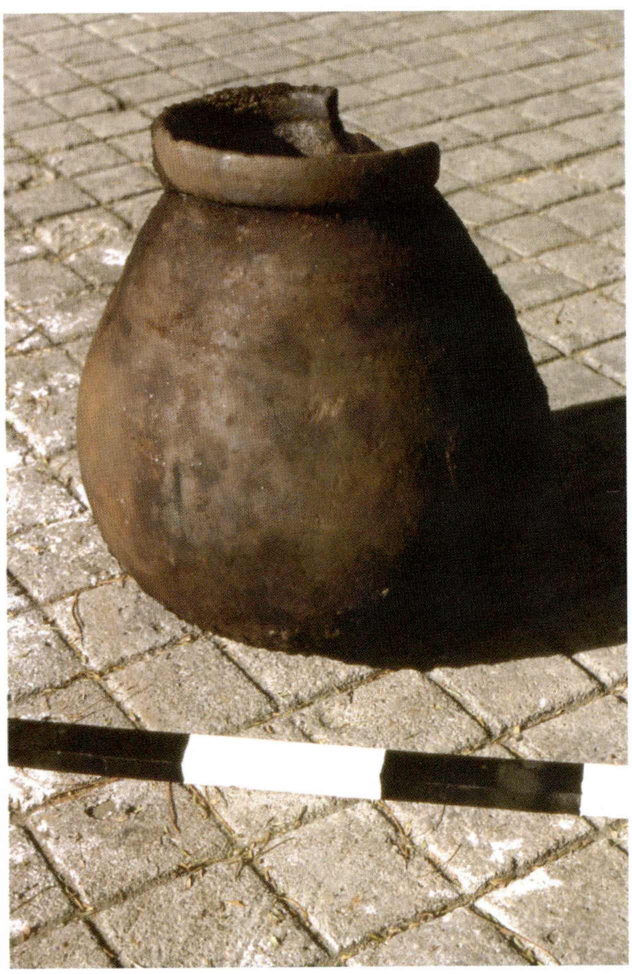

Plate 5.31 Zawar Mala mine: typical large ceramic vessel from long stope (cf. Fig. 5.12).

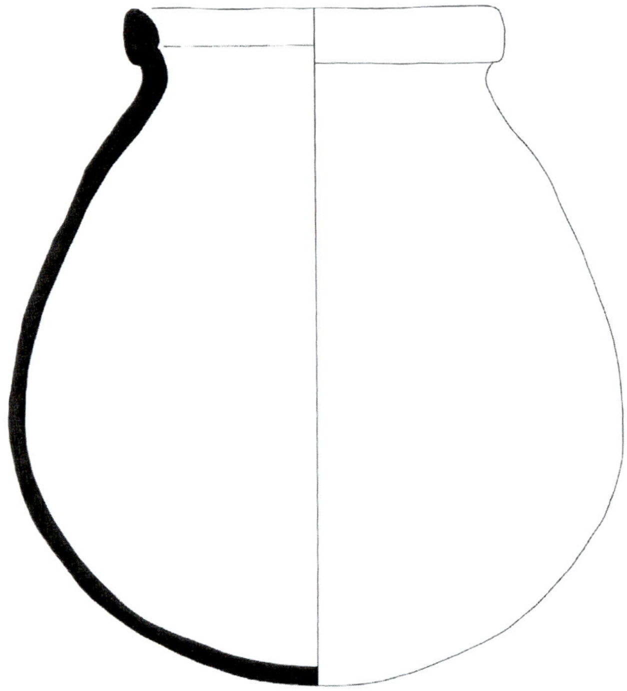

Fig. 5.12 Zawar Mala mine: typical ceramic jar from the long stope (cf. Plate 5.31 for scale).

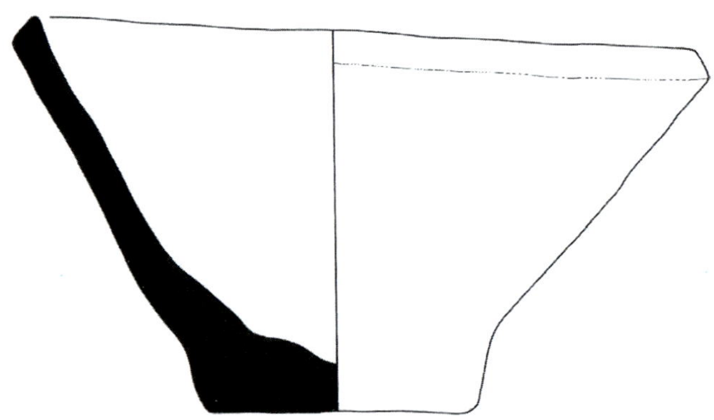

Fig. 5.13 Zawar Mala mine: ceramic lamp from near the bottom of the long stope. Note also the similarity with the vessel found at Dariba Site 10 (see Chapter 6, Fig. 6.56) (rim diameter: 15 cm).

Several pottery lamps were found (Fig. 5.13), resembling small plant pots, some of which originally formed the bases of larger vessels. These were still greasy to the touch, probably residues of the oil they once contained. Their positions on ledges or near the pathway suggest that lamps had semi-fixed locations rather than being carried by individual miners. Small jars, for water or oil, of about a litre capacity were also found, together with some larger vessels of perhaps three or four litres capacity. It is possible that the larger vessels were used for carrying water out of the mine.

The construction of good paths and ladderways, wide enough to allow workers carrying loads to pass each other, and the use of semi-fixed lamps illuminating the whole space rather than personal lamps, suggest the presence of a large labour force. There would have been a busy scene underground with perhaps hundreds of miners and workers, some carrying fuel or support timber in, and ore and rock waste out, probably using baskets carried on the head, with others conveying water in a similar way in jars (Fig. 5.11). Mining the deepest workings was perhaps considerably assisted by drought; after the four years of drought just suffered at Zawar when our surveys were conducted in the mid-1980s, inflows had diminished to almost nothing,

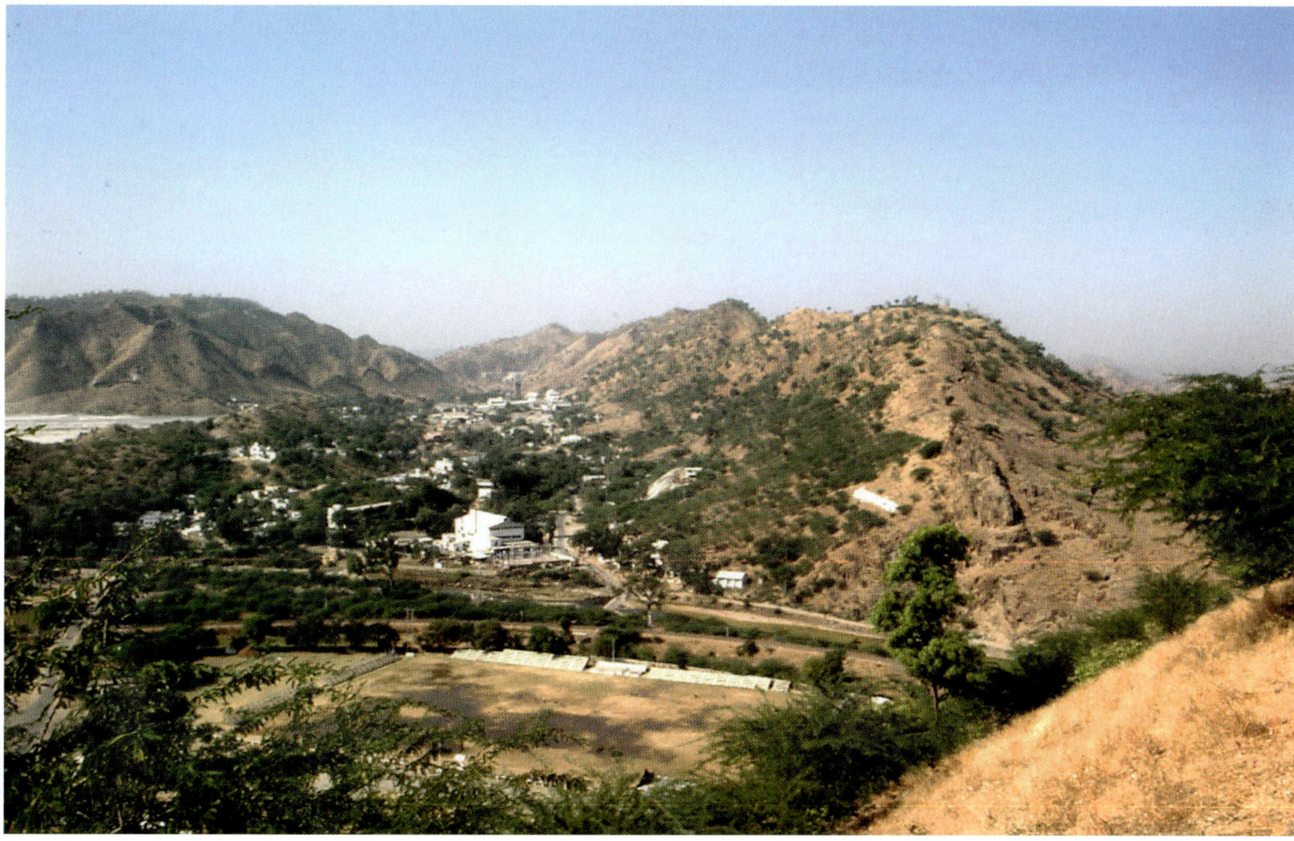

Plate 5.32 Zawar Mochia Magra from Balaria Magra looking north west. Compare with similar view (Chapter 3, Plate 3.18) taken 30 years previously.

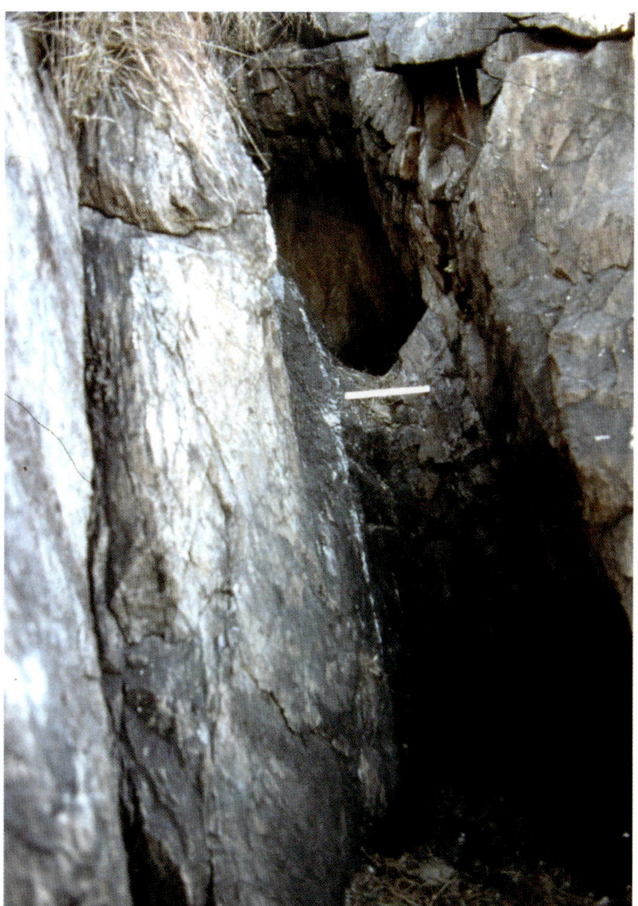

Plate 5.33 Zawar Mochia West: typical trench working at surface. Note the bridges left periodically in section.

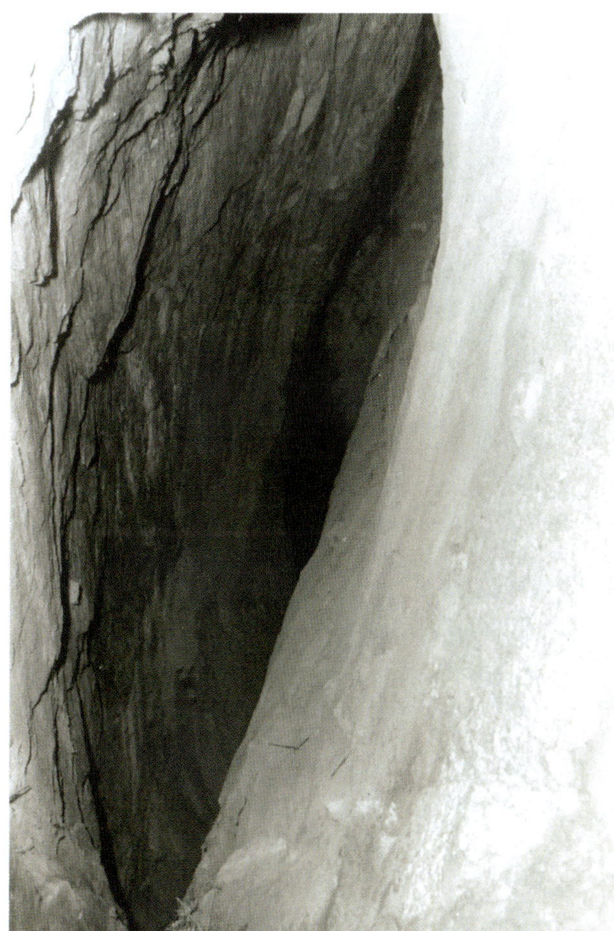

Plate 5.34 Zawar Mochia East: characteristic ovoid gallery section.

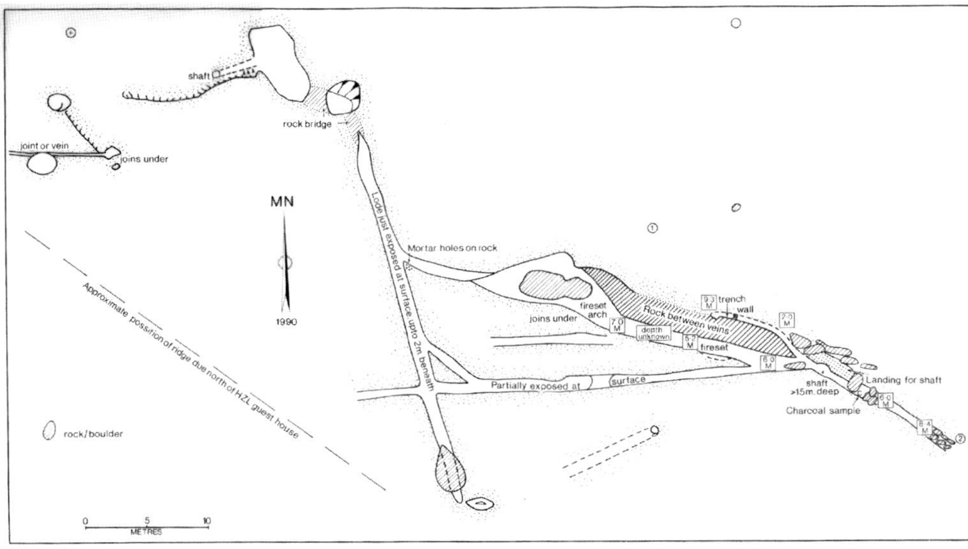

Fig. 5.14 Zawar Mochia east: plan of workings on the east slope towards the River Tiri.

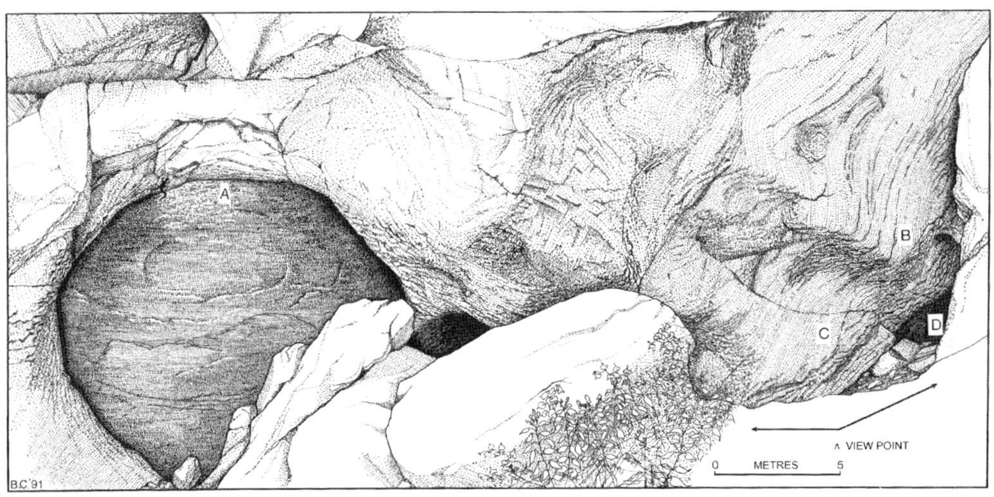

Fig. 5.15 Zawar Mochia east: post-medieval period opencast working using inclined benching which cuts into the earlier vertical stoping on two parallel veins: to the left (A) a large arch cuts into the remnants of the farther vein. To the right (B) an upper and earlier phase and (C) lower fireset bench have just cut through the nearer vein (extreme right [D]).

although 19th-century descriptions of the abandoned workings frequently described them as being flooded. Otherwise keeping the mines drained at depths well below the water table must have been a major undertaking and it would have made sense to work the deposit as quickly as possible with a large labour force. No drainage adits were found at any of the mines.

Firesetting was the major means of extraction (Chapter 8). The characteristic holes created by the fires suggest that the main impact was in the floor, with a reduced effect the higher above the fire the flames climbed. This produced the characteristic arched roof form. Sinking a decline was the preferential mode of attack while working up-dip was done by creating and driving forwards holes in the floor. As long as the fire was not extinguished by a large inflow of water, then the pore-water in the rock would expand as steam, creating rock-breaking forces parallel to the curved wall of the firehole. A typical depth affected by a single burn seems to have been about 5 cm. Many large pieces of rock, sometimes too heavy to lift, suggest that mechanical breaking methods were also used, probably employing iron tools and perhaps a battering ram. Gad and wedge marks were observed, mainly in narrow slits. However, investigation using a sharp pointed geological hammer indicated that tool marks do not show up well due to the sheared nature of the rock, which cleaves off very cleanly.

The nature of firesetting must have created major ventilation problems. Nowadays natural ventilation is good in most of the mine but is assisted by electric fans. Because of the vertical extent and heating of the air during firesetting, the overall original flow would probably also have been reasonable. However, this is not true for the finger stopes, and in the pillar and stall, or in the wide stope areas, where the air currents would have been dispersed. The ample roof height, as well as assisting human movement and carriage of materials, may have been necessary to allow for adequate air flows. It is noticeable that on many slopes the waste rock was not packed to the roof but considerable gaps of a metre or more were often left. It is possible that this was carefully judged to maintain the required air flow (Plate 5.22). It is not known how the firesetting was organised although the fireholes

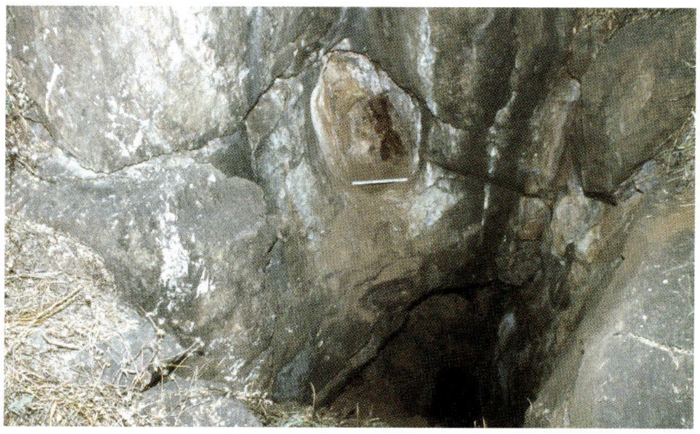

Plate 5.35 Zawar Mochia East: incline. Note the small fireset hollow in the side (scale: 30 cm) (cf. Fig. 5.17).

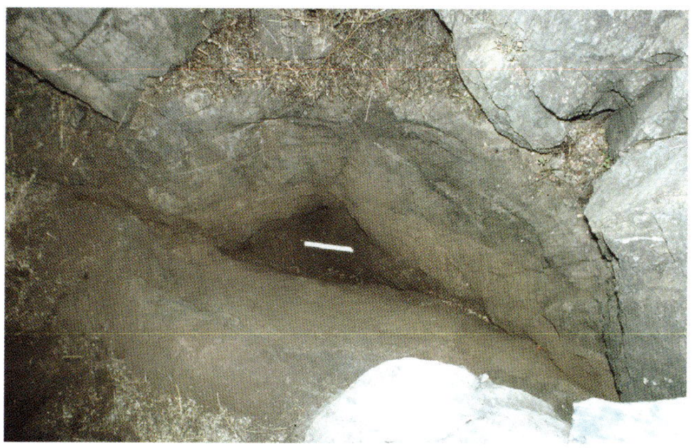

Plate 5.36 Zawar Mochia East: small fireset hollow (scale: 30 cm at bottom) with post-medieval radiocarbon date (cf. Fig. 5.18).

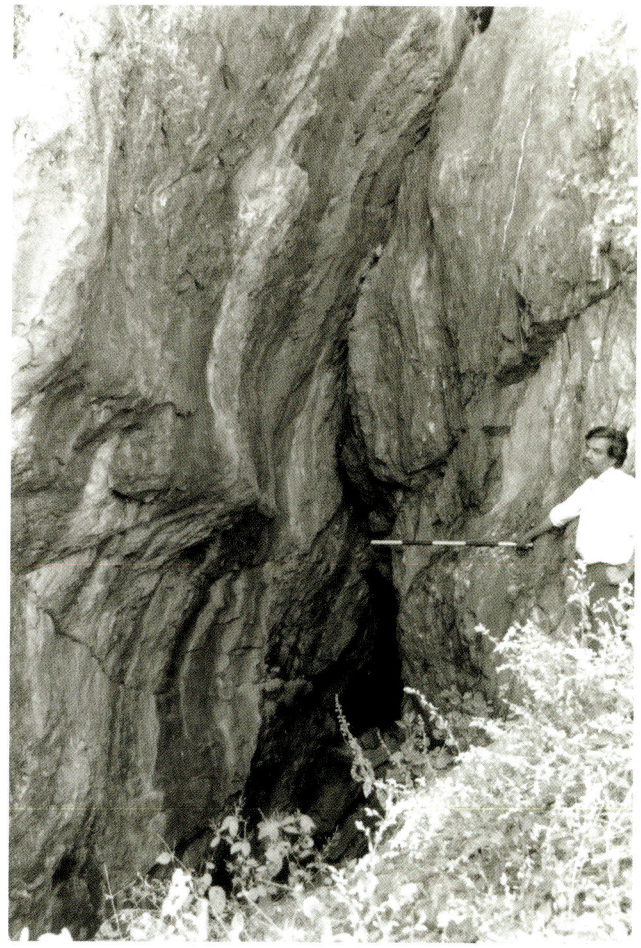

Plate 5.37 Zawar Mochia West: the intersection of two mined veins meeting one above the other.

are often small, suggesting that continuous fuelling may have been necessary during a burn. This raises questions of ventilation: if the air inside the mine became cooler than outside during daytime, reversal of ventilation might have taken place, allowing the fires to be approached while still burning and replenished. Fires for ventilation would also have been necessary as otherwise there would have been a substantial danger of carbon dioxide and monoxide build-up in the finger stopes and consequent suffocation. The combination of the smoke, together with the oxides of carbon and sulphur often coupled with high humidity, must have made working conditions difficult at best.

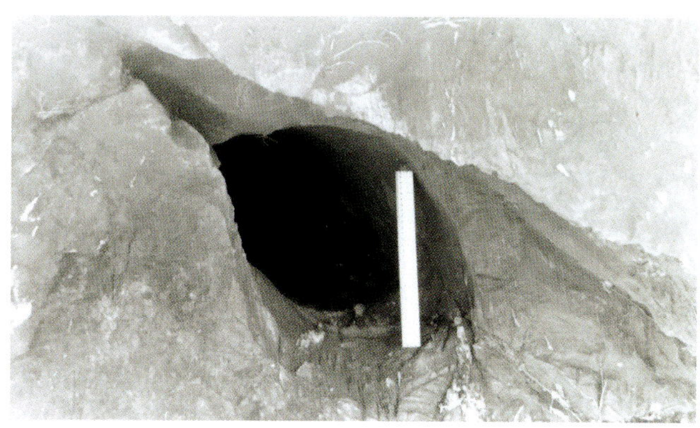

Plate 5.38 Zawar Mochia West: small fireset cross-cutting passage into an adjacent stope (scale: 30 cm).

Mochia Magra

Mochia Magra, separated by the gorge of the River Tiri from Balaria Magra (Plate 5.32) (see also Chapter 6, Fig. 6.2), has a series of parallel, almost vertical ore-bearing beds outcropping as a high ridge for a strike length of nearly 7 km (see Chapter 4, Fig. 4.5). The orebody is in general quite wide, and beds and ore shoots have been removed in the past across a total maximum width of around 30 m, but this varies considerably.

Most of the ridge is characterised by trench-type workings (Plate 5.33 and Figs 5.14 and 5.15), or by inclines and shallow shafts (Fig. 5.16), exploiting individual beds, presumably the richest. Use of round-pointed gads and wedge-ended chisels was common, and firesetting was also employed, producing the characteristic ovoid gallery sections (Plate 5.34) and rounded, hollow depressions in walls, roofs or gallery ends (Plate 5.35 and Fig. 5.17). Underground exploration in the near-surface workings during the survey was limited due both to time and the instability created by modern stopes, which have drawn off material below. Access seems mainly to have been by short inclines, perhaps because of the use of firesetting, or by steps cut in the rock or sides (Fig. 5.15D). Small timbers, known in English by their German name, *stemples* (see below), fitted in natural fissures or in 'egg and eye' holes, seem to have been used for climbing and for supporting waste overhead.

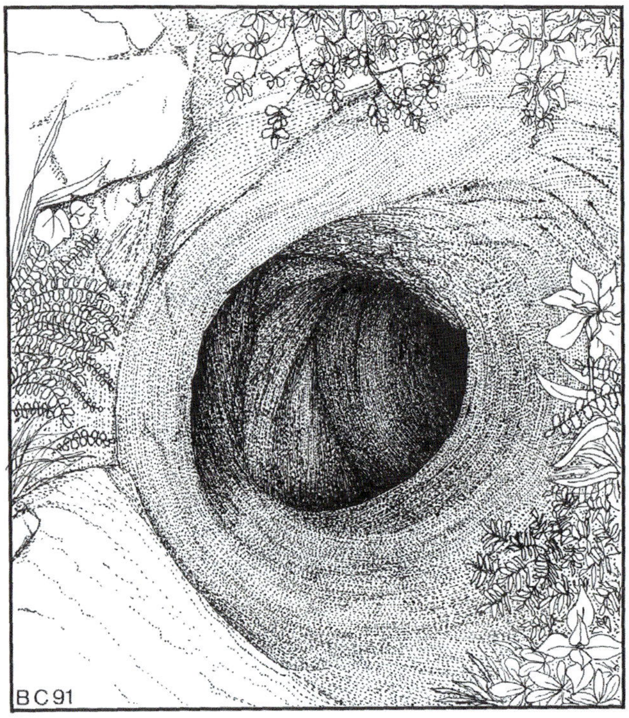

Fig. 5.16 Zawar Mochia east: fireset spiral shaft.

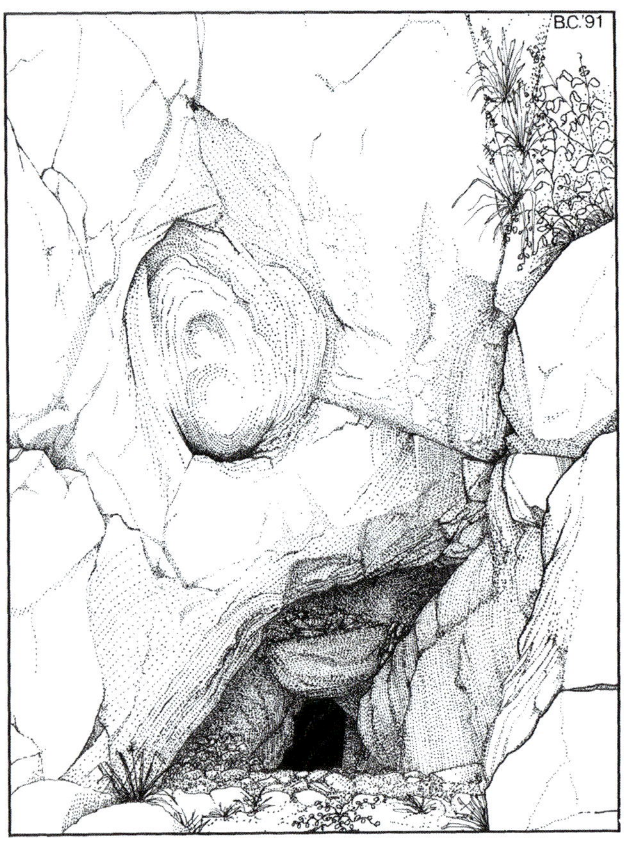

Fig. 5.17 Zawar Mochia east: fireset incline with small fireset hollow above (cf. Plate 5.35).

The tops of many shafts (Fig. 5.16) and some open stopes (Plates 5.36–5.38) occur along the ridge. Underground, ancient workings are found down almost to river level in what are now the upper levels of the modern Mochia mine, a depth of almost

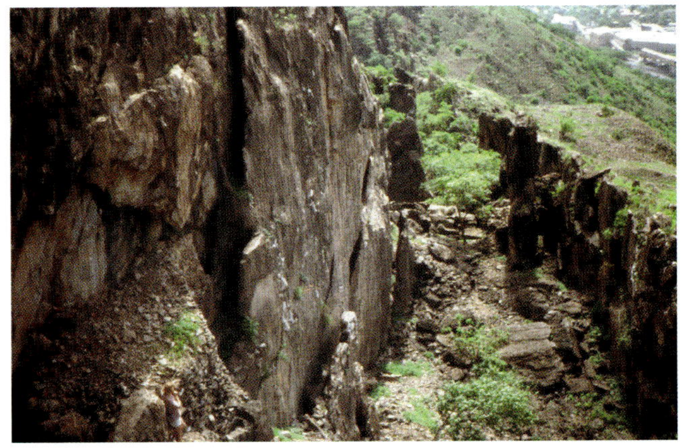

Plate 5.39 Zawar Mochia West: large quarry worked by firesetting.

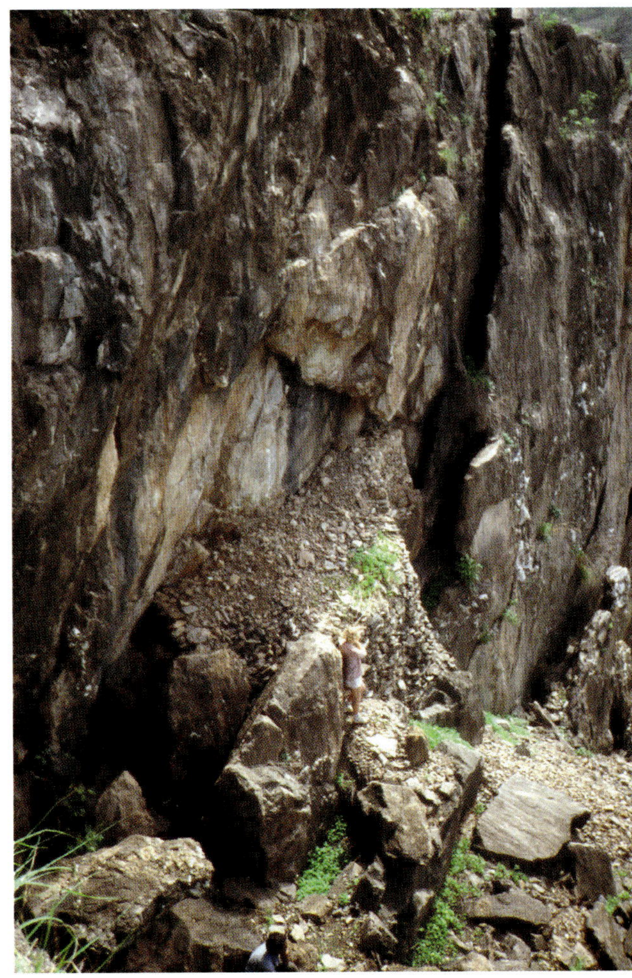

Plate 5.40 Zawar Mochia West: firesetting behind the wall to the left to bring down huge slabs of rock. Note the large slabs of fallen rock beneath.

200 m. Apart from a few holes visible at the top of inaccessible stopes or others infilled with debris, little or nothing can be made out of the detail. Iron gads, a draw hoe for filling baskets and two iron pestles (see Chapter 8, Plates 8.2 and 8.3), together with woven basketwork, rope and timbers, have been recovered during modern mining, but the locations were not recorded. Radiocarbon dating of a sample of wood from a pit prop gave a calibrated date of 770–680 BC, 660–630 BC or 600–360 BC (BM-2381).

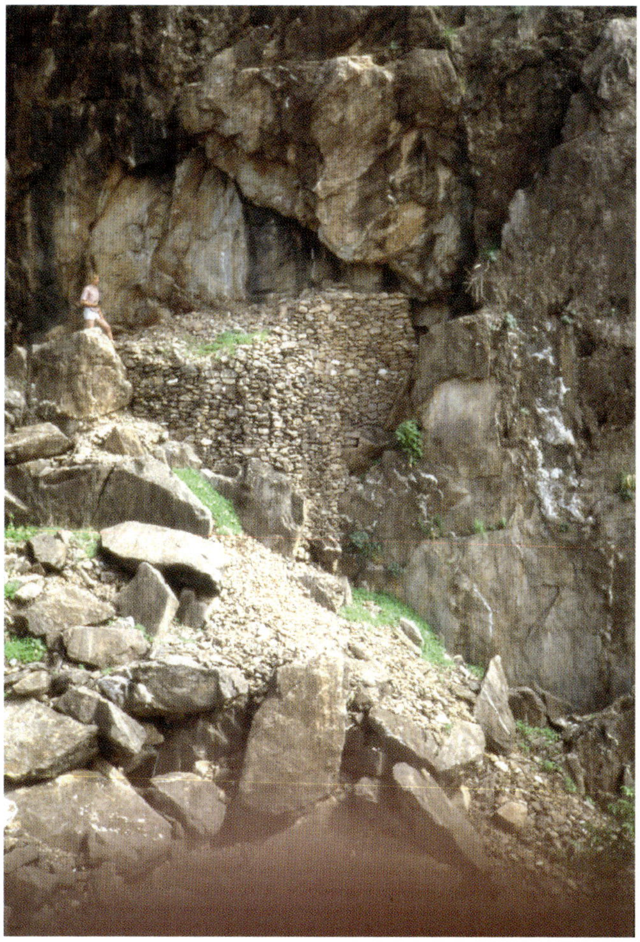

Plate 5.41 Zawar Mochia West: packs of stone rubble built to provide a platform for firesetting work.

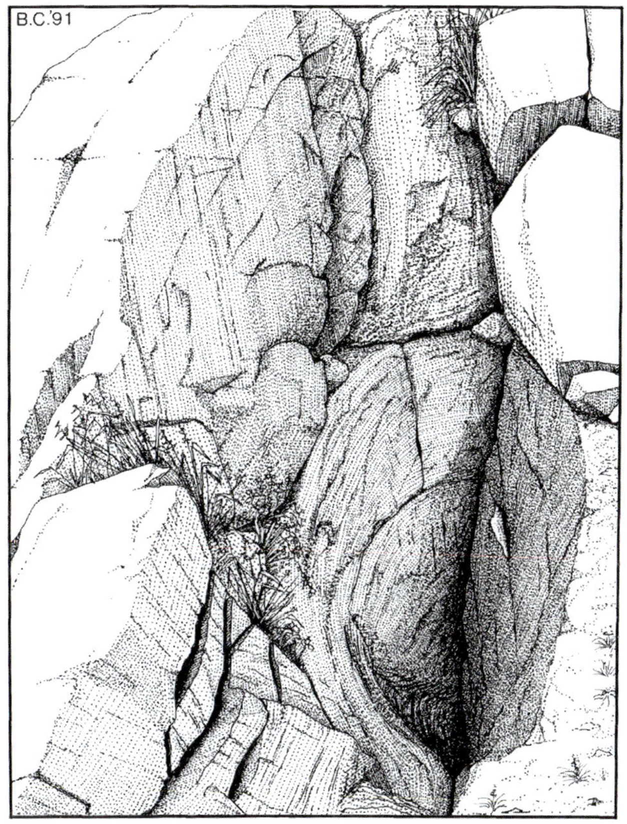

Fig. 5.18 Zawar Mochia east: small vertical working from the surface exploiting a joint (cf. Plate 5.36).

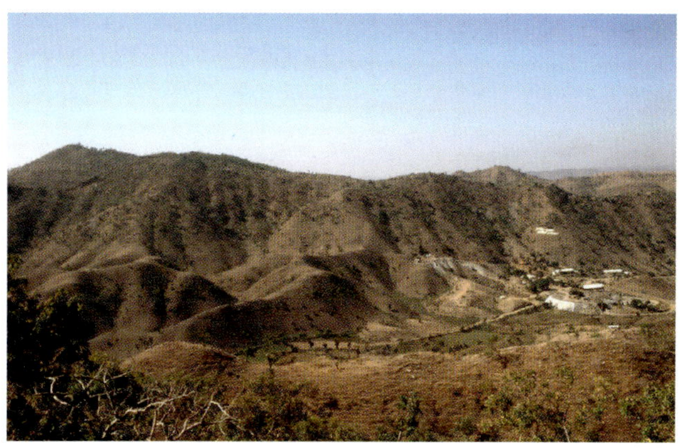

Plate 5.42 Zawar Baroi Magra seen from Zawar Mala.

At the western end of the open workings on Mochia, as well as smaller trench workings of the type noted above, a large opencast quarry has been developed, 30–40 m at its deepest (Plate 5.39). It is between 10 and 40 m wide and runs for at least 300 m along the strike, although the original extent is obscured by cave-ins from modern stoping. The quarry cuts into older workings and clearly postdates this earlier phase. The method of extraction seems to have been to cut upwards, probably from the original ancient underground workings, enlarging them to form the present large quarries. The individual beds were cut from behind causing them to fall to the quarry floor, no doubt assisting fragmentation (Plate 5.40). Platforms were constructed of stone and beds of brushwood upon which the fires could be built. In two places steep inclines of stone rubble were built to provide access to the last work platforms (Plate 5.41). Hammer and gad, or some form of pick, seem to have been used on the fire-weakened rock. Large quantities of benefication waste can be found downhill adjacent to the workings although part of this may have derived from earlier shaft-type workings.

The date of working of the trenches and opencast has not been fully determined. Charcoal from the firesetting debris in one small open working at the east end of Mochia (Plate 5.36 and Fig. 5.18) produced a calibrated date of AD 1430–1640 (BM-2666), a date supported by the post-medieval ceramics from the two excavated sites on Mochia East. It is presumed that some at least of the present workings are ancient, but it seems likely that much reworking took place in medieval and post-medieval times, if only as the source of ore for the smelting operations that generated the huge amounts of zinc smelting debris in this area.

Baroi Magra

Baroi Magra runs approximately north–south to the east of Zawar Mala (Plate 5.42) (see also Chapter 6, Fig. 6.2). This ridge has a number of shallow workings and deeper ones have been revealed by boreholes. However, the workings are badly weathered and partially collapsed. The problems of access and worthwhile examination are made even more difficult by quantities of wet bat excreta.

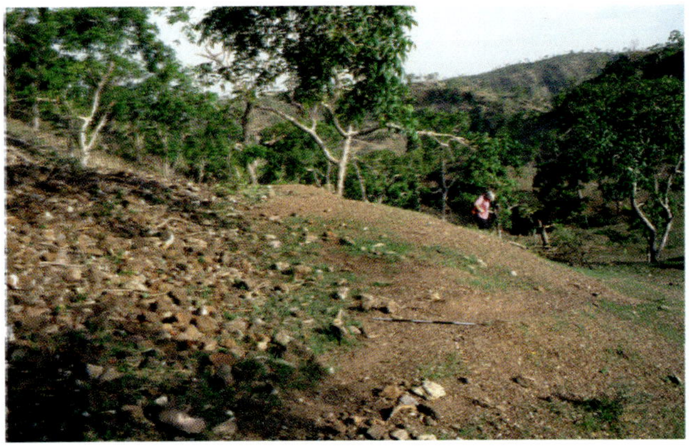

Plate 5.43 Zawar Baroi: heap of beneficiation debris. Note the smooth profile of the heap on the right-hand side, largely composed of pea-sized stones, partially overlain by more irregular mine debris on the left.

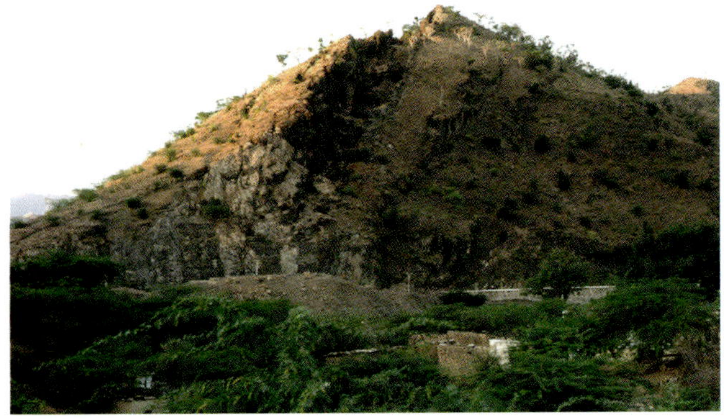

Plate 5.45 Zawar Balaria Magra seen from the east end of Mochia Magra across the River Tiri. The open workings on the west end of Balaria have collapsed into the underground workings below (cf. Plate 5.46).

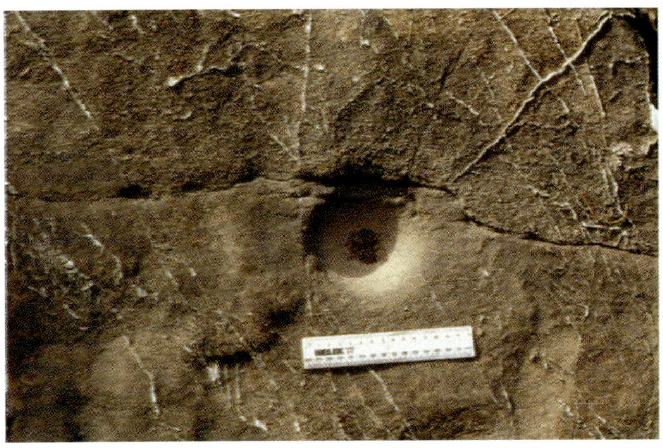

Plate 5.44 Zawar Baroi: typical small hollow worn in the rock surface by its use as a crushing surface (on either side of the 30 cm scale).

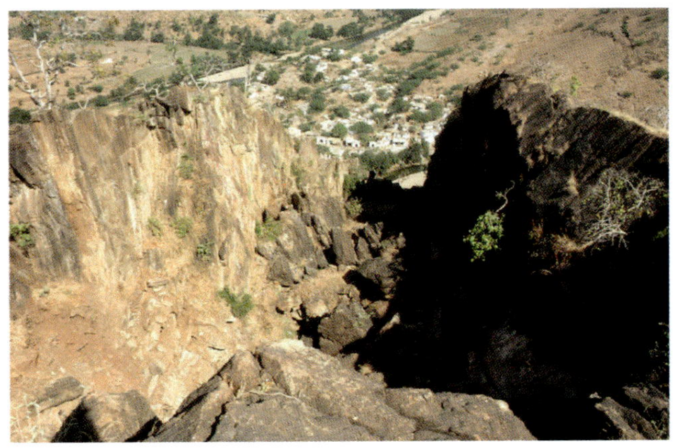

Plate 5.46 Zawar Balaria: looking down the open working that has collapsed into the workings beneath, towards Mochia East (cf. Plate 5.45).

At Baroi Magra, in contrast to the other Zawar mines, the ore is mainly of galena, probably worked primarily for its silver content, with only a little zinc, and the mines are believed to be post-medieval in date. In addition to the rough mine debris scattered around the mine entrances, but still within a short distance, are sizeable heaps of regular pea-sized rock fragments that are clearly the remains of controlled ore crushing (Plate 5.43). These heaps seem to be primary with no evidence of being reworked to recover ores not sought the first time around. Unlike the mine debris, these heaps are concentrated in certain areas on Baroi.

A few hammerstones of hard rock cobbles, typically weighing between 1 and 2 kg with clear marks on their ends evidencing their use as crushing tools, were found in the vicinity of the heaps on Baroi. There was no evidence of notching to suggest that the stones were originally hafted, and it is likely that they were handheld.

In many of the flat upper surfaces of rocks near to the heaps are small semicircular depressions approximately 20 cm wide at the top and about 10 cm deep (Plate 5.44). These are likely to have formed during the crushing of the ore. There are also a few larger, more regular mortars that do seem to have been deliberately carved into the rock, similar to those found at Dariba (Plate 5.51).

Washing the ore up in the hills at Zawar, if it took place at all, would seem impracticable given the location at some distance from any source of water apart from that brought out of the mines themselves as part of the drainage operations. It seems likely, therefore, that on Baroi the ore was crushed with the stone hammers and the mineral separated from the gangue by hand picking (as was the practice when the Mochia mine reopened at Zawar in the early 1950s) before further crushing and washing down in the valley.

Balaria Magra

The mines were not investigated underground due to safety concerns. Evidence for early workings exists in the form of shafts, trench workings and mine debris which run along the top of the ridge, such as are found at the other ridges. At the west end there was a major opencast working that seems to have collapsed into the underground workings below (Plates 5.45 and 5.46). A basket (see Chapter 8, Plate 8.6) and a selection of stemples were recovered from the mine (see Chapter 8, Plate 8.7), and a sample of wood from a mine prop gave a calibrated radiocarbon date of between AD 1650 and 1890 (BM-2338), that is sometime between the late 17th and early 19th century when the mines ceased production. Unfortunately, the locations of the props within the mine are not known.

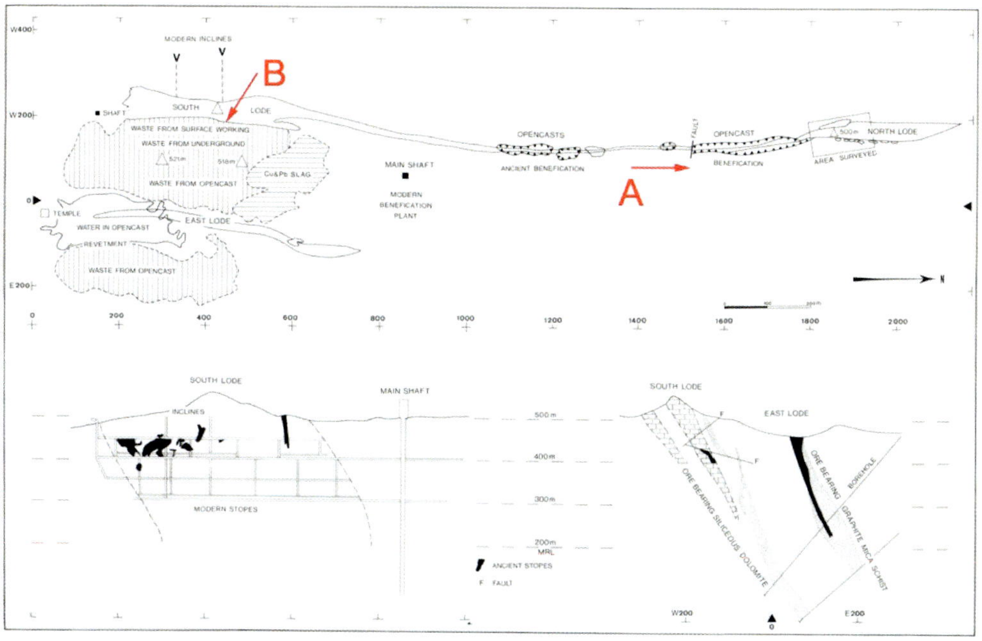

Fig. 5.19 Rajpura-Dariba: simplified plan and section of the principal deposits. The large numbers are the locations of the excavated surface sites described in Chapter 6 (the red arrows A and B indicate the position and direction from which Plates 5.47 and 5.59 were taken).

DARIBA MINES

Depth 64 mm layer 436	timber	TF-1117	2245± 100	410 to 170 BC	800 BC to 1 AD
East lode 4 to 5 mm depth, top of huge open cast mine	timber	NPL-208a	2140 ± 100	370 to 280 BC or 260 to 90 BC	400 BC to 60 AD
South Lode 100mm depth	*Aogeissus* sp	NPL-210	3040 ± 150	1450 to 1060 BC	1650 to 900 BC
South Lode 100mm depth	rope	teledyne	2100 ± 280	500 BC to 250 AD	900 BC to 600 AD
East Lode 263mm depth	bamboo	NPL-209	1790 ± 120	90 to 380 AD	50 BC to 550 AD

AGUCHA MINES

Old mine shaft	wood	PRL-930	2860 ± 100	760 to 680 BC or 660 to 630 BC or 600 to 580 BC or 560 to 390 BC	850 to 250 BC
Mine prop in gallery 30mm below surface	small timber *Terminalia* sp	BM-2356	2240 ± 60	390 to 350 BC or 320 to 210 BC	410 to 160 BC
RA/LW/87/13 south wall exposed by modern opencast	charcoal, *Anogeissus* sp	BM-2630	2320 ± 50	510 to 360 BC or 290 to 250 BC	800 to 700 BC or 550 to 200 BC
RA/LW/87/11 from basket found in old mine, site 11	wood (small branches)	BM-2631	2380 ± 130	770 to 380 BC	850 to 150 BC
RA/LW/87/10, from gallery	outer rings of timber post	BM-2632	2240 ± 50	390 to 350 BC or 310 to 210 BC	400 to 190 BC

Table 5.2 Radiocarbon dates for the Dariba and Agucha mines. Calibrated using OxCal v2. 18 (Bronk Ramsey 1995) and the calibration curve of Pearson and Stuiver (1986). References in the far right column are to the original publication of the figures in *Radiocarbon*.

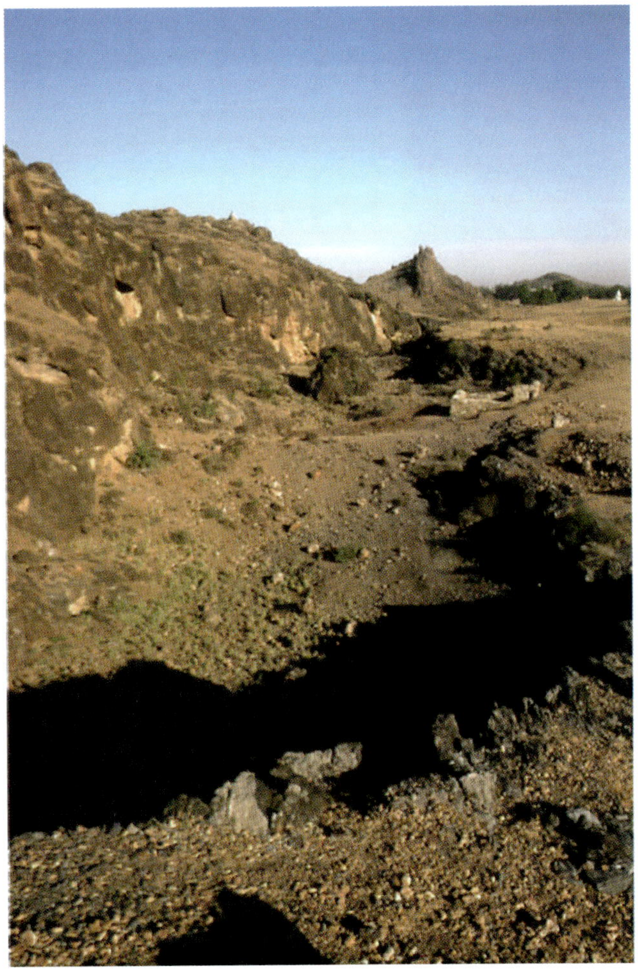

Plate 5.47 Dariba, North Lode: east side of open stoped area, looking north (cf. Figs 5.19 and 5.20). The location and direction of this photograph are indicated on Fig. 5.19, red arrow A.

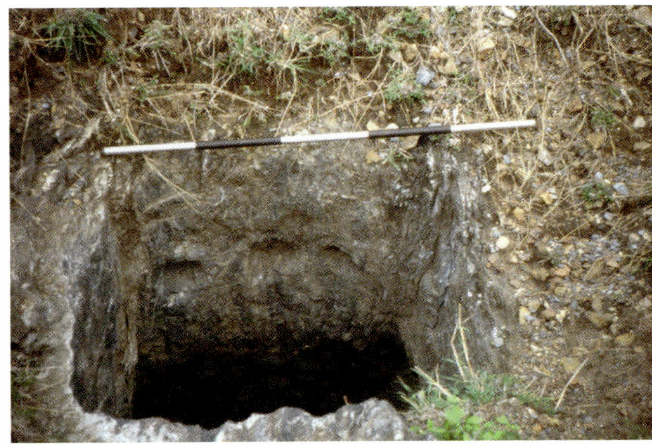

Plate 5.48 Dariba, North Lode: small climbing shaft. Note the footholds visible at the top.

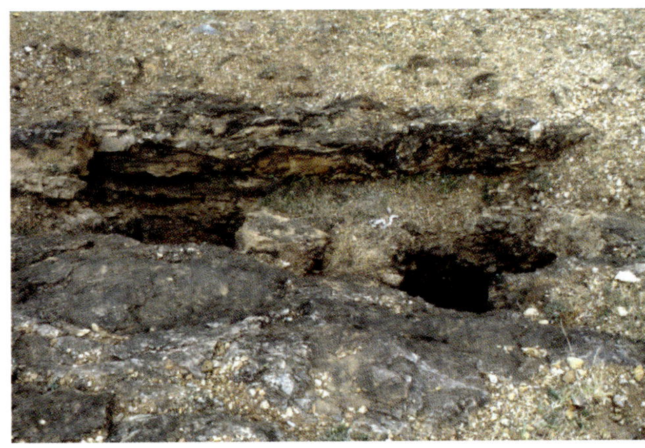

Plate 5.49 Dariba, North Lode: pair of shafts typical of North Lode workings.

Rajpura-Dariba

The site lies about 80 km north east of Udaipur on the Mewar Plain. Metamorphic calc-silicate rocks derived from mineralised sandy-dolomite rocks of Precambrian age form conspicuous ridges, with the plain formed from much greater thicknesses of more easily weathered graphite schist (Fig. 5.19) (see also Chapter 4, Figs 4.6 and 4.7). The main deposit worked at the time of our surveys was South Lode. At the surface there is a red gossan outcrop (see Chapter 6, Plate 6.48), which extends further north from the modern main shaft as the North Lode (Plate 5.47 and Fig. 5.20). The East Lode, which was not being worked at the time of our surveys, is a few hundred metres east and parallel to the South Lode. Its outcrop today is marked by a large ancient opencast working, flanked by extensive waste heaps. The opencast was usually filled with water until the recent past, although it has now been drained. Boreholes have penetrated ancient workings at a depth of 254 m beneath. Ancient workings are also found on South and North Lodes, and further north at Bethumi, where there are large opencast workings cutting earlier mine galleries. The major mines date from the first millennium BC (Table 5.2).

Although argentiferous lead was the chief product, there were extensive silver-rich fahl ores such as tetrahedrite and tennantite deposits worked in South Lode in antiquity. There was also some copper production, probably from the secondary mineralisation. This was worked in the post-medieval period and was still being mined and smelted on a small scale at the time of James Tod's visit in the second decade of the 19th century (Tod 1978, I: 618). Production probably continued for a little longer but was defunct by 1885. There is no evidence that the zinc in the ore was ever utilised in the past.

North Lode

This forms a series of conspicuous ridges rising to about 10 m above the general level of the plain, of metamorphic calc-silicate rock, dipping northward at a steep angle (Plate 5.47) (see also Chapter 4, Fig. 4.6). Ore is found in both the hard calc-silicate and in the adjacent softer graphite schist. At the surface and for a depth of 20–30 m the calc-silicate is within the oxidation zone. Although it is not conspicuously gossanised it is very well jointed, rather rubbly and, with difficulty, susceptible to the use of gad and wedge. Copper staining is apparent, but analysis suggests that the objective was mainly the silver-rich lead ore. The ore may have contained some anglesite and cerussite, but must have been primarily of galena.

Both opencast and underground mining methods are apparent (Plate 5.47, Figs 5.19 and 5.20). The earliest working

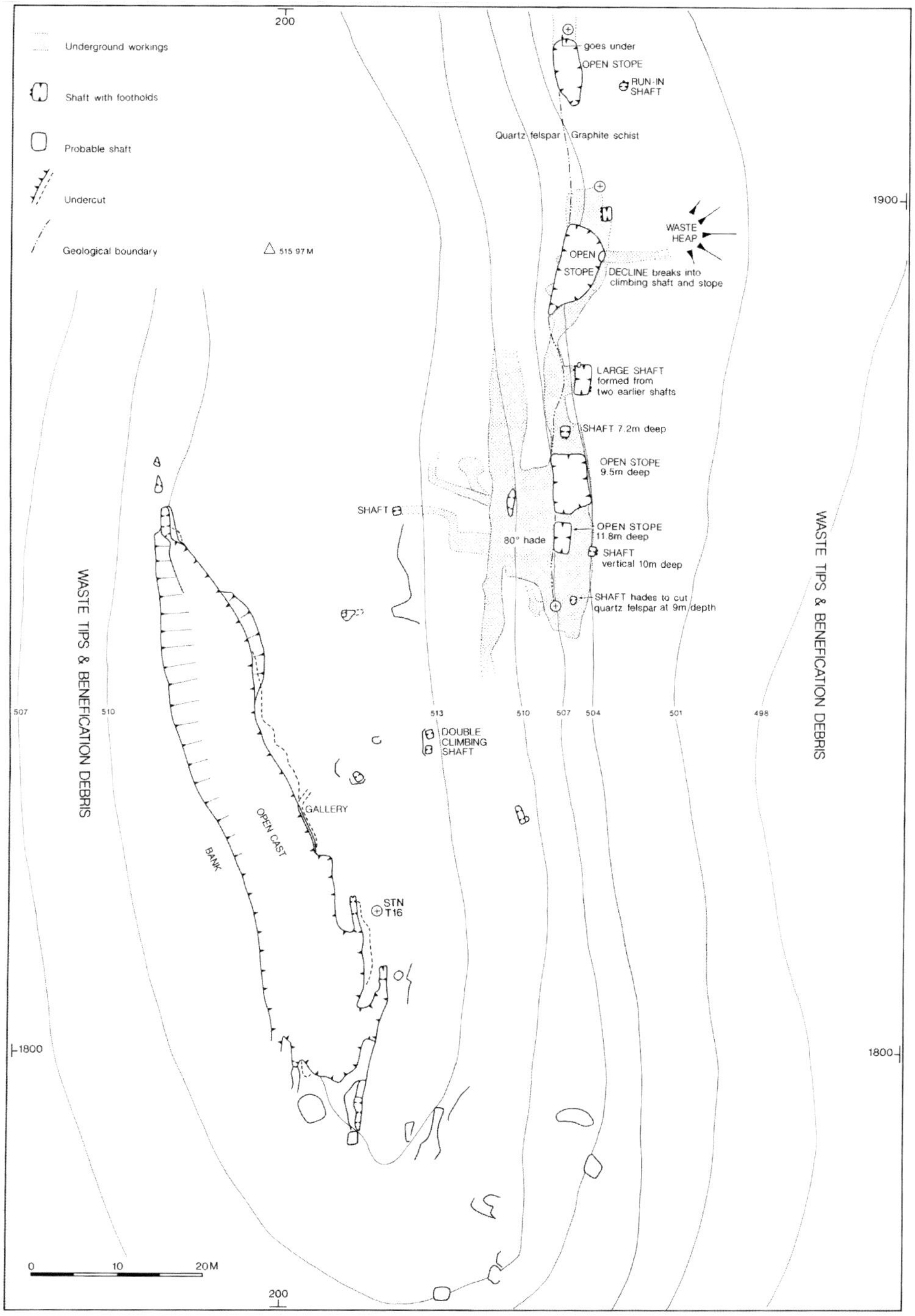

Fig. 5.20 Dariba North Lode: plan of the open stope area (cf. Fig. 5.19 and Plate 5.47).

appears to have been by narrow shafts sunk in ore-bearing beds, following the apparently irregular deposits. The shafts, often with climbing footholds (Plate 5.48), are usually in pairs (Plate 5.49) as at Agucha (see below), generally at intervals of only a few metres. Some shafts have been sunk in the softer graphite schist to meet the hanging wall of the calc-silicate at a depth of about 8–10 m.

On the hanging wall side of the lode in particular, some of the pairs of shafts have been joined, enlarged and fitted with large beam slots to form working and hauling platforms. These open out below into large stoped areas 10–12 m high and wide. There is at least one example of a cross-cut, driven probably to locate other ore-bearing beds a few metres away. Unfortunately large numbers of bats roost there, and as well as the resulting unpleasantness, the floors have been worked by locals in recent times for fertiliser, destroying hopes of locating dating material. On the footwall side, quarrying has cut into the old mine workings, with one quarry largely but not entirely in the graphite schist.

At the southern end of North Lode, a large section of the calc-silicate rock has been completely worked out to form the South quarry (Fig. 5.20). The rock was crushed to a fine gravel in the adjacent area. An outcrop further north, in which the original mine workings were cut into by another large quarry,

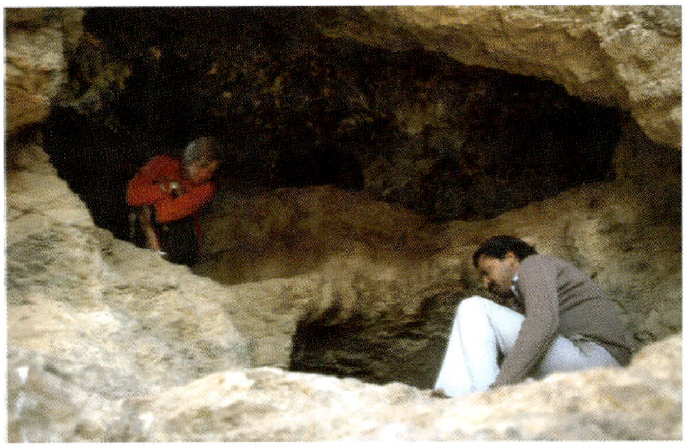

Plate 5.50 Dariba, North Lode: earlier workings exposed by the quarrying operations, reused by quarry miners for beneficiation operations. The crushing hollows are immediately in front of the team members (cf. Plate 5.51).

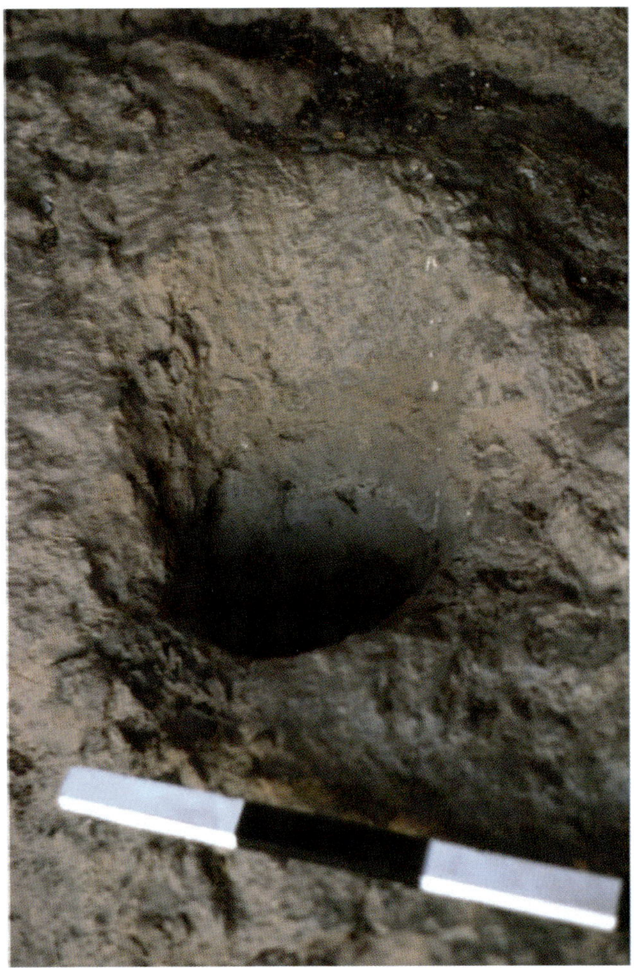

Plate 5.51 Dariba, North Lode: crushing hollow in old exposed mine workings (cf. Plate 5.50).

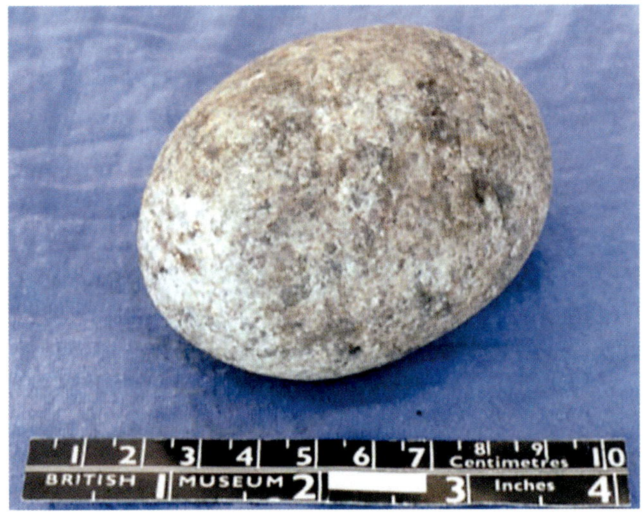

Plate 5.52 Lentoid hammerstone from the post-medieval heaps on South Lode, Dariba (cf. Fig. 5.21).

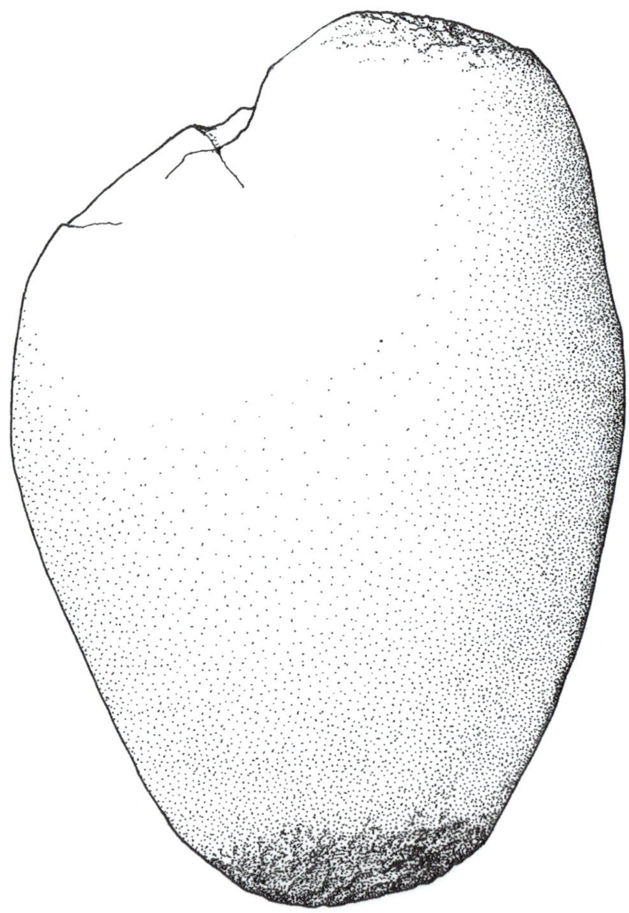

Fig. 5.21 Hammerstone from the North Lode area of Dariba (cf. Plate 5.52 and for scale).

had small but deep mortars formed in the calc-silicate rock just inside one of these mine openings (Plates 5.50 and 5.51). These postdate the original underground mining, but such locations would have been appreciated later as the all-important shade from the sun.

The ore had been crushed nearby as part of the beneficación process. The crushing was carried out with quartzite river cobbles, which had been used as hammers as they were damaged on their ends (Plate 5.52 and Fig. 5.21), against flat quartzite rocks (Fig. 5.22), known in English mining parlance as knocking stones (see Chapter 8, Plates 8.10 and 8.11). Many more might have been found had the area not been used as a source of road stone, gathered from the rock fragments that litter the plain below the mines. The waste product from

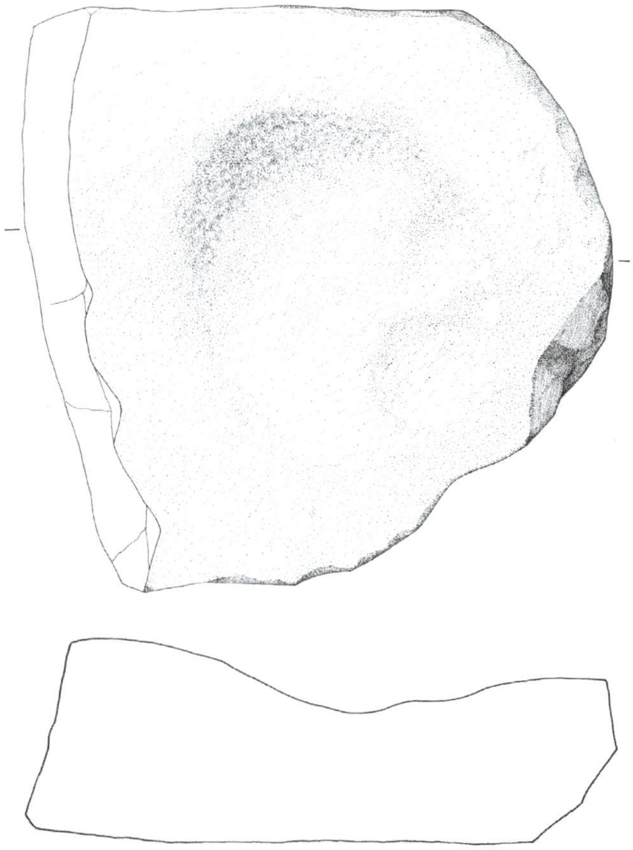

Fig. 5.22 Knocking stone anvil from the North Lode area of Dariba (see Chapter 8, Plate 8.10 and for scale).

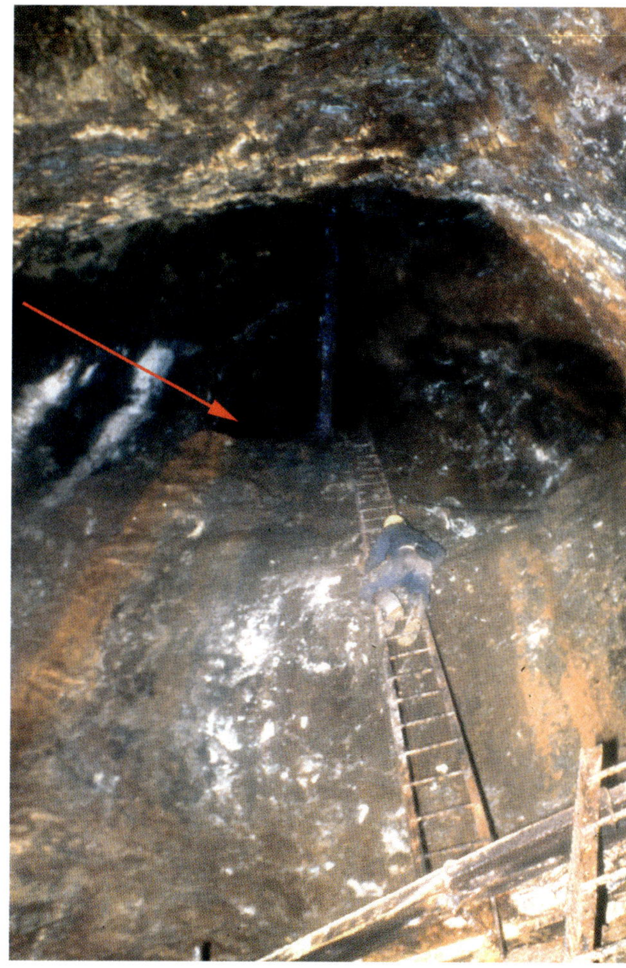

Plate 5.53 Dariba, South Lode: near vertical shaft-like stope with bailing pool at the top, indicated by the arrow (cf. Plate 5.55 and Fig. 5.25).

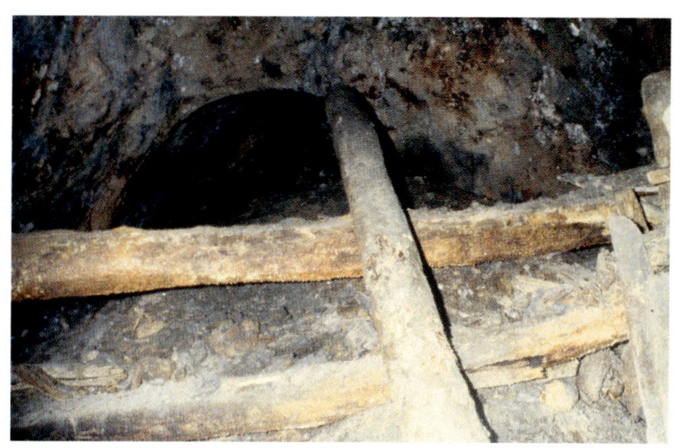

Plate 5.54 Dariba, South Lode: group of large timbers (cf. Fig. 5.24).

crushing was a fine gravel or coarse sand, found in very considerable quantities in adjacent heaps, or as outwash onto the plain. Iron pestles may have been used but the stone hammers found would have sufficed.

It is assumed that the larger scale stoping and the quarrying that cuts the older work is post-medieval – this might possibly account for the post-medieval metallurgical debris found in large quantities on the northern slopes of South Lode (Sites 13 and 14), the small slag scatters designated Sites 1–3 and 5 lying to the east of North Lode, and Site 15 lying to the west, the latter having a post-medieval radiocarbon date (see Chapter 6).

South Lode

The conspicuous gossan outcrop of South Lode (Fig. 5.19) (see also Chapter 6, Plate 6.48) has only one early mine opening in it, but the surface evidences many apparently infilled trenches. Much of the east side of the lode (hanging wall) has been quarried, and resultant heaps of waste overlap earlier debris from shaft mines sunk to meet the lode at depth. This quarrying is probably post-medieval as it is adjacent to extensive lead and copper slag heaps of this age (Sites 13 and 14) (see Chapter 6).

Underground, at depths around 100 m, the modern mine intersects a considerable number of ancient workings, mainly following shoots of lead ore in the footwall of the calc-silicate and the underlying graphite schist. These workings are predominantly fireset and mostly near-vertical (Plate 5.53 and Figs 5.23–5.25). In one small section a pile of timbers up to 25 cm in diameter and over 4 m long were found, partly worked on the ends by an adze to allow them to drop into slots (Plate 5.54 and Fig. 5.24). These timbers (stemples) may have been used either for heavy horizontal support or for spanning large vertical openings, as in the case where such a timber still spans a rectangular-section stope 3.5 × 2.8 m in size (Plate 5.55). A sample of wood from one such prop gave a radiocarbon date of between 380 and 100 BC (BM-2636). Three other radiocarbon dates on timbers from the deep mines of South Lode at the same levels described here, at

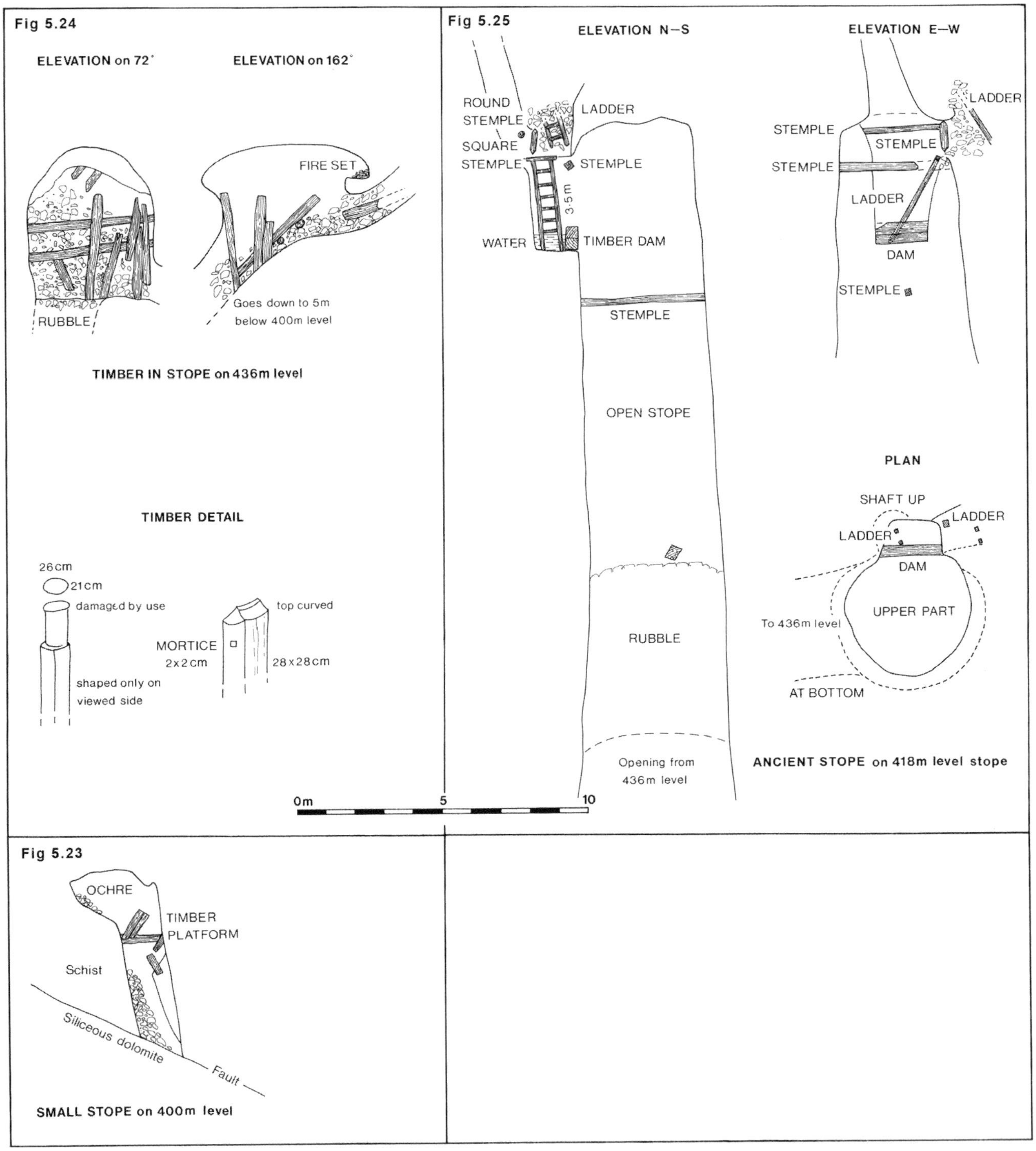

Figs 5.23–5.25. Fig. 5.23 Dariba, South Lode: profile of small stope. Fig. 5.24 Dariba, South Lode: large timbers in stope and details of the joints. Fig. 5.25 Dariba, South Lode: sketch plan and profile of near vertical stope and bailing system (cf. Plate 5.53).

approximately 100 m depth, have been reported ranging from the mid-second millennium BC to the mid-first millennium AD (NPL 210, TF 1117 and Teledyne, Table 5.2). Another chamber had a 3 m long ladder linking one level with another, with morticed rungs about 45 cm apart (Plate 5.56). Nearby firesetting used a timber platform to enable the wall to be attacked well above floor level.

The shaft-like stope mentioned above was some 20 m high, 7 m wide at the bottom and 5 m wide higher up (Plate 5.53). Just below the top a timber dam held back a pool of water in a recess cut into the wall (Plate 5.55 and Fig. 5.25). A ladder, again about 3 m long (Plate 5.57) with widely spaced rungs, linked it to another opening which, although infilled with rubble, had a further ladder just visible rising up a shaft (Plate 5.58). A second shaft rising directly above the dam was not so equipped, and unfortunately was blocked 10 m higher up, to which it had risen in two stages. The two shafts would have ensured air circulation, while the timber beam across the main part of the shaft was probably used for haulage, but could have been the sole survivor of others lower in the shaft that held shuttering to

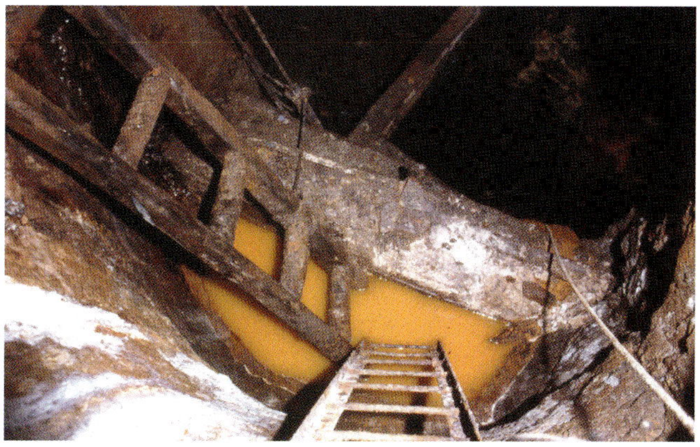

Plate 5.55 Dariba, South Lode: looking down on the bailing pool behind the timber dam at the top of the stope seen from above (cf. Plate 5.53 and Fig. 5.25). Note the large timber *in situ* across the top of the stope, probably used for hauling.

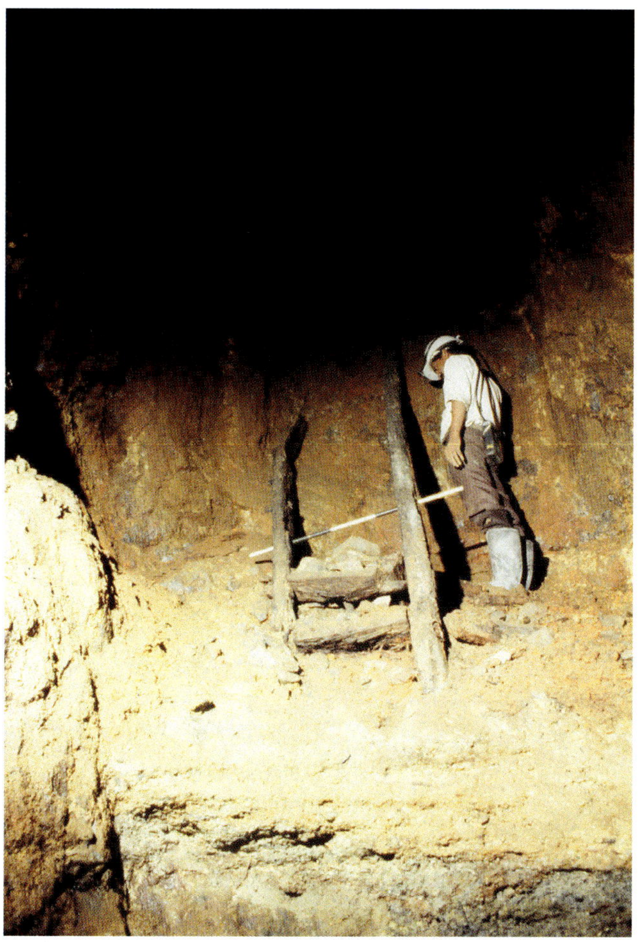

Plate 5.56 Dariba, South Lode: short ladder. Note the width suggesting it was used by people ascending and descending at the same time.

divide air flows as well as providing ladder stagings. The dam and ladder are clear evidence of the system of water removal, carried up in leather buckets or pot containers (see Chapter 8). A sample of charcoal from this complex was dated to 180 BC–AD 60 (BM-2637).

These workings continue below the water table for some 80–90 m at least. Inflow to the modern mine is considerable, but the separate vertical shafts – presumably like the Zawar finger stopes – permitted the water to be controlled to avoid

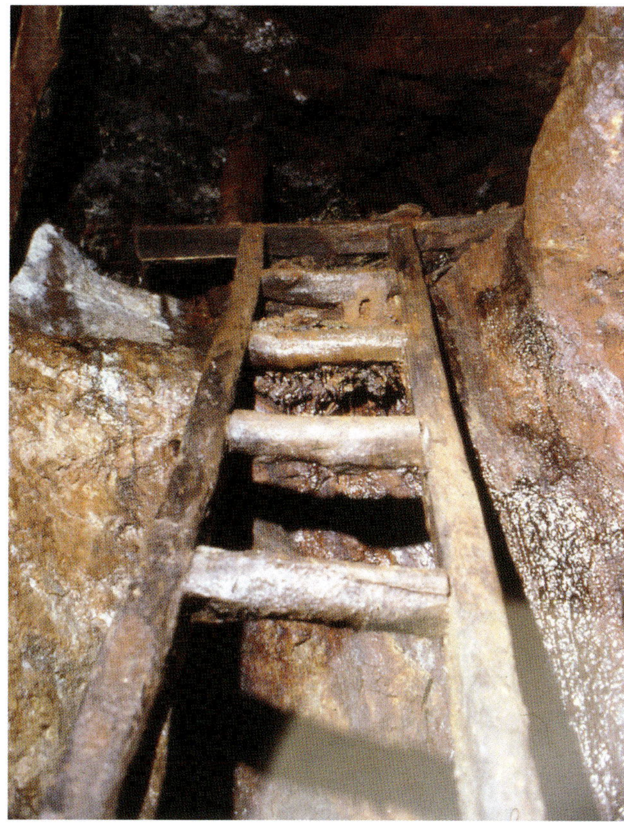

Plate 5.57 Dariba, South Lode: ladder of squared timber with rungs morticed into the sides, leading from the bailing pool up 3 m to a similar pool.

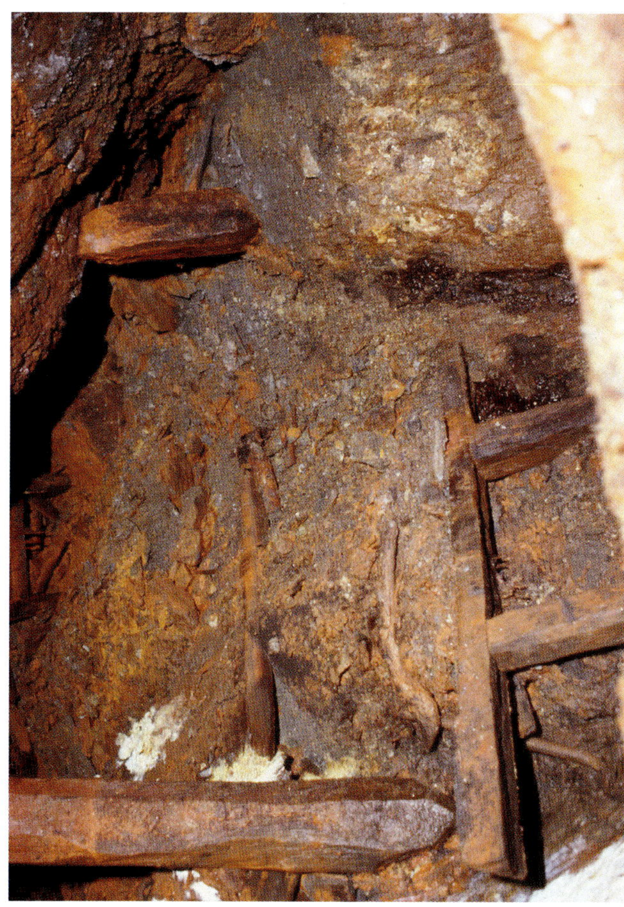

Plate 5.58 Dariba, South Lode: bottom of a ladder rising from the second stage or bailing pool.

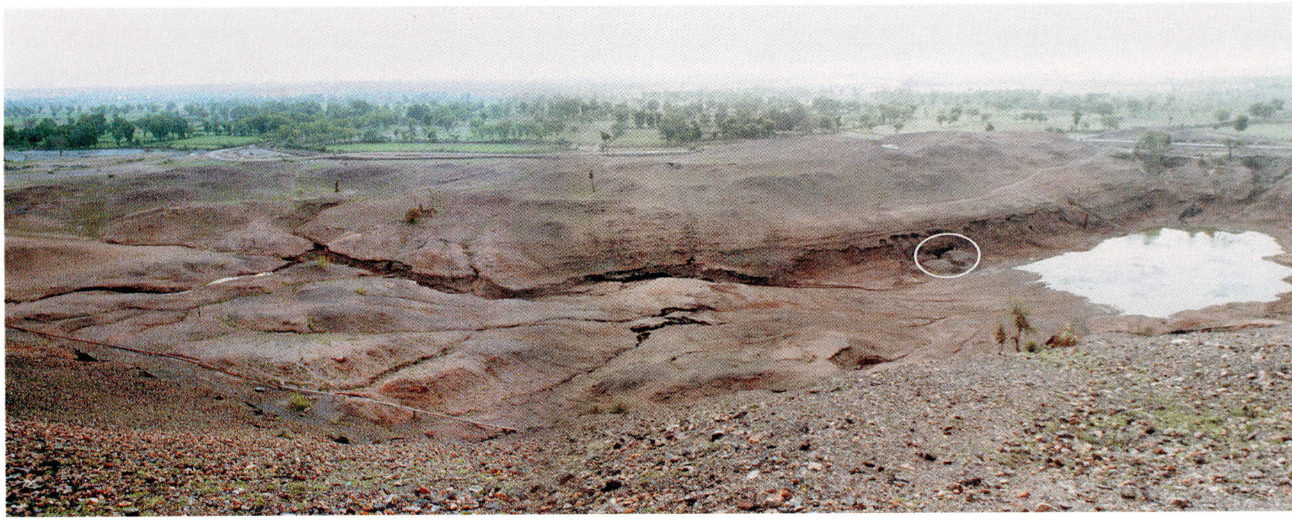

Plate 5.59 Dariba, East Lode: opencast looking south from the South Lode tips, partially drained in 1987 probably for the first time in centuries. Note the tall tips of mine debris dumped on the far east side that must in part have necessitated the need for the revetment. The excavated revetment site is circled. The location and direction of this photograph are indicated on Fig. 5.19, red arrow B.

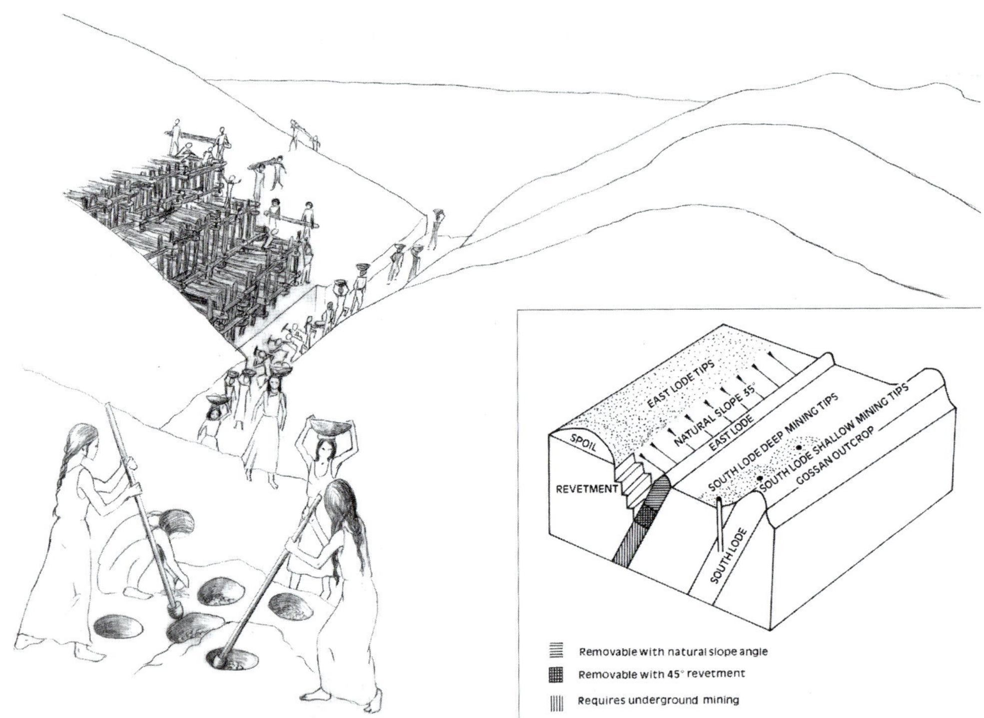

Fig. 5.26 Dariba, East Lode revetment: reconstruction of how a small section of the 300-m revetment may have appeared originally. The dip of the orebody suggests that not more than four steps could have been made before mining had to continue with shafts (see inset).

flooding the rest of the mine. Most of the workings appear at this depth to be in the graphite schist which would have been worked for the silver-rich galena that predominates near the footwall and is fairly watertight. This recalls an old local legend, recounted to the authors by Mr H.V. Paliwal. Durga, the goddess of the miners, instructed them not to be too greedy and remove all the ore but to leave some for her. This command was ignored and all the ore was removed, and as a punishment the mines flooded. This story could contain a memory of the consequences of working both the calc-silicate rock as well as the more impermeable graphite schist which had made the workings watertight.

As these workings are all below the surrounding plain there is no possibility of drainage by adits, but animals, presumably oxen or buffalo, may have been used on some of the very large shaft-like stopes to draw ore or water, or both, using an arrangement similar to that still used to draw water from the wells. The large size of the shaft stope would have enabled a zigzag-inclined ladder system to be constructed for those carrying materials in and out of the mine, in a manner analogous to some 19th-century Scottish coal mines. Remains of woven baskets and ropes about 2.5 cm in diameter have been found in the workings; these could have belonged to raising systems such as those outlined above.

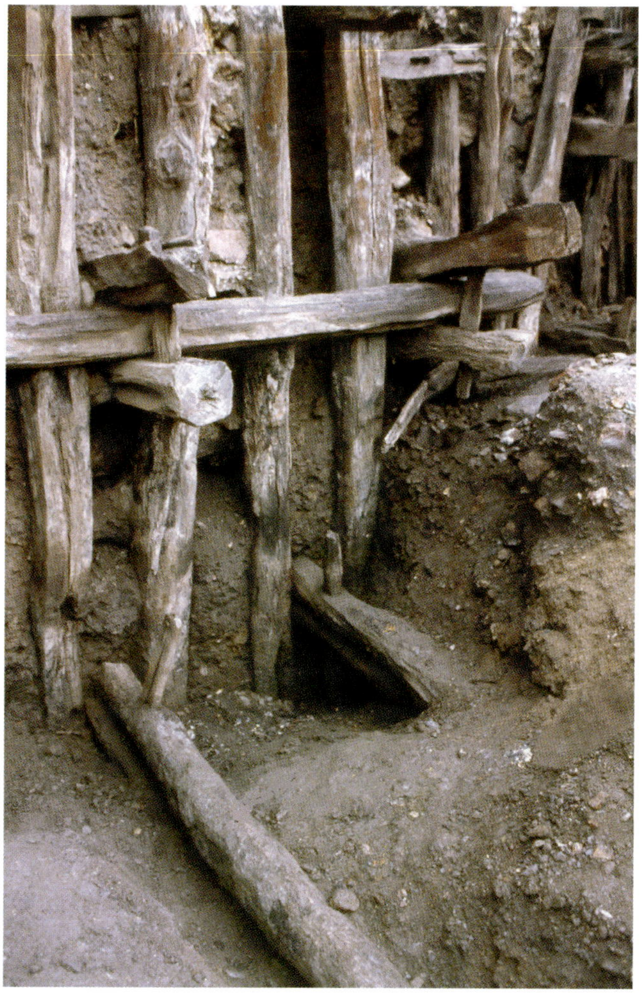

Plate 5.60 Dariba, East Lode revetment: front view. Note the reused ladder stiles (top right).

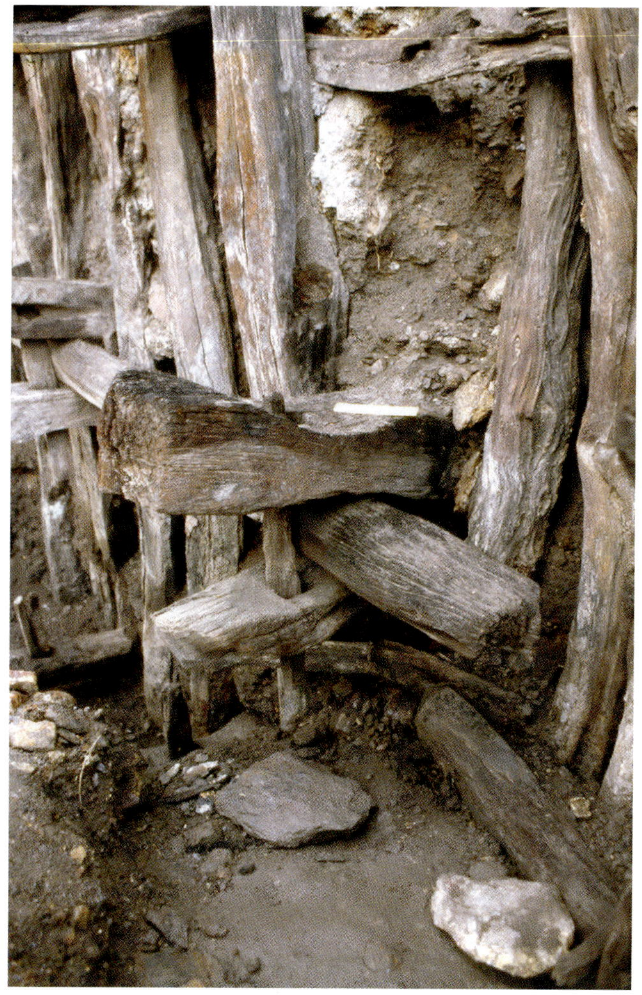

Plate 5.61 Dariba, East Lode revetment: detail of the arrangement for holding the front face horizontals in place, allowing them to be easily removed. Note the shaping of the ends of the beams (scale: 30 cm).

East Lode

Only the surface workings were examined since modern mining had not exploited the lode beneath at the time of our 1987 surveys. A sample of bamboo from a drill core at the 263 m level gave a radiocarbon date of between 50 BC and AD 550 (NPL 209), showing that the workings penetrate to much greater depths than any that have been investigated in recent times. The lode is some 30 m wide and inclines to the east (see Chapter 4, Fig. 4.6). Its footwall remains as low graphite schist cliffs, but the ore seems largely to have been in the calc-silicate rock, exposed at the north end as a worked-out stope 8 m wide. Otherwise no calc-silicate rock is visible as it has been worked completely away to leave a depression about 300 m long and 100 m or so wide (Plate 5.59 and Fig. 5.19). The opencast workings produced enormous quantities of waste, and here the ancient miners encountered a problem. Clearly such huge quantities would have to be dumped near the mine. The earliest excavated waste material is visible as a low tip at the south west end, but this is overlain by tip material from the deep shafts on South Lode and only limited further amounts could be dumped there because of the steep sides of South Lode, which sweep down almost to the edge of the opencast. Considerably more material has been dumped in very high tips along the length of the east side of the opencast, under which the lode inclines. However, the ground, although flat, is alluvial and very unstable such that without support the waste tips would have collapsed into the workings.

Until recently the depression was filled with water, 10 m or more deep, which for over a year provided the modern benefication plant with 600 tonnes of water a day without any substantial reduction in level. More recently, following four years of drought and a substantially increased inflow to the modern mine, the level dropped, revealing extensive timberwork on the east side (Fig. 5.26). This runs for at least 150 m and extends around the south east corner of the opencast. If it originally ran the full length of the east side as is likely, then it would have been over 300 m long. A small piece of wood was radiocarbon dated to between 400 and 170 BC (BM-2635). The wood of the main structure was identified as arjuna (*Terminalia arjuna*). A further radiocarbon date on a sample from the revetment timbers lay between 400 BC and AD 60 (NPL 208a).

A combination of recent landslips and some exploratory work by the company had revealed the exposed timbers to be part of a massive revetment. Our further, very limited excavation showed that the revetment was in three, probably four lifts each about 4 m high (Plates 5.60–5.65 and Figs 5.26 and 5.27). Because of the angle of the deposit there is unlikely to

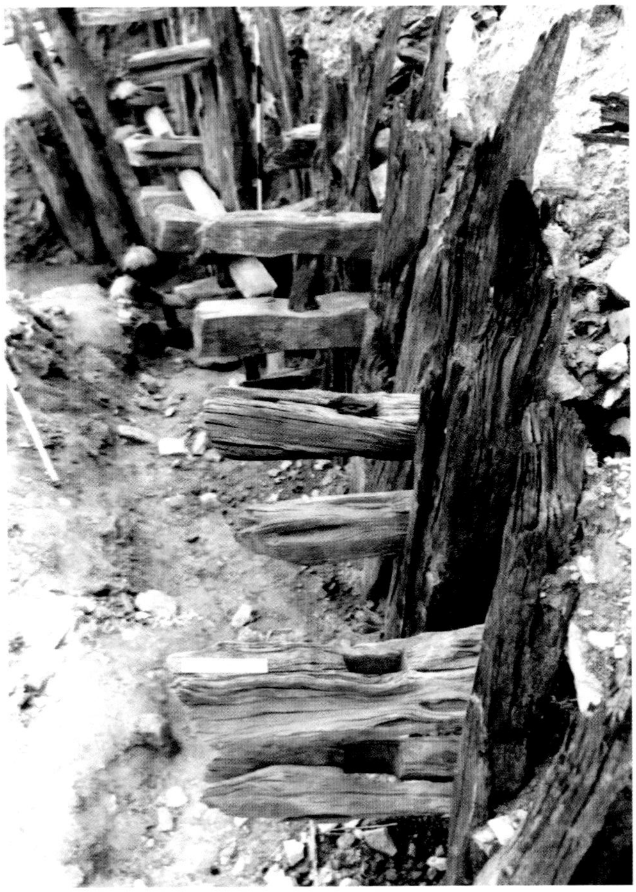

Plate 5.62 Dariba, East Lode revetment: view along an exposed section of the revetment.

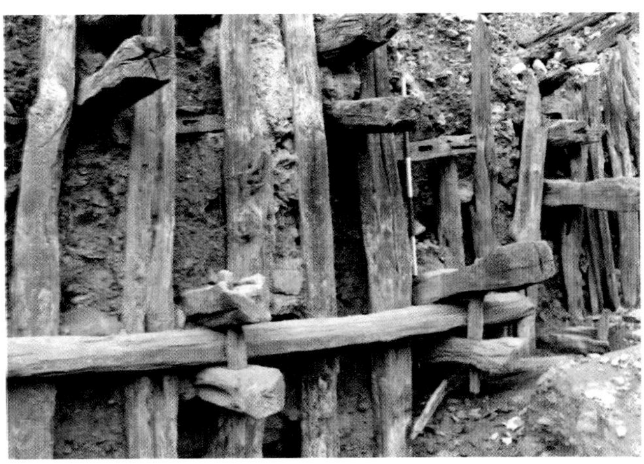

Plate 5.63 Dariba, East Lode revetment: detail of the face. Note the stile of an old ladder reused as infilling between the main timbers behind the scale rod.

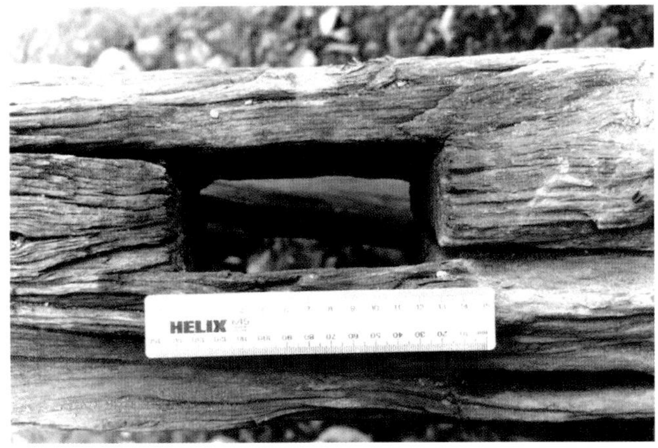

Plate 5.64 Dariba, East Lode revetment: typical precisely cut mortice joint.

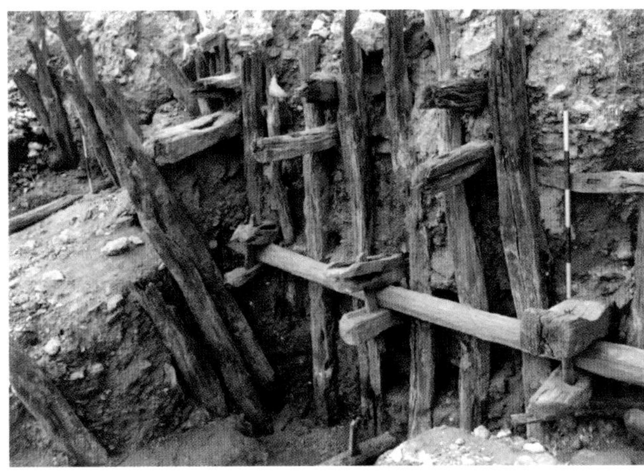

Plate 5.65 Dariba, East Lode revetment: pointed piles (left) driven into a bench to enable it to be mined.

have been more than four lifts before mining had to continue with shafts. Each lift consists of vertical squared posts, closely spaced, held back by three sets of horizontal pairs of timbers at right angles which extend back to the next lift (Plates 5.60 and 5.61), each pair pinned by long timbers across the front (Plates 5.62–5.65 and Fig. 5.26) that are themselves pinned at the back (Plate 5.60). Short lengths of timber passed vertically through mortices in the horizontal pairs to lock the long front timbers in place (Plates 5.63 and 5.64). Behind the vertical posts, smaller horizontals prevent material passing through the gaps, much of the timber being reused, including the stile of a ladder (Plate 5.63). The timbers were joined by simple but strong mortice and tenon joints as were also found on some of the underground timbers, with no evidence of pegging (Plate 5.64).

In one section, pointed piles had been driven into the bench to allow work on the face of the revetment behind (Plate 5.65 and Fig. 5.27 inset). The whole structure was infilled with mine waste. The revetment allowed the lode to be worked as an opencast without endangering the eastern side. Such extensive work can only have been worthwhile for a major economic mineral, almost certainly argentiferous lead, as evidenced by the principal contemporary slag heaps which lie to the south and east. The radiocarbon dates suggest the opencast postdates much of the deep mining, although not the deepest.

Removal of water from the opencast would have been an enormous task: it may have involved human labour, but the use of oxen for raising water in the area today suggests their use in the past. Observation of the operation of a modern well confirms that something like 40 pairs of oxen operating night and day (say 160 pairs in all) would probably have sufficed.

At the north end of the opencast a small calc-silicate rock outcrop has been used for crushing the mineral (Plate 5.66 and Fig. 5.28). The edge of the outcrop has a dozen or so small mortar holes similar to those found on North Lode, but more

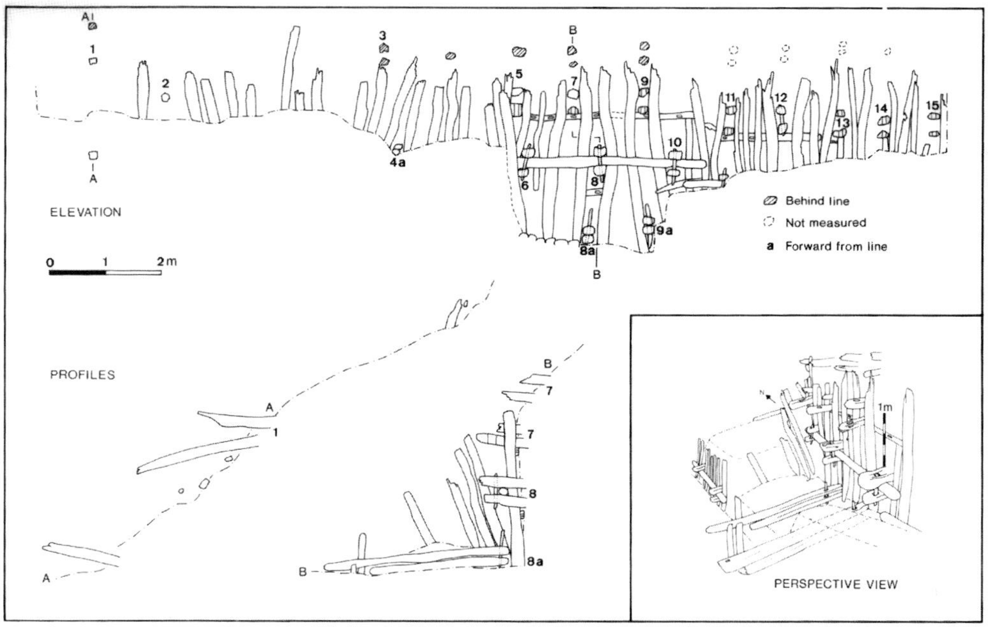

Fig. 5.27 Dariba, East Lode revetment: elevation and profiles at right angles to the timbers of one bench and lower left, perspective view through a bench. Note the pointed piles driven at right angles into the revetment to enable the bench itself to be worked.

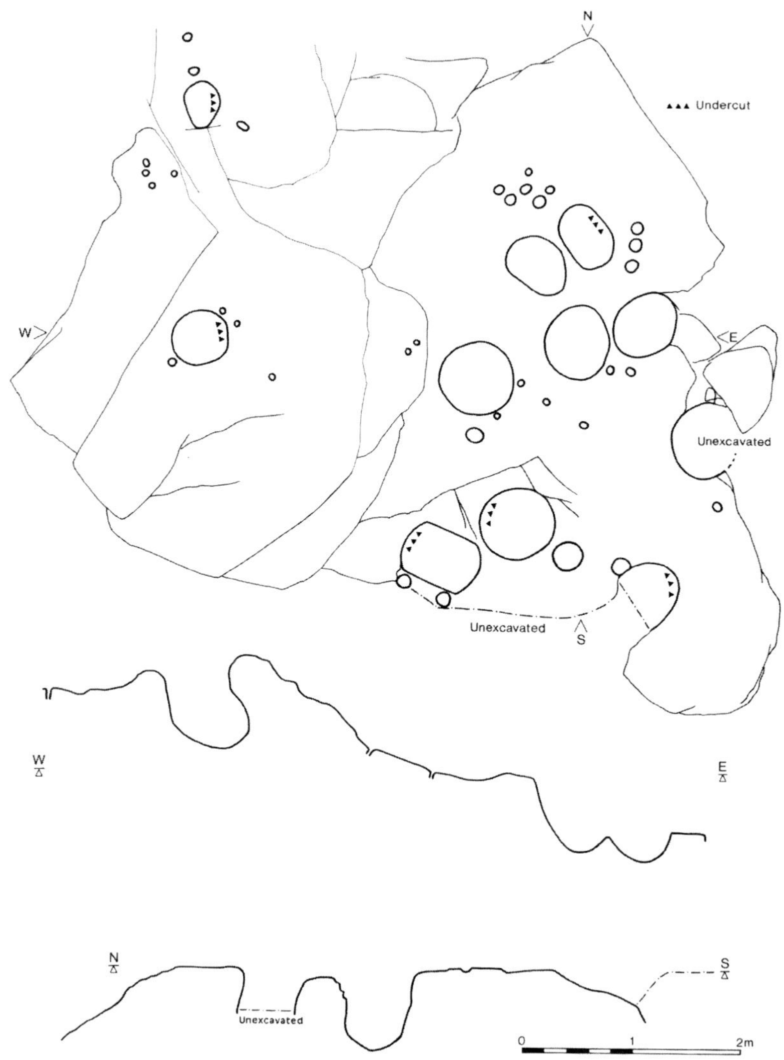

Fig. 5.28 Dariba, East Lode: plan and profiles of mortars carved in the hard calc-silicate rock. Note the undercutting on some.

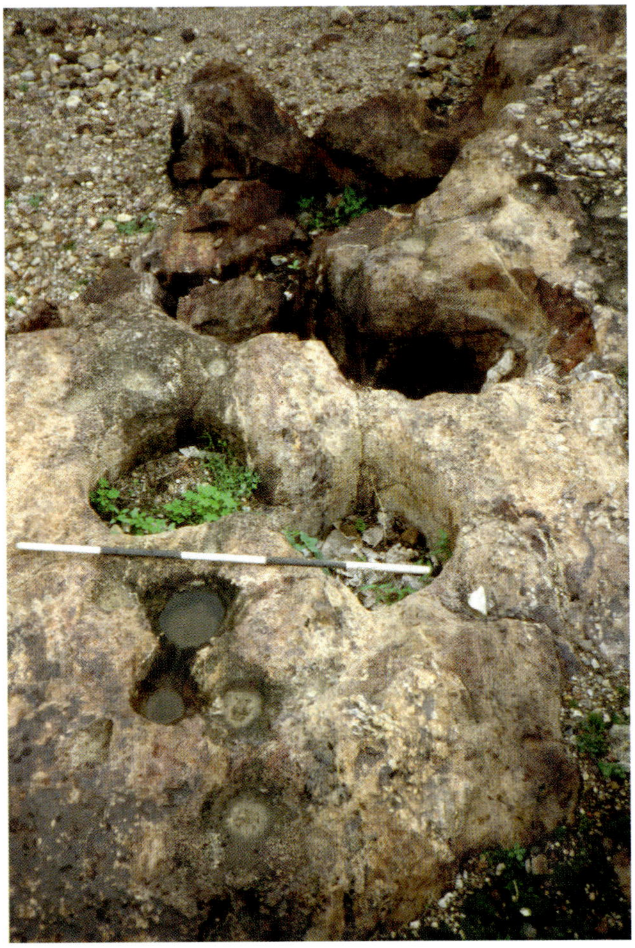

Plate 5.66 Dariba, East Lode ore crushing area: general view. Note the beneficiation debris piled up behind (cf. Fig. 5.28).

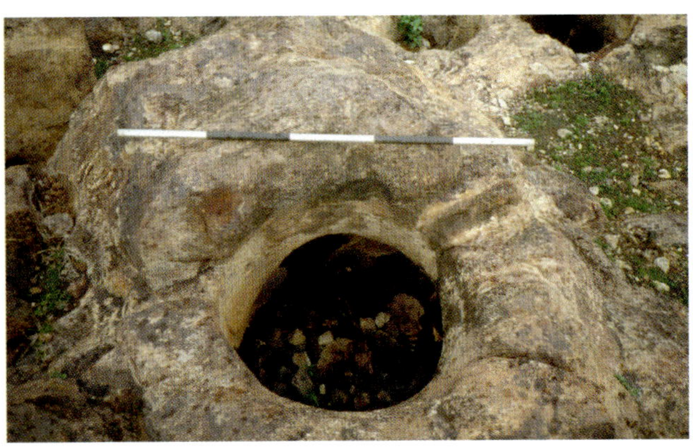

Plate 5.67 Dariba East Lode ore crushing area: large mortar.

conspicuous are ten large mortar holes with rounded bottoms (Plate 5.67). These have depths of about 50–100 cm and are approximately 50 cm in diameter at the surface. They were undoubtedly worked as stamps, with the operator standing, raising and dropping a timber shaft, presumably shod with iron, to pummel the ore. Most of the mortar holes are vertical, but the standing position of the operator meant that the stamping was done most conveniently at a slight angle, with the consequent effect on the inclination of the hole (Fig. 5.28).

Some holes have widened to cut into the next, possibly deliberately since the crushed ore would have been more easily

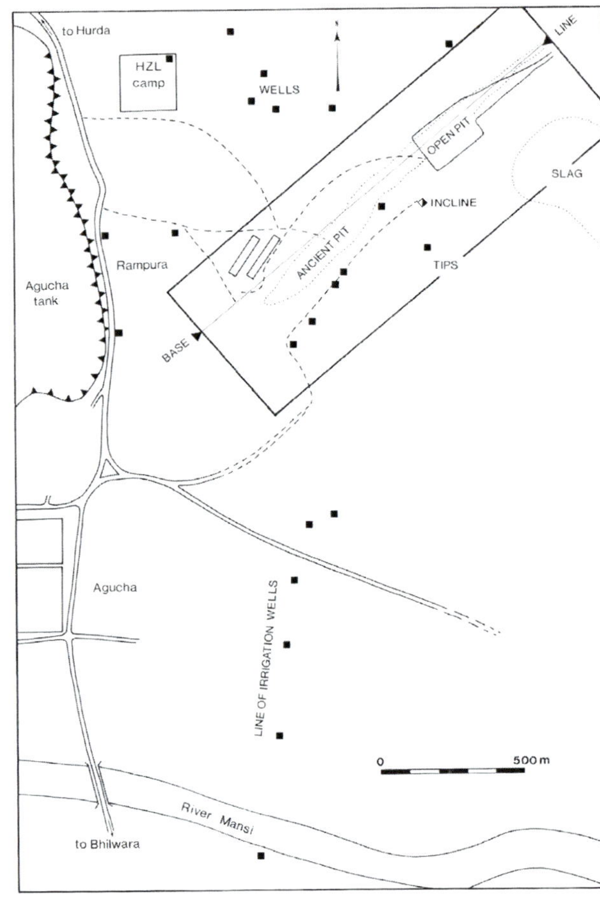

Fig. 5.29 Rampura-Agucha: general plan of the mine and surrounding area before the modern opencast working commenced.

removed. The close proximity of so many mortars no doubt reflects the paucity in the immediate area of suitable formations of hard quarzitic rock, but was also prompted by a desire to centralise operations as much as possible. They are undated but are almost certainly mortars where the ore from the deep mines of South and East Lodes and the East Lode opencast were crushed in the latter part of the first millennium BC. Crushed debris occurs in very considerable quantities in heaps in the area adjacent to the mortars and is clearly associated; much more has probably been washed down the slope towards and into the opencast.

Bethumni

Three large water-filled pits in the area are evidence for large-scale opencast mining in the past at Bethumni, which lies just to the north of the River Banas at the northern end of the Dariba–Bethumni ore belt (see Chapter 4, Fig. 4.7). The south western pit is 440 m long and 80 m wide in the central part, the middle pit is 160 m long and 50 m wide, and the eastern pit is 300 m long and 30 m wide. All these pits are irregular in shape and are parallel to the strike of the ore beds. Their maximum depth is not known but is probably between 10 and 15 m. There is no indication of their date at present, but by analogy with the large opencast pit above East Lode, they may also be Mauryan.

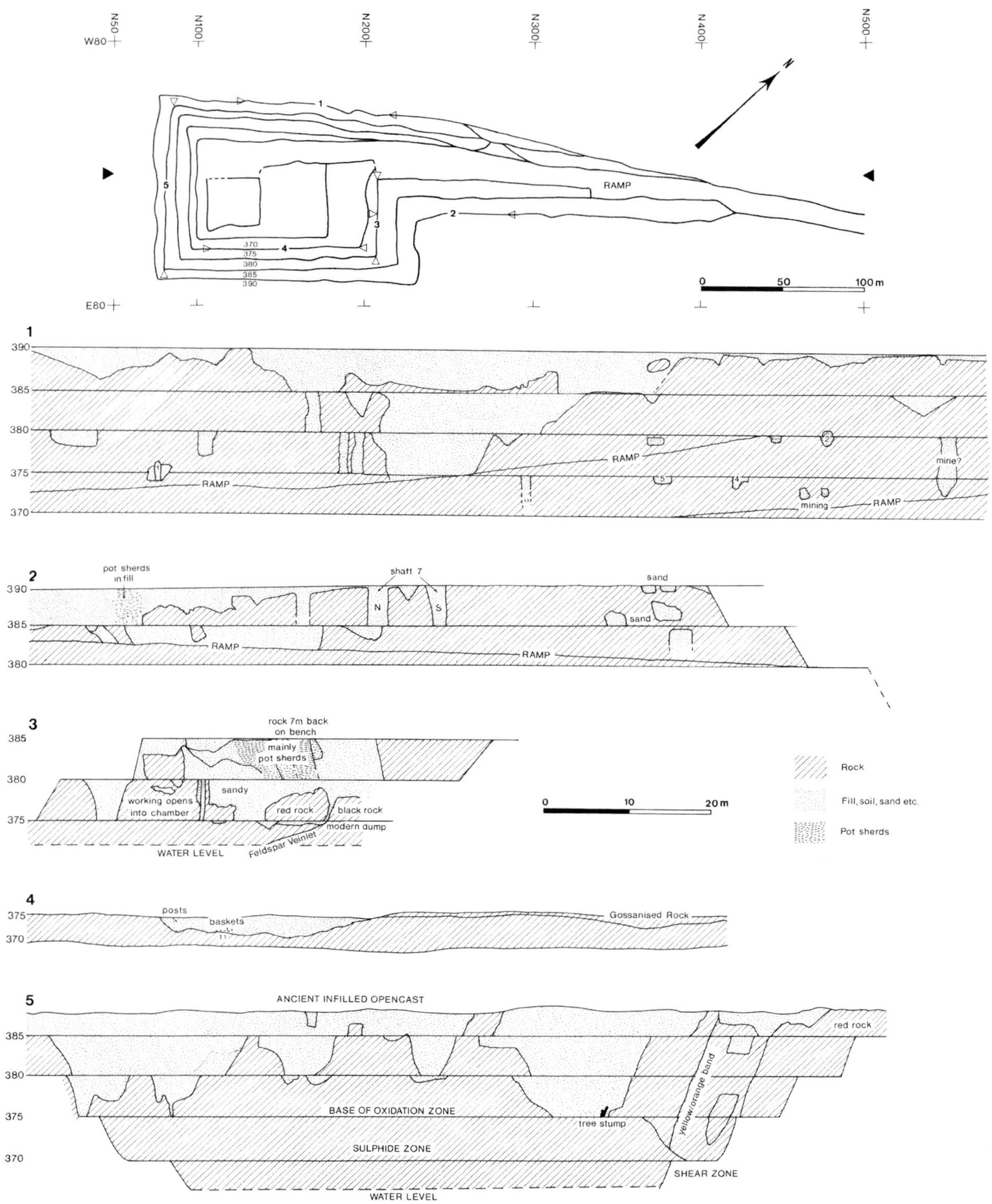

Fig. 5.30 Agucha: sketch sections of the four benched faces and the incline showing the fill of the ancient opencast (cf. Plate 5.68).

Cerussite, anglesite, hematite, goethite, siderite and calcite are found in the old mine dumps. No serious survey or excavation work has yet been carried out in this area, but the indications so far suggest that the ancient miners worked the argentiferous lead and possibly zinc ores by the open pit method. The minerals found in the dumps may well have originally been sulphidic when mined but have now weathered to hydroxycarbonates and oxides over the years of exposure to the atmosphere. No slag tips have been located in this area.

Rampura-Agucha

Exploration of the ancient mining activities at Agucha by this survey was far less extensive than that carried out at Zawar and Dariba, the main reason being that at the time of our visits (in the 1980s) the mine was just being reopened, in all probability some two millennia after it had last functioned as a mine and only about ten years after it had been realised that the shallow valley (see Chapter 3, Plate 3.19) with adjacent scatters of stone

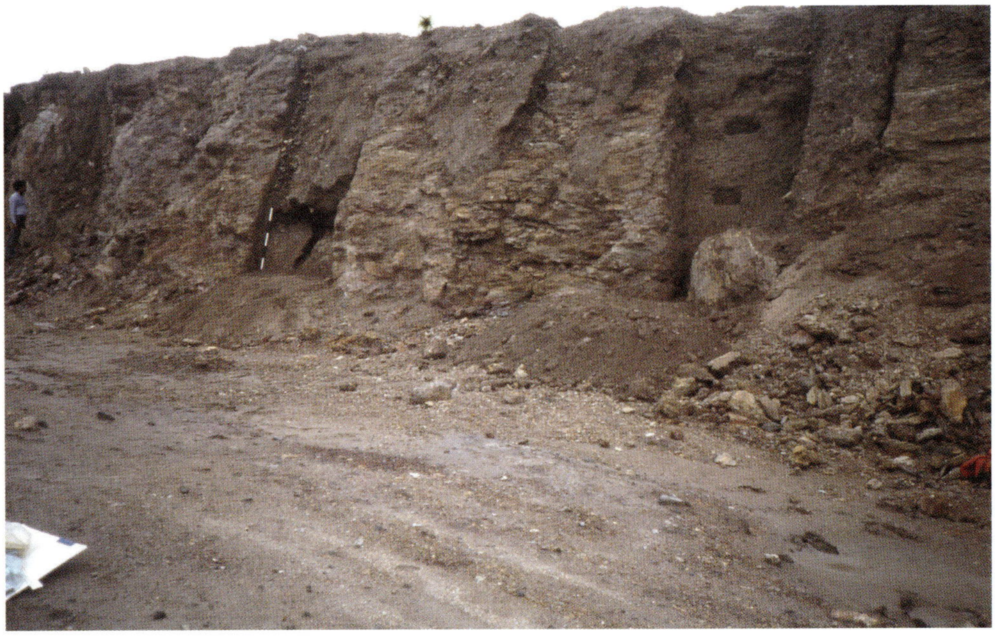

Plate 5.68 Rampura-Agucha: pair of shafts exposed on the east, hanging wall side of the 1987 incline ramp (Fig. 5.30). Note the climbing hand- and footholds in the shaft on the right.

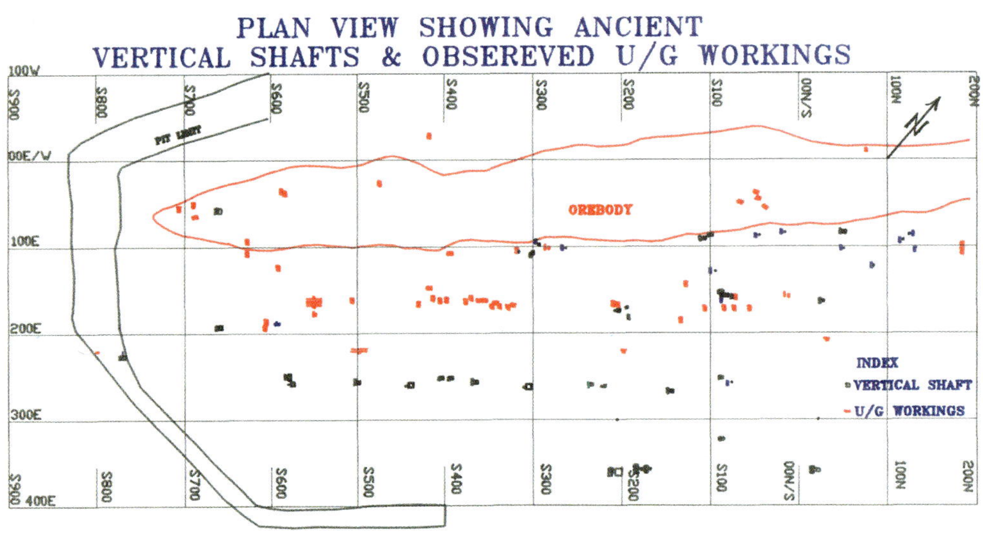

Fig. 5.31 Agucha: plan showing the U-shaped bund (to the left) and the ancient shafts and underground workings (from unpublished report, *Resource Identification at Agucha and Assessment through Interpretation of Ancient Mining*, by K.S. Choudhary).

and slag were the surface indications of a major early mine, and moreover one that contained huge reserves of zinc (Tiwari and Kavdia 1984; Gandhi *et al.* 1984). Therefore there was little in the way of extensive accessible underground workings exposed at that time. A major study of the mines was made in 1987, just as work at the modern opencast pit commenced (Fig. 5.29), during which some attempt was made to understand the mining sequence from the workings exposed in the faces of the benches of the new opencast (Willies 1987).

As the new mine developed through the 1990s more of the early underground workings were exposed. HZL, with its continued interest in the early history of its mine properties, ensured that records were made of the workings as they were exposed. These were brought together in an unpublished report, *Resource Identification at Agucha and Assessment through Interpretation of Ancient Mining* by K.S. Choudhary, the General Manager (Mining) in 1999. The description of the ancient mining presented here draws on both the 1987 and 1999 reports. However, even more than at Zawar or Dariba, much remains to be done.

Situation and geology

The host rocks of Precambrian age include graphite-mica-sillimanite-gneiss and a wide range of other schists and gneisses, with the footwall defined by shears and the hanging wall by a calc-silicate rock. The strike occupies a fairly level plain, extending for approximately 1.6 km, and varies in width from 10 to 110 m, being approximately lentoid in shape at the surface

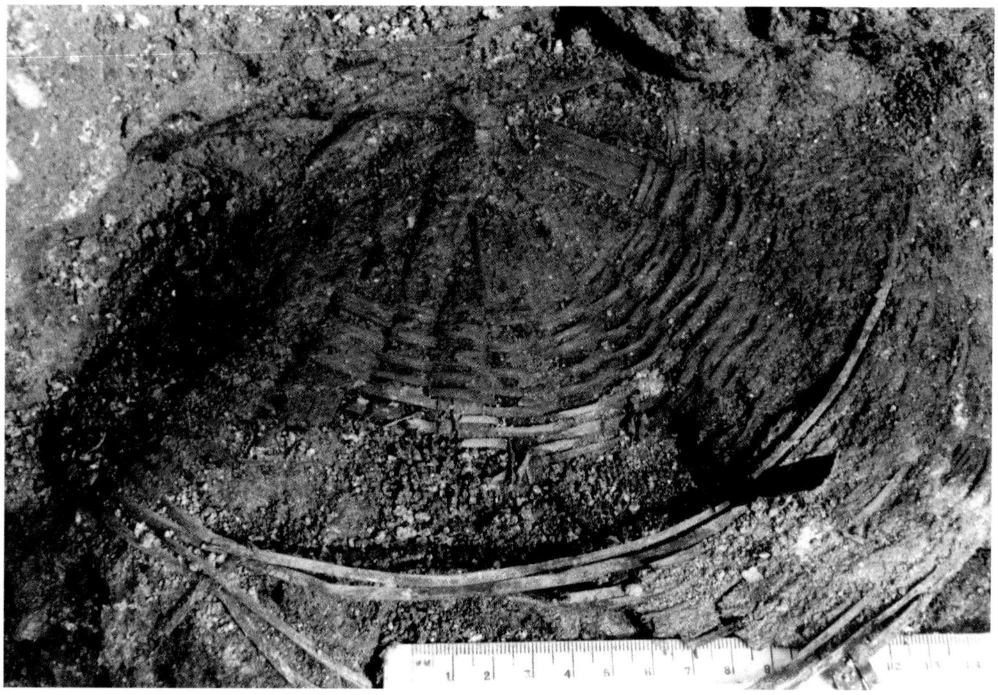

Plate 5.69 Agucha: basket exposed in the fill. One of many found abandoned at the contact zone, which was the floor of the old opencast. Site 11 east berm.

as indicated by the old workings (Fig. 5.31). The general strike is N 50° E–S 50° E with sharp wall contacts dipping between 55–65° due south east down to 130 m (see Chapter 4, Fig. 4.8). Most of the ancient mining has been on the hanging wall contact on the south eastern side of the opencast, ranging in width from 1 to 1.5 m, where recycled argentiferous lead veins and pockets, devoid of zinc, occur in fractures in the calc-silicate rocks within a zone up to 3 m from the ore contact (Plate 5.68). There was also some evidence of mining on the footwall contact on the north western side.

To the south of the opencast is an ancient bund running for approximately 440 m that is now typically approximately 4 m high and some 30–40 m wide at the base, tapering to 3 m wide at the top, made from mine waste (Fig. 5.31). The purpose of this bund was almost certainly to protect the workings from inundation during the monsoon by water emanating from the higher ground to the west, waters now held back by the walls of the present tank by the side of Agucha village.

From the presence of large quantities of gossan in the spoil tips it is likely that originally the presence of potentially profitable mineralisation was indicated by a gossan outcrop, much the same as that which survives at South Lode at Dariba although nowhere near as extensive or as prominent.

Samples of wood from four mine timbers gave calibrated radiocarbon dates around the mid-to-late first millennium BC (Table 5.2): 850–250 BC (PRL-930) 'at 20 m depth', i.e. about same depth as the workings investigated here; 800–700 BC or 550–200 BC from the opencast south wall (BM-2630); 400–190 BC from Site 11 (BM-2632); and 410–160 BC from a drill core at 30 m depth (BM-2356). A twig from a basket (Plate 5.69) from Site 11 gave a date of 850–150 BC (BM-2631). One of the major timbers, a mine prop, was identified as arjuna (*Terminalia arjuna*).

Much of the ancient mining activity seems to have been opencast working in the gossan judging from the contents of the dumps. These vary in size, extending for about 900 m on the south east, hanging wall side of the modern opencast as it was in 1987 at the time of our survey, but concentrated in the central parts between S 300 and N 200 (Fig. 5.30). Over the centuries the heaps have all eroded, rarely exceeding 10 m in height and very spread at their bases at the time of our surveys. Gandhi *et al.* (1984) identified three major heaps. One heap just to the north of Well 1, which Gandhi *et al.* estimated to have contained about 9500 m^3 of material, was surmounted by a banyan tree, forming a landmark in the otherwise very flat and featureless terrain and taken as base 0 in many recent surveys. Another heap lay east of the opencast along the 00 section line and was thought to comprise about 18,000 m^3 of material. The third major heap lay parallel to the opencast from S 120 to S 350 and was estimated to have contained about 13,000 m^3 of material. Together with some other smaller heaps it is estimated that there was approximately 44,000 m^3 of material dumped on the hanging wall side of the opencast. In addition, there were two sizeable heaps of regular pea-sized debris, probably the product of crushing operations on the northern footwall side between N 30 and N 480 that were estimated to have contained about 8400 m^3 of material. In contrast, only negligible dumps were associated with the enigmatic system of 'wells' running from the southern end of the opencast to the River Mansi and beyond.

The ancient opencast was not deep, with a maximum depth of about 15 m (Fig. 5.30). The principal foci of the open work are between S 450 and S 750 and again between S 200 to S 360, with some other smaller selected places (Fig. 5.31). From this it is clear that the gossan was only worked very selectively, with the argentiferous lead being sought while the very rich zinc mineralisation was ignored. Generally in the gossan zone of a zinc-lead-silver deposit, the zinc, being highly soluble as the sulphate or carbonate, is easily dissolved and leaches out or down. Lead is more resistant, whether as galena or as oxidised

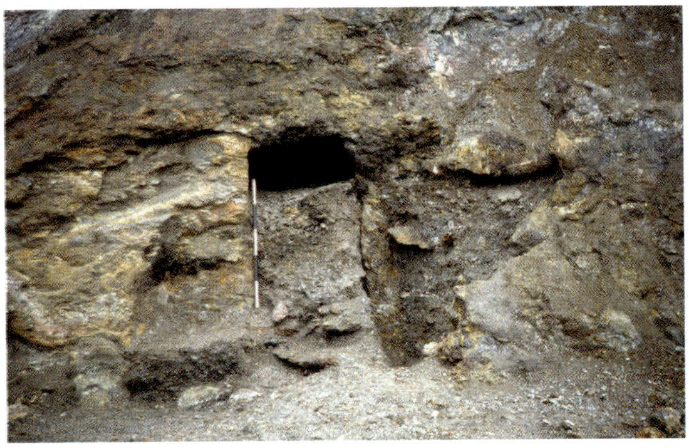

Plate 5.70 Agucha: small rectangular section gallery (left) and cross-cut (right) exposed in the west (footwall) side of the modern ramp. Note the gallery is 75% filled with debris.

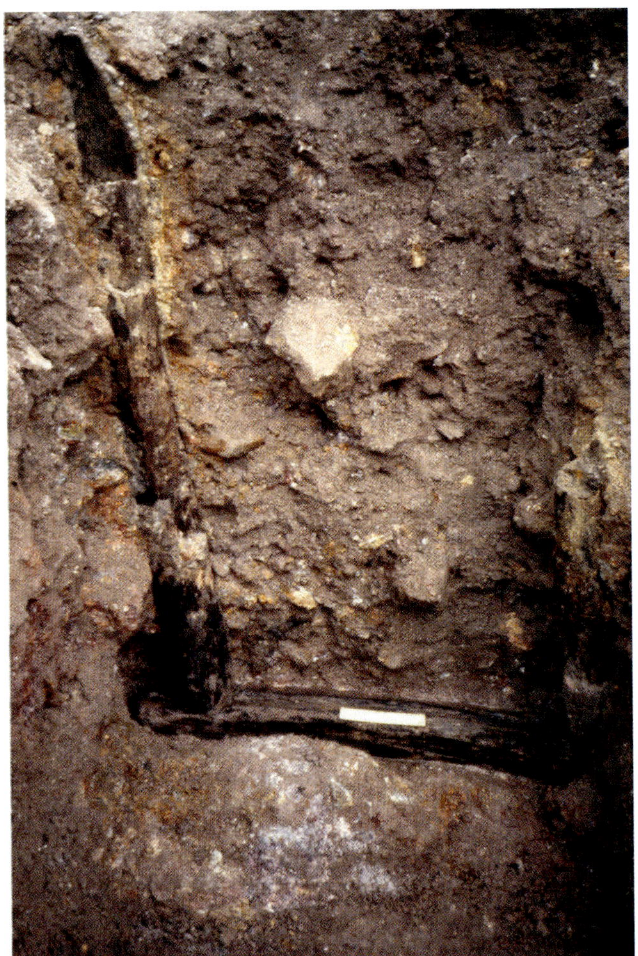

Plate 5.71 Agucha: lower parts of a small rectangular gallery driven through soft loose gossan and backfill necessitating timber supports. Site 11a on the east side of the berm.

cerussite or anglesite. Silver enrichment seems to occur, particularly at or just below the water table. Marmatite and pyrite ores are especially subject to leaching, and the limonitic filled boxwork structures which result make mining by ancient techniques much simpler. This was certainly the case at Agucha where the weathered gneiss is very soft and was easily and readily cut by wedge-ended chisels rather than gads.

Comparison with silver/lead deposits such as Mt. Stewart and Mt. Isa in Australia (Blanchard 1968), and Leadville in the USA (Kemp 1893), found in recent times and of similar size, age, rock types and semi-desert conditions, suggest a massive lead-silver deposit at Agucha in which the silver content can be very high: the pieces of lead metal from the well site (see Chapter 6) averaged about 1000 ppm of silver (see Chapter 7, Table 7.18).

The modern mine provided convenient sections across the ancient open pit, revealing the full extent of the large opencast working and exposing evidence of the ancient underground mining methods used. The ancient workings were mainly in the gossan from about 390 m down to the 375 m level (Fig. 5.30) but also in the only slightly altered sulphide zone below (Fig. 5.32). Most of the ancient workings uncovered by the end of the 1980s were opencast, penetrating down to the contact zone, with some 30–40% extraction on one section across the width (130 m) of the lode, to some 15 m depth (Fig. 5.30: 3 and 5). The sections exposed by the modern mining revealed several shafts and galleries penetrating the gossan and backfill (Plates 5.70 and 5.71). Limited excavations on the mine benches provided further information. On one section of the 375 m bench on the east side of the pit (Fig. 5.30), the remains of seven woven baskets were found (Plate 5.69). These would have been used for conveying ore and had been left on the bottom of the pit at the gossan/sulphide contact zone, where conditions favoured their preservation. One side of the modern pit cut through a layer several metres thick of red earthenware sherds, probably a heap of pottery wasters. These must have been dumped there contemporaneous with the ancient working since some were lying directly on the mined bedrock near the base (Fig. 5.30: 3).

In the same location timber supports, with verticals morticed into sole plates, remained from a gallery, although the tops had rotted away (Plate 5.71). This was probably an access to underground workings provided through the open-pit backfill. This suggests a long developmental history at Agucha with a partially backfilled and presumably abandoned opencast being penetrated by shafts and galleries to access more recently opened underground workings. In other areas, shafts with footholds had been sunk, working almost vertical shoots of ore in a joint or possibly a shear zone. Gossanisation was deepest at the shear zone, but this had not been uniformly exploited by the opencast, although a number of underground workings just penetrated it. The galleries examined seem to have been of walking height with the chisel marks in the usually soft rock clearly visible.

In order to follow the lead mineralisation, deeper deposits were worked by sinking a series of shafts from the floor of the opencast, near or through the calc-silicate hanging wall (Figs 5.31–5.33). There was also some working on the footwall contact. A series of seemingly regular main shafts was mostly situated between 140 and 160 m from the edge of the pit on the south east side, with only a few on the actual ore boundary, and some at about 60–70 m and one 250 m distant. The shafts were vertical and either square or rectangular with dimensions ranging from 2 × 2 m to 5 × 4 m with depths between 30 and 35 m. In the underground workings the drives, cross-cuts and vertical raises were also usually square or rectangular in section, but of much smaller dimensions than the main shafts. The dimensions varied between as small as 0.5 × 0.5 m to 0.75 × 0.75 m, with the main drives and cross-cuts having dimensions typically varying between about 1.2 × 1.2 m to 1.5 × 1.5 m. Some cross-cuts were

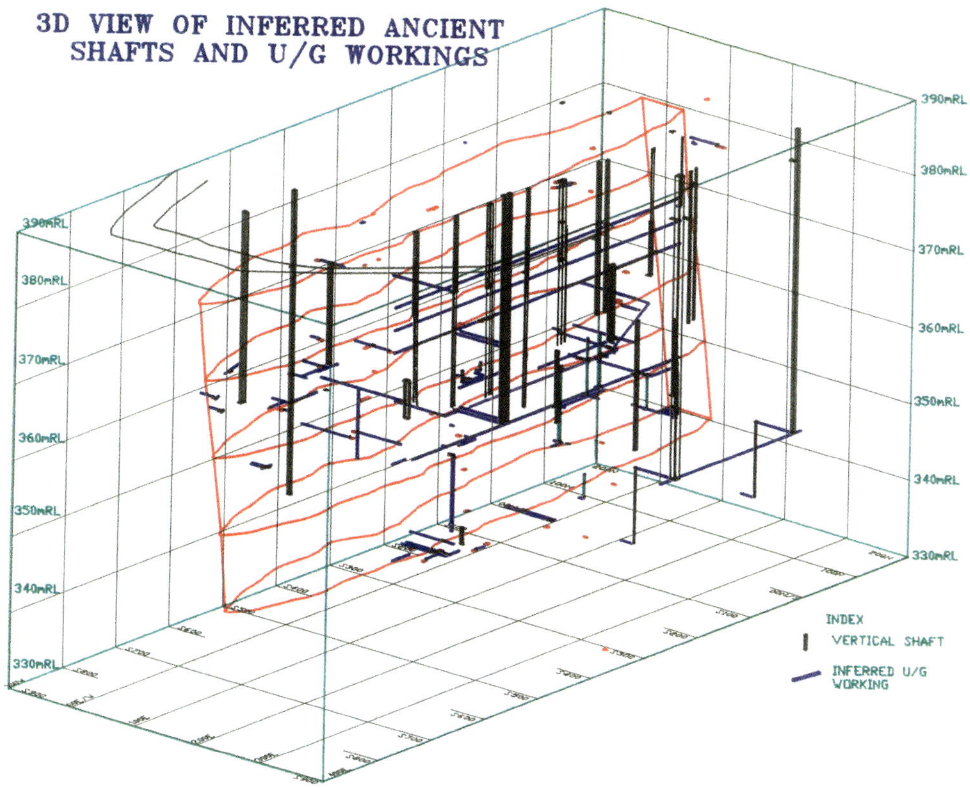

Fig. 5.32 Agucha: isometric view of inferred ancient shafts and underground workings (from unpublished report, *Resource Identification at Agucha and Assessment through Interpretation of Ancient Mining*, by K.S. Choudhary).

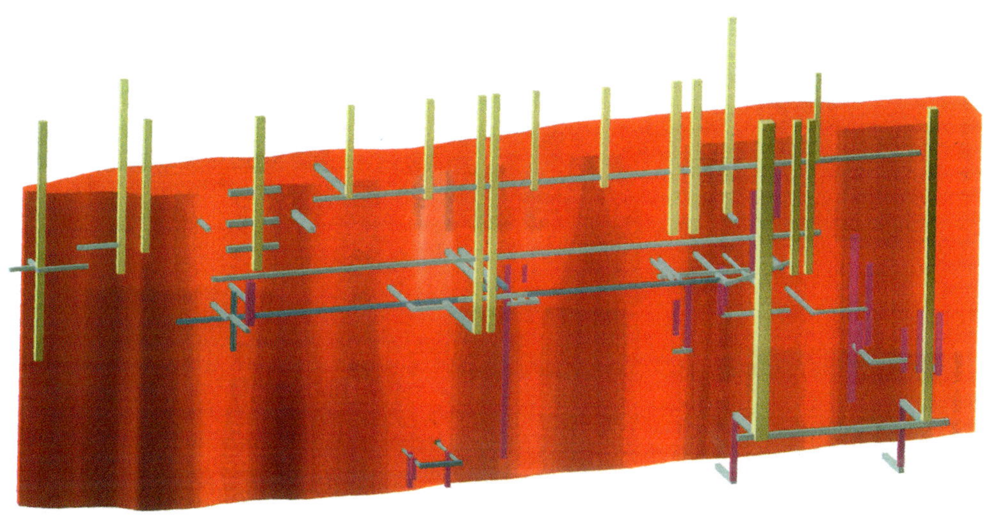

Fig. 5.33 Agucha: schematic view of the ancient mining scheme (simplified version of Fig. 5.32) (from unpublished report, *Resource Identification at Agucha and Assessment through Interpretation of Ancient Mining*, by K.S. Choudhary).

as large as 3 × 2 m and vertical raises of 1.2 × 1.2 m were not uncommon. There were some rising drives including one with a gradient of 1 in 20 running for some 60–100 m to the north. The stope blocks as defined by the cross-cuts and raises were typically 10 × 5 × 1.5 m, and these were selectively worked by overhand stoping following the thin and patchy veins of argentiferous lead. The underground surveying must have been of a high standard as evidenced by the regular series of right-angled interconnecting drives and cross-cuts over considerable distances.

The weathered nature of the rock meant that timber supports were necessary on the hanging wall and to lesser degree on the footwall. Possibly as a safety precaution the near-surface workings often had two exits, although these may have served as ventilation. Two parallel drives – one about 15 m from the ore contact and the other about 60–70 m – ran at about 30 m depth to link up with the main shafts. In places near the surface the loose sides of the ancient pit were held in position by horizontal timbers, which in turn were held in place

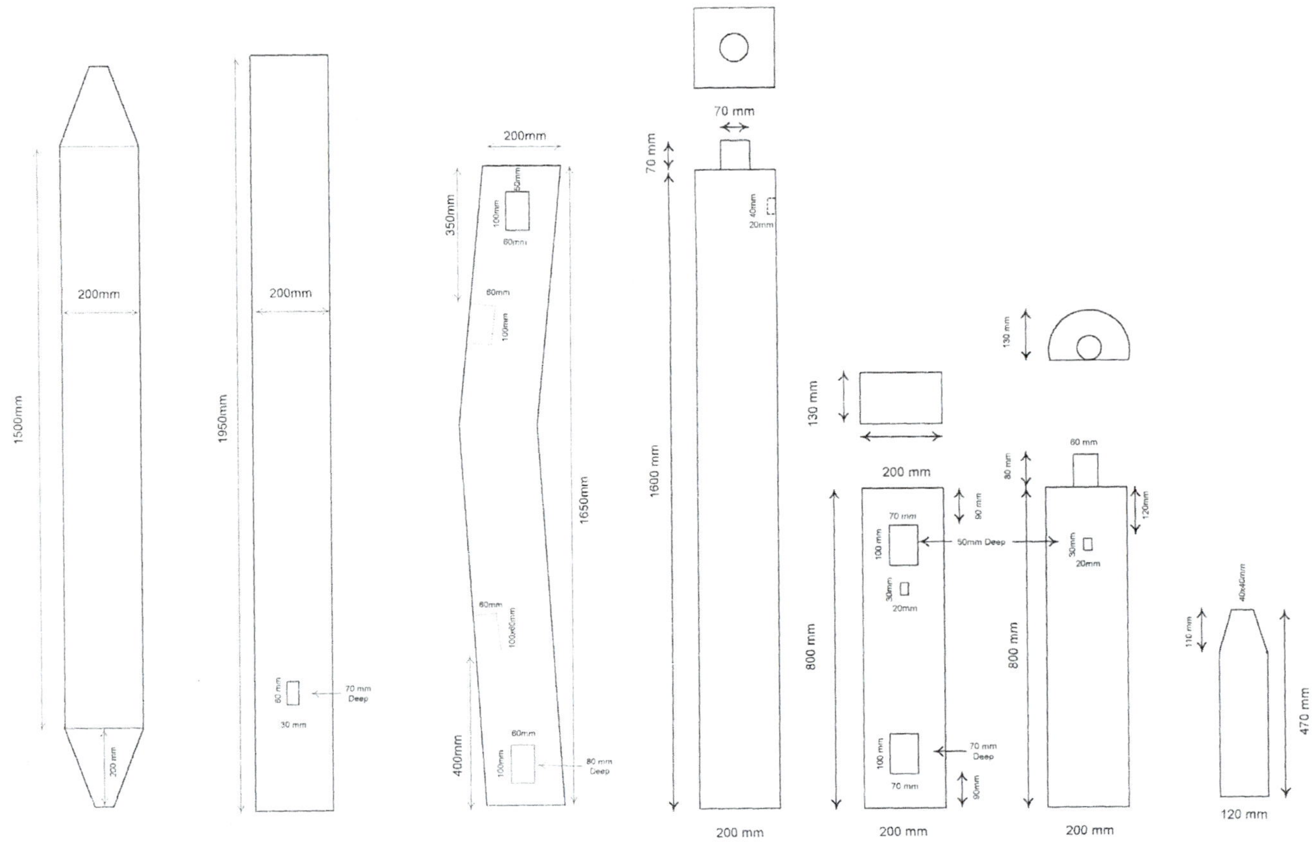

Fig. 5.34 Agucha: schematic illustration of the joints used on the wooden stemples.

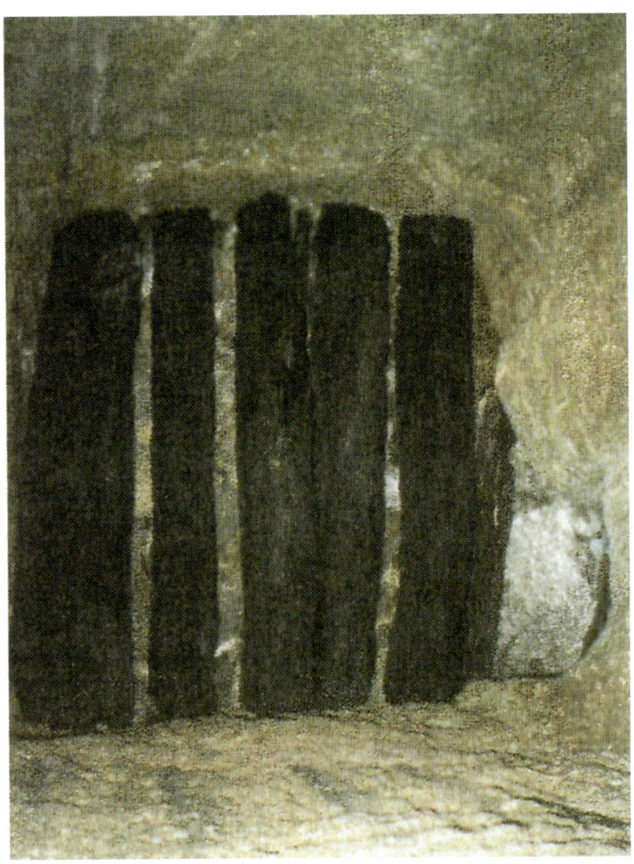

Plate 5.72 Agucha: a 2 × 2m square raise blocked off with logs probably to prevent inrushes of water during the monsoon.

by other horizontal timbers at right angles driven into the loose in a style reminiscent of the much larger system at East Lode, Dariba. Large flakes of the hanging wall had to be supported by carefully constructed timbers with mortice and tenon joints that showed by their sophisticated design that they had been made by miners with long experience of such problems (Fig. 5.34). Other timbers were used to seal off old shafts and raises both inside and outside the strike area (Plate 5.72), possibly to stop water from entering but also to maintain the required air flow through those parts of the mine that were still operational.

In addition to the main opencast and the associated shafts on the hanging wall, there is also a series of a further 14 or so 'wells' running from beyond the southern end of the opencast in a southerly direction at regular intervals for some 2 km to the River Mansi, and then a further three shafts on the far side of the river (Fig. 5.35). These wells are of square or rectangular section and many have passages, sometimes at more than one level, leading off from their sides which can still be followed for considerable distances into the rock (Fig. 5.36). In addition, in the immediate vicinity of several of the 'wells', there has been considerable subsidence, clear evidence that there must be extensive workings beneath. As these wells are away from and not on the same orientation as the main strike, as defined by the opencast, and as preliminary analyses failed to detect any significant evidence for lead or zinc, there has been considerable speculation as to the original function of the 'wells'. Were they:

- never anything more than a series of irrigation wells, possibly linked to the River Mansi but unconnected with the mines;

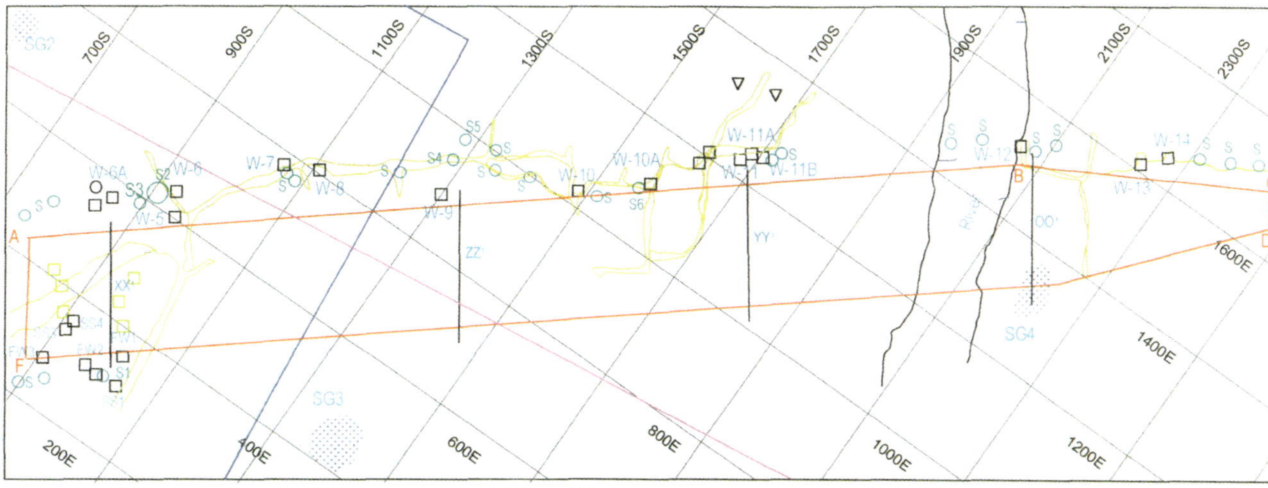

Fig. 5.35 Agucha: plan of 'wells', subsidence and other features between the ancient opencast mine and the River Mansi. Wells are indicated by a square, subsidence by a circle. Note that the 'wells' and subsidence continue on the far side of the river, ruling out mine drainage (from unpublished report, *Resource Identification at Agucha and Assessment through Interpretation of Ancient Mining*, by K.S. Choudhary).

- a very ambitious drainage system for draining the mines in the strike region for some 2 km through barren rock to the river;
- or was there originally further mineralisation in this region, and these were the main shafts associated with underground workings that were subsequently reused by the farmers for irrigation?

That they had just been used for irrigation seems unlikely. The soil is poor, seemingly not justifying the enormous effort required, and if irrigation was important why was there only a single line of wells rather than a pattern over the whole area? In addition, their layout and construction seem to have been too carefully planned and executed just to be the wells of simple farmers. They are in fact identical to the main shafts within the mined area and indeed the northern end of the chain is within the mined area. Could they then be part of a drainage adit driven through barren rock to take water from the mines to the River Mansi? The problem here is that the bottoms of the shafts are very nearly at the same depths as the bottoms of the main shafts in the mined area, which were not significantly above the bed of the river, giving a very poor flow. Also, some of the shafts have openings at several levels which is inexplicable if they were part of a drainage adit. Furthermore, during monsoon periods, when drainage was most necessary, the river level would rise and, if anything, flood the mines. Drilling near some of the wells close to the river located spaces presumed to be passages that were filled with clean river sand. Additionally, the continuance of the chain on the other side of the river away from the mined area would be totally unnecessary.

More recent geological and mineralogical work has shown little evidence for mineralisation in the vicinity of the well shafts, although it is possible that there was once a series of small but rich ore pods located and very thoroughly removed in antiquity leaving little evidence behind apart from the workings themselves. It is also possible that there is a change in the direction of the strike in the vicinity of the southern end of the main strike due to local folding and faulting, and the strata could well turn more sharply to the south and thus the 'wells' may lie over a continuation of the orebody. Therefore it is more than likely that the 'wells' were originally cut as the main shafts serving underground mining operations for argentiferous lead, and that in the recent past these were reopened for irrigation purposes (a conclusion also reached by K.S. Choudhary). This use would cause considerable movement of the ground waters in the flooded and filled ancient galleries, and this is probably responsible for the observed subsidence in the region of the reused shafts, some of which was clearly of very recent origin.

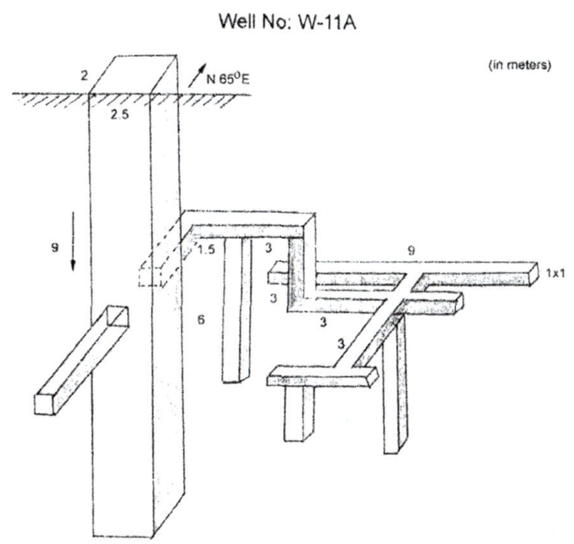

Fig. 5.36 Agucha: Well 11a (cf. Fig. 5.35). The complex series of small square shafts is similar to those found in the main working and totally unnecessary in a well system (from unpublished report, *Resource Identification at Agucha and Assessment through Interpretation of Ancient Mining*, by K.S. Choudhary).

Discussion

From the observed and inferred plan of the ancient workings it is possible to construct a tentative history of the development of the mine through the first millennium BC. First there could have been some exploratory tunnelling into the near-surface gossan deposits followed by selective opencasting. Then a series of fairly close-spaced shafts from the bottom of the opencut appears to increase the depth by about 10–12 m. This seems likely to indicate the first main phase of underground mining, some shafts penetrating possibly to a level substantially free of water in the dry season and, perhaps, at the deepest points, below this 'normal level' to somewhat greater depths after a succession of dry years. The presence of a 'double drift' – two roughly parallel levels at this horizon (usually driven to facilitate ventilation) – suggests a drainage adit at the base (Figs 5.32 and 5.33). Work below this to further shallow depths would have been likely in the dry season or in years of poor monsoons.

A further series of 'bottoms' seems to occur at a depth of around 20 m or a little deeper, with associated development of levels. One shaft (centrally placed in Fig. 5.32) appears to be a main shaft to this depth, which seems otherwise to have been served by underground shafts linked to those of an earlier date via the workings. The deepest shafts (shown on the right of Figs 5.32 and 5.33) and outside of the orebody, extend down to about 40 m deep. At this depth the two shafts are joined by a cross-cut and each has a level leading after a few metres to a further underground shaft or winze down to 50 m in depth. This is the lowest depth so far recorded of workings which could still be traced to the surface. However, a drill hole on the S 200 line encountered wood some 160 m below the surface, suggesting some very deep workings reminiscent of those at East Lode, Dariba.

There is a somewhat similar arrangement on the footwall side of the opencast (farther to the left on Figs 5.32 and 5.33) where another pair of shafts from within, or at least close to the orebody, descend to the same depth and are also connected by levels leading to another underground shaft down to the ultimate depth, obviously securing the air supply. This type of development seems to indicate two linked phases of operation requiring hoisting of water from the deep workings. There are also other underground shafts that appear to extend from the older workings on the footwall down to this same horizon. Earlier redundant shafts were sometimes sealed with heavy timbering, perhaps to prevent flooding of the workings during the monsoon and/or to facilitate changes in the ventilation system as the overall mine depth increased.

Although the duplication of deep shafts may simply have served for ventilation flows, it is at least as likely that the deepest shafts were sunk in order to isolate the workings from flooding each other, each deep shaft linking with shafts into old workings to form separate ventilation and drainage districts. This is a reasonably obvious practice used in more modern times, and perhaps employed also at both Zawar and Dariba where isolated workings seem to have been pursued at depth. From this analysis it is possible to suggest the following sequence of events:

- *Phase 1*: probably the development of shallow and very irregular tunnelling, selectively following richer veins or horizons of argentiferous lead ores within the upper oxidation zone.
- *Phase 2*: the realisation that the deposit was extensive enough with sufficiently rich levels to allow open working of the deposit, with the hanging wall supported by timber to allow maximum depths to be reached. From our observations this was likely also to be entirely within the secondary oxidation zone.
- *Phase 3*: possibly partly contemporary with the open working involving the sinking of shafts to work greater depths than could be economically worked by opencasting. This would continue until limited by water inflow, with the deepest workings only accessible during a series of dry years. It is possible that the double-drift drainage level was driven towards the end of this phase with, again, working below it in the drier years.
- *Phase 4*: sinking of a shaft from outside the openwork to a depth of some 40 m and, possibly, a new shaft from some way up the side of the openwork and deepening of others in a similar situation. The shafts would allow hoisting of water and, with the sealing of older shafts in the opencast, reduce the risk of flooding from the surface.
- *Phase 5*: deepening of the workings still further by underground shafts, possibly in separate ventilation and drainage districts, these new shafts unconnected to the main shafts to allow their isolation in case of local underground inundation. Workings at these depths seem much less extensive than at shallow depth, presumably selectively following only the richer ore shoots. Hoisting of water would need to be virtually continuous although with seasonal variations and possible disruptions.

After abandonment, it would have been almost impossible to reopen such deep workings without modern forms of power because of both accumulated water and damage to the workings.

Chapter 6 Survey and excavation of the surface sites at Zawar, Dariba and Agucha

P.T. Craddock

Zawar

As a preliminary stage of the fieldwork, surveys of the smelting sites were carried out both to get a feel for the overall scope and scale of the operations and to make a representative selection of sites for excavation. One member of the team, L.K. Gurjar, was, of course, already very familiar with these environs from his geological work for HZL and was able to highlight those areas with the most potential interest. However, all areas were carefully walked in surveys spread over several seasons. As noted in Chapter 3, we could find no evidence for any settlement or other occupation contemporary with the major mining activities of

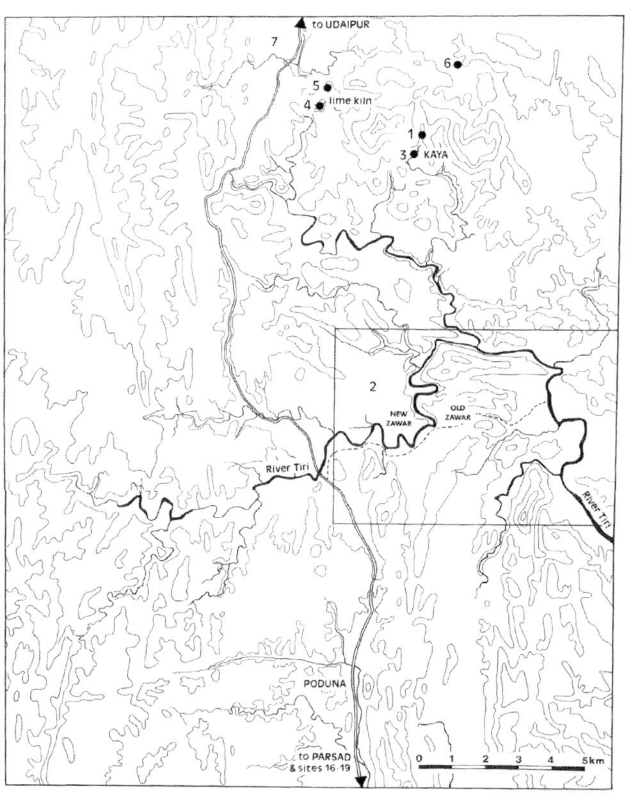

Fig. 6.1 Zawar: contour plan of the vicinity of the Zawar mines. 1. Old Kaya. 2. Zawar (inset Fig. 6.2). 3. Kaya zinc mines. 4. Spread of early copper tap slags. 5. Old copper mines. 6. Devimata copper mines (disused). 7. Present Kaya village.

the first millennium BC. We also failed to find any evidence of the pottery-making activities that must have taken place on a very considerable scale for the later zinc smelting activities. It is possible that, as in present-day Rajasthan, the majority of the ceramics were bonfire-fired, leaving little surface evidence beyond scatters of wasters (Kramer 1997: 67).

In the first season the majority of both the zinc and lead smelting sites were located (Figs 6.1–6.3). It was easy to differentiate between them, the surface evidence of the zinc production sites consisting mainly of retort debris and the lead production sites chiefly of slags; only in a very few instances were retort debris and slag found together (e.g. Site 5, where the two activities were probably not contemporary). Immediately after the first (1982) survey two zinc smelting sites, 30 and 37, were selected for excavation and collecting material for radiocarbon dating (Table 6.1) as it was felt they represented a good contrast. Site 30 was an isolated heap of small retorts up in a remote valley on Zawar Mala Magra (Plate 6.1) among the great system of mines for which dates of approximately 2000 BP had already been obtained (see Chapter 5, Table 5.1). Site 37 was in the main heaps of large retorts down in the Tiri Valley adjacent to the abandoned settlement of Old Zawar, among temples dated by inscription to the 15th century AD and located in places partially buried by the heaps. Subsequently an intermediate retort furnace (Site 33) was discovered exposed in a road cutting, also at Old Zawar, from which a plan and section could be obtained. In addition a number of fragments of very small retorts and handmade perforated plates were recovered from the fill of the great dam across the River Tiri (see Chapter 3, Plates 3.6 and 3.7) that appear to belong to a simpler, possibly earlier version of the process represented by the intact furnaces elsewhere on the site. The distinctive remains of zinc smelting seem to be confined to the immediate vicinity of Zawar itself, with just one site lying about 12 km to the north near to Kaya where there was some limited production (see below).

In the course of the first survey of slag, analysis of samples taken from the heaps that denoted the lead smelting sites indicated that they fell into two basic types: zinc-rich and iron-rich. It was suspected, correctly, that this could have chronological significance (Table 6.2) therefore the selection of slag heaps for investigation was determined partly on their geographic distribution over the whole site and partly on the composition of the slags. It is immediately clear from Fig. 6.2 that the

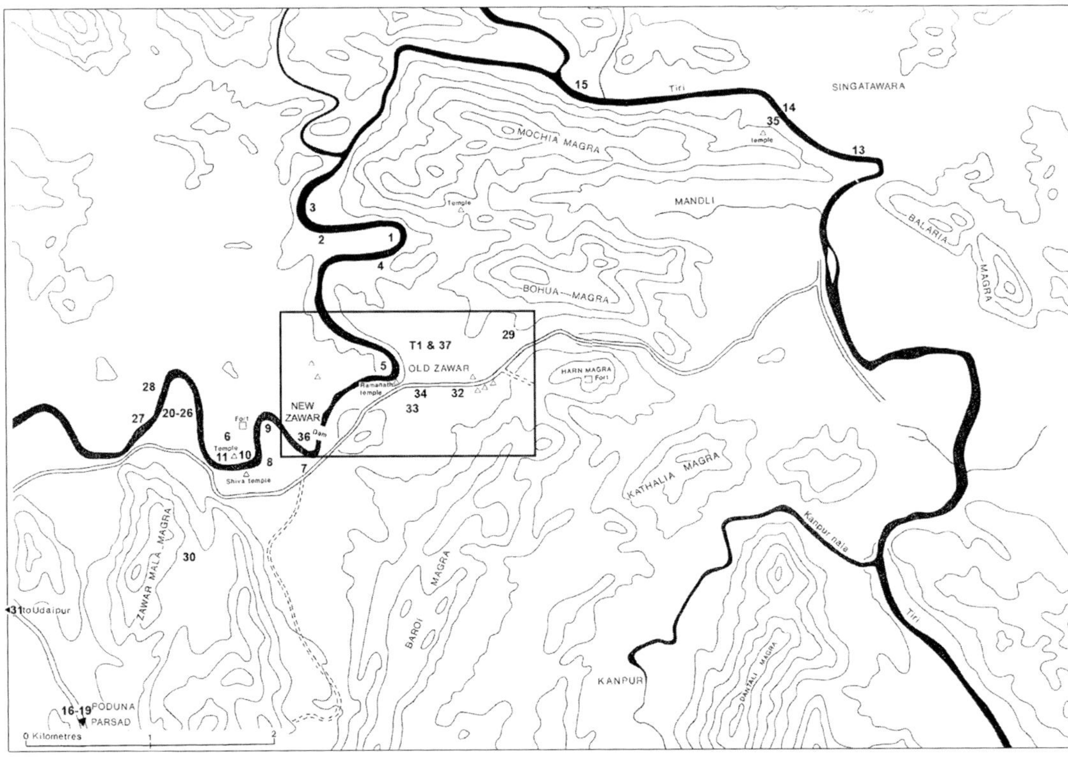

Fig. 6.2 Zawar: main area plan showing principal mines and smelting sites and Old Zawar (inset Fig. 6.3).

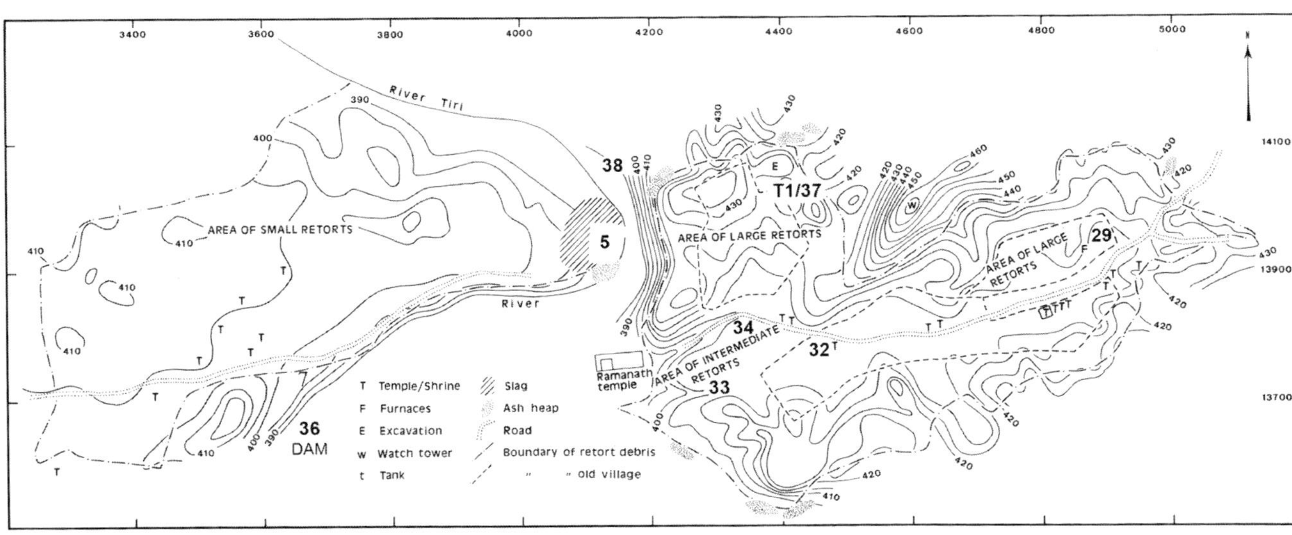

Fig. 6.3 Zawar: location of the sites excavated in the Old Zawar area.

overwhelming majority of the smelting sites are down by the River Tiri, with Site 30 being the only smelting site up in the hills, near to the mines.

One survey was made outside of the immediate area of the Zawar mines along the valley through which the main National Highway 8 between Udaipur and Ahmadabad now runs. The survey started about 20 km to the south at Parsad and worked back north towards the Tiri beneath Zawar Mala Magra. Numerous undated slag heaps were located (Sites 16–19). Analysis of the slags suggests that both copper and iron were being processed in addition to silver/lead. There was no evidence for zinc production (see the appendix). A further expedition was taken to the site of Kaya to the north west of Zawar where small retorts had been found (Fig. 6.1).

Zinc production operations

SITE 30

Function: Zinc production, small retorts.

Date range: Medieval, 530 ± 50 BP, BM 2223N.

Location: This is an isolated bank of retort debris (Plate 6.2) in a small valley in the heart of Zawar Mala Magra (Fig. 6.1) over 200 m above the general level of the Tiri Valley at 24° 20′ N, 73° 41′ E.

Description: The site has only poor communications with the main valley of the Tiri below. It is, however, located just below the main group of mines on Zawar Mala that were investigated as part of this project (see Chapter 5). These mines were worked principally over two thousand years ago, well over a thousand

ZAWAR ZINC SMELTING RETORT DUMPS

Ash dump north of Ramanath Temple (adjacent to Site 5)	charcoal	PRL-935	840 ± 130	1040 to 1270 AD	950 to 1400 AD	
Ash dump, Site 5	charcoal, *Terminalia* sp	BM-2782	modern			in press
Furnace, Site 30	charcoal *Artocarpus* sp	BM-2223N	530 ± 50	1320 to 1350 AD or 1390 to 1440 AD	1290 to 1450 AD	1985, **27**(3), 519
Retort heap, Site 5	charcoal	BM-2578	500 ± 50	1330 to 1340 AD or 1390 to 1450 AD	1300 to 1360 AD or 1380 to 1480 AD	1991, **33**(1), 61
Large retort heaps, Site 1, trench 1	charcoal	BM-2222R	240 ± 110	1490 to 1690 AD or 1730 to 1820 AD or 1920 AD to modern	1460 AD to modern	1985, **27**(3), 519
Large retorts, lower layers of Site 34, cut by road	charcoal	BM-2243R	350 ± 130	1420 to 1670 AD	1300 AD to modern	1985, **27**(3), 519
Mouth of furnace block, beneath coin hoard in furnace near Site 1	charcoal	BM-2638	Modern			1994, **36**(1), 108
Contents of retort from wall	charcoal	BM-2017R	110 ± 260	1530 to 1555 AD or 1635 AD to modern	1395 AD to modern	1984, **26**(1), 67
Contents of retort	charcoal	BM-2065R	modern			1984, **26**(1), 67

Table 6.1 Radiocarbon dates for the Zawar zinc retort sites. Calibrated using OxCal v2. 18 (Bronk Ramsey 1995) and the calibration curve of Pearson and Stuiver (1986). References in the far right column are to the original publication of the figures in *Radiocarbon*.

ZAWAR LEAD SMELTING SLAG HEAPS

Large slag heap, Site 5, layer 3	charcoal	BM-2484	100 ± 45	1690 to 1730 AD or 1810 to 1930 AD	1670 to 1755 AD or 1795 AD to modern	1989, **31**(1), 24
Large slag heap, Site 14, layer 3	charcoal	BM-2485	1950 ± 60	30 BC to 120 AD	110 BC to 210 AD	1989, **31**(1), 24
Small pit under layer of slag, Site 29, layer 2	charcoal	BM-2486	200 ± 35	1650 to 1680 AD or 1740 to 1800 AD	1640 to 1700 AD or 1720 to 1820 AD or 1850 to 1870 AD or 1920 AD to modern	1989, **31**(1), 24
Small slag heap, Site 2, trench 2	charcoal	BM-2487	1930 ± 80	40 BC to 180 AD	120 BC to 250 AD	1989, **31**(1), 24
Small slag heap, Site 7, trench 2	charcoal	BM-2488	1370 ± 80	590 to 770 AD	530 to 880 AD	1989, **31**(1), 24

Table 6.2 Radiocarbon dates for the Zawar lead smelting slag heaps. Calibrated using OxCal v2. 18 (Bronk Ramsey 1995) and the calibration curve of Pearson and Stuiver (1986). References in the far right column are to the original publication of the figures in *Radiocarbon*.

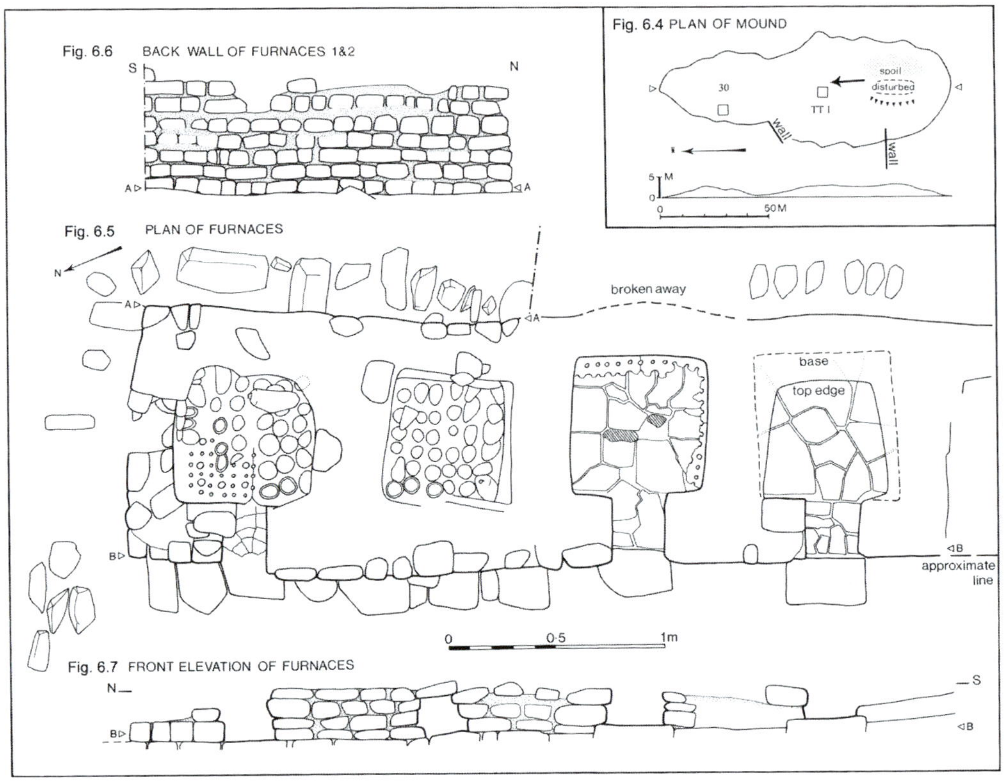

Figs 6.4–6.7. Fig. 6.4 Zawar, Site 30: plan of retort heap (the arrow shows the position from which Plate 6.2 was taken). Fig. 6.5 Zawar, Site 30: plan of the excavated portion of the furnace. Fig. 6.6 Zawar, Site 30: elevation of back wall behind furnaces 1 and 2. Fig. 6.7 Zawar, Site 30: elevation of the surviving front portion of the excavated portion of the furnace.

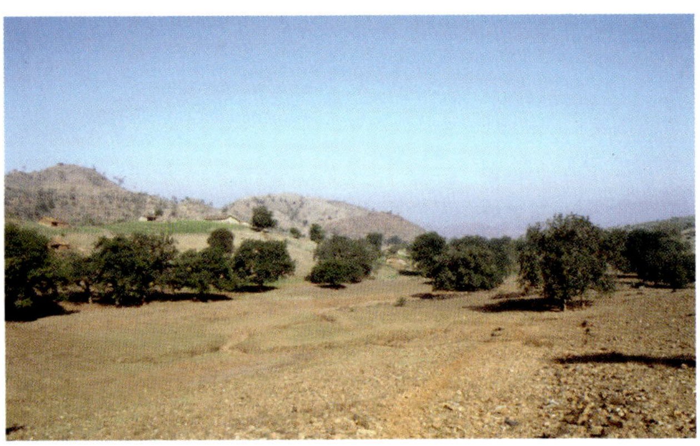

Plate 6.1 Small valley on Zawar Mala. Site 30 is in the trees in the left centre foreground.

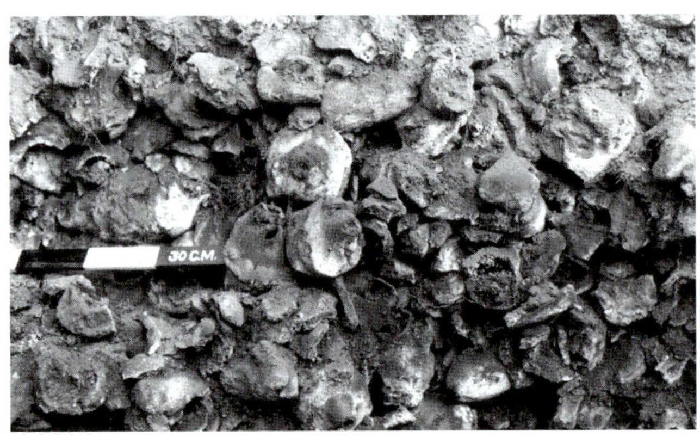

Plate 6.3 Zawar, Site 30: section of TT1 showing the uniform deposit of loosely packed retort fragments, quite typical of the heaps of retort debris.

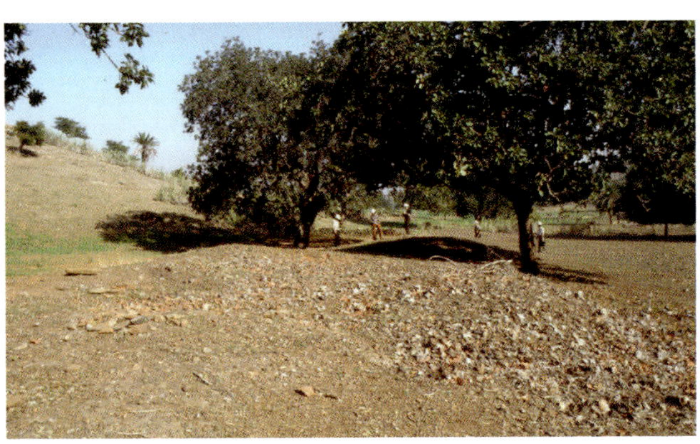

Plate 6.2 Zawar, Site 30: retort heaps looking to the north along the bank from the position marked on Fig. 6.4. The excavated furnaces were found beneath the large tree where people are standing.

years earlier than the zinc smelting operation described here. However, there is evidence of later quarrying cutting into the entrances of the earlier underground workings, at Pratap Khan for example.

The bank of retort debris is approximately 140 m long by 40 m wide and 1.5 m tall at the centre (Fig. 6.4 and Plate 6.2). Refractory material from it was scattered over quite a wide area, and it had been rather eroded by a small stream in the north west corner, which had helped to expose the furnaces (*koshthi*) described below. A small test area (TTI) 2 × 2 m was dug through the top of the heap towards the centre in order to investigate its nature and to take representative samples of refractory material. The first 20 cm was compacted soil, but

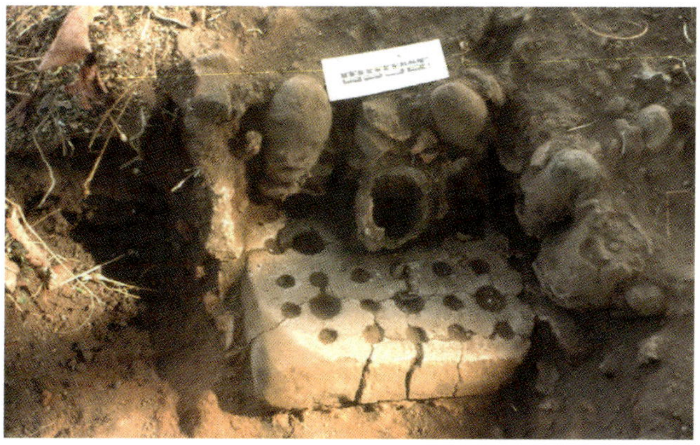

Plate 6.4 Zawar, Site 30: the first indication of the furnace block. The north west corner of the block is exposed here.

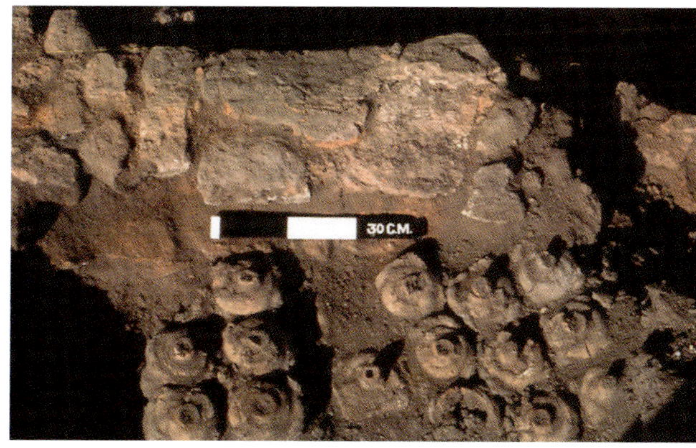

Plate 6.6 Zawar, Site 30: roughly stacked blocks of fused retorts, exposed in the area immediately behind furnace 4.

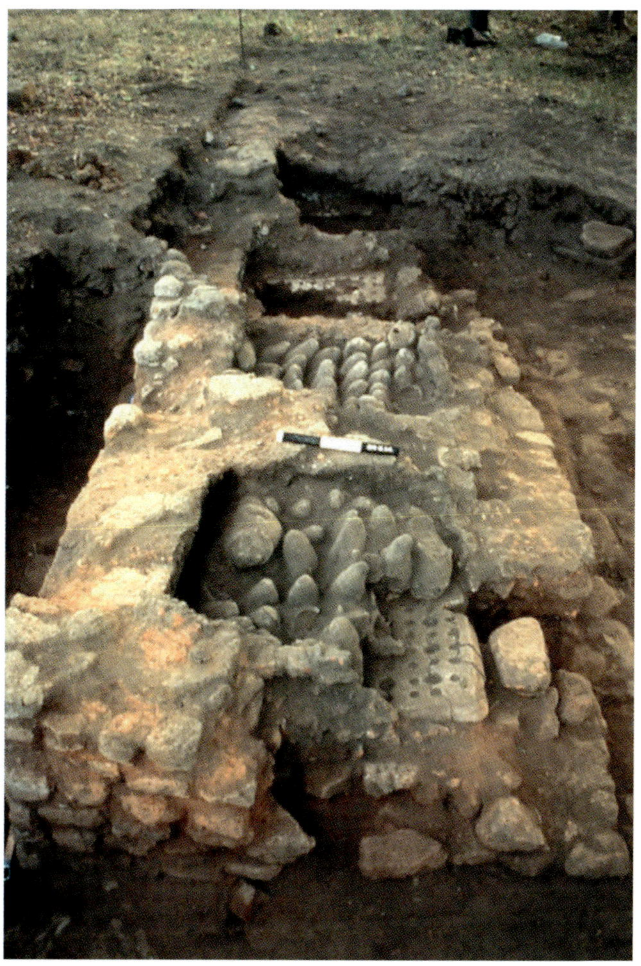

Plate 6.5 Zawar, Site 30: excavated portion of the furnace block looking south along the four excavated furnaces. The vertical iron rod at the top marks the end of the furnace block.

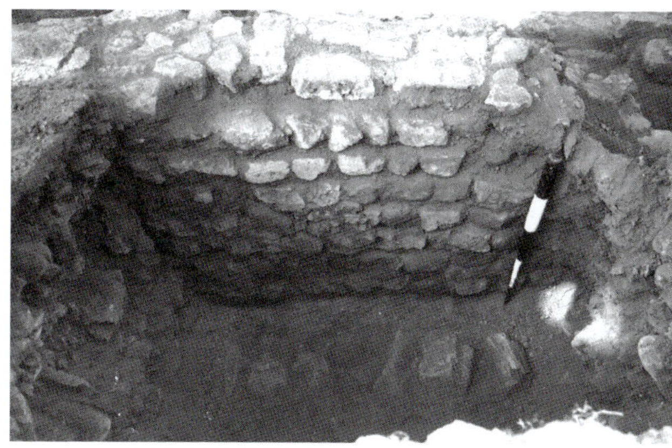

Plate 6.7 Zawar, Site 30: back wall of the furnace block behind furnace 4 where the wall survives to its full height of 60 cm. Note the exposed stone rubble plinth on which the wall stands.

from there down to the base the heap comprised very loosely packed retort fragments with a few pieces of furnace lining and perforated plate, but with no stratigraphy (Plate 6.3). The excavated material contained no mine debris, pottery or domestic debris at all. At the base there was a very loose brown soil that had presumably washed through the retort debris and this lay on the hard brown soil of the original land surface, which itself contained flecks of charcoal and red burnt clay. It is significant that although the present land surface all around the bank is covered with rock mine debris washed down from the mines above, the 4 m^2 of surface exposed by the excavation did not reveal a single fragment of broken rock, suggesting that either the fragments had been removed – broken rock was used in the foundations of the furnaces for example – or that the heap predates the appearance of the mine debris and therefore belongs to an earlier phase of working. The medieval date for the Site 30 furnace suggests that the former explanation is the more likely.

The furnaces

Only the form of the furnaces is described here; their function is described in Chapter 11. While this excavation was in progress a most important discovery was made in the north west corner of the bank: the edge of a perforated plate and the tops of several retorts were exposed, the latter seemingly in a regular pattern, much as they might have been when in the furnace. A limited excavation revealed that we had indeed located a substantially intact furnace still bearing its last load of retorts (Plate 6.4).

Initially it was thought that there was just the one furnace, but it rapidly became clear that it shared the south wall with a second installation. Accordingly a 2 × 2 m area was opened around the two furnaces, each of which was approximately 1 m^2. The surface of the heap in and around the area excavated had many large fragments of perforated plate. The first 10 cm was made up of compacted soil and retort debris above the loose fill of retort debris lying over and around the furnaces.

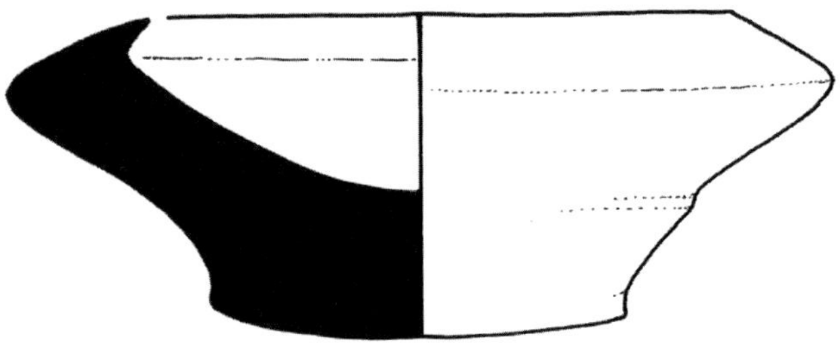

Fig. 6.8 Zawar, Site 30: small lamp found in retort debris (max. diameter 9.5 cm). This was one of the very few pieces of non-metallurgical ceramics found on the site and may be an indication of night working.

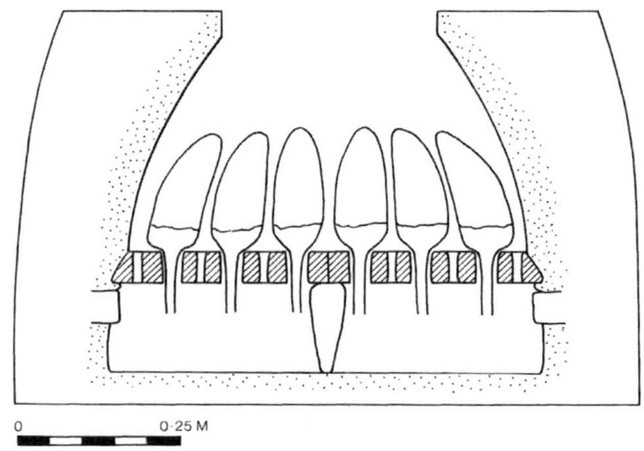

Fig. 6.9 Zawar, Site 30: schematic section through a typical furnace showing the retorts sitting on the perforated plates which divide it into two parts – the furnace proper and the cool chamber. Note the curve on the inner walls and the pronounced asymmetric tilt of the retorts at the edges (cf. Fig. 6.14). There would almost certainly have been individual vessels to collect the zinc as it dripped from the condenser ends, but as none were found they have been omitted.

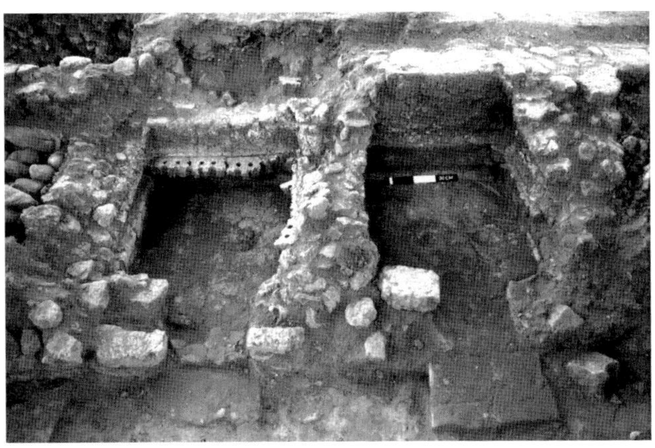

Plate 6.8 Zawar, Site 30, furnaces 3 and 4: note the large complete square ceramic tile in front of furnace 4.

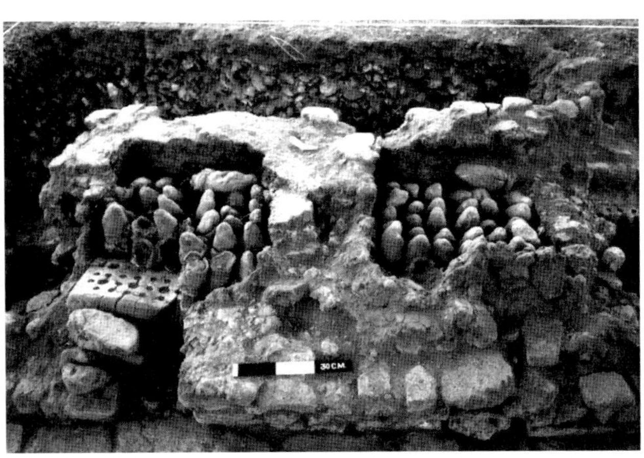

Plate 6.9 Zawar, Site 30, furnaces 1 and 2: the arrangement of retorts within the furnaces. Note also the two reused retorts placed horizontally across the others in identical positions in furnaces 1 and 2. They were slightly heat-fused to the other retorts in places.

It soon became apparent that the east, back wall of the furnaces ran farther to the south and it was followed for a total length of 6.14 m. This suggested that the furnace we had discovered was at one end of a block of seven furnaces, running approximately north–south along the axis of the main debris heap. The trench was therefore realigned to run parallel with the now-defined rear wall of the furnace block and the length extended to 5 m in order to accommodate four of the furnaces (numbered F1–F4, north to south) (Figs 6.5–6.7 and Plate 6.5). The furnaces had clearly rapidly become buried in retort debris in which there was little meaningful stratigraphy. At one point behind F4 the debris was composed of blocks of fused retorts roughly stacked from the ground surface directly against the wall up to its top (Plate 6.6). It could be argued that this represented the usual practice of dumping directly behind the furnaces but this seems inherently unlikely as the used retorts would have blocked direct access to the furnaces from the back in subsequent operations, such that if, for example, it was necessary to move a retort in the back row during the process, as must have often been the case, it would have involved leaning over the furnace from the front. However, unlike F1 and F2, F4's final load of retorts had been removed so it is very possible that the last load of used retorts was lifted straight out of F4 and stacked on the ground immediately against the wall in the knowledge that the furnaces were not going to be used again. This suggests that the usual practice was to dump the retorts just behind the furnaces while leaving some access to the back, and as soon as the furnaces were abandoned they rapidly became engulfed in the debris, most of which would have been from their own collapsing heaps but with some originating from their immediate successor(s) elsewhere on the site. The excellent preservation of the furnaces and the retorts also indicates rapid burial soon after their final use. The fill of the excavated area, as with TT1, was almost exclusively made up of loosely packed retort debris with some fragments of furnace lining and broken plates. Only one or two sherds of domestic pottery were found in the whole excavation plus one small lamp (Fig. 6.8). There was a considerable amount of charcoal in the upper portion of the fill, much of it in quite substantial pieces of several centimetres across, which could well represent the original fuel, although little was actually found *in situ* in the excavated furnaces. A sample for radiocarbon dating was collected from charcoal pieces found lying at the base of F4, possibly from the last firing of the furnace.

At the base of the retort dump there was loose brown soil among the debris, which lay on similar soft brown soil. The area (west) in front of the furnaces had originally been paved with large ceramic tiles laid directly on the soft ground surface. Once again, as with TT1, the soil around the furnace was essentially stone-free in contrast to the surrounding land surface around the heap, probably as part of a surface clearance before the furnaces were built.

A further block of furnaces was noted just to the south of the excavated group on approximately the same orientation. At the south end of the retort dump another concentration of perforated plate fragments and furnace lining was noted and led to the discovery of yet another block of furnaces, which was not investigated further.

From the uniformity of the debris it appeared that this was a standardised process therefore only four of the seven furnaces from the first furnace bank to be located were excavated. At the time of the excavation the possibility of conserving the furnace block was being actively considered so the furnaces were exposed rather than fully – and thereby destructively – excavated. For example, the last charge of retorts and perforated plates were not removed from furnaces 1 and 2 to investigate what lay beneath. Those furnaces were substantially complete but furnaces 3 and 4 had lost their charge of retorts and the plates, exposing the lower parts of the furnace, so very little information would have been obtained by excavating further. Unfortunately the subsequent discovery of a hoard of silver coins in another set of furnaces (see Chapter 3) ensured that these furnaces and those at Site 37 were completely ransacked by local treasure hunters after the excavations.

The furnace block

The furnace block, or *koshthi* as it was probably called, was oriented 22° west of north, lying along the axis of the valley therefore the openings to the furnaces were not oriented to the prevailing wind direction, which was up the valley. The later large and intermediate retort furnaces were sited along the edges of ridges, although not necessarily to direct the wind into the furnace openings.

The overall dimensions of the main brick superstructure of the block were 6.15 × 1.25 m at the base (Fig. 6.4), and stood at a height of 60 cm from the plinth as measured at the back wall of furnace 4, which stood to its full height (Plate 6.7). This height was also preserved for furnace 5, where the back wall was exposed but the furnace otherwise unexcavated. The back wall at furnace 2 was also preserved to full height in one place (Fig. 6.6), but otherwise in general the top two courses of the back wall were missing. Due to the contours of the heap the front wall was much less well preserved, and at furnaces 1 and 2 only survived to a height of four courses (25 cm) above the level of the tiled area (Fig. 6.7). The well-fired front inner lining of furnace 2 had survived to a considerable height, being only about 15 cm short of the full height surviving at the back of the furnace. This lining extended from the centre to the southern corner of the furnace and was featureless, suggesting that there were no fixed tuyeres or anything else set into the front furnace wall at a higher level which had not survived. This is confirmed by the retorts in furnaces 1 and 2, which showed no signs of manipulation or furnace control through the furnace walls.

The walls have a slight batter on their outer faces. As measured on the back wall at furnace 4, the top of the wall was 7.5 cm in from the base at a height of 61 cm. The furnace walls were built on a plinth of stone rubble extending irregularly for between 20 and 30 cm beyond the brick walls on the north and eastern sides, and presumably on the unexcavated southern

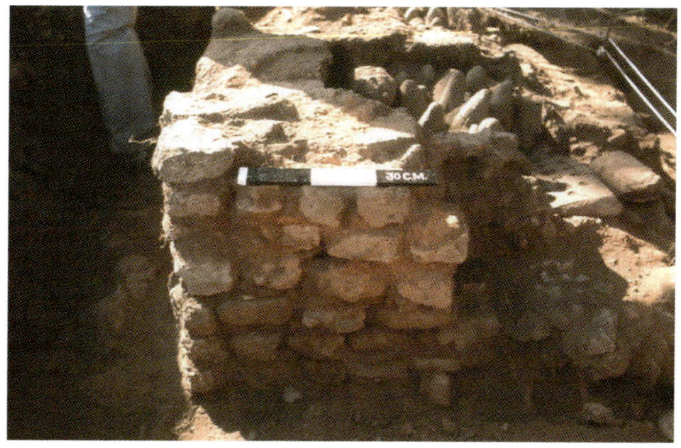

Plate 6.10 Zawar, Site 30: brick wall at the north end of the furnace block.

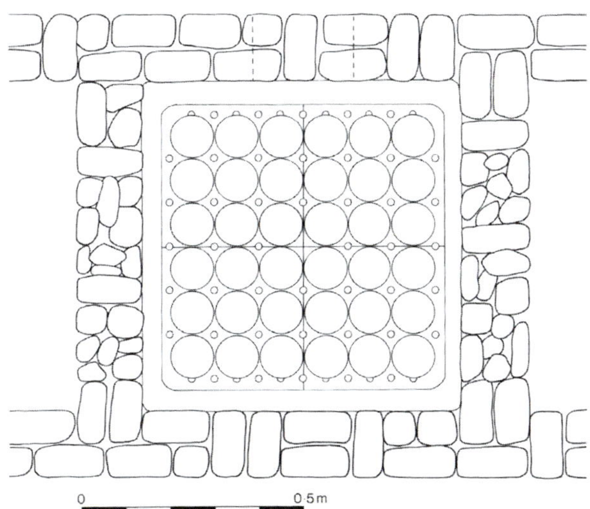

Fig. 6.10 Zawar, Site 30: schematic plan of a typical *koshthi* furnace at the level of the perforated plates (indicated as large circles) on which the retorts sat with their condensers protruding through the large holes into the cool chamber beneath.

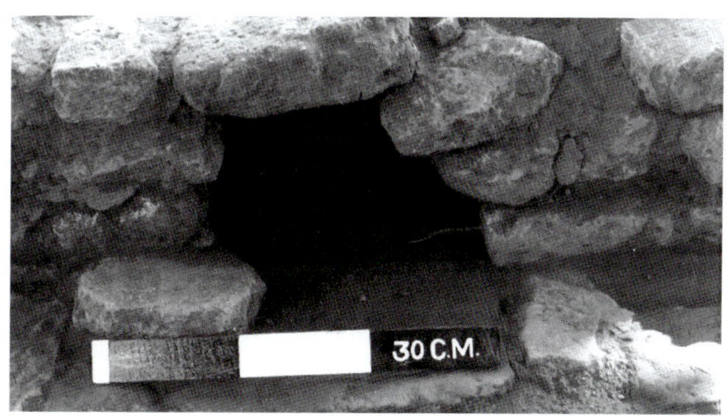

Plate 6.11 Zawar, Site 30: entrance to furnace 2, preserved intact. Note the slight corbelling.

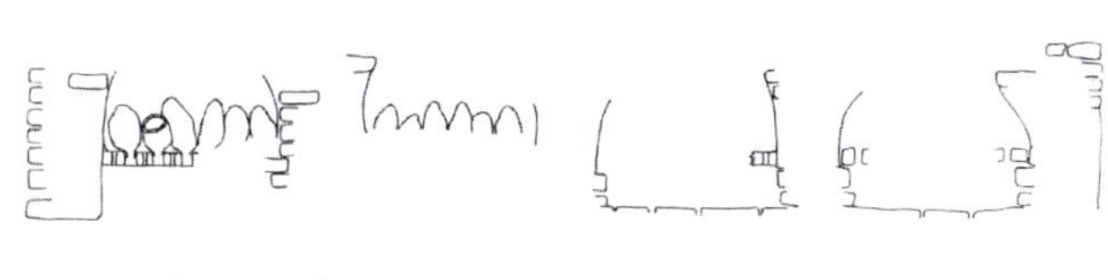

Fig. 6.11 Zawar, Site 30: detailed section through furnaces 1–4 showing the brick ledges for supporting the perforated plates and the curve of the upper furnace walls.

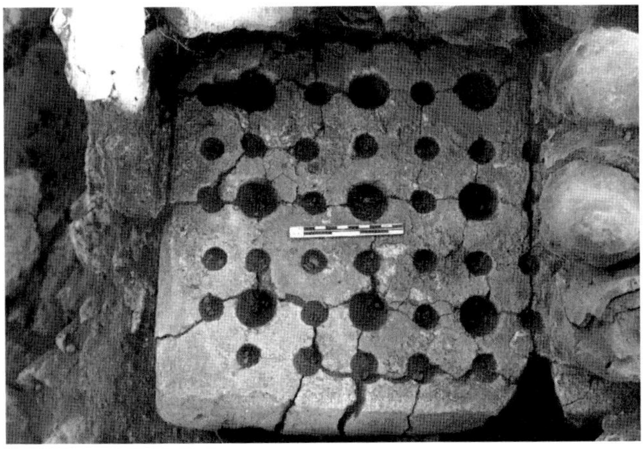

Plate 6.12 Zawar, Site 30: complete but badly cracked perforated plate in the north west corner of furnace 1. The nine large holes accommodated the condenser necks from the retorts and the smaller holes, some shared with the adjacent plates, were for the passage of air and ash.

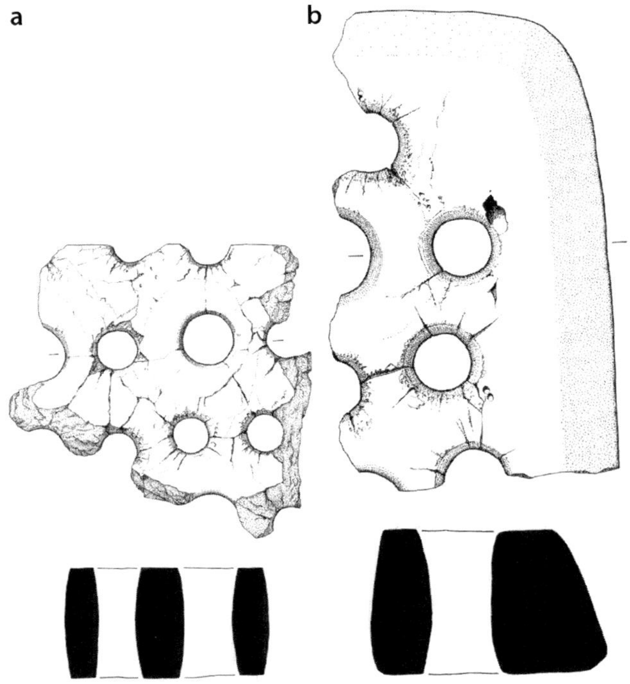

Fig. 6.12 Zawar, Site 5: large fragments of perforated plate from the Site 5 heaps, which contain material identical to that from Site 30 (max. dimensions: (a) 10.5 cm and (b) 13 cm).

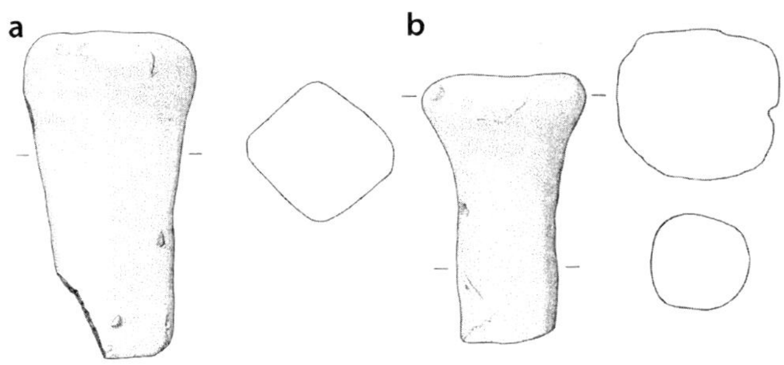

Fig. 6.13 Zawar. (a) Site 30: clay pillar found in the bottom of furnace 3 (length: 11 cm, cf. Plate 6.13); (b) Site 5: a more waisted example found in the Site 5 heaps (length: 8 cm).

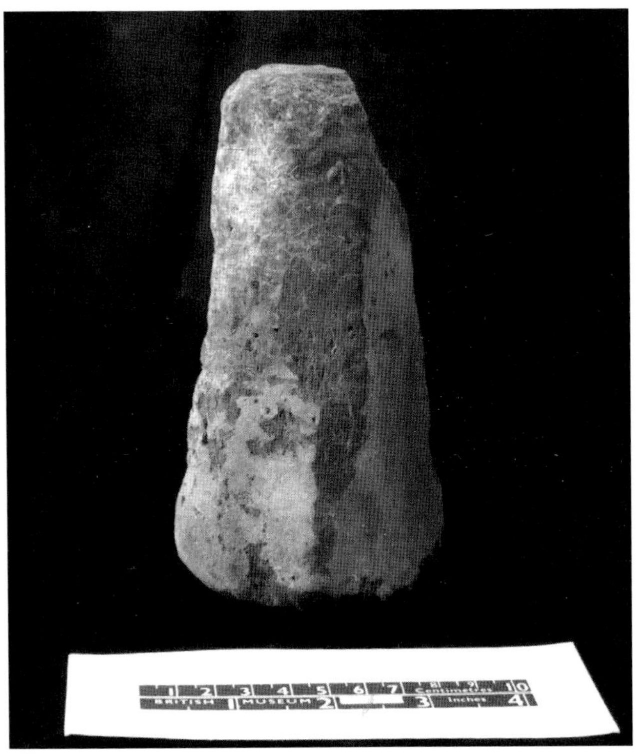

Plate 6.13 Zawar, Site 30: pillar found lying on the floor of furnace 3 (photographed standing upside down, cf. Plate 6.28).

end also. These undressed stones lay directly on the ground and were in effect the foundation of the furnace (Fig. 6.6 and Plate 6.7). This would have had the effect of raising the brick structure of the furnace slightly above the ground, helping to keep it dry. On the west side of the furnace, where the entrances to the lower chambers lay, the stone plinth was replaced by the large, square earthenware tiles. The only complete example excavated (in front of F4) measured 35 × 20 × 4 cm (Plate 6.8), but clearly some of the others had been fragmentary when laid and the floor must have been more of a 'crazy paving' of fragments than a regular pattern of whole tiles, the fragments probably reused from abandoned furnaces.

Internally each furnace consists of a hearth in a form approximating to a truncated pyramid, although the sides do have a pronounced and partially original curve. This is where the retorts were heated and then set over the cool, rectangular chamber where the descending zinc condensed (Fig. 6.9). The two parts of the furnace were separated by the perforated plates which supported the inverted retorts and also acted as the grate (Fig. 6.10 and Plate 6.9).

The furnaces were built substantially of rough, fired bricks, heavily mortared with mud, and originally stood eight courses high (Fig. 6.6 and Plates 6.7 and 6.10), although the coursing was very irregular and frequently bricks were replaced by pieces of refractory; in one place the spaces were just filled with mortar. The space dividing the individual furnaces between the now hard and highly fired furnace linings was made up of irregular refractory material and brick rubble in a mud mortar, with no attempt at coursing. In contrast, the walls of the lower, cool chamber of each furnace were built quite regularly of two courses of brick, the upper course projecting slightly to provide the ledge for seating the perforated plates (Fig. 6.11).

The jambs of the entrances to the lower chambers were made of longer, more regular bricks which were slightly corbelled,

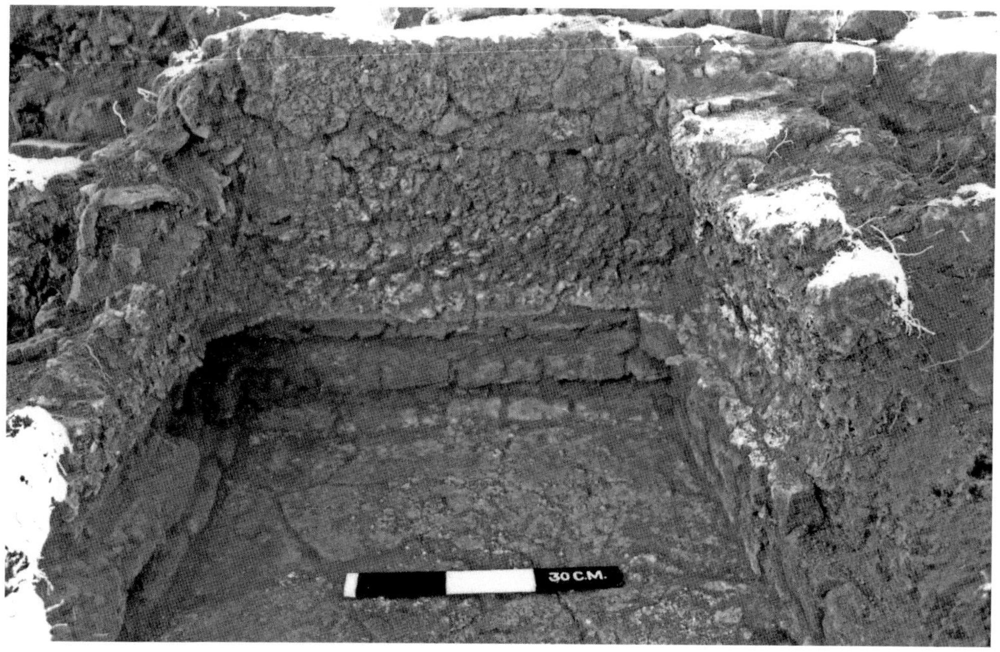

Plate 6.14 Zawar, Site 30: back wall of furnace 4. The hearth was in the form of a truncated pyramid.

supporting a brick lintel (Fig. 6.7 and Plate 6.11). The entrance to furnace 2 had a large brick set within it, possibly to control the air flow during the smelt. It would have been taken out at the end of the process while the zinc was removed from the cool chamber, and had presumably been replaced ready for the next smelt which, as it turned out, never happened. A similar arrangement was observed in the large retort furnace at Site 37.

The floors of the cool chamber in furnaces 3 and 4 were plastered with mortar about 2 cm thick, laid upon a surface of brick or tile. A little ash still lay on the floors, and the floor of furnace 4 was slightly glazed with zinc silicate, although the conditions should not have been hot enough in the cool condensing chamber to allow this to happen during the ordinary operation of the furnace. This material also contained some zinc[1] which must have dripped down directly from the condenser end while still very hot, somehow missing the collector vessel, and as such was the only metallic zinc found at Zawar. It is perhaps significant that, unusually, the last load of retorts and the plates had been removed from F4, suggesting that there had been some mishap in the collection chamber during the smelting and afterwards it had been necessary to gain direct access probably to scrape up zinc that had spilt onto the floor.

The floors extended through the entrances and slightly overlapped the tile paving outside (Plate 6.8). They were flat with no indication of any channelling through which molten zinc could have been tapped from the lower chamber as in some versions of the European Overpelt process, as suggested by Morgan (1985: 14). Instead, we believe the zinc must have been collected in individual collecting vessels beneath the condenser ends.

The upper and lower chamber of the furnace were divided by a grate made up of four perforated plates of refractory ceramic which had been moulded,[2] and which also supported the retorts (Figs 6.10 and 6.12a and b). The dimensions of the only complete perforated plate uncovered (on the north west corner of furnace 1) was 32.5 (north–south) × 34 (east–west) × 4 cm thick, although as Plate 6.12 shows, the plate was badly broken up. When multiplied by two the overall dimensions of the furnace grate work out at 65 (north–south) × 68 cm (east–west). These slight differences are not random, but are repeated in the dimensions of the empty furnaces 3 and 4 at the level of the bricks which would have originally supported the perforated plates. This suggests that the dimensions of these moulded bricks were regular and quite deliberate, and the furnace built to accommodate them.

Each plate has nine large cylindrical holes and 26 small cylindrical holes, together with the half sections of a further eight holes along the inner edges that align with similar holes in adjacent brick to form complete holes (Fig. 6.12a). There are no half holes in the centre edges as they would be blocked by the supporting pillar beneath. The central pillars, handmade of clay, have a roughly rhomboidal cross-section and taper towards the bottom, presumably to facilitate the withdrawal of the putative small collecting vessels through the front opening without having to dismantle the perforated plates. The pillar and the perforated plates in the Site 30 furnaces were essentially freestanding with no mortar. A support pillar was not excavated *in situ* at the Site 30 furnaces although presumably in furnaces 1 and 2 they still supported the plates. A pillar (Fig. 6.13a) was found lying in the fill of furnace 3, and another in the Site 5 heaps (Fig. 6.13b and Plate 6.13). The find spot of the former pillar, the only one found on the Site 30 excavation, suggests that it belonged to furnace 3. The excavated large furnace, Site 37, still had its pillar in place tapering towards the base (Plate 6.23). One of the later intermediate retort furnaces (Site 33) had the pillar firmly mortared to the floor (Plate 6.28), although the pillar in the adjacent furnace was freestanding.

The truncated pyramid forming the upper chamber had sides of 65 × 68 cm at the plate level as noted above, and about 45 cm (north–south) by an estimated 45 cm (east–west, the west wall now missing) at the top, 40 cm above the perforated plate as measured in furnace 4 (Plates 6.8 and 6.14). The back wall had an internal slope of about 70°, although the walls are noticeably concave internally as shown in Fig. 6.9. This is a regular feature continuing into the corners (although the corners are quite

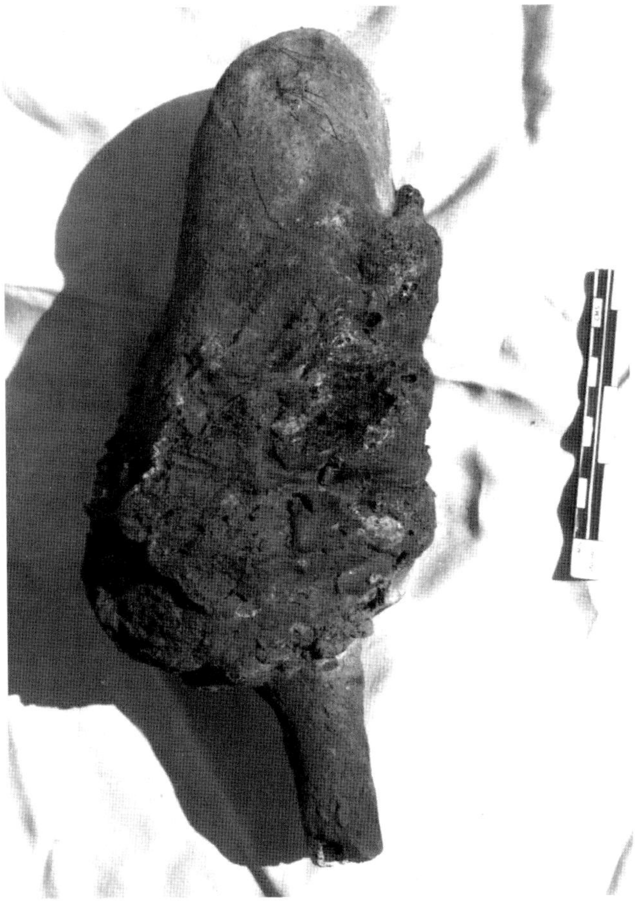

Plate 6.15 Zawar, Site 30: used retort from the Site 30 heaps showing the tilt necessary for the retorts adjacent to the sloping walls of the furnace. The condenser neck would have been vertical in use (cf. Fig. 6.14).

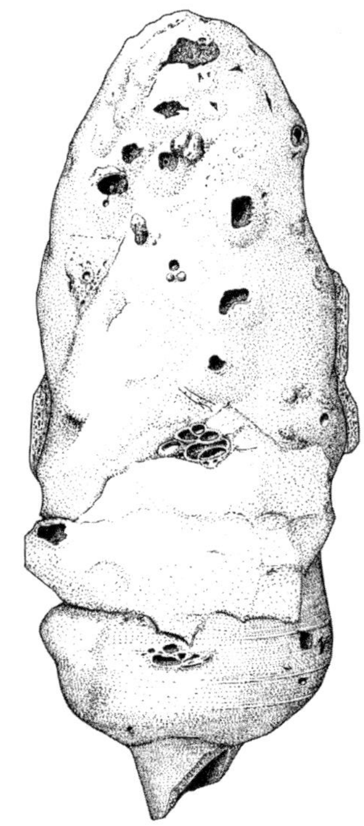

Fig. 6.15 Zawar, Site 5: small retort from the retort heaps. As is usual with almost all of the Zawar retorts the condenser end is missing (length to edge of luting: 22 cm). This retort is straight, presumably intended for the central regions of the furnace away from the sloping walls.

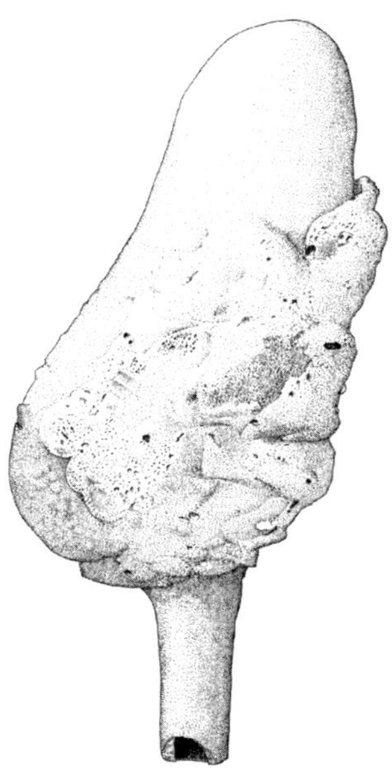

Fig. 6.14 Zawar, Site 30: small retort from the retort heap (overall length 28.5 cm, 21 cm to the retort base). Note the pronounced tilt to accommodate the retort to the slope of the furnace wall (cf. Fig. 6.9 and Plate 6.15).

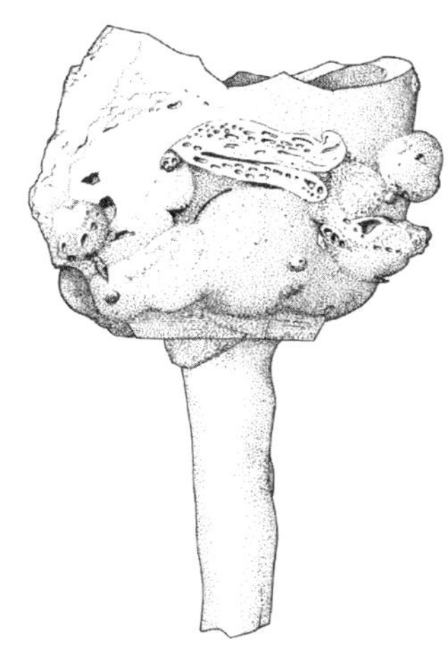

Fig. 6.16 Zawar, Site 12: condenser mouth from a small retort which still has several centimetres of the condenser tube attached (overall length: 12.5 cm) (cf. Chapter 11, Plate 11.1).

distinct) and seems to be a deliberate constructional feature rather than a result of vitrification and slumping of the furnace lining at the mid-levels where the temperature would have been at its maximum. The furnaces were undoubtedly reused and the linings would have been refettled by the application of fresh clay to the walls. The vertical fingermarks preserved in the lining of furnace 4 suggest that the furnace had been refettled with clay fairly recently before its final use. The preservation of the fingermarks shows the lining had not seriously eroded, again confirming that the curved profile was deliberate.

The top of the furnace back wall, as exposed at furnaces 4 and 5, was finished with a continuous layer of smoothed mortar over which the burning and vitrification had lapped continuously without a break at the inner edge. There was no indication that the top of the wall as found once bore a superstructure to create an updraught (contra Hegde *et al.* 1986: 254, fig. 24.6 and Hegde 1991: 82, fig. 35). The front wall has not

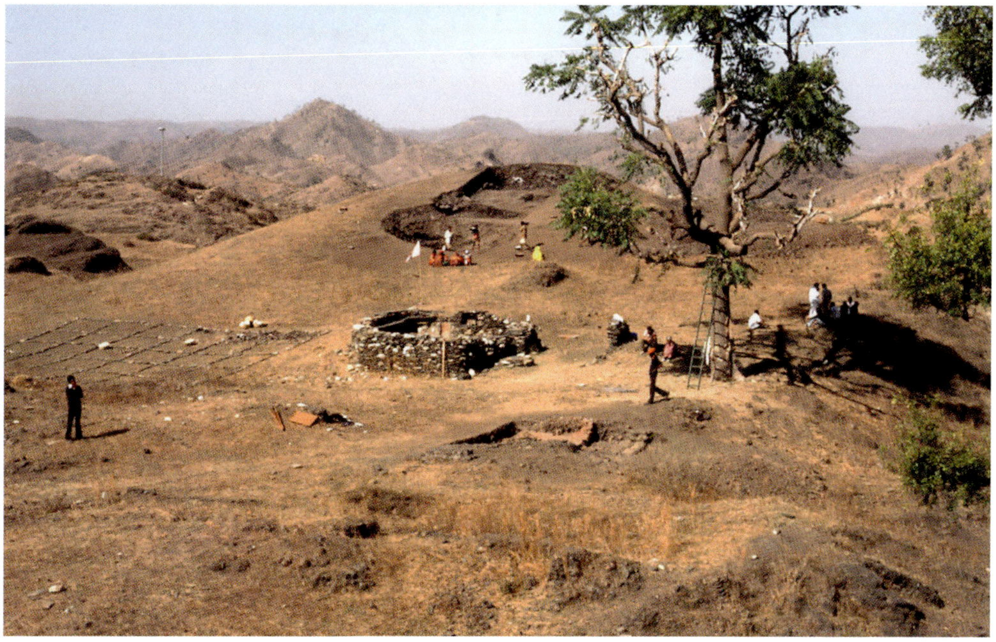

Plate 6.16 Zawar, Site 37: general environs of the large retort furnaces, which run along the edge of the old dumps to the right of the picture. The excavated furnace is in the centre foreground. Note the part-excavated tips associated with these furnaces in the background.

survived to its full height anywhere along the length excavated therefore the top must be conjecture, but the appearance of the partially vitrified retorts still *in situ* in furnaces 1 and 2 does not suggest that any fixed installation such as a bellows unit was mounted on or above the front wall; such an installation would only have hindered access to the retorts and the fire. The medieval Sri Lankan low iron smelting furnaces did not have a superstructure (Juleff 1996, 1998: 176–213), and the comparable traditional Chinese zinc smelting furnaces did not have any superstructure over the top of the furnace (Xu 1998: pls 4–6) therefore on balance, it seems most unlikely that the Indian furnaces would have been any different. If such a layer had existed then evidence of it would be expected on the vitrified walls of the *koshthi*, but in fact on the preserved back walls of the Site 30 furnace the vitrification is continuous up the wall and over onto the rim, strongly implying that the furnaces looked much as represented in the reconstruction drawing (frontispiece), and the fire was completely open at the top.

The retorts were carefully arranged in the upper chamber. Those in the outer two rows were asymmetric in shape such that when their condenser tubes were vertical in the holes of the perforated plate the retorts leaned inwards to accommodate the slope of the furnace walls (Fig. 6.14 and Plate 6.15); the other retorts in the inner rows were straight (Fig. 6.15). A number of sherds of retorts were found tightly wedged *in situ* between the retorts in the furnace, although none were actually vitrified in place. Furnaces 1 and 2 both had a complete used retort lying between the back rows of retorts (Fig. 6.5 and Plate 6.9). Although these retorts showed no obvious signs of having vitrified in their present position, neither could be easily moved so it must be assumed that they, together with the retort fragments, had been placed there during the course of the smelting operation to keep the retorts apart and to control the fire; the later furnaces had purpose-made spacers (Figs 6.22 and 6.23). It was noted that when the retorts in furnace 1 were removed to expose the perforated plate their necks were missing. It was also observed with the *in situ* retorts in the large and intermediate retort furnaces that the necks must have been broken off when the collecting vessels were removed. The very few necks that were recovered were filled with oxidised zinc and presumably were routinely recycled (see Chapter 7).

Only the area in the immediate vicinity of the furnaces was excavated – no attempt was made to excavate deep into the retort heap which covered the site in order to ascertain whether there had been a surrounding structure or roof. By analogy with the large retort furnaces at Site 37, which were in a structure, it is likely that the Site 30 furnaces were originally under some form of roof. It would have been necessary to keep the clay furnace dry, especially when not in use (see Chapter 9). No roof tiles were found at Site 30 such as occurred in large numbers at Site 37 therefore the putative roof would have been of thatch – like the Chinese furnaces (Xu 1998: pls 3–6) and the 19th-century South Asian copper and iron smelting furnaces (see Chapter 2, Fig. 2.3) – or conceivably wooden shingles. As depicted in Fig. 2.3, the structure would have been open to allow the fumes to dissipate and there seems to have been no wall to impede what would have been the growing heap of spent retorts behind the furnace. In contrast, the spent retorts from the large retort furnaces in their walled enclosures were not thrown behind but were moved some distance away.

Samples were taken for further study to ascertain the nature of the refractories and the heat regime to which they had been exposed during the smelting operations (see Chapter 7, Fig. 7.3). A series of samples was taken from the walls of furnace 4 and the sample positions recorded in the archive (see the appendix).

Site 37

Function: Zinc production, large retort process.
Date range: Post-medieval, 240 ± 110 BP, BM 2222R. Coin evidence from a similar furnace nearby suggests that the furnaces were defunct before the end of the 16th century AD (see Chapter 3).

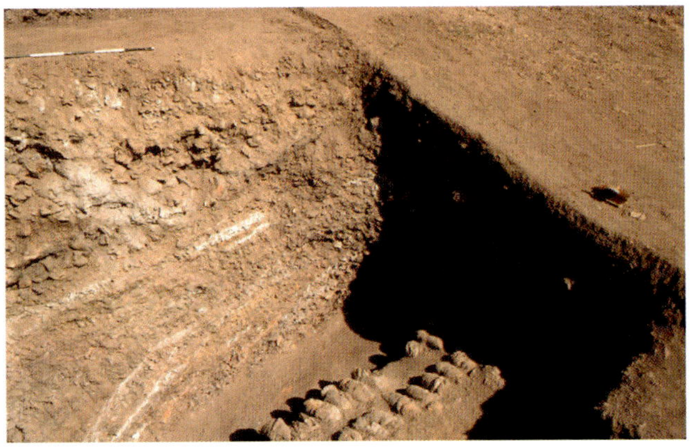

Plate 6.17 Zawar, Site 37 TT 1 Area C: excavation of the retort dump showing the wall of a reused retort and the north section. Note the major break between the upper retort heap and the earlier major heap, the tip lines of which are at a steep angle showing the heap was advancing forward by material tipped over the edge.

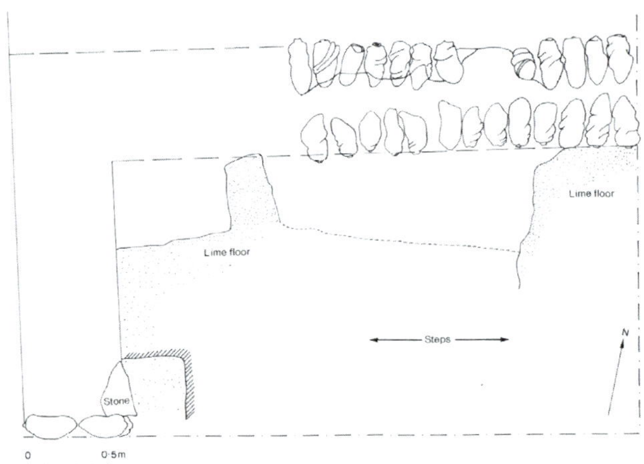

Fig. 6.17 Zawar, TT1 Site 37: walls of reused retorts enclosing a mortared floor beneath the heap of large retort debris from the Site 37 furnaces.

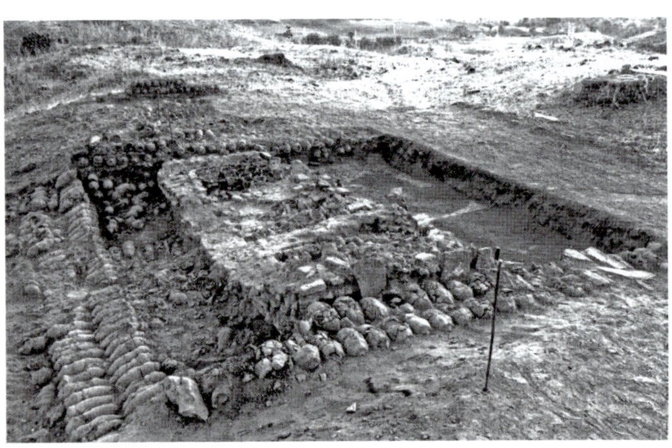

Plate 6.18 Zawar, Site 37: walls (cf. Fig. 6.19) built of a double thickness of mortared used retorts, forming the building and enclosure in which the furnaces were set (cf. Fig. 6.18 reconstruction).

Location: This site lies on the large banks of used retort and other debris that occur between Old Zawar to the east and the River Tiri to the west, at 24° 22′ N, 73° 42′ E (Plate 6.16 and Figs 6.2 and 6.3).

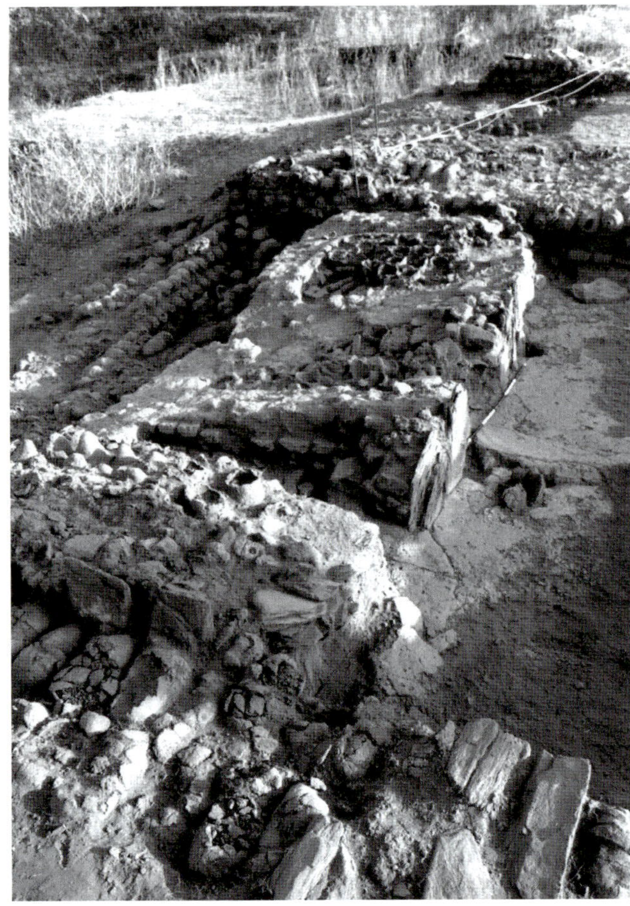

Plate 6.19 Zawar, Site 37: furnace block from the north showing the front with supporting stone slabs and plastered floor.

Description: The furnace block investigated was one of a group situated along the edge of one of the pre-existing banks of used retorts. Nearby lay a large heap of used retorts etc. apparently from these furnaces, lying on top of the existing mounds of debris. This heap was investigated first and therefore designated locally as Site 1.

A large area of approximately 5 × 12 m was laid out over the heap and divided into three primary areas, designated A, B and C, each 4 × 5 m; each of these areas was in turn subdivided. Excavation commenced in the west half of each of the areas. The heap was composed of very loose retort debris with some limited spreads of orange clay and grey and white ash (Plate 6.17). There were also fragments of roof tiles throughout the layers, perhaps from the roofs of the structures in which the large furnaces were housed. The tip lines in the heap were easily distinguished and in general had quite shallow slopes, broadly following the present contours of the heap. Thus it would appear that spent retorts and ashes from the furnaces which lay along the edge of the bank were brought out to the end of this promontory and dumped over the west end as depicted in Fig. 6.18. This gradually raised the level of the end to a maximum height of about 3 m above the pre-existing surface of the dump and a ramp developed on the south side for the convenience of those carrying the waste up onto the growing heap.

Excavation in areas A and B ceased after a depth of about 1 m. In areas B and C the lower layers contained limited amounts of domestic rubbish, indicated by some animal bone

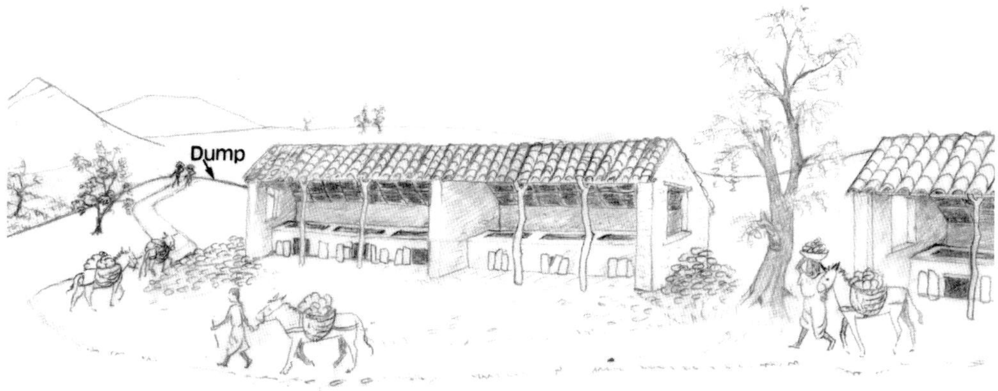

Fig. 6.18 Zawar, Site 37: reconstruction of the possible appearance of the large retort furnaces within open-fronted buildings set along the ridge of retort dumps.

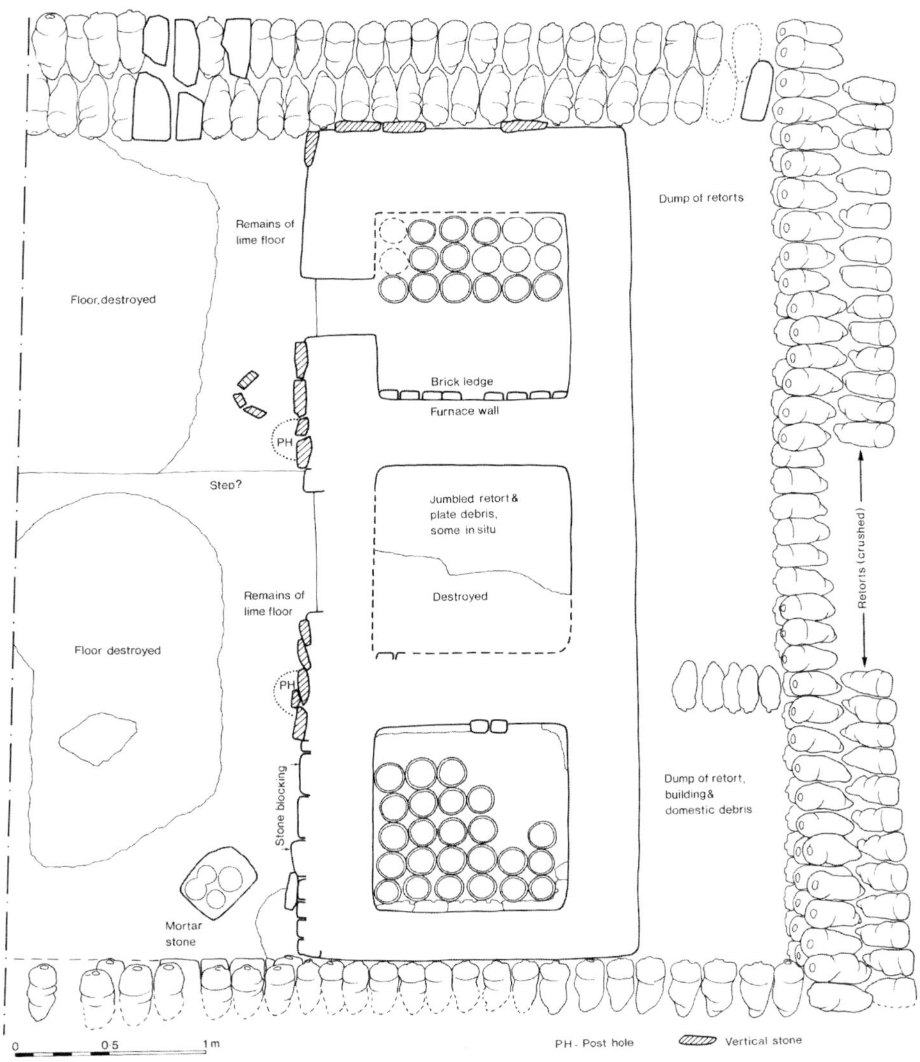

Fig. 6.19 Zawar, Site 37: plan of the excavated block of furnaces and its building. Note the substantial postholes in front of the furnaces and the large mortar stone.

(provisionally identified on site as belonging to sheep/goat, chicken and rodent), and domestic pottery. This lay directly on a hard continuous layer that was the stabilised ground surface of the mound upon which the last furnaces, exemplified by Site 37, were built and upon which the debris from them was dumped.

In area C the excavation continued and walls built of mortared retorts were encountered enclosing a mortared floor (Fig. 6.17). The lower courses of the wall were set in a trench which had cut down through the hard, continuous layer therefore the structure had been built sometime after the main mound had

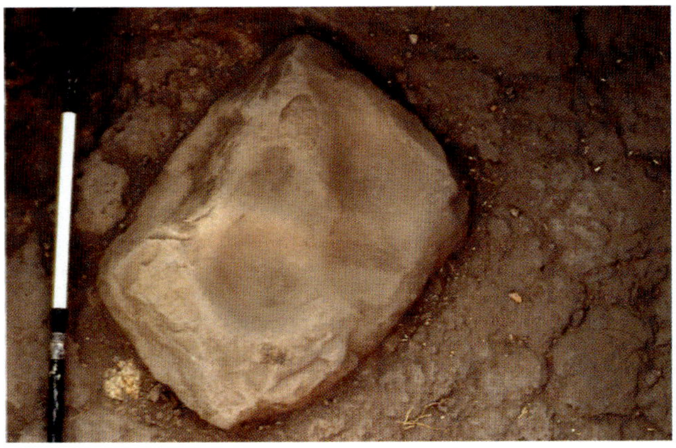

Plate 6.20 Zawar, Site 37: large mortar stone set in the floor before furnace 3.

stabilised and a surface formed, but before the deposition of the final heap which ultimately engulfed it. The walls were of a double thickness of mortared retorts (Plate 6.17), very similar to those surrounding the furnace blocks described below. However, there was no trace of a furnace structure and this together with the presence of contemporary domestic debris suggest they were probably the remains of a house.

The layers beneath the hard, continuous layer, which now form the main body of the mound, slope steeply down to the west and north, suggesting that the tipping advanced by material dumped over the edge from the south. The material in the upper layers of the main mound is also made up of large retort debris, interspersed with ashy layers or tip lines, and one major soil layer which clearly marked a hiatus in the development of the mound. It should be emphasised that although there is a considerable depth of material, and quite a complex stratigraphy, the whole mound probably built up over a relatively short time compared to ordinary domestic accumulations of debris, and the time scale should almost certainly be measured in decades rather than centuries. There were no small retorts in these heaps, although they were found in profusion at the Site 5 heaps just below alongside the Tiri.

The furnaces and their buildings

Very shortly after the excavation of the heaps began, evidence of *in situ* furnaces was noted in the present surface of the original retort heaps, and thereafter attention concentrated on these. The furnaces lay on the edge of the old mounds, some 30 m south of the excavated later heap, and it is highly likely that the retorts in that heap came from these furnaces. The evidence for the furnaces took the form of the remains of a group of large retorts standing upright in an apparent regular array, which on excavation proved to be the remains of the last batch of retorts still sitting in the furnace.

Preliminary cleaning of the surface revealed the remains of walls built of reused retorts surrounding the furnace block. An area of 5 × 3 m was laid out around the furnaces, aligned on the walls. Excavation exposed a block of three furnaces within the walled enclosure (Plate 6.18). Further clearing revealed evidence for an identical furnace block immediately to the south, on the same alignment, and also within a structure which had a common wall with the structure excavated. The presence of other furnace blocks within walls on a similar alignment still farther to the south suggests that there may have been a whole line of furnaces operating along the ridge (Fig. 6.18). The entrances to the furnaces face into the prevailing winds, and the position of the furnaces on the eastern edge of the retort dumps would tend to blow the fumes away from the settlement and industrial area. Only the one furnace block and its building were excavated, and the building will be considered first.

The building

The walls were built of a double thickness of reused large retorts set in mortar approximately 60 cm thick and survive to a height of over 1 m in the south wall (Fig. 6.19 and Plate 6.18). Here, the lower courses were partially of burnt brick, although apparently bricks were not in common use elsewhere on the site. The bricks may have been used to level the foundation to enable the retorts to be laid in horizontal courses. The walls formed three sides of a rectangular enclosure that was apparently open on the west side as no evidence of a wall was located there. This would have allowed access to the cool chambers of the furnace as their entrances were on this side. In addition the north and south enclosure walls were plastered internally on the west side of the furnaces. The many roof tiles in the associated heaps and the substantial postholes inside the enclosure indicate that the enclosure was roofed over the furnaces (Fig. 6.18). The form of this roof may be inferred from the location of the furnaces, which presumably were positioned to lie beneath the roof at its greatest height; they lie parallel to the back wall but, significantly, not against it, suggesting that the roof had a central ridge with gables at either end rather than a simple pentice roof sloping up or down from the east wall to the open front. If the latter had been the case then logically the furnaces would have been built against the east, rear wall. The roof must have been carried on the west (front) side by a beam supported by the side walls and by posts indicated by the two regularly spaced postholes of approximately 25 cm diameter just in front of the furnaces. Another posthole still farther to the west of approximately 30 cm diameter is probably a replacement. These postholes still retain some of their original stone packing. There are no corresponding holes on the east (rear) side and it must be supposed that this side of the roof rested on a timber wall plate laid on top of the east wall of the building.

It should be noted that as the furnace end walls reach right up to the north and south walls and there is no indication of an entrance in the east wall, the space, approximately 85 cm wide, between the backside of the furnace and the rear wall, was apparently unapproachable except by clambering over the furnaces themselves, which would have been quite impossible when they were in operation. However, there must have been some access to the space behind the furnaces, presumably through gaps in the east wall, so that the retorts on the east side could be adjusted while the process was in operation. Even so, as at the small retort furnaces previously, it is clear that the furnaces were controlled mainly from the front, and this is reinforced by the absence of any flooring or evidence of wall plastering on the east side.

The purpose of the roof was to keep the furnaces dry so it is unlikely that there would have been any major openings in the roof itself, and there is no evidence for chimneys. Even though the building was open on the front, west, side it does seem likely that the other walls would have had some openings to allow the noxious fumes, laden with lead and cadmium, to escape more easily.

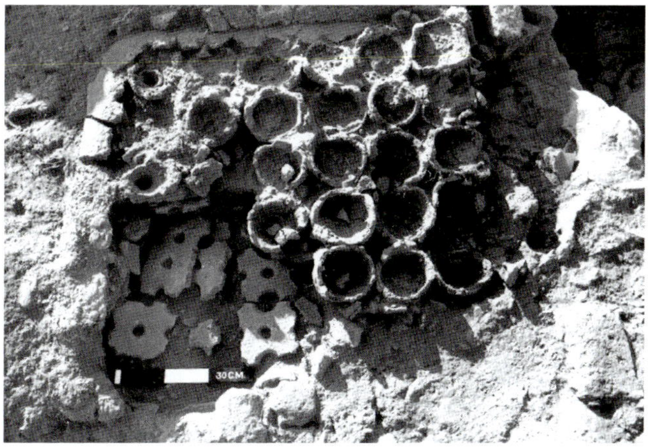

Plate 6.21 Zawar, Site 37: furnace 3 after cleaning. Little survives above the level of the perforated plates.

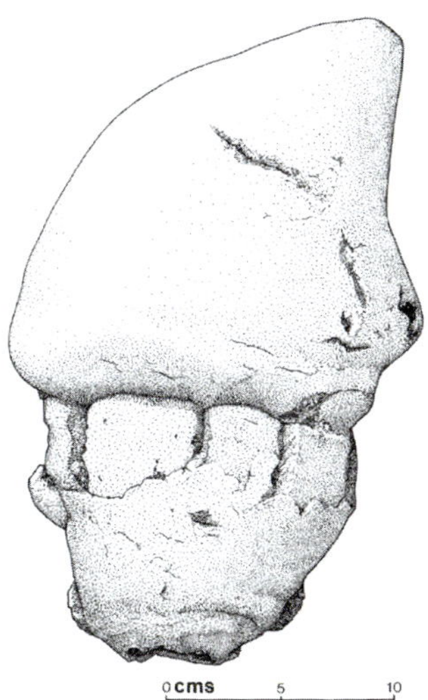

Fig. 6.20 Zawar, Site 37: used large retort from the main heaps. Note the asymmetry with one side straight and the other curved over at the top to enable the outer rows of retorts to be accommodated within the sloping walls of the furnace (cf. Plate 6.22).

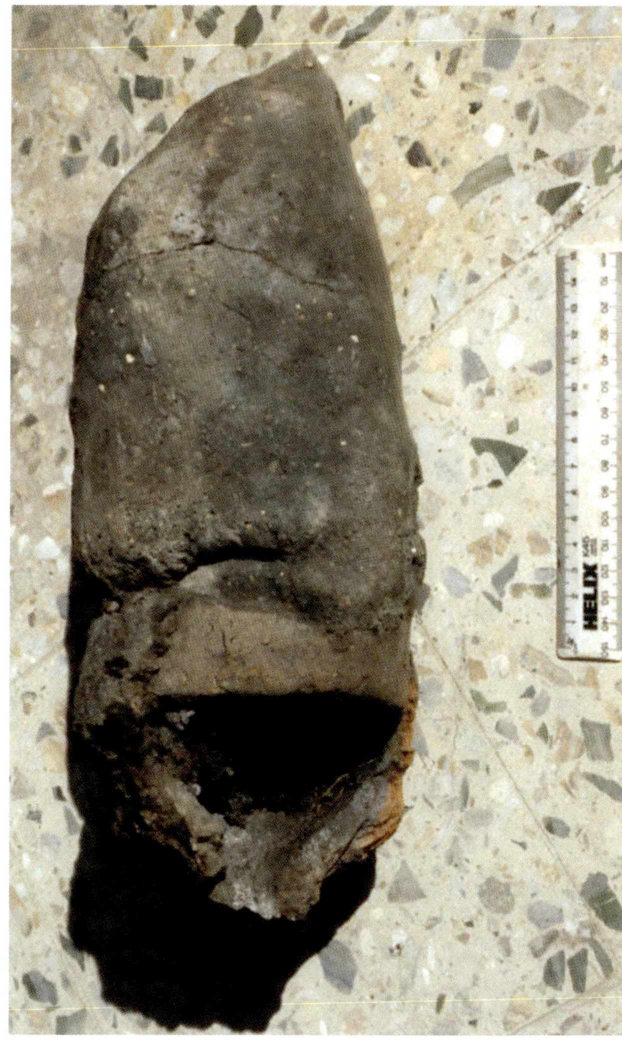

Plate 6.22 Zawar: undistorted large retort showing asymmetry at the top (right-hand side) necessary for the retorts in the outer rows to fit into the furnace (cf. Fig. 6.20).

The floor of the western, front part of the building was covered with a thick layer of plaster made very uneven by long use (Plate 6.19). A large mortar stone with many working hollows was set into the floor in front of the southernmost furnace (hereafter called furnace 3) (Plate 6.20). Its purpose is uncertain, but the general environment would surely have been too contaminated to permit food preparation, and the ore and other ingredients charged into the retorts are likely to have been prepared elsewhere. It may have been where the necks of the condensers were crushed to recover zinc-rich material for resmelting.

The furnace block

Only the form of the furnace is discussed here (its function is described in Chapter 11). In general the furnaces were very similar to the earlier, small retort furnaces excavated at Site 30. The large retort furnaces excavated here had not been buried under debris after use and thus were not as well preserved. Little survived above the level of the perforated plate (Plates 6.21 and 6.23) so there is not much information on the actual furnace chambers beyond their plan. The slight remains of the upper walls suggest they were once again in the form of a truncated pyramid, but the height is unknown. Some of the large retorts, found on the heap and which had not been distorted by heat, showed a slight but quite definite asymmetry (Fig. 6.20 and Plate 6.22), as did some of the small retorts (Fig. 6.14).

Analogy with those surviving *in situ* in the small retort furnaces would suggest that the asymmetric large retorts were located in the outer rows of the furnace leaning inwards to accommodate the sloping walls. The retorts were set in the perforated plates which are typically about 3 cm thick (Fig. 6.21). Many rough pieces of clay were found, approximately segmental in shape (Fig. 6.22), in the large retort dumps. Some of these were fused to the tops of large retorts (Fig. 6.23) and seem to have functioned as separators to control the spacing between retorts as they began to distort in the heat of the furnace.

The excavated furnace block was rectangular and measured 4.5 × 1.8 m. The exterior walls were built of regularly coursed fired brick, still standing some five courses high in places (Plates 6.24 and 6.25). A feature not found on the small or intermediate furnaces was the use of stone slabs set into the ground to hold the face and ends of the furnace block. On the furnace block we excavated the slabs were set along the north end between the furnace and enclosure walls and also along the front, west side (Plates 6.18 and 6.19). There was no indication that there had

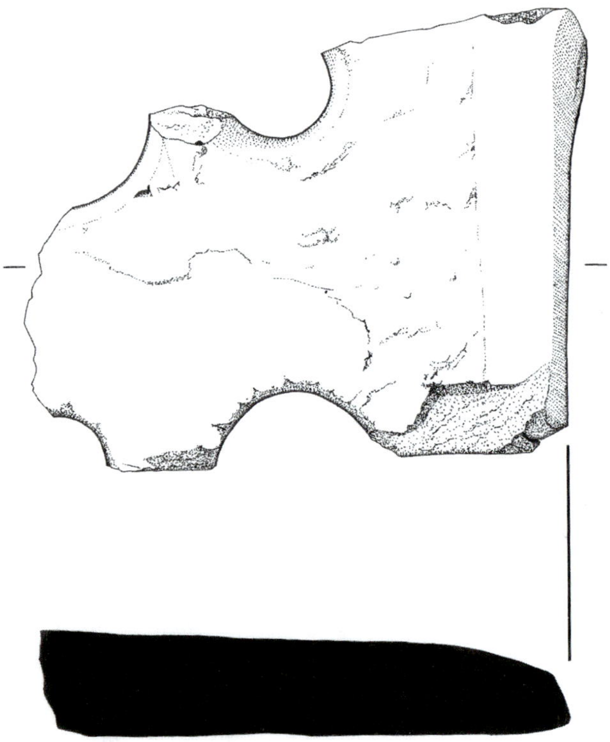

Fig. 6.21 Zawar, Site 37: fragment of a perforated plate (max. dimension: 9.5 cm). Note it is considerably thinner than the small retort plates (Fig. 6.12) but similar to the intermediate retort plates (Fig. 6.27).

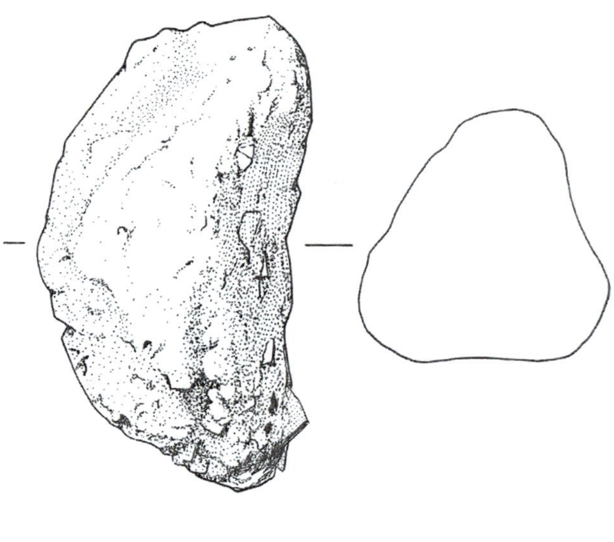

Fig. 6.22 Zawar: surface find near Site 29. Clay wedge used to keep the retorts apart during firing (length: 10 cm).

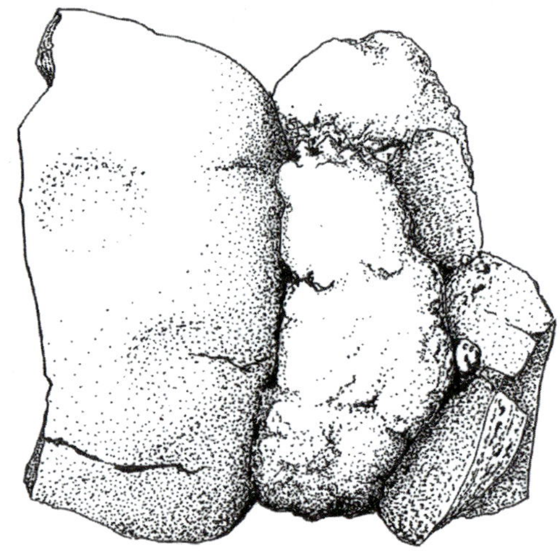

Fig. 6.23 Zawar, Site 5 retort heaps: clay wedge still in position between the fragments of the two large retorts it was separating (max. dimension of exposed wedge: 5 cm).

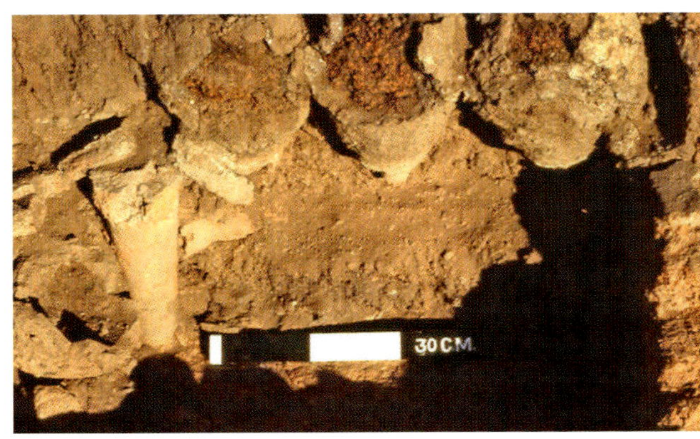

Plate 6.23 Zawar, Site 37, furnace 1: from the south side after excavation and half sectioning exposing the cool chamber. Note the supporting pillar is still in place.

ever been any set along the rear wall, and unusually there were none along the south end between the furnace and the wall of the building.

This setting of stones along the front of the furnace may be somewhat unusual for the large furnaces generally. With experience the team came to be able to recognise buried furnace blocks by the tops of the stone slabs set in pairs approximately 4.5 m apart and indicating either end of a furnace block. Their function was to provide additional lateral support. The area of the chambers of the large furnaces as represented by the perforated plates was approximately three times that of the small furnaces and double that of the intermediate furnaces, thus the tendency of the large furnaces to sag and spread during use must have been much greater. It is perhaps significant that the later intermediate furnaces did not need this additional support: possibly the large furnaces were found to be too unstable even with the additional support supplied by the stone slabs.

A long-established path passed through the remains of the central furnace (hereafter called furnace 2), rendering it more damaged than the other two. In the southern furnace the fragments of perforated plate lay on the floor of the cool chamber, still containing the remains of the last load of retorts. As the northernmost (hereafter called furnace 1) was better preserved it was selected for more detailed investigation (Fig. 6.19 and Plate 6.23). The retorts were in the same 6 × 6 arrangement as in the small furnaces, but being larger (typically 15 cm maximum diameters) required a larger furnace to accommodate them. At the perforated plate level furnace 1 had an internal width of 110 cm. Unfortunately, as the plates were completely shattered and the furnace walls were rather distorted and spread, the internal dimensions could not be measured with the same precision as on the Site 30 furnaces, therefore it is not known if the same slight but consistent differences existed between the

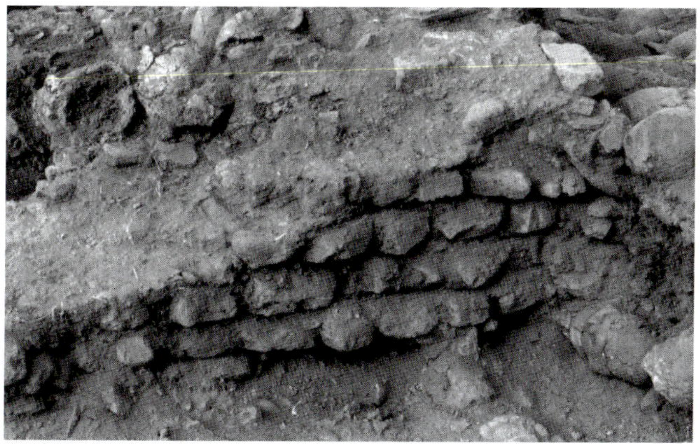

Plate 6.24 Zawar, Site 37, rear (east) of furnace 1: the brick wall still surviving five courses high against the north wall of the building.

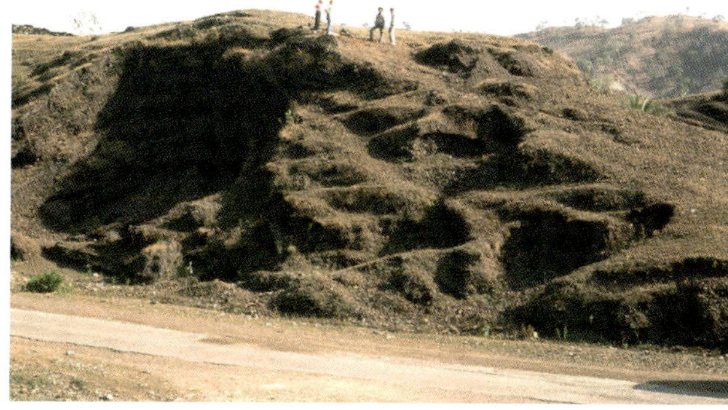

Plate 6.26 Zawar, Site 33: intermediate retort furnaces sited on the western edge of the large tips of small and large retort debris. The furnaces investigated (marked by the small group of people) were partially exposed by the modern road cutting.

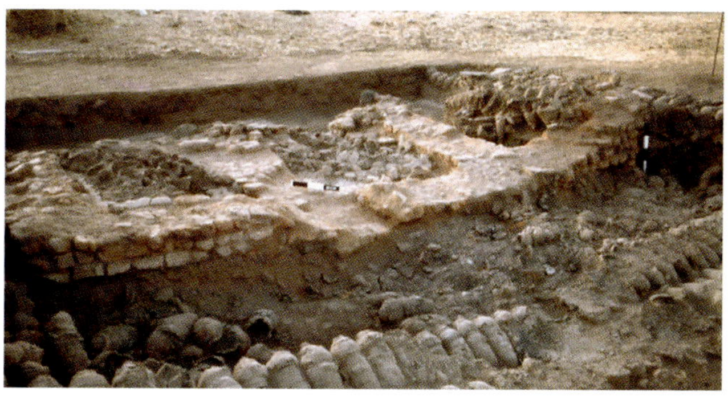

Plate 6.25 Zawar, Site 37: furnace block from the south east.

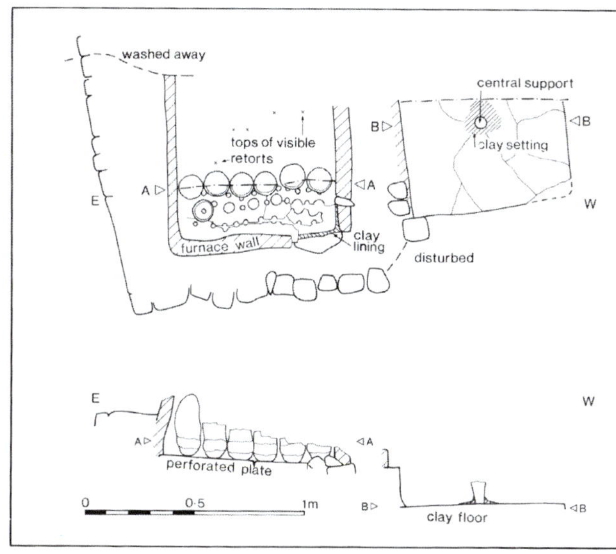

Fig. 6.24 Zawar, Site 33: plan and section of the intermediate retort furnace block after partial exposure.

north–south and east–west dimensions as found on the earlier furnaces.

Furnace 1 was half sectioned, exposing the central support pillar still in position (Plate 6.23). This was of circular section with a concave profile, very similar to those in the intermediate retort furnaces (Site 33) (Fig. 6.26 and Plate 6.28), and tapered towards the base. It was about 21 cm long, which gives the height of the cool chamber at the centre. As with the small furnaces, the walls between the furnaces were largely built of rubble set in mortar. The surviving walls of the cool chambers were much less regularly coursed than on the small furnaces, but did have a course of bricks to provide the ledge on which the perforated plates rested. This layer of bricks did not originally project beyond the face of the wall, but appeared to on excavation owing to erosion of the mud plaster facing of the walls.

The entrance to the cool chamber was of the order of 30 cm wide, although on a better preserved furnace a short distance away, which was subsequently excavated without proper record, the entrance was 45 cm wide (i.e. much wider than the entrances to the small furnaces) and about 30 cm tall, with the arch formed with a corbelled stone lintel. The entrance to furnace 3 was blocked with a large stone, recalling that at furnace 2, Site 30. Some samples were taken (see the appendix).

Sites 33 and 34

Function: Zinc smelting, intermediate retort process.

Location: On the south side of the Zawar road where it cuts through the western edge of the Old Zawar tips at 24° 22' N, 73° 42' E (Fig. 6.3).

Date range: Post-medieval.

Description: As a result of the fieldwork in the 1983 season it was noted that there were retorts intermediate in size between the large and small retorts (Fig. 6.28 and Plate 6.30). These intermediate retorts were concentrated in a limited area where the main road cuts through the major retort dumps to the west of Old Zawar, extending back for about 30 m behind the road on both sides for about 100 m along the road (Fig. 6.3). At the beginning of the 1985 season the remains of a block of intermediate retort furnaces were noted exposed in this area on top of a ridge (Plate 6.26). The block was obviously not complete and the already very full investigation programme precluded a full excavation of the furnace such as had been carried out for the large and small retort furnaces in 1983. Accordingly sufficient excavation was carried out to record the principal dimensions of a single furnace in the block.

The exact number of furnaces in the block was not ascertained, although there were at least three but probably not many more. The furnace block was sited on a ridge of debris from the small retort process above the road, and a further two furnace blocks were noted on the same east–west alignment farther along the

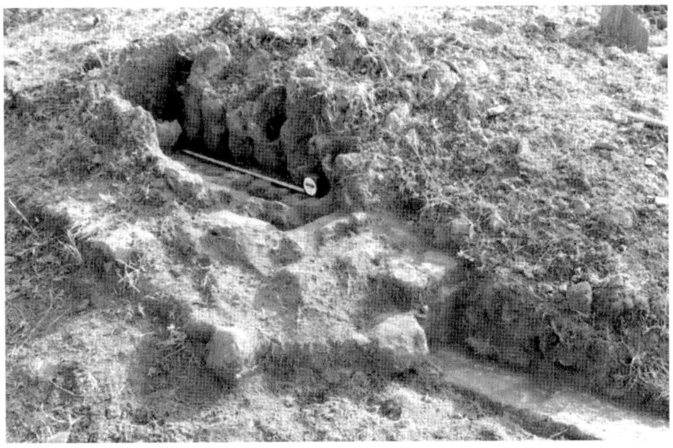

Plate 6.27 Zawar, Site 33: intermediate furnace block after partial exposure of two furnaces. Note the well-preserved in-curving rear wall. The tape is set at 50 cm.

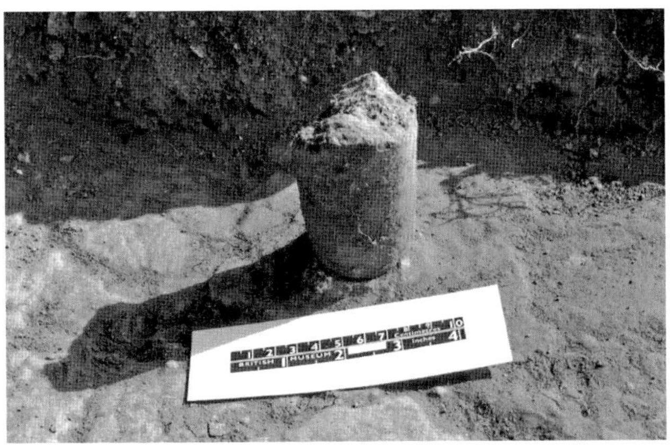

Plate 6.28 Zawar, Site 33: stump of the central support pillar of the westernmost furnace found firmly mortared in place.

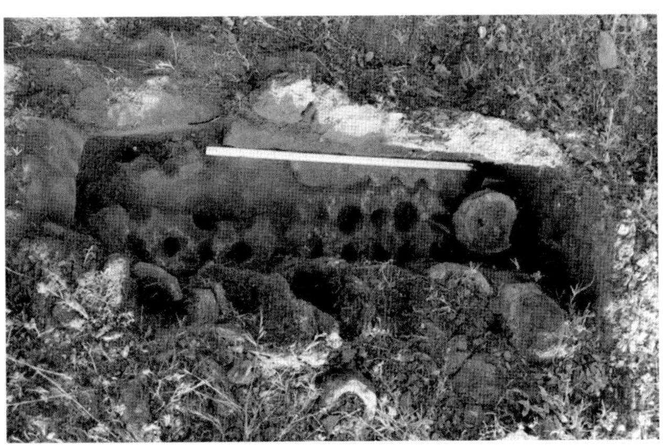

Plate 6.29 Zawar, Site 33: badly broken and distorted perforated plate exposed in the easternmost furnace. The tape is set at 50 cm.

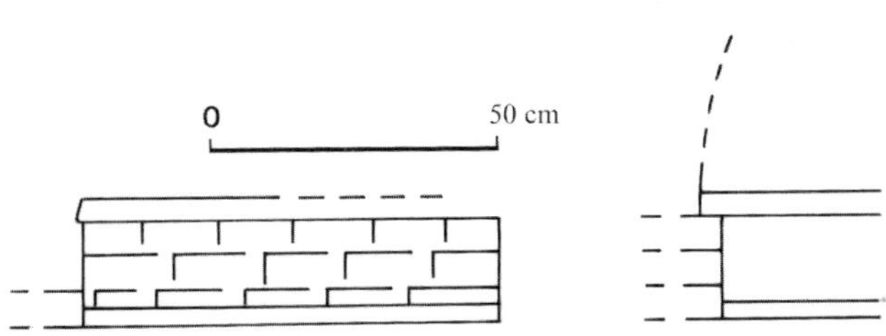

Fig. 6.25 Zawar, Site 33: schematic drawing of the brickwork in the central intermediate retort furnace showing how the perforated plates were supported (compare with Fig. 6.11).

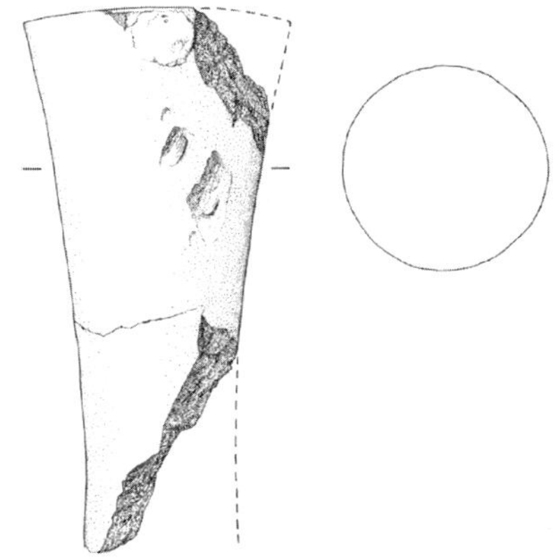

Fig. 6.26 Zawar, Site 33: fragment of a central support pillar from the cool chamber of the central furnace (surviving length: 13 cm).

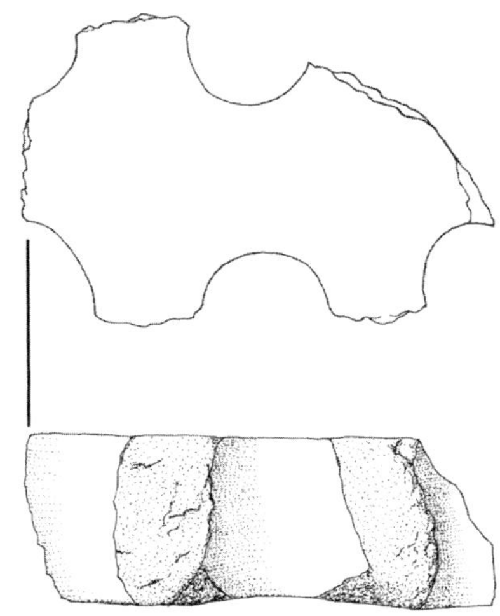

Fig. 6.27 Zawar, Site 33: fragment of a perforated plate (max. dimension: 9.5 cm).

ridge to the east, while a single chamber (Site 34) was recorded on the far west end of the heaps. No openings to the lower chamber were observed when the interior of the north side of the furnace block was cleared, suggesting that the entrance and 'front' must have been to the south, not in the prevailing wind direction. Possibly the furnaces were sited along the tops of the ridges to pick up the wind to help disperse the fumes rather than to blow the fire. No evidence was seen for a covering structure, and no tile fragments were observed in the surrounding heaps, but by analogy with the large retort furnaces it is likely that the furnaces were originally contained within a structure or at least under some form of roof. The date of the intermediate furnaces is not certain, but it was noted that the used intermediate retorts overlay both small and large retorts in the western heaps. Stratigraphically the intermediate process seems to represent the last phase of zinc smelting wherever it is found, probably of 17th–18th century date, although the large retort process could have continued contemporaneously at other parts of the site.

The intermediate furnaces were basically of the same materials and construction as the large and small furnaces (Fig. 6.24 and Plates 6.27–6.29). The outer walls were of brick with a rubble infill held in mud mortar. The insides of the lower, cool chambers were also of regularly coursed brickwork. However,

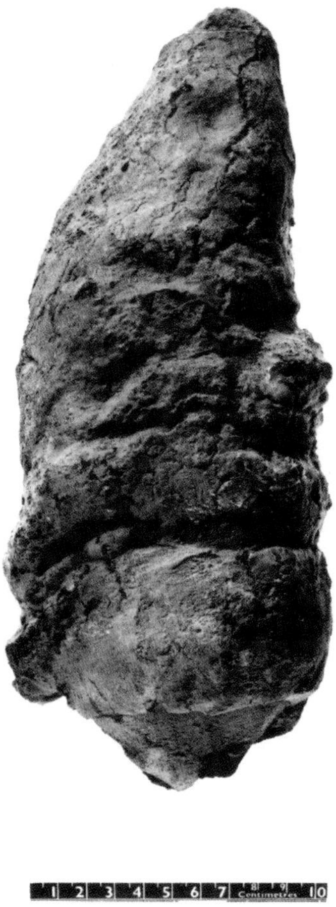

Plate 6.30 Zawar: undistorted intermediate retort showing asymmetry at the top necessary for the outer rows to fit into the furnace (cf. Fig. 6.28).

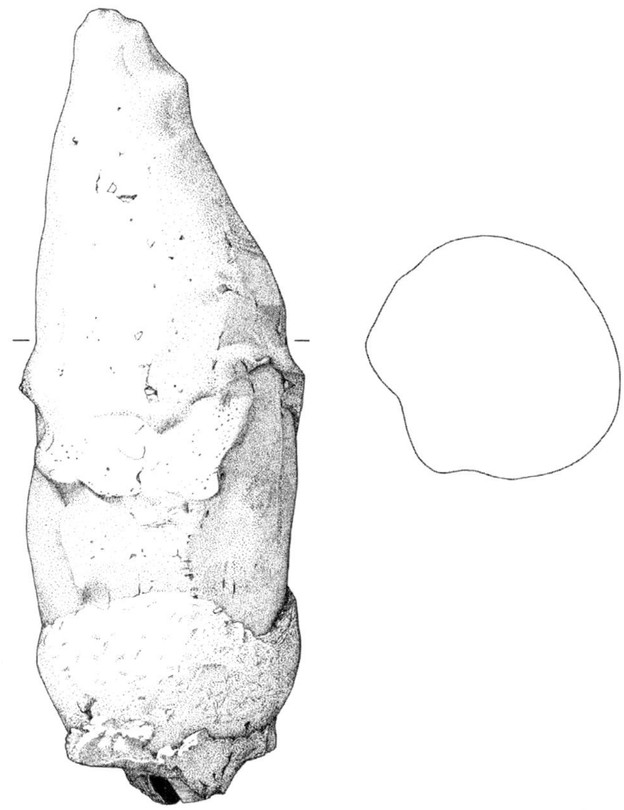

Fig. 6.28 Zawar, Site 33: intermediate retort from the surface of heaps adjacent to Site 33 (length: 29 cm). Note the asymmetry to enable the retorts to be set in the outer rows of the furnace against the sloping walls and also the evidence of hand fettling at the top (cf. Plate 6.30).

the arrangement for supporting the perforated plates was slightly different: in the intermediate furnaces the chamber wall was made up of three courses of brick, and the perforated plate sat on top of the third course of brickwork (Fig. 6.25), whereas in the small retort furnace the upper course of bricks in the chamber projected slightly to form the ledge (Fig. 6.11). The mortared floor of the lower chamber of the intermediate furnaces was exposed in at least two furnaces (Plate 6.27). They were very clearly comprised of a layer of mortar of irregular thickness varying between 1 and 3 cm which, as revealed by Site 34, sat directly on the surface of the debris heap, with no underlying foundation of brick such as was noted at Site 30.

The stump of the central pillar of the westerly of the Site 33 furnaces was firmly mortared in place, with the rest of the pillar lying alongside (Plate 6.28). This feature had not been noted in any of the furnaces previously investigated, and was not apparent on the pillar in the adjacent furnace which was found *in situ* when it was sectioned. The latter pillar was almost complete and 18 cm tall, corresponding to the height of the cool chamber as measured from the sides. It was of the same rather concave profile as the pillars from the large retort furnace (Fig. 6.26), unlike the small furnace pillars where the taper had a straight profile (Fig. 6.13 and Plate 6.13). Only the east–west dimension or 'width' of the lower chamber could be obtained, and this lay between 73 and 74 cm in both the furnaces investigated.

After recording the northern, or back, row of retorts from the central furnace, they were removed together with the badly broken fragments of perforated plate beneath (Fig. 6.27). Below this the fill of soft light brown soil was also removed to reveal the interior of the chamber, and the main plan and section were drawn across the furnace at the plate level (Fig. 6.24 and Plate 6.29).

Although the retorts were still sitting *in situ* they had all lost their condenser ends. It had already been noted that the retorts on the dump almost invariably lacked the condenser tubes, and these were only very rarely found separately (Figs 6.16 and 6.29), suggesting they that were deliberately removed and the contents recycled, the evidence here again suggesting that this must have happened when the putative collecting vessels were removed (see Chapter 11).

The furnaces were too severely damaged for any of the *in situ* but broken perforated plates to be completely exposed. Where this was partially done some measurements were recorded, the east–west width of the furnace at the perforated plate level height being approximately 75 cm. The plates rested on a ledge approximately 7 cm wide so they would have been about 44 cm in width. The small air holes of the plates varied between 2.5 and 2.7 cm in diameter and the large holes to hold the retorts measured between 4.25 and 4.5 cm. The plates were approximately 4.5 cm in thickness and had been made in a mould, a little thicker but otherwise quite similar to the plates in the large furnaces.

Very little survived of the upper walls of the furnace but what did showed the familiar features of the other furnaces,

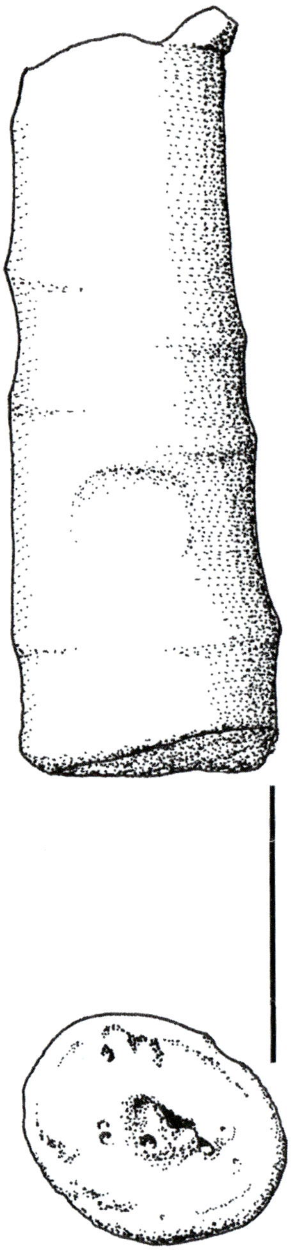

Fig. 6.29 Zawar, Site 33: condenser tube choked with zinc oxide and blue powder (length: 7 cm). Note the fingermarks (cf. Chapter 11, Plate 11.2).

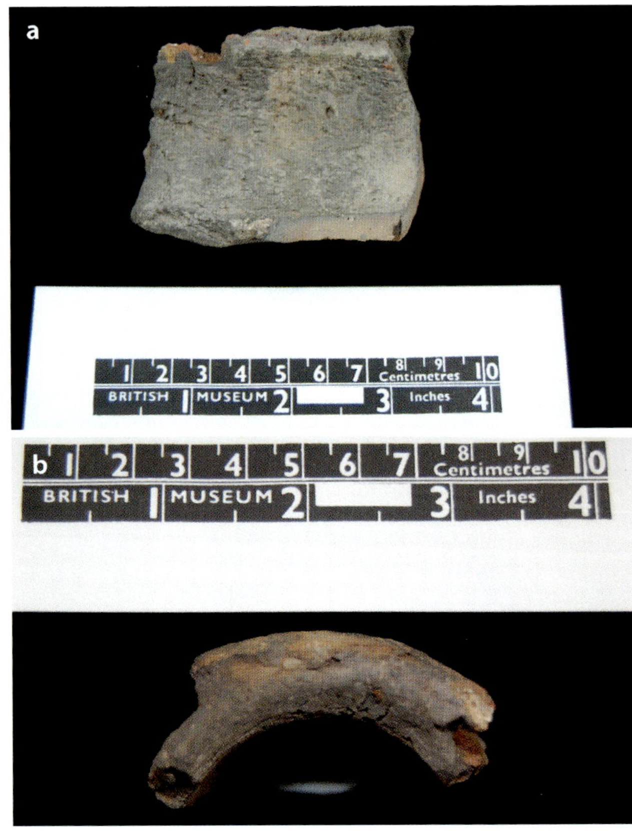

Plate 6.31 Zawar, Site 36: large fragment of a zinc smelting retort from the exposed section of the earth fill of the dam. These retorts were significantly smaller than the small retorts from Site 30. (a) The interior, the rim is to the left; (b) the profile (note the luting clay).

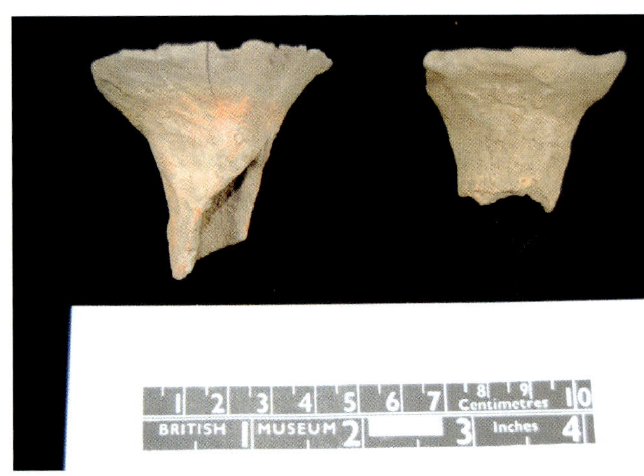

Plate 6.32 Zawar, Site 36: small handmade condenser mouths.

suggesting that the basic form was once again a truncated pyramid. Some of the intermediate retorts, where undistorted, displayed the same asymmetry as the small and large retorts to accommodate the outer rows into the inward sloping walls of the furnace (Fig. 6.28 and Plate 6.30). The outer walls of the intermediate furnaces were apparently brick-faced, with the centre filled with clay. At the plate level the outer back wall was about 27 cm thick – all but about 5 cm of this thickness was made up of regular brickwork and brick rubble. The clay furnace lining was typically about 3 cm thick (see the appendix).

Site 36: dam fill site

Function: Zinc smelting.
Date range: Pre-dam.
Location: 24° 21′ N, 73° 42′ E (Fig. 6.2)
Description: The great dam across the River Tiri is described in Chapter 3. The earth fill of the dam includes zinc smelting debris of retorts, condenser heads and perforated plates which are clearly related to the material excavated *in situ* at the other sites, but the retorts found in the fill are smaller (Fig. 6.30), and the refractories are generally more primitive (Plates 6.31a and b and 6.32, Figs 6.31 and 6.32a and b). The perforated plates were formed by hand rather than moulded. This is possibly of some significance suggesting that the building and operation of the

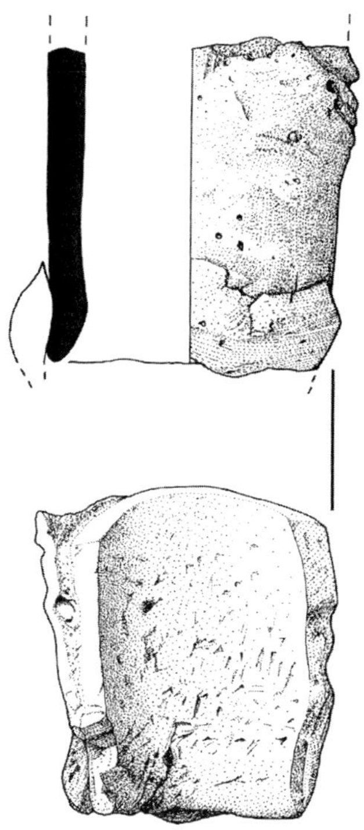

Fig. 6.30 Zawar, Site 36: substantial fragment of a very small retort with relatively undistorted neck showing internal diameter (estimated at about 5.5 cm).

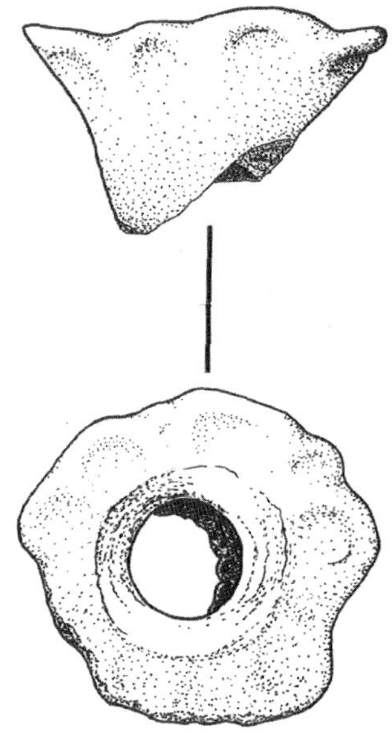

Fig. 6.31 Zawar, Site 36: small handmade condenser neck (max. diameter: 3.5 cm).

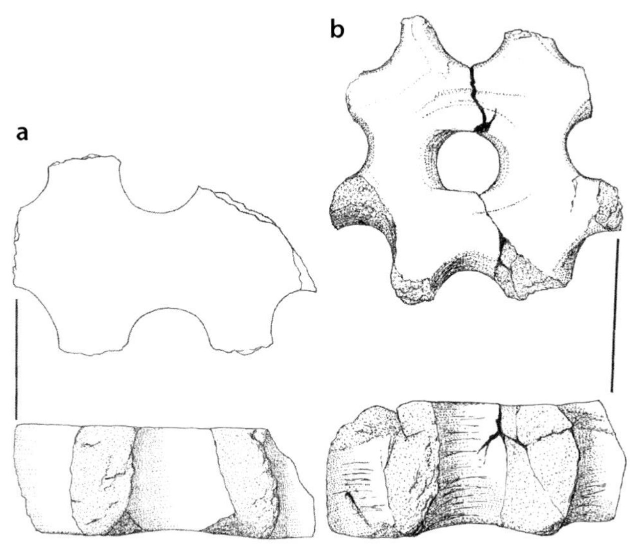

Fig. 6.32 Zawar, Site 36: fragments of handmade perforated plates (max. dimensions: (a) 6 cm and (b) 7 cm).

earlier furnaces was more independent and that later furnaces relied on supplies of components of regular dimensions presumably from a central supplier.

The material has no context, but the retort, plate and condenser fragments do seem to belong together. No slag was noted in the dam fill. Intensive field walking in the general vicinity of the dam failed to locate any more examples of these very small retorts that could indicate smelting sites surviving *in situ* (see the appendix).

SITE 38
Function: Ore roasting.
Date range: Apparently post-medieval.
Location: Immediately to the north of Site 37, just beyond the edge of the Old Zawar tips at 24° 22′ N, 73° 42′ E (Fig. 6.3).
Description: A number of mounds of distinctive white ashy material were noted lying just beyond the limits of the Old Zawar retort tips, particularly to the west. The mounds are often quite extensive, typically between 5 and 30 m in diameter and several metres in height at their centres. Some that lie directly beneath the Old Zawar escarpment have clearly been tipped from above, although the excavated example proved to have built up *in situ*.

One of the heaps, designated Site 38, was selected for excavation. It lies together with another heap just beyond the northern edge of the Old Zawar tips and is possibly associated with it. It is bow-shaped, 12 m in diameter and rising to a height of 2 m in the centre. A 6 × 2 m trench was dug from the centre to the edge of the mound. This exposed a series of layers made up principally of thick, continuous deposits of white ash separated by thin layers of charred twigs with crumbs of vitrified ceramic. The portion of the heap exposed in the cutting had been levelled on one occasion to create a more level surface (Fig. 6.33 and Plate 6.34).[3]

Another series of heaps lying at the edge of the Site 5 retort tips (Fig. 6.38) had a very similar pattern of material and deposition where exposed in sections created by traversing paths. Charcoal from the tip proved to be modern. No evidence of associated structures was observed at any of the sites, and there was nothing beneath the Site 38 mound itself. Nevertheless a 3 × 2 m area was excavated on a small area of flat land just to the north of the mound, but no features were uncovered.

The same white ashy material, also found in defined lenses within the large retort heaps (Plate 6.17), would appear to be broadly contemporary with the white ash heaps beyond the retort heaps. It is likely that the white mounds are made up of waste from the roasting operations in which the sulphidic ore was roasted to drive off the sulphur (see Chapters 7 and 11).

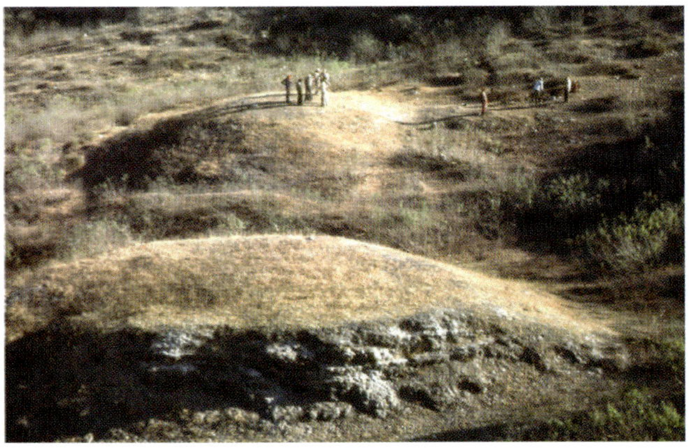

Plate 6.33 Zawar, Site 38: Two heaps of white ash. The excavated heap is in the background.

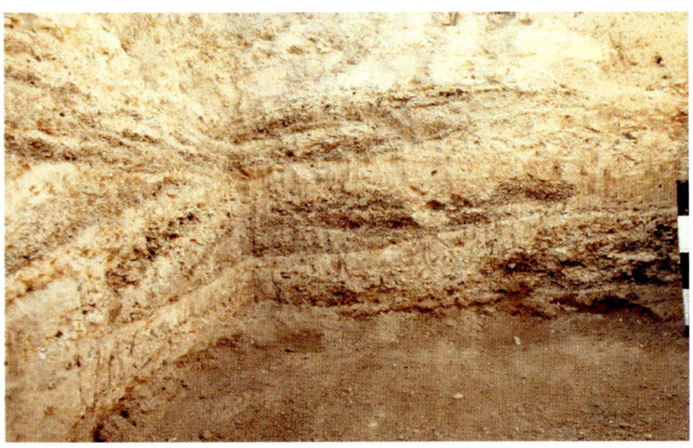

Plate 6.34 Zawar, Site 38: section through Site 38 showing discrete layers of ash and charcoal. Note the heap has been levelled at approximately halfway up the section (cf. Fig. 6.33).

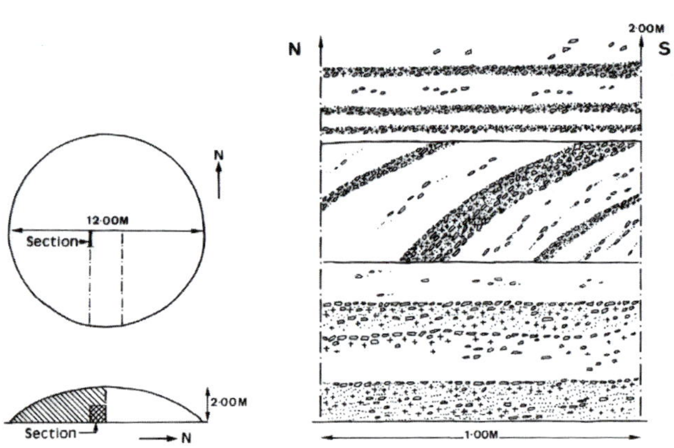

Fig. 6.33 Zawar, Site 38: plan and section through a white heap (cf. Plate 6.34).

The presence of crumbs of vitrified ceramic, seemingly from used retorts, is difficult to explain. It is possible that the Site 38 heap built up *in situ* from successive roasting operations apparently without any containing structures such as stalls. When the profile of the heap became too steep for convenient operation it was probably levelled to create a flat working area.

Other roasting operations had clearly taken place on top of the Old Zawar escarpment and the debris just shovelled over the edge. Roasting the beneficiated ore would have produced large quantities of sulphur dioxide and the ashes themselves would have had high calcium oxide contents and been very caustic when first produced. They also have high cadmium contents therefore it would have made sense to have kept them well away from the main body of workers tending the smelting furnaces.

Kaya

Function: Zinc smelting.
Date range: Small retort period.
Location: This site is set in remote and hilly country some 12 km north/north west of Zawar and about 5 km east of the village of Kaya at 24° 26′ N, 73° 41′ E (Fig. 6.1).
Description: The site was visited briefly in December 1988 and again in 1990. The area was not extensively field walked but there is abundant evidence of early metal mining and smelting activity in the vicinity. On the route through to the zinc smelting site from the main road just south east of the village of Kaya itself, several extensive spreads of copper smelting slags were observed in the vicinity of the lime kiln marked on the map, apparently the remains of much eroded heaps. Immediately to the north of these is a group of early but undated copper mines with at least 28 surviving shafts. There are also remains of extensive copper mining and smelting at Devimata which lies several kilometres to the north east of the Kaya zinc smelting site. Geologically the area is very similar to Zawar.

At the site itself there was only one farmstead, the other main activity in the vicinity being charcoal burning (described in Chapter 9). The hillock on which the site is situated is surrounded on three sides by very steep river valleys and had been terraced to create a well-defended settlement (Plate 6.35). Where visible the stone retaining walls of the terraces stand on stone-built berms and are up to 1 m in thickness. The only structures on the terraces are six circular walled enclosures, typically about 8–12 m in diameter. The walls are of dry stone, usually stand to about 1 m in height and have no entrances. This and their strategic positions around the terraces suggest they were strong points, possibly for defence against the local Bhil hill peoples or the Maharata incursions of the late 18th century. The fill of the terraces contains a great deal of stone rubble and roof tiles, as well as the zinc smelting debris discussed below, suggesting that there had once been structures on or around them. Although the terrace walls are substantial and well made it is all rubble work with no ashlar on the site, and no remains or record of a temple. There are more built terraces on the north side of the river, but these do not contain smelting debris. The local farmer knew nothing of the site, but said soldiers had once been there. The site is approached by a narrow terraced drystone trackway, now much overgrown, leading back eastwards over the surrounding ridge towards Kaya and the main road to Udaipur. There is no evidence of a built trackway leading south towards Zawar.

Evidence of extractive metallurgy

The area has not been properly geologically surveyed but there seem to be no mineral or mine workings in the immediate vicinity. There are, however, outcrops of dolomite containing sphalerite and galena, with indications of exploitation, on a hill ridge approximately 2 km to the south east at its nearest point and 1 km south of the terraced track. Samples taken from the walls of the old workings for analysis by HZL contained up to 2.4% zinc and 0.4% lead, which suggests that zinc was mined there. Crystalline galena with vein quartz in dolomite,

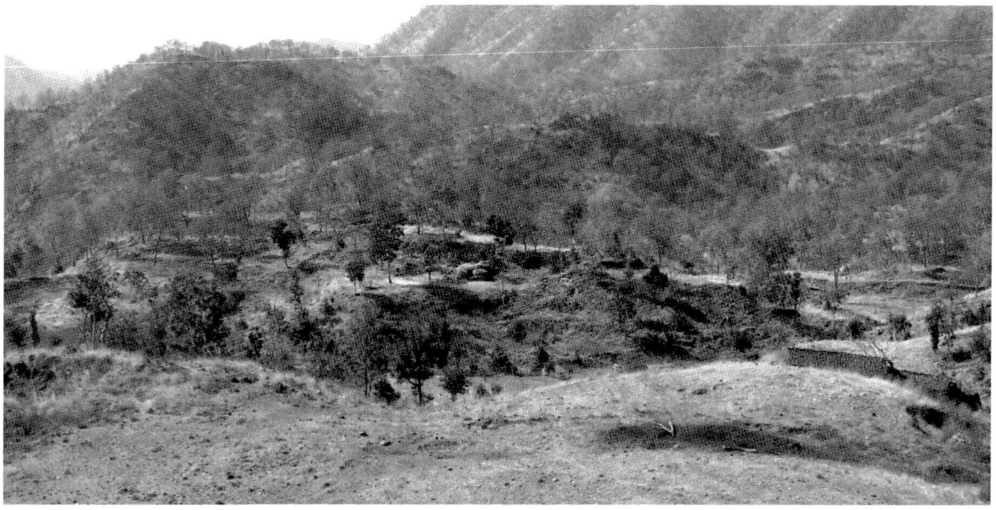

Plate 6.35 Kaya: this remote terraced settlement on the small, now densely wooded hillock in the foreground, lies beneath the hills but is separated from them by ravines on three sides.

together with quartz and dolomite boulders, is present in the stream beds.

The fill inside the terraced structure contains a great deal of loose soil and rubble as well as many retorts and condensers. The form and size of the retorts and of the handmade polygonal condenser heads are identical to the small retort debris from Zawar, the latter dated to the late 14th–15th century AD (see the appendix). This suggests that the Kaya site was established as an offshoot of the main Zawar enterprise during this period, when the mines were clearly flourishing. However, it seems that the Kaya operation was never very extensive and may well have been of rather short duration. The absence of a temple or a road for vehicular traffic also indicates that the site was never of any great importance. It is obviously difficult to assess the quantity of metallurgical debris, but there is little evidence of refractory material outside the terraced structure itself and no evidence of slags anywhere, suggesting that only zinc was processed. The present extant structures are likely to postdate the zinc smelting operations.

Slag smelting operations: Sites 2, 5, 7, 14 and 29

Site 2

Function: Silver/lead/zinc oxide production.
Date range: 1930 ± 80 BP, BM 2487.
Location: This is a rather isolated area of lead slags, lying on gently sloping ground just above the flood plain of the River Tiri (Fig. 6.2) at 24° 22' N, 73° 42' E.
Description: The main mound of slag is approximately circular, covering an area of 2000 m^2 rising to a height of about 1.5–2 m in the centre. A 2 m^2 area (T.1) was opened near the centre of the mound, but excavation stopped at about 1.25 m because the sides were extremely unstable and no section could be obtained. A second area (T.2) was opened about 25 m to the west, towards the edge of the mound, where the slags were more stable. The profile of the section through the mound obtained at T.2 is quite typical of a long-established heap (Plate 6.36). Beneath the soil of the surface the slag fragments were loosely piled, and all the small pieces, soil, refractories, charcoal etc. had been washed down to the base of the heap (Fig. 6.34). It was from this layer that the majority of the samples were taken, including the charcoal for dating. The refractory materials include furnace (Figs 6.35 and 6.37) and tuyere fragments (Fig. 6.36). Beneath this was the original land surface (see the appendix).

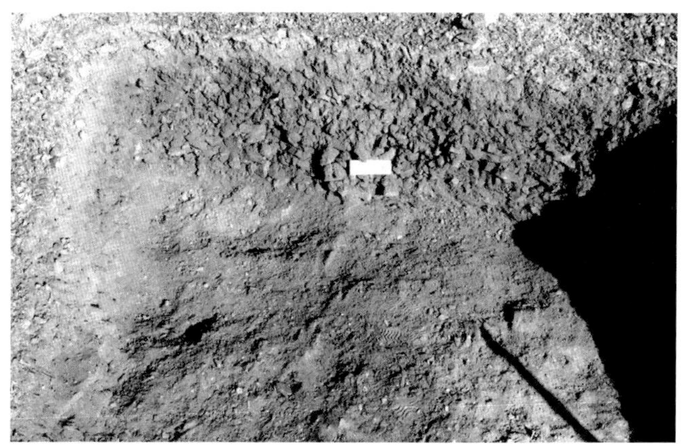

Plate 6.36 Zawar, Site 2: section (scale: 15 cm).

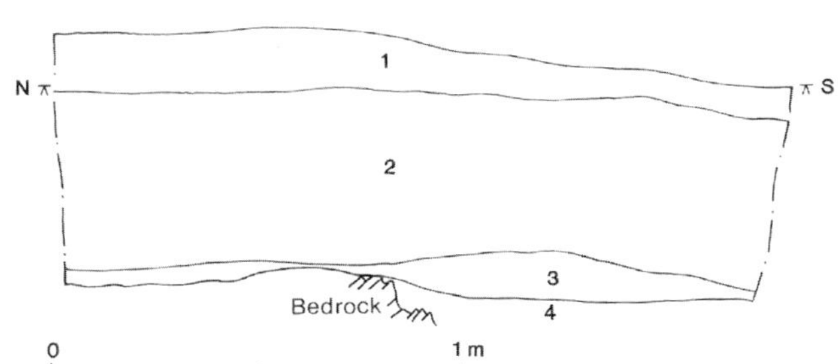

Fig. 6.34 Zawar, Site 2: west section of T.2.

Site 5

Function: Silver/lead production.
Date range: Post-medieval, 100 ± 45 BP, BM 2484.
Location: The site lies at 24° 22' N, 73° 42' E in a shallow bend of the River Tiri, adjacent to the Ramanath temple (see Chapter 3) below the steep scarp of the hills on which lie Old Zawar and the main large and intermediate zinc smelting heaps (Fig. 6.3).

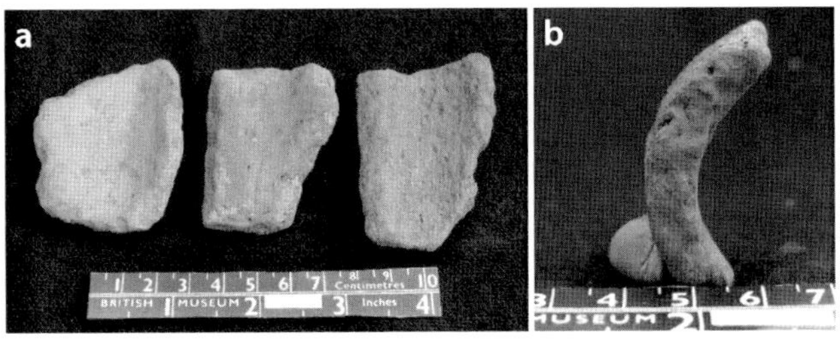

Plate 6.37 Zawar, Site 2: three tuyere fragments, thin walled but of quite considerable internal diameter. (a) Interior and (b) profile.

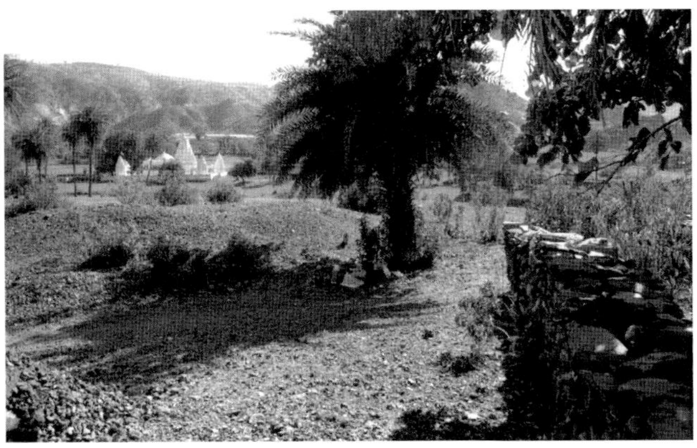

Plate 6.38 Zawar, Site 5: general view of the excavated slag heaps looking south towards the Ramanath temple in the background across the River Tiri.

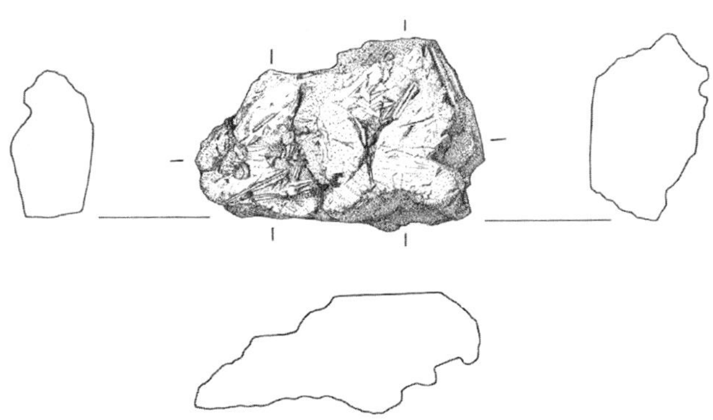

Fig. 6.35 Zawar, Site 2: furnace lining heavily tempered with straw. This lining was apparently applied to a flat furnace wall of stone (max. dimension: 7 cm).

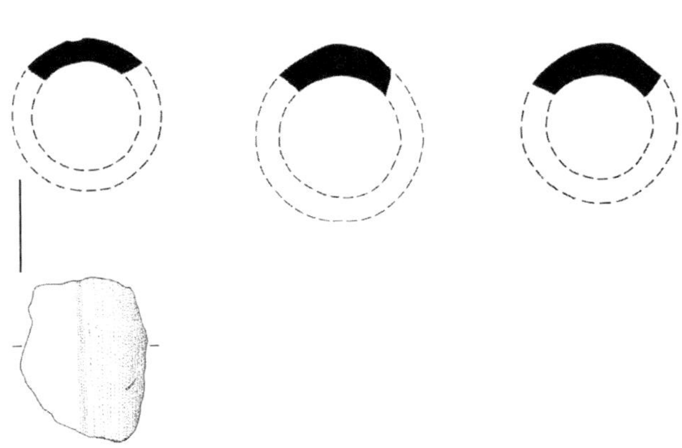

Fig. 6.36 Zawar, Site 2: fragments of a tuyere – thin-walled but of quite a wide internal diameter (estimated at 4 cm) (cf. Dariba, Site 10, Figs 6.57 and 6.58).

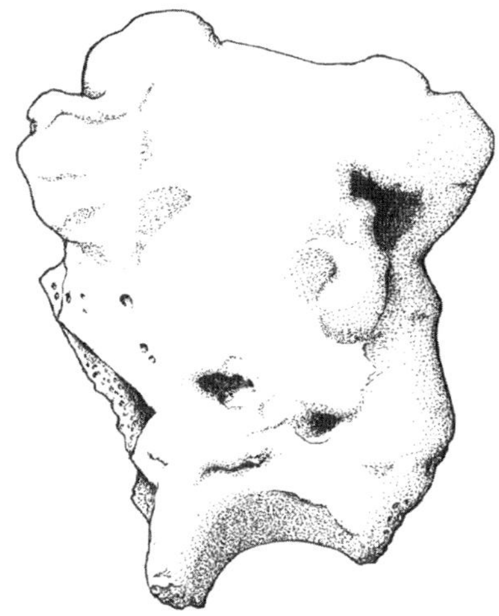

Fig. 6.37 Zawar, Site 2: furnace lining with an inclined hole for a tuyere (max. length: 9 cm).

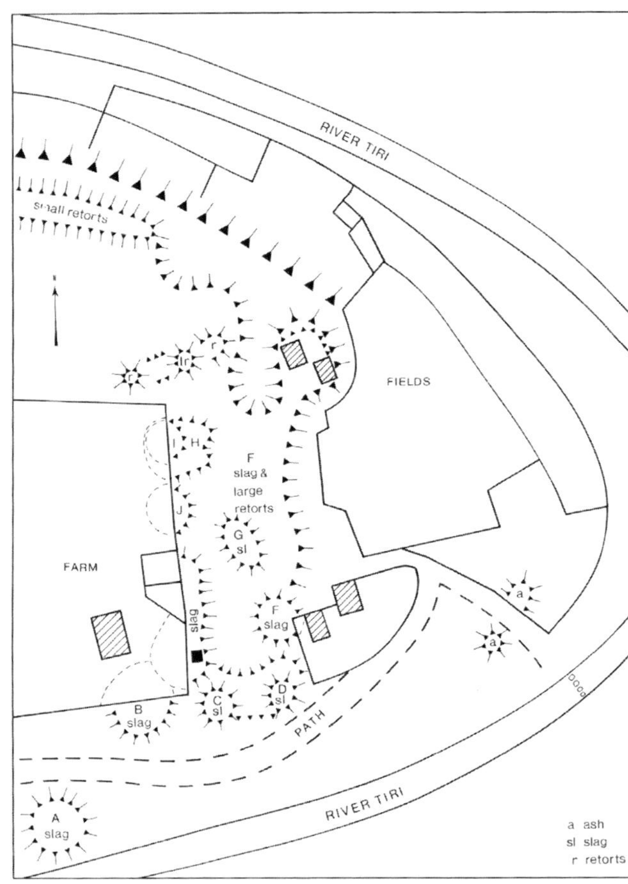

Fig. 6.38 Zawar, Site 5: sketch plan of the whole site which comprises both large and small retort heaps, white ash heaps as well as numerous slag heaps.

Description: The slag heap selected for excavation lies in an area of similar heaps (Fig. 6.38 and Plate 6.38), from which a comprehensive series of slag samples was taken. In addition there are also the white ash heaps from ore roasting (Site 38) and heaps of zinc smelting debris. A number of surface samples of interesting refractory fragments were collected from this area, including some fragments of perforated plates of types not encountered elsewhere, which were found on some of the

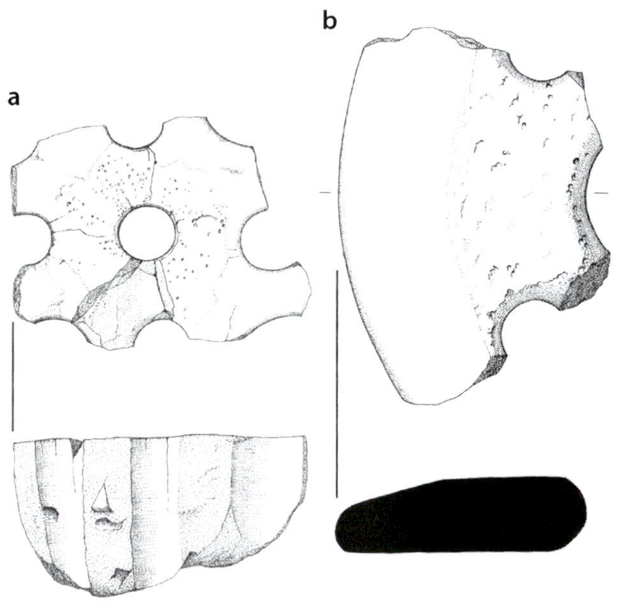

Fig. 6.39 Zawar, Site 5: unstratified fragments of perforated plates of types not encountered elsewhere. Note the substantial thickness of (a) (max. thickness: 4.5 cm), and the rounded profile of the edge of (b) (max. dimension: 11 cm), suggesting a circular furnace.

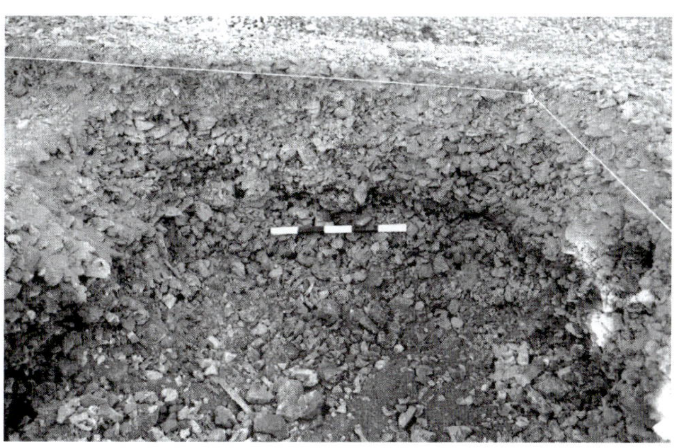

Plate 6.39 Zawar, Site 5: excavated area.

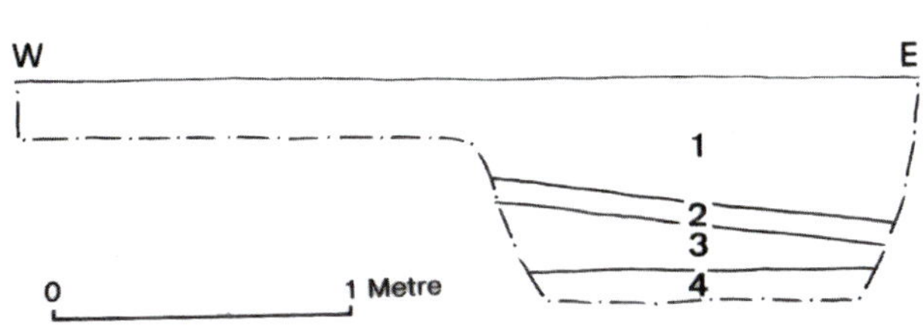

Fig. 6.40 Zawar, Site 5: north section of the excavated area. The slags were extremely loose and no proper face could be prepared.

heaps of small retort debris in the immediate vicinity, as well as heaps of large retort debris. They included plates similar to those found on small retort process sites, but almost twice as thick at approximately 6–7 cm (Fig. 6.39a), and other fragments apparently from a circular furnace of approximately 36 cm diameter (Fig. 6.39b). Thus this seems to have been an area of varied metallurgical activity probably extending over many hundreds of years. Behind this area to the west the small retort

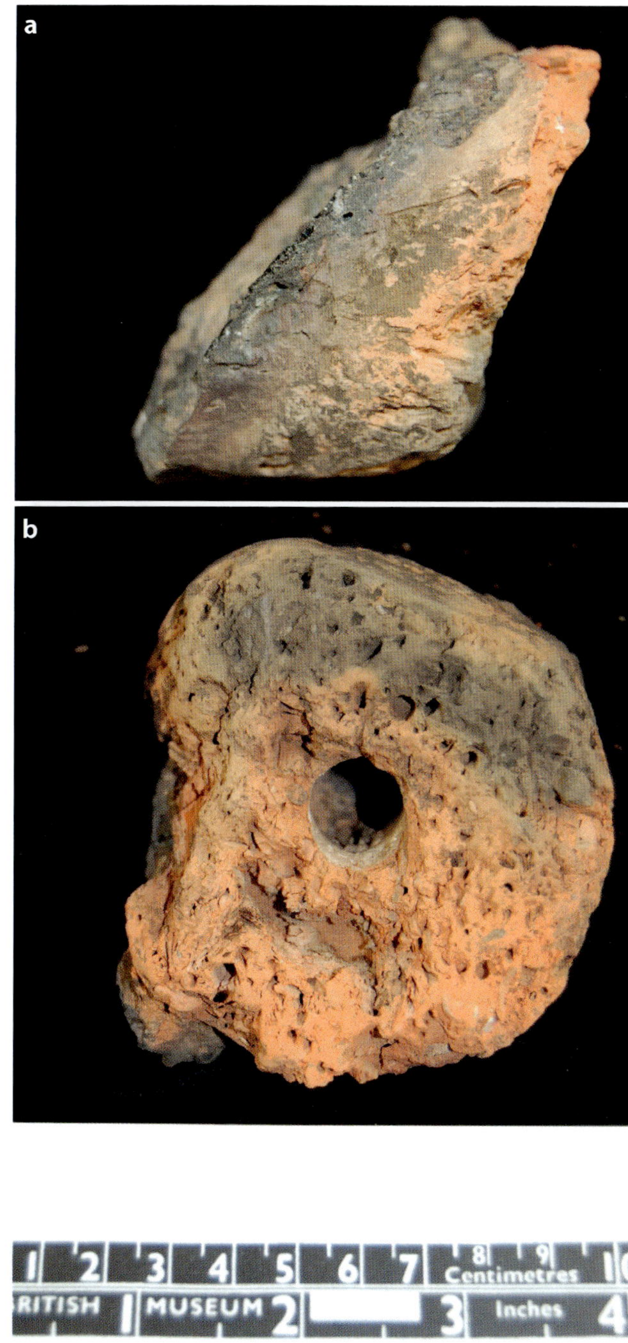

Plate 6.40 Zawar, Site 5: typical tuyere. (a) Side view – the slagged left face showing the angle of the tuyere to the furnace wall; (b) rear view – note the holes where straws of the plant tempering lying along the line of the tuyere have burnt out.

heaps are almost continuous, indicating that this was clearly the main area of zinc production at Zawar.

A 3 × 2 m area was opened in one of the heaps (designated Site 5) (Fig. 6.40 and Plate 6.39). The surface layer was loose slag identical to that underneath. The first 10 cm was designated Layer 1, the second Layer 2. After this the area was divided and digging continued in the eastern half. Loose, washed slags continued to a depth of about 50 cm and thereafter the slags had more soil washed down from above and charcoal etc. The excavation went down to a depth of approximately 1 m in the centre of the eastern area and ceased without reaching bedrock.

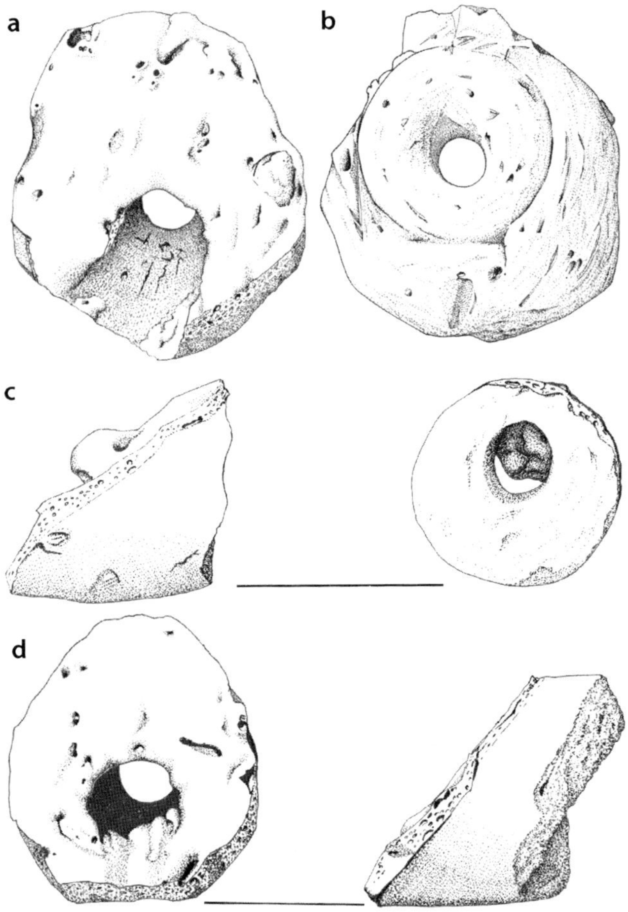

Fig. 6.41 Zawar, Site 5: selection of some of the numerous tuyeres excavated. (a) Front and (b) rear is complete with its added clay jacket (diameter: 9 cm); (c) profile and front (diameter: 5.5 cm) is partially blocked with slag; (d) front and profile (diameter: 5 cm).

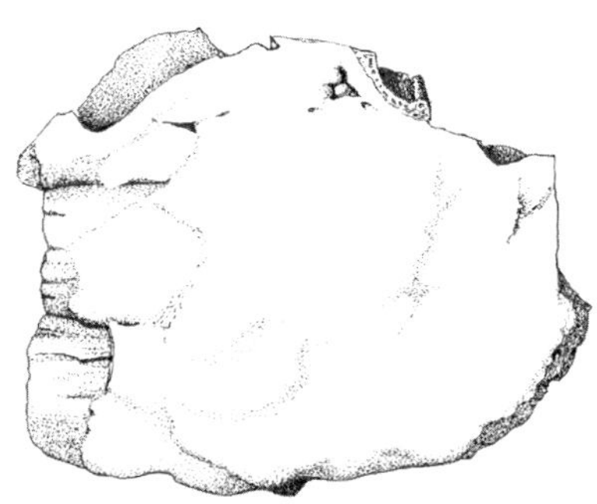

Fig. 6.42 Zawar, Site 5 L.3: outer face of a fragment of furnace wall showing the outer layer of applied clay (max. dimension: 8.5 cm).

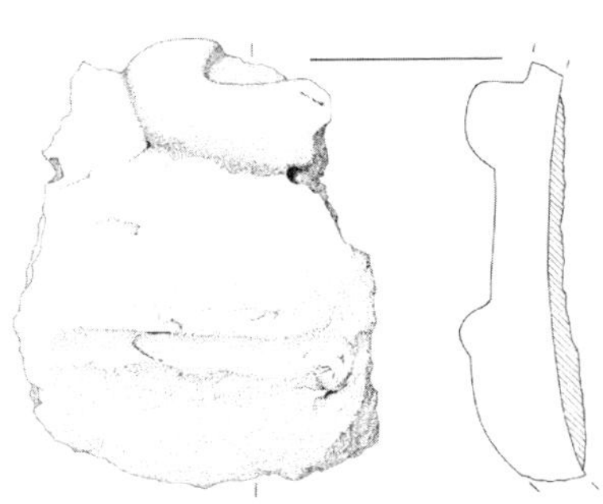

Fig. 6.43 Zawar, Site 5 L.3: face and section of a fragment of furnace wall. The wall at this point is very thin even though the slagging and vitrification show it is close to the reaction zone (max. dimension: 12 cm).

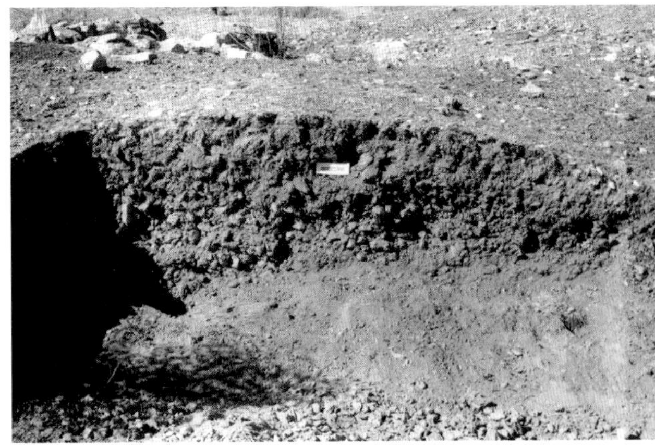

Plate 6.41 Zawar, Site 7: section in one of the small slag heaps exposed by recent removal of material for road metalling (scale: 10 cm).

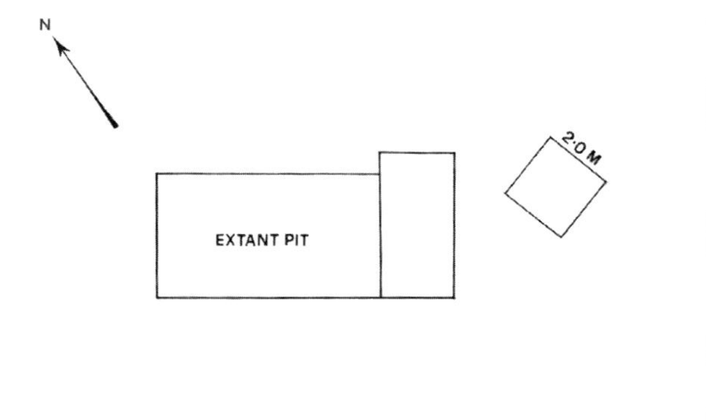

Fig. 6.44 Zawar, Site 7: sketch plan of the location of excavated area T.1 and the area exposed by recent digging.

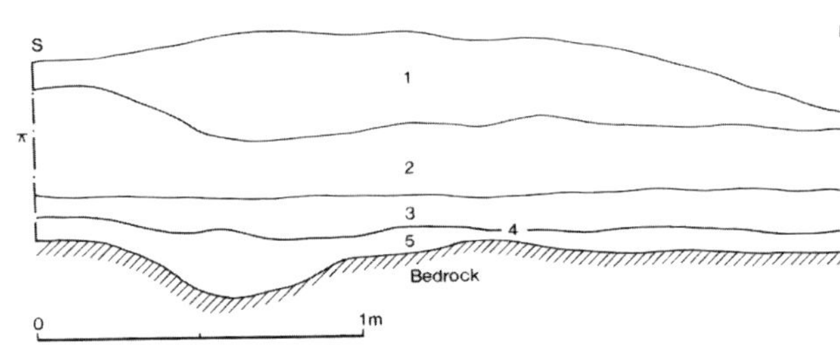

Fig. 6.45 Zawar, Site 7: south section exposed by recent digging. 1. Yellow/brown soil, occasional shale or slag. 2. Slag, mostly < 10 cm, some refractory and potsherds, no charcoal. 3. Yellow/brown soil, some shale and slag, pieces of burnt clay and pottery, and charcoal. 4. Compacted surface. 5. Weathered shale with brown soil.

The excavated area was rich in refractory fragments, and especially in tuyeres of which no less than 25 were recovered (Fig. 6.41a–d and Plate 6.40) (cf. Dariba, Site 14, Plate 6.62). There were also some fragments of coil-built furnace walls, some still retaining the outer layer of applied clay (Figs 6.42 and 6.43). In addition lead-rich slag, lead metal and burnt galena were found together with a piece of ceramic, heavily impregnated with metal oxides, which was provisionally identified as a cupel but is now believed to be a fragment of the neck of a condenser from the zinc smelting process (see Chapter 7) (see the appendix).

SITE 7

Function: Silver/lead/zinc oxide production.

Date range: 1370 ± 80 BP, BM 2487.

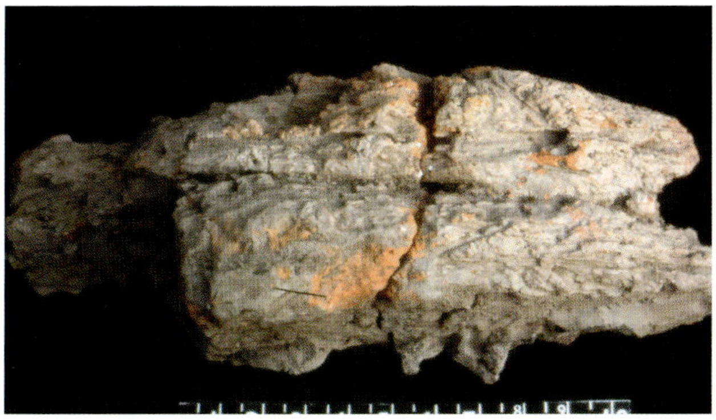

Plate 6.42 Zawar, Site 7: fragment of a furnace wall – the face showing two clay coils.

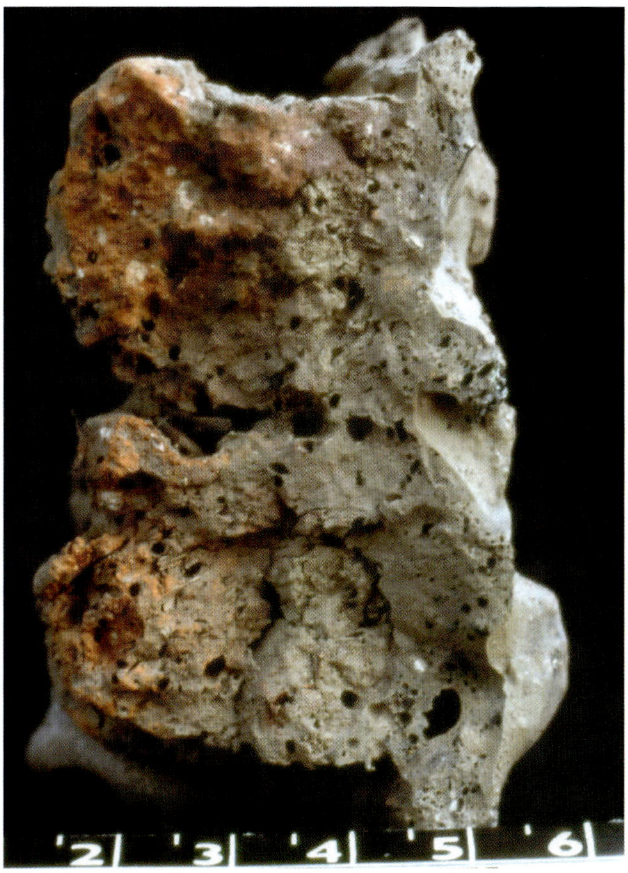

Plate 6.43 Zawar, Site 7: section through a furnace wall showing two rather squashed coils. Note the vitrification (right) extending deeply into the wall.

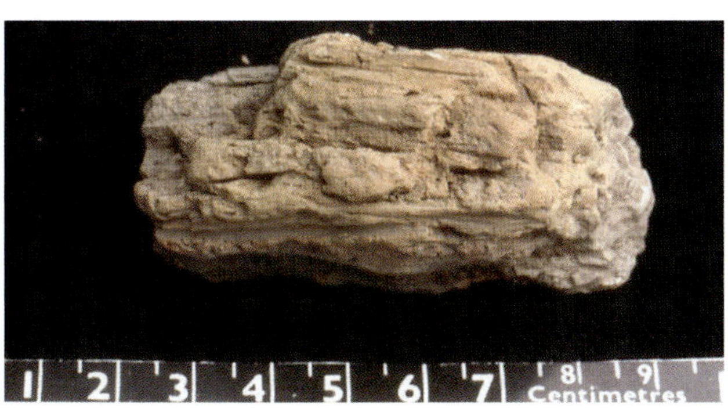

Plate 6.44 Zawar, Site 7: fragment of coil from a furnace wall. Note the straw impressions along the line of the coil.

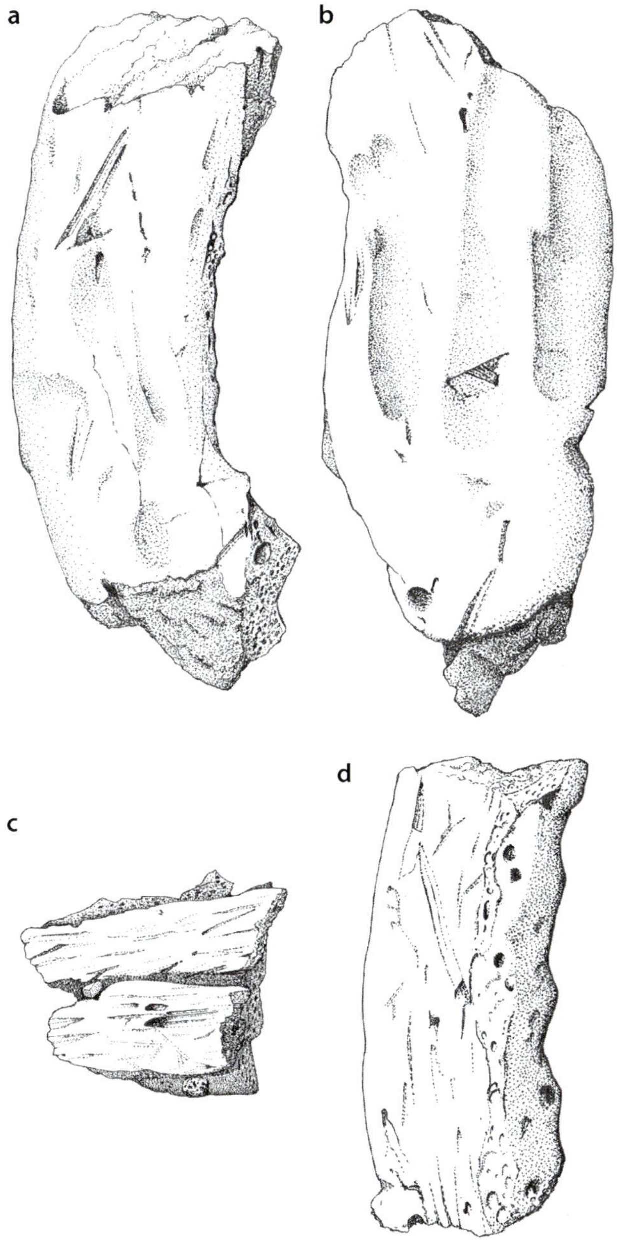

Fig. 6.46 Zawar, Site 7: selection of fragments of furnace wall of coiled clay, heavily tempered with straw. (a) and (b) are the section and face of the same piece (max. dimensions: (a) and (b) 13 cm, (c) and (d) 6 cm). Note they are all massively vitrified and slagged on the inner concave surfaces; (b) retains the original outer layer of clay with the fingermarks still clearly visible.

Location: The site is a small slag heap and scatter on the north side of the main Tiri-Zawar road, just outside New Zawar village and below Zawar Mala Magra at 24° 21′ N, 73° 42′ E (Fig. 6.2). *Description:* There are many similar slag heaps in the vicinity. The slags of this heap extend for about 400 m^2 and have a depth now of only about 30 cm as they have clearly been used in the recent past as a convenient source of road metal.

A 2-m square trench, T.1, was opened in the approximate centre of the mound. A 15-cm thick layer of loose, washed slag fragments overlay a brown soil which had some slag and fragments of refractory and charcoal, but little of interest. This in turn lay on bedrock. T.1 was abandoned and a second area, T.2, opened

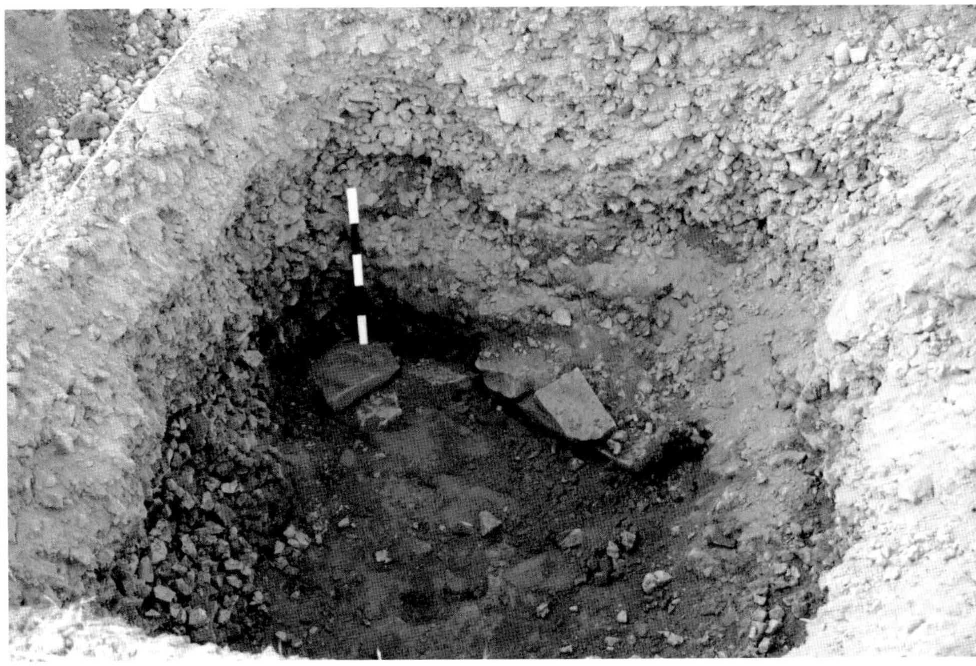

Plate 6.45 Zawar, Site 14: excavated area T.1 looking to the south east. Note the large stones at the bottom of the heap.

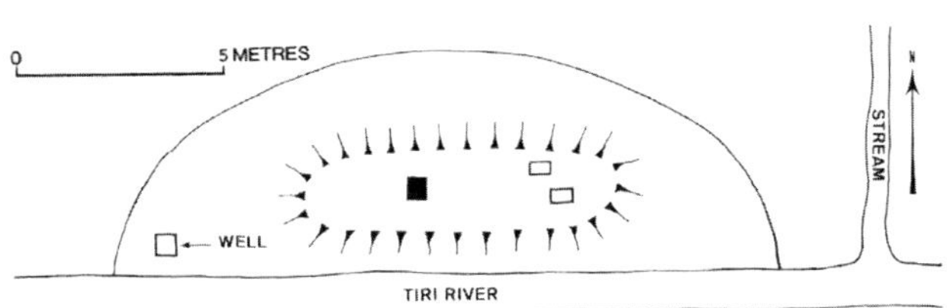

Fig. 6.47 Zawar, Site 14: sketch plan of the slag heaps by the River Tiri showing the location of the excavated area (black square).

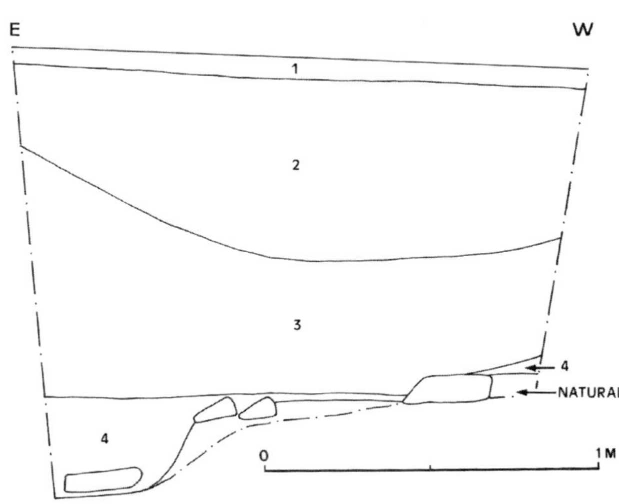

Fig. 6.48 Zawar, Site 14: north section of the excavated area.

in front of a section already exposed by the recent road metal diggings (Fig. 6.44 and Plate 6.41). The section was typical of an established, weathered slag heap (Fig. 6.45). A thin yellow/brown soil with occasional fragments of shale and slag (Layer 1) overlay the loosely piled slag fragments from which all the soil, charcoal, pottery, refractory etc. had been washed (Layer 2), down to the layer of yellow/brown soil (Layer 3) which lay on the original compacted surface (Layer 4) and beneath which was a brown soil with some shale but no evidence of slag or refractory (Layer 5). All the samples of refractory and pottery, together with the charcoal sample, were taken from Layer 4 and include furnace fragments of coiled clay (Fig. 6.46 and Plates 6.42–6.44), but no tuyere fragments were found. Note the date of this heap is coincident with that of the small workings in the top chamber of the nearby Zawar Mala mine (see Chapter 5). Also, possibly of some significance, the very small retorts from the nearby dam fill together suggest that this may be the area where the production of zinc metal developed from the existing zinc oxide production.

Several fragments were found of flat clay bases with sides of coiled clay and slagged on the inside. They have been interpreted as the bases of the small shaft furnaces (see the appendix).

Site 14

Function: Silver/lead/zinc oxide production.
Date range: 1950 ± 60 BP, BM 2485.
Location: The site is on the north bank of the River Tiri beneath Mochia Magra at 24° 22′ N, 73° 43′ E (Figs 6.2 and 6.47).
Description: There are almost continuous spreads of slag in the vicinity, and exposures in stream banks, well sides etc. show the slag layers are about 1 m in thickness. In one area the slags form a discrete low bank of approximately 40 × 20 m with a maximum height of about 1 m (Fig. 6.47).

A 3 × 2 m area was excavated into this bank (Fig. 6.48 and Plate 6.45). The first 50 cm of slags was mixed with soil, but below this the slag pieces were loose and washed. At the base of these a silty soil had developed on a bed of very eroded and fragmentary shale which still contained flecks of burnt material, charcoal and crumbs of pottery, and almost certainly represented the original land surface on which the slag was dumped. A series of large flat stones was uncovered on this surface, but with no evidence that they formed part of a structure.

Many pieces of refractory were found in the slag layers, most heavily tempered with plant material, and some have flat rear surfaces suggesting that they had been against a stone or brick structure. As at Site 2, no brick fragments were recovered, so presumably the furnaces must have been stone-built. Some of the pieces of lining have a wedge-shaped profile, which could have arisen if they formed the internal walls of a furnace that had a round internal section inside stone walls of square section. A heavily slagged tuyere was recovered similar to those from Site 2. The walls were approximately 1.5 cm thick and the internal diameter is estimated at between 3 and 4 cm. The tube was at an angle of about 45° to the slagged surface. Some rough domestic pottery was also excavated (see the appendix).

Site 29
Function: Silver/lead production.
Date range: 200 ± 35 BP, BM 2486.
Location: The site lies on a steep rocky area just outside the eastern limits of the zinc retort dumps of Old Zawar at 24° 21′ N, 73° 42′ E (Fig. 6.2 and Plate 6.46).
Description: A platform, approximately 11 × 6.5 m (Fig. 6.49 and Plate 6.47) had been cleared on an outcrop, with evidence for a structure at one end and apparent smelting activities at the other. The front face of the platform was retained by a low wall built of reused retorts and contained a plastered basin or small cistern. The southern end had a considerable scatter of slags, which also spilled onto the ground some 3 m below. An area 2 × 2 m was opened around the latter area, but bedrock was encountered within 10 cm. A slag sample was taken together with some typical coils or plates of clay furnace lining, and the area was designated Site 29 (lower). A 3 × 2 m area was excavated on the platform above, around the slag which was concentrated at the south end of the platform and designated Site 29 (upper). The slags were concentrated in the surface layers. Excavation revealed bedrock in the eastern part of the area just below the surface, but the upper part of the western part was made up of very loose broken rock with some soil and slag. Beneath this, at a depth of about 20 cm, was a thick localised spread of charcoal against the rock face, with a diameter of about 25 cm. A charcoal sample was taken for radiocarbon dating from this feature which was probably a hearth. The remainder of the area was left unexcavated. Clearly the slag lay over and in the rocky fill which covered the hearth. On the other side of the rocky outcrop were some slag scatters above the level of the platform. Thus it is likely that lead smelting had taken place near the summit of the outcrop, although no evidence remained of a smelting installation, and the slags had fallen down to the platform and areas below (see the appendix).

Survey and excavation of the surface sites at Rajpura-Dariba

The field surveys focused on the extensive near-surface mine workings of the South and North Lodes. These were followed for approximately 5 km north of the main modern workings as far as Bethumni (see Chapter 5, Fig. 5.19). At Dariba the main evidence of smelting activity lies to the south east of the large East Lode opencast (see Chapter 5, Fig. 5.19). Slags and other smelting evidence cover the ground over approximately 50 ha (Plate 6.48), although due to periodic monsoon flooding and consequent spreading and burying of material it is difficult to be certain of the true extent from the surface alone. Wells

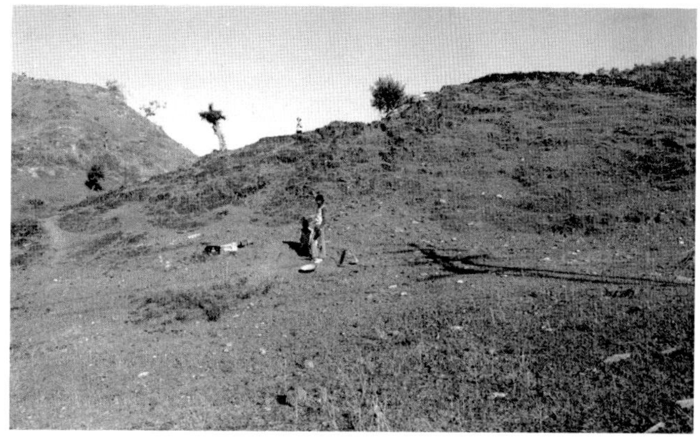

Plate 6.46 Zawar, Site 29 from the north east corner of the Old Zawar retort dumps: the figure in the foreground is standing on 29 (lower), the figure on the skyline on 29 (upper). The actual furnace site was probably just above 29 (upper) on the right.

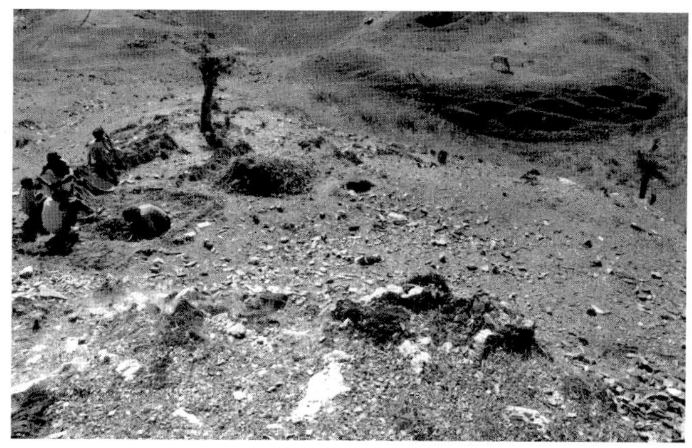

Plate 6.47 Zawar, Site 29: looking down from the summit of the hillock to the west. The upper area is being excavated to the left, and the levelled platform is in the centre of the picture (the arrow on Fig. 6.49 shows the position from which this photo was taken). Note the edges of the smooth heaps of retort debris in the background.

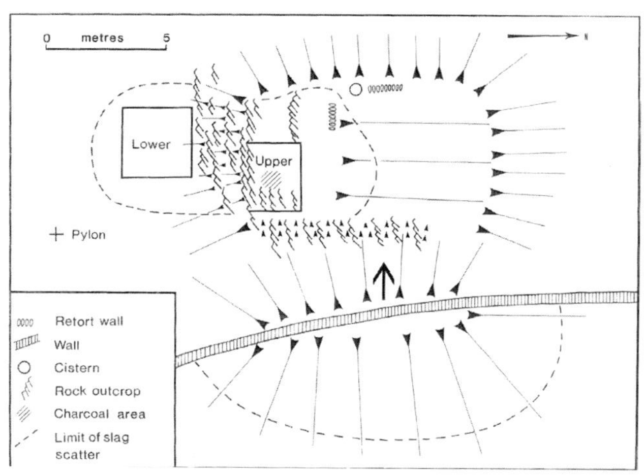

Fig. 6.49 Zawar, Site 29: sketch of the general area of the slag spills in which the excavated areas were located (the arrow shows the position from which Plate 6.47 was taken).

have been sunk in this area in the recent past and inspection of their sides show that the metallurgical debris reaches a depth of several metres in some places (Plate 6.49).

To the immediate south east of the opencast are many hectares of heaps made up of large loose fragments of tap slags.

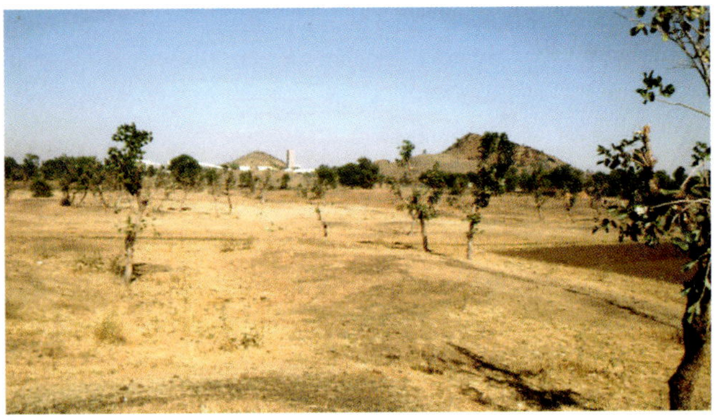

Plate 6.48 Dariba: general view over the area of deeply stratified metallurgical debris which covers an area of approximately 50 ha. The photo was taken from Site 9 looking north west back to South Lode and the East Lode headstock seen in the distance. Note the rather weathered appearance of the whole area.

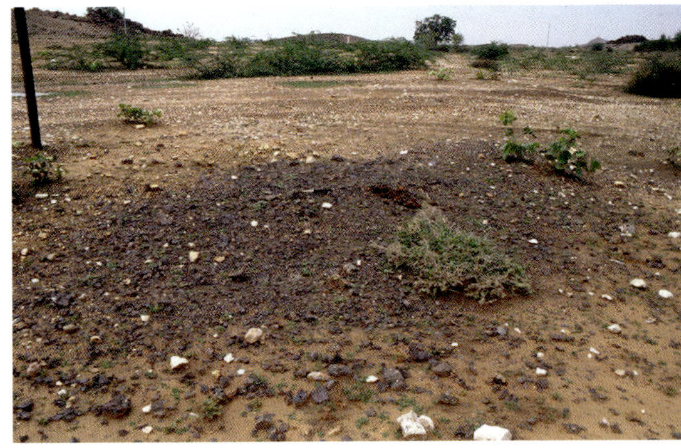

Plate 6.50 Dariba, Site 15: the largest of the small scatters of slag to the north of the main ancient workings suggest that only a little smelting activity took place here.

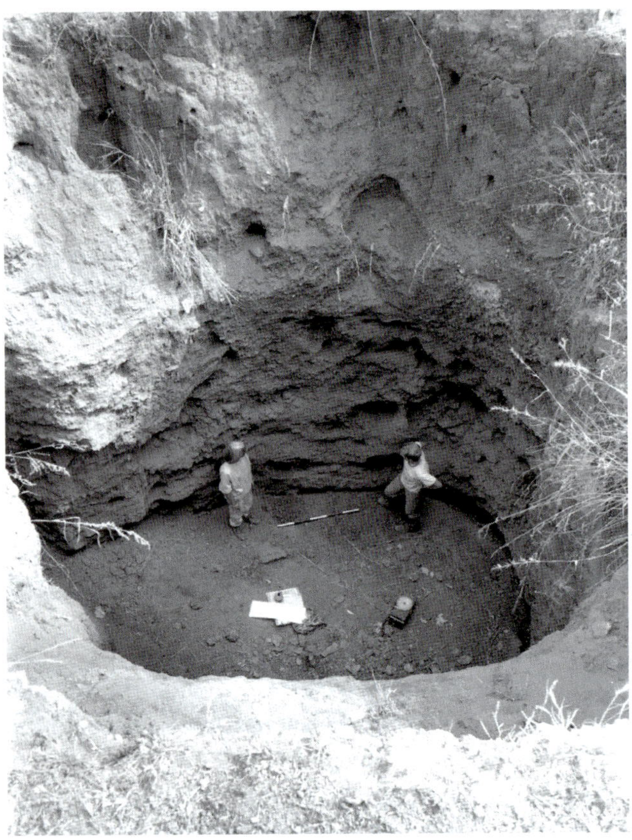

Plate 6.49 Dariba, Site 11: a considerable depth of occupation material including metallurgical debris had built up to the south east of the mines at Dariba as exposed in the sides of later wells such as this.

This area of slag dumps seems to have originally extended into the cultivated fields west of the canal, but peter out quite quickly as evidenced by the well sections. Beyond these heaps and continuing to the south east, after a gap of between 50 and 100 m of flat open ground, is a series of low mounds of occupation and metallurgical debris. Some of these proved to be the much-eroded remains of stone and mud brick structures when investigated, and similar heaps were also located in the main area of the slag heaps. One of the slag heaps (Site 7) and two of the low mounds – one with evidence of bricks and stone (Site 8) and one covered with metallurgical debris (Site 9) – were selected for excavation.

The sides of three of the wells were cleaned to reveal the nature and depth of the occupation debris through which they were sunk (Sites 10, 11 and 12). The sides of Site 10 were very rich in slags and well-preserved refractory material through the 3.5 m that was exposed. Site 11, at the southern edge of the mounds area, had some slag in the top layers but continued down for over 5 m of occupation debris. Site 12 in the area of the slag mounds also contained some metallurgical debris. The lower slopes of the eastern side of the South Lode outcrop are also covered with slags (Sites 13 and 14), some of which had been stacked against the mine waste from the opencast.

Farther to the north there is some evidence of smelting – just a very few minor spreads to the north east of North Lode (Sites 1–3, 5 and 6) (Plate 6.50) (see also Chapter 5, Fig. 5.19) and a small heap to the west of North Lode (Site 15), strongly suggesting that at all periods the bulk of the smelting took place in the vicinity of the East Lode opencast or to the south east. The radiocarbon dates from the surface sites at Dariba broadly follow the mine dates through the later part of the first millennium BC for the ancient Mauryan activity (Table 6.3).

Site 7

Function: Silver/lead production.
Date range: 1940 ± 70 BP, BM 2492.
Location: To the south east of the large opencast is an area of continuous slag heaps covering several hectares (Plate 6.51) (see also Chapter 5, Fig. 5.19).
Description: The individual heaps are of the order of 20 m in diameter and typically rise to between 1 m and 2 m above the present ground surface. One of the heaps was selected for excavation and a 3 × 2 m trench dug into the side. The heap was composed entirely of loosely piled fragments of tap slags, many of which were quite large – fist-sized and above. There was very little soil or other material in the mound therefore the sides of the excavation were very weak, necessitating a pronounced batter to the section (Plate 6.52). Some steep tip lines were observable in the slag section (Fig. 6.50), but these represent only very brief interruptions in the build-up of the heap, which must effectively be regarded as a single event.

At the base of the heap (Layer 3) the slag pieces lie more horizontally, and immediately beneath (Layer 4) was a layer of soft brown soil with a great deal of cindery material, the refractory

DARIBA SILVER/LEAD SMELTING SLAG HEAP

Area of extensive silver production, Site 9, layer 4	charcoal	BM-2490	2340± 60	750 to 730 BC or 530 to 370 BC or 280 to 260 BC	800 to 200 BC	1989, **31**(1), 25
Lower layers in massive buildup of debris, Site 10, layer 23	charcoal	BM-2491	2170 ± 45	360 to 280 BC or 260 to 170 BC	370 to 110 BC	1989, **31**(1), 25
One of number of slag heaps, Site 7, layer 4	charcoal	BM-2492	1940 ± 70	40 BC to 130 AD	110 BC to 230 AD	1989, **31**(1), 25
Site 15, west of North Lode opencast	charcoal	BM-2640	160 ± 50	1660 to 1700 AD or 1720 to 1820 AD or 1850 to 1880 AD or 1920 AD to modern	1650 to modern	1994, **36**(1), 106

AGUCHA SILVER/LEAD SMELTING SITES

Well site layers 3 and 4	charcoal	BM-2489	2380 ± 40	510 to 380 BC	760 to 690 BC or 540 to 370 BC	1989, **31**(1), 25
Below slag heap, Site 1, layer 3,	charcoal	BM-2633	2140 ± 50	360 to 310 BC or 240 to 100 BC	370 to 50 BC	1994, **36**(1), 107

Table 6.3 Radiocarbon dates for the Dariba and Agucha smelting and cupellation sites. Calibrated using OxCal v2. 18 (Bronk Ramsey 1995) and the calibration curve of Pearson and Stuiver (1986). References in the far right column are to the original publication of the figures in *Radiocarbon*.

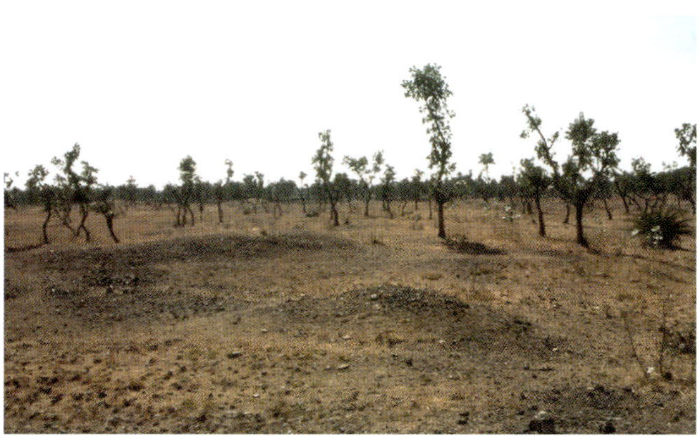

Plate 6.51 Dariba, Site 7: slag heaps to the immediate south east of the large East Lode opencast (cf. Chapter 5, Fig. 5.19).

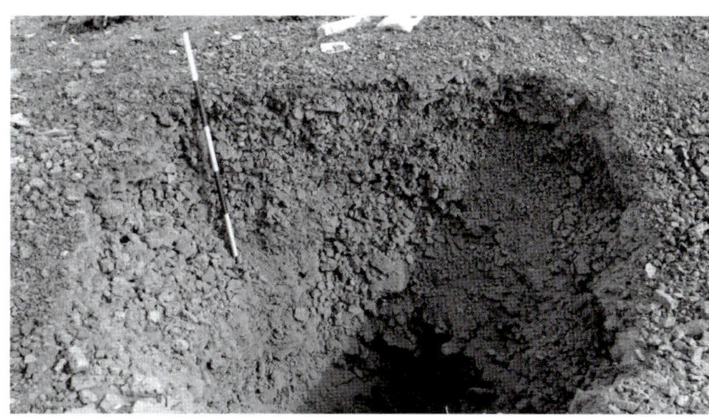

Plate 6.52 Dariba, Site 7: excavated area through a heap of loose tap slags.

material sampled, charcoal and flecks of burnt material, together with rather smaller slag fragments, all of which had presumably washed down from the heap above although there were no large slag pieces. The charcoal sample was taken from this layer. Beneath is a much more compacted and harder soil that probably represents the original surface upon which the slag was dumped, although Layer 4 may be the original surface, in which case the radiocarbon date and the refractory samples do not necessarily belong with the slag heap (see the appendix).

Site 8

Function: Occupation.

Date range: Possibly Mauryan.

Location: This is one of the small tells that lie at the south end of the slag heaps represented by Site 7 (see Chapter 5, Fig. 5.19).

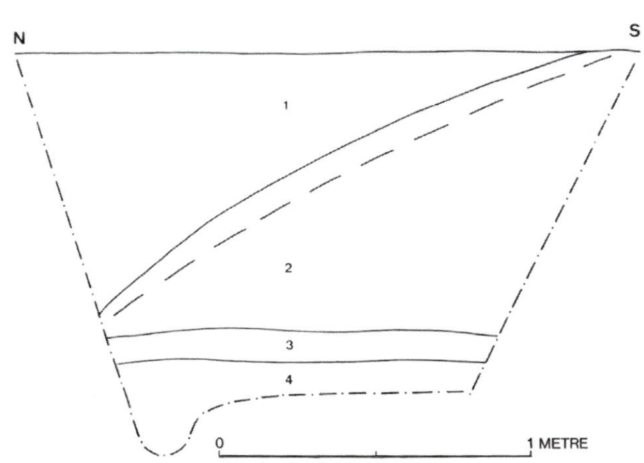

Fig. 6.50 Dariba, Site 7: east section of area dug through extremely loose slags.

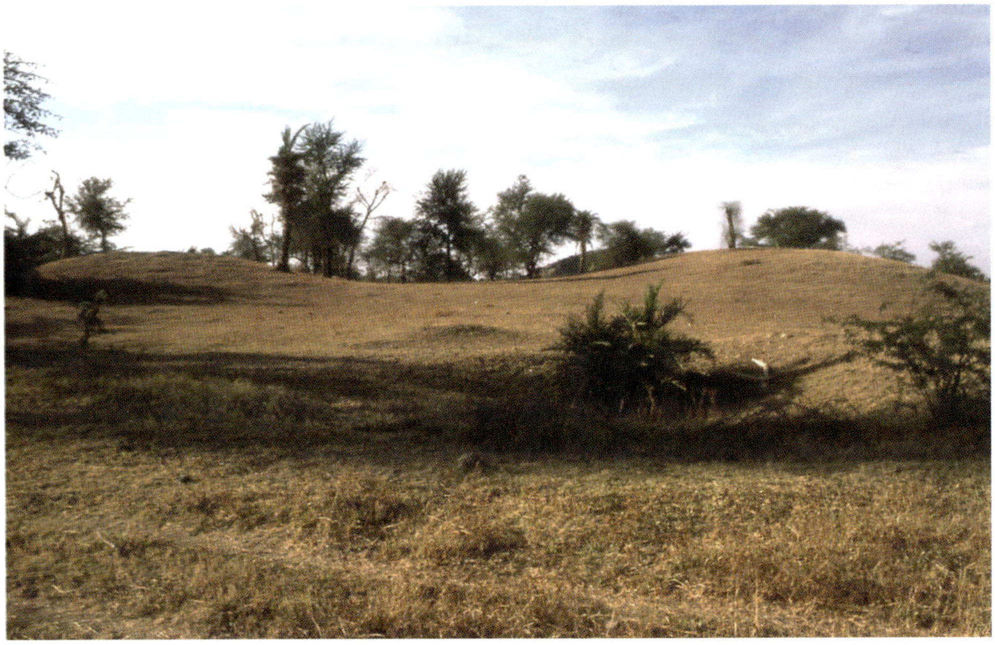

Plate 6.53 Dariba, Site 8: very substantial but eroded banks containing mud bricks forming a square enclosure open at the corners.

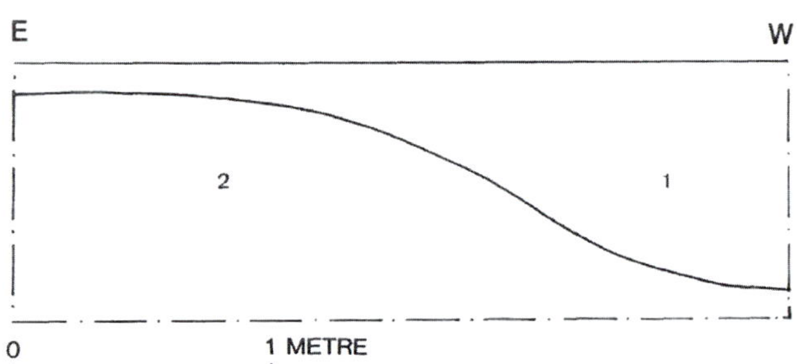

Fig. 6.51 Dariba, Site 8: section, facing south, dug into the outer face of a decayed mud brick bank.

Description: Site 8 is specifically an area of four conjoined banks that together approximate to a square with sides of about 20 m, although with no proper corners (Plate 6.53). Some dressed stone and large coarse bricks project from these banks, suggesting that they may originally have formed a walled enclosure and thus apparently a building of some status. The layout recalls the instructions in the *Arthaśāstra* (2.13.1) (Kangle 1965–1972: 110) for a workshop for the purification and working of precious metals: 'The Superintendent of Gold should cause to be built a workshop with a court-yard having four work-halls, without inter-communication [and] with a single door, for the manufacture of gold and silver.'

The four halls would be arranged in a square with a central courtyard, did not intercommunicate, and had a single door to the outside world. Copious quantities of metalworking refractories were exposed in the nearby well section (Site 10). A small trench 3 m in length was cut into the north west end of the most northerly of the banks to produce a section that was about 1 m tall (Fig. 6.51). Layer 1 was a light soft soil with a considerable quantity of much eroded domestic ceramics and a few small slag fragments (see the appendix). Layer 2 comprised more compacted clay, again with some small eroded domestic ceramics and small slag fragments, but no charcoal. This was probably the remains of a mud brick structure. However, as this was clearly not an area of primary metal production debris no further excavation took place.

Site 9

Function: Principally silver/lead production with some evidence of copper and iron production.
Date range: (Layer 4) 2340 ± 60 BP, BM 2490.
Location: This was a small eroded mound rising about 50 cm above the general level of the plain towards the far south east of the area of metallurgical debris (see Chapter 5, Fig. 5.19), therefore farthest from the mines (Plate 6.48). Immediately to the south lay the last of the small tells of building debris.
Description: The site is crossed by a minor unmetalled road with flanking side ditches which revealed metallurgical debris in their sides. A small trench, 2 × 1.5 m, was excavated adjacent to the track (Fig. 6.52 and Plate 6.54) and produced abundant evidence of metallurgical activity, principally in the form of charcoal and fragments of ceramic refractories. Some of these were clearly ordinary potsherds that had been utilised as such, but nearly all were heavily slagged. There was only a little primary smelting slag in the area. The first layer was made up of gravelly topsoil, which was a water-lain flood deposit. Beneath this was Layer 2, a very dark soft soil containing a great deal of small charcoal (sampled for radiocarbon dating). This distinctive layer was followed for some distance along the section exposed by the side ditches and suggests that a charcoal store was located either here or nearby. The charcoal would have been sieved before use and these are small pieces and dust, known as 'fines', which had been rejected. Layer 3 was soft brown soil but with a great deal of refractory debris, including a tuyere stand (Fig. 6.53). Much of this was apparently oxidised-fired and some of the pieces were covered with a lead glaze (Fig. 6.54 and Plate 6.53). Once again the layer was

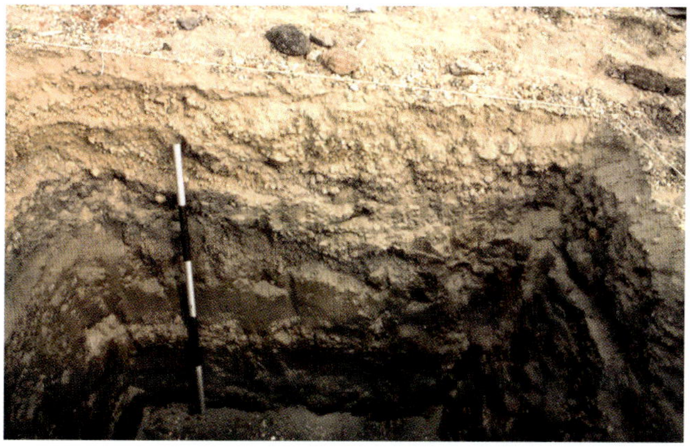

Plate 6.54 Dariba, Site 9: excavated area looking to the east.

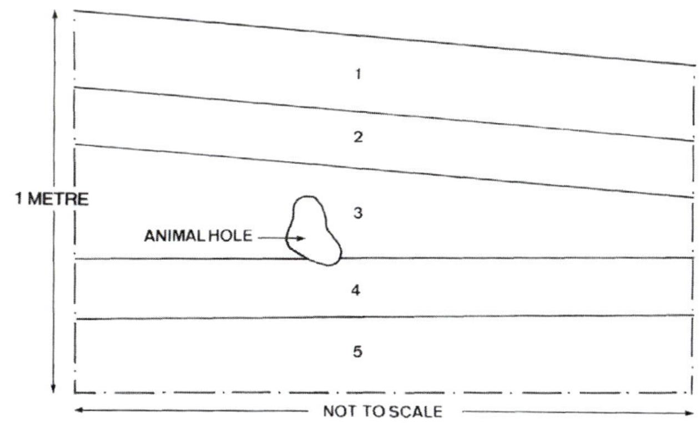

Fig. 6.52 Dariba, Site 9: east section.

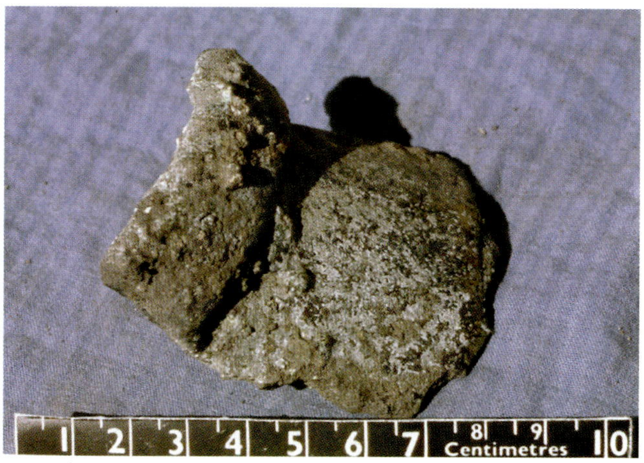

Plate 6.55 Dariba, Site 9: rim sherd massively lead-glazed on both sides and over the breaks, suggesting that it was reused in the cupellation process (cf. Fig. 6.54).

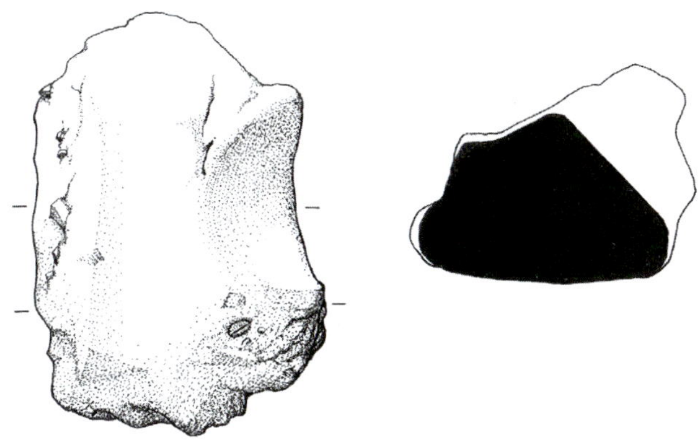

Fig. 6.53 Dariba, Site 9 L.4: fragment of a clay tuyere support, similar to those from Agucha (cf. Fig. 6.73). Profiles as indicated (outline at max. width: 4.5 cm).

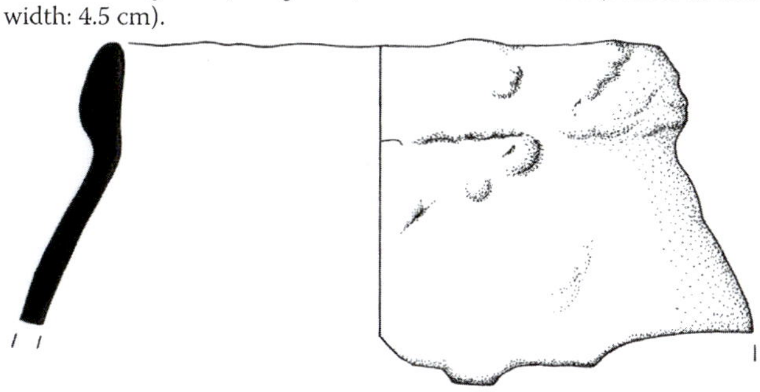

Fig. 6.54 Dariba, Site 9 L.4: sherd, massively glazed all over including the breaks, probably as a result of having been used in the cupellation process (max. dimension: 5.5 cm) (cf. Plate 6.55).

traced farther away in the road ditch section and a great deal of the same lead-glazed refractory was observed.

Beneath Layer 4 was a harder brown soil containing a little charcoal and only minor flecks of burnt ceramic. The composition of the slag from this layer is very different from the other slag samples collected from Dariba, but is typical of direct-process iron smelting slags. As the well sections to the north west demonstrate, occupation debris with evidence of some metallurgical activity possibly continued on down. However, Site 9 is located very much towards the edge of the area of metallurgical debris and the first 20 cm of Layer 5 contained no further evidence of metallurgy so excavation was terminated (see the appendix).

SITE 10

Function: Debris from a variety of metallurgical processes and some silver/lead slags.

Date range: (Layer 23) 2170 ± 45 BP, BM 2491.

Location: Well section site south of the large opencast (see Chapter 5, Fig. 5.19), close to Site 8.

Description: This is a well section site where the top of the well has cut through the occupation layers that form the present ground surface of the plain. Additionally a dyke had already cut through the early deposits and then become filled with tap slags (Layers 1–6) long before it was itself cut by the well (Fig. 6.55 and Plate 6.56). From the relatively small amount of non-slag debris in the bottom of the dyke it is suggested that it filled with slag relatively soon after its construction, or at least after the last clearing out. The slag is part of the western edge of the major slag dump exemplified by Site 7. The underlying occupation/metallurgical debris built up in approximately horizontal bands, interspersed with pits. The section created by the well was cleaned and straightened, but no attempt was made to discover the depth of the occupation material by excavation into the material filling the remainder of the well.

The slags from the dyke fill and those from the layers below are all associated with silver/lead smelting and are similar to those from Sites 7 and 9. The main interest in the section lay in the huge concentration of refractory material, especially of crucibles, in the centre of the exposed section (notably Layers 20–26 and 45)

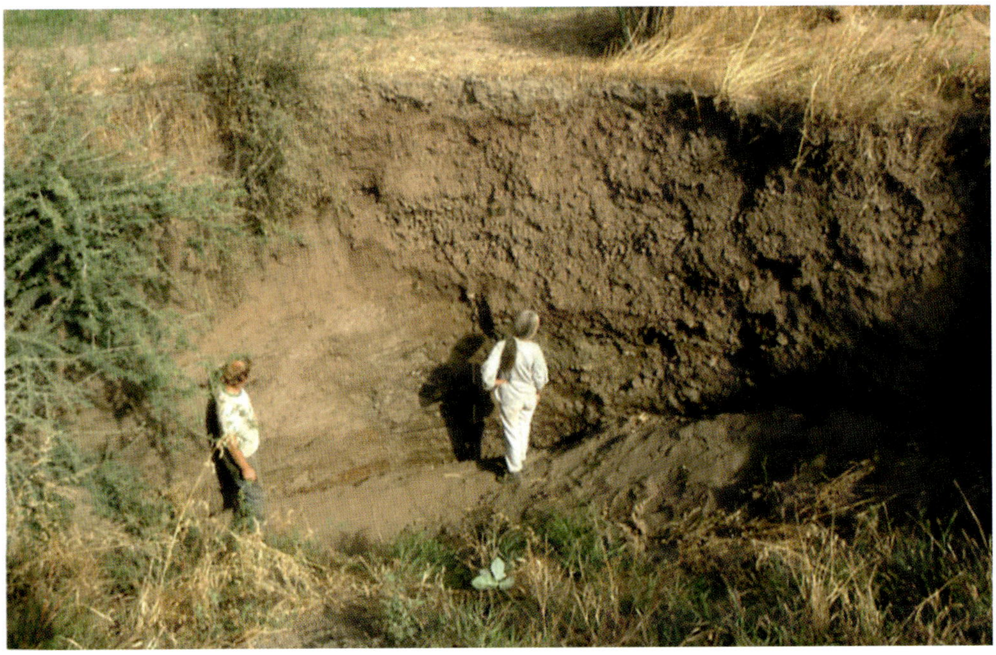

Plate 6.56 Dariba, Site 10: section looking to the north east before cleaning and sampling. The slag-filled dyke is to the right.

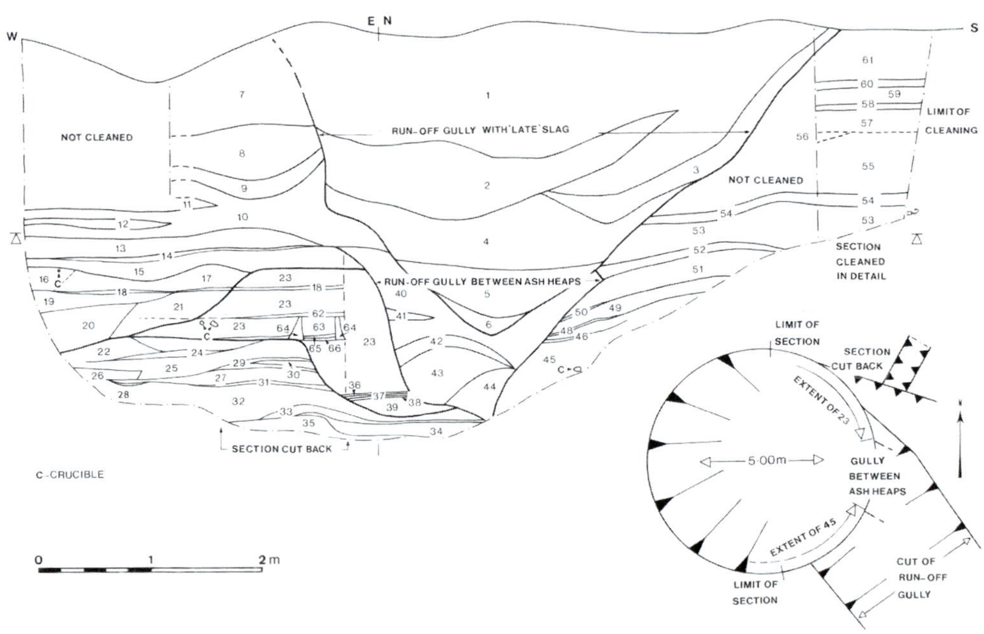

Fig. 6.55 Dariba, Site 10: section exposed in the north side of a well. Note the slag-filled channel cutting through the horizontal layers.

and many samples were taken (see the appendix). The material making up the layers was often ashy in nature with a far greater preponderance of burnt material and charcoal than is found in ordinary occupation debris. There was some slag throughout the layers but no great concentrations such as occurred in the dyke fill. Some of the layers could have been water-lain from the monsoon flash floods, but the considerable quantities of very fragile refractory together with complete crucibles and pots (Fig 6.56 and Plate 6.57), all in pristine condition, must be *in situ*, and had probably been dumped immediately after use into pits or channels where they were protected from further damage.

The wide range in shape and size of the various refractories and other debris suggested that a variety of metallurgical activities was performed there, however, qualitative analysis of these (see Chapter 7) detected tin in the contact surfaces of all of them, suggesting that they were used to melt copper and bronze. Therefore they were not from primary metal smelting operations but rather were associated with the alloying and purification of bronze, possibly casting, although no mould material was recognised. This is supported by the fragments of copper alloy metal recovered from Dariba which were also of bronze. There were some slags and furnace lining fragments

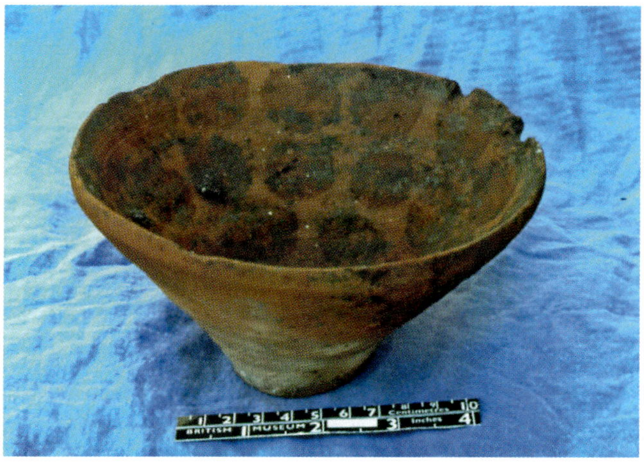

Plate 6.57 Dariba, Site 10 L.45: one of several complete pots recovered from these contexts. Note the painted decoration of large dots on the inside (cf. Fig. 6.56).

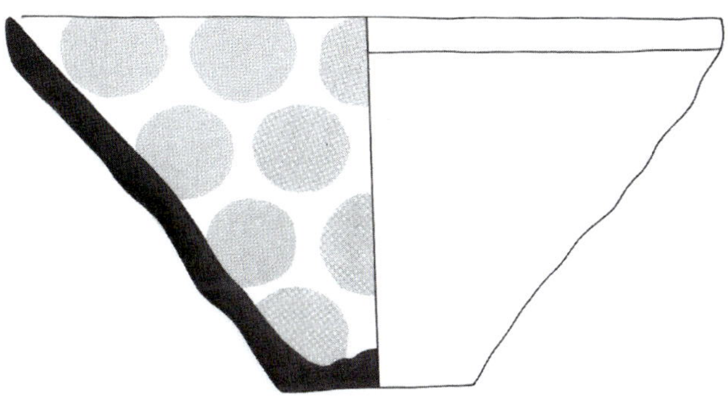

Fig. 6.56 Dariba, Site 10 L.23: complete domestic pot (diameter: 12 cm). Note the internal painted decoration (dark brown) (cf. Plate 6.57) and also the similarity with the vessel found underground at the Zawar Mala mine (cf. Chapter 5, Fig. 5.13).

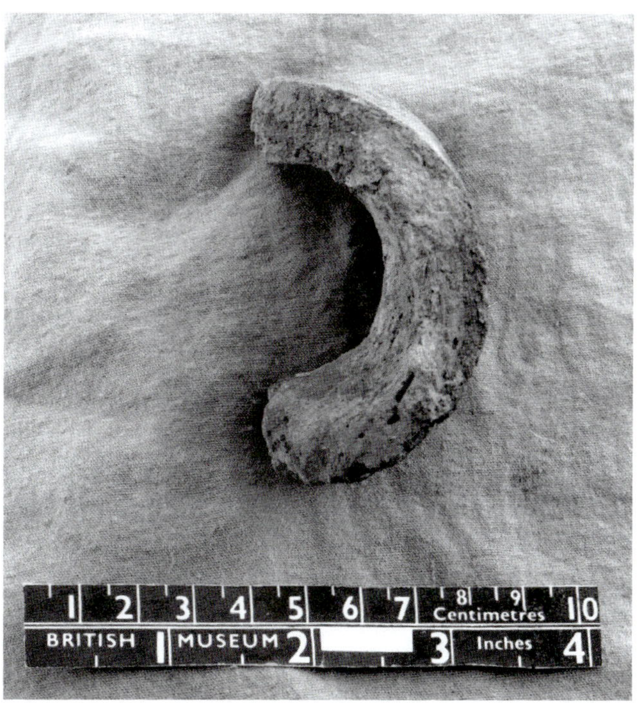

Plate 6.58 Dariba, Site 10 L.43: fragment of a large diameter tuyere tube of fine clay.

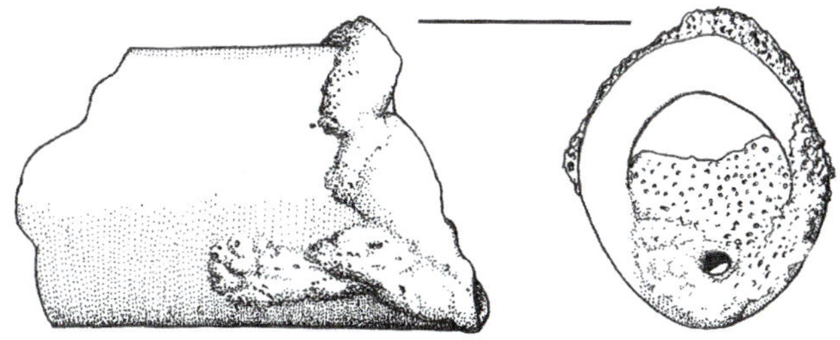

Fig. 6.57 Dariba, Site 10 L.23: complete tuyere end (length: 5 cm) (cf. Zawar, Site 2, Fig. 6.36).

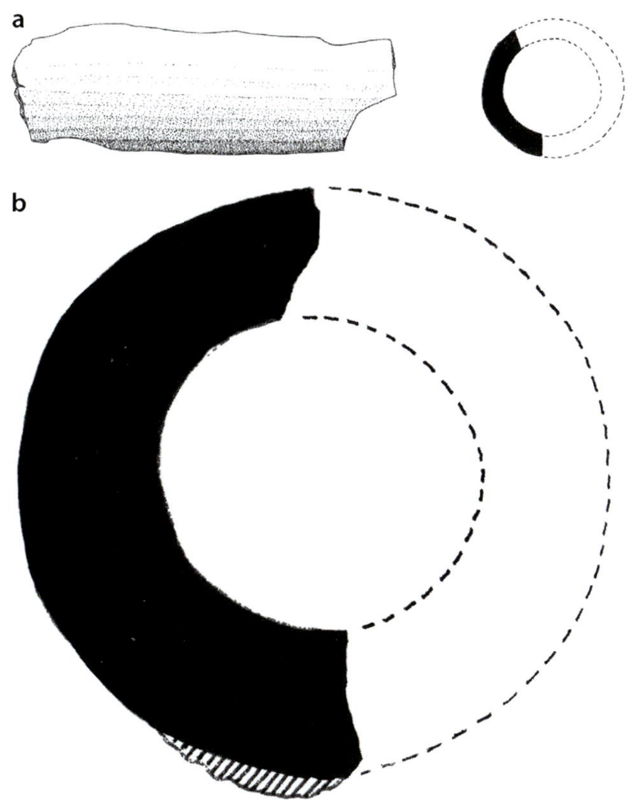

Fig. 6.58 Dariba, Site 10 L. 23 and 43: fragments of a tuyere with thin walls made from fine clay. (a) Length: 8 cm, outer diameter 3 cm; (b): outer diameter 5 cm (cf. Zawar, Site 2, Fig. 6.36).

that could be associated with smelting operations, although these were not necessarily taking place in the immediate vicinity. The tuyere fragments are rather thin-walled and of a fine clay (Figs 6.57 and 6.58 and Plate 6.58), reminiscent of those from Zawar Sites 2 and 14 (Fig. 6.36) and the well site at Agucha (Fig. 6.74), all of approximately the same age.

There were many large fragments of heavily burnt refractory, either the edges of heat-buckled flat tiles or of hearth linings. Other pieces are less substantial but have a more regular profile; these are probably the rims of large open vessels, although they were clearly also involved in metallurgical processes at high temperature. The majority of the refractory material at Site 10 seems to have been associated with the melting of bronze, as evidenced by the often very well-preserved crucibles (Fig. 6.59 and Plate 6.59), some with lids (Fig. 6.60). Many of these have pronounced nipples on their bases and the vitrification shows that they stood in burning charcoal. The nipples would have served to raise the

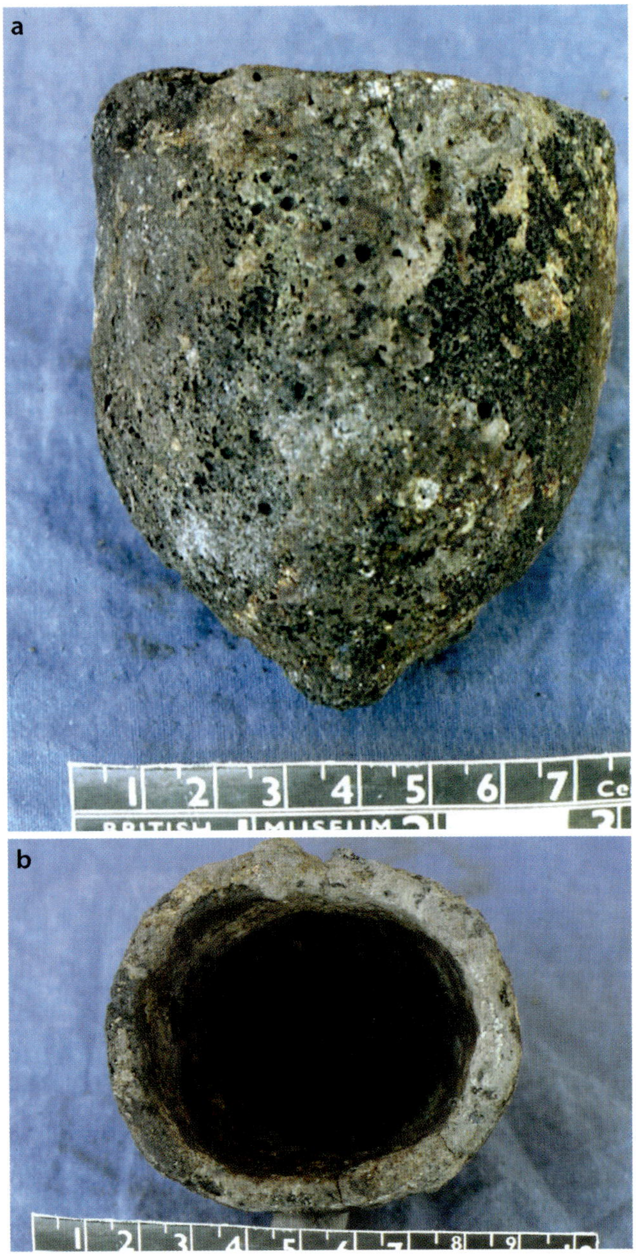

Plate 6.59 Dariba, Site 10 L.23, complete crucible: (a) profile and (b) top (cf. Fig. 6.59b).

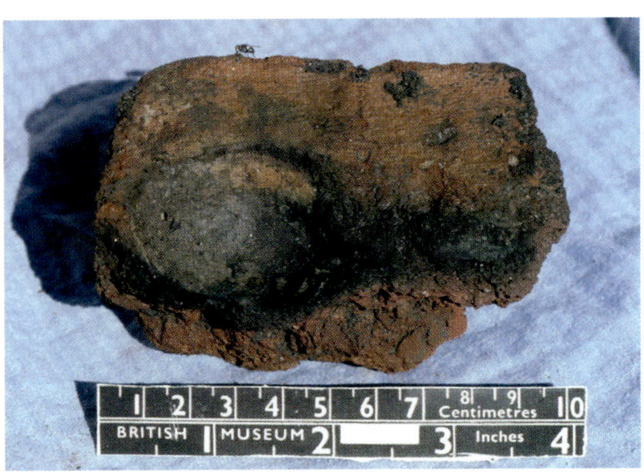

Plate 6.60 Dariba, Site 10 L.23: corner of possible ceramic ingot mould (cf. Fig. 6.61).

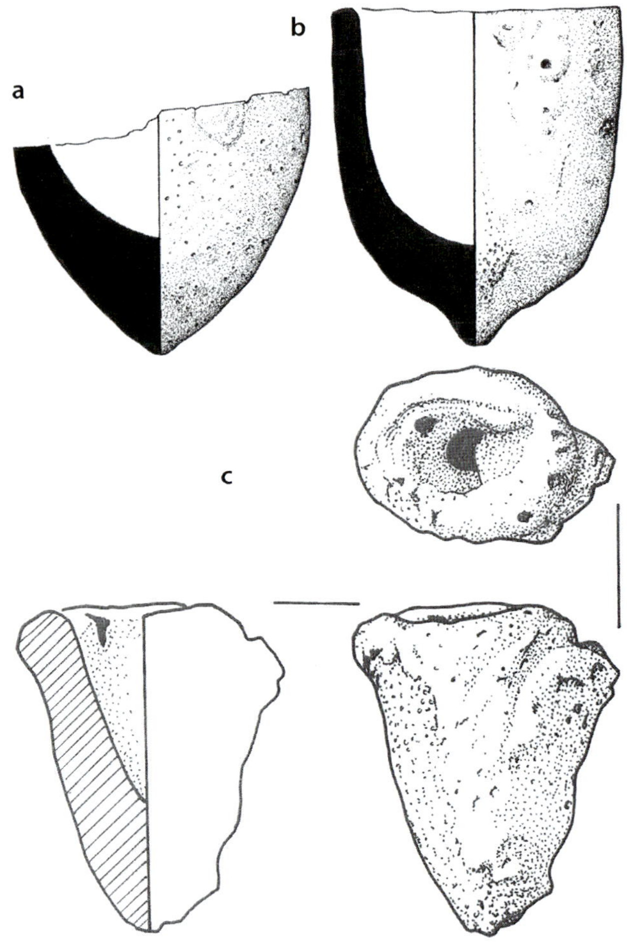

Fig. 6.59 Dariba, Site 10 L.23 and 50: selection of crucibles. (a) Max. outer diameter: 7 cm; (b) outer diameter at top: 5.5 cm; (c) outer diameter at top: 3.5 cm.

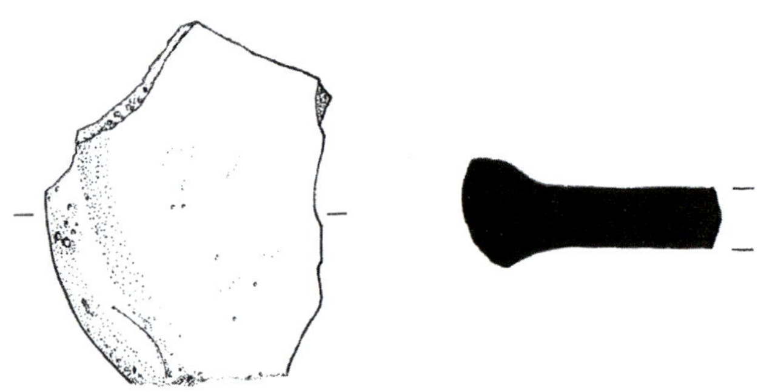

Fig. 6.60 Dariba, Site 10 L.23: possible crucible lid (max. dimension: 4 cm).

base slightly to ensure that the burning charcoal was located all around and beneath the crucible. This is quite a common feature on early crucibles worldwide and can be paralleled in India by crucibles from Chandraketugarh in the North 24 Parganas District of West Bengal, dated to between the 2nd century BC and the 2nd century AD (Chattopadhyay 2004: 143, plate 1A). Ingot moulds were also found (Plate 6.60 and Figs 6.61 and 6.62).

Sites 11 and 12

Function: Site 11 produced some evidence of smelting in the upper part, but mainly occupation debris. Site 12 also produced some evidence of smelting activity, lead slags, etc.

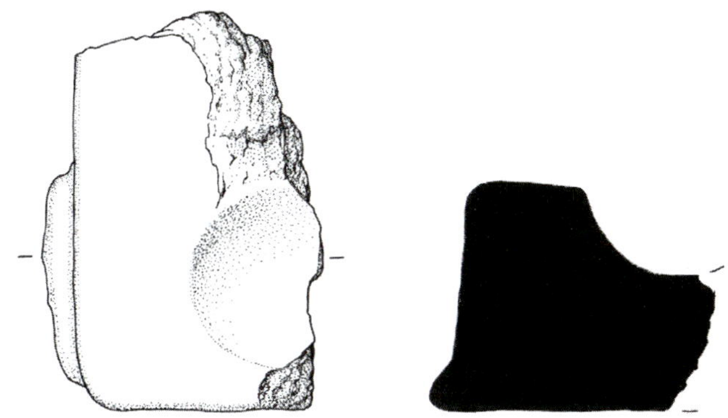

Fig. 6.61 Dariba, Site 10 L.23: corner of a possible assay block or ingot mould (max. dimension: 6.5 cm).

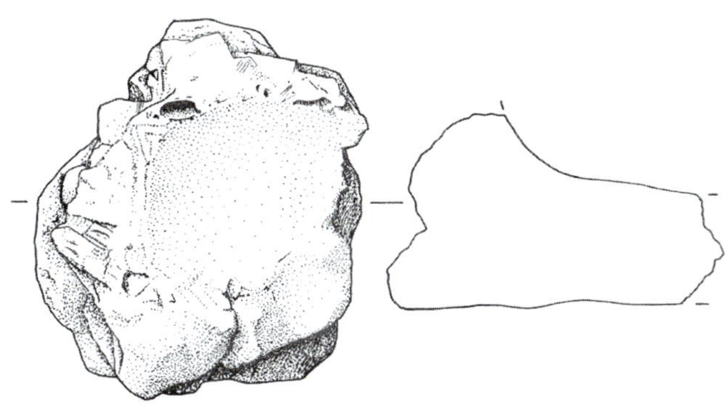

Fig. 6.62 Dariba, Site 10 L.23: fragment of possible assay block or ingot mould (max. dimension 7 cm).

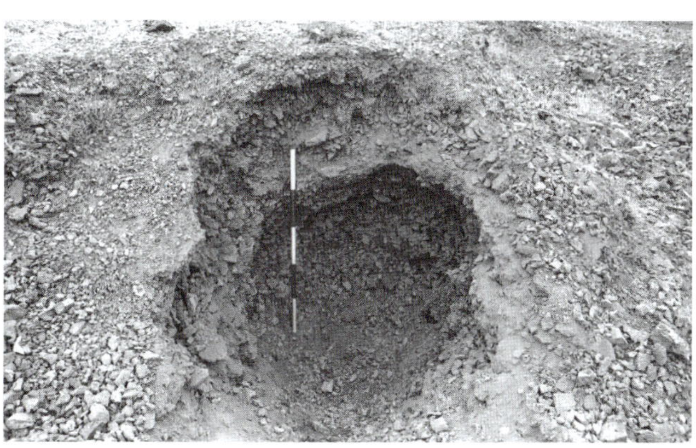

Plate 6.61 Dariba, Site 14: rough section into the heap of very loose slags. The refractory material was concentrated in the upper layers (cf. Plate 6.62).

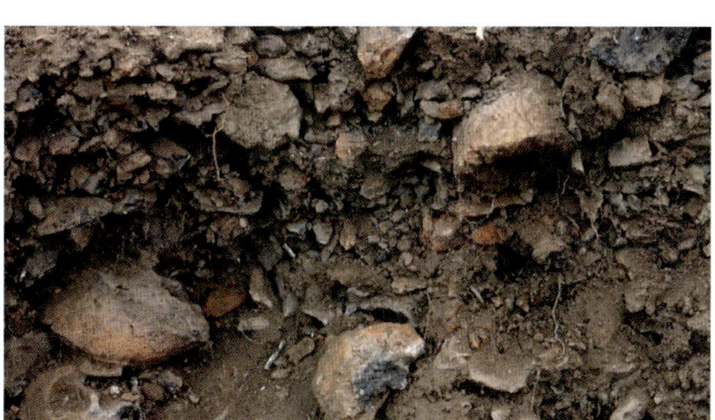

Plate 6.62 Dariba, Site 14 L.2: concentration of tuyeres. Three are clearly visible here.

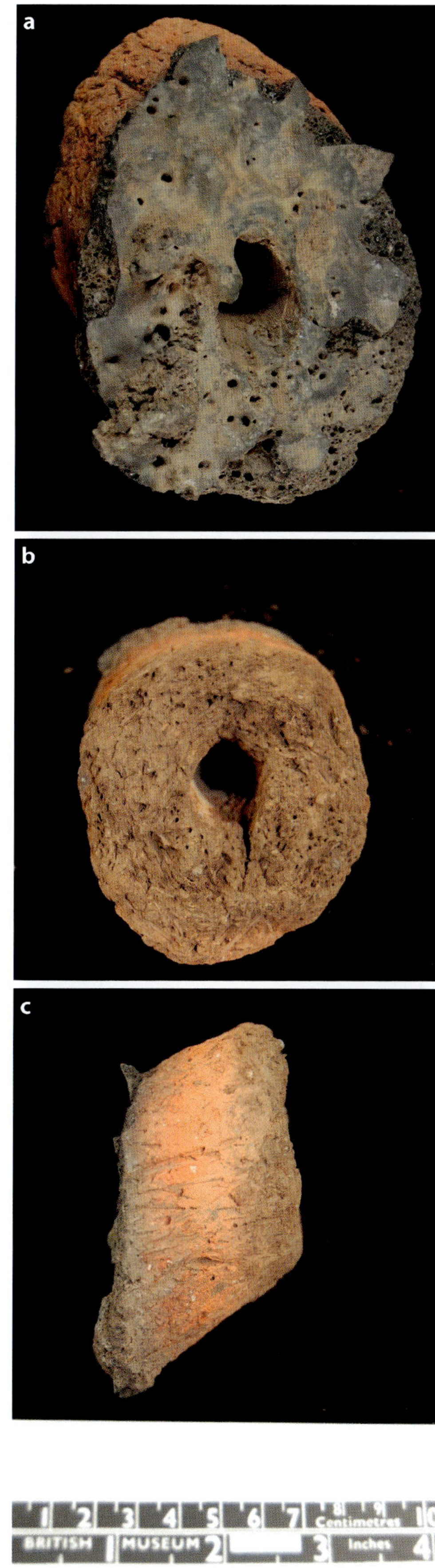

Plate 6.63 Dariba, Site 14 L.2, typical tuyere: (a) slagged front; (b) rear showing the many small holes where straw once lay along the line of the tuyere; (c) side view showing the angle of the tuyere to the furnace wall.

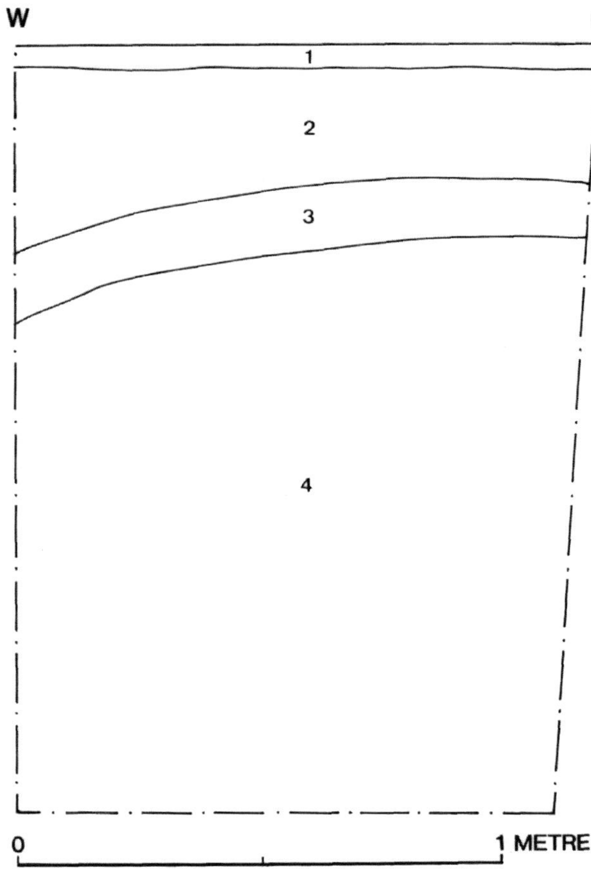

Fig. 6.63 Dariba, Site 14: rough section of the sampled face in the very loose slags. The refractories came mainly from L.2.

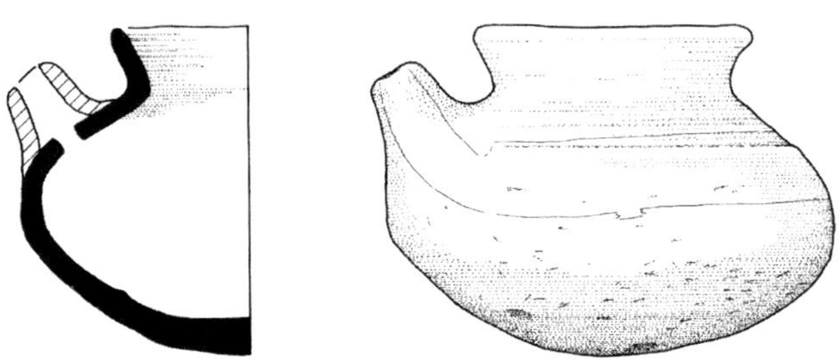

Fig. 6.64 Dariba, Site 14 L.2: small complete water pot (max. diameter: approx. 10 cm).

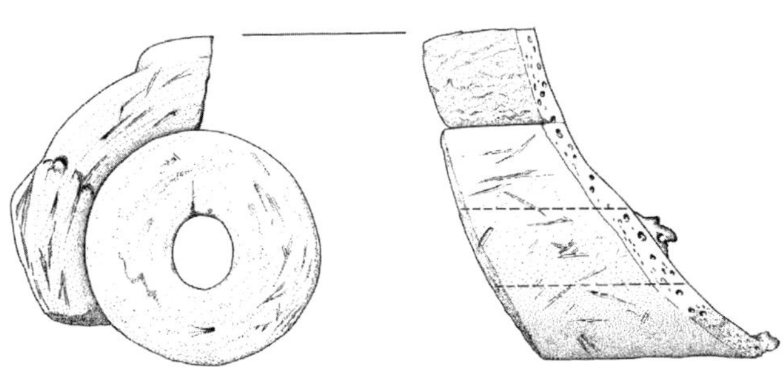

Fig. 6.65 Dariba, Site 14 L.2: typical tuyere showing inner tube and a fragment of the outer collar (outer diameter of tube: 5 cm).

Date range: Undated, but both are probably broadly contemporary with Sites 7, 9 and 10.
Location: Site 11 is at the southern end of the slag heaps, Site 12 is some 200 m north of Site 8.

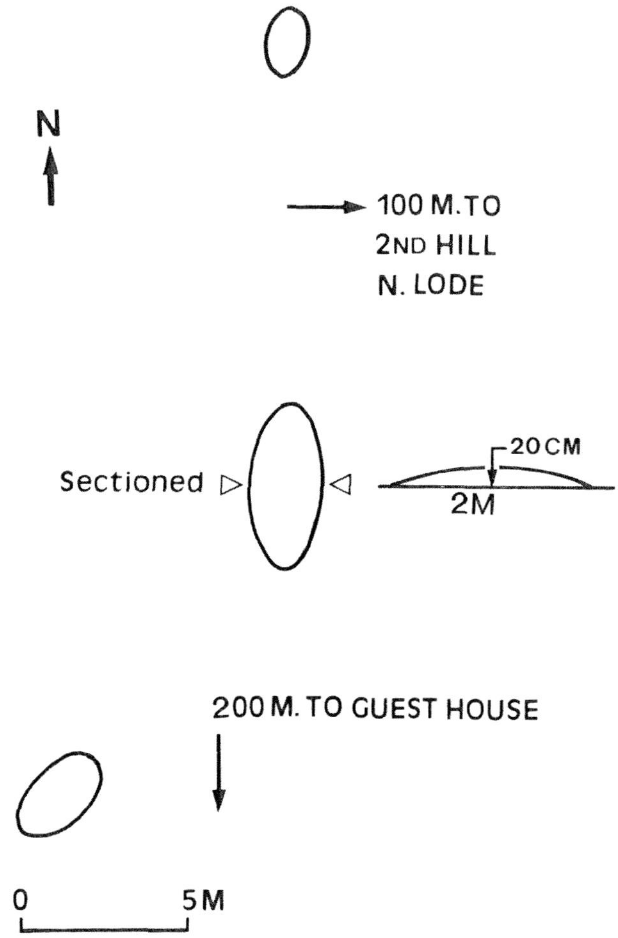

Fig. 6.66 Dariba, Site 15: sketch plan of the three small scatters of slags (cf. Chapter 5, Fig. 5.19).

Description: Site 11 is a large well at the southern end of the heaps and spreads of large slags (Plate 6.49). At a depth of about 80 cm was a layer about 80 cm thick containing tap slag. Beneath this a further 3.5 m of the well side was exposed and contained occupation material down to the top of the present filling of the well. There was some crushed slag throughout, but little direct evidence for extractive metallurgy having taken place in the immediate vicinity. Site 12 is a well of rather small diameter, and the sides were only exposed for about a metre in depth. In one place a pit containing domestic pottery and slag was revealed (see the appendix).

Sites 13 and 14
Function: Copper and lead smelting.
Date range: Recent, as indicated by the pottery (Fig. 6.64).
Location: Sites 13 and 14 are in the large heaps of slag that lie stacked against the east face of South Lode (see Chapter 5, Fig. 5.19). Site 13 lay some 50 m east of mine survey point 814.
Description: Site 13: retaining walls of slag blocks had been built to stabilise the slag on the rather steep slope. Previous exploration of these slag tips had located large concentrations of refractory, especially tuyeres and some spills of lead metal. A 3 × 2 m area was opened on a reasonably flat surface. The top 20 cm of the slag (Layer 1) contained a great deal of soil and ordinary domestic pottery sherds not dissimilar to the local country wares currently in use. The next 20 cm (Layer 2) of slag still contained a

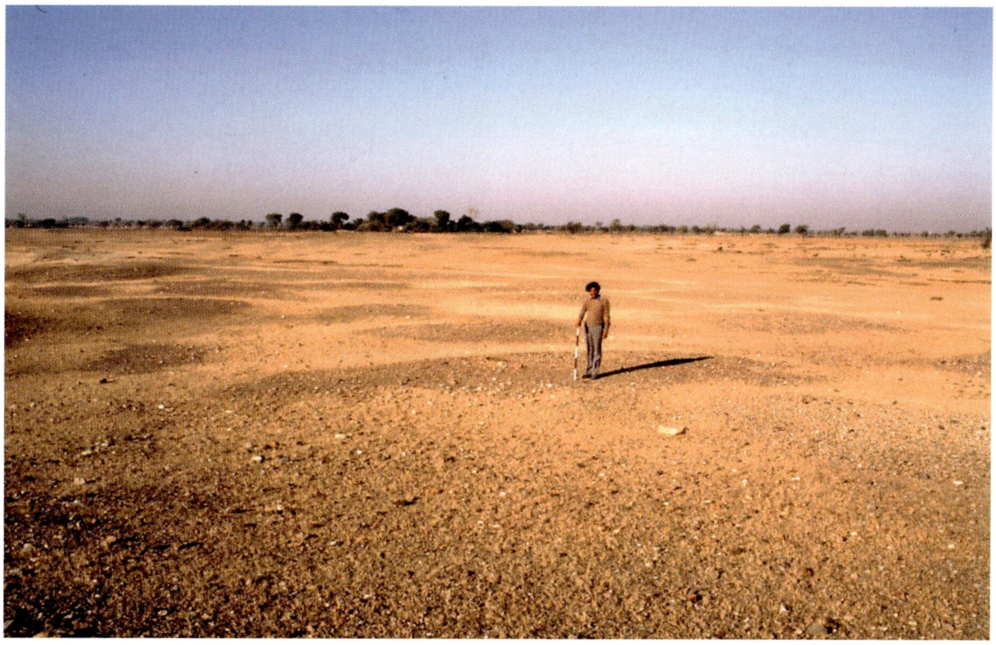

Plate 6.64 Agucha, Site 1: heap selected for excavation showing the generally eroded landscape to the east of the opencast mine.

considerable amount of domestic pottery including roof tile fragments, but only rather small fragments of refractory. There was still a great deal of soil and ash dust around the slag, suggesting there had not been sufficient time for it to be washed through; certainly it was very different from the heaps represented by Site 7. Further excavation at Site 13 was discontinued after the discovery of the nearby exposed section that became Site 14.

At Site 14, about 6 m to the east of Site 13, a cutting in the slag was observed to be very rich in well-preserved pottery and refractory material (Plate 6.62). It was not possible to establish the stratigraphic relationship between the two sites on the steep slope of South Lode, with exposed bedrock, but clearly the two sites are broadly contemporary. This north-facing exposure was cleaned as much as possible given the very loose nature of the heap, and sampled (Fig. 6.63 and Plate 6.61). The refractory was concentrated just below the surface in Layer 2 and included fragments of furnace wall and tuyeres. Large quantities of domestic pottery sherds in excellent condition and one complete vessel (Fig. 6.64) were also recovered from Layer 2. In one small area alone no less than 12 tuyeres were found (Fig. 6.65 and Plate 6.62) (see the appendix).

The slag layers throughout the section were permeated with soft soil and dust, again suggesting that little weathering has taken place, and the abundant pottery is identical to the wares used by the local country people to this day. Recalling the mention of the 'tin and copper mines of Dureba' by Tod (1978, I: 618) as operational at the time of his residency in the early 19th century, it seems likely that these dumps belong to the last phase of the traditional industry and confirmed by the radiocarbon date of the nearby Site 15 (see below), therefore no radiocarbon determinations were attempted.

Site 15
Function: Copper smelting.
Date range: Post-medieval, 160 ± 50 BP, BM 2640.
Location: Approximately 150 m west of North Lode and 200 m north of the mine guest house.
Description: This was one of a series of small scatters of slag (Sites 1–3, 5, 6, 15 and 16) lying to the east and west of the North Lode opencast workings. Site 15 was the only site where the slag was sufficiently concentrated to constitute a small heap that offered a chance of excavating a meaningful charcoal sample from an undisturbed context in order to obtain a date for the northern sites (Fig. 6.66). Analysis of the slag from Sites 14, 15 and 16 (see Chapter 7, Table 7.15) shows that they were engaged in copper smelting. The site comprised three small scatters or heaps, each about 5–10 m in diameter, and rising no more than 20 cm above the general ground surface. The largest heap was selected for excavation and half sectioned. The slags were in a soft soil full of burnt clay with some rough refractory and large amounts of charcoal. Many of the charcoal pieces were quite sizeable, typically up to walnut size, which together with the presence of soil still among the slags showed that the heap was little eroded and thus likely to be of no great age. This sat on soft yellow silt which is the present land surface all around. The slag scatters which constitute Sites 1–3, 5, 6 and 16 could be smaller, less well-defined heaps of the same age, or possibly eroded heaps of much greater age (see the appendix).

Surface sites at Rampura-Agucha

Survey and excavations

Extensive field surveys were carried out in 1983 and again in 1985 for about 1 km on either side of the main trench mine in advance of the development of the new mine (Agrawal 1999) (see Chapter 3). The ground surface to the immediate north of the trench mine was covered by eroded heaps of rather uniform-sized crushed rock that had the appearance of beneficiation debris, and to the immediate south were several

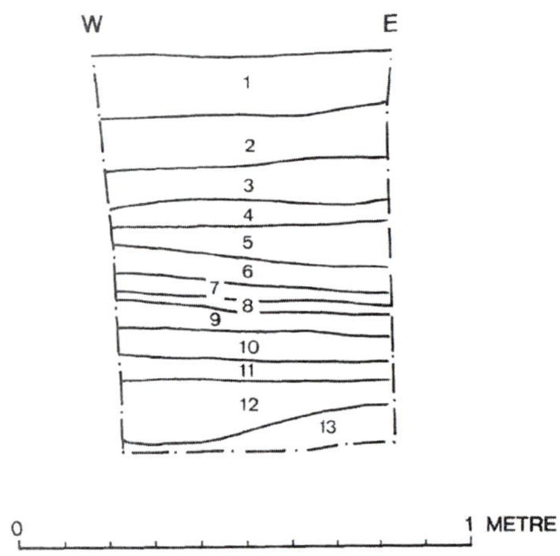

Fig. 6.67 Agucha, Site 1: south section.

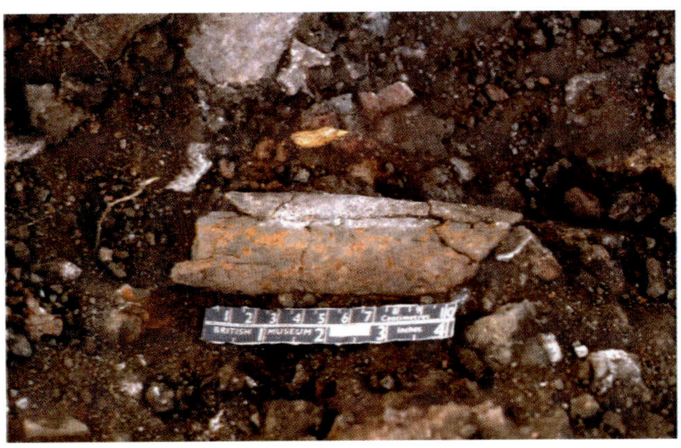

Plate 6.65 Agucha, well site, 'tuyere area': fragmentary tuyere typical of the wide bore, thin-walled tuyeres found here and at contemporary sites at Zawar and Dariba.

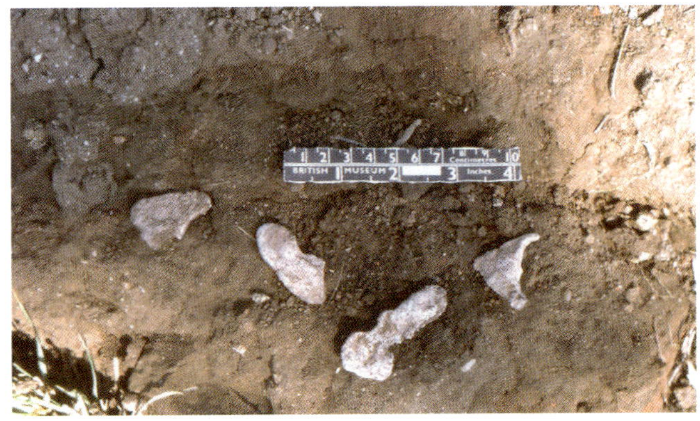

Plate 6.66 Agucha, well site, 'lead area': spills of argentiferous lead (cf. Chapter 7, Table 7.18) as uncovered.

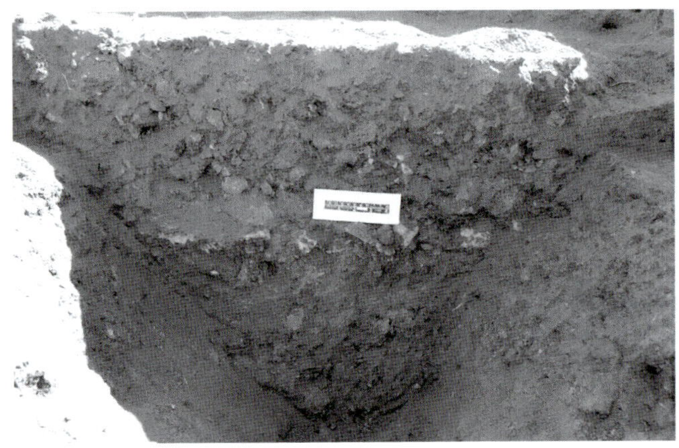

Plate 6.67 Agucha, well site, 'tuyere area': east section across the pit. The fill is L.4 which produced most of the well-preserved refractories.

large eroded heaps of mine debris. The whole ground surface in the area between the immediate south east of the mine and the village of Bheru Kera appeared to be made up of slags and much-eroded refractory material covering approximately 30 ha (Plate 6.64). The true extent of the surface metallurgical operations was difficult to estimate due to the undoubted spreading of material over the centuries. The slags sometimes formed low, spread-out mounds that gave every indication of having been heavily eroded and levelled, caused in part no doubt by periodic monsoon flooding which is prevalent in the area. The radiocarbon dates broadly follow the mine dates through the later part of the first millennium BC (Table 6.3).

In such documentary records that survive concerning Agucha there is no mention of mining or metal production so it must be assumed that there was no production throughout the medieval and post-medieval periods up until the rediscovery of the deposits in the latter part of the 20th century. The areas for investigation were selected purely on the evidence of material exposed in the sides of the recent wells.

SITE 1

Function: Silver/lead smelting.

Date range: 2140 ± 50 BP, BM 2633.

Location: 25° 50′ N, 74° 44′ E. The site lay approximately 70 m north/north east of the well section sites described below.

Description: The heap selected for excavation was approximately 20 m in diameter and rose about 30 cm above the general level of the plain (Plate 6.64). A 1 × 2 m trench was excavated down to a depth of 90 cm (Fig. 6.67), and the fill comprised three main events. The surface slags (Layers 1–3) lay on a series of thick water-lain compact soil layers containing a few small charcoal pieces and eroded refractory (Layers 4–10). Beneath these was a horizon containing crushed slag (Layer 11) that lay on a layer containing slag pieces, refractory and charcoal (Layer 12) (see the appendix). None of the refractory fragments had any recognisable features or form. Layer 12 in turn lay on a buried soil (Layer 13) which contained a few flecks of charcoal but no slag or refractory. This layer was taken as representing the base of the metallurgical debris, although it still contained some occupation debris. Therefore there would seem to be two periods of slag deposition separated by a period of abandonment. Charcoal was taken from first and third horizons, and the radiocarbon date obtained on the latter.

WELL SITE

Function: Silver/lead smelting, silver cupellation.

Date range: 2350 ± 40 BP, BM 2489, Layers 3/4, tuyere area.

Location: 25° 50′ N, 74° 44′ E. Approximately 300 m east of mine datum point 00 (see Chapter 5, Fig. 5.31).

Description: The site was a modern well that was now choked towards the top, dug through the ancient occupation and metallurgical debris (see the appendix). In the exposed section a buried surface of hard red clay was observed at about 40 cm

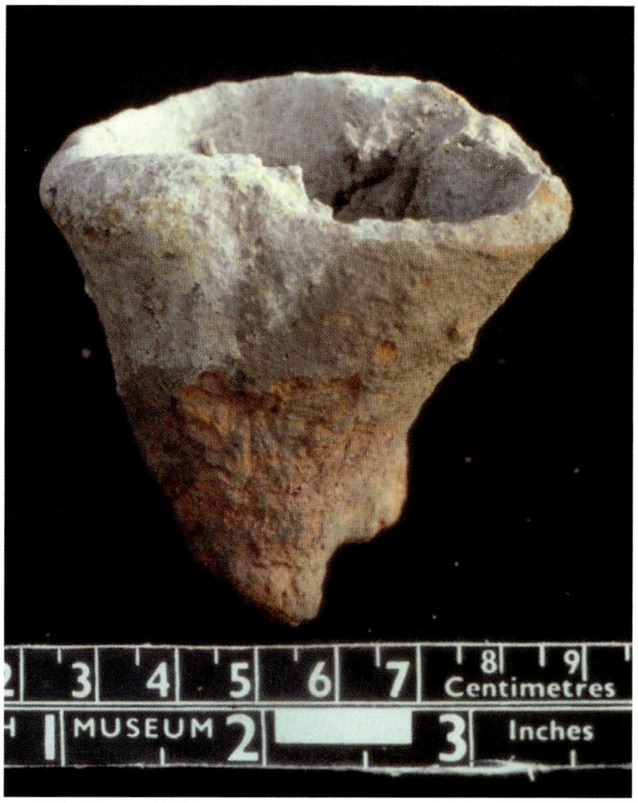

Plate 6.68 Agucha, well site, 'tuyere area' L.4: complete cupel. Note the line between the glazed (white) and unglazed (red) areas showing the angle at which the cupel lay in the fire (cf. Plate 6.72 and Fig. 6.72d).

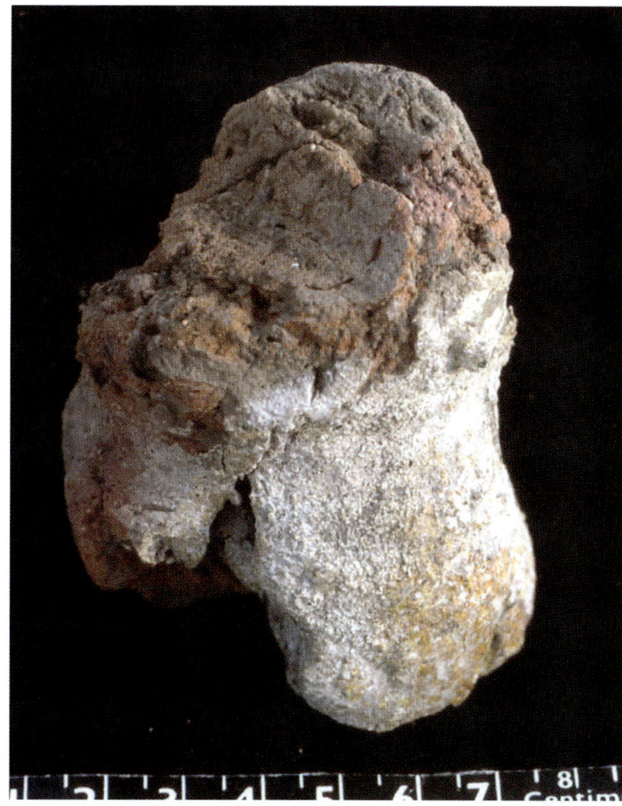

Plate 6.69 Agucha, well site, 'tuyere area' L.4: tuyere support showing the curve at the top to accommodate the tuyere.

depth, together with some refractory. The bedrock of amphibolite continued irregularly from about 1 m down. Although this was not a deep stratigraphy, it was of potential interest because substantial pieces of refractory were visible in the sides (Plate 6.65).

A small area, 1 × 1.5 m, was opened around these refractory fragments, which became known as the 'tuyere area'. A further small excavation was made on the north side of the pit and became known as the 'lead area' because of the metal spills found there (Fig. 6.68 and Plate 6.66). In fact the two areas revealed similar features and material: both contained very substantial amounts of refractory in a generally excellent state of preservation, all apparently connected with the cupellation process. Only a small amount of slag was found; some at least was inevitable given that the site lay within the general area of the slag tips. None of the abundant refractory material seemed to relate to the primary smelting of ore to metal. It appears that the modern well had cut through an area of pits which contained cupellation debris, but from two quite distinct activities: the assay of ores to determine their silver content and the regular production of silver from argentiferous lead. The first activity is evidenced by the numerous small cupels (Plate 6.68), some with their tuyeres still attached (Plate 6.72) and their separate pedestals (Plate 6.69), and the small pieces of argentiferous cerrusite, which together suggest that the mine assay office was in the vicinity (see reconstruction Fig. 6.69). The second activity is evidenced by a segment of a litharge cake, the massively vitrified and glazed ceramic furnace or hearth fragments (Plate 6.71), and a large tuyere fragment set in clay

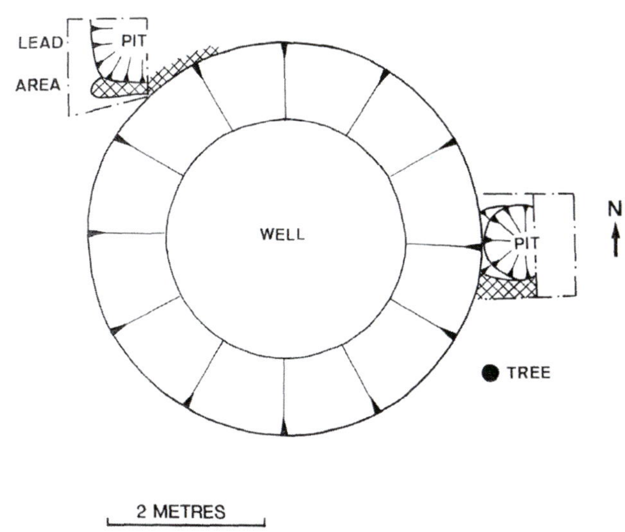

Fig. 6.68 Agucha, well site: sketch plan showing the location of the 'tuyere' and 'lead areas'.

(Plate 6.70 and Fig. 6.74), which indicate the actual production of silver from argentiferous lead (Fig. 6.70).

Tuyere area

An area was opened on the east side of the well (Figs 6.68 and 6.71 and Plate 6.67). Beneath the upcast from the well digging (Layer 1) was a compact brown soil (Layer 2). Underneath this surface the brown soil continued (Layers 3–5), but was much softer and contained a great deal of refractory material including many cupels, tuyeres and pedestals (Figs 6.72 and 6.73)

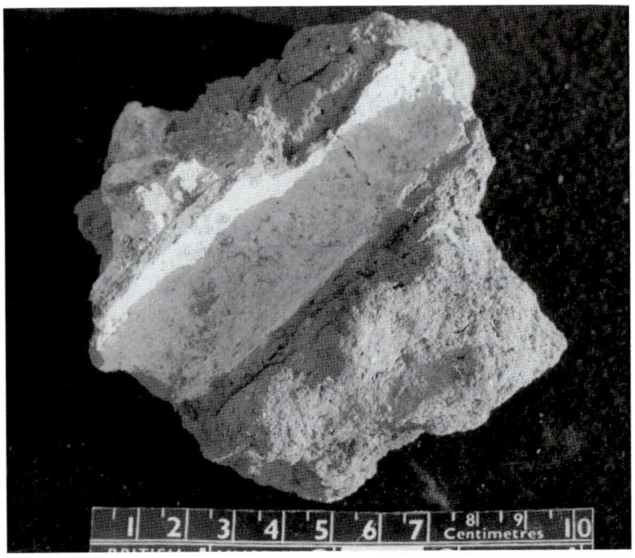

Plate 6.70 Agucha, well site, 'tuyere area' L.4: fragment of a thin-walled tuyere set in clay with a flat underside (probably from a cupellation hearth, cf. Figs 6.70 and 6.74).

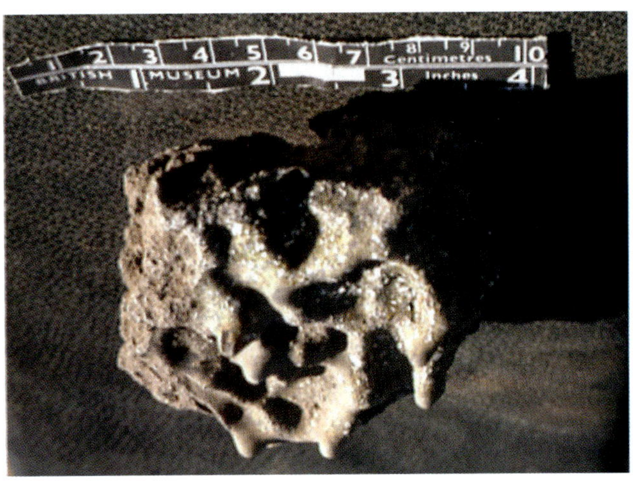

Plate 6.71 Agucha, well site 'tuyere area': typical heavily vitrified and lead-glazed ceramic fragment from a cupellation hood (cf. Fig. 6.75) (shown upside down to represent its original position, cf. Fig. 6.70).

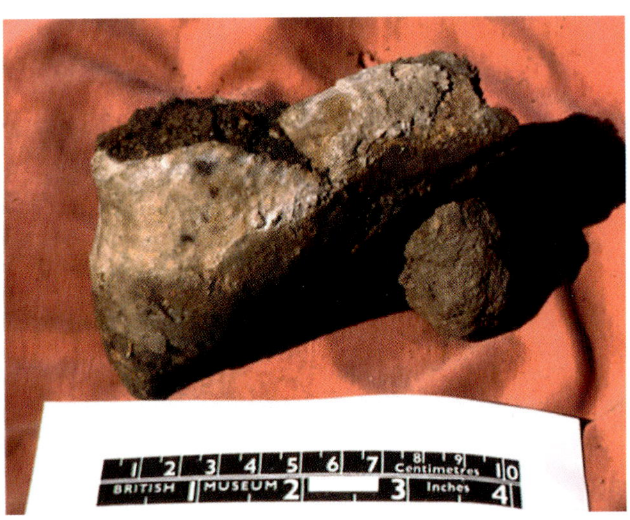

Plate 6.72 Agucha, well site, 'lead area': complete cupel with tuyere still attached (cf. Fig. 6.72C) placed upon a suitably sized pedestal for this photograph (cf. Fig. 6.69). Note the soil filling had not been removed at this stage.

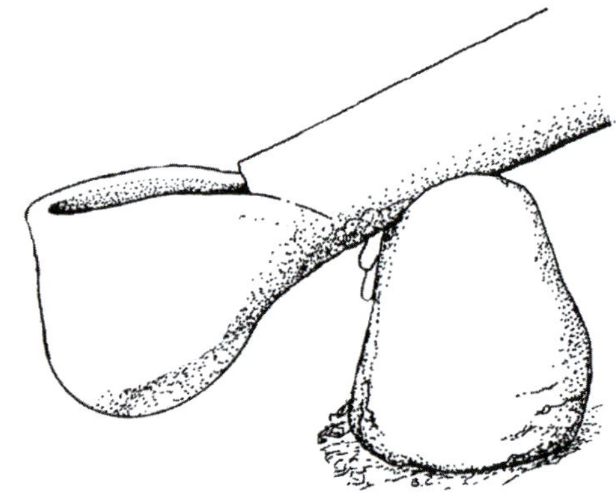

Fig. 6.69 Agucha: reconstruction of the cupels, especially Fig. 6.73c, tuyeres and their stands.

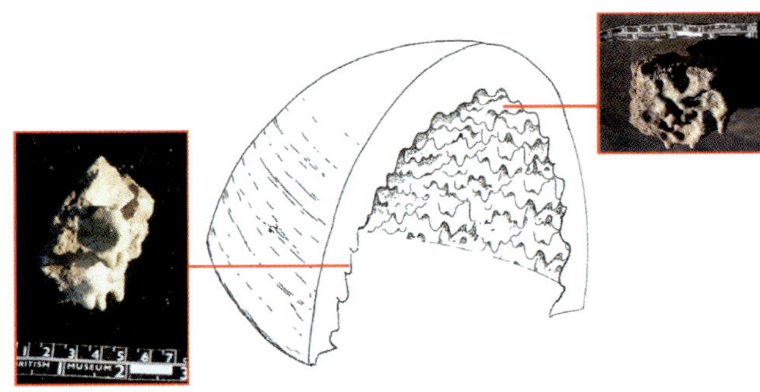

Fig. 6.70 Agucha: reconstruction of the form of the ceramic hood based on the drip pattern (image: A. Simpson).

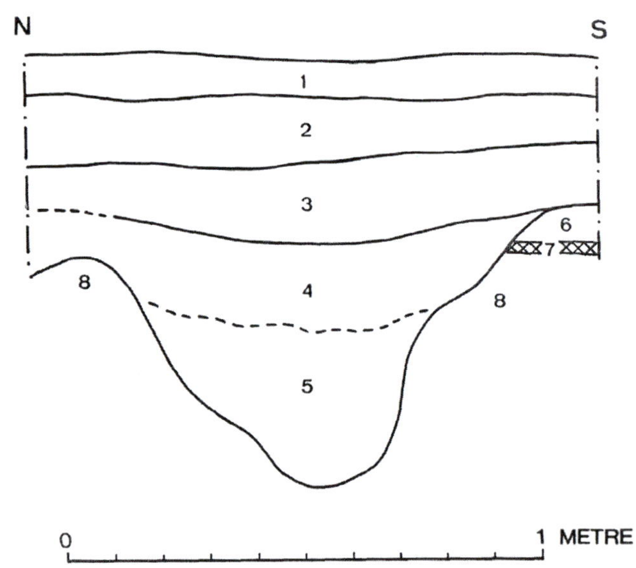

Fig. 6.71 Agucha, well section, 'tuyere area': east section.

that are interpreted as tuyere supports – indeed refractory material is estimated to have formed approximately 40% of the volume of Layer 3 (see the appendix). Much of this very fragile refractory was substantially complete and it soon became obvious that it had survived intact because the bulk of it was in a pit. The brown soil of Layer 3 permeated throughout the pit fill, but the concentration of refractory declined in Layers 4 and 5 which were otherwise similar. The edges of the pit dipped down sharply away from the edge of the well to a depth

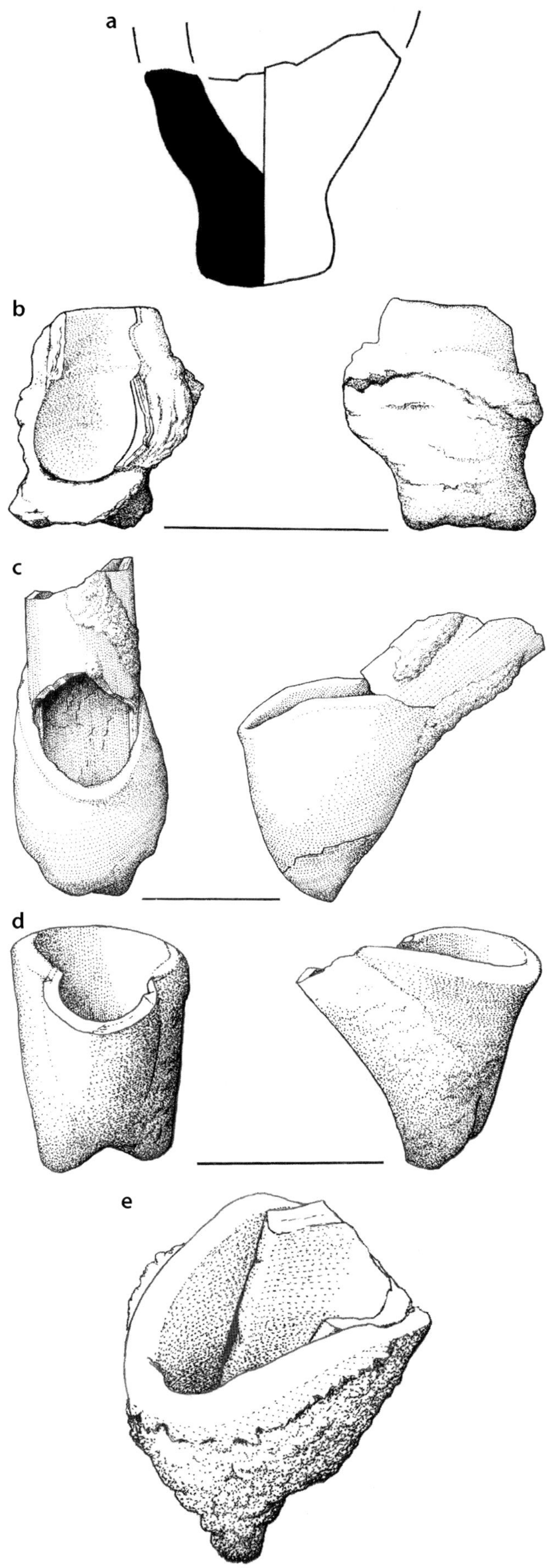

Fig. 6.72 Agucha, well section: selection of complete cupels and fragments. (c) Max. dimension: 8 cm; (d) 5 cm tall; (e) max. dimension across: 5 cm.

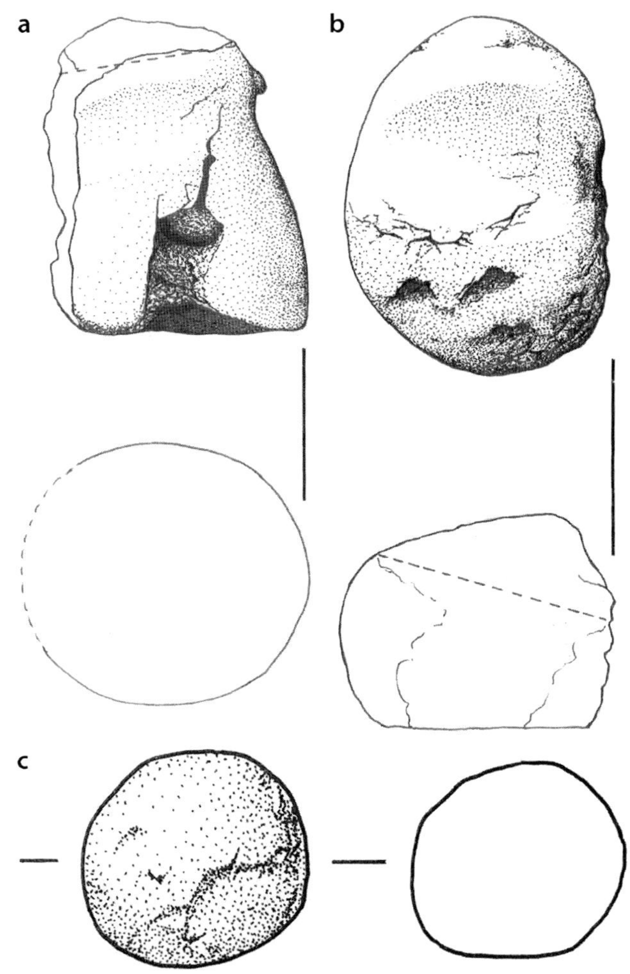

Fig. 6.73 Agucha, well section: selection of tuyere stands. (a) 5 cm tall; (b): width 4 cm; (c): diameter: 2.5 cm (cf. Fig. 6.53).

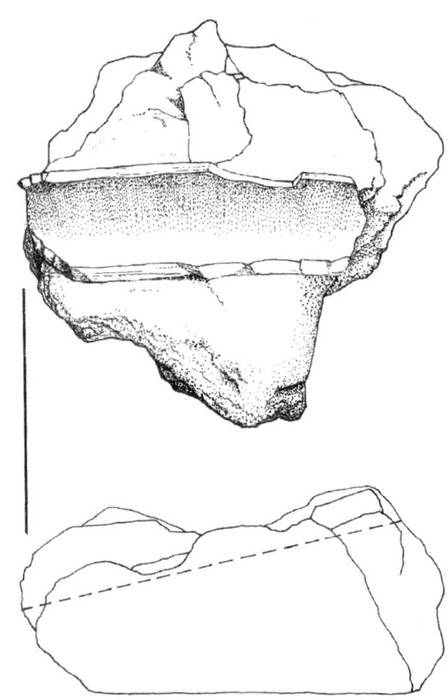

Fig. 6.74 Agucha, well section, 'tuyere area' L.4: fragment of a thin-walled tuyere set in clay with a flat underside, probably associated with the cupellation hearths (surviving length along the tuyere tube: 7 cm) (cf. Fig. 6.70).

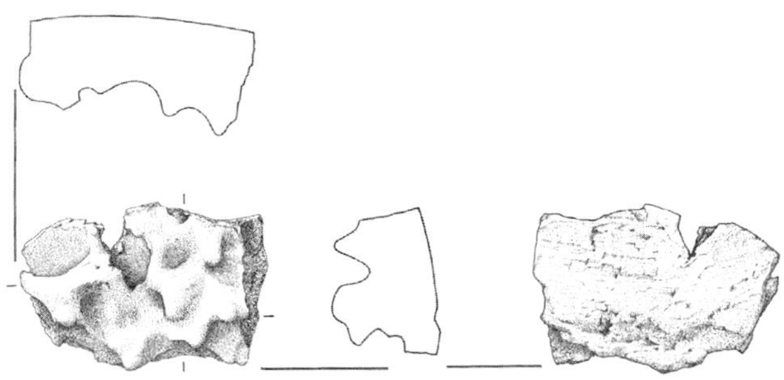

Fig. 6.75 Agucha, well section: fragments of clay massively attacked by lead oxide to form the drips of molten lead silicate. This was probably part of the walls of a hood over the cupellation hearth (max. dimension: 5 cm) (cf. Fig. 6.70).

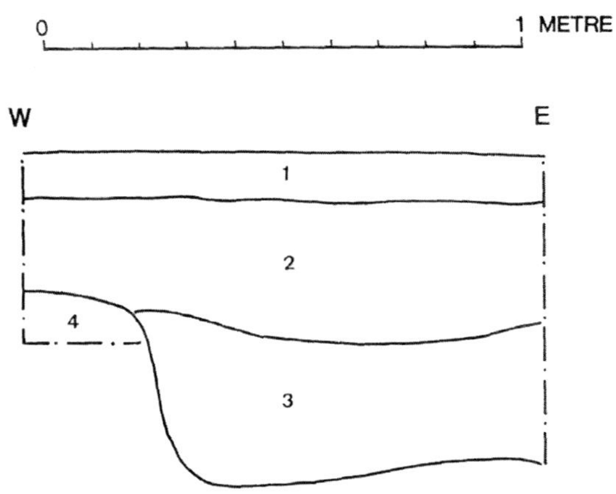

Fig. 6.76 Agucha, well section, 'lead area': north section.

of about 50 cm, and probably only a very little of the pit edge was cut by the well. On the south side the pit had cut through a compacted yellow/brown soil (Layer 6). Immediately below was the hard red surface of clay (Layer 7) noted initially in the well section. The pit was principally cut down into a layer of brown soil that contained occasional flecks of charcoal, but was apparently free of metallurgical debris (Layer 8). This soil was continuous with Layer 4 in the 'lead area' (see below), similar and probably analogous to Layer 13 at Site 1, approximately 70 m away.

Lead area

An area approximately 1.3 m square was opened on the north side of the well (Figs 6.68 and 6.76). Below the upcast from the well (Layer 1) was the present land surface and just beneath that the ground was full of substantial pieces of refractory, many of which were very heavily glazed (Fig. 6.75), together with large pieces of charcoal (Layers 2 and 3) (see the appendix). This was observed to be the upper part of a pit which cut through the hard surface of red clay (Layer 7) noted in the 'tuyere area' and elsewhere in the well sides. In this area the hard red clay only occurred on the south side of the pit, adjacent to the well at the north of the area. Where the section was drawn, the pit cut directly into the pale brown occupation soil (Layer 4) noted at the base of the 'tuyere area' (Layer 8), and very likely similar to Layer 13, Site 1. Once again the fill of the pit, although similar throughout, contained markedly less refractory material in the lower regions.

Appendix: List of samples taken

Survey of surface sites at Zawar (Figs 6.2 and 6.3)

The samples here were taken in the field survey from sites not subsequently investigated further. All are from the 1983 survey unless stated otherwise.

- Site 1: *Slag*: 21905.
- Site 4: *Slag*: 21910-12.
- Site 6: *Slag*: 21957 and 58; *Refractory*: Furnace lining: 21959 and 21960.
- Site 8: *Slag*: 24131.
- Site 9: *Slag*: 21917.
- Site 10: *Slag*: 24133.
- Site 11: *Slag*: 21916.
- Site 12: Condenser mouth with remains of a tube (Fig. 6.16).
- Site 13: 1983: *Slag*: 21918-23. 1985: *Slag*: 26067.
- Site 15: *Slag*: 21928 and 30.
- Site 16: *Slag:* 21949 and 50, 52, 54 and 55; *Refractory*: Furnace lining: 21953.
- Site 17: *Slag*: 21932-34.
- Site 18: *Slag*: 21935-38.
- Site 19: *Slag*: 21939-42.
- Site 31: *Slag*: 21947 and 48, 5048-51; *Refractory*: Furnace lining: 21952; *Mineral*: Dark red material (principally iron oxide): 21951.
- Site 32: *Refractory*: Condenser head from large retort: 5048-52; Perforated plate samples: 27326, 24129 and 30.

Samples collected from Zawar, Site 30

1983: Unstratified surface material

- *Refractory*: Complete retorts: 5048-33 and 34 (Figs 6.14 and 6.15 and Plate 6.15).
- Condenser: 5048-30.
- Perforated plate pieces: 5048-31, -35 and -36.
- Pillar: 5048-37 (Fig. 6.13a and Plate 6.13).
- Fragment of furnace wall from top of furnace: 5048-38.

1983: Excavation samples

- *Refractory*: Furnace wall and lining: S.2, 5048-13; S.3, -14; S.7, -15; S.8, -16; S.18, -22; S.19, -23; S.20, -24; S.21, -25; S.22, -26; S.23a, -27; S.23, -28; S.25, -29.
- Furnace floor with zinc: 52582.
- Retort fragments: S.9, 5048-17; S.10, -18; S.11, -19.
- Perforated plate fragments; S.12, 5048-12; S.13, -20.
- Retort contents: S.15, 5048-21 and -32.
- Lamp: Fig. 6.8.

Samples collected from Zawar, Site 37

These are just the samples retained for study at the British Museum from the 1983 excavations. The main sample collection is in Baroda. The positions of the sample numbers, suffixed S. are included in the archive.

Excavated samples

- *Refractory*: Furnace lining: S.1, 23987; S.6, 5048-42; S.7, -43; S.8. -44.
- Retort samples: S2. 5048-39; S.3, -40; S.4, -41.
- Perforated plate samples: S.5, 23986.

Surface samples

- *Refractory*: Perforated plate: 24503, 27324 and 5048-48.

Samples collected from Zawar, Site 33

1983: Surface collection

- *Refractory*: Retort samples: 23865-67 and Sample 28, 5048-46.
- Furnace wall: 23863.
- Perforated plate: 23864 and 5048-6.
- Retort spacer: 5048-11.
- Retort contents: 5048-6 and -9.

1985: Excavated material

- *Refractory*: Intermediate retorts: 27333, 27355, 27356, 27357.
- Condenser tube: (Fig. 6.29): 27358.
- Perforated plate (Fig. 6.27): 27367, 27325, 27360.
- Clay pillar (Fig. 6.26): 27365.
- Furnace lining: 27366S, 27364.
- Clay retort separator: 5048-6 (Fig. 6.22).

Samples collected from Zawar, Site 34 (all surface collections)

- *Refractory*: Complete intermediate process retort: 5048-7 (Fig. 6.28 and Plate 6.30).
- Retort contents: 5048-8 and -9.
- Other intermediate process retorts: half-sectioned retort (see Chapter 11, Plate 11.3) 5048-49 (collected by S.W.K. Morgan, donated by Bristol University). Retort with intact mouth (see Chapter 11, Plate 11.5) 5048-50.

Samples collected from Zawar, Site 36

- *Refractory*: Retort (Fig. 6.30 and Plate 6.31a and b), 37920S: fragment preserving approximately a third of the circumference of the retort, and although fired, the shape was not distorted. When complete this would have had an internal diameter of about 5.5 cm.
- Two retort ends fused together and heavily vitrified: 37921.
- Two retort heads: 27316, 27317 and 2718.
- Condenser mouths: (Fig. 6.31 and Plate 6.32): 27135 and 27314, 37928-32. Handmade. Rather irregular but outer diameter usually approximately 5.5 cm; one example had a minimum diameter of approximately 4.5 cm.
- Perforated plates (Fig. 6.32a and b): 27320, 37922, -23, -24, -25, -26. Basically similar to those from other sites but handmade. There are apparently two thicknesses, represented by four pieces that are about 4–5 cm thick and associated with evidence of reduced firing, and two pieces that are only about 3 cm thick associated with evidence of the usual oxidising firing condition. It is possible that the two thinner pieces are flakes of the thicker plates.

Samples collected from Kaya

- *Refractory:* Fragmentary retorts: 37918, 37917.
- Condenser head: 37916.
- *Other materials:* Charcoal from the present-day charcoal burning operations. 37919. Identified as neem (*Azadirachta indica*).

Samples collected from Zawar, Site 2

- *Slags*: Mostly glassy, some vesicular. 1983 surface: 21906-08. 1985: Layer 1. 26066.
- *Refractory*: Furnace lining: 27389-91 and -94, all heavily tempered with plant material. The pieces varied in thickness between 1 and 3 cm (Fig. 6.35) and had clearly been applied to a flat surface, suggesting they were the lining of a stone or possibly brick furnace (although no brick fragments were found on the site). The clay is generally in an oxidised condition and not heavily vitrified, although slagged on the exposed face. No estimate could be made of the internal dimension of the furnace of which they formed a part. One piece of furnace lining had the remains of a circular hole of approximately 4.5 cm diameter through it at an angle of about 45° to accommodate a tuyere (Fig. 6.37) of similar diameter to those of the tuyere fragments recovered from the site.
- *Tuyere*: A number of fragments of tuyere were recovered (Fig. 6.36 and Plate 6.37a and b) and one, 27392, was retained for study. The pieces were of hard, fine red ceramic and only one piece showed any trace of vitrification. The inside diameter is estimated as being about 3–4 cm, and the thickness of the pieces recovered is about 0.8–1.0 cm. Similar pieces were found on Zawar Site 14, Dariba Site 10 (Figs 6.57 and 6.58), and at the Agucha well site (Fig. 6.74), which are all of approximately the same age.[4]
- *Ceramic:* Small sherd of domestic pottery: 27387, for comparison with the refractories.
- *Mineral:* Two pieces of dense tabular mineral, both rich in zinc and lead: 27388, three conjoined pieces, 9 × 5 cm and 0.5 cm thick, weight 80 g; 27394, 4 × 3 cm, and 0.9 cm thick, weight 55 g (see Chapter 7, Plates 7.4–7.6, and Chapter 9).
- Red iron-rich mineral, 27393, vein stuff or possibly flux.

Samples collected from Zawar, Site 5

Silver/lead smelting debris: 1983 surface collection

- *Slags*: 21899-904 and 21961, 24502.

Zinc smelting debris: 1983 surface collection

- *Refractory:* Small retorts, complete: 5048-2, -3 (Fig. 6.15); Fragments: 23856 and 58, 24502.

- Perforated plate fragment: 23857 (Fig. 6.39a).
- Pedestal: 5048-4.
- Furnace brick: 5048-5.
- Small retort furnace lining fragments: 23860-62.
- Retort filling: 23859.

1985: Surface collection

- *Refractory*: Small retort fragments: 27334-36. Small retort process perforated plate fragment: 5048-1. Large retort fragments: 27321, -28-30, -32.

1985: Excavation collection

- *Slags*: L.3: 26085-87. Survey: a-26088, b-89, c-90, d-91, e -92, f-93, g-94, h-95, i-96, j-97.

Zinc smelting debris

- *Refractory*: Fragment of condenser head: 27352 (see Chapter 7).

Silver/lead smelting debris

- *Refractory*: Complete tuyere: 27339 (Fig. 6.41a and b, Plate 6.40a and b). Fragments: 27337, 27338.
- Flat fragments of burnt tile: 27342, 27343, 27345 and 27346-8.
- Fragments of coil-built furnace: 27341, 27344 (Figs 6.42 and 6.43).
- *Partially burnt minerals and metal*:[5] Molten slagged galena, 27349 and 50. Burnt galena, 27353. Droplets of argentiferous lead metal 27351.

Samples collected from Zawar, Site 7

- *Slag*: 1983: 24132. 1985: 26071 and 26082.
- *Refractory:* Coiled furnace lining (Fig. 6.46 and Plates 6.42–6.44), 27377, 27377, 27385-6. The clay coils are about 2.5–3 cm thick, and were originally plastered with clay on the outside and inside. Samples 27385 and 27386 (see Chapter 7, Plate 7.7) retain their original internal clay lining. The fingermarks on the external plaster show that this was actually the outer face of the furnace wall and not just a lining. Thus at this point the wall was quite thin, no more than 4 cm thick, although from its slagged and vitrified condition the fragment must have lain close to the reaction zone. From the curvature of the surviving fragments the internal diameter of the furnaces is estimated to have been between 40 and 55 cm.
- Furnace base of slagged red clay: 27384.
- One base sherd of a very roughly made vessel: 27395.

Samples collected from Zawar, Site 14

- *Slag*: 1983: Surface: 21924, -25 and -26.
- 1985 Surface: 26078. L.2: 26081. L.3 26080 L.3/4 26078 -79.
- *Refractory*: Flat pieces of heavily plant-tempered furnace lining, L.3: 27371-75, 276376, 27278-79. Samples 27372 and 27376 both have remains of a circular hole, of about 1 cm diameter, in their fractured edges. These do not seem to be contemporary with the use of the furnace as they go through the slag and vitrification, but could be pick marks from the demolition of the furnace. The usual thickness of the refractory lining where it is flat is between 1 and 2 cm.
- Large lump of totally vitrified ceramic, L.3: 27375.
- *Mineral*: L.3: dense white burnt material with layered structure, 27382, 4.8 × 4. 0 cm, and 0.7–0.8 cm thick, weight 41.3 g, similar to 27388 and 27394, Site 2 (see Chapter 7).
- Lighter, white rock, probably also calcined, 27380.

Samples collected from Zawar, Site 29

Zinc smelting

- *Refractory*: Surface, near excavation: clay wedge spacers from large retort process, 2736-62 (Fig. 6.22.).[6] Retort fragments: 27363-69.
- Furnace wall 27370, L.3, but intrusive from nearby zinc smelting operations, for comparison with furnace lining 27368, and found in the same layer.

Silver/lead production

- *Slag*: 1983: Surface: 21943-47. 1985: L.1 26069 and -76.
- *Refractory*: Slagged flat furnace lining, L.3: 27368.

Samples collected from Dariba, Site 7

- *Slags*: 26037, 26049-50, 32272-77.
- *Refractory*: Lining: 27286, 33948-50, numerous pieces of plant-tempered clay that are only lightly burnt and little slagged. These pieces are typically between 1.5 and 3 cm thick, and have slightly curved profiles in all directions. The concave faces have been exposed to the heat, which suggests they are the clay linings from either hemispherical bowl furnaces or hearths with a diameter of about 30 cm.
- 27311-12 Two small, more heavily vitrified fragments of ceramic.
- Utilised pottery sherd: 27294, from the surface adjacent to the heap; a sherd that has become heavily lead glazed with evidence of the glazing running over the breaks in one place, showing that it was reused as a sherd probably in the cupellation process (see Chapter 10).
- *Mineral*: 26100.

Samples collected from Dariba, Site 8

- *Slag*: 26038 and 26039, 26051 and 26052.

Samples collected from Dariba, Site 9

- *Slags:* 26040 and 32278,-79, silver/lead smelting; L.4 26041, 32280 -81, iron smelting (see Chapter 7, Tables 7.9 and 7.15).
- *Refractory* (all from Layer 4): Curved clay furnace linings either from a domed bowl furnace or hearth. The fabrics

are heavily tempered with plant material or dung, slagged on the inside but not massively burnt: 27274, approximately 1.5 cm thick, and the curvature would suggest an internal diameter of approximately 40–45 cm.[7] 27276: fragment of large crucible or curved hearth lining of 25 cm diameter. Rough, heavily plant/dung tempered outer layer and an inner layer of fine clay. There is a distinct slag/vitrification rim on the inside which SEM examination showed to be largely composed of a fayalite slag containing iron-rich copper, suggesting that either copper smelting or refining had taken place within it (see Chapter 7, Plate 7.17).
- Pottery sherds with a lead glaze, probably as a result of incorporation in a smelting hearth: 27294-27302. 27298 is a rim sherd where the vitrified glaze is running away from the edge. That is, the rim was uppermost during the operation and the ceramic must have formed a hearth lining rather than inverted as a bowl furnace. 27295 is totally glazed all over including the breaks (Fig. 6.54 and Plate 6.55), showing it was used in the process as a sherd, probably in a cupellation process (see Chapter 7).
- Fragment of tuyere stand 27275 (Fig. 6.53): this is a piece of coarse, roughly shaped clay with many ill-sorted inclusions, some quite large. It has the curved impression of where a tuyere was once positioned at an angle of about 45°, and is a more or less intact rough clay stand to hold the tuyere at the correct angle into a smelting hearth. Similar stands from Agucha (Fig. 6.73 and Plate 6.69) are associated with cupellation debris; this example could have directed the blast either into a cupellation hollow or a smelting hearth. The tuyere would have had an external diameter of about 5–5.5 cm and the thin-walled tuyeres recovered from Site 10, 27253 (Fig. 6.57) or 27249 (Fig. 6.58a), for example, would have fitted in perfectly.

Samples collected from Dariba, Site 10

- *Slags*: L.2: 26053, 32282-83. L.10: 26054, 32284-85. L.55: 26055, 32286-87. L.58: 26056, 32288-89.
- *Metals*: Copper stains/fragments, L.23: 27245 and 27247 (fragment of sheet), L.45: 27246 (several small fragments), all identified as being of bronze.

Refractories: crucibles

- Complete crucible + contents, L.45: 26337 (Fig. 6.59b and Plate 6.59).
- Crucible, L.15: 26336.
- 16 crucible fragments of similar dimensions and fabric (26320-35).
- Base of large crucible 26319.
- Large crucible, L.23: 26338.
- Small crucible L.15: 26339 (Fig. 6.59c).
- Various crucible fragments, L.23: 26340-44.
- Crucible rims, L.23: 276214-22.
- Crucible rims, L.45: 27237-39.
- Large crucible base and side, L.23: 26318.
- Crucible lids, L.23 (Fig. 6.60): 27235 and 27236.
- Decorated sherd cut down to form a lid.

Tuyeres

- Complete tuyere end, L.23: 27249, oval, internal dimensions 3 × 2 cm. Wall 6 mm thick, but varies (Fig. 6.57 and cf. Fig. 6.36).
- Fragments of large tuyere, L.43: 27251. 5.5 cm internal diameter, 7.5 cm external. Wall thickness typically about 1.5 cm, but varies (Fig. 6.58b).
- Long fragment of small diameter tuyere, L.23: 27253. Internal diameter 3.0 cm, wall thickness 5 mm (Fig. 6.58a).
- Small fragments same body as the tuyeres, but apparently flat, L.23: 27250.

Furnace lining

- L.23. Large friable fragment: 27229 (see Chapter 7). Possibly relined as there is vitrification in the centre.
- Heavily burnt ceramic, furnace wall: 27252. Some copper staining.
- Brick fragment, L.23: 27226. Possibly from domestic use.
- Two fragments of burnt brick (?), L.10: 27228 and 27227. Very heavily tempered with plant material compared to 27226.
- Fragment of furnace wall, L.64: 27268. Rather unburnt.

Heavily burnt rims of large vessels/tiles

- L.23. Large, thick, wedge-shaped piece, but with some curvature, heavily burnt: 27230.
- Very large, rough, clay rim (?): 27232. Outer diameter 45 cm, some burning and glazing.
- Large rim or tile fragment: 27223. Inner diameter 40 cm, intermediate between 27230 and 27224, burnt and vitrified.
- Rim sherds from large vessels with similar burning/glazing to the above: 27225, 27224 and 27231.
- This group all seem to be rim fragments of large bowls or open dishes that had been exposed to high temperatures.

Clay ingot moulds (?)

- L.23. Corner fragment: 27233. 5 cm thick, 9 × 5.5 cm, containing about half of one hemispherical depression, 1.5 cm deep × 4 cm diameter. It is of very rough, clay, heavily tempered with plant material, and seems to have the impression of a potsherd on one side. It was made by pressing the wet clay into a rectangular mould. The clay is burnt and blackened, and copper and tin were detected only in the depression. There are slight remains of an apparent linking channel near the base of the depression, and the fragment could be the remains of a single valve ingot mould for casting small hemispherical bronze ingots (Fig. 6.61 and Plate 6.60).
- Ingot mould: 27234. This is more poorly preserved. It has a vitrified base, suggesting it was either heated from beneath or possibly reused in a furnace wall. 5 cm thick. The depression is about 2.5 cm deep × 4.5 cm diameter, with an apparent channel running into it. Clay heavily tempered with plant material, macroscopic quartz etc. (Fig. 6.62).

- Possible fragment of ingot mould: 27241. This piece of burnt clay has a shaped hemispherical depression in one side of about the same dimensions as those in 27233 and 27234.

Miscellaneous samples

- Heavily burnt and vitrified clays, containing copper, L.23: 27242 (possibly crucible slag), 27252-54; L.25: 27240. 27255 was a roll of clay about 10 cm in length and about 4 cm diameter, of which about half survives split along its length. One end is flattened and it apparently stood as a small pillar. The clay contains much grit and the surviving surface contains many particles of crushed white quartz as if the wet clay had been rolled in it. The surface is now vitrified.
- Mortar and ash samples: 27266.
- Crucible contents: 27267.
- Utilised potsherd from neck of vessel, exposed to heat on convex side, L.23: 27248.
- *Ceramic*: Potsherds for comparison with refractories, L.23: 27256-60.
- *Mineral:* L.45: 27243 (too light to be ore).
- L.23: 27 244, gypsum.

Samples collected from Dariba, Sites 11 and 12

- Site 12: Slag 26042, 32300-01.

Samples collected from Dariba, Site 13

- *Refractory:* Furnace wall and lining: 27277-79, various small fragments of red clay coil-built furnace wall, heavily tempered with plant material/dung. The coils are about 2.5 cm diameter, from a small cylindrical furnace approximately 25–30 cm in diameter. 27278: a fragment of the rim of a furnace, apparently little heat distorted, diameter difficult to estimate. 27277 is piece of coil-built furnace wall.
- *Tuyere*: 5046-1-Y with one small piece of a tuyere collar attached, 15 cm outer diameter.
- *Ceramic:* Potsherds: 27308-10.

Samples collected from Dariba, Site 14

- *Slag*: 26057Y, 32290T-1R. 28499Y, 32292P (see Chapter 7, Table 7. 15).
- *Refractory:* Furnace wall: 27270-72, 33947, lumps of coil-built furnace wall heavily tempered with plant material or dung. The curves appear to be just in the plane of the coils, suggesting a small cylindrical shaft furnace. The total thickness of the walls is about 4 cm with the individual coils being about 2.5 cm in diameter. The diameter of the furnace would have been very small, about 20–25 cm. One piece coated with clay on the outside. 27280-82 are very similar pieces. 27269 is a fragment of a similar clay coil, but the flat rear surface suggests it was a lining against a wall of brick or stone. Lining about 1.5 cm thick, and burnt but not slagged, plant material/dung temper.
- *Tuyeres:* 27283, complete tuyere tube about 6.5 cm diameter, internal diameter 1.5 cm, surviving length 3 cm. Inclined at an angle of about 45–40° to the furnace wall (Plate 6.63c). Slagged but only lightly vitrified. Heavily tempered with plant material/dung. 27284 comprises numerous tuyere fragments. 27285 is a complete tuyere with about one-third of the outer collar recently broken off in three joining fragments (Fig. 6.65). The tuyere tube is about 6.8–7 cm external diameter, the inner diameter is about 1.5 cm, and the surviving length is approximately 4.5 cm. As with 27283, it is lightly vitrified and slagged. Tuyere collar, about 2–2.5 cm thick. It is inclined at an angle of 45° to the furnace wall. Several essentially complete tuyere/lining assemblies were found (see, for example, Fig. 6.65). Examination of some of the furnace fragments suggests that they were built up from coils, rather than the slabs as recorded in the 19th century at Khetri (see Chapter 2, Fig 2.4). Successive coils have been roughly smoothed together and then a clay lining applied to the inner surface. The morphology of the tuyeres indicates that they were positioned in the furnace wall so that the air blast entered the furnace at an angle of about 45°, blowing down onto the charge in the heart of the furnace. A similar method for the construction of furnace walls was in use at Zawar at this period.

Samples collected from Dariba, Sites 1–6, 15 and 16

- Site 1: *Slags*: 26033 and 26044.
- Site 2: *Slags*: 26084 and 26045.
- Site 3: *Slags*: 26035.
- Site 4: *Slags*: 26036.
- Site 5: *Slags*: 26047.
- Site 6: *Slags*: 26048.
- Site 15: *Slags*: 32296 and 97, 32293-5 (see Chapter 7, Table 7.15). *Refractory*: Several pieces of heavily slagged and vitrified furnace wall with no recognisable features: 33951.
- Site 16: *Slags*: 32298 and 32299 (see Chapter 7, Table 7.15).

Samples collected from Agucha, Site 1

- *Slags*: Surface 26061V, 32076-77; L.3 26058, 32070-71; L. 4 26059, 32072-73; L.5 26060, 32074 (see Chapter 7, Plate 7.26), 32075 (see Chapter 7, Table 7.17).
- *Ceramic*: Small body sherds, L.3: 26268 and 26269.

Samples collected from the surface of the Agucha well site

- *Slags:* 26062 and 32078.

Samples collected from the Agucha well site, 'tuyere area'

- *Slags:* L. 4: 26063, 32080 and -81 (see Chapter 7, Plate 7.27 and Table 7.17).
- *Refractories:* Cupels, L. 4: a complete cupel, capacity 20 ml (Fig. 6.72d and Plate 6.68), 26306.
- A cupel base: 26305.
- Six cupel fragments, L. 3-5: 26252 to -57 (Fig. 6.72a–e).

- Pedestals or tuyere supports, L.3-5: 26273, 26274, 5045-1 (Fig. 6.73a–c and Plate 6.69).
- Various tuyere/cupel fragments, L.3: 26279-87, -26288 (see Chapter 7, Plates 7.28–7.30).
- Fragment of thin-walled tuyere set in clay with flat bottom (Plate 6.70 and Fig. 6.74).
- Furnace wall and linings, L.3-5: 25988, 26245-50. Various fragments of heavily glazed refractory: 26271 and 26272 (Fig. 6.75 and Plate 6.71) (see also Chapter 7, Plate 7.31). A flat-faced fragment: 26275.
- *Ceramic:* Body sherds, L.3-5: 26251T, 33957-59.
- *Mineral*: Small piece of cerussite, L.3: 26277 (see Chapter 7, Table 7.19).

Samples collected from the Agucha well site, 'lead area'

- *Slags:* L.2: 26064, 32082-83. L.4: 26065 and 32084.
- *Metal and litharge:* L.3 and 5: 26278, 43612-15P, spills and droplets of lead total weight 109 g (Plate 6.66).
- Litharge: L.3: (33960), segment from a plano-convex cake of litharge, approximately 6 cm diameter and 3 cm thick at the centre, and weighing 80 g.[8]
- *Refractory*: Cupels: L.2 and 3: 26300-03. All fragmentary.
- Cupels: L.3: 26304, with tuyere in place. Capacity 30 ml (Fig. 6.72c and Plate 6.72).
- L.2: 265989-99.
- Furnace lining: L.2: 26258.

Notes

1. See Chapter 7 (p. 149) for analysis.
2. See Chapter 7 (p. 145) for a petrological description.
3. See Chapter 7, Table 7.13 for analyses.
4. See Chapter 9 (p. 189) for the probable method of manufacture and more on tuyeres generally.
5. See Chapter 7, Table 7.11 for analyses.
6. See Chapter 6, Fig. 6.23 (p. 100), Site 33 for their use.
7. Compare with Zawar, Sites 2 and 14 (pp. 108 and 113).
8. For analysis see Chapter 7, Table 7.19.

Chapter 7 The scientific examination of industrial materials from Zawar, Dariba and Agucha

A.P. Middleton, C.R. Cartwright, P.T. Craddock, K. Eckstein, I.C. Freestone, D.R. Hook and N.D. Meeks

General introduction

This chapter describes the scientific investigation of the industrial materials recovered from the excavations at Zawar, Dariba and Agucha. The scientific methods of investigation are outlined in Chapter 1 and the broader metallurgical interpretations are presented in Chapters 9, 10 and 11. The materials studied here fall into a number of categories including:

- *Ceramics:* the zinc, lead and copper smelting installations from Zawar and Dariba, the large, small and intermediate zinc smelting retorts from Zawar, and the cupellation material from Agucha, together with their tuyeres. The report also includes refractories from the metalworking area at Site 10 and pottery sherds (some apparently reused in metallurgical processes) excavated at Dariba.
- *Fillers:* the residues in the zinc retorts at Zawar.
- *Slags:* the lead or copper smelting slags found at all three sites.
- *Ashes:* the zinc ore roasting debris from Zawar, some of which was used as a mortar in the walls built at Old Zawar.[1]
- *Zinc-rich crusts:* plate-like deposits rich in zinc associated with earlier lead smelting sites in Zawar, some adhering to ceramic materials as a fume.
- *Metals and minerals:* zinc metal from Zawar, lead from Agucha, lead minerals from Zawar and Agucha.

Some preliminary scientific reports were published while the investigations were still in progress (Craddock *et al.* 1985, 1989, 1998a and 1998b, 2013; Craddock and Gurjar 1991; Craddock and Eckstein 2003; Eckstein *et al.* 1996; Freestone 1989; Freestone *et al.* 1985a and 1985b, 1991). While the actual data are reliable, some of the interpretations may have changed as a result of the ongoing investigations, discussed below. Lead isotope work on early Indian zinc coins is reported in Chapter 13 (Fig. 13.1) and Craddock *et al.* (2015), and on some early silver punch-marked coins in Chapter 2 (Fig. 2.7).

Industrial materials from Zawar

Introduction

The scale and diversity of the metallurgical remains at Zawar dictated a problem-oriented approach to the analysis of the residue materials. Throughout the programme of analysis, specific issues served as a focus, with a view to determining:

- The nature of the raw materials used
- Preparation processes
- Temperatures and durations of the metallurgical processes
- The nature of the products
- Chronological and locational variations in process
- Recycling of materials

At Zawar the process debris indicate that very different operations were being carried out in the early, Mauryan and post-Mauryan periods and in the later, medieval and post-medieval periods. The early sites are represented by slag heaps; the later sites by heaps of zinc smelting debris as well as by slag heaps. The slags fall into two groups – zinc-rich (Type I) and iron-rich (Type II) (Tables 7.1 and 7.2) – which occur separately and may be distinguished in the field by their appearance.

The zinc-rich slags occur in the more peripheral smelting areas away from the central Old Zawar retort dumps and, where dated, belong to the early centuries AD. They are found with the distinctive encrusted furnace linings and the plates of solidified lead/zinc fumes. The iron-rich slags occur principally in and around the central smelting area just below Old Zawar, designated Site 5, dated to the 17th–18th century AD, and often contain or are associated with galena. All the Zawar slags are homogenous, showing that the slag was fully molten in the furnace, unlike the slags from early Dariba and Agucha (see below).

Table 7.1 Composition of Zawar Type I and II slags and their petrography.

BMRL no.	Site	Type	SiO_2	Al_2O_3	FeO	MnO	MgO	CaO	BaO	Na_2O	K_2O	PbO	ZnO	Cu	Ag	S	Method	Petrography
21905	1	I	45.1	7.1	15.2	0.5	3.6	12.7	ND	0.5	2.0	3.8	9.6	0.0	ND	ND	AAS	gl (xtl)
26083	1	I	41.2	6.72	12.9	0.5	4.9	14.7	0.2	1.6	2.0	5.3	9.9	0.1	0.002	ND	ICP	ND
32099	1	I	43.2	5.49	11.2	0.5	4.5	14.1	0.2	1.3	2.3	6.0	11.3	0.0	0.001	ND	ICP	ND
21906	2	I	45.6	5.12	11.8	0.8	6.7	17.8	ND	0.4	1.6	3.4	6.8	0.0	ND	ND	AAS	cpx, gl (gal, wu)
21907	2	I	46.7	5.23	11.6	0.8	6.6	16.6	ND	0.3	1.7	3.4	7.1	0.0	ND	ND	AAS	cpx, gl (gal, wu)
21908	2	I	40.5	5.71	12.2	0.8	6.8	18.4	ND	2.5	1.9	3.6	7.5	0.0	ND	ND	AAS	cpx, gl
26072	3	I	38.4	5.81	8.5	0.6	3.6	8.4	0.1	0.4	1.6	9.1	23.6	0.0	0.002	ND	ICP	ND
32093	3	I	35.9	3.71	10.8	0.5	3.5	11.2	0.1	0.3	1.2	12.8	20.0	0.0	0.001	ND	ICP	ND
21909	4	I	33.3	5.95	10.8	0.9	6.2	18.3	ND	2.4	2.1	8.1	12.0	0.0	ND	ND	AAS	gl (xtl)
21910	4	I	34.2	6.16	11.5	0.9	5.9	17.4	ND	1.2	2.5	8.1	12.2	0.0	ND	ND	AAS	gl (ol, mel, cpx)
21911	4	I	38.3	5.65	9.7	0.8	5.7	15.3	ND	0.5	1.8	9.4	12.8	0.0	ND	<0.2	AAS	ol, mel, gl (cpx)
21912	4	I	38.9	5.52	10.6	0.7	5.2	15.4	ND	0.4	1.8	9.9	11.7	0.0	ND	<0.2	AAS	cpx, gl (ol)
21914	4	I	38.9	5.56	9.9	0.8	5.6	16.4	ND	0.4	1.9	8.6	12.0	0.0	ND	ND	AAS	gl (cpx)
21957	6	I	44.8	6.88	16.1	1.0	3.0	12.5	ND	0.5	2.4	3.4	9.5	0.0	ND	ND	AAS	gl (cpx)
21958	6	I	50.6	9.33	6.8	0.4	3.7	12.4	ND	0.7	2.5	8.9	4.7	0.0	ND	ND	AAS	gl (xtl)
26071	7	I	48.8	9.24	8.9	0.4	2.9	8.4	0.1	0.8	2.6	13.2	4.7	0.0	0.000	ND	ICP	ND
26082	7	I	41.5	6.97	13.2	0.3	2.8	11.6	0.1	0.4	1.9	11.4	9.9	0.0	0.001	ND	ICP	ND
32097	7	I	45.3	5.43	12.8	0.3	3.1	11.4	0.1	0.4	2.0	8.1	11.1	0.0	0.000	ND	ICP	ND
32098	7	I	45.4	5.24	12.6	0.3	3.2	11.5	0.1	0.5	2.0	8.0	11.1	0.0	0.001	<0.2	ICP	ND
24131	8	I	42.1	6.97	13.1	0.6	2.8	9.3	0.0	0.4	1.9	11.9	10.9	0.0	0.001	ND	ICP	ND
26070	8	I	45.1	5.83	16.1	1.1	5.2	12.0	0.1	0.2	2.1	7.0	5.2	0.0	0.002	ND	ICP	ND
26077	8	I	61.9	15.54	8.5	0.3	1.9	3.1	0.1	0.9	2.9	4.5	0.5	0.0	0.000	ND	ICP	ND
32094	8	I	60.9	11.21	7.1	0.2	1.8	2.4	0.0	0.6	2.6	4.6	8.7	0.0	0.001	ND	ICP	ND
32095	8	I	41.8	5.91	22.7	0.5	2.6	9.2	0.1	0.5	1.6	5.6	9.4	0.0	0.000	ND	ICP	ND
32096	8	I	47.8	5.05	15.4	0.7	4.0	18.9	0.1	0.5	2.8	3.9	1.0	0.0	0.001	ND	ICP	ND
21917	9	I	44.9	8.01	10.9	0.5	3.1	11.1	ND	0.5	2.0	11.1	7.9	0.0	ND	ND	AAS	gl (xtl)
26068	10	I	41.8	8.16	8.3	0.3	3.2	8.6	0.1	0.4	2.0	18.4	8.7	0.0	0.001	ND	ICP	ND
32090	10	I	44.9	6.02	10.1	0.4	4.1	12.5	0.1	0.4	2.1	6.1	13.3	0.0	0.001	ND	ICP	ND
21915	11	I	43.5	7.48	14.8	1.0	2.5	9.2	ND	0.5	2.1	9.0	10.1	0.0	ND	0.3	AAS	gl (xtl)
21916	11	I	44.5	7.37	11.8	0.8	2.9	10.2	ND	0.5	2.1	10.7	9.1	0.0	ND	ND	AAS	gl (xtl)
26073	11	I	40.2	6.32	16.8	0.3	4.3	17.8	0.1	0.4	2.4	3.7	7.7	0.0	0.001	ND	ICP	ND
21918	13	I	46.8	4.18	14.8	0.6	5.8	15.4	ND	0.3	1.5	3.5	7.1	0.0	ND	0.4	AAS	xtls, gl (gal)
21919	13	I	38.2	4.04	17.0	0.7	5.5	14.5	ND	0.3	1.4	2.9	15.5	0.0	ND	0.5	AAS	ol, gl (mel)
21920	13	I	42.4	5.46	9.3	0.7	5.1	11.0	ND	0.4	1.9	7.2	16.6	0.0	ND	ND	AAS	gl
21921	13	I	42.9	4.69	11.2	0.8	6.7	16.5	ND	0.3	1.6	4.6	10.8	0.0	ND	0.2	AAS	cpx, gl (ol)
21922	13	I	39.5	4.46	11.7	0.8	7.4	17.3	ND	0.3	1.8	3.5	13.3	0.0	ND	ND	AAS	ol, mel, gl
21923	13	I	40.4	5.1	10.2	0.6	4.7	12.6	ND	0.4	1.8	7.2	17.0	0.1	ND	ND	AAS	xtls, gl

21924	14	I	41.4	4.1	10.6	0.7	7.6	17.1	ND	0.3	1.6	3.9	12.7	0.0	ND	ND	AAS	ol, gl
21925	14	I	43.3	4.36	11.5	0.7	6.7	14.5	ND	0.3	1.5	4.0	13.2	0.0	ND	ND	AAS	xtls, gl
21926	14	I	40.6	4.1	13.9	0.7	6.2	13.8	ND	0.3	1.4	6.0	13.0	0.0	ND	0.1	AAS	cpx, gl (ol)
21928	15	I	36.5	5.5	8.8	0.6	4.7	15.8	ND	0.3	1.7	8.5	17.7	0.0	ND	ND	AAS	mel, will, cpx, sp, (gl, gal)
21930	15	I	39.5	5.9	11.3	0.7	4.6	13.7	ND	0.4	1.7	6.8	15.4	0.0	ND	<0.2	AAS	gl
21931	15	I	34	4.61	10.6	0.6	4.6	12.3	ND	0.4	1.3	9.8	21.9	0.0	ND	ND	AAS	will, cpx, mel, sp (gal)
21947	31	I	42.3	5.16	13.6	0.7	6.2	14.7	ND	0.5	1.4	3.9	11.4	0.1	ND	ND	AAS	ol (gl)
21948	31	I	45.3	6.39	13.3	0.7	4.3	12.0	ND	0.4	1.6	4.7	11.2	0.0	ND	ND	AAS	ol, gl, (rel qutz)
21949	31	I	45.6	6.12	13.1	0.7	4.3	13.4	ND	0.4	1.6	3.9	10.9	0.0	ND	ND	AAS	ol, gl (gal, rel qutz)
21950	31	I	44.9	5.82	14.2	0.8	4.5	14.5	ND	0.3	1.7	3.6	9.8	0.0	ND	ND	AAS	ol, gl (gal)
21899	5	II	36	3.99	34.2	0.5	1.1	6.5	ND	0.4	1.3	11.4	4.6	0.0	ND	0.7	AAS	ol, cpx, gl, gal, wu
21900	5	II	38.5	5.31	37.6	0.6	1.4	6.5	ND	1.0	1.9	2.5	4.7	0.0	ND	1	AAS	ol, gl (gal, wu, CuXS, FeAs)
21901	5	II	46.8	6.73	20.8	0.6	2.6	14.8	ND	0.4	2.5	2.2	2.5	0.0	ND	0.4	AAS	cpx, gl
21902	5	II	40.1	5.9	26.1	0.6	2.8	14.0	ND	0.6	2.5	3.0	4.5	0.0	ND	0.7	AAS	ol, gl (gal)
21903	5	II	33.9	4.52	44.0	0.2	0.9	3.4	ND	0.9	1.5	8.3	2.4	0.0	ND	0.6	AAS	ol, gl (gal)
21904	5	II	41.3	5.08	24.3	0.7	2.3	11.8	ND	0.5	2.0	6.9	5.3	0.0	ND	ND	AAS	ol, gl (qutz, gal)
26087	5	II	45	4.68	34.2	0.3	1.6	5.7	0.1	0.2	1.6	3.9	2.9	0.0	0.001	ND	ICP	ND
26088	5	II	42.3	5.2	32.6	0.6	1.7	7.6	0.2	0.2	1.9	4.0	3.9	0.0	0.008	ND	ICP	ND
26089	5	II	38.3	3.4	39.2	0.7	1.3	5.2	0.1	0.2	1.2	5.1	5.5	0.0	0.002	ND	ICP	ND
26096	5	II	39	3.9	37.9	0.4	1.0	4.1	0.1	0.2	1.4	7.5	4.5	0.0	0.005	ND	ICP	ND
32100	5	II	39.1	4.92	41.5	0.4	1.0	3.5	0.1	0.9	2.3	4.7	1.5	0.0	0.003	1	ICP	ND
32101	5	II	40.5	3.25	36.9	0.4	2.2	9.0	1.3	0.2	1.4	1.8	3.0	0.0	0.001	ND	ICP	ND
32102	5	II	41.5	5.68	23.9	0.5	3.1	13.4	0.1	0.6	2.5	5.0	3.6	0.0	0.001	0.5	ICP	ND
32103	5	II	36.2	2.93	41.7	0.3	1.0	2.6	0.0	0.2	1.4	6.7	6.8	0.0	0.007	ND	ICP	ND
32104	5	II	42.8	3.43	36.5	0.2	1.7	5.3	0.0	0.3	1.5	4.7	3.6	0.0	0.001	ND	ICP	ND
32105	5	II	38.8	3.56	39.7	0.8	1.7	3.4	0.0	0.1	1.4	7.8	2.7	0.0	0.003	ND	ICP	ND
32106	5	II	41.1	3.55	31.8	0.6	1.5	8.3	0.1	0.4	1.6	3.0	8.1	0.0	0.002	1.8	ICP	ND
32107	5	II	37.8	3.65	30.2	0.8	1.7	11.4	0.1	0.1	1.9	5.6	6.9	0.0	0.002	ND	ICP	ND
21943	29	II	33.5	4.84	41.9	0.9	1.3	7.1	ND	0.3	2.2	4.0	4.0	0.0	ND	0.7	AAS	cpx, ol, Fe ox, gl (gal)
21944	29	II	40.9	6.27	30.8	1.5	1.3	7.3	ND	0.3	1.9	5.1	4.7	0.0	ND	ND	AAS	ol, gl (gal)
21945	29	II	32.3	5.2	41.0	0.9	1.5	8.5	ND	0.4	1.6	2.8	5.9	0.0	ND	0.9	AAS	cpx, ol, Fe ox, gl (gal)
21946	29	II	31	4.65	40.5	0.7	1.6	9.3	ND	0.3	1.7	7.2	3.3	0.0	ND	ND	AAS	cpx, ol, Fe ox, gl (gal)
26069	29	II	34.2	4.11	41.5	0.7	1.4	7.3	0.1	0.2	1.3	4.0	5.1	0.0	0.001	ND	ICP	ND
32091	29	II	38.7	5.1	37.8	0.6	1.2	3.5	0.1	0.2	2.7	4.8	5.3	0.0	0.002	ND	ICP	ND
32092	29	II	37.4	4.87	34.8	0.6	2.0	4.5	0.1	0.1	2.8	6.9	6.0	0.0	0.002	ND	ICP	ND
24129	32	II	35	3.59	39.4	0.4	2.0	5.2	0.0	0.2	1.3	4.8	8.3	0.0	0.002	ND	ICP	ND
24130	32	II	36.8	3.8	34.2	0.6	1.5	4.5	0.1	0.1	1.7	10.5	6.2	0.0	0.002	ND	ICP	ND

ND = not determined; S by large area SEM-EDXA; petrography by transmitted and reflected polarised light microscopy plus SEM-EDXA of selected samples

Key: ol = olivine; cpx = clinopyroxene; mel = melilite (hardystonite); gl = glass; Fe ox = iron oxide; wu = wurtzite; will = willemite; xtal = fine unidentified silicates (spherulitic); parentheses signify minor phase

Table 7.2 Composition of silicate phases in Zawar slags.

Phase	Slag no.	Method	Site	Slag type	SiO_2	TiO_2	Al_2O_3	FeOt	MnO	MgO	CaO	ZnO	PbO	Total
Clinopyroxene	21908	BMNH	2	I	51.7	0.2	3.2	8.5	0.3	14.0	23.0	2.8	<0.3	103.6
Clinopyroxene	21908	BMNH	2	I	52.5	<0.2	2.1	7.1	0.4	13.5	23.0	2.6	<0.3	101.2
Clinopyroxene	21908	BMNH	2	I	52.1	0.2	2.4	7.0	0.6	13.9	23.2	2.7	<0.3	101.9
Clinopyroxene	21911	BMNH	4	I	49.4	0.5	3.7	5.7	0.3	12.7	23.8	2.5	<0.3	98.7
Clinopyroxene	21911	BMNH	4	I	46.1	0.7	6.3	9.7	0.5	9.6	23.8	3.5	<0.3	100.2
Clinopyroxene	21911	BMNH	4	I	46.8	0.4	5.4	9.2	0.3	9.7	22.9	3.1	<0.3	97.8
Clinopyroxene	21910	BMB	4	I	43.0	0.5	6.2	10.4	0.3	10.0	22.2	4.0	<0.3	96.6
Clinopyroxene	21914	BMNH	4	I	47.3	0.2	6.1	9.6	0.6	10.3	22.7	3.6	<0.3	100.3
Clinopyroxene	21912	BMNH	4	I	47.3	0.8	6.2	9.4	0.3	10.7	23.6	3.6	<0.3	101.8
Clinopyroxene	21912	BMNH	4	I	45.9	0.5	6.3	10.8	0.4	9.8	23.2	3.9	<0.3	100.7
Clinopyroxene	21912	BMNH	4	I	49.4	0.3	4.9	8.0	0.5	11.6	24.1	4.0	<0.3	102.7
Clinopyroxene	21912	BMNH	4	I	47.4	0.2	6.1	10.1	0.3	10.7	23.4	4.3	<0.3	102.5
Clinopyroxene	21918	BMC	13	I	49.6	0.3	4.6	14.2	0.4	9.2	19.6	3.4	<0.3	101.3
Clinopyroxene	21931	BMNH	15	I	42.0	0.4	8.1	14.4	0.2	8.0	23.0	4.1	<0.3	100.0
Clinopyroxene	21931	BMNH	15	I	40.8	0.4	6.8	16.4	0.3	6.0	23.2	4.8	<0.3	98.7
Clinopyroxene	21931	BMNH	15	I	41.0	0.2	6.7	16.5	0.7	5.8	22.9	5.2	<0.3	99.1
Clinopyroxene	21928	BMB	15	I	42.5	0.7	8.7	13.9	0.6	7.4	21.8	3.4	<0.3	99.0
Clinopyroxene	21901	BMNH	5	II	49.1	0.4	1.7	15.0	0.4	9.0	21.3	0.9	<0.3	97.9
Clinopyroxene	32102	BMC	5	II	44.1	0.5	6.0	21.4	0.5	4.5	20.0	1.6	<0.3	98.5
Olivine	21911	BMNH	4	I	36.6	<0.2	<0.2	19.9	1.2	26.2	2.3	13.7	<0.3	99.9
Olivine	21911	BMNH	4	I	36.6	<0.2	<0.2	17.9	1.1	23.2	4.4	13.1	<0.3	96.2
Olivine	21911	BMNH	4	I	36.6	<0.2	<0.2	17.2	0.8	31.6	1.6	13.5	<0.3	101.2
Olivine	21912	BMNH	4	I	35.6	<0.2	0.7	25.3	1.3	24.2	1.4	12.9	<0.3	101.4
Olivine	21912	BMNH	4	I	36.6	<0.2	<0.2	19.3	1.1	31.2	1.8	12.5	<0.3	102.5
Olivine	21910	BMB	4	I	37.0	<0.2	<0.2	12.2	1.6	33.2	2.6	12.1	<0.3	98.5
Olivine	21921	BMA	13	I	31.6	<0.2	<0.2	19.2	1.0	26.8	1.3	19.5	<0.3	99.4
Olivine	21919	BMA	13	I	31.2	<0.2	<0.2	26.6	0.7	18.6	5.0	17.5	<0.3	99.6
Olivine	21919	BMC	13	I	33.5	<0.2	<0.2	28.7	1.1	18.7	5.7	12.7	<0.3	100.4
Olivine	21919	BMC	13	I	35.6	<0.2	<0.2	17.7	0.9	29.7	1.6	15.3	<0.3	100.7
Olivine	21919	BMC	13	I	34.3	<0.2	<0.2	26.6	1.2	20.8	6.2	12.5	<0.3	101.5
Olivine	21926	BMC	14	I	32.3	<0.2	<0.2	32.1	1.2	15.5	1.5	16.3	<0.3	98.9
Olivine	21899	BMA	5	II	29.1	<0.2	<0.2	58.3	0.7	2.6	1.3	7.5	<0.3	99.5
Olivine	21899	BMA	5	II	29.1	<0.2	<0.2	58.2	0.7	1.6	1.4	8.5	<0.3	99.5
Olivine	21902	BMA	5	II	28.7	<0.2	<0.2	64.7	0.5	1.8	0.3	3.0	<0.3	99.0
Olivine	21904	BMA	5	II	29.8	<0.2	<0.2	49.9	1.2	7.7	1.9	9.0	<0.3	99.5
Olivine	21900	BMNH	5	II	30.6	<0.2	<0.2	56.2	0.7	5.4	0.6	4.8	<0.3	98.2
Olivine	21900	BMNH	5	II	31.0	<0.2	0.3	57.0	0.6	4.1	0.7	5.3	<0.3	99.0
Olivine	21900	BMNH	5	II	30.6	<0.2	<0.2	56.5	0.7	4.8	0.7	4.6	<0.3	97.9
Olivine	32100	BMC	5	II	29.6	<0.2	<0.2	65.4	0.7	2.0	0.4	1.3	<0.3	99.4
Olivine	21943	BMA	29	II	28.8	<0.2	<0.2	58.6	1.2	4.0	1.6	5.4	<0.3	99.6
Willemite	21928	BMNH	15	I	31.3	<0.2	<0.2	7.1	0.5	10.9	0.5	49.1	<0.3	99.4
Willemite	21928	BMNH	15	I	32.1	<0.2	<0.2	7.2	0.4	11.4	1.1	47.4	<0.3	99.6
Willemite	21931	BMNH	15	I	31.1	<0.2	<0.2	6.1	0.2	11.7	<0.2	50.8	<0.3	99.9
Willemite	21931	BMNH	15	I	31.3	<0.2	<0.2	7.5	0.4	11.5	<0.2	49.4	<0.3	100.1
Willemite	21928	BMB	15	I	31.3	<0.2	<0.2	8.1	0.6	12.4	0.2	49.4	<0.3	102.1
Orthosilicate	21919	BMC	13	I	32.8	<0.2	0.2	24.6	0.8	8.3	22.3	10.7	<0.3	99.7
Orthosilicate	21919	BMA	13	I	32.8	<0.2	<0.3	21.8	1.0	8.0	23.5	14.7	<0.3	101.6
Orthosilicate	21919	BMC	13	I	32.8	<0.2	0.3	26.0	0.9	6.7	21.3	13.7	<0.3	101.7
Orthosilicate	21919	BMC	13	I	32.8	<0.2	0.3	24.0	0.9	16.0	13.2	10.9	<0.3	98.0
Hardystonite	21911	BMNH	4	I	32.8	<0.2	1.6	2.0	0.2	5.3	36.3	13.0	2.2	93.3
Hardystonite	21911	BMNH	4	I	32.8	<0.2	2.0	2.3	0.1	4.1	34.9	12.2	3.1	91.6
Hardystonite	21911	BMNH	4	I	32.8	<0.2	1.2	1.8	0.4	5.5	36.6	13.4	1.8	93.5
Hardystonite	21911	BMNH	4	I	32.8	<0.2	1.5	1.8	0.2	5.0	35.9	14.8	2.5	94.5
Hardystonite	21910	BMB	4	I	32.8	<0.2	1.5	2.0	<0.2	4.7	34.9	12.3	2.0	90.1
Hardystonite	21910	BMB	4	I	32.8	<0.2	1.3	2.0	<0.2	5.6	36.8	12.3	1.7	92.5
Hardystonite	21928	BMNH	15	I	32.8	<0.2	2.5	1.8	0.2	3.6	35.1	15.6	2.6	94.1
Hardystonite	21928	BMNH	15	I	32.8	<0.2	1.2	1.5	0.1	3.5	36.2	15.9	3.1	94.3
Hardystonite	21931	BMNH	15	I	32.8	<0.2	3.1	2.5	<0.2	3.3	35.1	14.6	2.4	93.9
Hardystonite	21928	BMB	15	I	32.8	<0.2	2.2	1.6	<0.2	3.5	34.1	15.1	2.0	91.2

Method: BMNH – EDXA, Geoscan, Natural History Museum (IF); BMA – EDXA British Museum (NDM), normalised to 100%;
BMB & BMC – EDXA, British Museum (ICF)

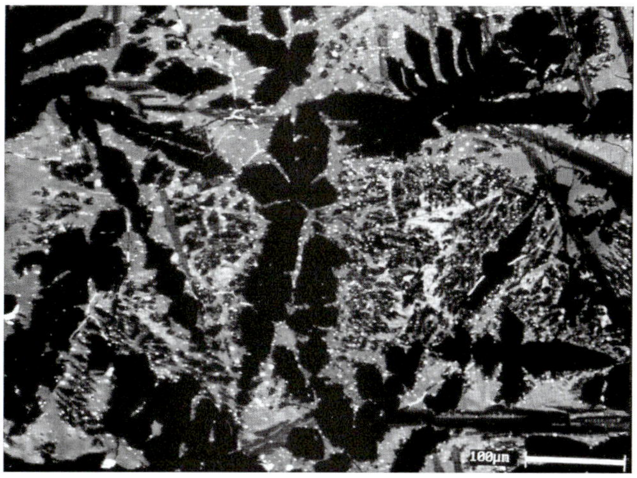

Plate 7.1 SEM photomicrograph of Type I slag 21926 from Zawar Site 14 showing clinopyroxene crystals (black) in a glassy matrix. An unusually large number of tiny lead oxide inclusions (white) are present in this section (scale bar: 100 µm).

Early zinc oxide/lead smelting at Zawar

Type I slags

These are either glassy in the hand specimen, rather like obsidian, or alternatively may be very fine-grained, matt and stony. They have clearly solidified from a homogeneous liquid after tapping from the furnace. Although their phase composition is varied, in addition to glass they typically contain minor amounts of an iron-poor olivine, a zincian melilite and a clinopyroxene (Table 7.1 and Plate 7.1). The less glassy varieties appear to reflect a higher zinc concentration and to have crystallised to melilite (hardystonite), a calcium-zinc silicate, and to willemite, a zinc orthosilicate (Table 7.2). Examples of the oxide compositions are given in Table 7.1 where it is observed that the slags are complex, with silica contents of around 40%, calcite, iron, lead and zinc oxide contents in the range 5–20% and around 5% magnesia and alumina. The complex chemistry of the slags probably served to lower the melting temperatures and facilitate smelting as in some cases the slags appear to be close to the cotectic crystallisation of the pyroxene, olivine and melilite phases, and is reflected in their glassy nature.

Sulphides are uncommon in these slags; in particular sulphides are absent from the very small and occasional inclusions of lead that are now present as the carbonate but would originally have occurred as the metal. The lead that was being smelted may have been further treated to recover the silver values but due to their small size and degradation it was not possible to estimate their silver content. However, no evidence of cupellation debris was found in our admittedly limited excavations at Zawar. It is believed that the Type I early slags result from smelting the roasted mixed sphalerite/galena ore to produce zinc oxide and lead metal (see below and Chapter 11).

Furnace linings

The early slag heaps (Sites 2, 14 and 7) produced quantities of slagged furnace lining (see Chapter 6). These have large amounts of coarse filler in the form of quartz, plant remains and also fragments of crushed furnace lining, recognisable by the slagging, as a form of grog filler.

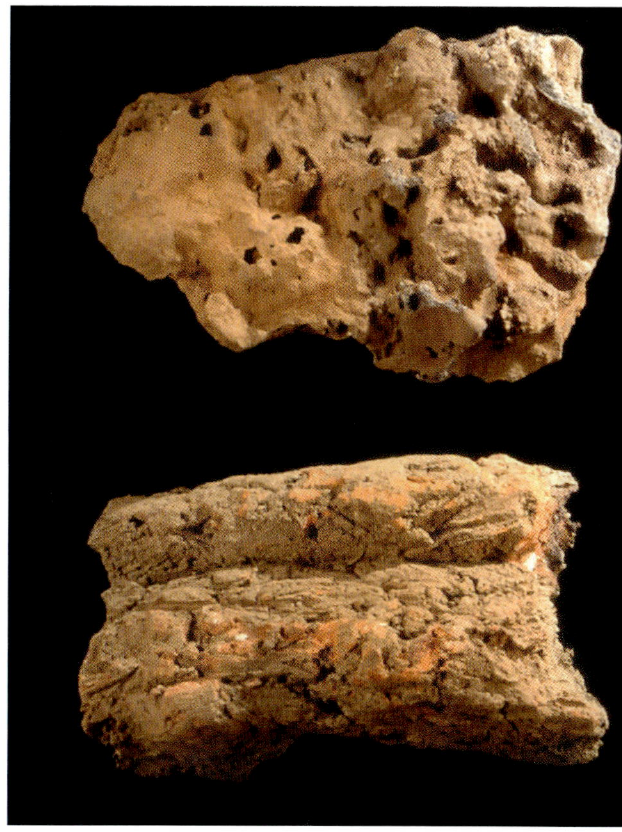

Plate 7.2 Two furnace rim fragments 27389 (lower) and 27391(upper) from Zawar Site 2, heavily impregnated with zinc oxide. The lower piece is 5 cm across.

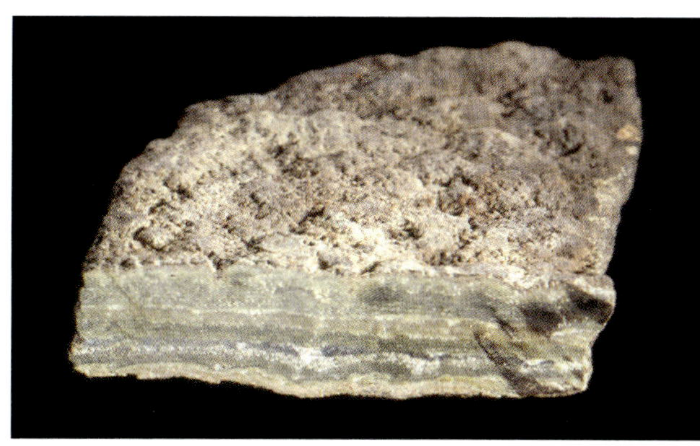

Plate 7.3 Flat layered plate of zinc oxide 27394B from Zawar Site 2. The piece is approximately 4 × 3 cm and 0.9 cm thick.

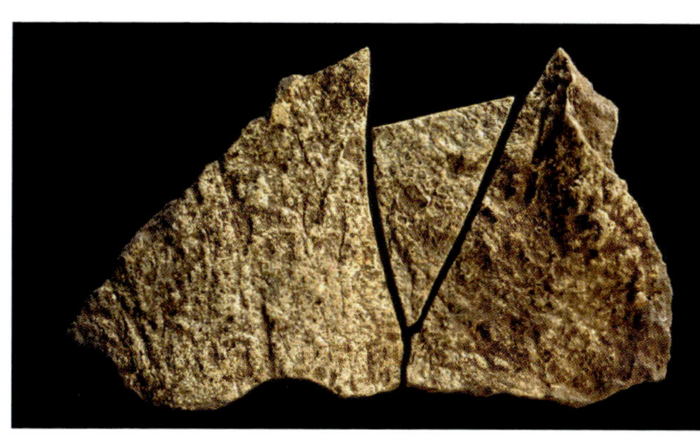

Plate 7.4 Flat layered plate of zinc oxide 27388 from Zawar Site 2. The three conjoining fragments are approximately 9 × 5 cm and 0.5 cm thick.

Table 7.3 Composition (wt.%) of materials from Zawar, Site 2, determined by SEM-EDXA.

BMRL no.	SiO_2	Al_2O_3	FeO	MnO	MgO	CaO	Na_2O	K_2O	PbO	ZnO	TiO_2	SO_3	comments
Loose ZnO - crusts with layered structure (type II)													
27388									20.6	79.4			bulk composition
27394B									18.7	81.4			bulk composition
									52.6	47.4			lead rich layer
									10.0	90.0			zinc rich layer
Roasting crusts with sulphides (type I)													
27394A	14.9	0.7	0.9		0.8	0.7			44.3	29.2		8.3	roasting crust, bulk composition
	65.5	17.3	5.3		0.7	1.5		4.6		4.5	0.8		ceramic, bloated
	27394 A and B were two pieces in only one bag with the same sample number. SiO_2 in roasting crust only secondary.												
27391	17.6	0.7	0.7		0.4	0.4		0.2	42.1	29.4		8.4	roasting crust, bulk composition
	63.2	16.2	6.8		1.9	0.7	2.1	3.1	2.2	3.1	0.7		ceramic, bloated
	SiO_2 in roasting crust only secondary.												
furnace lining, slagged													
27389	67.1	17.1	7.3		1.6	0.9	1.0	3.5			1.0		lining, straw tempered, remote from slag, bloated
	68.8	17.0	6.4		1.7	1.1		3.9			0.4		dito
	70.1	13.8	5.3		1.1	1.3	1.5	4.0	1.0	1.2	0.6		lining, closer to slag,already molten
	38.4	5.0	7.9		5.2	12.7		2.2	10.7	15.8			slag layer, area analysis
	35.0	6.6	6.1		3.6	9.6	4.8	2.4	22.8	8.3	0.4		dito
	30.2	5.9	10.8	0.8	0.5	6.9		2.5	29.1	11.0			glass in slag layer
	29.55		7.1	0.5	9.9	0.3				52.6			willemite, ca. $(Zn_{1.3}Mg_{0.5}Fe_{0.2})$ $[SiO_4]$
	13.65		*2.8*	*0.2*	*6.8*	*0.1*				*18.0*			*dito, atomic % of the elements, rest to 100: oxygen*
	The slag layer contains predominantely willemite, or willemitic olivine with inclusions of glass, further salite ca. $Ca_{0.4}Mg_{0.4}Fe_{0.1}Zn_{0.1}[Si_2O_6]$ (with zinc and iron-rich cores) and leucite ca. $K[AlSi_2O_6]$ as well as an intergrowth leucite-willemite. Ceramic is bloated already in those parts which have the greatest distance to the slag on the section. Closer to the slag, it was entirely molten (now glassy with bubbles).												
slag													
21908	40.5	5.7	12.2	0.8	1.4	6.8	2.5	1.9	3.6	7.5			slag (ICP - analysis)
	43.1	6.2	16.5	0.9	2.1	10.7		3.7	8.0	8.5	0.4		glassy matrix of slag, SEM
	41.6	7.7	17.6	0.7	1.7	8.4		4.9	10.1	5.3	0.4		dito
	42.5	7.6	17.1	n.d.	1.4	8.5		4.8	10.0	8.1			dito
	39.7	6.9	15.8	1.1	0.5	5.5		4.5	13.6	11.8	0.5		dito
	40.8	5.1	17.2	0.9		10.6		3.6	6.5	15.3			glassy matrix (centre of the polaroid)
	35.9	6.8	17.2	0.9		6.3		4.6	11.1	17.4			glassy matrix
	38.0	5.5	15.5	0.6	0.9	13.2		3.4	15.4	7.0	0.5		glass, inhomogeneous
	37.3	6.0	14.5	0.6	0.5	9.1		4.9	17.0	9.9	0.3		dito
	37.7		16.0	0.7	4.9					40.7			zinc-melilithe ca. $Zn_2(Fe_{0.6}Mg_{0.4})[Si_2O_7]$
	17.0		*6.1*	*0.3*	*3.3*					*13.6*			*dito, atomic % of the elements, rest to 100: oxygen*
	51.9	2.5	11.9	0.7	9.9	20.9				2.2			pyroxene (salite) ca. Ca $(Mg_{0.6}Fe_{0.4})[Si_2O_6]$
	20.2	*1.2*	*3.9*	*0.2*	*5.7*	*8.7*				*0.6*			*dito, atomic % of the elements, rest to 100: oxygen*
	38.2		27.2	1.1	22.9	1.6				8.9			olivine ca. $Mg(Fe_{0.8}Zn_{0.2})SiO_4$
	16.0		*23.4*	*0.4*	*14.3*	*0.7*				*2.8*			*dito, atomic % of the elements, rest to 100: oxygen*

Table 7.4 Composition (wt.%) of materials from Zawar, Site 7, determined by SEM-EDXA.

BMRL no.	SiO_2	Al_2O_3	FeO	MnO	MgO	CaO	Na_2O	K_2O	PbO	ZnO	TiO_2	comments
ZnO-crust from smelting (type II) adhering to furnace lining (coil)												
27386	5.0								22.6	72.4		crust, main part, area analysis
	30.7					0.5			22.3	46.4		crust, outer surface, area analysis
	24.8								2.3	72.9		Zn_2SiO_4 in outer surface of the crust
	13.5									*29.4*		*dito, atomic % of the elements, rest to 100%: oxygen*
	15.3	1.1			0.9				61.0	21.7		Larsenite, ca. $PbSZnSiO_4$, in outer surface
	13.4	*1.2*			*1.2*				*14.4*	*14.1*		*dito, atomic % of the elements, rest to 100%: oxygen*
	39.3		1.3		0.5	0.9			3.0	54.9		transition zone to ceramic, ca. $ZnSiO_3$ (darker on polaroid)
	19.0		*0.5*		*0.4*	*0.5*			*0.4*	*19.7*		*dito, atomic % of the elements, rest to 100%: oxygen*
	26.3								3.0	70.7		transition zone to ceramic, ca.Zn_2SiO_4 (brighter on polaroid)
	14.2								*0.4*	*28.2*		*dito, atomic % of the elements, rest to 100%: oxygen*
	30.6	3.6	3.1			3.8			1.3	57.3	0.3	ceramic, behind transition zone, quartz temper not included
	72.4	4.1	1.8		0.3	0.5		0.8	0.4	19.7		ceramic, close to coil boundary
	68.5	18.8	6.1		1.4	0.7		4.5				ceramic, hardly chemically influenced
	72.1	14.1	5.5		1.1	1.7		3.1		2.0	0.5	dito
	The ceramic contains small particles of used furnace lining. No bloating but ZnO-impregnated. Transition zone, 40 microns, of $ZnSiO_3$ and Zn2SiO_4 only. Within the crust, quartz particles with reaction zones of Zn_3SiO_3, same phenomenon around temper in ceramics.											
Slags												
26071	48.8	9.2	8.9	0.4	2.9	8.4		2.6	13.2	4.7		ICP - analysis
	48.6	8.9	9.8	0.5	2.6	7.9		2.7	12.6	5.4		area analysis SEM, slag entirely glassy
26082	41.5	7.0	13.2	0.3	2.8	11.6	0.4	1,87	11.4	9.9		ICP - analysis
	38.6	6.5	15.8		2.5	11.4		1.9	10.5	11.5		area analysis SEM, slag entirely glassy

Table 7.5 Composition (wt.%) of materials from Zawar, Site 14, determined by SEM-EDXA.

BMRL no.	SiO_2	Al_2O_3	FeO	MnO	MgO	CaO	Na_2O	K_2O	PbO	ZnO	TiO_2	comments
ZnO- crust, layered structure (type II) but on both sides adhering to ceramic												
27382	3.7	0.6	0.6			0.3			10.1	84.7		crust, bulk composition without ceramics
	41.7	1.7	4.1		7.6	37.4		0.7		6.6		ceramic, eutectic structure in molten part, bright in SEM
	53.9	23.4	0.4					21.7		0.7		ceramic, eutectic structure in molten part, dark in SEM (leucite)
	65.6	12.7	5.7		1.4	3.4		4.9	1.7	4.7		ceramic, molten part with Pb- and Zn-rich inclusions
	60.5	11.8	6.0		1.5	3.8		4.6	2.9	8.2	0.6	ceramic, with increasing distance from the crust
	63.0	14.6	5.0		0.9	3.2		4.4	3.2	4.9	0.7	dito
	52.6	12.6	11.8		4.6	6.8		3.5	2.2	4.8	0.6	dito, without coarse quartz grains
	74.5	9.5	3.8		0.9	0.5		2.2	0.9	7.8		ceramic, opposite site of the crust, hardly influenced, fading into crust
	On one side heavily bloated, already molten ceramic, in the middle of the crust and on the other side ceramic hardly influenced. Quartz grains in crust not influenced like those from site 7.											
Slag												
27926	40.6	4.1	13.9		6.2	13.8	0.3	1.4	6.0	13.0		slag (ICP - analysis)
	39.4	4.9	15.9	0.6	5.9	12.7		1.9	4.5	14.1		area analysis, centre of the section, crystallised region
	33.6	6.6	13.1	0.5		9.8		3.4	13.3	19.7		glass, brighter in SEM
	40.6	5.2	12.0	0.4	1.9	14.5		2.6	8.1	14.7		glass, in crystallised area, darker
	32.2		30.3	1.3	18.7	1.6				16.1		olivine, ca. $(Mg_{0.8}Fe_{0.8}Zn_{0.4})[SiO_4]$ (brighter crystals)
	13.5		*10.7*	*0.5*	*11.7*	*0.7*				*5.0*		*dito, atomic composition of the elements, rest to 100%: oxygen*
	48.9	3.5	10.2	0.5	11.5	21.1				4.3		salite, ca. $Ca(Mg_{0.7}Fe_{0.3})[Si_2O_6]$ (darker in SEM)
	17.8	*1.5*	*3.1*	*0.2*	*6.3*	*8.2*				*1.2*		*dito, atomic composition of the elements, rest to 100%: oxygen*
	46.9	4.1	14.0	0.7	8.1	20.2				6.0		ferrosalite, ca. $Ca(Fe_{0.5}Mg_{0.5})[Si_2O_6]$
	17.6	*1.8*	*4.4*	*0.2*	*4.6*	*8.1*				*1.7*		*dito, atomic composition of the elements, rest to 100%: oxygen*
	45.7	16.4	6.0			0.7		16.3	1.3	13.7		eutectic structure, dark part
	31.8	2.1	12.2			10.8		1.0	14.2	28.1		eutectic structure, bright part

Zinc-rich crusts

These are fragments of dense material principally of zinc and lead oxides (Tables 7.3–7.5). They fall into two types: the first often contains sulphides (now usually oxidised to sulphates), and are always found adhering to a furnace lining (designated Type I, Plate 7.2). These linings were identical to the previously mentioned slagged furnace linings but came from above the slag level, and suggest that the zinc and lead-rich material had been deposited either during charging or as a fume. The second type is of purely oxidic material either adhering to ceramic or apparently as free plates (designated Type II, Plates 7.3 and 7.4).[2]

All the samples are considerably weathered with much of the lead oxide now in the form of lead carbonate ($PbCO_3$) and the zinc oxide now largely zinc carbonate ($ZnCO_3$). Some zinc silicate (Zn_2SiO_4) is also present. It could be argued that the Type II plates are simply the by-products of smelting or roasting within a furnace or hearth, but no slag minerals were present, indicating that they were not in direct contact with the smelt, and no quartz or other gangue minerals were found, showing that they were not formed during a roasting operation. The general purity, especially the near absence of iron (Tables 7.3–7.5), strongly suggests that they formed predominantly from the deposition of zinc and lead oxide fume.

Type I

The Site 2 samples, 27391 and 27394A, are very weathered although the large grains may still contain unweathered sulphides, sulphates or oxides in their core (Plate 7.5). Zinc sulphide usually occurs inside the lead carbonate together with small rounded particles of zinc oxide. Sample 27394A is especially rich in lead sulphide, although much of this has now weathered to lead sulphate but sometimes still with some residual lead sulphide in the centre. There is also some original unweathered lead oxide (Table 7.3). On Sample 27394A the furnace lining was covered by a thin continuous layer of lead sulphide less than 1 mm thick (white), now partially oxidised (Plate 7.6). With increasing distance from the ceramic body the material becomes more porous, weathered and the zinc content increases It seems that both of these samples (Plates 7.5 and 7.6) represent the original deposition of lead sulphide, coating but not penetrating or reacting with the ceramic upon which it was deposited. This is surprising as the ceramic in both samples is bloated, suggesting temperatures well in excess of 1000°C. If the ceramic had been at this temperature when the galena was deposited or subsequently the galena would have oxidised.[3]

The material from Site 7 includes a fume formed against a coil of ceramic furnace wall (27386). The bulk of the weathered deposit (Table 7.4) is composed of zinc and lead carbonates and silicates with traces of larsenite ($PbZn(SiO_4)$) and some quartz grains (Plate 7.7). No sulphides were present. The penetration of the zinc oxides into the ceramic body and their reaction with the quartz and other silicates in the body show that the zinc oxide had deposited as a fume while the ceramic was hot. The absence of sulphides suggests that in this case the furnace was used for a single operation, and that the sulphidic ore having been roasted elsewhere had been transferred to a fresh coil-built furnace, represented by the present sample.

Type II

Both of the Type II free plates, 27394B and 27388, have layered structures (Plate 7.3), with some difference in composition (Table 7.3). There are no sulphides present. The green layers tend to be rich in zinc and the white layers rich in lead. The zinc oxide is in particles with diameters between 1 and 5 µm on Sample 27391. There are some fine charcoal particles on and between the layers on 27382. Once again little else is present, suggesting that these have been separated from the main body of the smelt as a fume. Sample 27394B also has charcoal dust, not only on its surfaces but in between the layers, suggesting that the temperatures where the deposition took place did not exceed 1000°C, and that this deposition must have taken place while charcoal was being charged to the furnace.

The material from Site 14 included a flat weathered plate, 27382[4] (Plate 7.8) with a very thin layer of unbloated ceramic

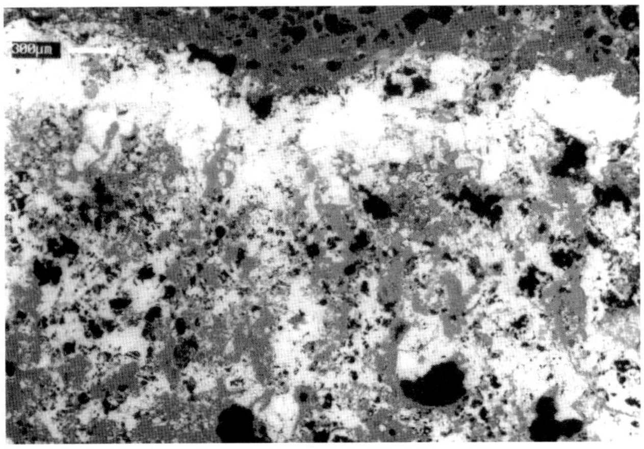

Plate 7.5 SEM photomicrograph of Sample 27394A from Zawar Site 2 showing the crust (lower part) adhering to the ceramic furnace lining (dark upper part). The furnace lining is covered with a thin layer (< 1 mm) of PbS (white) now partially oxidised to $PbCO_3$.

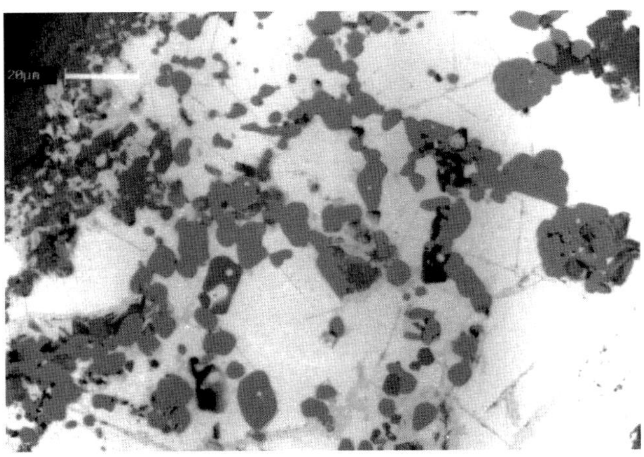

Plate 7.6 SEM photomicrograph detail of Sample 27394A (Plate 7.5) showing PbS grains with octahedral cracks (white) that have weathered along the grain boundaries (pale grey). Rounded droplets of ZnO (dark grey) sit in the PbS along with some primary Zn_2SiO_4 (dark grey, partially dendritic).

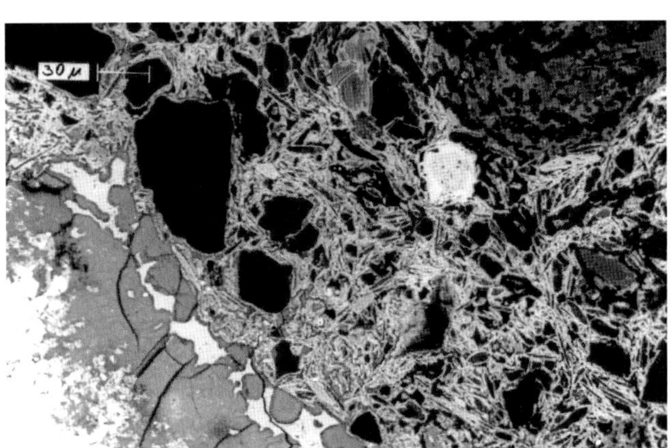

Plate 7.7 SEM photomicrograph of impregnated furnace lining 27386 from Zawar Site 7. The main part of the print is the ceramic heavily impregnated with ZnO particles (white) and the surfaces of the quartz temper (black) have reacted to form Zn_3SiO_3. The area of zinc oxide deposition (lower left) is of Zn_2SiO_4 (white) while the darker cracked material contains Zn_3SiO_3 (*note:* little or no lead or sulphidic material is present).

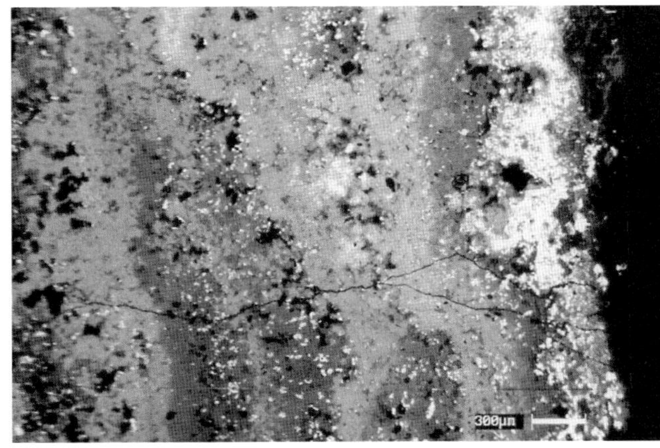

Plate 7.8 SEM photomicrograph of zinc oxide fume 27382 from Zawar Site 14. The grey areas on the centre right are agglomerated $ZnCO_3$ particles (diameter approx. 5 µm); the white area to the right is $PbCO_3$. The black particles are mainly holes and quartz grains. The dark area on the far right is ceramic material.

on one side and a discontinuous layer of bloated ceramic on the other. The latter had clearly been subjected to a very high temperature, but not apparently while the zinc and lead oxides were being deposited as there is no evidence of reaction. The bulk of the plate is made up of layers of zinc and lead oxides (Table 7.5), now often carbonates, apart from a few quartz grains and a very few tiny inclusions of lead sulphide against the unbloated ceramic. There is no evidence of reaction at the surfaces of the quartz grains in the zinc oxide, suggesting that temperatures were not very high. It is likely that the zinc and lead had deposited on the thin layer of ceramic, but the bloated ceramic adhering to the other side is anomalous and could suggest that the sequence of events was more like the samples from Site 2, where the zinc oxide had deposited against a lining that previously had been much hotter.[5]

Summary

These materials are likely to be the debris from a two-stage process of producing zinc oxide, together with some metallic lead as a secondary product (see Chapter 11, Fig. 11.3). First the sphalerite/galena ore was roasted, followed by smelting under reducing conditions – sometimes in the same furnace as at Site 2 or in separate furnaces at the later Site 7 – to produce a vapour of zinc with some lead which promptly reoxidised and deposited as a fume. The bulk of the lead would have drained through the slag and collected at the base of the furnace.

Medieval/post-medieval zinc smelting

Ashes from the roasting heaps

Ash deposits occur in heaps particularly around the periphery of the retort dumps at Old Zawar where the ore was roasted (see Chapter 6, Plate 6.33). Partially burnt twigs, charcoal and ceramic fragments are visible in discreet layers in the grey-white friable matrix (see Chapter 6, Plate 6.34 and Fig. 6.33 and Chapter 11). The ash heaps were not sampled for analysis in a fully systematic fashion but the available data are sufficient to suggest their origins. All the ashes are rich in lime and magnesia (Table 7.6) found mainly as carbonates. They are thus broadly similar in composition to the spent charge

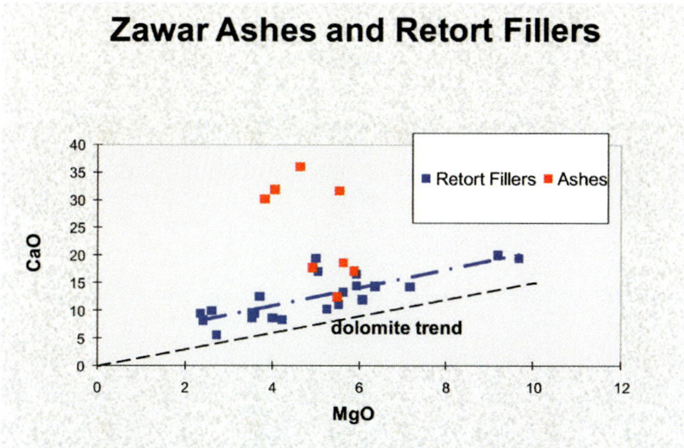

Fig. 7.1 CaO vs MgO in retort residues and ashes. The retort residues show a marked correlation and similar slope to that of dolomite. The average CaO content of the ashes is higher than in the residues.

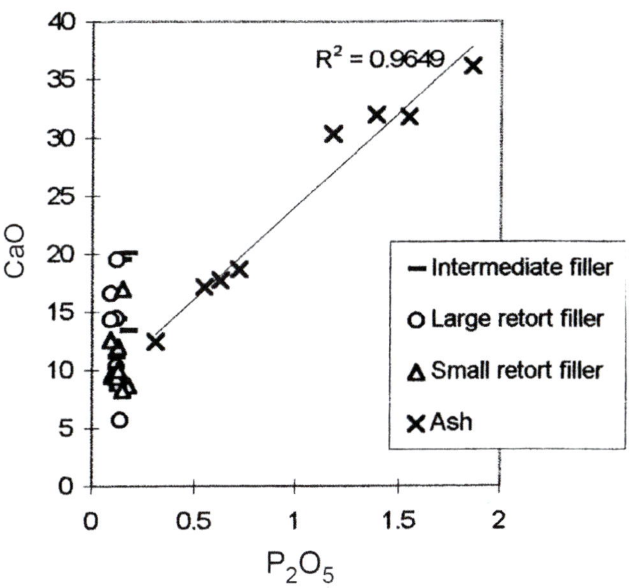

Fig. 7.2 CaO vs P_2O_5 in retort residues and ashes showing a marked correlation for the ashes and the distinction of the two groups.

residues in the zinc smelting retorts (see below and compare with Table 7.7). However, the ashes are typically richer in lime for a given magnesia content than the retort residues (Fig. 7.1). Furthermore, the ashes from the heaps have higher phosphate contents with a strong correlation between the lime and phosphorous oxide (Fig. 7.2). Lime and phosphate are two of the characteristic components of the ashes of both dung and wood, so the association of these minerals is to be expected. These characteristics, and the presence of charred wood and twigs in the heaps, indicate that a large component of the ashes is derived from the burning of wood.

The zinc oxide contents of the ashes average 2–3%, strongly suggesting a relationship with the zinc smelting process. The lead oxide contents are typically less than 0.5%, whereas those of the retort residues average 1.1%. Similarly, silver contents of the ashes are low, with a maximum of only 32 ppm, while those of the retort residues average 67 ppm. In contrast the cadmium contents of the ashes are much higher – four measurements yielding values between 133 and 214 ppm – while the mean cadmium content of the retort residues is 15 ppm. Cadmium is more volatile (BP 766°C) and would have vaporised preferentially during the preliminary roasting process. In recent industrial processes zinc ores typically lost about 60% of their cadmium on roasting (Mellors 1969: 420) therefore the heaps are likely to be located where the sphalerite ore was roasted to produce the zinc oxide for the retort charges.[6]

Furnaces and retorts: petrography and clay composition

Field observations, using the naked eye and a ×10 hand lens, indicated marked variations in the fabrics of the various components of the furnaces and retorts. In particular, there is a variation in the coarseness of the fabrics. The ceramics associated with the large retort process exhibit coarse fragments of phyllite/slate in the furnace walls, finer fragments of phyllite and/or quartzite in the moulded perforated plates and relatively

Table 7.6 Composition (wt.%) of ash from the white heaps etc. at Zawar.

BMRL no.	Material	Lab	Site/description	P_2O_5	CaO	MgO	PbO	ZnO	FeO	Ag	Cd	MnO	Cu
						wt%						ppm	
27410	Ash	BM	Heap nr site 5	0.72	18.65	5.62	na	na	na	na	na	na	na
41123	Ash	BM	Heap nr site 5, gritty	0.55	17.15	5.88	na	na	na	na	na	na	na
41124	Ash	BM	Heap nr site 5, upper part	0.31	12.44	5.49	na	na	na	na	na	na	na
41128	Ash	BM	Heap nr Site 5	1.86	36.11	4.62	na	na	na	na	na	na	na
41126	Ash	BM	Site 5, white, powdery ash	0.63	17.76	4.91	na	na	na	na	na	na	na
41125	Ash	BM	Heap nr site 31	1.39	31.94	4.04	na	na	na	na	na	na	na
41127	Ash	BM	Old Zawar East heap, road cutting	1.55	31.74	5.53	na	na	na	na	na	na	na
41129	Ash	BM	Heap nr. Site 3	1.18	30.29	3.81	na	na	na	na	na	na	na
14	Ash A	HZL	heap below zawar I	na	15.53	6.80	0.32	2.05	7.59	32	na	0.43	48
15	Ash B	HZL	heap below Zawar I	na	22.66	7.63	0.22	1.74	2.19	15	na	0.19	64
21	ash	HZL	Layer 1, Zawar I	na	19.87	5.80	0.16	0.25	5.40	18	na	0.74	52
LKG1	ash	HZL	Zawar	na	na	na	0.23	1.63	1.93	10	133	na	na
LKG2	ash	HZL	Zawar	na	na	na	0.66	3.42	3.47	22	214	na	na
LKG£	ash	HZL	Zawar	na	na	na	0.71	4.16	4.89	24	167	na	na
LKG4	ash	HZL	Zawar	na	na	na	0.15	1.93	2.06	12	198	na	na

Analyses by AAS (HZL) and ICP-AES (BM)

Table 7.7 Composition of selected components in the zinc smelting retort residues.

			wt. %					ppm			
BMRL no.	Retort type	Site	CaO	MgO	PbO	ZnO	FeO	Ag	Cd	MnO	Cu
27412	Intermediate	Old Zawar, West end	11.12	5.52	0.41	2.71	14.8	88	11	na	na
27413	Intermediate	Old Zawar, West end	8.32	4.21	0.70	3.74	13.1	58	10	na	na
27414	Intermediate	Old Zawar, West end	19.51	4.99	0.61	3.71	13.0	63	9	na	na
27468	Intermediate	Old Zawar, West end	13.38	5.62	1.08	3.51	10.3	52	16	na	na
28500	Intermediate	Old Zawar, West end	14.38	6.36	1.88	4.86	12.9	88	17	na	na
28501	Intermediate	Old Zawar, West end	20.05	9.19	0.52	2.32	11.2	83	10	na	na
	Mean		**14.46**	**5.98**	**0.87**	**3.47**	**12.54**	**72**	**12**		
	s.d.		***4.22***	***1.58***	***0.50***	***0.81***	***1.45***	***15***	***3***		
27469	Large	Old Zawar, East end	16.59	5.93	5.39	3.29	13.9	73	15	na	na
27470	Large	Old Zawar, East end	5.65	2.72	bd	2.54	24.6	30	9	na	na
27471	Large	Old Zawar, East end	8.63	4.00	0.30	3.88	28.8	59	19	na	na
28472	Large	Old Zawar, East end	14.44	5.94	0.25	10.58	10.5	26	17	na	na
28473	Large	Old Zawar, East end	19.47	9.67	1.03	3.44	11.2	104	25	na	na
28474	Large	Old Zawar, East end	14.33	7.17	0.97	3.21	12.3	67	11	na	na
28475	Large	Old Zawar, East end	10.28	5.25	0.61	4.23	17.1	46	10	na	na
	Mean		**12.77**	**5.81**	**1.43**	**4.45**	**16.92**	**58**	**15**		
	s.d.		***4.45***	***2.07***	***1.80***	***2.55***	***6.58***	***25***	***5***		
28476	Small	Site 5	9.48	3.59	0.69	3.29	13.1	46	10	na	na
28477	Small	Site 5	17.00	5.04	1.94	7.49	9.8	85	25	na	na
28478	Small	Site 5	11.98	6.07	1.68	16.56	9.3	126	31	na	na
28479	Small	Site 5	9.61	3.53	0.74	14.69	16.2	108	14	na	na
28480	Small	Site 5	12.57	3.71	0.70	4.98	10.5	42	13	na	na
28480	Small	Site 5	8.66	3.53	2.64	5.79	9.1	85	38	na	na
28480	Small	Site 5	8.28	2.41	0.64	3.71	17.0	46	10	na	na
28480	Small	Site 5	9.42	2.35	0.70	4.31	15.9	63	10	na	na
28480	Small	Site 5	10.00	2.61	0.54	4.78	16.1	35	9	na	na
	Mean		**10.78**	**3.65**	**1.14**	**7.29**	**13.00**	**71**	**18**		
	s.d.		***2.73***	***1.23***	***0.75***	***4.90***	***3.35***	***32***	***11***		
LKG 16	Retort 2 front	Old Zawar	9.79	6.47	2.41	7.99	9.26	92	na	0.45	76
LKG 17	Retort 2 centre	Old Zawar	14.69	5.14	2.78	0.15	10.42	119	na	0.45	65
LKG 18	Retort 2 back	Old Zawar	14.41	5.31	2.58	0.00	12.09	264	na	0.46	68
LKG 19	Retort 1 back	Old Zawar	10.91	5.80	0.54	0.15	14.15	57	na	0.68	32
LKG 20	Retort 1 front	Old Zawar	9.79	7.79	0.92	3.26	15.69	58	na	0.69	65

Analysis for Ca, Mg by ICP-AES (BM), phosphorus analysed but below 0.5%
Analysis for Pb, Zn, Fe, Ag, Cd by AAS (HZL)
Except LKG16-20, where all elements by AAS (HZL)

fine, silty fabrics selected for the moulded condenser heads. The clay wedges that held the large retorts apart contain abundant fragments of phyllite and a little quartzite.

In general, the ceramics associated with the small retort process are finer than those associated with the large and intermediate retort processes. For example, observations of the Site 5 small retort material indicate an absence of coarse inclusions in the perforated plates, as well as a lack of quartzite from the retort walls, whereas coarse fragments of crushed, angular quartzite were added to the fabrics of the large and intermediate retorts. In thin section, the ceramics display a number of common characteristics: silty materials consisting of sub-angular quartz and phyllite fragments ranging from coarse silt to fine sand grade, set in a matrix comprising micaceous minerals and fine silt-grade quartz.

Several ceramics from the excavated large retort furnace, Site 37, were analysed for major and minor elements by SEM-EDXA using the method described in Freestone and Tite (1986) to determine the compositions of the fine-grained matrix and the bulk of the ceramic (Table 7.8). The bulk compositions are similar and are fairly typical of common clays, such as those used to produce house bricks, rather than material which would be considered to possess particular refractory properties by a modern ceramicist (Freestone and Tite 1986; Freestone 1989). The ceramic matrices are typically richer in alumina and poorer in silica, which is consistent with the presence of coarse particles of quartz in the fabric. The furnace linings are composed largely of fragments of phyllite/slate with a low proportion of matrix material (Plate 7.9), while the perforated plates and the retorts contain phyllite/slate and quartz fragments with more abundant matrix material (Plate 7.10). Analysis of the phyllite/slate fragments gave a composition very close to that of the bulk material.

These observations suggest that the clays used for the various components of the zinc smelting process were all closely related and were derived essentially from the weathering of the local argillaceous phyllites. They represent the major local clay resources and do not have particular refractory properties.

Table 7.8 Compositions of zinc-smelting ceramic materials determined by SEM-EDXA.

BMRL no.	23871: Retort		23987: Furnace wall			23986: Perforated plate		Surface glaze	
	Bulk	Matrix	Bulk	Matrix	Phyllite	Bulk	Matrix	Bulk	Matrix
SiO_2	68.7	55.1	68.9	64.4	68.8	68.2	56.7	64.9	58.0
TiO_2	0.8	0.7	0.8	0.4	1.1	0.7	0.7	0.4	0.6
Al_2O_3	16.4	24.0	17.1	19.5	16.1	16.7	24.0	13.8	14.0
FeO	6.3	8.8	6.0	6.6	7.2	6.3	8.8	7.5	15.4
MnO	0.2	0.3	<0.2	<0.2	0.2	0.2	<0.2	<0.2	<0.2
MgO	1.6	2.4	1.4	1.7	1.4	1.2	2.5	0.9	1.4
CaO	1.1	2.6	0.5	0.7	0.3	0.8	1.7	0.4	0.7
Na_2O	0.9	1.0	0.6	0.9	0.8	0.5	0.6	0.7	0.6
K_2O	3.4	3.9	4.2	0.9	3.9	4.4	4.9	9.9	7.5
PbO	<0.4	<0.4	<0.4	<0.4	<0.4	0.4	<0.4	<0.4	<0.4
ZnO	<0.3	<0.3	<0.3	<0.3	<0.3	<0.3	<0.3	0.6	0.6
Cl	0.1	0.2	0.1	0.1	<0.1	0.1	0.2	<0.1	<0.1

Also analysed but not detected, P_2O_5 (<0.2), SO_3 (<0.2), CuO (<0.3)

Even so, it would appear likely that they were carefully selected and prepared to achieve optimal performance for the required purposes. Clays for the bricks of the furnace walls and their linings required little plasticity hence a relatively low proportion of clay matrix, but needed dimensional stability at moderate temperatures, which would have been provided by the relatively large quantities of phyllite fragments. The rounded nature of these fragments indicates that they are the result of weathering rather than crushing (Plate 7.9). The high proportion of phyllite suggests that the clays used for the bricks of the furnace walls and linings were residual or colluvial. The clays used to make the perforated plates and the retort walls were required to be more plastic, so had a higher proportion of plastic matrix to inclusions (Plate 7.10). These clays may have come from alluvial deposits from the local river valleys, fluvial transport having removed the coarser particles. In particular, the deposits of clay that built up behind the major dam across the River Tiri (see Chapter 3, Plate 3.6 and Chapter 6, Fig. 6.2) may have been exploited. In order to provide the retorts with the additional stability such as resistance to slumping that would be required when exposed to very high temperatures, they were tempered with coarse fragments of crushed, refractory quartzite. The funnel-shaped condenser heads carried the zinc vapour into the cooler condensing chamber beneath the furnace. They are usually unvitrified but because a key requirement would have been low porosity to minimise penetration by zinc vapour and consequent loss of yield, they were prepared using the finest clays.

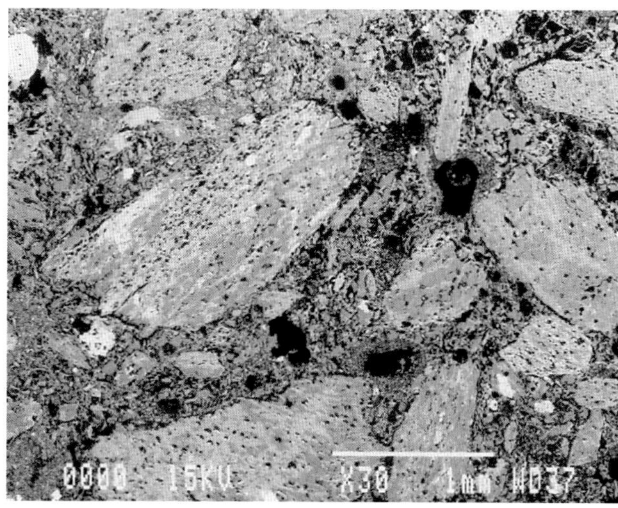

Plate 7.9 SEM photomicrograph of furnace lining Sample 23987 showing abundant lozenge-shaped fragments of weathered phyllite in a fine clay matrix. Backscattered electron image (BEI).

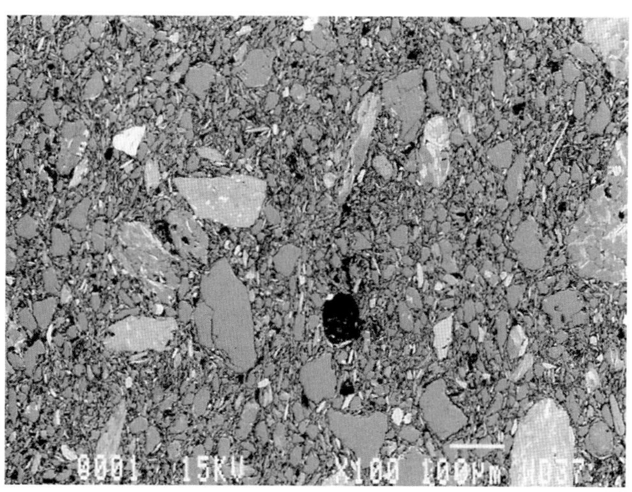

Plate 7.10 SEM photomicrograph of retort Sample 23871 showing fragments of quartz (grey) and phyllite (pale grey with pale micaceous inclusions) in a fine-grained matrix. Note the fillers are much finer than in the furnace lining (cf. Plate 7.9).

Furnace regime

The conditions of smelting may be inferred from the furnace and retort ceramics. The temperatures to which the ceramics were exposed clearly varied considerably. The outer parts of the furnace linings, in the corners for example, are soft and hardly vitrified, whereas the retorts are sometimes highly vitrified and bloated, and slumped in a concertina-like fashion. In some cases the ceramic locally became completely fluid and ran in a trail down the outside of the retort (Plate 7.11).

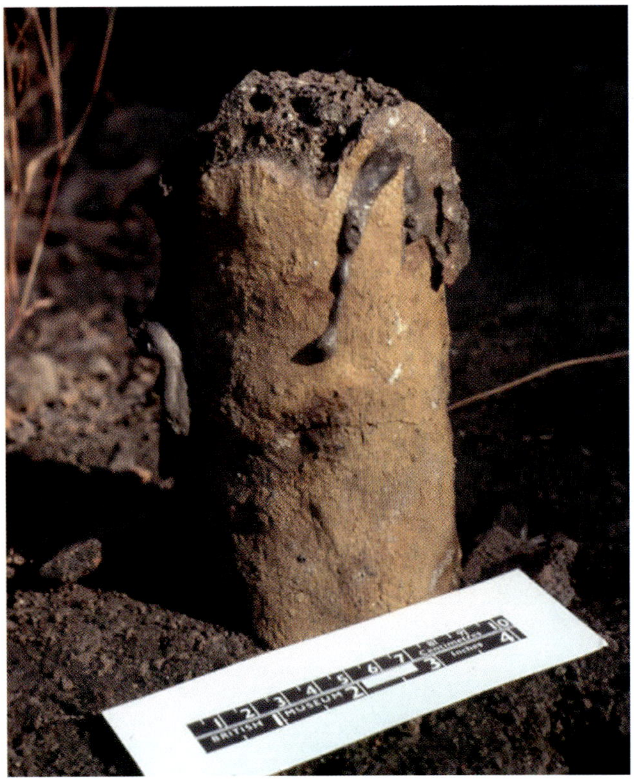

Plate 7.11 Trail of molten ceramic running down the side of a small retort from the heaps at Old Zawar cut by the road.

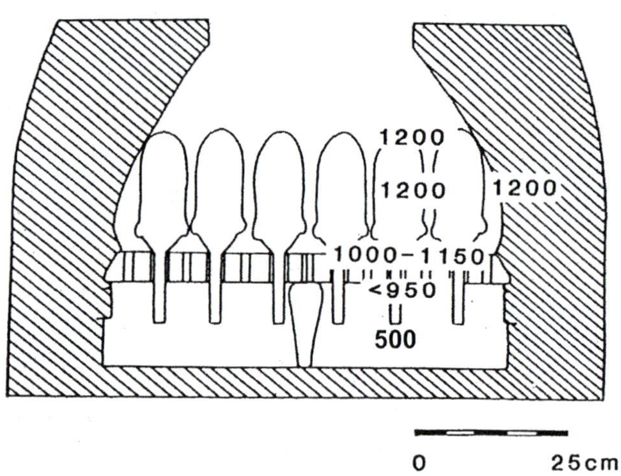

Fig. 7.3 Schematic cross-section showing the distribution of temperatures in a furnace of the type used with large retorts in Old Zawar (cf. Table 7.9).

Table 7.9 Firing temperatures and durations of zinc-smelting ceramics deduced from examination of their microstructures in the SEM (cf. Fig. 7.3).

Large Retort BMRL no.	**Site**	**Description**	**Max temp (degreees C)**	**Duration (hours)**
23986	37	Perforated plate (upper)	1150	2 to 3
24503	37	Perforated plate (lower)	1000-1050	<= 2
23987	37	Furnace wall	1200	>=5
23985	37	Retort tail	1200	n.d.
23984	37	Retort centre	1200	n.d.
23983	37	Condenser cone	<<1100*	n.d.
23870	37	Condenser cone	<<1100*	n.d.
Small Retort				
23874	30	Perforated plate, near large hole (upper)	<<1100*	n.d.
23874	30	Perforated plate, near small hole (lower)	1150	c. 0.5
24502	30	Perforated plate (upper)	1050	c. 0.5
23875	30	Furnace wall	1200	<=5
23871	30	Retort (underfired/ unused)	<<1100*	n.d.
23872	30	Retort condenser junction (i)	1000-1050	n.d.
23872	30	Retort condenser junction (ii)	1200	n.d.
23873	30	Retort tail	1050	n.d.

* These samples do not appear to have been exposed during use to temperatures higher than their original firing temperatures, which were probably below 1000°C

Estimates of the temperatures at which the ceramics had been fired were made using the approach of Maniatis and Tite (1981), as described in Chapter 1. Polished section or polished thin sections of a ceramic sample from Zawar examined in the SEM demonstrate the clear changes in the microstructure when fired in the laboratory at temperatures between 1000°C and 1100°C. Two groups of surface finds were included in the study: one from Old Zawar, associated with large retorts and the other from Zawar Mala (Site 30), associated with small retorts. Where there was a vitrification gradient across a sample a fragment of ceramic was removed from an area of low vitrification and refired as described in Chapter 1. SEM images of all of the refired samples showed continuous vitrification with medium bloating pores (10–40 μm diameter), while two of the refired samples from Zawar Mala (furnace lining 23875 and perforated plate 23874) showed slightly coarser and more rounded pores, suggesting that they were slightly less refractory, perhaps because of their finer grain size. These results are consistent with the petrographic and analytical results outlined above, indicating that all of the clays used had very similar properties.

The refiring experiments were used as a calibration in order to estimate the maximum firing temperatures and times at maximum temperature for the selected samples (Table 7.9). The maximum temperature that the ceramics can record unambiguously is 1200°C. At this point the ceramic is on the verge of failure and shows continuous vitrification with coarse bloating pores. The microstructure would look the same in the SEM however if the ceramic attained higher temperatures such as 1250°C or 1300°C for a short period. A further potential uncertainty occurs because a thin potassium-rich glaze layer has formed on some retorts.[7] The fluxing of the surface of the ceramic by fuel-derived potash could in principle lead to an overestimate of the surface temperature.

The results for the large retort process are summarised schematically in Fig. 7.3. The highest temperatures in the region of 1200°C occurred in the chamber around the retorts themselves where the fuel was burned (and against the furnace lining in their immediate vicinity). Field observations indicate that maximum vitrification (bloating or glazing) in the large retort process began 50–100 mm above the front, condenser, end of the retort, and did not always extend to the tail. The perforated plates below the retorts only attained temperatures of 1000–1150°C on their upper surfaces due to shielding by the closely packed retorts.

It was possible to estimate the temperature attained by the underside of the perforated plates from the measured vitrification gradients. They indicate that the maximum temperatures there were in the region of 750–950°C. These temperatures, and those obtained for the condenser cones, suggest that the plates acted as very effective heat barriers, allowing the temperature in the lower chamber to be maintained at around 500°C, which was necessary in practice for zinc vapour to condense quickly to a liquid (see Chapter 11). Maximum temperatures in the large retort process were maintained for about five hours.

The temperatures and process durations estimated for the small retort process appear to be somewhat lower, and less uniform, than those for the large retort process. The very short durations recorded at the upper surfaces of the perforated plates (Table 7.9) may only be an indication of the period before the plates became shielded by the retorts, which appear to have slumped down during firing and then flattened and spread laterally against the plates. Unfortunately, it is not possible to deduce the temperatures at which the ceramics were originally fired during fabrication before they were deployed in the furnaces.

The necessary reducing conditions within the retorts can be estimated by examining their residues (Freestone *et al.* 1985b). Taking mole fractions of 0.05 FeO and 0.1 ZnO in the glassy phases, assuming that the metallic zinc and iron phases were pure and making some thermodynamic simplifications, such as ideal mixing of the oxide components in silicate liquids, the reducing conditions can be estimated. At the estimated temperatures of about 1200°C the partial oxygen pressure would have been of the order of 10^{-20} (Fig. 7.4).

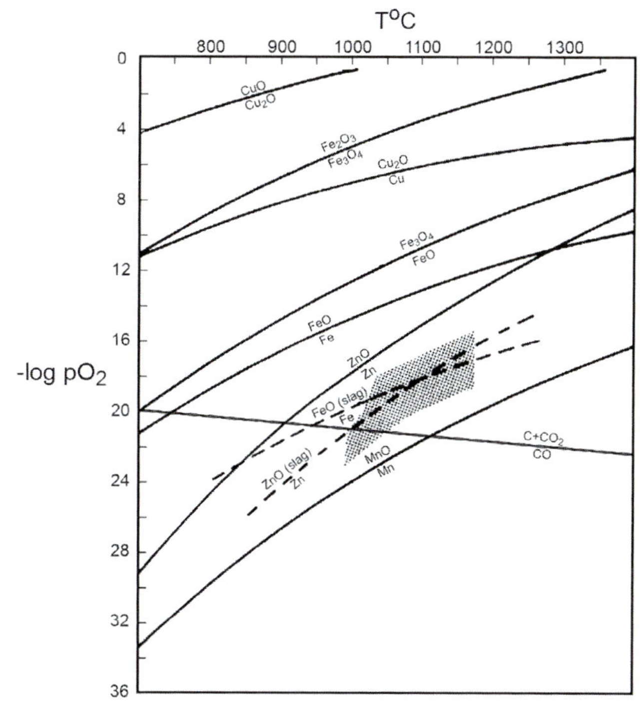

Fig. 7.4 Plot of partial oxygen pressure (–log pO_2) against temperature. The shaded area represents the likely reducing conditions in the retorts at the estimated smelting temperature of about 1200°C. The intersection of the $2C + O_2 - 2CO$ and the $2Zn + O_2 - 2ZnO$ lines shows that no reaction is possible below 900°C therefore the reaction must proceed in the vapour phase.

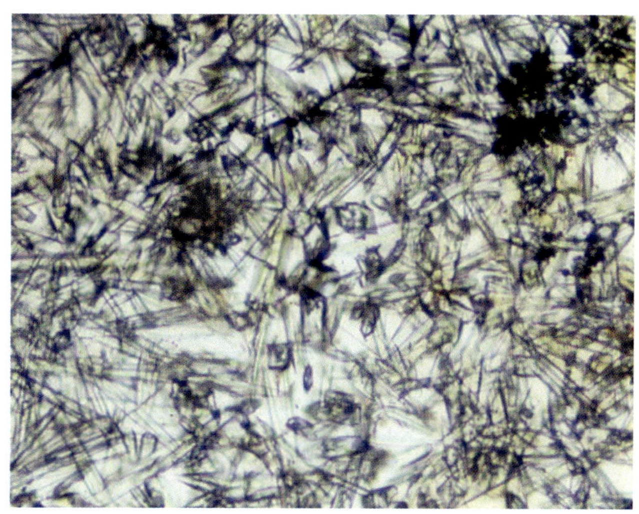

Plate 7.12 Thin section of glassy residue material in a retort showing abundant laths of diopside in a glass matrix.

The charge

The residue inside the retorts consists of a mixture of irregular scoriaceous (cindery) particles, typically around 10 mm diameter, and an abundant, friable fine-grained fraction. The walls of the retorts are commonly glazed internally. The scoria adheres more commonly to the central high temperature zone and less commonly to the inside of the cooler condenser head or the tail, particularly in the large retorts. In the large and intermediate retorts, the spent charge is generally tightly packed and sintered so that the impression of the stick inserted through the mouth of the retort may still be seen (see Chapter 11, Plate 11.6 and Fig. 11.5). The residues in a large retort typically weigh around 1 kg. In the small retorts, however, the charge appears not to have sintered and the retorts are usually empty.

In thin section, the coarse scoria fragments comprise a glassy vesicular skeleton lined with calcite. The calcite presumably represents free lime which formed at high temperature and has since recarbonated. Some goethite (hydrated iron oxide) is also present, associated with the calcite linings. The glassy fraction consists of abundant lath-like crystals of diopside ($CaMgSi_2O_6$) in a matrix of colourless glass (Plate 7.12). Rounded grains of metallic iron occur relatively frequently in the glass, since iron oxide would reduce under the extreme conditions that reduce zinc oxide to zinc metal. The goethite is likely to have formed as a result of the weathering of metallic iron originally present.

Compositions of a number of scoria fragments and of the fine (<2 mm) fraction from one large retort, Sample 50439, are given in Table 7.10. The major components of the scoria are lime, silica, magnesia, iron oxide and alumina. The fine fraction is significantly enriched in silica and alumina, and depleted in lime and magnesia, relative to the scoria. In addition, selected

Table 7.10 Composition of charge residues from large retort 50439, analysed by AAS.

	Scoria (1)	Scoria (1)	Scoria (1)	<2 mm fraction
SiO_2	20.8	13.6	24.3	40.8
TiO_2	0.2	0.1	0.1	0.3
Al_2O_3	3.6	1.6	1.9	6.5
Fe_2O_3	9.2	7.6	9.6	10.0
MnO	1.6	1.6	1.2	1.0
MgO	10.2	7.8	9.7	6.1
CaO	24.6	33.5	24.3	15.3
Na_2O	0.4	0.4	0.2	0.8
K_2O	0.5	0.3	0.3	1.1
PbO	0.4	0.0	0.4	0.3
ZnO	0.6	0.7	0.3	1.2
BaO	0.1	0.1	0.1	0.4
CuO (ppm)	110	54	57	88
Ag (ppm)	104	45	41	41

elements were analysed for a range of large, intermediate and small retorts (see Table 7.7). The results for lime, magnesia and iron oxide are broadly consistent for all retort types and also consistent with the more complete analytical data for the <2 mm fraction, presented in Table 7.10.

The high silica and alumina contents indicate the presence of a significant argillaceous (clay-rich or clay-derived material, such as phyllite) fraction in the charge. In the original report (Freestone *et al.* 1985b), it was considered that this may have been ceramic which had spalled off the retort walls during firing. However, there is no physical evidence for the large amount of spalling that would have had to have taken place. An alternative source for the high levels of silica and alumina is possibly the recycled condenser tubes (see Chapter 6, Fig. 6.29). Varying amounts of zinc condensed in the ceramic condenser tubes either as the oxide or as an intimate mixture of metallic zinc and the oxide, known in English as 'blue powder' (Ingalls 1906: 526–30). These tubes were usually broken off the ends of the retorts, and it is likely that they were collected, crushed and added to a new batch of charge in order to reclaim the zinc; only a few empty tubes were found on the excavation. One tube, 10 cm long, which cannot originally have been more than 12 cm and still fitted into the cool chamber, weighed 60 g. Even this empty tube had some zinc oxide adsorbed in the walls and thus the weight of ceramic contributed to each charge was probably of the order of 50 g. One tube that was found choked with zinc oxide and blue powder weighed 250 g of which about 200 g would have been zinc salts. This must have been an extreme case that would have terminated the operation of the retort. Typically, between 10 and 200 g of zinc minerals was returned to each smelting charge with the crushed tubes. However they do not contribute to calculations on the likely overall yield as presumably on average a comparable amount would be lost again in the condenser tube during the smelt to which the crushed tube had been added.

Animal dung is also rich in silica and if, as is likely, the crushed ore was made into balls with dung and charcoal, then this would be an additional source (see Chapter 9). If the putative collecting crucibles were also added regularly to the retort charge this would be an additional contribution. It is possible that extra clay was added but no direct evidence for this was observed.

The host rock of the orebodies at Zawar is dolomite ($CaMg(CO_3)_2$), probably the major source of the lime and magnesia found in the spent charges, although not the only one. The dolomite host rock was analysed by the HZL laboratories and typically found to contain 30% lime and 20% magnesia. Thus there is excess lime in the residues over and above that likely to have been contributed by the dolomite alone, which would have been of the order of 40 g of lime per retort. Lime, and to a lesser extent, magnesia are the major components of wood ashes (Misra *et al.* 1993) (see Chapter 9). However, this is likely to have made only a small contribution to the charge residues. A large retort can hold about 500 g of roughly crushed charcoal. Charcoal typically contains 3–4% ash (Thorpe 1922: 259), so even if the retort was completely filled with charcoal and the ash was composed of 100% lime, this would only have added 15–20 g lime. It seems, therefore, that additional lime must have been added. Many ancient metallurgical recipes suggest the addition of lime (usually as limestone) to the charge, and it is possible that the additional lime here may have been added as an initiator (see Chapter 11).

The locally elevated amounts of chlorine in the glasses associated with the silicate fraction of the retort residues has been noted previously (Freestone *et al.* 1985a). It was observed that these might reflect the addition of salt to the charge, a component that would have been beneficial in the process and which was specified as an additive in Indian iatrochemical texts (see Chapter 11). It was also noted, however, that the chlorine might have derived from a mineral in the gangue. These observations are still valid so it is not possible to state unequivocally that salt was added to the charge.

The iron is assumed to have derived mainly from the ore where it is present as marcasite (Fe_2S) in solid solution in the sphalerite (ZnS) and also as pyrite (FeS_2) with perhaps 2–3% derived from the recycled ceramic condenser tubes. Silver (and lead) would have derived entirely from the ore. Taking the means and standard deviations of the concentrations of iron, lead, zinc and silver in the residues from 22 retorts (Table 7.7), and comparing them with the modern zinc concentrate at Zawar, some interesting parallels can be made. The averages for the retort residues are 11% iron, 1.1% lead, 4.3% zinc and 70 ppm silver, and for the modern zinc concentrate the figures are 8% iron, 1.4–2% lead, 50% zinc and 140 ppm silver.

With the exception of zinc, which volatilised from the charge, the ratios of the other components are roughly the same in both the residues and the modern zinc concentrate, particularly if a little of the iron in the residues was derived from recycled ceramic. This suggests that the beneficiated ore added to the charges in the ancient process was similar to the modern ore concentrate. Selection of a rich ore and careful hand picking might be expected to produce a similar grade of concentrate as the modern beneficiation process. It can therefore be deduced that the amount of zinc oxide (derived from the ore) added to the charge was around six times the amount of iron. The large retorts contain about 1 kg of spent charge, of which iron typically contributes about 110 g; from this the original charge added to the large retorts can be estimated. Allowing for the minor additions of iron from the ceramic, the amount of beneficiated and roasted zinc ore added to each retort would therefore probably have been in the region of 1300 g of which approximately 650 g would have been zinc oxide producing 500 g of zinc. A small amount of zinc would have burnt off at the end of the condenser with the excess carbon monoxide.[8]

Zinc metal

XRF analyses carried out across the sintered contents of two large retorts that had broken in half along their length showed the distribution of lead and zinc contents (Fig. 7.5). In both instances, although the actual contents of the two retorts were quite different, the lead content remained approximately constant throughout and there was a progressive increase in the zinc content towards the condenser outlet. That is, the zinc was exiting towards the bottom of the retorts while the lead, being less mobile, was largely retained within the retort.[9]

The excavations produced only one sample of metallic zinc: a small spill (52582) in a localised zinc-glazed surface on the floor of Site 30 furnace 4 (see Chapter 6). SEM examination showed the zinc to be heavily corroded to zinc oxide or carbonate especially at the grain boundaries, although some metallic zinc was still present in the centres of the grains, together with globules of lead that had precipitated at the former grain boundaries. Semi-quantitative SEM-EDX microanalysis of the surviving uncorroded zinc metal (reported in Craddock *et al.* 1998b) estimated that the zinc contained about 0.35% lead; other metals such as cadmium were below the detection limit (estimated at 0.02 % for the cadmium). A more recent SEM-EDX area analysis of the whole area including the corroded zinc in which the lead globules deposited at the grain boundaries were included suggested that the original overall composition had been approximately 98% zinc and 2% lead.

The ratio of the zinc to lead in the beneficiated and roasted ore is likely to have been between 96% and 97% zinc and 3–4% lead if similar to current beneficiated ore. It can be seen that overall through the combined retort–condenser system the proportion of lead in the zinc is progressively reduced, even while apparently remaining fairly constant through the retort itself. Thus, the residues from the tail of the retort contained approximately 2 parts zinc to 1 of lead, while in the top of the condenser the metal probably contained approximately 10 parts zinc to 1 of lead. The metal that left the end of the condenser contained approximately 98% zinc and only about 2% lead. Under the prevailing temperature and reducing conditions much of the lead was retained within the retort due to its lower volatility.

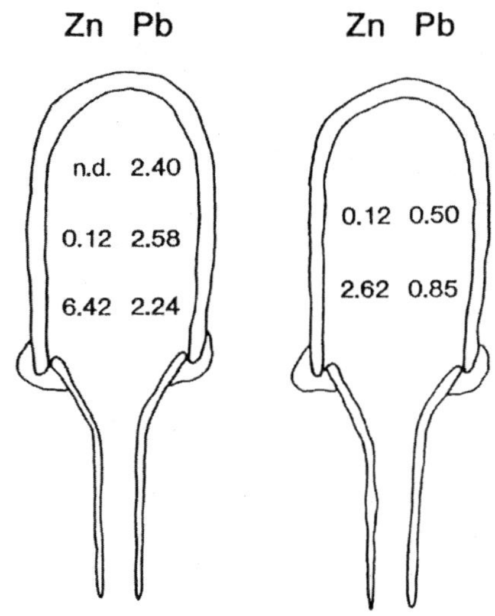

Fig. 7.5 Distribution of lead and zinc in the sintered contents of two large retorts. Note the relative consistency of the lead and the strong enrichment in the zinc content towards the bottom of the retorts.

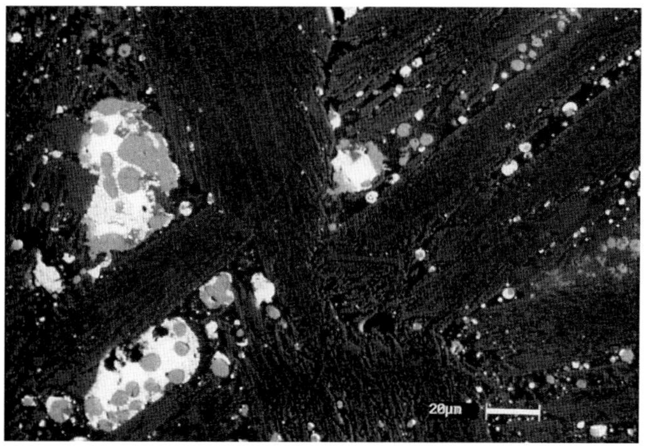

Plate 7.13 SEM photomicrograph of Type II, post-medieval slag, 32103 from Zawar Site 5 showing fayalite crystals with numerous inclusions of lead sulphide (white) and zinc-iron sulphides (grey).

Post-medieval lead/silver smelting

Slags

The later, Type II, slags, contemporary with the zinc production, may have glassy rims, but are predominantly crystalline and have a slight sheen. Their mineralogy is completely different from the earlier, Type I slags. It is dominated by the iron-rich olivine fayalite and they typically contain subordinate iron-rich clinopyroxene and iron oxide minerals. Minor sulphide phases of lead, zinc and iron are common (Plate 7.13). There are also a few inclusions of either nickel arsenide or iron-nickel arsenide speiss.

These are typical fayalite slags found on traditional silver/lead smelting sites around the world (Bachmann 1982), and more specifically at the broadly contemporary post-medieval sites at Dariba (see below). Their chemistry is correspondingly simple with around 40% silica and 35% iron oxide; the oxides of lead and zinc, calcite and alumina are present at around 5% (Table 7.1). They result from the direct smelting of galena ore as evidenced at Site 5 by the galena that had solidified within some of the slags and also in cracks in the furnace wall (Table 7.11). Sample 27353 contained 61% galena and 600 ppm silver, and Sample 27350 contained 140 ppm silver in 18.8% lead oxide, both of which, if all the silver is associated with the lead, equates to silver contents of approximately 1000 ppm, a rich ore. A spill of lead metal (27351) from the same context had 490 ppm silver.

Ceramics

Many of the fragments were heavily tempered with plant remains, which have burned out leaving numerous voids running through the ceramic bodies (see Chapter 6, Plate 6.40b). This showed that stalks several centimetres in length must have been aligned along the length of the clay cylinder used to form the furnace coil or tuyere tube (see Chapter 6, Plates 6.42 and 6.44 and Chapter 9).

Table 7.11 Composition of molten galena and lead metal from Zawar, Site 5.

BMRL no.	**Ag***	**Al_2O_3**	**(As_2O_5)**	**Au***	**[BaO]**	**Bi***	**CaO**	**(CdO)**	**CoO**	**CuO**	**FeO**	**(HgO)**	**[K_2O*]**	**MgO**	**MnO**	**[Na_2O*]**	**NiO**	**P_2O_5**	**{PbO*}**	**(S)**	**(Sb_2O_5)**	**SiO_2****	**SnO_2**	**TiO_2**	**(ZnO)**
26001 Mochia	0.038	0.46	<0.04	<0.003	0.19	<0.02	1.00	<0.005	<0.005	0.007	0.27	<0.009	0.13	0.36	0.020	0.62	<0.008	<0.07	84.6	0.02	0.03	11.3	<0.01	<0.02	0.91
27349 galena	<0.009	3.14	0.04	0.006	0.18	<0.02	4.42	0.005	0.034	0.009	40.4	<0.009	1.27	1.06	0.456	0.74	<0.008	0.18	9.61	0.03	<0.02	38.2	<0.01	0.17	0.05
27350 galena	0.014	3.26	<0.04	0.006	0.17	<0.02	3.94	0.006	0.010	0.023	40.3	<0.009	1.42	1.48	0.351	0.76	0.014	0.47	18.8	<0.02	<0.02	24.8	<0.01	0.20	3.96
27353 galena	0.060	2.24	0.06	<0.003	0.19	<0.02	1.54	<0.005	<0.005	0.019	10.0	<0.009	0.57	0.41	0.073	0.74	<0.008	<0.07	61.0	<0.02	0.05	20.2	<0.01	0.11	2.74
Lead metal:	**Ag***	**As***	**Bi***	**Cd***	**Cu***	**Fe***	**Pb***	**Sb***	**Sn***	**Zn***															
27351V	0.049	-0.08	0.01	<0.001	<0.02	<0.01	99.0	0.07	0.07	0.003															

* denotes analysis by AAS, ** SiO_2 by difference, < denotes less than the value quoted (i.e. the detection limit), round brackets ( ) denote some of analyte may have been lost during fusion, square brackets [] indicate probable contamination from glass vials/distilled water, braces { } indicate low levels of lead may be cross-contamination from sample preparation, italics denote Pb expressed as sulphide, bold denotes metal as carbonate. Precisions: c.±2-5% for major elements above 10%, ±10% for minors between 1-10%, deteriorating to ±50% at the detection limit.

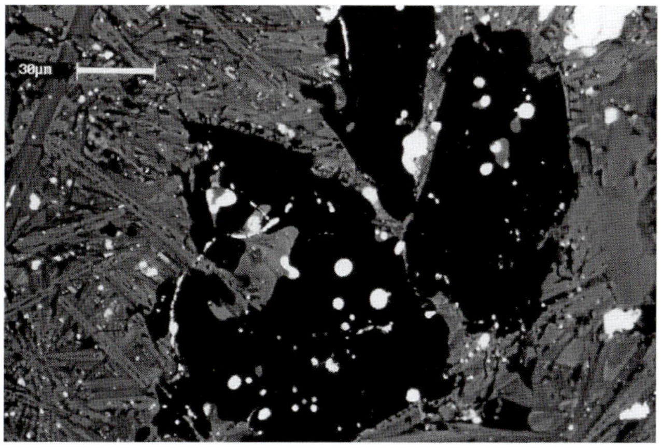

Plate 7.14 SEM photomicrograph of slag 32274 from Dariba Site 7 showing large relict sanidine with Zn-Fe-S and PbS inclusions surrounded by olivine (dendrites) and hyalophane.

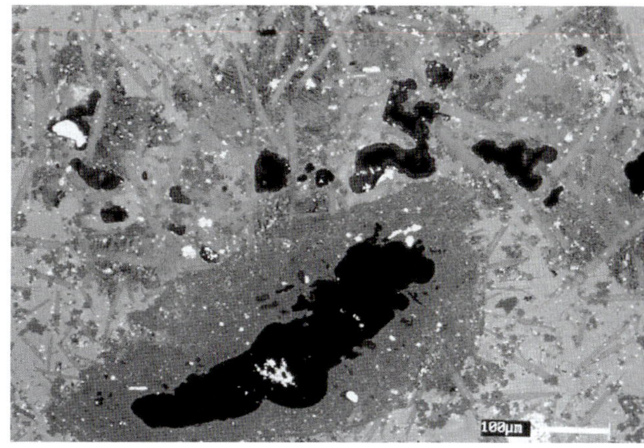

Plate 7.15 SEM photomicrograph of slag 26040 from Dariba Site 9 showing relict of metamorphic host rock with $PbCO_3$ and PbS inclusions surrounded by hyalophane.

Industrial materials from Dariba

The process debris excavated from Dariba included lead/silver production from the Mauryan period together with some evidence for copper and iron smelting. In addition, very well-preserved refractories from a foundry operation were found. The post-medieval sites investigated were all producing copper, although lead was also produced.

Early silver/lead smelting at Dariba

Industrial materials related to the extraction of lead and silver were recovered at Dariba principally from Sites 7 and 9 (Chapter 6). It is thought that these sites were closely related and the materials from the two sites are therefore considered together here.

Slags

The most striking aspect of the early Dariba silver/lead slags is their heterogeneity (Plates 7.14 and 7.15). It would appear that many of the slags had never been fully molten in the smelting installation, due very largely to their composition. In this they are similar to the contemporary slags from Agucha, but very different from those at Zawar. Overall they do have higher copper contents than those from Zawar or Agucha, although still well below what would be expected in an early copper smelting slag, as exemplified by the post-medieval slags from Sites 13–15 (see below). This elevated copper content almost certainly reflects copper in the argentiferous lead ore.

Unreacted fragments of quartz are present from the original charge together with sanidine ($(K,Na)AlSi_3O_8$) and partially dissolved aluminium silicates (Plates 7.14 and 7.15). The component that was fully molten solidified into a glass, but this can contain some crystalline phases including olivine, hyalophane, sanidine and spinel. The inclusions are mainly of sulphides including lead, iron-zinc and copper-iron sulphides, now often weathered to the sulphate. Occasional inclusions of pyrrhotite and chalcopyrite occur. The slags have a high barium and aluminium content originating from the minerals in the graphite schists, including the sanidine, which would have been part of the host rock. The high alumina content

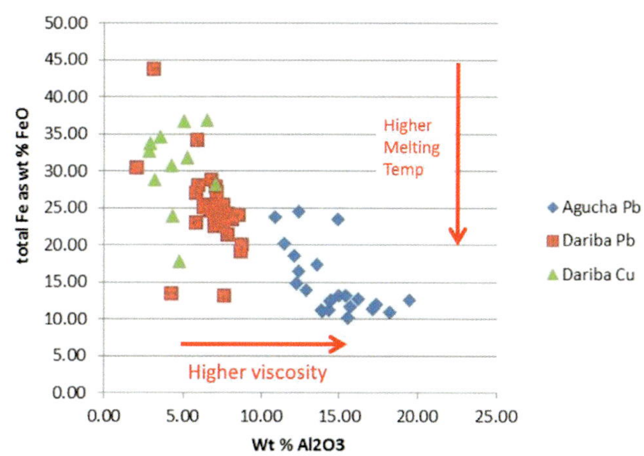

Fig. 7.6 Plot of Al_2O_3 against FeO for the Dariba and Agucha slags. The Dariba copper slags are post-medieval and their aluminium contents are lower than those found in the earlier silver-lead slags as their ores tend to come from the calc-silicate rather than from the graphite schists.

would have given the slags an elevated melting point and rendered them very viscous at the estimated smelting temperatures of between 1000 and 1300°C (Fig. 7.6). Despite this high viscosity, the low incidence of lead-rich blebs shows that the argentiferous lead was removed successfully (see Chapter 9).

Four of the early Dariba slags have silver contents over 100 ppm (Site 10 26055, 26040 and 32288, and Site 9 32278) (Table 7.12) associated with lead antimonide and arsenide speiss inclusions. The propensity for speiss to absorb silver is well documented (Sisco and Smith 1951: 48–9ff.; Tafel and Wagenmann 1953) and its formation avoided in modern silver production. Argentiferous speiss is not uncommon on ancient silver production sites as exemplified by Rio Tinto in Spain (Craddock *et al.* 1985b, 1987; Craddock 1995: 219–21) and Rammelsberg, Goslar in Germany (Hoover and Hoover 1912: 408). In general the majority of the silver is contained in the lead, but in the Dariba slags the lead inclusions are few and tiny, such that the silver content of the individual inclusions could not be determined. Overall, the slags exhibit a close correlation between the lead and silver contents (e.g. Samples 32288 and 26040).

Table 7.12 Composition of silver/lead slags from Dariba, Sites 7, 9, 10 and 12.

BMRL no.	Context	Slag type	SiO_2	Al_2O_3	FeO	MnO	CaO	BaO	MgO	PbO	ZnO	Na_2O	K_2O	Ag	Cu	Cd	Inclusions
26055	S1/L55	Ag/Pb I	30.25	5.83	34.13	0.31	8.23	4.99	1.79	2.69	4.25	1.00	2.56	0.011	0.140	0.001	PbS, Fe(Zn)S, Zn(FeS)
26054	S10/L10	Ag/Pb I	36.63	7.16	27.14	0.21	8.06	5.02	1.86	3.01	4.62	0.78	2.35	0.003	0.117	0.001	PbS, Zn(Fe)S
32284	S10/L10	Ag/Pb I	36.47	5.78	27.05	0.23	9.15	5.29	2.11	2.73	4.69	0.84	2.51	0.002	0.107	0.001	NA
32285	S10/L10	Ag/Pb I	36.10	5.95	28.09	0.22	8.42	5.01	1.80	3.11	4.84	0.76	2.43	0.003	0.117	0.001	NA
26053	S10/L2	Ag/Pb I	39.71	8.51	24.02	0.26	10.06	4.20	1.89	1.48	3.35	0.84	2.90	0.002	0.078	0.000	NA
32282	S10/L2	Ag/Pb I	38.91	7.78	23.60	0.26	10.48	4.22	1.83	1.74	3.77	1.49	3.17	0.001	0.084	0.000	PbS
32283	S10/L2	Ag/Pb I	43.29	6.98	22.57	0.25	9.48	4.26	1.72	1.52	3.84	0.81	2.65	0.002	0.099	0.000	NA
32286	S10/L55	Ag/Pb I	38.92	2.06	30.45	1.80	7.17	5.43	1.77	4.07	3.99	0.26	0.51	0.002	0.140	0.001	PbS, Fe(Zn)S
32287	S10/L55	Ag/Pb I	35.69	3.10	43.75	1.10	2.86	3.40	0.47	3.75	0.37	0.10	0.28	0.025	0.215	0.000	NA
26056	S10/L58	Ag/Pb I	37.24	7.87	24.12	0.34	9.54	4.55	1.81	3.21	5.42	0.79	2.28	0.003	0.120	0.001	NA
32288	S10/L58	Ag/Pb I	48.56	4.26	13.38	0.58	7.54	3.39	1.94	14.86	1.98	0.39	0.98	0.015	0.524	0.002	NA
32289	S10/L58	Ag/Pb I	42.67	5.76	22.96	0.85	8.83	6.18	2.23	3.06	2.78	0.70	1.30	0.003	0.098	0.001	NA
26037	S7	Ag/Pb I	46.02	7.66	13.20	0.65	17.37	4.61	2.13	0.89	3.14	0.70	2.08	0.002	0.073	0.000	NA
26049	S7	Ag/Pb I	38.33	8.11	23.50	0.45	10.27	4.78	1.92	1.77	4.85	0.75	2.53	0.003	0.104	0.000	NA
26050	S7	Ag/Pb I	38.39	7.54	25.31	0.40	10.06	4.41	1.79	1.73	4.16	0.75	2.53	0.002	0.090	0.000	?Pb & Pbs
32272	S7	Ag/Pb I	39.01	6.42	25.31	0.41	10.02	4.54	1.93	1.30	4.66	0.81	2.68	0.001	0.081	-0.000	?Pb & PbS, speiss
32273	S7	Ag/Pb I	40.21	7.58	22.63	0.41	10.15	4.48	2.17	1.44	4.13	1.29	2.86	0.001	0.101	0.000	NA
32274	S7	Ag/Pb I	36.08	7.14	25.43	0.35	10.57	4.44	2.18	1.81	3.86	2.26	2.94	0.003	0.095	0.000	NA
32275	S7	Ag/Pb I	39.67	6.31	25.06	0.42	9.94	4.60	2.05	1.68	4.23	0.71	2.42	0.002	0.091	0.000	NA
32276	S7	Ag/Pb I	37.37	6.67	24.65	0.50	10.56	5.02	1.86	1.82	5.25	1.10	2.33	0.002	0.106	0.000	NA
32277	S7	Ag/Pb I	36.47	7.00	23.96	0.41	10.89	4.84	2.03	2.17	4.30	2.63	2.49	0.002	0.098	0.000	NA
32300	S12	Ag/Pb II	39.73	7.84	21.31	0.30	9.83	3.41	2.20	5.45	2.85	2.31	2.20	0.003	0.167	0.001	NA
32301	S12	Ag/Pb II	46.07	8.76	19.93	0.19	7.67	2.62	1.87	4.05	2.13	1.36	2.76	0.009	0.300	0.001	NA
26040	S9	Ag/Pb II	39.17	8.66	19.14	1.44	6.18	4.14	1.24	6.91	9.08	0.50	1.29	0.023	0.092	0.006	PbS
32278	S9	Ag/Pb II	35.12	6.75	28.78	1.16	4.71	3.96	1.27	5.76	7.01	0.56	1.46	0.016	0.206	0.003	NA
32279	S9	Ag/Pb II	35.30	7.03	27.91	0.29	8.09	4.43	1.57	2.42	6.22	1.11	2.33	0.006	0.159	0.001	NA

Analyses by ICP-AES
Inclusions by SEM-EDXA

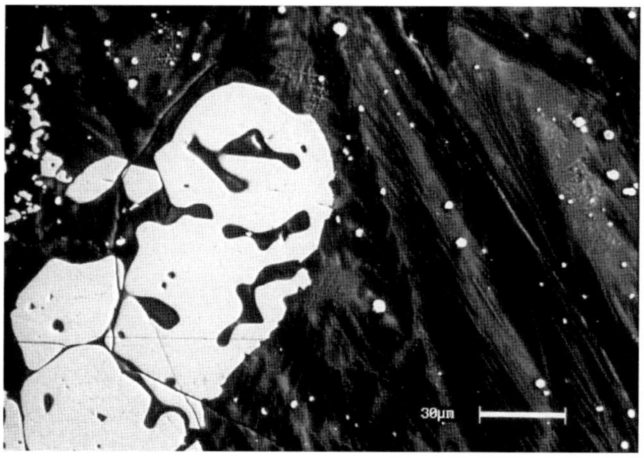

Plate 7.16 SEM photomicrograph of slagged crucible or hearth lining 27276 from Dariba Site 9 showing a large magnetite crystal (white) in a fayalite slag containing numerous prills of iron-rich copper (cf. Plate 7.17).

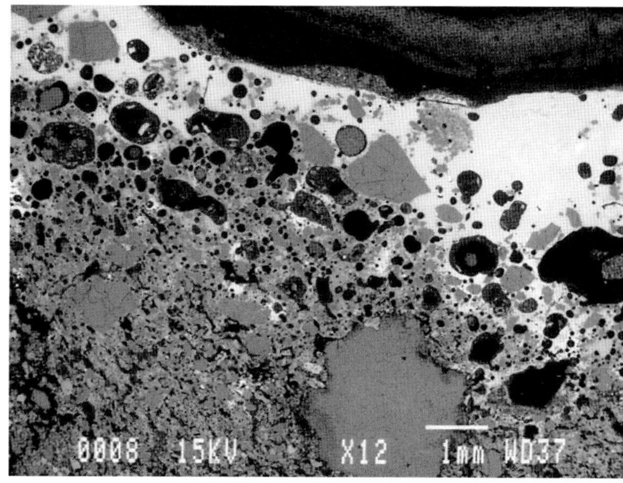

Plate 7.17 SEM photomicrograph of a polished section of hearth lining 27276 (Dariba Site 9) associated with copper production showing the iron-rich fayalitic slagging. Note the diffuse boundary with the main body of the ceramic suggestive of a relatively long process, i.e. smelting rather than melting (cf. Plate 7.16) (BEI) (scale bar: 1 mm).

Table 7.13 Composition of 'bodies' and vitrified surfaces on some ceramics from Sites 7 and 9 at Dariba.

	1	2	3	4	5	6	7	8	9	10	11	12	13
BMRL no.	**27286Y**	**33948S**	**33948S**	**33950T**	**27274V**	**27274V**	**27275T**	**27276R**	**27276R**	**27296U**	**27296U**	**27295W**	**27295W**
	furnace lining, body	furnace lining, body	furnace lining, vitrified surface	furnace lining, body	?crucible, body	?crucible, vitrified surface	tuyere stand, body	furnace lining, body	furnace lining, vitrified surface	?crucible, body	?crucible, vitrified surface	'glazed' sherd, body	'glazed' sherd, 'glaze'
SiO_2	65.3	57.8	27.3	55.5	61.1	30.9	61.3	61.1	47.5	62.2	40.0	61.0	15.8
TiO_2	0.9	1.0	0.3	0.7	0.6	0.5	0.9	nd	0.4	1.0	0.6	0.9	nd
Al_2O_3	16.4	16.3	5.4	18.3	15.7	7.1	17.1	23.4	5.9	17.0	8.0	17.3	3.9
FeO	6.5	9.2	21.5	8.4	8.0	7.8	8.2	1.0	18.7	8.3	4.1	8.3	1.7
MnO	nd	nd	0.7	0.3	nd	nd	nd	nd	0.3	nd	nd	nd	nd
MgO	2.6	3.5	1.7	4.4	2.7	1.6	3.6	1.0	3.3	3.4	1.9	3.0	0.9
ZnO	nd	nd	7.8	nd	1.1	4.8	nd	nd	1.6	nd	4.4	nd	nd
CaO	1.0	7.1	9.9	4.9	2.4	2.3	1.5	2.0	14.0	3.1	3.3	3.6	1.2
CuO	nd	nd	nd	nd	nd	nd	nd	nd	1.7	nd	nd	nd	nd
BaO	nd	nd	4.5	nd	nd	nd	nd	nd	0.7	nd	0.8	nd	nd
Na_2O	2.1	3.0	1.9	3.3	1.2	1.8	2.5	3.1	1.0	1.0	2.1	1.7	nd
K_2O	3.8	1.8	3.4	3.3	2.0	1.4	3.8	6.2	1.8	3.0	1.3	3.0	0.4
PbO	nd	nd	15.1	0.5	4.3	41.5	0.8	2.1	2.6	0.8	33.1	0.8	75.8
P_2O_5	nd	nd	0.4	nd	nd	nd	nd	nd	0.5	nd	nd	nd	nd
SO_3	1.2	nd	nd	0.3	0.5	nd	nd	nd	nd	nd	nd	nd	nd
Cl	0.1	0.2	nd	nd	0.3	0.2	0.1	nd	nd	nd	0.2	0.2	0.3
	99.9	99.9	99.9	99.9	99.9	99.9	99.8	99.9	100.0	99.8	99.8	99.8	100.0

Three slags, all from Site 9, layer 4 (26041, 32280 and 32281), have very different compositions (Table 7.12) and mineralogies from the other contemporary Dariba slags. They possess very high iron contents with low lead and copper contents, and much lower silica and alumina contents, resulting in a much lower melting temperature than the contemporary silver/lead smelting slags. They are homogeneous having solidified from a completely molten slag, and their mineralogy is of fayalite (Fe_2SiO_4) with some wustite (FeO). There can be no doubt that these are iron smelting slags produced by the direct, so-called bloomery process. It is just possible that these slags were brought in from elsewhere to flux the lead or copper smelting as was practised in India in medieval and more recent times (see Chapters 2 and 9), but their prevalence in Layer 4 makes it more likely that they are the product of iron smelting being carried out in the immediate vicinity.

Sample 27276, from Site 9, is a heavily slagged and vitrified ceramic from a large crucible or hearth lining (see below). The micrograph of the slag component (Plate 7.16) shows a large area of magnetite (Fe_3O_4) set in a fayalite slag which contains many inclusions of ferruginous copper. The vitrified layer is about 2 mm thick but not well defined, having a rather diffuse boundary with the underlying ceramic (Plate 7.17). XRF analysis showed that it contains appreciable quantities of copper together with small amounts of lead and zinc (Table 7.13, column 9).[10]

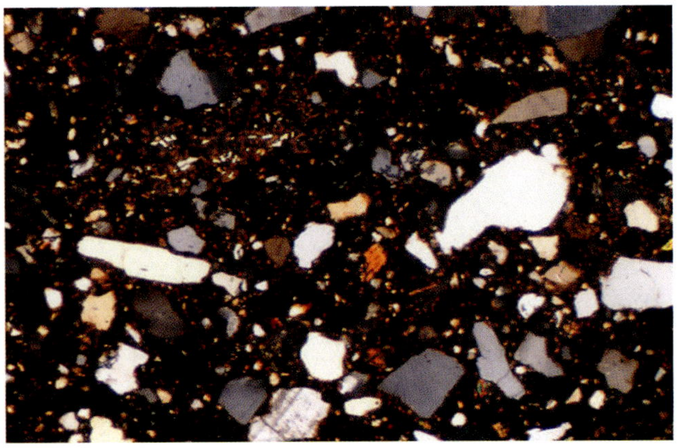

Plate 7.18 Photomicrograph of thin section of hearth lining 27274 (Dariba Site 9). Plane polars, width of field c.5 mm.

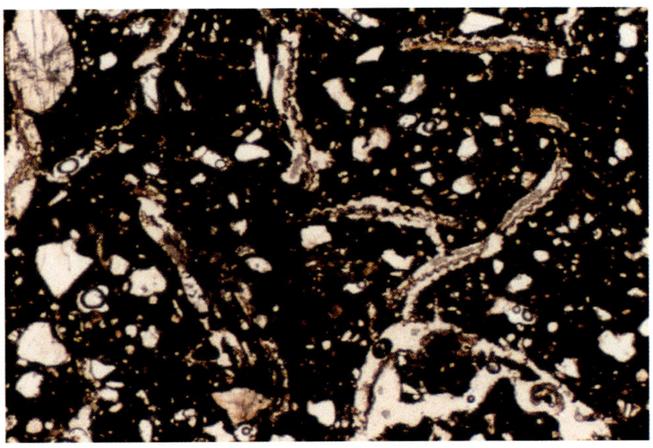

Plate 7.19 Photomicrograph of thin section of furnace lining 27286 illustrating the fabric typical of many of the ceramics found at Sites 7 and 9 at Dariba. Note the plant temper (cf. Plate 7.20). Crossed polars, width of field c.5 mm.

Ceramics

The ceramic material recovered from among the slag that dominates Site 7 included six broken pieces identified tentatively as fragments of curved hearth or furnace lining. All are very weathered and irregular in shape, with no diagnostic features that might indicate any particular function; they are of variable thickness, ranging from c.10 mm to several centimetres. To some extent this may be a function of fragmentation and abrasion after discard and during burial under the slag. Most have vitrification on their concave surfaces, which are mainly dark grey-black in colour (i.e. they were exposed to a reducing atmosphere during use), but may grade into red, oxidised material.

Several examples from Site 9 (27294, 27295, 27297, 27299–27301) appear to be pottery sherds reused in the smelting process. With the exception of 27295, these are from relatively large diameter vessels, with wall thicknesses typically approaching 10 mm; several exhibit coarse ribbing on their outer (concave) surfaces. The inner, concave surfaces of these sherds are usually heavily vitrified. A single example (27295) (see Chapter 6, Fig. 6.54 and Plate 6.55) has thinner walls (c.5 mm below the thickened rim) and a 'glaze' that extends over both surfaces as well as the broken edges, indicating that it was broken before or during its use for some metallurgical purpose. The composition is also very different to the other 'glazed' sherds (see further discussion below). Another of the sherds from Site 7 (27294) also exhibits a 'glaze' that extends over one of the break surfaces.

A group of samples from Site 9 (27274, 27296 and 27298) share a similar appearance to the fragments of hearth or furnace lining described above from Site 7. One sample (27298) has a well-formed rim from which the vitrification clearly runs away. This shows that the rim was above the main operation and thus that the ceramic was in a bowl position, that is, acting as a hearth lining during the smelt rather than inverted as a hood. The other pieces, typically c.15–20 mm thick, may also be fragments of hearth linings.

XRF analysis confirmed that the surfaces of all of the furnace/hearth linings and reused sherds (with the exception of 27276, discussed above) are ferruginous with lead and zinc. The surface of the unvitrified tuyere stand (27275) is also ferruginous and was found to contain a small proportion of lead. Microanalysis in the SEM affirmed the XRF finding that the vitrified layers on these ceramics contain variable amounts of lead (with the exception of 27295 which has massive amounts), with a few percent also of zinc, while copper was below the detection limit (with the exception of 27296) (Table 7.13, columns 3, 6 and 11).

Thin sections for petrographic examination were prepared from a selection of the samples, supplemented by observations made in the SEM, discussed more fully in the section on process conditions below. The fabrics of the ceramics from the two sites are generally similar to one another: all are made from coarsely tempered, ferruginous clays (Plate 7.18). The angular temper ranges in size from silt to grains several millimetres across and are likely to have been obtained from a sedimentary deposit. These aplastic grains are mainly quartz and feldspar but there is also a suite of minerals more clearly diagnostic of derivation from metamorphic rocks; these may include staurolite, garnet, amphibole and kyanite (and occasional rock fragments). A few fragments of slag were observed, probably fortuitous additions.

Most sections include regions of vitrified and bloated ceramic. In non-vitrified regions, not exposed to such high temperatures, the fine matrix may be orange-brown in colour and birefringent. In some of the samples from Site 9 there are some rather irregular areas of calcite, which appears mainly to fill voids and may be a post-depositional feature.

The ceramics from Site 7 have a few elongated voids indicative of chaff from wheat (*Triticum* sp.) and barley (*Hordeum* sp.) incorporated into the mix. In general the ceramics from Site 9 have fewer such vegetal voids, except for the hearth linings 27274, 27296 and 27298 which have more abundant voids from sparse to common plant remains used for temper (Plate 7.19). Some vestigial structures from the plant remains survive. These have been identified as chaff from wheat, barley, sorghum (*Sorghum* sp.) millet (*Pennisetum* sp. and *Setaria* sp.) and rice (*Oryza* sp.), as well as fragments of chickpea (*Cicer* sp.), rice and sorghum seeds/husks. This fabric is similar to that used for the crucibles recovered from Site 10 (see below). It is interesting to note the highly comminuted nature of the wheat and rice stem fragments and especially the wheat, rice and sorghum husk fragments from 27296. The condition and fragmentation of these plant remains appear typical of dung, unlike Samples 27274, 27298, 27275 (Site 9) and 27286 (Site 7) in which the chaff and husk fragments were added in the form of threshing or crop processing waste. This is clearly important from a technological perspective (see below) but it also implies a close relationship with and proximity to arable and pastoral farmers.[11]

The 'glazed' potsherd (27295) from Site 9 has a much finer fabric than the other ceramics, containing only sparse aplastic grains coarser than silt grade, mainly quartz and feldspar, but with very occasional metamorphic minerals. It is characterised by common plant remains temper, identified as highly abraded and fragmented wheat and rice stem chaff, and wheat, rice and sorghum husk fragments, probably added in the form of dung.

To summarise, these observations suggest that these ceramics, thought to have been used in the metallurgical operations, were made from local clays derived from the underlying metamorphic rocks (Chapter 4). Plant remains temper was added to all the clays (and as dung in some instances), and this too must have been obtained locally. The ceramics were produced from clays very similar or identical to those used for domestic pottery. At both Site 7 and Site 9 sherds of household pottery appear to have been used in the metallurgical processes (the use of these sherds is also discussed in the section on cupellation below).

Process conditions

Analysis of the bodies of those ceramics thought to have been associated with the smelting of lead (Table 7.13, columns 1, 2, 4, 5, 7, and 10) showed the 'clay pastes' (i.e. excluding grains coarser than about 50 μm) to be ferruginous, with low to moderate amounts of soda and potash and rather variable amounts of calcium. They are, therefore, not particularly refractory and comparison of the vitrified textures with those of fired materials of similar refractoriness indicates exposure to maximum temperatures in the region of 1150°C.

The vitrified and bloated textures of the furnace lining fragments close to their 'hot' slagged surfaces grade sharply back into an unvitrified texture in which the aplastic grains can be seen, set in a silty matrix. This indicates a relatively steep thermal gradient during use, suggesting that the process for which the installation was used was of short duration, probably not more than about an hour. As expected the fabric of the tuyere support (27275) showed that it had not been exposed to the full heat of the fire.

The 'glazed' sherds from Sites 7 and 9 were initially thought to have been associated with cupellation or to the related process of scorification (see Chapter 10). However, with the exception of 27295, minor amounts of zinc were found in their slagged and vitrified surfaces which could not have survived such processes, but instead demonstrates that they were associated with the primary smelting of the argentiferous lead ore, possibly serving as hearth linings.[12] The short duration is surprising, but is supported by the survival of large inclusions in the slags, such as Sample 32274 (Plate 7.14) for example, which are likely to have survived from the original ore.

Cupellation

Although it seems likely that lead was being smelted not so much for its own value as for the silver which it contained, evidence for the recovery of silver by cupellation of lead metal at Dariba is limited to the tuyere support, 27275 (see Chapter 6, Fig. 6.53), which is very similar to those found at Agucha (see below), and one pottery sherd (27295) (see Chapter 6, Plate 6.55 and Fig. 6.54). This reused sherd from Site 9 now bears a thin glaze that extends over the break surfaces. The vesicular glaze layer is of variable thickness but typically c.250 μm. It is mostly pale yellow in colour but there are a few patches of pale turquoise. The boundary with the underlying body is sharp. The vesicles in the glaze and also many of the voids within the body of the sherd are filled by a crystalline carbonate mineral, which originally was probably lead oxide. XRF analysis of the surfaces of this sherd showed that the glaze is lead-rich and this was confirmed by EDX microanalysis in the SEM (Table 7.13, column 13). The body is vitrified throughout, with fine to medium bloating pores, indicating that it was exposed to a temperature in the region of 900–1000°C, significantly lower than the hearth/furnace lining fragments. The uniform nature of the vitrification texture of the glazed sherd precludes estimation of the period for which it was used. These observations, in particular the lack of zinc and very high lead content, are consistent with, but do not confirm, its association with a process of cupellation.

Copper alloy processing

At Site 10 a wide range of metallurgical debris was recovered from the excavation of the well section, including slags and a considerable quantity of ceramic artefacts. The slags from the site were all from lead smelting operations, identical to those from Sites 7 and 9 (and included in Table 7.12). They came either from the layers above those producing the refractory materials or as a minor component within those layers, and are likely to be intrusive from the smelting activities in the immediate vicinity therefore they will not be considered separately here.

Ceramics

The ceramic materials included hearth or furnace linings, tuyeres, numerous crucibles, discs that may have functioned as crucible lids, fired clay blocks that may have been used for melting small quantities of metal, and potsherds. Qualitative analysis using XRF showed that all of these samples have tin concentrated in their working surfaces, along with copper. Tin, which is not found in any of the ores, was not detected in any other samples from Dariba and was not present in the background material from Site 10. This clearly indicates that the refractories were all associated with the processing of bronze.

Examination in the petrographic microscope and the SEM showed that most of the ceramics from Site 10 were made from clays similar to the majority of those used at other sites at Dariba. Typically, the clay paste of the crucibles (Plate 7.20) contains relatively common silt/fine sand, which is angular and consists mainly of quartz and feldspar. In addition, there is sparse coarser sand up to c.1 mm, which includes some feldspar and rare grains of metamorphic minerals such as staurolite and garnet. There is some fine-grained carbonate (probably calcite) in some sections occurring mainly as linings or infillings of voids and around the margins of the sherds that is likely to have entered during burial. Bright orange infillings of the hydrated zinc silicate mineral, allophane, seen in some of the sections, are also thought to be secondary in nature. In most examples, the matrix is vitrified and bloated, but one example (27235, a crucible lid) includes a non-vitrified region, which can be seen to be micaceous. The elemental composition of the ceramics used for the crucibles and tuyeres was assessed by EDX analysis

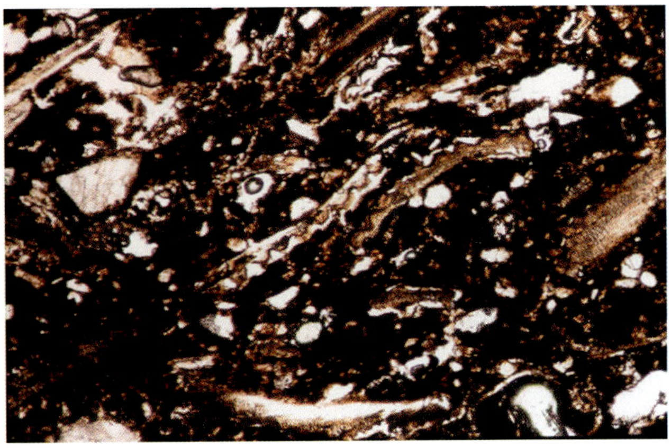

Plate 7.20 Photomicrograph of thin section of fragment of crucible 26319 (Dariba Site 10) illustrating the vegetal fibre-rich fabric of the crucibles. Plane polars, width of field c.2 mm.

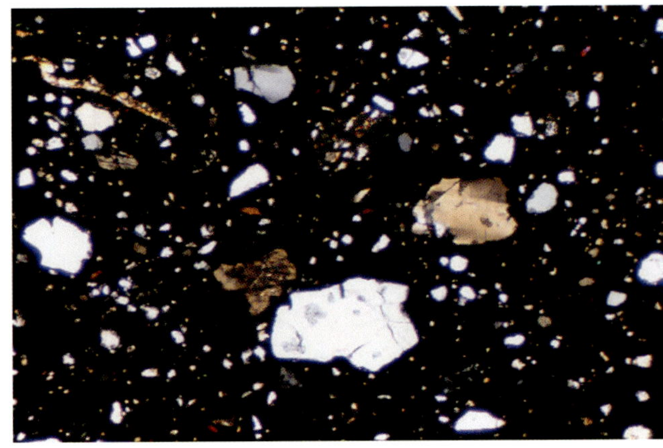

Plate 7.21 Photomicrograph of thin section of fragment of tuyere 27249 (Dariba Site 10), Plane polars, width of field c.2 mm.

Table 7.14 Composition of some of the ceramics examined from Site 10 at Dariba.

	1	2	3	4	5	6
BMRL no.	27235Y	27323Y	27253U	27249X	26322P	26322P
	crucible lid	crucible	tuyere	tuyere	crucible	crucible
	bulk	bulk	bulk	bulk	glassy phase	glassy phase, between iron spherules
	4	4	3	3	5	3
SiO_2	63.3	70.6	61.4	60.4	64.0	54.2
TiO_2	0.9	1.0	0.7	1.1	0.6	0.6
Al_2O_3	15.9	13.0	15.1	15.9	17.2	29.1
FeO	7.3	5.9	6.9	7.9	8.3	3.5
MnO	nd	nd	nd	nd	0.1*	0.2*
MgO	2.9	3.1	3.5	4.4	2.7	3.4
ZnO	nd	nd	nd	nd	nd	nd
CaO	1.3	1.5	5.1	2.0	1.2	2.9
CuO	nd	nd	nd	nd	0.2*	nd
BaO	na	na	na	na	na	na
Na_2O	2.3	1.0	2.8	2.3	1.8	2.4
K_2O	5.5	2.7	3.4	4.6	3.3	3.3
PbO	nd	nd	nd	nd	nd	nd
$P2O_5$	nd	0.6	0.3	nd	0.3	nd
SO_3	0.2	0.2	0.4	0.7	0.2	nd
Cl	0.3	0.4	0.3	0.6	0.3	0.2*
	99.9	100.0	99.9	99.9	99.9	99.4

* not present in all analyses

of areas selected to exclude aplastic (mainly quartz) grains coarser than c.50 μm (Table 7.14, columns 1–4). The analyses showed that the clays used are ferruginous (c.7% FeO) and that the total alkali content is rather variable (c.4–8%), probably reflecting the inhomogeneous distribution of mica. All the crucible sherds contain abundant plant remains temper, identified as short 'chopped' lengths of rice and sorghum stems, rice and sorghum husk fragments and short lengths of wheat and barley straw. Sherds of several other ceramic artefacts (large rim vessels or tiles, tuyeres) from the site were in a similar fabric, again with a micaceous clay paste and plant remains temper (mainly wheat and barley chaff) but in smaller proportions and with more sand temper (Plate 7.21).

The two heating blocks – a clay wedge, a tuyere and several other artefacts, including all of the pottery sherds – were made from a similar clay paste but contain little (the bricks and heating blocks) or no (the pottery sherds) plant remains temper. The plant remains temper in the heating/assay block 27234 consisted of abraded rice and sorghum stem and husk fragments, most probably added in the form of dung. The single example of furnace/hearth lining examined, 27229, is heavily vitrified and bloated, so that microscopical observation was difficult. It appears to be rather similar to the fabrics described above except that it was very coarsely tempered with quartzose fragments up to several millimetres in diameter. The small amount of plant remains temper present included wheat and barley stem chaff.

The clay used for the brick, 27227, is banded in lighter and darker brown, but is distinctive mainly in the present context because it contains very little aplastic material. There

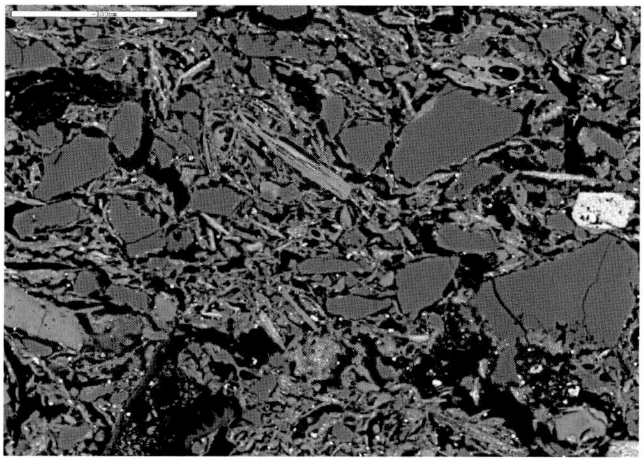

Plate 7.22 SEM photomicrograph of polished section of crucible fragment 26323 (Dariba Site 10) showing less vitrified portion of the ceramic (BEI) (scale bar: 100 μm).

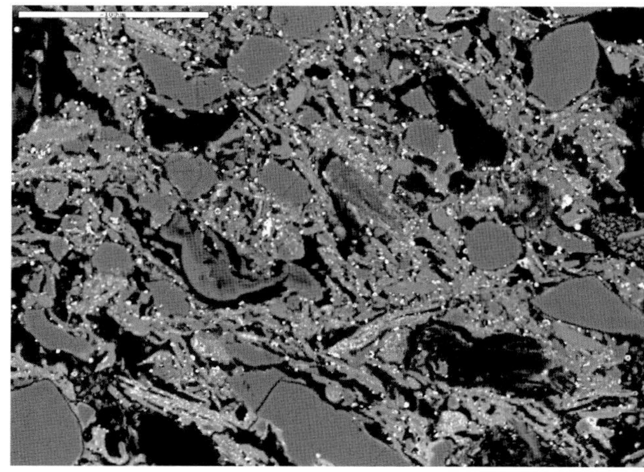

Plate 7.23 SEM photomicrograph of polished section of crucible fragment 26323 showing vitrified texture and fine sub-spherical (white), iron-rich particles (BEI) (scale bar: 100 μm).

is rare-sparse silt and occasional aplastic grains up to c.0.5 mm. The fine matrix contains abundant mica and the limited areas of fine-grained carbonate are thought to be secondary. Once again, petrographic examination has demonstrated that local clays, derived from underlying metamorphic rocks, were used to make the industrial ceramics. Although similar 'base clays' were used for all materials, it appears that these were modified by the addition of coarse quartz and of plant remains temper, according to the intended purpose of the ceramic. In particular, it is evident that the crucibles and, to a lesser extent, the tuyeres used for post-extractive metallurgical processing at Site 10, were made in a distinctive fabric containing an exceptionally large proportion of plant remains temper (cf. the furnace/hearth ceramics from Site 9 discussed above). The addition of plant remains temper to modify the behaviour of pottery clays during forming, firing and use is well known in India and elsewhere (London 1981; Gaimster 1986; Rice 1987: 407; Tomber *et al.* 2011), and there can be little doubt that in this instance the clay paste was modified in the knowledge that this would enhance the metallurgical performance of the crucibles. The medieval Indian *silpasastra* treatises on casting operations frequently specify the addition of plant material to the clays forming crucibles and other foundry ceramics (Reeves 1962).

During use the crucibles would have been embedded in a bed of hot charcoal, which might itself have been sufficient to produce reducing conditions within their fabric. However, the very high levels of plant remains temper included in the clays used to make the crucibles would also have contributed to this reducing atmosphere and to a general enhancement of their refractory performance (Freestone and Tite 1986 and Freestone 1989, who discussed the analogous use of rice husk temper in crucibles used in the production of 'wootz' steel in India). During the initial firing of the crucibles only a small proportion of the carbon associated with the abundant, plant material temper would have been burnt out. When used subsequently for metal melting, the remaining carbon would have acted to reduce the iron oxides in the clay paste to the metallic state. Evidence for this process is now seen as the fine spherules of iron observed in the more heavily vitrified (i.e. hotter) parts of the fabric, closer to the convex external surfaces (Plate 7.23). This reduction of the iron in the fabric had the effect of reducing by more than half the amount of iron available to combine with the silica, thereby reducing vitrification (compare columns 5 and 6 in Table 7.14).

As noted by Freestone, the use of organic temper under these conditions would have been important in helping to maintain a reducing atmosphere within the bowl of the crucible and the resulting voids would also have conferred improvements in other properties, such as heat transfer, shrinkage and thermal shock resistance (Skibo *et al.* 1989; Tite *et al.* 2001). Ceramics with a good thermal shock resistance should be low-fired and contain large quantities of platy or fibrous temper. Cracks developing in the low-fired ceramics were of lower energy and more easily dissipated, and also rapidly ran into a void left by burnt-out plant remains. Quantities of quartz temper (typically more than 20% and in the size range of 100–700 μm) also lead to a significant increase in thermal shock resistance because of the differential thermal expansion on firing and cooling: the clay expands and contracts much more than the quartz-creating regions of microcracks in the immediate vicinity of the quartz particles. It must be borne in mind that although low firing temperatures and large quantities of temper are good for thermal shock resistance, they leave the ceramic physically weak, which can be a problem especially for crucibles that must contain a load of heavy metal. This is one of the main reasons why crucibles before the post-medieval period tend to be small.

Process conditions

Polished sections from two of the crucible fragments, 26322 and 26323, two of the tuyeres, 27249 and 27253, and also from the crucible lid, 27235 (see Chapter 6, Fig. 6.60) were examined in the SEM and analysed by EDX to provide evidence of the conditions under which they were used. All exhibit a variation in texture from a continuously vitrified, sometimes fine-medium bloated texture, to a vitrified texture with little bloating (Plate 7.23) in regions exposed to less severe conditions. The vitrification textures observed in the sections of the tuyeres suggest that they were exposed to temperatures in the region of 950°C. The ceramics of the tuyeres are red, indicating the oxidising conditions expected in the tubes supplying air, whereas the dark grey-black colour of the crucible fabrics suggests that these were situated

Table 7.15 Composition of copper and iron slags from Dariba, Sites 14, 15 and 16 (copper), Site 9 (iron).

BMRL no.	Context	Slag type	SiO_2	Al_2O_3	FeO	MnO	CaO	BaO	MgO	PbO	ZnO	Na_2O	K_2O	Ag	Cu	Cd	Inclusions
26057	S14	Cu	44.03	7.03	28.18	0.30	10.11	0.59	3.92	0.09	0.11	0.63	1.52	0.001	0.270	nd	Cux(Fe)S
28499	S14	Cu	34.99	6.49	36.93	0.32	11.66	0.21	3.11	0.19	0.14	0.28	0.79	0.001	0.628	nd	NA
32290	S14	Cu	43.39	4.24	30.75	0.24	10.27	1.75	3.21	0.29	0.19	0.76	0.93	0.001	0.451	0.001	NA
32291	S14	Cu	41.86	5.22	31.86	0.24	9.03	0.70	3.83	0.10	0.16	1.67	1.35	0.000	0.351	nd	Cu(Fe)S
32292	S14	Cu	36.42	5.02	36.73	0.30	11.95	0.21	3.02	0.18	0.17	0.33	0.85	0.000	0.586	nd	Cu_2S
32293	S15	Cu	38.39	2.82	32.67	0.47	12.76	2.02	4.37	0.59	0.64	0.48	0.63	0.000	0.419	0.000	CuxS & PbS
32294	S15	Cu	41.84	3.54	34.59	0.39	7.85	2.27	2.66	0.63	0.79	0.27	0.75	0.001	0.442	nd	$Cu(?Pb)_2S$
32295	S15	Cu	44.78	3.18	28.81	0.40	11.13	2.11	3.78	0.53	0.44	0.36	0.81	0.001	0.363	nd	Cu_2S, $Cu(?Pb)_2S$
32296	S15	Cu	40.52	2.86	33.67	0.40	9.85	1.75	3.58	0.71	0.57	0.39	0.60	0.002	1.073	0.000	NA
32298	S16	Cu	42.97	4.74	17.79	0.18	17.32	0.12	10.08	0.11	0.06	0.71	0.68	0.001	2.608	0.000	NA
32299	S16	Cu	41.76	4.32	23.86	0.15	16.47	0.20	7.81	0.05	0.11	0.95	1.07	0.000	0.463	nd	NA
26041	S9/L4	Fe	20.78	3.24	62.51	0.22	3.99	0.14	1.15	0.21	0.11	0.29	0.39	0.001	0.005	0.004	NA
32280	S9/L4	Fe	20.41	4.23	58.60	1.07	4.45	0.37	1.61	0.15	0.13	1.68	0.76	0.000	0.010	0.001	NA
32281	S9/L4	Fe	22.30	3.04	60.47	0.22	4.73	0.15	1.21	0.23	0.19	0.31	0.41	0.000	0.020	0.011	NA

Analyses by ICP-AES
Inclusions by SEM-EDXA

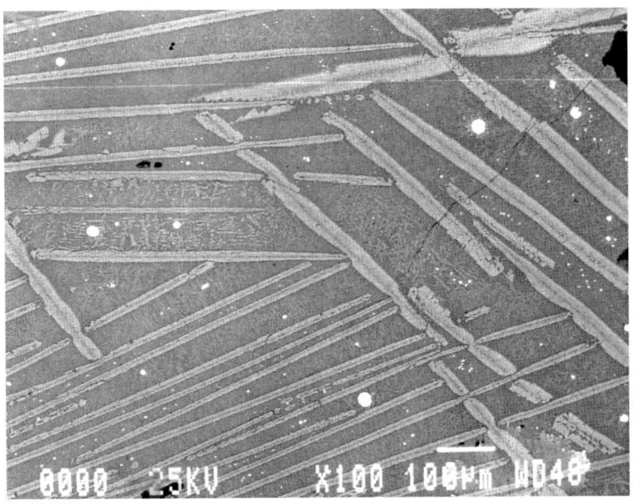

Plate 7.24 SEM photomicrograph of slag 32294 from Dariba Site 15 showing laths of fayalite and inclusions of copper iron sulphide.

in a part of the furnace where the supply of air was more restricted. After making some allowance for this effect it is estimated from the degree of vitrification of the crucibles that they were heated in a charcoal bed that raised the crucible to a temperature in the region of 1150°C. As the crucibles were used to melt copper and bronze, temperatures in excess of 1100°C would have been required. The crucible fabrics were vitrified throughout therefore it was not possible to estimate the duration of the process.

The hotter surfaces of the crucibles, which exhibit the most developed bloating, correspond to the convex, exterior surfaces. The small nipples on the bases of the crucibles would have served to lift the bases slightly thereby improving airflow and heat circulation beneath the crucible. This further demonstrates that the fire was around and beneath these rather deep crucibles rather than above them, as is often the case with shallower vessels.

Post-medieval copper smelting

The archaeological evidence and radiocarbon dating indicate that Sites 13, 14 and 15 are contemporaneous and were active at a much later date than the other metallurgical activities investigated at Dariba. All three sites yielded both slag and ceramic 'refractories'.

Slag

The high copper content and very low lead content (Table 7.15) strongly suggest that the slags from Sites 13, 14 and 15 are from copper smelting operations. The mineralogy of these slags is very different from the earlier silver/lead slags. Typically they are homogeneous and have a large quantity of fayalite, with copper being mainly concentrated in the inclusions of either copper metal or copper iron sulphide (Plate 7.24).

Ceramics

The samples of ceramic debris include fragments of furnace wall and lining, and also tuyeres and the clay linings that appear to have been used to lute them into the furnace wall (see Chapter 6). Many of the fragments, particularly the tuyeres, were heavily tempered with plant remains, which have burned out leaving numerous voids running through the ceramic bodies. This showed that stalks several centimetres in length must have been aligned along the length of the clay cylinder used to form the furnace coil or tuyere tube (see Chapter 6, Plate 6.63b and c and Chapter 9). The impressions left by the organic material were characteristic of rice, wheat and sorghum chaff fragments. In the rim coil from a furnace from Site 13 (27278) the chaff fragments are severely abraded, typical of dung. Most of the samples bear dark, almost black, slaggy deposits, and the ceramic in the vicinity of these deposits is often bloated and dark grey-black in colour, presumably in response to the hot reducing atmosphere of the furnace.

Petrographic examination confirmed the indication from the macroscopic examination that all the furnace components were made from local ferruginous clays, similar to those used at the

Table 7.16 Composition of some furnace linings and tuyeres from Sites 13, 14 and 15 at Dariba.

	1	2	3	4	5	6	7	8
BMRL no.	27278	27279	27270	27271	27284	33946	33947	33951
	furnace lining	furnace lining	furnace lining	furnace lining	tuyere	tuyere	tuyere	furnace lining
	Site 13	Site 13	Site 14	Site 14	Site 14	Site 14	Site 14	Site 15
SiO_2	63.0	63.6	71.6	68.5	67.6	70.1	70.5	59.5
TiO_2	1.1	0.7	0.6	0.4	0.8	0.6	0.6	0.7
Al_2O_3	16.6	14.5	10.7	13.0	13.3	12.9	12.3	17.8
FeO	7.8	10.2	7.0	9.3	8.9	6.3	7.1	9.4
MnO	nd	nd	nd	nd	0.2	0.2	0.1	nd
MgO	3.2	3.8	1.9	2.7	1.9	2.3	1.9	3.2
ZnO	nd	1.6	0.7	nd	0.2	na	na	nd
CaO	2.5	0.8	1.5	1.3	2.0	1.9	2.4	2.1
CuO	nd	nd	nd	nd	nd	nd	nd	nd
BaO	nd	nd	nd	nd	na	na	na	nd
Na_2O	2.3	nd	0.9	1.0	1.4	1.4	0.6	2.9
K_2O	3.1	2.4	2.2	2.2	2.6	2.0	2.5	3.1
PbO	nd	1.6	1.2	1.0	nd	1.0	nd	0.3
$P2O_5$	nd	nd	nd	nd	0.7	0.9	1.5	nd
SO_3	nd	0.4	0.4	nd	0.1	nd	0.4	0.2
Cl	0.1	0.2	0.4	0.2	0.2	0.2	0.2	0.4
	99.7	99.8	99.1	99.6	99.9	99.8	100.1	99.6

earlier installations at Sites 7, 9 and 10. All the samples from Sites 13, 14 and 15 contain common to abundant angular silt (mainly quartz), together with sparse to common coarser grains (up to several millimetres across). These coarser grains include metamorphic rocks and minerals such as mica schist, quartzite, meta-siltstone and amphibole, and also occasional fragments of slag. The clay matrix is sparsely micaceous. The chemical composition of the clay pastes, determined on areas selected to exclude most aplastic grains coarser than about 50 μm, are summarised in Table 7.16. These show the ceramics to be of only moderately refractory composition. No clear distinction was noted between the fabrics of the furnace linings, coils and the tuyeres. In many of the pieces examined there were vitrification and bloating of the fine matrix in those areas closest to the slaggy layers. The slaggy material has de-vitrified in parts with the development of fine spinifex crystals of fayalite.

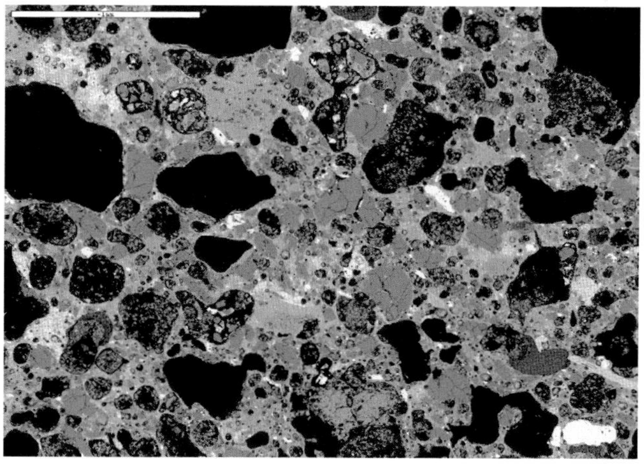

Plate 7.25 SEM photomicrograph of polished section of tuyere fragment 33947 (Dariba Site 14), close to the slagged 'hot' face of the tuyere, showing the vitrified and bloated texture of the ceramic body (BEI) (scale bar: 1 mm).

Process conditions

The vitrification textures exhibited by the polished sections in the SEM were all rather similar. Close to the slaggy, 'hot' face, the fine matrix is continuously vitrified, with the development of very coarse bloating pores (Plate 7.25). These textures grade back to areas in which the individual flakes of clay/mica can be discerned but still show some indication of vitrification at the grain boundaries. Based upon the composition and degree of vitrification observed, it appears that the smelting process undertaken at these sites was carried out at temperatures in the region of 1150–1200°C and that the process was operated at this temperature for a relatively short period of about two hours, that is the lower end of expectations for a smelting process. The contemporary 19th-century descriptions (see Chapter 2) suggest smelting times of between 6 and 12 hours for a typical copper smelt. The estimated times obtained from the furnace linings for the latter process imply that the quoted figures must have referred to the total duration of the smelt including the preheat of the furnace prior to the actual smelt, and that the process was only running at the high temperature required to melt the copper and form a molten fayalite slag for much shorter periods (see Chapter 9).

Industrial materials from Agucha

Slags found at the two sites excavated at Agucha are thought to have been by-products from the smelting of argentiferous lead with some freibergite. Most of the materials recovered from the excavated well section site are industrial ceramics that appear to have been associated with ore assaying and silver production of silver from argentiferous lead by cupellation.

Table 7.17 Composition of slags from Agucha.

BMRL no.	Context	SiO_2	Al_2O_3	FeO	MnO	CaO	BaO	MgO	PbO	ZnO	Na_2O	K_2O	Ag	Cu
32071	S1/L3	49.53	16.18	12.65	0.20	7.04	0.07	4.36	2.97	0.36	1.57	3.65	0.003	0.018
26058	S1/L3	48.24	15.01	13.06	0.23	7.84	0.08	4.18	3.62	0.12	1.60	3.69	0.003	0.014
32070	S1/L3	55.64	14.49	12.55	0.14	5.71	0.06	3.01	2.30	0.20	1.13	3.35	0.003	0.031
32072	S1/L4	44.55	13.60	17.33	0.46	8.99	0.08	3.65	4.25	0.20	2.07	2.87	0.002	0.015
26059	S1/L4	48.63	12.87	13.92	0.23	8.57	0.06	3.99	5.65	0.25	1.47	2.76	0.003	0.029
32073	S1/L4	40.93	12.08	18.54	0.38	8.11	0.07	3.72	10.15	0.39	1.29	2.25	0.006	0.018
26060	S1/L5	48.11	15.43	13.16	0.24	7.88	0.05	4.44	3.52	0.95	1.52	3.22	0.003	0.015
32075	S1/L5	40.28	11.42	20.13	0.41	7.60	0.09	3.39	10.50	0.25	1.53	2.14	0.005	0.015
32074	S1/L5	40.60	10.88	23.70	0.38	7.08	0.08	3.12	8.31	0.43	1.42	1.33	0.006	0.019
32076	surface	44.61	12.23	14.79	0.23	11.02	0.17	3.69	6.04	0.29	2.67	2.57	0.002	0.040
32077	surface	45.77	12.33	16.36	0.54	10.71	0.19	3.97	3.28	1.11	1.59	2.30	0.002	0.025
26061	surface	39.64	12.37	24.45	0.50	8.27	0.08	3.27	3.98	0.15	1.34	3.22	0.004	0.020
32079	well	47.29	15.70	11.72	0.16	7.63	0.25	3.98	5.56	0.24	2.39	3.76	0.003	0.019
32078	well	53.24	15.55	10.10	0.12	6.65	0.07	3.61	3.55	0.16	1.70	4.12	0.004	0.010
26062	well	38.70	14.92	23.48	0.45	7.38	0.07	4.25	5.30	0.10	1.09	1.61	0.004	0.027
32083	well/pb/L2	50.22	13.93	11.01	0.20	9.78	0.13	3.83	4.87	0.25	1.43	3.06	0.002	0.036
32082	well/pb/L2	48.54	13.87	11.16	0.19	9.88	0.09	3.80	6.35	0.13	1.59	3.10	0.005	0.048
26064	well/pb/L2	50.14	14.31	11.14	0.19	8.40	0.07	3.82	5.63	0.32	1.50	3.20	0.005	0.018
32084	well/pb/L4	51.04	14.44	12.35	0.13	6.65	0.10	3.90	4.05	0.09	1.45	4.40	0.006	0.013
26065	well/pb/L4	48.80	17.28	11.91	0.13	6.40	0.10	4.08	3.99	0.11	1.42	4.44	0.007	0.013
32081	well/tuyere/L4	49.55	17.08	11.37	0.15	6.96	0.08	4.05	3.51	0.20	1.67	4.11	0.003	0.013
26063	well/tuyere/L4	49.62	18.18	10.84	0.12	6.31	0.06	4.59	3.43	0.33	1.61	3.66	0.005	0.027
32080	well/tuyere/L4	37.79	19.44	12.59	0.18	8.82	0.14	4.01	4.51	0.25	4.66	6.19	0.007	0.017

Analysis by ICP-AES

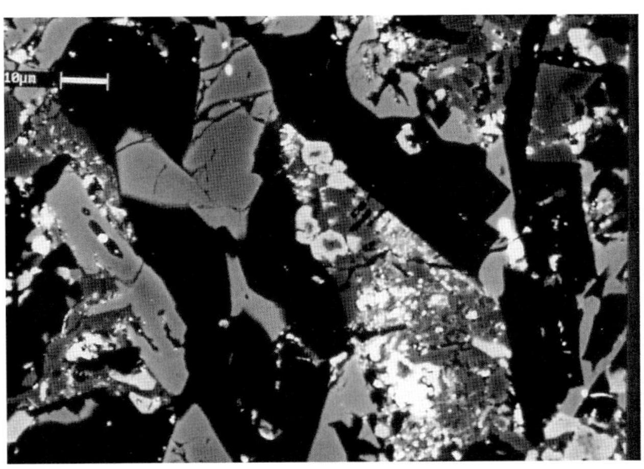

Plate 7.26 SEM photomicrograph of slag with refractory 32074 from Agucha, Site 1 showing well-crystallised slag with olivine (white large), bytownite (dark large) and spinels with zonal growth.

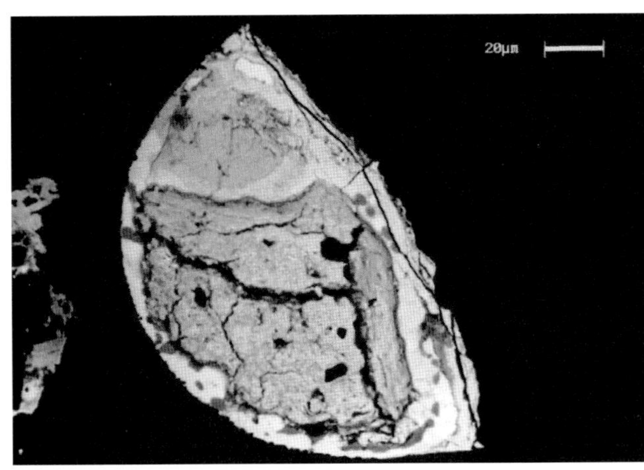

Plate 7.27 SEM photomicrograph of slag 32080 from Agucha well section. Inclusion containing $PbCO_3$, $PbSO_4$, Fe-Ni arsenides, Ag and Ag antimonides adhering to the left side of a gas bubble.

Table 7.18 Composition of lead metal from Agucha well section sites.

ICP-AES analyses of lead samples

Sam No.	Wt/mg	Ag	As	Bi	Cd	Co	Cu	Fe	Pb	S	Sb	Sn	Zn
Agucha:													
46312	9.67	0.107	0.654	0.024	<0.001	<0.001	0.032	<0.01	98.8	<0.03	0.407	0.058	<0.002
46313	9.93	0.089	0.629	0.023	<0.001	<0.001	0.032	<0.01	98.9	<0.03	0.386	0.025	<0.002
46315	9.87	0.100	0.634	0.023	<0.001	<0.001	0.032	0.013	98.8	<0.03	0.395	<0.02	<0.002
26278	9.78	0.170	0.520	0.017	<0.001	<0.001	0.014	<0.01	99.0	<0.03	0.382	<0.02	<0.002
46314	9.23	0.079	0.525	0.030	<0.001	<0.001	0.026	<0.01	99.0	<0.03	0.371	<0.02	<0.002
46314 ii	9.63	0.093	0.525	0.026	<0.001	<0.001	0.030	<0.01	99.0	<0.03	0.386	<0.02	<0.002

Notes:
The results have been expressed as weight per cent and have been normalised to 100%.
"<" denotes less than the figure quoted, i.e. beneath the detection limit.
Precision should be *c*. ±1-2% for the major elements, *c*. ±5-10% for the minor elements, deteriorating to ±50% at the detection limit.

Table 7.19 Composition of cerussite and litharge from Agucha well section sites.

BMRL no.	Ag*	Al_2O_3	(As_2O_5)	Au*	[BaO]	Bi*	CaO	(CdO)	CoO	CuO	FeO	(HgO)	$[K_2O^*]$	MgO	MnO	$[Na_2O^*]$	NiO	P_2O_5	{PbO*}	(S)	(Sb_2O_5)	SiO_2**	SnO_2	TiO_2	(ZnO)
26277 **cerussite**	0.017		<0.04	<0.003		<0.02		<0.005	<0.005		1.03	<0.009		0.27	0.044		<0.008		88.9	0.09	0.10	9.5	<0.01		0.05
33960 **litharge**	0.012	1.23	<0.04	<0.003	0.19	<0.02	4.65	<0.005	<0.005	0.071	0.36	<0.009	0.24	0.42	0.009	0.59	0.009	0.27	69.9	0.02	0.34	21.6	<0.01	0.05	0.09

* denotes analysis by AAS, ** SiO_2 by difference, < denotes less than the value quoted (i.e. the detection limit), round brackets ( ) denote some of analyte may have been lost during fusion, square brackets [] indicate probable contamination from glass vials/distilled water, braces { } indicate low levels of lead may be cross-contamination from sample preparation, italics denote Pb expressed as sulphide, bold denotes metal as carbonate. Precisions: c.±2-5% for major elements above 10%, ±10% for minors between 1-10%, deteriorating to ±50% at the detection limit.

Table 7.20 Composition of fine matrix and selected phases in some ceramics from Agucha.

	1	2	3	4	5	6
	cupel	tuyere	tuyere	"hood"	"hood"	"hood"
BMRL no.	26573	26279	26279	26248	26258	26258
	matrix	matrix	kalsilite	matrix	matrix	"glaze"
SiO_2	57.4	56.2	37.0	62.4	70.6	33.7
TiO_2	1.3	1.1	nd	0.7	0.7	0.3
Al_2O_3	21.0	18.4	31.1	18.0	13.3	7.3
FeO	9.0	11.1	0.6	8.4	5.3	2.6
MnO	nd	nd	nd	nd	nd	nd
MgO	3.7	3.6	0.8	2.4	1.9	1.1
CaO	1.3	2.4	0.3	1.8	1.8	1.2
Na_2O	1.8	0.9	1.2	1.4	1.5	1.0
K_2O	3.3	2.5	28.0	2.8	2.8	1.8
PbO	1.1	3.4	1.0	1.4	1.7	51.0
P_2O_5	nd	nd	nd	nd	nd	nd
SO_3	nd	nd	nd	0.2	nd	nd
Cl	0.1	0.4	nd	0.4	0.3	nd
	100.0	100.0	100.0	99.9	99.9	100.0

Means of two analyses, normalised

Silver/lead smelting

Slags

The slags are heterogeneous which is reflected in their composition (Table 7.17) and the presence of numerous fragments of host rock from the ore, as at contemporary Dariba. These again suggest a short process although, unlike Dariba, no significant fragments of hearth/furnace lining were found at Site 1 to study further. The Agucha slags are rich in both silica and alumina, but poor in iron oxides compared to the contemporary slags from Dariba and Zawar (Fig. 7.6). Their high aluminium content in particular, deriving from the graphite-schist deposits in which the silver-rich ores were located, would give them a high melting point and a high viscosity. The Agucha slags are especially rich in fragments of quartz which could have been added as a flux, in which case they would have been tapped together with the slag before they had time to react with the other mineral species. The molten component has usually solidified as a glass, with minor amounts of crystalline phases such as bytownite ($(Ca,Na)Al(Al,Si) Si_2O_8$) and olivine (Plate 7.26). The majority of the inclusions were of lead sulphide or metallic lead, the latter now usually weathered to lead carbonate. Occasionally there are inclusions of pyrrhotite and iron-nickel arsenide and antimonide speiss. The mineralogical study of the Agucha ores carried out by Höller and Gandhi (1995) suggested that much of the silver was likely to have been associated with the galena and antimony-rich freibergite ($(Ag, Cu, Fe)_{12}Sb_4As_4S_{13}$), and the presence of galena and antimonides in the slag inclusions supports their hypothesis. There are very few inclusions of copper-silver sulphide within some of the lead carbonate inclusions, but the bulk of the silver is in the former lead metal inclusions or in the antimonide speiss (Plate 7.27). The silver content of the lead in the inclusions can be as high as 1.1%, showing beyond question that these slags are from the smelting of argentiferous lead.

The slags contain quite small quantities of lead in the form of very small dispersed blobs either of galena or now-corroded lead metal, and only occasional inclusions of speiss. This indicates that, as at Dariba, the bulk of the argentiferous lead had been successfully removed from the slags despite their high viscosities (see Chapter 9).

Ore and metal

The lead droplets from the well site (Table 7.18) contain of the order of 0.5% each of antimony and arsenic, which would have been expected to have been largely lost in a long rigorous smelt. The silver content is variable but averages out at about 1000 ppm, a rich ore. The fragment of cerussite ($PbCO_3$) (26277), which may have originally been of galena, contained only approximately 200 ppm silver (Table 7.19), and this is possibly why it was rejected.

Cupellation

Most of the material described here was excavated at the well section site (see Chapter 6), and is thought to have been associated with the extraction of silver from lead. The ceramic materials included several cupels, tuyeres, and pedestals which are interpreted as being the supports for the minute cupel/tuyere assemblies (see Chapter 6, Fig. 6.69). These items were all used in a small-scale cupellation probably to assay ore samples. In addition, a larger-scale industrial process is thought to be represented by the fragments of coarse, heavily vitrified ceramic material (Table 7.20). These may have formed parts of specialised furnace linings in the shape of hoods which surrounded the cupellation hearths used for the large-scale extraction of silver. This is also evidenced by a large fragment from a disc of litharge (33960) and the large square-section tuyere (see Chapter 6, Fig. 6.74) suitable for supplying air to the hooded hearths (see Chapter 6, Fig. 6.70).

The clay pastes used for the cupels, tuyeres and stands vary somewhat but all are characterised by an abundance of poorly sorted, angular grains, typically 0.1–0.5 mm in diameter but ranging in size from silt grade (< 60 μm) to coarse sand (up to c.1 mm). The relative proportions of finer and coarser aplastic grains vary. Typically they are composed mainly of quartz and feldspars, along with green-brown pleochroic amphibole, garnet, and occasional fragments of metamorphic rocks. The proportions of amphibole and garnet vary from sherd to sherd. The fine clay matrix, which contains rare opaque grains and flakes of white mica, has oxidised to a red-brown colour towards the 'outer' surfaces that would have been exposed during firing or use, but the 'cores' are almost opaque in the thin sections. A little burnt-out organic material was observed in some sherds. The pedestal, 27273, included very sparse wheat and barley chaff fragments, visible mainly in the hand specimen. Pale yellow, isotropic glass can often be seen at the edges and penetrating along fissures and cracks in the fragments of cupels, tuyeres and pedestals.

The hood fragments have a distinctive 'glaze' on their concave surfaces, with 'drips' frozen in place (see Chapter 6, Plate 6.71). The morphology of the drips of glaze suggests that the various fragments recovered were positioned at different angles during use, from almost vertical to essentially horizontal; on the latter the frozen drips are roughly perpendicular to the surface of the slab of ceramic. There is some indication that the underside surfaces may originally have been dimpled to encourage the forming lead silicate to drip vertically rather than run down the sides of the hood. By inferring the orientation of the numerous fragments recovered, it was possible to make a tentative reconstruction of the original form of the hood (see Chapter 6, Fig. 6.70 and Chapter 10). The 'back' faces of some fragments, such as 26247 and 26249, appear to have been pressed against a relatively smooth, concave surface while still wet. Fragment 26250 seems to have been pressed against a more undulating surface. The pieces are vitrified that appears to extend from the 'hot' (glazed) face through to the 'cold' face; the thickest sample is 26246 at c.50 mm. Some of the 'cold' faces bear a whitish deposit. Their analysis and similar white or pale yellow surface deposits visible on several other examples of the refractory materials by XRD showed that they consist mainly of cerussite, sometimes mixed with litharge (PbO).

Visual inspection of the fabric of the hood fragments shows that it is coarser than the others described above. Petrographic examination indicates that it contains abundant, very poorly sorted aplastic inclusions, which range in size from silt to very coarse sand grade; the grains are typically sub-angular but some are sub-rounded. The size distribution of the aplastic grains is bimodal, with common very fine sand/silt (< 0.1 mm) and common medium-coarse sand (> 0.25 mm). In other respects, the fabric of the hood fragments is very similar to the other ceramics, including the presence of very sparse wheat and barley chaff fragments, visible mainly in the hand specimens (for example 25988). The differences between the clays/tempers used for the cupels, pedestals and tuyeres were probably fortuitous; they may reflect the use of clays and temper from different sources but all within

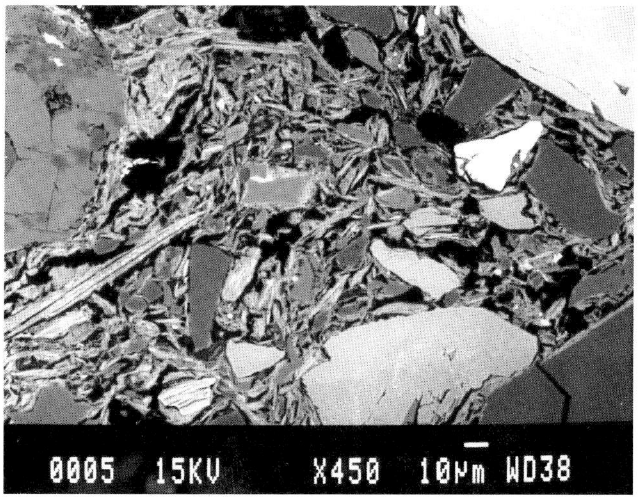

Plate 7.28 SEM photomicrograph of polished thin section of tuyere 26279 showing the essentially unvitrified texture of the 'clay' matrix. Note the preservation of the lath-like crystals of mica/clay. The larger aplastic grains include quartz (dark grey), feldspar (mid-grey) and amphibole (light grey) (BEI) (scale bar: 10 μm).

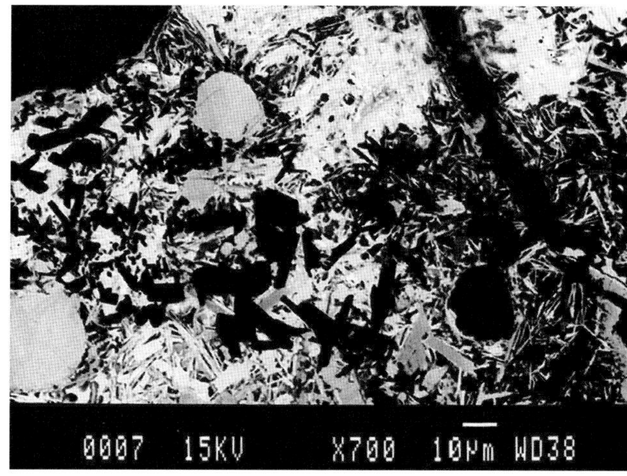

Plate 7.30 SEM photomicrograph: detail of polished thin section of tuyere 26279 showing the texture of the lead-rich layer. The dark prismatic crystals are kalsilite thought to have been formed by reaction between the silicate minerals of the ceramic and the lead oxide deposited on the surface during use (BEI) (scale bar: 10 μm).

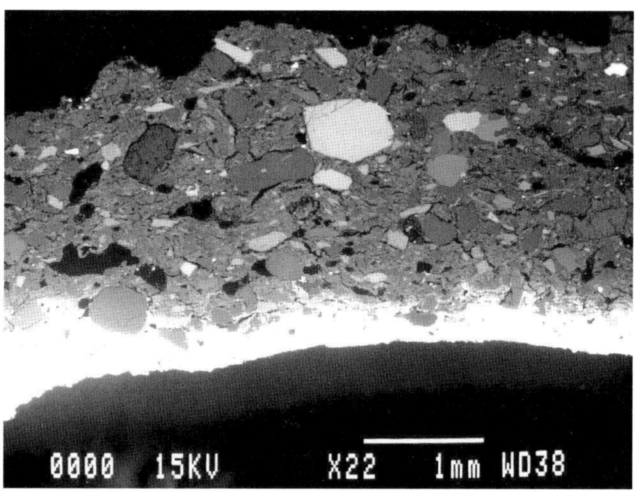

Plate 7.29 SEM photomicrograph of polished thin section of tuyere 26279 showing the general texture of the ceramic with particles of quartz (dark grey), feldspar (mid-grey) and amphibole (light grey); a large grain of garnet can be seen towards the top centre of the image. The bright white layer along the lower margin is a lead-rich deposit (cf. Plate 7.30 for more detail at higher magnification) (BEI) (scale bar: 1 mm).

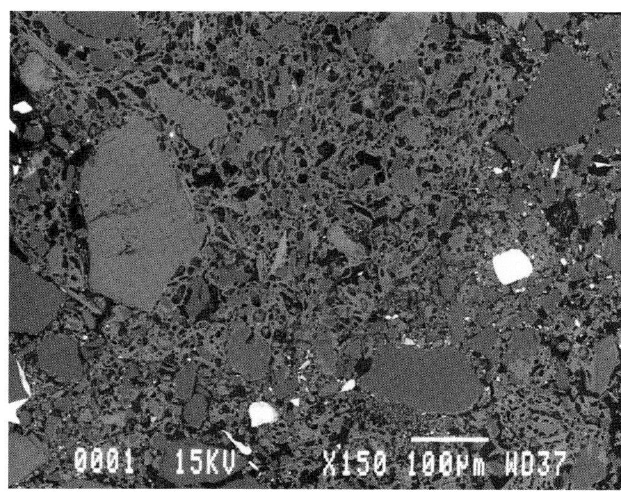

Plate 7.31 SEM photomicrograph of polished thin section of ceramic hood 26248 showing the continuously vitrified and bloated texture of the ceramic with particles of quartz (dark grey) and feldspar (mid-grey); the bright particle is a fragment of lead-rich 'glaze', probably introduced during sample preparation (BEI) (scale bar: 100 μm).

the same local area. However, the use of a coarser clay fabric for the hoods seems likely to have been deliberate, probably because these were much thicker and were exposed to more severe thermal conditions.

The litharge cake (33960) is actually a mixture of lead oxide and lead silicate, as with early litharge cakes generally (Bayley 1992). It contains 120 ppm of silver, which is very low compared, for example, to the quantities found in the litharge cakes at the slightly earlier Lydian gold refinery at Sardis in Turkey, which varied between about 200 ppm and several percent of silver (Ramage and Craddock 2000: 160, table 6.2).

Examination in the SEM of the polished sections of the cupel, tuyere and pedestal samples confirmed the presence of quartz, garnet (almandine), amphibole and feldspar (typically andesine), in a fine-grained matrix. The observations showed that the fine-grained matrix exhibits only very limited vitrification, with lath-like particles still clearly visible (for example, Plate 7.28). SEM-EDX analysis revealed that the clays were non-calcareous and not particularly refractory, containing approximately 15% or more of iron, calcium, potassium and sodium oxides (Table 7.20, columns 1 and 2). The analyses also confirmed a few percent of lead that is almost certainly a result of contamination during use, but which would nevertheless have contributed towards a reduction in the temperature at which vitrification occurred.

A lead-rich layer, c.200 μm thick can be seen on the surface of tuyere 26279 (Plate 7.29). More detailed examination showed that this layer consisted of a mixture aggregates of fine elongated prismatic crystals and more compact areas (although seen to be composed of interlocking fibrous crystals when observed at higher magnification), with spherical grains scattered throughout. Variations in the electron backscatter properties across this layer (seen as differences in grey level in the photomicrograph), coupled with EDX analysis, suggest that this lead-rich layer is composed of an intimate mixture of lead carbonate and lead oxide. In addition to these lead-rich phases, it also contains a small proportion of silicate material, which in some regions has recrystallised to form prismatic crystals (Table 7.20, column 3) (Plate 7.30) of potassium aluminium silicate, kalsilite ($K_6Al_6Si_6O_{24}$).

The two fragments of hood examined in the SEM were vitrified and bloated throughout (Plate 7.31) with, as expected, a very different texture from the cupels, tuyeres and pedestals. The ceramic of the hood samples, like the cupel, tuyere and pedestal, appears to be composed of a non-calcareous, low-refractory clay with abundant aplastic inclusions of quartz and silicate minerals. The texture and composition vary within each of the hood samples and also between the two samples examined (Table 7.20, columns 4 and 5). This is thought to reflect the inhomogeneous nature of the coarsely tempered 'clay'. Both hood samples have higher silica:alumina ratios than the cupel and tuyeres, reflecting the inclusion of a higher proportion of quartz in the areas analysed. There was no evidence on either sample for any systematic variation in the degree of vitrification from the glazed surface to the 'back' of the sample. On both samples examined, the glaze is of very variable thickness but typically up to several hundred micrometres. This glaze is essentially a lead silicate, with minor amounts of alumina, iron, magnesia, soda and potash (Table 7.20, column 6).

Process conditions

In order to assess the maximum temperature at which the cupels were used, apparently low-fired fragments of one of the pedestals, 26273, were refired at temperatures from 950° to 1100°C in an oxidising atmosphere. The samples refired to 950°C have a higher degree of vitrification than any of the used cupel, tuyere and pedestal samples examined. This suggests that the cupel/tuyere/pedestal assemblies were unlikely to have been exposed to temperatures exceeding about 900–950°C for any prolonged period. However, although the degree of vitrification observed in the refired samples is typical of that observed for non-calcareous, low refractory clays examined by Maniatis and Tite (1981), indicating that the present observations are reliable, some caution is necessary because of the possible fluxing effects of lead absorbed by the ceramics during use. The complete tuyere/cupel assembly (26304) (see Chapter 6, Fig. 6.69 and Plate 6.72) suggests that the charge in the cupel was heated from above, with the highest temperatures being achieved at the surface of the charge within the opening of the cupel. The bases of the cupels were embedded in loose material in the cupellation hearth, protecting them from the lead-rich fume and shielding them from the heat applied from above. The estimates of the temperatures at which the ceramics were used are towards the minimum that would have been required to complete the cupellation process. At least near the end of the process, it would have been necessary to raise the temperature to about 1000°C, sufficient to melt the silver (MP 961°C).

Comparison of the vitrification textures of the hood samples with those of the refired pedestal samples suggests that the hoods were exposed to temperatures in the region of 1100°C, confirming the indication from the macroscopical observations that these materials were exposed to a significantly higher temperature regime than the cupel assemblies. However, as noted already, the ceramic materials contain a small proportion of lead (Table 7.20, columns 4 and 5), which would have had the effect of reducing slightly the vitrification temperature of the ceramic, leading to an overestimate. Further confirmation was therefore sought by the refiring of fragments from the hood, 26258; no discernible changes in the texture of the ceramic were noted in samples refired at temperatures up to 1100°C, suggesting that the hood really had been exposed to temperatures of at least this order during use.

The relatively uniform vitrification throughout the thickness of the hood fragments, up to a maximum observed thickness of 50 mm, suggests that the thermal gradient from the 'hot' face of the hood to the 'cold' face was very low. This in turn indicates that the process time was of the order of several tens of hours (Tite *et al.* 1982a). The reconstruction of these cupellation operations is investigated more fully in Chapter 10.

Notes

1. See Chapter 3 (p. 21). No analyses of these mortars are given here but are reported in Freestone *et al.* (1985a).
2. Similar plates have been reported by Hezarkhani-Zolgharnian *et al.* (1994) (see Chapter 11, p. 216).
3. It is possible that had the furnace been reused through a number of cycles (see Chapter 11, Fig. 11.3) then following a previous high temperature zinc smelt, in which the ceramic had vitrified, it was charged again with more crushed ore for roasting at a lower temperature including galena, some of which coated the now cooler ceramic, and this never became as hot as in a previous cycle enabling the galena to survive. Seemingly at a later, smelting, stage the fume began to deposit on the galena becoming more oxidic and was increasingly composed of zinc rather than lead although both are represented (Table 7.5).
4. The plate is 4.8 × 4.0 cm, and 0.7–0.8 cm thick, and weighs 41.3 g, but with broken edges all around would originally have been much larger.
5. See note 3 above.
6. This was stated in the first publications (Craddock *et al.* 1985a) but then subsequently doubted before a complete reassessment was made considering the context (Craddock *et al.* 1998; Freestone *et al.* 1985a).
7. An earlier report (Freestone *et al.* 1985a) interpreted this glaze as a partial melt, segregated from the ceramic. Subsequent experience in the investigation of high fired stoneware ceramics suggests that the enhanced potassium, up to 10% (Table 7.8), was probably derived from the fuel.
8. The overall daily yields of the furnace blocks are estimated in Chapter 11, p. 229.
9. A fragment from Site 5, 27352, from the top of a condenser, is massively impregnated with metal, now totally corroded. SEM-EDX analysis showed a ratio of about 90% zinc and 10% lead.
10. This piece is either from the smelting or refining of copper. In the matte process of smelting copper-iron sulphides under moderately reducing conditions (Chapter 10), magnetite and fayalite can form and co-exist in the slags (Bachmann 1982: 9). In the refining process, the iron-rich copper is melted under oxidising conditions to remove the iron as magnetite, some of which will react with the silica of the crucible or with sand that may be sprinkled onto the molten metal. These slags usually include minerals such as delafossite ($CuFeO_2$), rather than just fayalite. The refining process is also of relatively short duration compared to the smelting process and the deep penetration of the slag into the ceramic suggests an extended period at high temperature. Thus on balance Sample 27276 is likely to have been generated by smelting rather than refining. Whichever method was used, it provides strong evidence for primary copper production at Dariba back in Mauryan times.
11. Donkey dung was used as a temper by Rajasthani potters in the recent past (Kramer 1997: 56).
12. With the exception of sherd 27276 (see above), and of the cupellation sherd 27295 (Chapter 10, p. 202).

Chapter 8 Mining and beneficiation

P.T. Craddock and L. Willies

In this chapter and those that follow it the processes by which the metals were mined and extracted from their ores at the three major mines investigated will be reconstructed as far as possible, based on the evidences of archaeology, scientific examination and such records as exist of the mining and smelting processes in use in South Asia and its environs prior to the 20th century.[1]

With four metals, at three major production sites spread over three millennia, the evidence for the process metallurgy could be presented in a number of different ways: by technology, chronology or metal. It seems more logical to adopt a basic structure of technological progression from the mining of the ores through some of the processes common to all the metals. Within that framework the chronological progressions and, more probably, regressions will be discussed (and again in more detail in Chapter 12). For the actual smelting the copper, silver/lead and zinc processes will be considered separately as the processes for each metal are so completely different from each other in many aspects.

Mining

It is necessary to consider the background to the great mines reported here, both technically and economically. Through the first millennium BC in northern India there was a significant increase in intellectual activity with a flourishing literature, of which many works, both religious and secular, are extant. From these it is clear that mathematics, medicine and other scientific subjects were being studied systematically (Allchin 1995: 84–9). This new phenomenon of literate secular intellectual activity was appearing throughout the Old World from Greece to China. The great mining systems that were developing at the same time across this enormous area are a direct and tangible result of this vastly increased knowledge. The development of prodigious mines such as Rio Tinto in Spain, Laurion in Greece, Dariba in Rajasthan and Tonglushan in China at much the same time as the increase and formulation of technical knowledge is not coincidental (Craddock 1995: 69–92). Mining systems of the size and complexity encountered from this time required at least an empirical knowledge of mathematics, mechanics, hydraulics and pneumatics (Aralere 2016). The exact relationship between practical need and theoretical possibilities is as uncertain in this instance as in later times (Musson and Robinson 1969). To paraphrase the well-known truism from the European industrial revolution concerning science and steam: did early science owe more to the mines than the mines owed to early science? A serious discussion of these arguments is beyond the scope of this volume, but the association between the development of intellectually sophisticated urban societies and large-scale deep mining in South Asia and elsewhere throughout the first millennium BC should be appreciated as the link is both real and causal.

In South Asia, as elsewhere, one of the economic driving forces behind the development of the silver mines – and to a lesser extent of the gold and copper mines – is likely to have been the demands of coinage. Punch-marked coins of silver with some copper (Rao *et al.* 1990; Agrawal *et al.* 1990) were produced in huge numbers, initially in the north but then throughout South Asia (Gupta 1969). The sources of the silver used for the punch-marked and later coins have long been debated by Indian numismatists with hitherto no major recognised sources in South Asia (Bhardwaj 2000: 152). The early history of coinage in South Asia is still very uncertain. Some scholars maintain that the first silver coins of South Asia should be dated to the late 7th century BC (Gupta and Hardaker 1985; Hardaker 1992) contemporary with the earliest coins in the West – the Lydian electrum coins at Sardis. However, the consensus would place an origin sometime around the mid-first millennium BC with general use following rapidly throughout the 5th century BC (Cribb 1985, 2003). The dates for the major exploitation at the silver mines at Agucha and Dariba would seem to support the later dates (see Chapter 5, Tables 5.1 and 5.2).

The very complete and extensive early mining systems surviving at Zawar, Dariba and Agucha allow the mining operations to be understood in some detail. The geology and mineralogy of the mines are broadly similar, with quite thick deposits that usually dip at steep angles and continue unbroken to great depths. This similarity is reflected in many aspects of the mining strategies adopted at the mines, enabling several of their features to be considered together. The overall impression is of a well-organised mines administration working the ore deposits in a systematic and ordered manner on a considerable scale with a large workforce. This is evidenced by the planned sequence of mining at Zawar Mala, Zawar, and by the ability to control drainage and ventilation hundreds of metres below ground and the water table, as well as by major works such as the timber revetments of the opencast mines at both Dariba and Agucha. This evidence accords with the description of the organisation of mines given in the *Arthaśāstra*. However, as with many state-run enterprises in antiquity, there was perhaps a tendency to rely on a large labour force that was readily available at the expense of technical innovation. Therefore there is little evidence of devices for raising materials or water within the mines, or for ore crushing above ground, that are found in some contemporary Roman mines (Craddock 2008). Overall there is not much evidence for mechanical geared devices or for the application of animal or water power to mining and metallurgy. This was to be a feature of the whole region until recent times and stands in strong contrast to both Europe (Agricola: Hoover and Hoover 1912) and the Far East (Golas 2000).

Organisation

Although the Aravalli mines are apparently never mentioned specifically in any of the ancient sources, the *Arthaśāstra* contains a great deal of information on the organisation of mining generally that is very relevant to the understanding of the operation and control of the Aravalli mines in the broader sense.[2]

The *Arthaśāstra* is a major work dealing primarily with state administration assumed to be based on that pertaining in the Mauryan Empire. It is ascribed primarily to Kautīlya, the chief minister of the first Mauryan Emperor Chandragupta, and thus should have been compiled in the 4th century BC. Although there are doubts over the authorship and dates of the work in its present form, it does seem feasible that much of it was compiled during the period of the Mauryan Empire in the last centuries of the first millennium BC,[3] therefore it is likely that the text is broadly contemporary with the main period of operations at Agucha and some of those at Dariba and Zawar. It is interesting to compare the *Arthaśāstra* – which contains so much almost firsthand information on ores and detailed instructions on mine organisation and the complex revenues associated with mining – with the later *Nitisara* compiled by Kamandakiya in the 8th century AD (Dutt 1896: 38–9, 60) when the mines in the Aravallis at least were largely inoperative. This latter treatise broadly covers the same subjects of governance and considers mining to be one of the sources of revenue of the monarch, but provides no instruction as to their organisation or administration. It is significant, as Sarma (2001: 41–2) points out, that the *Arthaśāstra* provides a great deal of information on the bureaucracy of the mines but almost nothing on the actual underground mining technology or basic smelting technology, activities carried out in the field far away from civilisation. Conversely subjects such as precious metal refining and alloying, activities that would normally be carried on in urban centres, are covered in some detail.

In the second book of the *Arthaśāstra* the duties and activities of the various heads of departments of the government are described.[4] In order to be more realistically comprehensive, the situation is envisaged in which a newly acquired territory is being opened up and peopled. This pioneering spirit is particularly apposite to the Mauryan administration that was probably establishing, supplying and protecting the mines in what must have been the very remote territories in the Aravalli Hills.

It is clear from the *Arthaśāstra* that most economic and political activity was (ideally at least) to be centrally controlled. As was (and is) common throughout much of the world, the mineral resources of the earth were regarded as state property. Certainly the major mines for precious metals were owned and usually worked by the state in the Roman Empire (Craddock 2008: 99–100) and in the Islamic world (Hill 1993: 211–15). Alternatively the state might develop the mine, sink the ventilation and climbing shafts, install and maintain drainage etc. but contract out the actual working of the deposits under supervision. It seems that in the Mauryan Empire mining was also largely a state activity so in all probability the major mines would have been viewed as planted settlements with every aspect of life and support, including agriculture, developed under central control (Kangle 1965–72, III: 182–4; Chakrabarti and Lahiri 1996: 140–42). Chapter 6 (2.6.1: 74) deals specifically with the setting up of revenue collection (mines inevitably being one of the major sources) while chapter 12 (2.12.1: 105) concentrates on the establishment of mines:

> The Director of Mines, being conversant with the science of [metal] veins in the earth and metallurgy, the art of smelting and the art of colouring gems, or having the assistance of experts in these, and fully equipped with workmen skilled in the work with implements, should inspect an old mine by the marks of dross, crucibles, coal and ashes, or a new mine, where there are ores in the earth, in rocks or in liquid form, with excessive colour and heaviness and with a strong smell and taste.

Note that gemstones and metals are frequently discussed in the same sections and this is not entirely unreasonable – the Aravalli Hills are, after all, an established source of gemstones (Krishnaswamy 1988: 462–66). The metal ores at Agucha, for example, were rediscovered by gem prospectors looking for garnets. The reference to ores in a liquid form is interesting. Described as flowing out of rock-cut caves etc. they are variously yellow or red and turn the vegetation near where they flow the same colour. This description would fit the acidic solutions of ferric sulphate formed by the oxidation and dissolution of the iron pyrites in the veins. These are frequently found flowing out of old workings, poisoning the vegetation around them, and would be a good indicator of a metalliferous deposit.

From the first passage it is clear that the *Arthaśāstra* regarded the prospection of mines as a state activity, although the Director of Mines could employ other, presumably non-state experts. Similarly the mines would seem to have been state run unless there were problems when they could be leased to others to work:

> He should let for part-share or on lease a mine that is burden-some in point of expenses or working; a light one, he should work himself (2.12.22: 108).

Dishonest and unauthorised mining was clearly a problem:

> He should make a miner who robs pay eight-fold, except in the case of jewels. The thief and the person who lives [by mining] without permission, he should bind and

force to work [in the mines], also the person who offers personal labour in place of a fine (that is someone who cannot afford their fine) (2.12.20–21: 108).

The question of the sources of the workforce in early mines is frequently discussed with especially gruesome accounts of slave labour and those condemned to the mines in the Classical world (Healy 1978: 132–8).[5] The *Arthaśāstra* has little to say concerning the labour force for mines, but makes clear that the state could conscript men not just for the army but also for labouring work (1.4.1–2: 9; III: 167). This would seem to have been some sort of *corvée* system of enforced service from labourers who were otherwise free, similar to the *m'ita* system used by the Inca state in South America to operate its mines. In addition there were workers who had in effect mortgaged themselves into temporary slavery (Sharma 1966). Expert (apparently free) miners could be called upon to help in the establishment of mines but in all likelihood the usual labour force was made up of slaves and labourers as in other state enterprises; this probably also applied to the private mines. However, mining and extractive metallurgy would have required a considerable expertise which could only have been provided by a core professional staff: firesetters, timbermen, supervisors etc. Clearly there would be problems recruiting a large labour force in remote locations, and the situation was probably not hugely dissimilar to that faced when the Zawar mines were being reopened after the Second World War when a mixture of local labour and professional miners from elsewhere was recruited (see Chapter 3).

In earlier times, prisoners would be serving out their sentences in the state mines. For example, there are references in the *Arthaśāstra* (1.13.21: 29) to dissidents being sent to the mines so clearly there was an element of coercion among at least parts of the labour force.

The *Arthaśāstra* also makes clear that some of the broader economic and logistical problems of large-scale distant mining had been carefully thought through. There is a discussion on the need to establish good communications and the rival merits of small deposits of rare, valuable commodities and larger deposits of less valuable materials:

> Of two mines, he who causes a mine to be opened, yielding plenty of valuable ores, with roads not inaccessible, and operated with small expenditure, over-reaches. Even in that case, as between few things of high value and plenty of things of small value 'The few things of high value are to be preferred, for, diamonds, rubies, pearls, corals, and gold and silver ores swallow up (that is surpass in value) the many objects of small value by their excessively high value,' say the teachers. 'No,' says Kautilya. A purchaser for an object of high value is rare and (only) found after a long time, while there are plenty for objects of small value because of a constant demand (2.7.13–17: 360).

In this passage, which seems to refer to mines, the need for good communications is stressed, and then Kautilya argues that a steady trade in utilitarian commodities is preferable to the more chancy dealings in more valuable products. By extension to mines, a large-scale operation producing a regular supply of metal is to be preferred to a smaller operation, potentially high yielding but less predictable. This, of course, is a good justification for what must have the very considerable investment needed to establish the Aravalli mines. This theme is continued when the discussion moves on to various methods of transport where (7.12.22–25: 360–61) it is argued that trade routes, such as existed to the east and west for low value bulk commodities for which there was a ready and continuous market, were to be preferred to the low-cost route such as that to the south for valuable commodities. It is stated (7.13.27: 361) that a road for wheeled vehicles is preferable to the cheaper path for pedestrians or pack-animals only, as the former enables large-scale undertakings to take place, advice that the directors of the disastrous 19th-century Kumaon iron smelting venture (see Chapter 12) would have done well to heed two thousand years later!

In chapter 22, which deals with tariffs and tolls, as well as elsewhere in chapter 12, it is clear that the distribution of the metal products and the metal manufacturers were also tightly controlled and to some extent state run under a Director of Metals:

> And no sale of commodities [shall be allowed] in the places of their origin. For taking metal goods from mines the penalty is six hundred [*paṇas*] (2.22.9–10: 146–7).

> What is produced from ores he (the Director of Metals) should put to use in factories for the respective metals. He should establish trade in manufactured goods in a single place, and [lay down] a penalty for those who manufacture, purchase or sell elsewhere (2.12.18–19: 108).

> The Director of Metals should establish factories for copper, lead, tin, *vaikṛntaka*,[6] brass (*arakuta*), steel, bronze (*kamsaya*), bell-metal and iron, also [establish] trade in metal-ware (2.12.23: 108).

In section 13 more extensive descriptions follow for the production and testing of gold and silver and the minting of silver coins (see Chapter 10). It does seem that the intention was to restrict production of the metal and the manufacture of artefacts to a limited number of locations so that they could be more easily controlled. This could explain the unusual situation at Site 10, Dariba, where bronze, and possibly brass alloys, was being produced in quantity but apparently nothing was made from them locally. Possibly Dariba at this period was a designated metal producer, with the tin being brought in from elsewhere and the metals then sent away for fabrication.

One of the best-known pronouncements in the *Arthaśāstra* makes clear the importance of mines to the supply of the treasury and ultimately the power of the state:

> The treasury has its source in the mines; from the treasury the army comes into being. With the treasury and the army, the earth is obtained with the treasury as its ornament (2.12.37: 110).

The *Arthaśāstra* stipulates in more detail the various taxes and dues levied on mines and metals than perhaps for any other commodity, reflecting this importance:

> Thus from the mines he (the Director of Metals) should collect the price, the share, the surcharge, the monopoly tax, penalty, duty, compensation, fine, inspection fee,

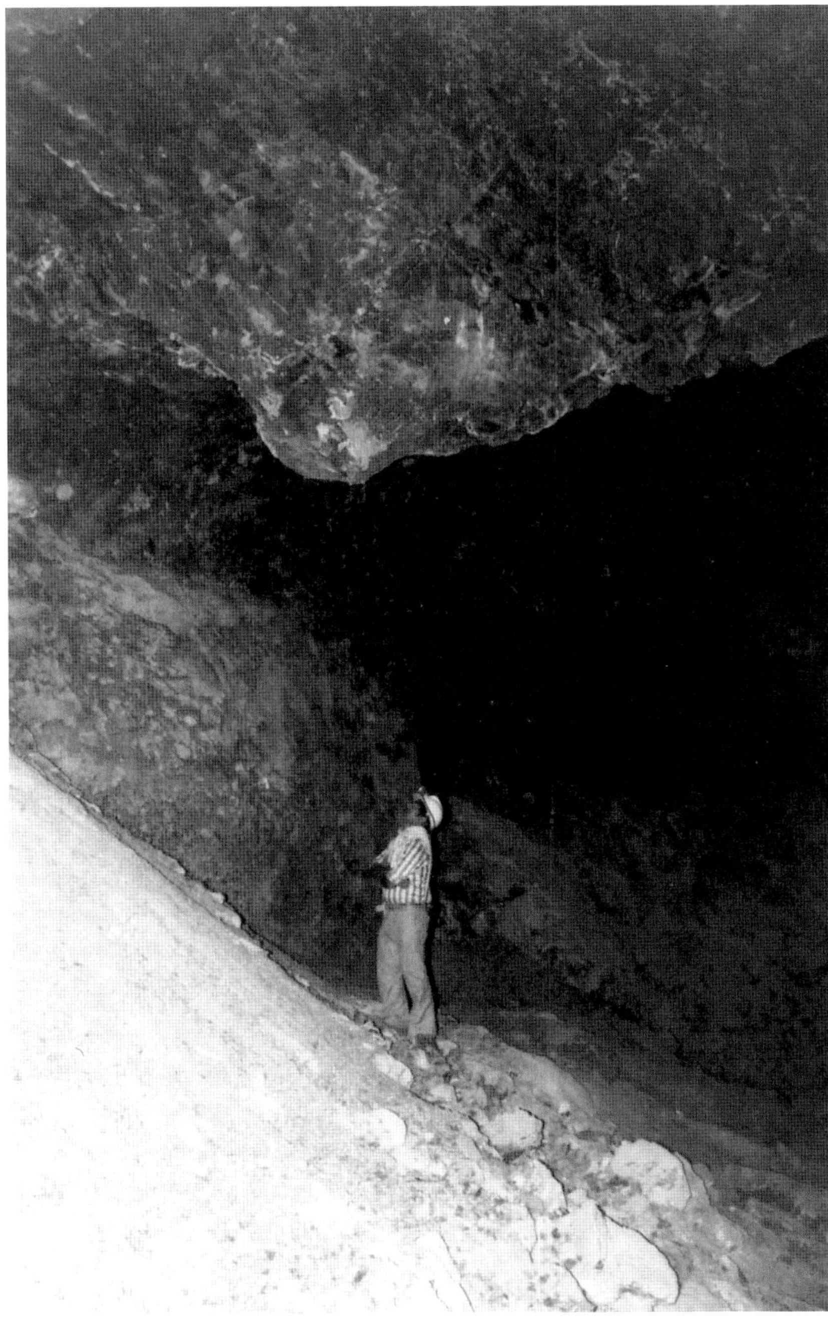

Plate 8.1 A 'pendule' in the roof of a gallery in the Kehtri-Singhana mines (photo: L. Willies).

> and manufacturing charges as well as the twelve kinds of metals and commodities [made from them]. In this way he should fix the collection [of income] under various heads in the case of all commodities (2.12.35–36: 110).

Kangle (1965–72, III: 183) attempts to interpret these as follows:

- the price is presumably the money from the sale of the metal;
- the share, monies from a shared enterprise;
- the surcharge, a sort of sales tax;
- the monopoly tax, a kind of protective duty;
- the penalty, money from fines levied against those who broke the state monopoly;
- the duty, a custom or excise duty;
- the compensation, paid to the state for the transfer of some of its mining rights to a private concern;
- the fine, probably also monies collected for infringement of various state regulations;
- the inspection fee and manufacturing charges having more obvious meanings.

It is difficult to ascribe these various levies exactly, but they do serve to demonstrate the very close involvement of the state at all stages in the production of metals.

The *Arthaśāstra* sets out that:

> On *kṣauma, dukūla*, silk, yarn, armours, yellow orpiment, red arsenic, antimony, vermilion, metals of various kinds and ores, on sandal-wood, aloe, spices, fermentation, and

minor substances, on skins, ivory, bed-spreads, coverings and silk cloth, and on products of goats and rams, [the duty to be charged is] one-tenth part or one fifteenth part (2.22.6: 145).

It is interesting that both metals and ores are mentioned suggesting that not all ores were smelted at the mines but that some were apparently traded as ores, which could be significant for the putative early production of zinc in the form of zinc oxide rather than as metal at Zawar.

Latterly the system of state ownership and direct involvement in the running of the mines was operating in the Hindu Rajput kingdoms of Rajasthan as well as in the Moslem Mughal Empire as was made clear by Tod (1978, I: 10):

the mines are royalties; their produce a monopoly, increasing the personal revenue of their prince. '*Án-Dán-Kán*' is a triple figurative expression, which comprehends the sum of sovereign rights in Rajast'han, being *allegiance, commercial duties, mines*.

However by the 19th century much of this control and even meaningful ownership had ceased together with the loss of authority of the titular rulers generally. Thus in Brooke's description of the Khetri and Singhana mines it is made clear that the ownership of the mines was divided between the miners themselves and local businesses. The extent of the local ruler's involvement was limited to extracting a tax as described by Lockett (1831),[7] but could provide neither leadership nor the funds for carrying out the repair and maintenance that by the 19th century were clearly necessary if the mines were to continue operating. The system of bidding for short-term concessions for working the mine, not dissimilar to the system current in many British mines – although probably a very efficient way of achieving the maximum revenue from an established mine in the short term – also discouraged long-term investment or large-scale development.

One notable absence from within the early mines is any evidence for shrines, and the 19th-century records of primitive mining make no mention of this aspect of life underground. This is rather surprising as shrines are common in modern mines the world over, typically to St Barbara in Catholic communities, and in even the most modern mines in India there are shrines, usually to the goddess Durga. She is an ambivalent deity, being both dangerous, her wrath having to be placated, but at the same time upholding safety and stability in the world, not unimportant considerations in a mine.

Prospection

In section 2.12.1 of the *Arthaśāstra* (quoted above) instructions are given for locating metal ores, either by inspecting old abandoned workings or by examining new ore prospects. In the following sections detailed descriptions are given of many ores and how they may be recognised.[8] For example, 'In the case of all ores, when there is an increase in heaviness there is increase in metal content' (2.12.7: 106). Deep underground with only poor lighting, weight is indeed of prime importance in assessing an ore's potential value. The *Arthaśāstra* is one of the very few ancient works to contain instructions on how ores should be located.

At Zawar and Dariba the orebodies would have been exposed at the surface at intervals all along the crests of the ridges formed by the hard metamorphic host rocks, and the quartz veins and distinctive metallic ores would have been easily recognised. The ore minerals at Zawar seem to have been relatively little oxidised and thus would have been readily identifiable, but even where oxidised their density would have revealed their potential. At Dariba there is a considerable and very distinctive gossan deposit formed of oxidised iron minerals from the veins and this has been taken as an indicator of the presence of other metals the world over. At Agucha it is very likely that even though the gossanised ore deposits were near the surface they were totally covered by alluvium and probably by vegetation as well. There are two possibilities: the ore could have been encountered during well sinking, which as mentioned above was partially responsible for its recent rediscovery, or the vegetation itself could have provided the evidence of heavy metals contained beneath. This method of ore detection apparently has a long history both in Europe and the Far East (Cannon 1960; Craddock 1995: 30–31; Buchanan 1992; Needham 1959: 675–80; Golas 2000: 219–20). The vegetation generally has a rather dark and withered appearance and there are also specific plants that are tolerant of heavy metals and flourish just in those areas. Studies by Aery and Tiagi (1986) on the waste dumps at both Zawar and at the old copper mines at Khetri found that a wild variety of *Impatiens balsamina* (the familiar garden balsam, called *dopati* in India) flourished and as such could be regarded as a good indicator.

Mining strategy

In common with most early mines the strategy seems to have been to follow the ore from where it outcropped usually by digging down into the rock beneath (underhand stoping), as explicitly stated for example by Drummond (1841) in the mines he observed in northern Afghanistan. Where the ore broadened out into a thick deposit and was not at too steep an angle then it would be taken out in stepped benches, as exemplified at Zawar Mala and also as suggested by Conophagos at Laurion (1980: 192, fig. 9-37). As the mines became deeper some shaft sinking through the country rock was needed, principally for ventilation but also for climbing, but overall there was a marked reluctance to mine anything except the ore and so access was usually through the previously mined areas (see Chapter 5, Figs 5.8 and 5.16, Plate 5.5).

In addition, there was no attempt to dig drainage adits beneath the workings into the hillsides at Zawar such as were dug by the Romans, for example, at Rio Tinto in Spain (Palmer 1926/7; Salkield 1987; Willies 1997). This was true of Indian mines generally. Stephens (1901–2: 393–414) commented on the mines at Garhwal and Kumaon that:

Absence of ventilation and the want of adits seem to have stopped mining operations. It is inexplicable why these old miners did not grasp the advantage of adits. A cross-cut about 600 ft (183 m) in length would unwater all these formations nearly 180 ft (55 m) below the old workings.

The miners seem to have taken the richest ore first from simple opencast pits and then followed it down until it was lost by faulting or the drainage/ventilation problems became insuperable. At North Lode, Dariba, there were some cross-cuts,

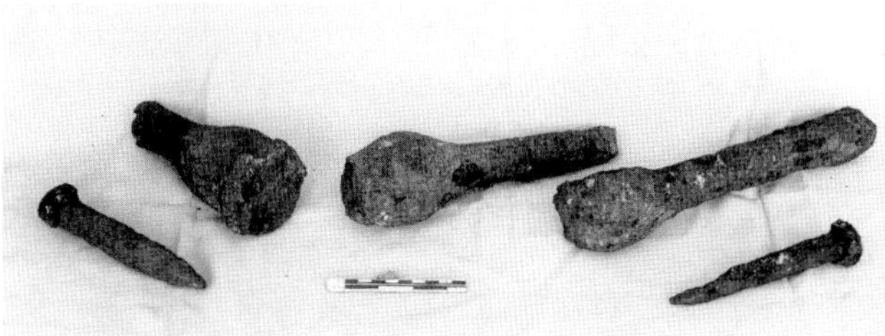

Plate 8.2 Two iron gads with three iron pestles found separately inside the mines on Mochia Magra, Zawar.

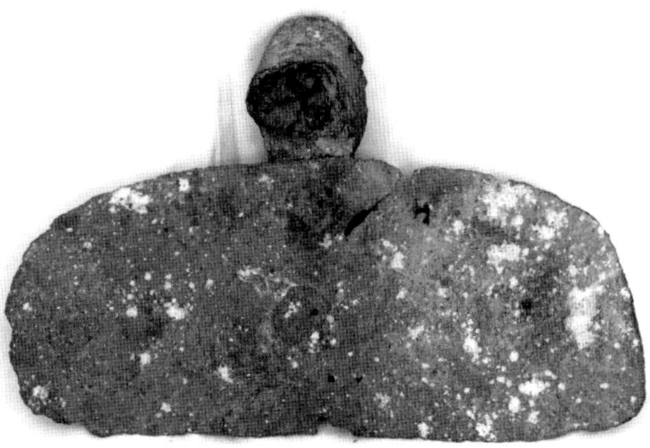

Plate 8.3 The head of an iron hoe used for raking ore or gangue into baskets. One of a group of undated iron tools from Mochia Magra mine, Zawar.

probably made to try to locate new orebodies (see Chapter 5). Then the rest of the orebody was systematically removed by taking out the less rich ore from the walls and roof by overhand stoping; finally, retreat mining removed the pillars of ore that had been previously left to support the roof (Plate 8.1). These stages in the exploitation of the deposits seem to have been fairly close together in time, but many of the deposits have opencast workings at their top which often, as at North Lode, Dariba, cut through the existing gallery system (see Chapter 5, Plate 5.50). Such dating evidence as exists suggests that with the notable exceptions of the East Lode opencast at Dariba and that at Agucha, the opencasts are considerably later workings. There is also the possibility that a return to orebodies previously abandoned may have taken place because different metals were being sought, as exemplified by the possible reworking of the old silver/lead workings for copper at North Lode, Dariba. However, in general, it just seems to reflect a strategy of aiming for the richest ore rather than the simplest technology, even if this meant leaving parts of the deposit unmined. A similar policy of taking the richest first seems to have been adopted by the lead miners at Ajmer in more recent times (Dixon 1831) (see Chapter 2).[9]

Mining methods

Rock stability and support

At Zawar the problem of roof support and rock stability was fairly simple as the rock is reasonably uniform in quality and of high competency. It is noticeable that at depth, where the rock is much less weathered, the arched openings are much larger than

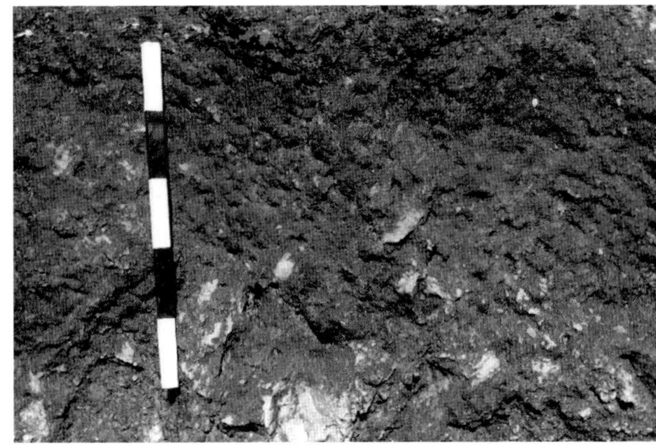

Plate 8.4 Rock surface covered with rounded pick or gad marks at North Lode, Dariba.

the earlier higher workings, with spans of flat roof of up to 20 or 30 m, which is not dissimilar to spans in the modern carefully calculated stopes. Many pillars seem to have been removed by retreat working and it is tempting to consider this as systematic, allowing the maximum depth to be achieved, with less risk of rock instability causing inflows of water to increase. At Dariba the weakness of the rock was, and remains, a greater stability problem, although it eased the problems of excavation. Here large timbers functioned for both wall and roof supports (see below). At all three sites some stability problems were overcome at the surface by quarrying, which obviously required sufficient quantities of ore for this to be an economically viable option.

Haulage

Mining involves the movement of very considerable quantities of material: ore, waste, timber for supports and firesetting and water. As much waste material as possible would be stacked as deads in the worked-out stopes but much remained that had to be removed from the mine. There is little evidence of haulage arrangements or any form of trucks, wheelbarrows or even sledges such as are attested in some Roman and Chinese mines. Instead, in the early Aravalli mines, there were well-developed paths that were frequently lit by lamps set into niches in their vicinity. The paths sometimes zigzag to accommodate steep slopes but where this was not possible they were provided with series of wooden or stone steps which are sometimes cut into quite gentle slopes (see Chapter 5, Plates 5.16 and 5.17). These steps are up to a metre in width enabling two lines of miners to pass easily carrying their loads. Therefore the best evidence for the movement of materials in the mine is that it was by direct human haulage. In the constricted spaces of some of the 19th-century mines it was more efficient for stationary chains of miners to pass the materials one to another, as described by Captain Dixon at the lead mines at Taragarh, near Ajmer (see Chapter 2), or just dragged along the floor on wooden trays or even skins as recorded by Drummond (1838). Together with the baskets and pots found in the mines (see below) it is possible to envisage almost continuous lines of miners carrying ore and rock in baskets and water in pots out of the mine, and returning with timber etc. The height of the passages on the main routes would have allowed loads to be carried in baskets on the head. Today this is done mainly by women, as in the building and road construction industries and even in the early days of HZL (see

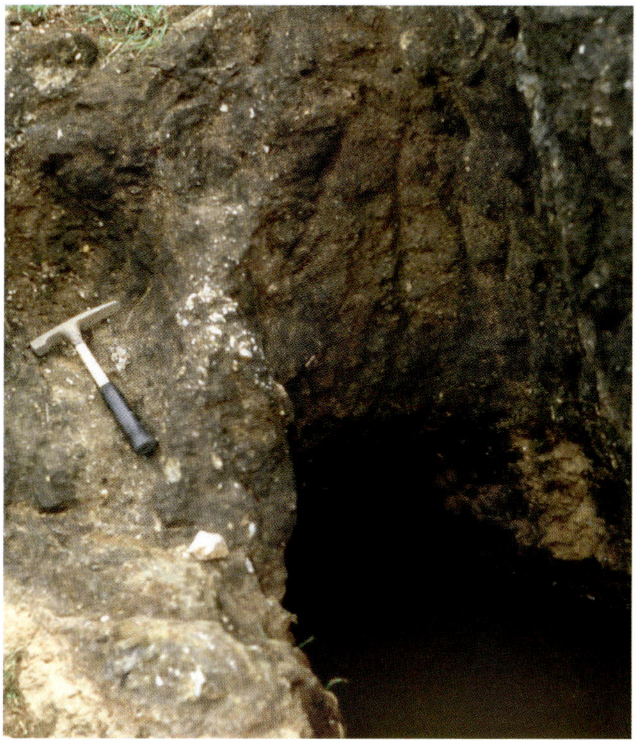

Plate 8.5 Long vertical chisel or pick marks. Mochia Magra mine, Zawar.

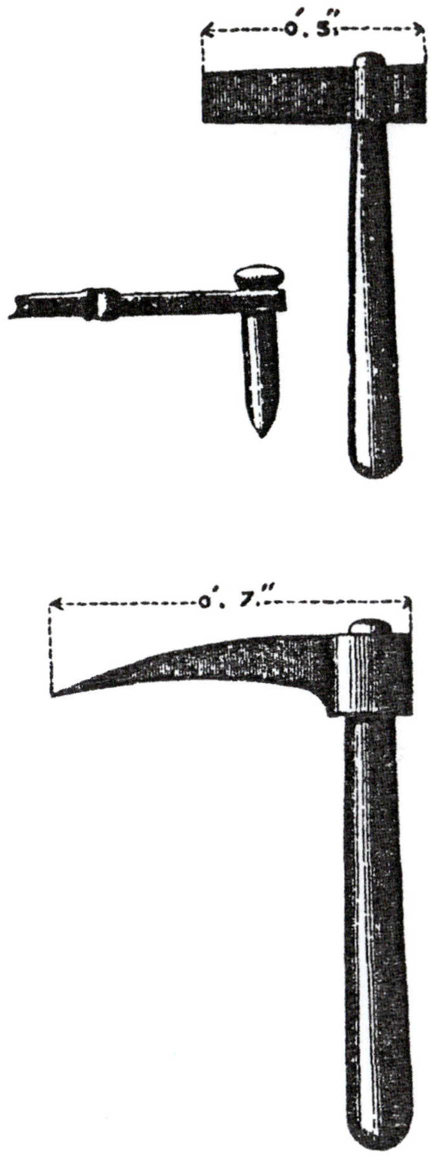

Fig. 8.1 Mining tools from Sikkim recorded by Blandford in the mid-19th century (from Percy 1861: 388).

Chapter 3, Plate 3.16). The loads are filled by men and lifted on and off the women's heads at the beginning and end of the journey, as depicted in the reconstruction drawing (see Chapter 5, Fig. 5.11).

At the three mines investigated many of the smaller vertical shafts linking the inclined workings to the surface have footholds cut into them so must be considered as climbing shafts as well as serving for ventilation. None had pairs of holes cut into the rock at the top to take the verticals for a windlass so there is no evidence for this or for any other winding gear to raise material, such as Munn (1936) found at the Hutti gold mines, together with rocks with grooves worn by ropes. Presumably all the materials were carried out of the inclined workings on the stepped paths.

Tools

The dates of even the earliest mine workings in this project fall well within the usage of iron in India; no stone mining hammers were encountered. It would seem, therefore, that iron was the normal material employed for the mining tools, in contrast to the crushing tools utilised for beneficiation where stone cobbles seem to have been used for much longer (although the late usage could be a recession, see below). Very few iron tools have survived with the exception of some undated examples recovered from the Mochia mine at Zawar at various times. These include two gads (see below) and three heavy iron pestles that clearly were used for crushing or grinding rather than primary mining (Plate 8.2), and an iron draw hoe (Plate 8.3) that would have been ideal for pulling the ore and debris back from the mine face and into the baskets that were used in large numbers in the mines.

From a study of the tool marks on the rock faces it is possible to get some idea of the tools used. These seem to have been round-pointed tools known as gads, which would have been hammered into cracks and the face of the rock exploiting weaknesses exposed by weathering, leaving round depressions in the rock (Plate 8.4). Wedge-shaped chisels or mining picks were also used which created long cuts on the rock face (Plate 8.5). Blandford (as republished by Percy 1861: 338) illustrated the iron tools used at a 19th-century copper mine in Sikkim, between Nepal and Bhutan in the southern Himalayas (Fig. 8.1).

In the Timber Chamber of the Zawar Mala mines were the remains of two bipods in the vicinity of a pillar that had been removed in the last phase of work in the mine (see Chapter 5, Plates 5.29 and 5.30, Fig. 5.9). They were simply formed from two posts, one pegged through a socket in the other. It is possible that heavy iron-shod posts were suspended from these bipods and swung as battering rams against the fire-weakened pillars. However there does not seem to be any surviving method of attaching the necessary ropes and it is possible that they were used in haulage of material up the slopes.

Baskets were found underground at both Zawar and Dariba and in quantity at Agucha (see Chapter 5, Plate 5.69), suggesting

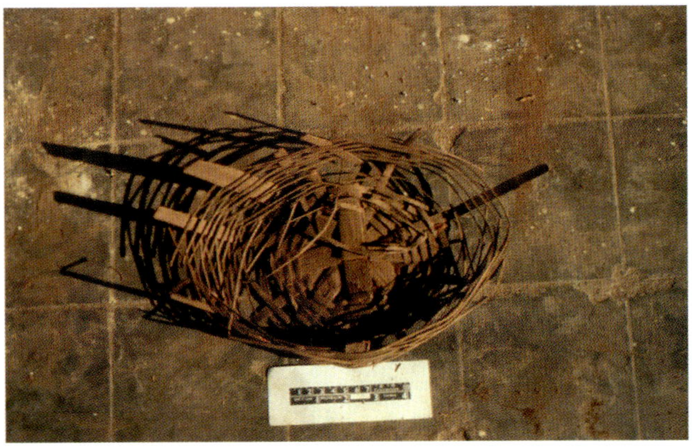

Plate 8.6 Undated basket from Balaria mine, Zawar.

that they were the main means of carrying the ore and waste out of the mines (Plate 8.6). Baskets were used for this purpose more recently as recorded by Brooke (1864) at Singhana and by Stephens (1901–2: 406–10) and Mallet (1875) in the Himalayas, recorded in Ball (1881: 276–7) for example.

Numerous pots of a suitable size for carrying water were found in the workings (see Chapter 5, Plate 5.31 and Fig. 5.12). Some of these would have contained drinking water but 19th-century descriptions of mine drainage using pots to carry water from the mines suggest that this may have been prevalent in earlier times (see below). There is no evidence for wooden vessels, and leather would probably be proscribed to Hindu miners, especially those belonging to the Jain faith, who seem to have been active in the medieval Zawar mines as evidenced by the temples.

Very few purpose-made lamps were found during the excavations (see Chapter 5, Fig. 5.13), but underground the bases from broken pots that could have served as lamps were discovered in several instances still sitting upon ledges from where they could have shed some light on the adjacent workings and pathways. Several of these were still perceptibly greasy to the touch. In more recent times a good steady illumination was achieved using a simple dish filled with cotton waste soaked in oil. No evidence was found for rush lights on the walls of the mines, such as Stephens (1901–2: 406–10) recorded, and we have no information as to whether or not the miners carried individual illumination as was sometimes the case in later mines (Boileau 1831; Dixon 1831).

Firesetting

The more weathered and fissured rock near the surface could be successfully mined directly with hand tools such as those described in the previous section, but the hard rock beneath would have required weakening beforehand. This was achieved by firesetting,[10] a technique that was in use everywhere from the inception of mining until replaced by explosives in the post-medieval period (Fig. 8.2). An early Indian reference to firesetting is contained in the *Brhat Samhita* compiled by Varahamihira in the 5th century AD (Bhat 1981), which states that if the rock resisted splitting, a fire was to be lit against it of ebony or bulia wood, and when the rock was red hot water mixed with lime was to be sprinkled over it, causing it to burst. More recently firesetting was recorded as being used both at the lead mines near Ajmer and the copper mines at Singhana-Khetri, in Rajasthan in the 19th century (see Chapter 2).

As observed in Willies (1994), firesetting was not just a simple bonfire technology. The practice of firesetting generally has been described in the articles listed above and will only be outlined here before detailing the more specific evidence for its use in the Aravalli mines. The heat generated by the fire acts in a number of ways on the rock including the vaporising of the free water in the rock as well as much of the water contained in chemical combination in hydrated minerals (decrepitation). In addition the heat creates tension due to the differential thermal expansion of the various different rocks and to the temperature gradient in the rock. All of these result in huge strains that a soft rather than a hard rock is more able to accommodate, usually causing the latter to split and craze (Willies 1994).

Firesetting underground causes a variety of problems and hazards depending on the particular task envisaged, but the Indian miners clearly used the technique with considerable skill and versatility. An obvious problem is the safe removal of the smoke and fumes generated, the latter being principally carbon dioxide with smaller quantities of the more lethal carbon monoxide and sulphur dioxide. In the Aravalli mines with large stopes it was probably possible to keep small fires burning while work continued around them, which also meant that they could be replenished while still burning – and even on occasion moved from one location to another while developing major cracks, as is current practice in the granite quarries of southern India (Collins 1892–3; Vikram 1984: 33–4; Craddock 1996). The relatively small quantities of warm smoke and fumes would rise to the roofs of the often tall chambers and galleries and continue up the steeply rising galleries until either the surface or a vertical shaft was reached, allowing the miners to move about freely beneath. Large quantities of soot coat the roofs of many of the ancient galleries (see Chapter 5, Plate 5.14). In large shafts there may have been some form of partitioning to effectively divide the shaft into two – the fire beneath on one side driving the smoke and fumes upwards while fresh air was drawn down the other side. When large firesetting operations were conducted it is likely that sufficient fuel for the entire operation was stacked in the appropriate manner, lit, and then the mine evacuated until such time as it was deemed safe to return and resume work. After a long fireset it is unlikely that dousing the surface of the by-now deeply fissured rock would have produced any further cracking, but it would have been necessary to cool the rock sufficiently for mining to start. Firesetting required considerable amounts of timber as fuel, typically of the order of between about 1:1 and 1:1.5 of timber to rock removed (Dubois 1996, table 1) but Willies (1984) reckoned 1:2 (or more likely 1:5). Therefore the collection, preparation and transport of the fuel represent a considerable amount of labour in the overall mining activity. In this project considerable quantities of half-burnt twigs and minor branches were observed which were presumably used as kindling, but in addition larger logs that would burn for hours must have been required. The large quantities of mature timber necessary would have required dedicated woodland and forestry. This may well have been worked together with the charcoal production, and in the Mauryan phase of the mines operation would probably have been under the control of the Director of Forest Produce (as discussed in Chapter 9).

Heat rises and a passage advanced into rock by firesetting will also tend to rise unless measures are taken to control the direction (as shown in Dubois 1996, fig. 4, for example). In the

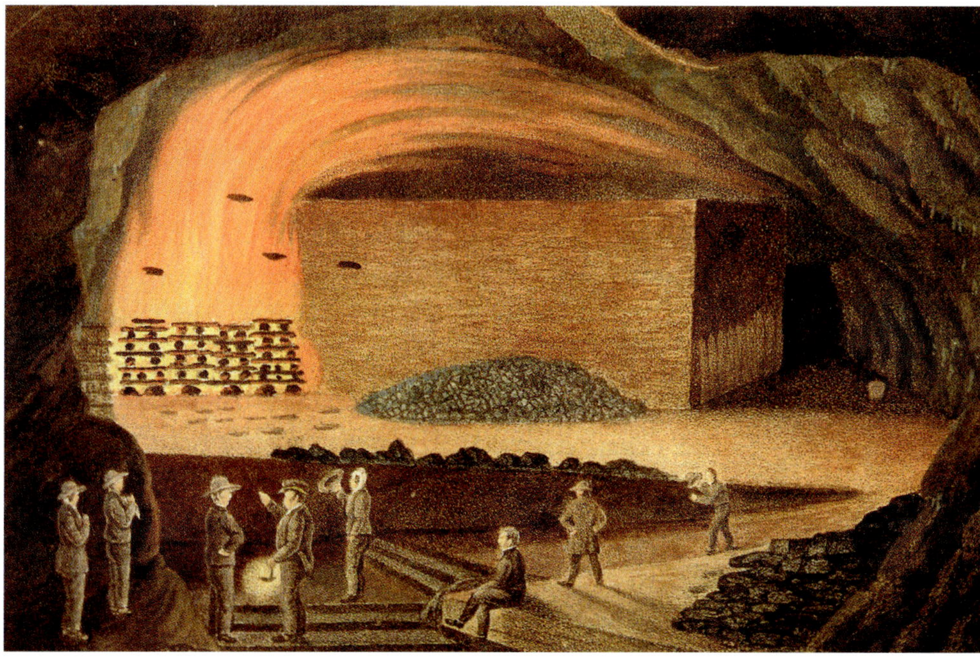

Fig. 8.2 Lithograph of firesetting at Rammelsberg in 1837. Fires on this scale must have burnt in the Aravalli mines. Note how the flames seem to stick to the walls, creating the very smooth sinuous walls that are such a feature of the Indian mines. Note that because of the size of the workings and the control of the fumes and smoke it seems that a party was allowed to remain underground and watch the fire. This suggests that it might have been possible for work to have continued in the Indian mines during firesetting.

Aravalli and other Indian mines firesetting was used successfully to sink vertical shafts and to drive passages horizontally as well as upwards (inclines) and down (declines). Similarly when mining the ore, the firesetting could advance the work by overhand stoping in descending drives and, conversely, could then remove further ore from the sides, floor or roof by underhand stoping back up the drive in retreat mining. Although similar arched pillar and stall areas of firesetting are observable from recent centuries in countries such as Norway, Sweden and Germany, the use of fireholes in the benching of floors and sinking of inclines has not previously been reported in detail. Clearly, the Indian miners could use firesetting to advance the work in any direction they desired. A large part of this control would have been in the design and layout of the fire itself, of which by its nature little or no evidence of its original form survives. It is likely that most shaft and decline sinking was carried out by a series of small fires creating hemispherical depressions in the floor beneath the fire, such as those recorded at Zawar Mala (see Chapter 5, Fig. 5.6 and Plate 5.11), and it is estimated that a thickness of about 5 cm of rock could have been removed after each fireset. Similar intersecting hemispherical fireholes were observed by Llewellyn Jones (Collins 1892–3) in the Kolar Gold Mines. A typical firehole was excavated in the small fireset depression on top of Mochia at Zawar, which could have been the start of an intended shaft that was subsequently abandoned (see Chapter 5, Plate 5.36 and Fig. 5.18). It is possible that the fuel of the fire was arranged in such a manner that much of the radiant heat was reflected back onto the floor, or alternatively that charcoal rather than wood was used to give a more locally intense heat that would conduct into the rock beneath. Charcoal fires would have the added advantage of producing very little smoke. McCarthy (1882) described the sinking of shafts about 1 × 2 m across through quartzite in the native gold mines of West Africa. This was done with a fire made up of bundles of faggots of wood, replenished at intervals for two days, followed by controlled dousing over another two days before the cracked rock was prised out using iron chisels

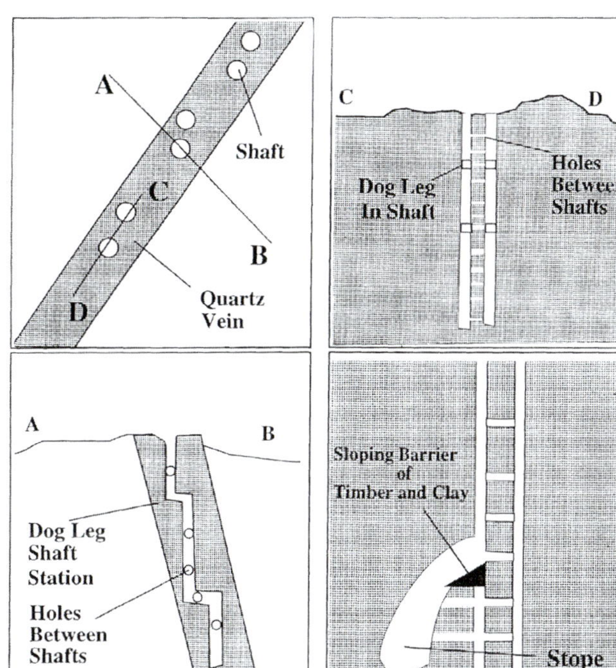

Fig. 8.3 Shaft sinking by firesetting at Manso Kkran in the Aadansi East District of Ghana. Top left is a schematic plan of shafts sunk in the vein. The shafts, sunk in pairs top right and bottom left, are stepped for safety and to provide platforms for hauling waste out of the mine. Bottom right shows underhand stoping by fire (from Laidler and Sharp 2000).

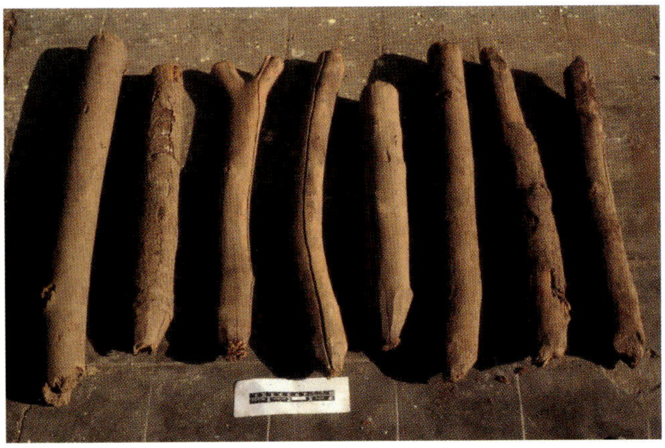

Plate 8.7 Selection of timbers from Balaria, Zawar mine: stemples probably all of post-medieval date. Note they are adze-sharpened at the ends to enable them to slip easily into cracks or hollows in the rock face but are otherwise unworked (scale bar: 10 cm).

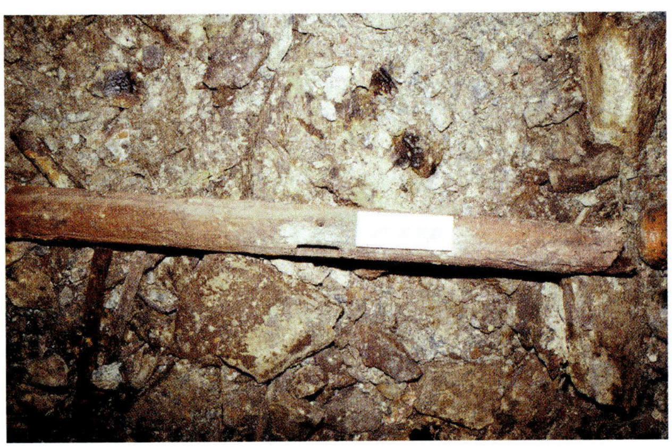

Plate 8.8 Fragment of a ladder from Zawar Mala showing a well-cut mortice and hole for a now missing peg.

on a metre-long handle. Laidler and Sharp (2000) give a more recent detailed description of how vertical shafts up to 40 m in depth were driven through quartz veins in the gold mines in Ghana (Fig. 8.3). Two small shafts were excavated and an opening made between them. A fire was lit in the bottom of one shaft and the smoke rose through the other. Although the fuel was brushwood, care was taken to ensure that the fire was largely composed of glowing coals so that much of the heat was directed downwards into the rock. The fire was kept burning all night and then doused suddenly with several hundred litres of water. This reduced the quartz to a sugary powder to a depth of about 1–1.5 m and a further 0.5–0.8 m could easily be removed with hand tools. The work progressed stepwise from one shaft to the other, stopping up the openings as the depth increased. Larger shafts such as those in the Aravalli mines could have been made by taking out the dividing wall, which would have been quite easy, and by enlarging the sides until the hole was of the desired width and then repeating the operation.

Driving declines would also have required the same control of the fire to counter its natural tendency to rise. The smoke and fumes would escape along the rising roof of the passage. With an incline the problems would have been reversed – the fire would naturally tend to rise, but there would have been severe ventilation problems in a rising chamber, and somehow the smoke and fumes had to be induced back down the roof of the incline. Where such work was encountered in the Aravalli mines the area was normally fairly broad and by controlling the inflow of air to the fire it could be possible to force the smoke to descend at least for short distances. In the large stopes typical of the Aravalli mines it would be relatively easy to control the fires and safely remove the smoke and fumes. In the sloping deposits, the underhand stoping would be advanced downwards by a series of stepped benches such as were preserved at Zawar Mala (see Chapter 5, Fig. 5.8). Subsequently the stope could be enlarged by firesetting at the lowest point and then continued by working upwards. This could lead to problems positioning the fire against the higher walls and roof and necessitate the building of stone and timber platforms like those preserved at Zawar Mochia (see Chapter 5, Plates 5.40 and 5.41) as well as at other mines in the Aravallis (see Chapter 2, Plate 2.3). Similarly pillars were removed by firesetting during retreat mining often leaving pendules hanging from the roof (Plate 8.1).

Timbering[11]

Wood preservation in the underground workings is quite good in the Aravalli mines, as is common in the damp, rather poisonous conditions that exist in many mines where non-ferrous metals were extracted. All the common mine timbers are represented: the uprights to support the roofs in the chambers or to stabilise the waste material stacked on the sloping floor of the worked-out stopes; the horizontal timbers to support the sides known as stemples (Plate 8.7); stairways and ladders; the hollowed-out tree trunk drainage launders; the various scaffolds needed when attacking the higher walls and roofs; various uprights at the tops of stopes probably associated with haulage up the slopes; as well as some more unusual items such the bipods found in the Timber Chamber in Zawar Mala (see Chapter 5, Plates 5.17, 5.21–5.24, 5.27, 5.29, 5.55–5.58).

The timbers are usually unshaped although occasionally squared. The stemples have some sharpening at one end (Plate 8.7) (see also Chapter 5, Fig. 5.34a) to aid their firm fixing into fissures or holes chiselled into the rock face, and the other end was cut off at a slight angle to enable it to be dropped into place against the opposing face. Drummond (1838) commented on the absence of sawn timbers in the Indian mines he visited. The major squared timbers in the early Aravalli mines are likely to have been adzed and then planed, but sawn across their grain. Where timbers are joined one timber is usually morticed into the other such as on the bipods at Zawar, the ladders at Zawar (Plate 8.8) and at Dariba, and the frames at Agucha The mortice and tenon joins were carefully chiselled (see Chapter 5, Plates 5.30, 5.63 and 5.64, Fig. 5.34b).

The rock at Zawar is very stable and relatively little support was necessary other than to prevent rock flakes from peeling off the roof. In the more gossanised deposits that were worked at Dariba and at Agucha, much more support was necessary and in places the workings required timber frameworks to hold back the soft wall material. At Agucha, on one such frame, the vertical supports, morticed into the sole plates on the floor, survived although the tops had rotted away (see Chapter 5, Plates 5.21 and 5.71).

Ventilation

Ventilation in mines is always a most pernicious problem. Foul gases such as carbon dioxide and sulphur dioxide, which are heavy and accumulate at the lowest points of workings, were produced in huge amounts during the firesetting of the dolomite and the pyritic and other sulphidic host rocks. One of the major advances in early mining technology was the ability to control the air supply underground. Before that advance in technology the work could proceed no more than a few metres from the shaft. Hubert (1974), for example, estimated that a miner could work no more than a maximum of 4 m from the shaft in the European Neolithic flint mines. The Bronze Age mines at Timna in Israel provide an excellent example of the consequences of this, where thousands of shafts no more than a few metres apart were required to supply air to the maze of small galleries (Conrad and Rothenberg 1980). These were necessary as the technology did not then exist to control the movement of air, and must have constituted a major constraint on the development of a mine, especially where the deposit dipped steeply. Such a primitive close array of shafts was observed by Haughton (1854) sunk into the placer deposits at the gold mines in Singhbhum, although the weak nature of the host rock may have been partly responsible.

By the period of major activity in the Aravalli mines, work could proceed over 100 m from the nearest shaft, although this must have required careful organisation and monitoring to ensure that all parts of the mine system were being adequately supplied and that the smoke and fumes from the numerous small firesets were being safely removed. To some degree the smoke would have been a useful indicator of the direction of the air flow. The situation would never have been constant as without powered air blowers the flow is influenced by temperature differences between the mine and the outside. Thus at night it is likely that the air inside the mine could be warmer than outside, but during the day the temperature could rapidly become much hotter outside, possibly leading to reversal of the air flow; the ever-changing pattern of firesetting work must also have affected the air flow. Even so, in the large galleries and stopes of the Aravalli mines it must have been possible to maintain a movement of air relatively easily, with hot air rising to the roof and cooler fresh air entering at floor level. Traditionally in complex mining systems the air flow was controlled by doors and shutters between different parts of the workings, as well as by linking shafts at strategic points. Air can be moved in a number of ways: simple fans made of sheets of cloth operated on the *punkah* system are recorded in Roman mines (Pliny, *Natural History* 31.49; Jones 1963: 408–9) for example and illustrated in *De re metallica* (Hoover and Hoover 1912: 212). Air can also be moved by the judicious use of fire to warm it, causing it to rise. At its very simplest an underground working could have a shaft at each end with a fire at the base of one shaft causing the air to rise up it, sucking more air down the other shaft and through the workings. Murray (1940) believed that some of the shafts in the old copper mines at Ramanchandra Pahar in Singhbhum were for ventilation and Maclaren (1906, quoted in Allchin 1962) suggested that the close proximity of certain shafts in the Gadag Gold Field, now in the southern Indian state of Karnataka, indicated that they were used primarily for fire-induced ventilation. Ventilation can be locally improved in the mine by separating and directing the air flows, and there is evidence in the paired shafts and linked finger stopes that this was done at Dariba and at Zawar Mala, as also suggested by Singh (1997). Paired shafts have also been found at Agucha (see Chapter 5, Figs 5.32 and 5.33). An advance on this would be for one shaft to be made to do the work of two by carefully partitioning it vertically to become two halves sealed from each other so that the only intercommunication would be through the workings. Vertical grooves in the sides of the shafts to hold the bratticing survive in some ancient mines but have not been positively identified in the Aravallis. Wulff (1966: 15) described primitive copper mines in Anarak (Iran) with evidence of much earlier working where fires had been lit under one shaft to draw the air through the mine for ventilation. If there was only one shaft then a corner of the shaft was cut off with stone slabs to make in effect a triangular chimney under which lamps were set to draw the air out; alternatively earthenware pipes could be installed for the same purpose.

In the Aravalli mines it is likely that the prime function of many of the smaller vertical shafts was ventilation as there is no evidence that they were used for ore raising or climbing. It is also very probable that the whole profile of the major workings underground were carefully shaped by judicious stacking of the debris (deads) supported by timber uprights (see Chapter 5, Plate 5.22). If the only concern was to dispose of as much waste as possible in the space available then clearly they would have been packed to the roof, but this is manifestly not the case: there is normally a gap of at least a metre between the roof and the tops of the deads. Additionally, in some places major rock flakes have been carefully supported by timbering, yet there would seem to be no reason why they should not have been allowed to fall the few metres onto the deads beneath other than to maintain the smooth profile and thereby preserve the pattern of air movement within the mine.

Mine drainage

Water has always been a major problem, and whereas other difficulties may cause risk, or uneconomic operations, if the water cannot be removed then it will stop work entirely in underground workings. Ideally a mine should have the water problem tackled both strategically and tactically. Strategies include surface drainage, mining in order to prevent inrushes, separated areas underground to stop total flooding, and the provision of adequate drainage systems. Tactics might include the use of barriers to prevent the entry of water, channels, launders and dams to prevent it descending any deeper than necessary, and emergency systems to cope with actual inrushes. The great depths reached at Dariba and Agucha in particular are evidence that not only were the problems appreciated, but also that practical and economic systems of water control had evolved, some of which have been noted above. More surprising perhaps is the apparent lack of drainage adits, even where, as at Zawar, they would have been technically feasible, driven in from the sides of the deep valleys.

The actual drainage of the mines seems rather simple and to have relied largely on manpower. At Zawar Mala there are hollowed-out tree trunks that clearly channelled the water to various collection points or ponds. The water must have been carried up to the surface manually from the mine, probably in the large ceramic pots that were frequently encountered (see Chapter 5, Plates 5.23, 5.24, 5.31 and Fig. 5.12). Where the passages were more tortuous or steeper it was more economical to employ a human chain to pass the containers from one to another, as recorded by Brooke (1864) at the Kulihan mine near Khetri in the mid-19th century (see Chapter 2). Curle (1899: 210) interpreted the large number of pots and evidence of standing places as indicating that this method had been used in the Kolar Gold Mines in antiquity.

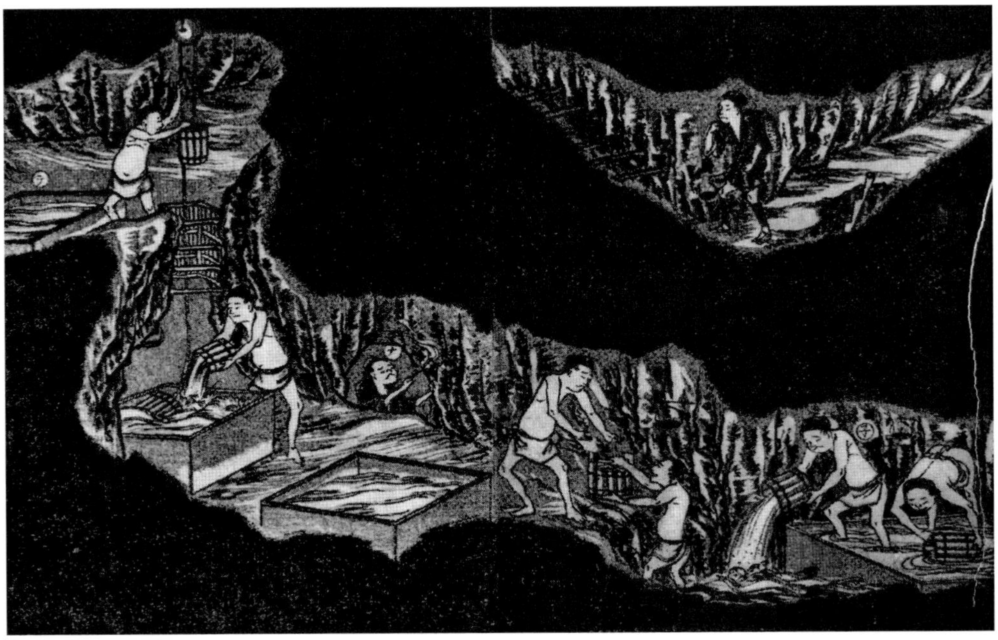

Fig. 8.4 Bailing system (19th century) at a Japanese mine. Note that although the water can be easily bucketed from the lower to the upper dam, in most cases the farthest left dam is too high for this to be done and a simple rope and pulley has to be used to raise the bucket (from Gowland 1899).

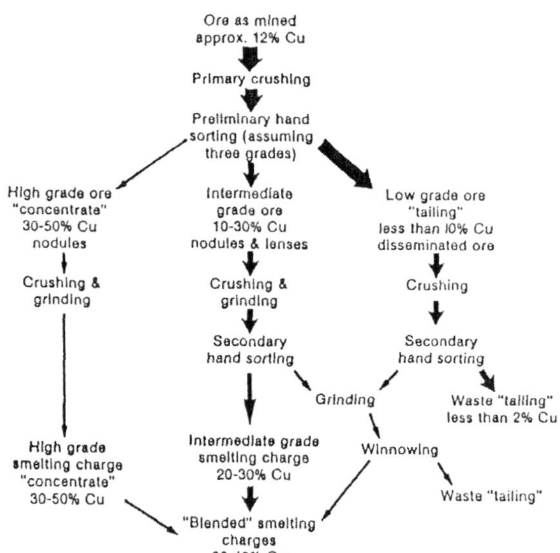

Fig. 8.5 A beneficiation sequence postulated for the copper ores from the Late Bronze Age/Early Iron Age mine at Timna, Israel. There is every reason to suppose that the sequence in the Aravalli mines was equally complex (from Merkel 1985).

Where the way was both tortuous and steep, bailing was another alternative. At Dariba there was the small but well-preserved section of a system of small dammed bailing ponds (see Chapter 5, Plates 5.53, 5.55, 5.57 and 5.58, Figs 5.23–5.25), probably similar to those recorded in Japan in the 19th century (Fig. 8.4) and also by Drummond (1838) at the Sheera mine at Pergunah in Kumaon. The preserved section was apparently quite steep with the upper pond being about 3.5 m above the lower so it would not have been possible simply to have bailed from one to the other – the bailer would have had to use the ladder unless the bucket was raised using a simple pulley as depicted in Fig. 8.4, or a system of hand pumps as shown in some Japanese and Chinese illustrations of mining. However there is no evidence for the use of pumps in the Indian mines at this period.

In contrast to the Classical mines (Healy 1978: 93–100; Craddock 1995: 76–81), and to some degree their Islamic successors (al Hassan and Hill 1986; Hill 1993), the Indian mines do not seem to have employed water wheels, the Archimedes screw or any other water-raising device inside the mines, although it is possible that animals may have been used to raise the water from the opencast workings in a similar fashion to that used today from the deep wells. The *Arthaśāstra* (2.24.18) lists the different dues charged for taking water for irrigation from the various canals and reservoirs, including by hand, using animals and by wheel, indicating that more sophisticated means of water raising existed even if they were not used in mines. Similarly, the bucket chain *noria*, so prevalent for water raising in Indian agriculture since medieval times, has not been encountered in a mine.

Beneficiation

The nature and extent of the beneficiation[12] processes have a profound influence on the smelting process as well as on the composition of the resulting metal (Merkel 1985) (Fig. 8.5). For successful smelting operations the furnace operator would require an ore charge of consistent composition and of uniform size – achieving this preliminary stage in metal production is known as beneficiation. It was also sometimes necessary to separate mixed metal ores, as at Zawar, where the lead and zinc ores were separated manually after crushing. Since good ore is notably heavy it could immediately be sorted underground into waste and ore by a competent miner and sent to the surface or waste heap as appropriate. A large proportion, perhaps at least half, of the excavated material was sent out of the mine judging by the spaces left behind.

Where possible it is likely that the good ore was extracted while in lump form and useless waste gangue discarded. The bulk of the ore would have to be crushed to facilitate the physical separation of as much of the gangue as possible. The ore would also have to be crushed to uniform size ready either for charging directly into the furnace or for roasting and briquetting.

Ore crushing

The use of water power in general was very unusual in medieval and even post-medieval South Asia (Habib 1986; Khan 1986), although water power was harnessed for grinding corn etc. throughout much of the Islamic world, especially in the lands bordering the Mediterranean where water mills had a long history (Lewis 1997; al Hassan and Hill 1986: 52–5). Throughout South Asia ores and minerals were almost always crushed by hand.[13] Brooke 1864 (see Chapter 2, Fig. 2.2) recorded the rather unusual two-handled drop hammers used at Khetri to crush copper ore, and with no anvil other than the floor these would leave very little archaeological evidence. Another method of carrying out primary crushing recorded in India in the 19th century and later, mainly for crushing gold-bearing rock, was using a boulder. This could either be pivoted on an arm to form a primitive stamp (Simonin 1869: 445, fig. 151), or simply rolled over the ore contained in the dish-shaped depression which formed in the rock surface beneath (Munn 1936, fig. 2). No evidence was found at the Aravalli mines for either utilised boulders or depressions such as have been reported from elsewhere in India (Medlicott and Blandford 1893: 49).

The evidence from Zawar, Dariba and Agucha suggests that originally the ore was crushed in large mortars with vertical hand-operated stamps, but that latterly handheld hammers of iron or stone were used. Stone hammers are usually regarded as indicators of great antiquity, and for the hammers employed for the actual mining this is probably true, but they remained in use for crushing much longer. In fact they are still employed in some parts of the world, apparently not as part of a continuing tradition but in recently established low-tech enterprises (Worthington and Craddock 1996). Undated crushing hammers are frequently found at old mines in India, and they were certainly in use in India in the 19th century. Mallet, recorded in Ball (1881: 277), described ore-crushing stone hammers held in a forked stick, which he observed being used at copper mines in Sikkim in the 1870s.

Our surveys in different parts of India encountered many crushing hammers, in particular on the post-medieval workings and heaps at South and North Lodes at Dariba (see Chapter 5). The hammers are all unmodified cobbles of hard rock typically between about 500 g and 2 kg. They fall into two categories: those generally of lentoid shape that are worn on their ends (see Chapter 5, Plate 5.52 and Fig. 5.21) and those of a more cubic shape where the striking contact was in the centres of the flat faces resulting in small pits. These are not to be confused with the anvils, known as knocking stones (described below) as they are generally far too light for that function. The very different wear patterns observed on the two types of hammer would seem to be due to their shape rather than to their function.

At the mines of Baroi at Zawar and at South and North Lodes at Dariba there are small hollows in the rock surfaces and stone hammers near to heaps of beneficiation debris. It is often thought that these hollows, found near old mine sites worldwide, were deliberately prepared as mortars for the reception of the ore to be crushed and some, such as those at the stone quarries in Lower Swat dated to the Gupta period that still

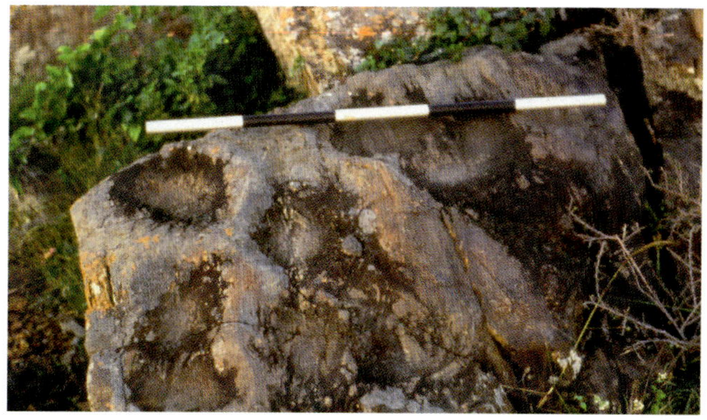

Plate 8.9 Series of undated crushing hollows at the Kolar Gold Field. Note one has broken into the side of another suggesting that their shape and depth are the result of use and not deliberate.

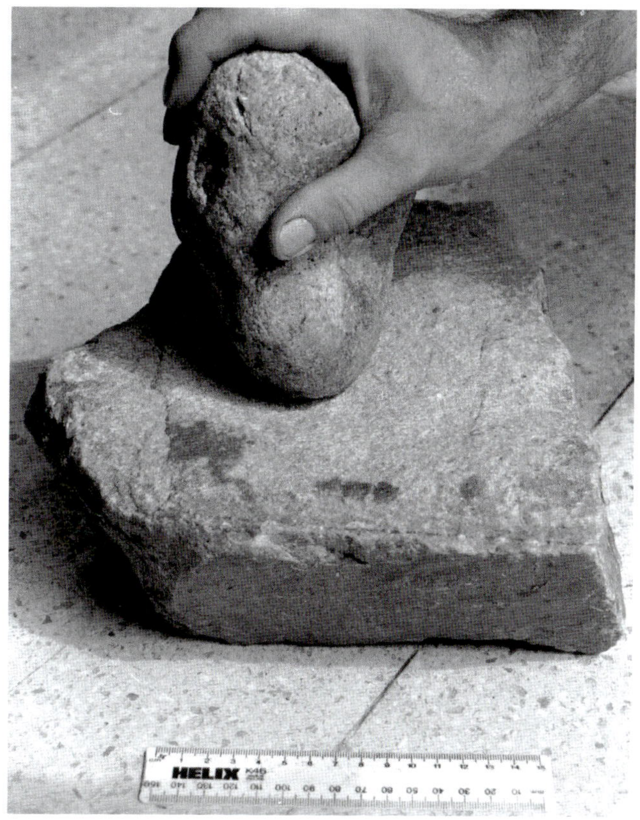

Plate 8.10 Stone hammer and knocking stone, both from North Lode, Dariba.

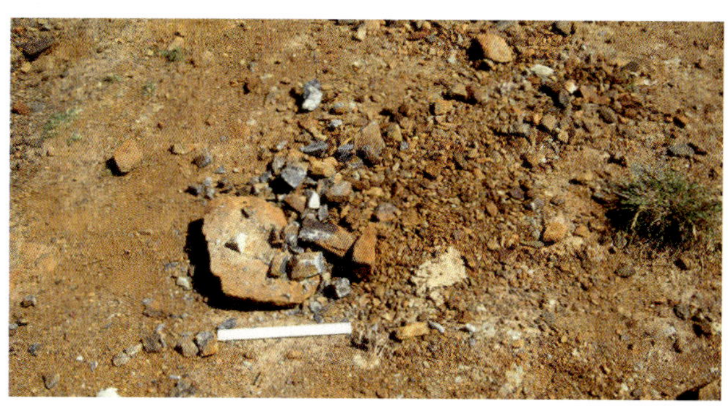

Plate 8.11 Ore-crushing area associated with small-scale unofficial gold-washing operations using a steel hammer and a knocking stone at the Kolar Gold Field, reminiscent of those found at North Lode, Dariba (scale bar: 30 cm).

carry chisel marks, are clearly deliberate creations (Di Florio *et al.* 1993), exemplified in this project by one mortar from Baroi, Zawar and another from North Lode, Dariba (see Chapter 5, Plate 5.51). However, it is likely that the majority formed as a consequence of the ore-crushing operations on previously flat surfaces (Craddock 1995: 157–61). For a while these forming depressions could be quite useful to contain the ore while it was being crushed, but in soft rock they would develop and deepen quite quickly, and once a depth of about 10 cm had been reached there was a danger of the operative holding the hammerstone crushing his, or more likely her, fingers between the hammerstone and the hollow. This helps to explain why sometimes there are so many of them, side by side (Plate 8.9). They should therefore be regarded as worn-out crushing surfaces rather than deliberately made mortars, and should not be confused with the multi-dimpled surfaces of stone washing tables (see below). Alternatively the ore could be crushed against flat plates of hard rock, known as knocking stones (Plate 8.10). Numerous knocking stones were found at North Lode, Dariba (see Chapter 5, Fig. 5.22) and were observed in use at 19th-century copper mines in Sikkim by Blandford (recorded in Percy 1861: 388) and also at Singhana in northern Rajasthan (see Chapter 2). Very similar stones were observed in use on our 1990 survey at the Kolar Gold Field in unofficial workings where material from the mine tip was crushed on small stone anvils using iron hammers (Plate 8.11).

In traditional European mining practice the large lumps of ore were broken with heavy long-handled hammers (sledging), followed by further breaking with long-handled small hammers (spalling), and final breaking up with small short-handled hammers (cobbing), prior to hand picking or washing (as described by Agricola, see Hoover and Hoover 1912: 268–73). Recognisably similar processes were observed in India by Taggart (1927: 618–24), and at Singhana (see Chapter 2, Fig. 2.2) almost 400 years later.

Ore concentration by hand picking and washing

Having crushed the ore to fine gravel, with only a small fraction larger than about 2 cm diameter, the gangue or waste material could be removed and the various metal ores separated by hand picking. Alternatively the heavier metalliferous minerals could be concentrated by agitating the ore in moving water, causing the lighter material to move further than the heavier mineral. Many systems for washing the ore are known; the method employed depending to some degree on the availability of water and the characteristics of the ore (Craddock 1995: 163–7; Hoover and Hoover 1912: 300–348). At the lead mines near Ajmer, recorded by Dixon in the 1830s, where there was a lack of continuous water supplies, large pits were dug in which to wash the crushed ore, the water being supplied from wells (see Chapter 2). Similar washing tanks are also described at Agnigundala in the Guntur District of Andhra Pradesh in the early 19th century by Heyne (Chakrabarti and Lahiri 1996: 20). In India some gold, iron and tin ores were traditionally washed from sand and gravel as part of the mining process and it is likely that washing was a regular feature for most ores (Babu 2003). Little direct evidence of washing was found at any of the sites excavated as part of this project, certainly nothing like the complex contemporary arrangements that survive for washing the argentiferous lead ores at Laurion in Greece (Conophagos 1980: 213–73; Healy 1978: 144–8; Jones 1984, 1988) possibly because the wooden washing tables etc., such as were used for gold and for copper, are unlikely to have survived. Further beneficiation could be carried out by winnowing the crushed ore for which there is considerable evidence in India. It was recorded by Pāṇini in the late first millennium BC in the *Astādhyāyī* (8.4.34,Vasu 1899), it was still in widespread use in the traditional iron smelting industry (Tripathi and Tripathi 1994) and was observed at the tin workings at Hazibagh and Ranchi in Bihar in the 19th century (Chakrabarti 1979). The usual procedure was to crush, hand pick and then winnow the ore.

At some sites in India, the surfaces of rocks lying near streams are covered in a regular array of small dimples, no more than 1–2 cm in depth and 3–5 cm in diameter. These may well have been deliberately worked to make a table for washing the ore to separate the heavy mineral from the lighter gangue. A rock surface of this type is illustrated by Hegde and Ericson (1985: pl. 1), and was said to be typical of many found during their survey of early copper workings in the Aravalli Hills. However, such prepared washing stones were not found on or near to any of the streams at the sites investigated as part of this project.

The locations of the beneficiation operations, as marked by the evidence of crushing activities at the three mines relative to the mining and smelting operations, are of some interest. At both Dariba and Agucha the main mines were established in reasonably flat terrain with the smelting operations taking place nearby. At mountainous Zawar mining and smelting were generally physically well separated. There is little direct evidence for the early periods of operation, but it is apparent that in the later periods, the beneficiation processes started close to the mines. Mactear (1894–5) recorded a similar situation at Samān, in Iran, where the copper ore was hand dressed at the mine before being carried some 3 km to the settlement for smelting. The beneficiation evidence from the three mines will be considered separately.

Zawar

As has already been made clear, at Zawar the mine entrances are all up on the ridges where the ore outcrops, well away from the smelting operations in the valleys below. Immediately adjacent to the mine entrances heaps of rough, unsorted mine debris are often found containing rock fragments of all sizes from fire-shattered slabs weighing many kilograms down to dust (see Chapter 5, Plate 5.43). There is no evidence for later reworking of these heaps for another mineral. At Baroi the crushing hollows, hammerstones and large heaps of rejected debris of regular size are located farther away but still within the vicinity of the mine openings. The hand-picked ore would then have been carried down to the valley for smelting. It is perhaps significant that these rather primitive and small-scale crushing activities seem to be mainly associated with the one mine group at Zawar that is considered to be post-medieval.

In the early phases of operations at Zawar the ore could have been brought down from the mines to be beneficiated in the valley. No stone crushing hammers were found so perhaps the ore was crushed with large handheld stamps such as are envisaged for the early workings at Dariba (see below). After crushing the ore may have been washed in the River Tiri, but possibly was concentrated by hand picking after roasting (Chapter 9). Again no direct evidence was found for washing operations, such as the rock surfaces covered with dimples noted elsewhere in the Aravallis as mentioned above. However, both in India and worldwide, the more usual practice was the construction of wooden washing tables, known as tyes, such as those described by Blandford in Sikkim (recorded in Percy 1861: 389) which would not have survived. The finely crushed rock debris was probably washed away by the Tiri during the monsoon season which might explain the absence of the large quantities of beneficiation debris at Zawar of the type found at contemporary Dariba and Agucha that do not have a river running through the ore preparation and smelting areas. Therefore, overall at Zawar in the early phases, it seems that the beneficiation of the ore is likely to have taken place down in the valleys by crushing with iron-shod hammers, followed by concentration by hand picking aided by washing, through most of the mine's operation. In the later phases the silver/lead ore mined on Baroi, near to the mines, was crushed with small hammers, some of which were of stone.

As a precursor to the extraction of zinc by distillation, the zinc and lead minerals were separated by hand picking from the crushed ore. This must have taken place before the roasting stage as the debris of the zinc roasting operations has a low lead content and the contemporary lead smelting slags are rich in galena (see Chapter 10).

Dariba

At Dariba there is more surviving evidence of the ancient beneficiation processes such as the large heaps of crushed rock at the north end of the major opencast mine at East Lode. Analysis of samples from the heaps showed the average grade to be 1.26% lead, 0.82% zinc, 0.20% copper and 4.81% iron. This suggests that the ore had been sorted and some beneficiation carried out, producing a concentrate for smelting. In the immediate vicinity of the heaps is a group of mortars cut into an outcrop of the hard calc-silicate rock (see Chapter 5, Plates 5.66 and 5.67, Fig. 5.28). Such deliberate rock-cut mortars are not found either at Zawar or Agucha due to the absence of suitable calc-silicate or other hard rocks. Cylindrical pits, about 30 cm in diameter and 40 cm in depth or somewhat larger ones, 60 cm in diameter and 70 cm in depth, used for ore crushing, were noted by Hegde and Ericson in their survey of the Aravallis (1985) and Hegde (1991: 17) at the foot of the hills containing traditional copper mines. They assumed that these remains were Chalcolithic, but in reality they are undated and were probably in use over long periods of time.

No hammerstones were found on the adjacent spoil heaps at Dariba, and it seems that the ore was crushed by stamping, probably with iron-shod wooden posts repeatedly raised and dropped onto the ore in the mortars. These would have needed to be fairly heavy. No actual physical remains of stamps or of battering rams have survived at Dariba or at Agucha and Zawar. The postulated iron-shod posts were probably operated manually: this may seem self-evident but water-powered, cam-operated stamp mills were used in China from the Han period (Needham 1965: 82–3) and also by the Greeks and Romans in the West, probably from the 3rd or 2nd century BC (Lewis 1997). At the Roman gold mines of Três Minas, in the north of Portugal (Bachmann 1993a; Wahl 1993: 141, fig. 19), Fresnedo, Pola de Allande in the Asturias region of Spain (Sánchez-Palencia 1989, fig. 5) and at Dolaucothi-Pumsaint in central Wales (Craddock 2002; Burnham and Burnham 2004: 281–4), for example, the remains of the distinctive anvil stones for cam-operated multiple stamps, worked by water or animal power, have now been recognised (Craddock 1995: 161), all dating to the 1st century AD. These stones are very similar to the so-called Cornish stamps of the post-medieval period in Europe. Small cam-operated stamps are used to crush coffee or spices in the markets throughout southern India to the present day, and may have a long history. If the postulated stamps at Dariba had been power-operated in some way then they would have had to have been held in a quite substantial frame. This, in turn, probably had its supporting posts set in holes in the rock around the mortars. There are several small holes around the mortars but these are the typical shallow basin-shaped depressions of the small crushing hollows and could not have held a post securely. They do seem to cluster around some of the mortars and at least two are actually in the top of the mortar, suggesting that the large mortars may have provided convenient sitting or squatting places for later small-scale crushing operations such as abound on the post-medieval workings on the nearby South Lode. Also the stamps would necessarily have dropped vertically creating little undercutting, whereas manually operated stamps could have been angled to crush the ore at the edges leading to the observed undercutting of the mortars in certain directions (as indicated in Chapter 5, Fig. 5.28); this can be quite pronounced as the sectioned mortar on the far left shows. This further suggests that the workmen stood in preferred spots on the rims of the mortars.

Another possibility is that the mortars were operated similarly to the traditional Indian sugar cane presses. Large stone mortars, not dissimilar in shape and size to those at Dariba, held a large wooden post that was turned at an angle by a couple of oxen. However, the very irregular rock outcrop in which the Dariba rock-cut mortars are set would have made it difficult for the oxen to walk in a circle around the mortars unless some staging was built for them and there is no surviving evidence for this, nor is there any evidence of scratches running around the walls of the mortars as would be expected. This combined, albeit rather negative, evidence suggests that at Dariba the ore was crushed in the large mortars with manually operated stamps. The comparison between the large-scale organised crushing installations in use over 2000 years ago at East Lode, with the very minor and primitive beneficiation methods of the more recent past, is quite striking.

Agucha

No stone hammers or small crushing hollows were observed anywhere at Agucha, suggesting that the ore may have been crushed in a manner similar to that practised at East Lode, Dariba. The absence of stone tools and small mortars at Agucha, which unlike Zawar and Dariba has no recent history of working, supports the impression that the use of stone tools on unprepared rock surfaces is relatively recent.

Overall, throughout the three mines there is a strong suggestion of regression of scale from early large volume stamping to the smaller scale stone hammer crushing in use in the later phases.

Notes

1. Principal sources: Hoover and Hoover 1912 (Agricola); Percy 1861, 1870, 1880; Ball 1881; Kangle 1965–72 (Kautīlya); Willies 1989, 1992a and b; Craddock 1995: 23–92.
2. Kautīlya: Kangle 1965–72; Sarma 2001.

3. Kautīlya's *Arthaśāstra* describes in some detail the establishment of mines in remote areas, together with their defence and management (Chapters 8 and 9). One of the main arguments previously advanced for dating the *Arthaśāstra* after the Mauryans (discussed and refuted by Kangle 1965–72, III and Mital 2000, esp. p. 32) was that they had no experience of mining. It can now be appreciated that these mines in the Aravalli Hills were Mauryan enterprises, set up far from their main centres, and it is not impossible that a very real Kautīlya had these operations in mind when he compiled his treatise on governance in the 4th century BC.
4. As outlined in the *Arthaśāstra* the Mauryan administration was highly organised and strictly controlled. Administration was split between towns and countryside. The rural administrator, the *samāhartṛ*, had ultimate control over various heads of departments, the *adhyakṣas*, responsible for all aspects of the productive economy, including those for mines, the *ākarādhyaksa*. Another important department was that of stores, whose function went well beyond just warehousing. Basically none of the producing departments could sell their produce independently; instead it had to be delivered to the *saṁnidhātr*, the Director of Stores. For example, the Department of Forest Produce charged with supplying charcoal and straw to the mines could not do this directly, but only through the Department of Stores. By this means the central administration retained complete control of the economy. The Aravalli mines would have sent their products away for finishing, fabrication and distribution. This could explain why there is no evidence of brass production at Zawar, and why the Site 10 foundry at Dariba was producing items in copper alloy for internal use, rather than in the silver which they were producing.
5. Megasthenes (Chapter 2, p. 8) noted that organised slavery such as was prevalent in the Hellenistic world did not exist in India. Kangle (1965–72, III: 187) discusses this and concludes that most of the mining force would have comprised free *śūdra* labourers.
6. The identity of *vaikṛntaka* is uncertain. At 2.12.16 it states that its ore is the colour of a crow's egg or of a birch leaf, both green. Therefore it would seem probable that a copper or iron ore was meant (Biswas and Biswas 1996, II: 48). At this time the Bactrian kingdoms to the north of the Mauryan Empire were issuing cupro-nickel coins; the alloy is believed to have been made by co-smelting mixed copper nickel ores (Needham 1974: 237). Alternatively a zinc ore was intended here: willemite (zinc silicate) is green, in which case *vaikṛntaka* could be zinc oxide, although there is no evidence for deposits of willemite at Zawar. In recent centuries *vaikṛntaka* was a term associated with a variety of minerals used in Ayurvedic medicine, including zinc minerals.
7. Quoted in Chapter 2, p. 14.
8. Those of lead, silver or zinc are described in detail in Chapter 10, p. 201 and Chapter 11, p. 212.
9. At most mines in antiquity once the vein had been located at the surface it was worked simply by digging down into it. The modern concept of identifying the extent of an orebody and then sinking a straight shaft by the side so it can be attacked from beneath dates only from the European Renaissance. As Biringuccio critically remarked in his *Pirotechnica* in 1540: 'The Ancients ... instead of beginning from the bottom, as moderns do, they began their mine in the upper part where the ore appeared in the light of day at the surface, and digging down as in wells they followed it to the bottom, now here, now there' (Smith and Gnudi 1942: 19).
10. Dixon 1831; Boileau 1831; Collins 1892–3; Craddock 1992, 1995: 33–7; Willies 1994; Dubois 1996; Weisgerber and Willies 2000; Golas 2000: 300–306.
11. The major timber revetment of the South Lode opencast at Dariba is considered separately in Chapter 5.
12. General: Merkel 1985; Craddock 1995: 156–67). Classical antiquity: Conophagos 1980: 213–72; Healy 1978: 139–48. Europe: Hoover and Hoover 1912: 279–351. China: Golas 2000: 352–69.
13. Atkinson (1882) noted that water power was sometimes used in the crushing of copper ores at the Kumaon mines, now in Uttarakhand State, northern India, possibly utilising the small indigenous horizontal water mills that used the streams of the foothills of Himalayas, but at this late date water power in a mine could have been a European introduction.

Chapter 9 Smelting

P.T. Craddock

Calcining, roasting and briquetting[1]

It was the usual practice in the past to heat metal ores prior to their actual reduction in the smelting stage in order to drive off the water of crystallisation they may have had as well as to decompose carbonates to oxides and to render them physically more friable. This is known as calcination in metallurgical terminology. In post-medieval Europe zinc carbonate ores were usually calcined to convert them to zinc oxide before charging into the retorts (Almond 1998: 166). In practice, almost all such operations would have taken place with free access to air. If they were undertaken to decompose sulphidic ores, they are known as roasting operations (Maréchal 1985). Most sulphidic ores, with the usual exception of galena (lead sulphide (PbS)), were roasted to at least partially oxidise the sulphides. Sulphidic copper ores certainly require a separate roasting operation (see below). The debris from the roasting of copper ores as observed at Singhana in Rajasthan (see Chapter 2, Plate 2.4) is quite distinctive: cracked, reddened and very obviously oxidised and burnt, but not generally fused. It is quite different, therefore, from the once-molten smelting slags that were created under reducing conditions and are thus generally grey or black, but certainly not red.[2]

At its simplest, in the traditional heap-roasting process, the sulphidic ore was mixed with a little brushwood, usually alternating layers of brushwood and ore, to form heaps and fired (Peters 1907: 115–25). Sulphidic ores usually contain some pyrite (ferric sulphide (FeS_2)) which burns freely, generating immense heat. The burning timber ignited the pyrites which then burnt slowly, converting much of the ore from sulphide to oxide. The white heaps at the edges of the main smelting areas of Old Zawar (see Chapter 6, Plates 6.33 and 6.34) are almost certainly where the zinc sphalerite ores were roasted in heaps prior to their reduction (see Chapter 11), otherwise little or no evidence of roasting was encountered at the three sites. This is mainly because, apart from zinc at Zawar, the bulk of the ore smelted was galena which, as noted above, does not require roasting prior to smelting. The lead ores at Ajmer, for example, were smelted directly as recorded by Dixon in the 19th century (see Chapter 2).

In the remote past when a gentle heat was needed, the Indian alchemists usually specified cow dung as the fuel (Roy and Subbarayappa 1976: 86) and, as with the smelting of the ores, it was often specified that the materials should be ground up and made into cakes with cow dung prior to roasting (Rây 1956: 220). The mixing of the crushed materials with cow dung was clearly widely used in Indian metal production in more recent times, both in the roasting and smelting stages. The usual practice as recorded in 19th-century European descriptions of smelting of non-ferrous metals in India and neighbouring countries was for the crushed ore, whether roasted or not, to be made into small cakes or sausages mixed with dung, and sometimes also clay, ready for the furnace (see Chapter 2). Recent replication copper smelting experiments using the technique based on the 19th-century descriptions worked well in small traditional furnaces or hearths (see below). The high alkali content of the dung promoted the sintering of the silica in the clay or ore, creating an ideal open structure sinter. The practice of making the ore into sintered balls or sausages continued into the 20th century as recorded by Wulff (1966: 16–17) at the copper and lead smelters of Anārak in central Iran, for example, where the dust fraction of the crushed ore was made into balls with clay and charcoal dust and then roasted to form a sinter in a small separate furnace before joining the rest of the ore in the main smelting furnace.

This was, in effect, an early form of sintering and briquetting the ore, which is now standard practice in many modern smelting processes (Stillman1923).[3] As with roasting operations generally, this aspect of ore preparation has not received sufficient attention from archaeometallurgists seeking to reconstruct ancient and traditional smelting processes elsewhere. Rehder (1999, 2000: 104–8) noted the similarity of these 19th-century Indian copper smelting activities to some aspects of his carbothermic smelting process, which he believed could have been widely practised in antiquity. In this process, the roasted ore is intimately mixed with charcoal dust and strongly heated in a small contained environment such as a crucible. There is no evidence – either archaeological or from the more recent detailed historical descriptions from India – for these small containers for copper smelting. However, there are medieval descriptions of the smelting of galena with iron in crucibles (Dube 2006) (see below) and crucibles were certainly used in India for the more sophisticated smelting processes necessary for crucible steel production by *in situ* carburisation (Bronson 1986; Craddock 1998b) and for the smelting of zinc, the contained environments being either the crucible or the retort.

Fuel[4]

There are many accounts of traditional charcoal burning: Rehder (2000: 55–62); Constantine (1975) in Australia; Armstrong (1978) and Kelley (1986), in Britain; Horne (1982)

Table 9.1 Wood ash compositions (from Misra *et al.* 1993).

Element	Pine	Aspen	Poplar	R. Oak	W. Oak	W. Oak Bark	D.F. Bark
Calcium	29.05	21.17	25.67	36.58	31.35	36.14	34.26
Potassium	16.24	11.25	7.93	6.08	10.25	0.97	2.78
Magnesium	7.03	3.55	9.09	5.20	7.57	0.34	0.37
Sulphur	1.07	0.70	1.02	1.80	1.21	0.40	0.52
Phosphorus	0.84	1.18	0.95	1.56	0.56	0.08	0.51
Manganese	4.04	0.14	0.45	1.49	0.14	0.16	0.37
Zinc	0.36	0.34	0.04	0.22	0.08	0.05	0.07
Iron	0.58	0.26	0.32	nd	0.09	0.01	0.26
Aluminium	0.47	0.14	0.35	0.68	<0.03	<0.03	0.59
Sodium	0.06	0.06	2.30	0.08	<0.06	<0.06	<0.06
Silicon	nd	0.11	nd	nd	0.13	0.12	0.24
Boron	0.06	0.05	0.05	0.08	0.04	0.007	0.07
Copper	0.04	0.03	0.03	0.07	0.02	<0.002	0.02

nd = not determined

in contemporary Iran; Wagner (1985: 33–7) in 20th-century China; Craddock (1995: 189–96). In addition, some descriptions of charcoal burning in 19th-and 20th-century India are given below.

Wood and dung

Wood, dung and charcoal were used variously and extensively in the mines and in the smelting of the ores, and their sustained procurement must have required careful logistical planning at major operations. Wood was required for the firesetting operations at all the mines; some charcoal was probably also needed for more specialised firesetting tasks such as shaft sinking (see Chapter 8). As already noted, brushwood was also probably used in the roasting operations where the oxidation was exothermic and the reaction only required initiation. The small twig charcoal that was found in quantity in the white heaps at Zawar probably resulted from burning brushwood.

The dung of herbivores, usually containing considerable quantities of plant remains, was used for a wide range of metallurgical processes in the past but the extent of its use is probably not fully appreciated generally. In India, however, and to a lesser degree in Iran, it is well documented. Cow dung was employed as a filler in the refractories, as a support material in the briquetted ore, and as a source of alkali in some refining processes of precious metals (see Chapter 10), as well as serving as a fuel. Many of the 19th-century descriptions of ore roasting state that cakes of cow dung were used as the fuel, and it is very possible that cow dung was employed in conjunction with brushwood in the zinc ore roasting process at Zawar. In addition, the roasted ore was routinely kneaded with cow dung to make small cakes or sausages as described above.

The heat content of well-dried animal dung is about 17 to 18 megajoules per kilogram (Winterhalder *et al.* 1974), which is only a little more than that of grasses or low resin wood, but appreciably less than softwood charcoal, which produces about 27–30 megajoules per kilogram. For dense hardwood charcoal, the heat content per unit volume is even greater, an important consideration where space is restricted. Both materials are postulated to have been used as fuels in various early metallurgical operations, usually without supporting evidence, so it would be useful to be able to distinguish between the use of dung and charcoal from their surviving ash. Cow dung has a high alkali content and when it burns at high temperature a considerable quantity of very distinctive light vitrified material is created by reaction partly with the straw and other siliceous material it contains and partly with any other clay material in the vicinity (Evans and Tylecote 1967). This is in sharp contrast to charcoal, which creates just a little powdery ash. A familiar example of the products of burning cow dung is the large ancient burnt mounds of cow dung that are found in some areas of southern India (Allchin 1963) which were studied by Zeuner (1959), who concentrated on the vitrified material. Samples of this were found to melt between about 1200 and 1230°C, and one sample was analysed and found to contain 67.4% SiO_2; 9.47% combined Fe_2O_3, Al_2O_3 and Mn_2O_3; 4.38% CaO; 2.98% MgO; 6.24% K_2O; 1.9% Na_2O and 3.4% P_2O_5. Samples of vitrified material from other burnt mounds were found to contain 1.17% and 1.57% P_2O_5, and a sample of cow dung ash from England contained 56.3% SiO_2 and 2.46% P_2O_5.

In contrast, wood ash (Sanderson and Hunter 1981) and charcoal ash (Tylecote *et al.* 1977) contain only small quantities of silica but more calcium. Vitrified wood ash analysed by Evans and Tylecote (1967) was found to contain little silica and only traces of phosphate. The ash of vitrified straw as represented by rice hulls seems to be relatively pure, containing silica with about 1% potash and only about 0.4% phosphates (Biek 1970; Houston 1972). Misra *et al.* (1993) studied the composition of wood ash as a function of the temperature of combustion. The main components of wood ash after combustion at 600°C are given in Table 9.1; it can be seen that calcium is the principal cation present. On heating progressively to 1300°C the main losses are in potassium (63–90%) and sulphur (7–55%).

The more detailed analyses of Misra *et al.* (1993), while confirming the low silica content of wood ash, do not suggest that there is a significant difference in the phosphorus content between dung and wood, so it would seem that it is only possible to chemically distinguish vitrified cow dung from wood or charcoal ash by the silicon or calcium content.[5]

Charcoal

All the traditional and 19th-century descriptions of metal smelting, in India as elsewhere, specify that apart from exceptional circumstances charcoal was the only fuel used in the actual smelting process, in contrast to the ancillary roasting and refining operations mentioned above. For smelting in small installations, charcoal is an ideal fuel: it contains very few harmful elements to contaminate the forming metal such as sulphur, which bedevils all fossil fuels, and it has a high calorific value therefore it can generate high temperatures in small volumes. The study of the materials from the slag heaps at all three of the major mines and at all periods, tells the same story (see Chapter 7). Slags that must have been molten at temperatures well in excess of 1000°C were produced in hearths or furnaces – reconstruction from their fragments shows that they had internal diameters typically of no more than 20–40 cm. Similarly the space in the *koshthi* furnaces for smelting zinc was very restricted. It is simply not possible to generate those temperatures and reducing conditions with wood or dung in such small volumes; charcoal is the only feasible fuel.

It is, however, possible to smelt lead with wood. Collins (1893–4) recorded that dry wood was used for smelting lead at Ghorband in Afghanistan and kiln-dried wood was sometimes used for lead smelting in post-medieval Britain, where it was known as white coal (Willies 1991b: 98). However, the smelting installation needed to be larger, whether a furnace or a hearth, as exemplified by the medieval/post-medieval lead smelting 'boles' of western Europe (Craddock 1995: 207–9).

Charcoal size

Charcoal is a fragile material and during transport invariably breaks and abrades, creating quantities of dust known in English as fines. It is important to remove these (usually by shovelling the charcoal against a grid) prior to charging the charcoal lumps to the furnace to avoid restricting the passage of the various reaction gases. The surviving archaeological evidence for this practice are the large areas of fines such as those excavated at Dariba, Site 9, where they form much of layer 2 (see Chapter 6). It is also important that the charcoal should be reasonably consistent in size otherwise the combustion gases will tend to follow certain paths adjacent to the larger lumps rather than moving uniformly through the whole mass, leading to differing conditions within the smelt (Rehder 1990, 2000: appendix 2; Merkel 1985: 88).

The size of the individual charcoal pieces is an important parameter in the overall conditions during the operation of the furnace: the larger the pieces the greater the air flow through the fire so the conditions will tend to be hotter because the rate of combustion is faster. The conditions are more oxidising due to the smaller surface area of charcoal exposed and shorter contact time to reduce the carbon dioxide formed by the initial combustion in the tuyere zone to carbon monoxide (see below). Also, the larger the charcoal size the greater the risk of the ore and flux falling unchanged through the reaction zone and into the slag layer beneath.

In small traditional shaft furnaces for smelting metals, the spaces between the furnace wall, whether curved or flat, and the charcoal are greater than the spaces between adjacent lumps of charcoal (Rehder 2000: 16–17). There should be a propensity for the combustion gases to pass up the spaces adjacent to the wall and thereby fail to enter the main central reaction zone. By experience it has been found that in a furnace of circular cross-section, the wall effect only becomes serious when the charcoal diameter is more than about 8–10% of the furnace diameter; the combustion rate tends to be greatest towards the centre.

As charcoal is very fragile it is rarely preserved intact on old smelting sites, making it difficult to determine the original preferred size. There are some exceptions: at the Roman mines at Rio Tinto, in southern Spain, for example, a stone-built charcoal store was excavated and the well-preserved lumps found were typically between 3 and 10 cm in maximum dimension (Craddock 1995: 192). The contemporary silver smelting furnaces were about 60 cm in diameter (Craddock *et al.* 1985b) which equates well with the 8–10% rule of the wall effect size-limiting factor. At the Bronze Age copper smelting sites at Timna (Israel), a charcoal dump was found in which the lump size was typically about 3.5 cm in diameter, adjacent to furnaces that were around 30 cm in diameter (Rothenberg 1990b: 85), again a good approximation to the rule. The copper smelting experiments carried out by the Early Mines Research Group (EMRG) (Timberlake 2005; Craddock *et al.* 2005), using a small, low-shaft furnace of about 35 cm diameter, found that the best results were obtained with charcoal that was approximately 3–5 cm in diameter. Thus the furnace size can perhaps give some indication of the optimum lump size for the charcoal.

No well-preserved charcoal stores were found on the excavations reported here, but some complete lumps of charcoal were occasionally found in association with a variety of installations. At the post-medieval copper smelting Site 15, at Dariba, the charcoal lumps were typically about 2–3 cm in diameter, and the fragments of the smelting furnaces at the nearby contemporary Sites 13 and 14 suggested an internal diameter of only about 20 or 25 cm. The charcoal associated with the medieval zinc smelter excavated at Site 30 at Zawar (see Chapter 6) was also between 2 and 3 cm in diameter, but although the furnaces were considerably larger than 20 or 30 cm across, they also contained the retorts, creating all sorts of local wall effects. In addition the charcoal had to fit into very confined spaces.

Another important function of the fuel is to support the ore charge which must be held as a solid in the main reaction zone just above the tuyere long enough to be reduced to metal. Clearly, there is an optimum size for this, dependent in part on the size of the ore charged to the furnace. For the small briquetted cakes recorded in the recent past, of the order of about 10 cm diameter or length and weighing no more than a few grams, the EMRG smelting experiments suggest that charcoal pieces of about 2–5 cm diameter are preferable (Timberlake 2005).

Charcoal production

There are several descriptions of simple charcoal production in South Asia. As in the rest of the world both pit and heap methods of charcoal burning were employed. In the following extracts, original text is in [ ] and added comments here in ( ).

Ball (1881: 382) recorded the method used by the Lohars, itinerant iron smelters of the Rampur region (now in the Sambalpur and Sundargarh Districts of Orissa, in central India), who remained in one location for a few months only until the timber suitable for charcoal had been exhausted:

> Although *sál* (*Shorea robusta*) is the wood most commonly used for making charcoal, the bijasal (*Dipterocarpus marsupium*) seemed to be preferred by some. Bamboo, although abundant, is never used. The wood is cut into logs, about 3½ feet long (110 cm), or rather more, and is burnt in holes dug into the ground, which are about 4 feet square (125 cm^2) and 18 inches deep (45 cm). Small branches are not used, only the very best of the timber.

This is not dissimilar to the pit method which the present team recorded in the vicinity of Kaya, near Zawar in 1990.

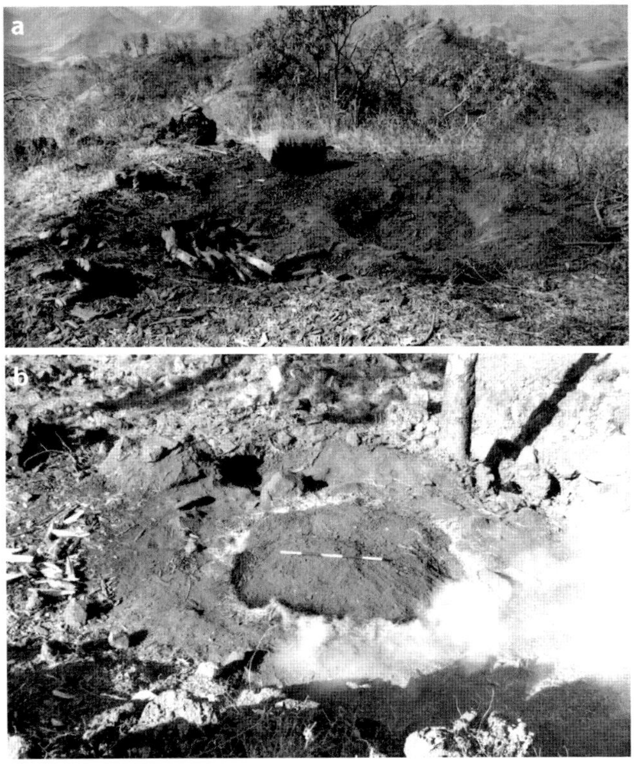

Plate 9.1 Present-day charcoal burning by the pit method in reserved forest at Kaya, near Zawar: (a) the pit is empty with some logs to the left; (b) the process is under way beneath a layer of ash.

Franklin (Dharampal 1971: 229), writing in the 1820s of charcoal production for iron smelting in central India, recorded the heap method:

> They are fully aware of the effects of certain kinds of wood on the quality of the iron and know from experience those which are best suited for their purpose; but as they cannot always obtain the trees they prefer – they use a mixture for their smelting process, excluding only those that are noxiously pernicious; but in their refineries they use exclusively *teak*, *mowa* or *bamboo*, to the last of which they give the preference; in their preparation of it, they are expert from habit and no men are able to make better charcoal; they usually allow a month for drying the wood after felling, and their method of piling it, for burning, is in conical heaps; the remainder of the process being exactly similar to the practice of Europe.

Oldham (1856: 6) recorded the production of charcoal by a rather unusual variant of the heap method used by the inhabitants of the Tributary Mehals, in what is now the Dhenkanal and Cuttack Districts of Orissa:

> Only logs of one size are employed, namely, about 12–18 inches (30–45 cm) in diameter. Trees of larger dimensions cannot be broken up by the tools the natives possess, and the trunks are left to lie in the forest. The smaller branches are perhaps used for firewood; they are never, as far as I am aware, used for making charcoal. The logs are cut into pieces about 5–6 feet (150–180 cm) long which are piled on each other in layers crosswise, and then the small stack of logs, without any other preparation, is set fire to. When well lighted, water is thrown on, and the crusts of charcoal formed are knocked off each log.

Oldham was surprised by the amount rejected and could not resist the usual European pejorative jibe against the local technology, this time to the effect that:

> They seem to exceed all others that I have heard of, in the ingenuity with which they seem to have adopted a process for obtaining the very smallest possible amount of useful fuel from the greatest amount of timber.

Dousing burning wood is a quick but wasteful method of making charcoal, but the careful selection of a reasonably uniform wood size and age is desirable in order to achieve uniform combustion. The methods of charcoal production used by the Kamar iron smelters at the village of Pentabahal in the very rural and undeveloped areas behind Cuttack in the Sambalpur District of rural Orissa were recorded in the 1961 census (Ahmed 1963: 18) as follows:

> 15. The Kamars are expert charcoal burners. They make it usually from *sal* (*Shorea robusta*) or *jaman* (*Syzygium cumini*) trees, which yield charcoal suitable for conditions of combustion and blast as obtaining inside the smelting furnace. They move about in the forest trying to locate dead and dried up trees which are approximately a foot and a half (45 cm) in circumference. Trees which are thinner, and twigs and small branches of trees are not used for production of charcoal, because they may either get completely burnt into ashes or get reduced into too small a size of charcoal-pieces which are uneconomical for use in the furnace or the forge. The dried up trees of suitable girth are cut into logs 4 to 5 feet (1.2 to 1.5 m) long. Half a dozen of such logs are piled up and a quantity of dried leaves or twigs are inserted in the openings to take fire from a straw torch which is carried from the village for setting fire to the pile of logs. It takes about 2½ to 3 hours for these logs to burn into charcoal. The well-burnt portion of the charcoal is collected into baskets, leaving the unburnt or half-burnt portion at the site as useless for their purpose. At times, this discarded material is used up in the next burning process but generally the inclination is to try fresh logs. After charcoal is collected in baskets, it is transported either by women in head-baskets or by men in *bhars* (two baskets slung on a pole and carried over the shoulder, also used to carry the ore) ... Two days a week, namely Wednesday and Thursday, are set apart for burning charcoal and transporting ore and charcoal to the site of the furnace.

It is inherently unlikely that dead timber was used as it does not make good charcoal: either the description was confused, or more likely, the Kamars were trying to give the impression to the government recording officials that they were not cutting down good, sound timber from the protected forests. The combustion time of only 2.5–3 hours seems very short. The considerable amount of waste should be borne in mind when making estimates of the timber requirements for charcoal production. In Singhbhum charcoal was produced for copper smelting in the recent past in small clay kilns approximately

1.5 m high and 2.5 m in diameter (Shrivastava 2006: 70 where they are erroneously described as smelting furnaces).

Modern charcoal burners favour young wood not more than 10 or 20 years old, and between 10 and 15 cm in diameter, usually from pollarded timber (Constantine 1975). Dead wood or old trunk wood is considered too dry, and tends to burn rather than pyrolyse. Many instances occur in the literature of preferred species of wood for particular operations or in metal production in particular areas, and the smelter's insistence on a specific charcoal can seem unwarranted or even capricious, especially as such a wide range of species has been 'preferred' throughout the world (see above and Elwin 1942: 121–3, for the preferences of the Agaria, the traditional itinerant iron smelters of India, and Mishra 2003: 112–13, for some of the species favoured by the present day tribal iron smelters of Chotanagpur). In fact the properties of charcoal do vary greatly from one species to another. For example, there are huge variations in density that directly affect the time taken to burn, which in turn affects the time of exposure of the ore to the hot reducing gases in the furnace. It would therefore seem logical that a smith would only use charcoal from the limited number of species with whose properties he was familiar in order to ensure that the all-important consistency of the overall furnace charge was maintained in the furnace operation.

Horne (1982) described the production of charcoal by the pit method in the desert regions of contemporary Iran. This seems to have been fairly inefficient with a weight conversion of only 14% wood to charcoal, although Rehder (2000: 56) suggests between 10 and 15% for antiquity generally, compared to between 20 and 30% for modern European charcoal-burning kilns. These figures must be considered when attempting to estimate the wood requirements for the large Aravalli mines.

Unregulated charcoal firing by the pit method was observed by the author during surveys in the reserved forests around Kaya to the north of Zawar in 1990 (Plate 9.1). The pits were about a metre in depth and 1.5 m in diameter, and the combustion was controlled by soil placed over the top. As the cutting was unofficial there was no systematic selection of timber; anything, including dead wood, was claimed to be used (although much of the wood we saw being prepared was green), however the preference was for small trunks of approximately 10 cm diameter. A firing lasted for about 24 hours with straw to aid the combustion of the green wood. A pit would be loaded with about a tonne of timber which produced between 60 and 70 kg of charcoal (that is a conversion rate of only about 6–7%). This is highly inefficient and may be due largely to the random selection of timber used – the dead wood probably just burned. Traditionally the hard timber, neem (*Azadirachta indica*) was preferred at Zawar in the recent past for general and domestic use, but *silari* or *siris* (*Albizia*) is often used now due to depletion of the better timber. The present-day process probably relates more to the practice of the local tribal peoples and the minor metallurgical operations still carried on at Dariba and at Zawar in the 19th century rather than the industrial production that must have gone on in the region in the more remote past.

The supply of charcoal and the impact on the landscape

The amount of wood needed to produce the charcoal required to smelt metal, its procurement and the effect of wood cutting for charcoal on the landscape in the vicinity of metal mines and production sites generally has been considered by several authors (Craddock 1995: 193–4; Sim and Ridge 2002: 35–42; Rehder 2000: ch. 14).

A regular supply of charcoal must always have been an important consideration for the major smelters, especially in the more arid areas around the northern Aravallis. Many 19th- and 20th-century descriptions of small-scale smelting operations, especially in desert areas, emphasise the problems of obtaining charcoal; in some instances the metal production was only sporadic because of supply difficulties. For example, the itinerant iron smelters had to move on to new locations once the timber suitable for charcoal in their current area had been exhausted (Ball 1881: 382). Elwin (1942: 236) recorded that the Agaria required about 30 acres of woodland per furnace. Such recent accounts have been used to suggest that this is likely to reflect the situation in the more remote past. In the case of the major established mines this seems inherently unlikely. The large stopes and galleries at Zawar and the great opencasts at Dariba and at Agucha would only have been viable if they were producing tons of ore per day on a continuous basis, and this required tons of charcoal to smelt, which in turn needed many more tons of suitable timber.

Realistically, charcoal can only have been supplied in quantity on a regular day-to-day basis from managed woodlands which are well recorded in the post-medieval period in Britain, for example (Fell 1908; Hammersley 1973), and in the United States in the 19th century (Temin 1964: 65).[6] Stands of coppiced timber from plantations providing large quantities of fairly uniform logs would have been vital for the regular bulk production of charcoal. Furthermore, charcoal made from coppiced timber has been recognised at some Roman iron smelting sites in Britain (Figuerial in Fulford and Allen 1992). Given the evidence of the organisation of ore production underground, the state operators of the major mines in India in antiquity almost certainly ensured the establishment of managed woodlands to guarantee their fuel supply.

Wagner (1985: 33–7) records that in mid-20th-century China, charcoal was produced in southern Henan by specialised groups of four–six persons who could achieve between 600 and 700 kg of charcoal every two days from managed woodland. He suggested that there may have been about 400 such groups each producing about 100 tonnes of charcoal per annum from plantations of a few hundred hectares each. Based on various estimates of charcoal production and consumption he estimated that approximately 180,000 tonnes were consumed per annum to produce about 14,000 tonnes of iron in the traditional iron smelting industry at Dabieshan in southern Henan in the first half of the 20th century, and that this could have been produced from about 500–700 km^2 of well-managed woodland.

Something of this controlled supply is implicit in the previously mentioned treatise on state administration, the *Arthaśāstra* (Chapter 8). The major mines were under central state control which included woodlands managed by a Director of Forest Produce (2.17.1; Kangle 1965–72, III: 129–31), responsible for its products including charcoal (2.17.16). The actual distribution of the charcoal would probably have been the responsibility of the *saṁnidhātr*, the Superintendent of the Magazine (2.15.1), who dealt with the collection and distribution of agricultural produce generally. Presumably there were local store houses serving each of the major mines as it specifically states that: 'He should cause charcoal and husks to be taken to metal workshops and for plastering walls' (2.15.60). The instruction to also deliver husks to the metalworkers is interesting – they were probably to be used as temper for the clay in the production of the refractories such as furnace linings, crucibles, etc.

The woodlands necessary to supply the Aravalli mines at the height of their production must have been extensive. Calculations of the fuel supply at contemporary mines in other parts of the world, such as the Roman mines at Rio Tinto in southern Spain, have shown that there was no need for the farthest plantation to have been more than 40 km distant, even with ample allowances for land set aside for food production, such as is attested in the organic tempers of the refractories at the mines excavated in this project (see Chapter 7). Large plantations can be envisaged at Agucha and Dariba amid the farmed estates, and at Zawar managed woodlands could have clothed the hills, away from the immediate vicinity of the smelter pollution. Therefore, together with the large amounts of arable land necessary to produce the food for the workforce, overall the mines could be seen as long-term establishments, developing the landscape rather than denuding it.

Hooker (1854 quoted in Percy 1864: 265) described the environment of the iron smelting communities in the Nonkreem Valley in the Khāsi Hills of Meghalaya in eastern India. According to Percy's thoughts and Hooker's description:

> One would naturally suppose that voracious iron-works, consuming as they do almost every living plant or tree in their vicinity, and changing a beautiful landscape into a hideous desert, would be regarded with peculiar abhorrence by an ardent and illustrious botanist like my friend Dr. Hooker. But, so far from this, the metallurgist will be gratified to learn that Dr. Hooker, with a catholicity of scientific feeling which is somewhat surprising, indited the following passage even in view of the gorgeous scenery of the Himalayas: 'few houses were visible, but the curling smoke from the valleys betrayed their hiding-places, whilst the tinkling sound of the hammers from the distant forges on all sides was singularly musical and pleasing.'

Perhaps Zawar might have looked a little like that from a distance in centuries past.

Air supply

Bellows[7]

The silver/lead and copper smelting operations of all periods excavated as part of this project had a forced draught supplied by bellows. The air would have been delivered to the interior of the furnace or hearth through the clay bellows pipes known as tuyeres to protect the bellows' nozzle from the intense heat. Incomplete examples of these survive in large numbers at all three sites, but no direct evidence was found for the bellows themselves so we have to speculate on the types likely to have been used at various times in the mines' histories.

There is no evidence for water power being harnessed to drive the bellows of furnaces in either Central or South Asia, unlike contemporary China and medieval Europe, a fact commented on by Drummond (1841). The bellows were all manually operated which was the main constraint on the furnace size. It becomes very difficult to supply sufficient air from manually operated bellows with the pressure necessary to penetrate to the middle of the reaction zone once the diameter of the furnace or hearth exceeds about 50 or 60 cm. It should be appreciated, therefore, that regardless of the scale of the mining operations and the size of the slag heaps, the actual individual smelting installations were very small and transient.

The main types of manually operated bellows used in the past have been studied by Weisgerber and Roden (1985, 1986) who suggested that bag bellows were likely to be the earliest, followed by pot or cylinder bellows, with concertina bellows introduced probably around the mid-first millennium BC in the Mediterranean world. The familiar hinged bellows were introduced only in the first millennium AD. Tylecote (1981), followed by Merkel (1990: 109–10) carried out experiments on the efficiency of the various bellows systems. A pair of bag bellows with flap valves operating together delivered about 150 litres of air per minute. A pair of foot-operated pot bellows produced about 200 litres per minute, although Merkel believed that theoretically 360 litres per minute should have been possible. Hand-operated hinged concertina bellows delivered between 400 and 500 litres per minute.

The usual form of bellows used in South and Central Asian smelting processes in the 18th, 19th and 20th centuries were quite simple bag bellows, although sometimes of prodigious size, and cylinder or pot bellows using ceramic or wooden pots. Percy (1864: 255–6) gave a detailed description of the construction and mode of operation of the simple bag bellows used by the iron smelters in South Asia, based on the reports of Howard Blackwell, an experienced manager of ironworks in England, who conducted a survey of traditional iron smelting for the Indian government in the mid-19th century:

> The blast is produced by bellows very unlike those of Europe. The bellows in common use consists of the skin of a kid or goat, taken off the animal by opening the hinder part only; the holes corresponding to the legs are sewn up; at the neck-end is inserted a nozzle of bamboo, and the tail-end is cut transversely, so as to form, when the edges are brought together, a long, straight slit for the admission of air. For a considerable length, but not to the whole extent, of each of the edges or lips composing this slit, a piece of split bamboo is firmly tied on the exterior. By this means the slit may be readily closed or opened, and made to act the part of a valve. The skin should be rendered very supple by rubbing it with oil or buttermilk. Each furnace is provided with a least two such bellows. One man is required to work them, and he sits cross-legged on the ground between them, moving them alternatively in order to keep up a continuous and regular blast. A leathern thong passes from one of the lips of the valve round his hand, and to fill the bellows he drops the lower lip and raises the upper one, when air enters, swelling out the skin into the shape of a conical bag. He then quickly catches up the lower lip, closes the valve, and bears with his weight on the inflated skin, forcing out the contained air through the nozzle into the furnace. Large bellows of similar construction are made of buffalo-hide, with the exception that the skin is sewn up along the line

> of the belly, and the bamboo strips are fixed together at one end of the valve, and can be separated only at the other, where they are prolonged beyond the skin, so as to form convenient handles. Such a bellows requires one man to work it.

As noted the bellows were used in pairs with the air entering the furnace either through a common tuyere or paired tuyeres lying side by side. The bellows were worked alternately to provide the continuous blast necessary for iron smelting.

Campbell (1843) provided a slightly different description of the bag bellows based on his detailed observations of iron smelting in the south of India where the arrangements of bellows and tuyeres were somewhat different to those described above. The air was introduced through a single cylindrical tuyere 14 inches (35 cm) long, four inches (10 cm) thick and with an internal diameter of one inch (3 cm). This was placed in the blocked front opening of the furnace. His account of the bellows and smelting is reproduced in Percy (1864: 256) and Dharampal (1971: 256–8):

> The bellows are two goat skins taken off the animal, by opening the hinder part only. The orifices at the legs are sewn up, and a piece of bamboo is inserted, and tied tightly into the neck of each skin, and these bamboos being inserted into the outer part of the blast pipe, which is made conical for that purpose, the vacant openings are then stopped with wet clay. The open end of the skin is finished by folding the edge of one side, as a flap, about 4 inches (10 cm) over the other edge, and sewing up the upper and lower corners, so as to leave a part of both flaps open for about 9 inches (20 cm). When the skin is filled with wind, and pressed, the inner flap closes therefore against the outer, and stops the passage. Each skin is managed by one man, who places it in his lap, and squeezes it down with the elbow and lower part of the right arm, grasping at the same time a projecting sort of handle of leather, formed at about the part where the tail of the animal might have been. To enable the blower to fill the skin again with wind, a piece of string is attached to the lower corner of the posterior part, which is tied to a peg driven into the ground about a foot behind the man's elbow, and keeps the skin distended to its full extent as it lies in his lap: and a loop of leather is also fastened to the outer flap, through which the arm is passed, by which the opening into the skin is distended upon raising the elbow; and the skin being pulled out horizontally by the neck in one direction, and the string and the peg in the other, upon pulling up the middle part vertically by the leather grasped in the hand, the skin is opened out into a triangular shape, and fills with wind through the open flap. While squeezing the skin in blowing also, by pressing the hand forward so as to pull against the string attached to the posterior part, the valve-flaps are made to close fairly and properly, so as to allow hardly any wind to escape. The left hand is employed in assisting the right, or in squeezing the distended parts of the skin upon the side. It will be observed that, as both the necks of the skins open into the blast-pipe, a portion of the wind expelled from one skin passes back again into the other, as they are worked alternately; a defect which might have been remedied in a most simple manner, by attaching little hanging door-valves to the ends of the pipes.

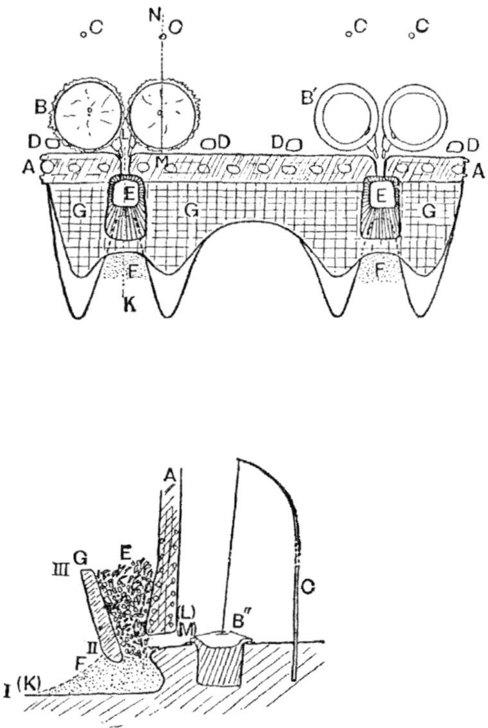

Fig. 9.1 Plan (upper) and section (lower) of 19th-century clay iron smelting furnace (G) at Hatarabage, Balangoda in Sri Lanka. The plan is at tuyere level depicting the interior (E) as being rectangular with rounded corners at that point, with the slag opening closed with sand (F). Note the interior as shown here is nowhere near as complex as that recorded recently in a contemporary surviving furnace (the plan of the furnace block is approximately 2.75 m along). The section through the furnace at K, L, M, N shows the arrangement of the pot bellows (B and B') with the sticks to act as springs for the bellows (C), and support sticks (D). Note also the wall (A), described as being of wattle and daub, protecting the bellows operator from the heat and fumes of the furnace (from Coomaraswamy 1908: 191).

In the Balangoda District of Sri Lanka, Coomaraswamy (1908: 191–2) described the pot bellows and their operation as follows (Fig. 9.1):

> The bellows, from which a continuous blast is kept for three hours or more, are worked by two other men (i.e. other than the third man who operated the furnace), who share the work by turns: they are of a primitive but ingenious type. Behind the furnace are embedded vertically two logs of wood, about 17 inches (43 cm) in diameter, and hollowed out above to a depth of three or four inches (8 to 10 cm); over each of these a piece of deer skin is tied, having a small hole near its centre, and at the centre a piece of cord attached, which latter is fastened six or seven feet (1.5 to 2m) above to a springy stick, the lower end of which is embedded in the ground. From each of these two sections of the bellows a small pipe conducts the blast into the hole leading into the well of the furnace. The bellows blower stands with his feet, one each on each of the deer skins, facing the wall and lifting and putting down his feet in turn, drives a

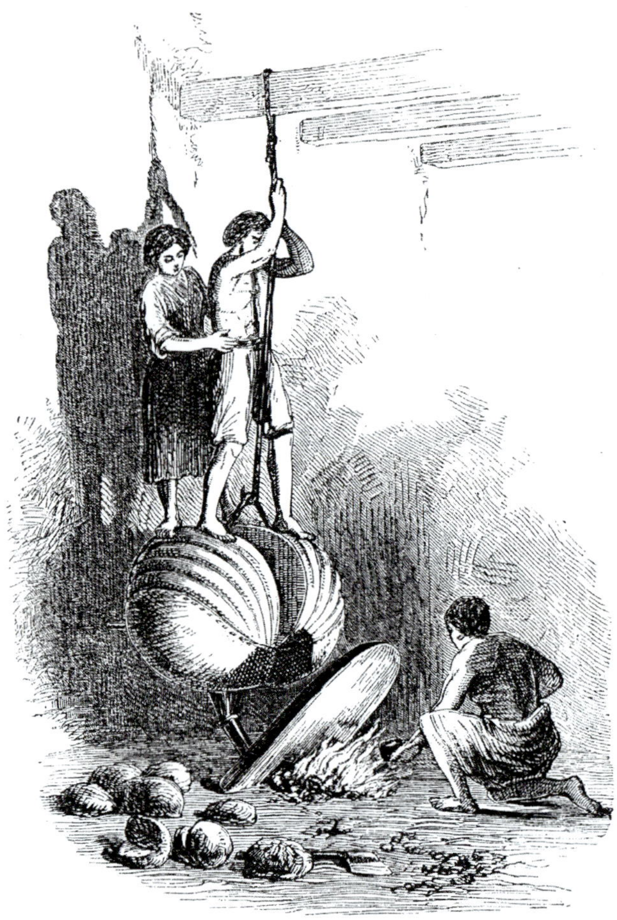

Fig. 9.2 An almost unique example of the use of hinged bellows in 19th-century India at the iron smelting furnaces in the Nonkreem Valley in the Khāsi Hills. The unusual mode of operating suggests they were an indigenous rather than a European introduction (from Hooker 1832, reproduced in Percy 1864: 264).

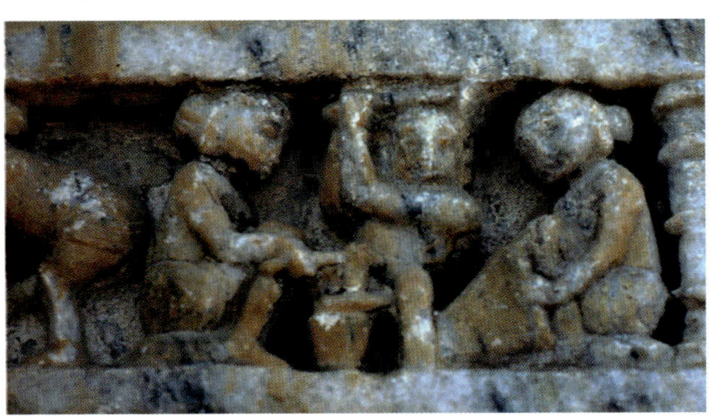

Plate 9.2 Stone carving on a 10th-century AD shrine at Ahar, Udaipur, showing a lively scene at a forge. The blacksmiths are apparently cutting a flat piece of metal lying on the anvil by hammering a wedge held in tongs. To the right another smith is operating a bellows. Note the assistant's right hand seems to be holding a handle or lever protruding to the right of the bag as would be expected if this was a hinged concertina bellows but not if it was a bag bellows, which would require both hands to operate.

> continuous blast of air into the furnace; for as he puts down his foot to press out the air from one skin, he covers the hole in it with the sole of his foot; meantime the other skin is filled with air through the hole, which is uncovered when he lifts his foot, the skin being pulled up by the tension of the stick upon the string. On either side of the blower is a stake which he grips hold of, and to which a strap is fastened that passes round behind him and supports his back.

The arrangements for supplying air to the smelting furnaces of the Kamars at the time of the 1960 Census in Orissa (Ahmed 1963: 19–21) were as follows:

> 18. The bellows used in the smelting shop are distinctive and very much different from the hand (bag)-bellows of country smiths or rotary blowers of modernized shops. The bellows are made of a section of trunk of *gambhari* tree (*Gmelina arborea*) shaped so as to be 11 inches (28 cm) in diameter on the upper side and 10 inches (25 cm) on the bottom side, with a height of 5 inches (13 cm) in all. The inside of this piece of wood is dug out so as to make a cavity 7½ inches (20 cm) in diameter on the upper side, leaving the bottom intact and a hole is made on one side to take the bamboo pipe which is meant to carry the blast to the furnace. A piece of hide of *sambar* (*Cervus unicolor*), deer or goat is stretched across the top in the fashion of a drum and tied down strongly with seven rounds of *siali* rope (a leguminous liana, *Bauhinia vahilii*) so as to make it air-tight. In the centre of this cover a hole is punctured and a rope is passed through and held in place by a twig or a nail. The other end of the rope is tied to a springing bamboo called *dang* (?), one end of which is buried in the ground.
>
> 19. In this manner a pair of bellows lie side by side in front of the furnace, fitted air-tight with bamboo pipes 2½ feet (75 cm) long leading from the drum to the nozzle of the furnace (the tuyere). Before the bellows are worked, the hide-cover is soaked in water to make it sufficiently pliable. When the pair of bellows are ready for operation; a Kamar woman, generally the wife or the daughter-in-law, rides over them with one foot on each of the bellows and starts depressing and releasing them alternately. The holes in the centre of the covering hide act as valves. As the foot of the bellows-worker depresses, her heel serves as a stopper to the valve and forces the air through the bamboo pipe into the nozzle of the furnace. When the foot is taken off, the bellows get released and the bamboo stick springs lifting the hide-cover. At this stage the hole in the cover is open and the bellows are filled with air by such action. She holds the springing bamboo for support but occasionally holds a stick in her hand. The two bamboo pipes are kept under a piece of heavy stone to remain stable during the bellowing operation. Otherwise the pipes are liable to be thrown out with the pressure of air coming from the drum.

A pair of foot bellows was estimated to have a life of about a month (Elwin 1942: 216). It is significant that the recorder stresses that the bellows used by the Karmars for their smelting operations were different from the simpler bag bellows still used by the country blacksmiths for their forges.

The only record of hinged concertina bellows were those made by Cracroft (1832) and Hooker (1854: 310 reproduced in Percy 1864: 262–5) of the iron smelting operations carried out in the Nonkreem Valley in the Khāsi Hills of Meghalaya State in north eastern India in the first half of the 19th century. Hooker says little of the construction of the bellows but does provide a rather remarkable woodcut of them in operation (Fig. 9.2).

Percy (1864: 264) attempted to interpret the bellows operation from the illustration and summarised Cracroft's description of their operation as follows:

> As shown in the engraving, they seem capable of being worked by the feet alone; for, by means of the attachment between them, as the air would be expelled by pressing obliquely with the foot on one side, the other by the same action become inflated. The two sides may thus be alternately filled and emptied, so as to produce a continuous blast. The attractive damsel figured by Dr. Hooker will naturally excite the sympathy of the reader.

Percy (1864: 265) then quotes the description of a similar iron smelting operation in the same hills in an illustrated article by Cracroft (1832):

> Two large double bellows, with the nozzles pointed downwards, are set up on one side of the apartment, on the upper part of which a man stands with one foot on each, his back supported by two planks. He holds a stick in his left hand, which is suspended from the roof, and has two strings attached to it below, connected with the two bellows: these are worked quickly by a wriggling motion of the loins, and the strength of the leg. The nozzles of the bellows unite in a tube which leads underground, from a sort of wind-chest, to the hearth about 4 ft. in front of them.

Percy commented that 'The bellows represented in the illustrative etching seem to be composed of wood and skin or leather and closely resemble ordinary European bellows'. The bellows as illustrated by both Cracroft and Hooker do indeed appear to be hinged concertina bellows, but their arrangement and operation are very different from any known European practice, suggesting that they were likely to have been indigenous. A carving on a shrine at Ahar, now in the outskirts of Udaipur, dated to the 11th century AD, seems to show a hinged concertina bellows being operated to blow a blacksmith's hearth (Plate 9.2).

Thus in South Asia over the past two centuries bag and pot bellows predominate, yet the more advanced hinged concertina bellows are apparently depicted much earlier. Is this yet another indication of technical regression by indigenous industries in recent centuries?

Tuyeres[8]

The air from the bellows usually entered the furnace in clay pipes known in English by the French term, *tuyeres* (often referred to as *twyers* in Percy and other 19th-century English publications). There were two arrangements for supplying the air from the bellows to the tuyere, depending to some degree on the type of bellows used. The end of the bellows could be tied directly to the tuyere thus at the end of each stroke pushing the air into the tuyere, the bellow's valve would have to be opened to prevent gases being sucked back from the furnace into the bellows. The alternative was to sit the end(s) of the bellows in the rear of the tuyere protruding from the furnace. The ends were usually expanded to accommodate them, but were otherwise open. The air movement in this arrangement was a little more complicated (Fig. 9.3). On the blowing stroke the air entering the tuyere would actually drag in more air from the surroundings and on the reverse stroke if the bellows did not have a valve then air would be sucked into the bellows. From the limited descriptions and illustrations of 19th-century practice in India it seems that the nozzle from each bag bellows in the multi-tuyere arrangements, such as are found at the copper smelting furnaces at Singhana, was described as being sealed to the rear of the tuyere with clay.[9] Sometimes the nozzles from the pairs of pot bellows used most often for iron smelting were just lodged in the rear of the tuyere. Unfortunately, no complete tuyeres or recognisable rear portions of tuyeres were found at any of the sites excavated because they were not fired before use. However it is likely that the early thin tuyere tubes were pre-fired and as none of the fragments show any signs of flaring, this suggests that they were sealed to the bellows tube.

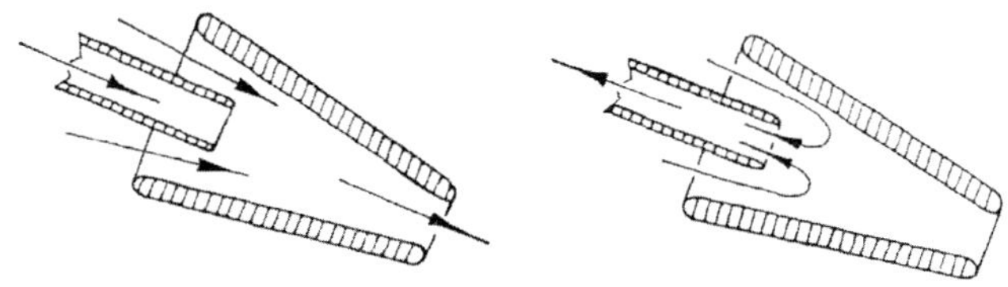

Fig. 9.3 Flow of air at the junction of a bellows end just lodged in the rear of the tuyere (adapted from Hodges 1964: 67).

The tuyeres used at the three Aravalli mines and their changes through the ages seem to be very similar. In the early periods up to the 7th century AD the tuyeres are thin-walled (typically about 0.5 cm) with internal diameters of between 1.5 and 3.5 cm, as evidenced by the well sites at Agucha (see Chapter 6, Fig. 6.74), Sites 2, 14 and 7 at Zawar (Fig. 6.36), and Dariba, Site 10 (Fig. 6.57). The post-medieval tuyeres from Zawar and Dariba have much thicker walls, typically about 2 cm but with smaller internal diameters of about 1.5–2 cm, than previously (Zawar, Site 5, Fig. 6.41; Dariba, Site 14, Fig. 6.65, for example). The tuyeres could either pass directly through the furnace wall or be set in a clay collar of greater diameter that was itself inserted into the wall. The later tuyeres all seem to have been set in collars, but there is no evidence for this on any of the earlier fragments. The earlier tuyeres seem just to have penetrated through the furnace wall as exemplified by the furnace fragment at Zawar, Site 2 (Fig. 6.37). The general early absence of collars at Dariba and at Agucha may be further evidence that the ores were smelted in hearths there rather than furnaces.

The tuyeres are likely to have been made by wrapping clay[10] around a rod of the appropriate thickness and then withdrawing it, creating a tube. Alternatively the stick could have been inserted into a ball of clay and then rolled out as recorded by Elwin (1942: 186, pl. 24):

> On a reversed bellows, the wooden bottom of which makes a smooth and convenient platform, a woman works up the clay (*nari-māti*). She beats it with a special rounded stick called the *thondi-beth* and then makes it into a ball. In the middle of this she shapes a hole, inserts the stick, rolls out the clay along it with her hands, presses it out and smoothes it by rolling it on the platform, gradually splays out one end, removes the stick,

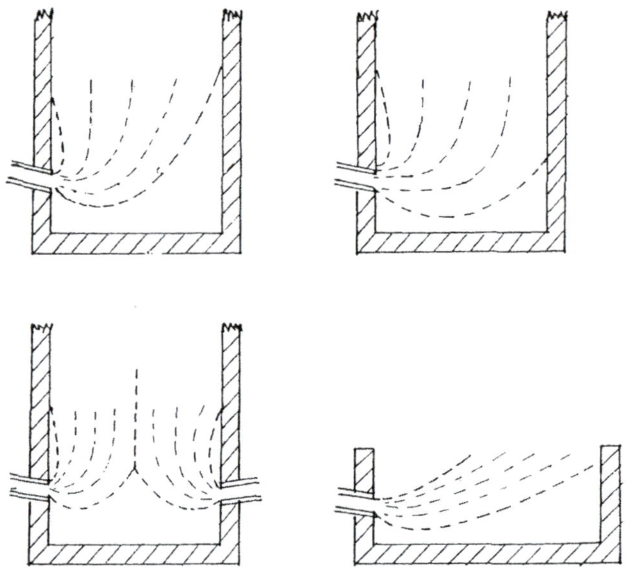

Fig. 9.4 The spread of the blast through a deep charcoal bed in a small shaft furnace with one or two tuyeres and in a hearth. Top left: lower air velocity; upper right: higher air velocity; lower left: opposed tuyeres; lower right: flow in hearth (from Rehder 2000: 68, fig. 7).

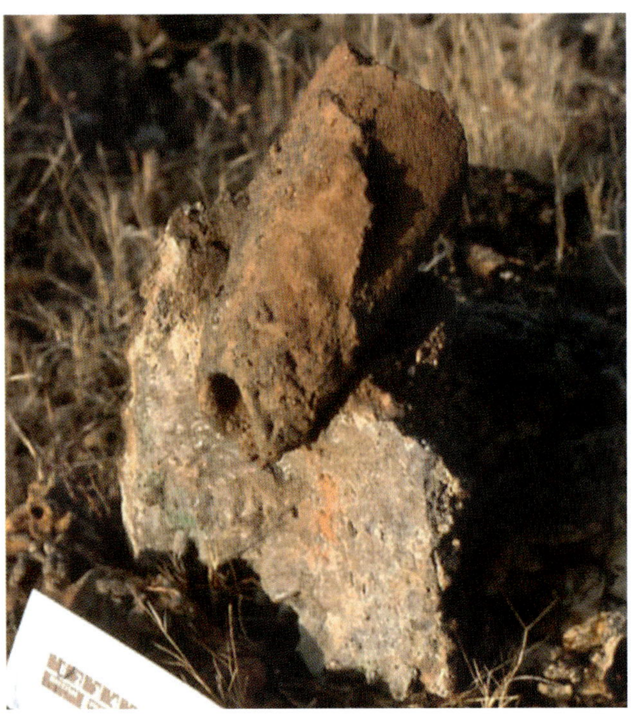

Plate 9.3 Tuyere set at a steep angle in a clay surround, collected by Hegde and Ericson from a copper smelting site in the Aravallis.

> and carefully shapes and finishes the cylinder with her hands, finally placing it to dry standing on its broad end in the sun. Half-a-dozen are made at a time.

The evidence of how the tuyeres were made in the Aravalli mines is clearest on the thin-walled early tuyeres. The sheets of clay were wrapped around a rod. Their thickness is quite regular except for a slight but real thickening which runs along the length (Zawar, Site 2, Fig. 6.36), presumably the seam where the edges of the clay overlapped and were joined around the rod. They have parallel striations on the inside probably where the rod was pulled out. On the outside of the tube there are often quite pronounced and regular parallel ridges, suggesting that the outer wall of the tuyere was contained or at least shaped in a mould or former to make the tuyere more regular in section and thickness (Fig. 6.58a).

The later, much thicker and coarser tuyeres are also likely to have been made by rolling clay around a rod, but there is no evidence for moulding the exterior of the tube. Only the portions exposed to the heat of the furnaces survive – the rear portions are never found, including those on sites such as Site 14 at Dariba which is of no great age and where the earthenware local pottery survives perfectly. This suggests that these tuyeres were just sun-baked rather than properly fired before use, similar to those of the Agaria (see above). This, together with the open structure of the ceramic, would enable them to accommodate the large thermal gradients while in use without breaking, but they would disintegrate rapidly when exposed to weathering on the spoil tips.

The larger internal diameter of the early tuyeres would allow a greater volume of air to enter the furnace, but required a higher flow rate to maintain the pressure necessary for the air to penetrate through the furnace charge to the centre of the reaction zone. For the early furnaces and hearths this would be about 15–20 cm from the tuyere end (assuming that the tuyere end was flush with the furnace wall, which it certainly would be by the time the smelting operation was well under way through erosion).

The angle at which the tuyere enters the furnace largely determines the flow of the blast through the furnace as shown in Fig. 9.4. Shires (1960, reported in Rehder 2000: 67) showed that the inclination of the tuyere by about 15° to the horizontal gave the most effective penetration of the blast into the fuel bed. It has proved possible to estimate the angle at which the tuyeres entered the furnaces or hearths in a number of instances at Zawar and Dariba, either where the hole is preserved in a fragment of wall (Zawar, Site 2, Fig. 6.37) or where the slag-eroded ends of the tuyeres survive. Assuming that the furnace wall was near vertical the angle between the face and air passage can be measured. In the early furnaces, Zawar, Site 2 (Fig. 6.37) and Dariba, Site 10 (Fig. 6.57) for example, the angles are quite shallow, of the order of 15°. In the later furnaces, Zawar, Site 5 (Fig. 6.41) and Dariba, Site 14 (Fig. 6.65), the angle increases to about 45°. The steep angle is most clearly seen on a tuyere still set in a square jacket collected from the slag heaps by Hegde and Ericsson (1985), apparently from Ambaji at the southern end of the Aravallis (Plate 9.3). It is probably similar to the 'nozzles connected with clay to form a little circular wall' in the manner described by Boileau (see Chapter 2, Figs 2.3 and 2.4), upon which the horizontal furnace rings sat. The tuyere pipe shown on the section of the Singhana furnace illustrated in Subbarayappa (1971) was apparently almost horizontal as it entered the furnace, but the separate passages in the tuyeres, made in order to admit a rigid poker to be pushed into the furnace itself to keep the air passage clear of slag, were at an angle of about 45°, suggesting that the rest of the passage must also have been at that angle.

Another important process parameter is the number and arrangement of the tuyeres in the furnace, together with their mode of operation. Two or more tuyeres will give a much more powerful and uniform flow of gases through the furnace as shown in Fig. 9.4, provided that they are operated simultaneously. Determining the number and arrangements of tuyeres is a perennial problem with early furnaces where little is preserved *in situ* above the base and the tuyeres would typically have entered the furnace a few centimetres above.

There are two reliable and detailed descriptions of the operation of small shaft furnaces in the Aravallis for non-ferrous smelting: Dixon for lead smelting at Ajmer (Percy 1870: 94–6) and Boileau for copper smelting at Singhana (Percy 1861: 394–6) (see Chapter 2, Figs 2.3–2.5). Both describe furnaces served by three bellows each with a tuyere set into the furnace. Percy assumes that the tuyeres would have been set regularly at approximately 120° to each other at Ajmer as Dixon does not specify. However, at Singhana, Boileau described the 'nozzles' or tuyeres as being set in moist clay at 90° to each other such that they could form one half of a ring, the other half being left free for the slag to be tapped.

The furnaces at Zawar and Dariba were of different construction from the contemporary furnaces in the accounts listed above, but it would seem more likely that they were also served by three bellows each with its own tuyere. Another important factor is whether there was a constant or intermittent blast, that is, were the bellows operated simultaneously or alternately? It appears from the early records (see Chapter 2) that many of the single tuyere systems, supplied by two bellows, were operated alternately to give a reasonably constant blast. Indeed, songs were sometimes sung or drumbeats made to establish a rhythm in order to keep the bellows operatives in synchrony. This would allow something approaching equilibrium conditions to be established in the reaction zone in the furnace. It is perhaps significant that these systems, supplying a constant blast, were intended for the solid state smelting of iron which remains immobile in the reaction zone after reduction, and where it was important to maintain the reducing conditions (see below).

For the smelting of the non-ferrous metals where the product is a liquid which coalesces and falls out of the reaction zone almost immediately on reduction this is not necessary. Dixon (1831) (see Chapter 2) recorded that the Ajmer lead smelters used a pulsed blast. It should be noted that because of the intermittent nature of the blast, the smelting operation was always in a dynamic state, never achieving equilibrium. Air delivered as an intermittent, pulsed blast when the bellows are operated simultaneously can be much more effective than a constant supply, as copper smelting replication experiments have demonstrated (Merkel 1990: 102–3).

Jaikishan (2009: 105) recorded an old iron smelter from the Telanga District of Andhra Pradesh and stated that he preferred the old-fashioned bellows to the modern blowers quite specifically because they did produce a pulsed blast:

> In this modern age, we are using only metal blowers which are comparatively inferior to the buffalo skin blowers. The old buffalo skin blowers are preferred to modern metal blowers because the old buffalo skin blowers could push larger quantity of win (wind?) into the furnace at once, which helped in heating the material soon. The repeated blowing with short intervals was of grate (*sic*) use for sooner heating of the metal. The modern blowers which blew wind continuously give some cooling effect at the Tuyeres inside so the required heat may not be provided by the wind. Only the blowing intervals are effective, I think so.

Different arrangements were necessary where the tuyere supplied a hearth or some other open installation without a wall to hold and direct it, as exemplified by the Agucha remains. The majority of the tuyere fragments are of similar construction and diameter to those from the early sites at Zawar found with shaft furnaces, but at Agucha they were associated with the small cupels and clay supports which presumably held them at the correct angle. The Dariba, Site 9 tuyere stand (Fig. 6.53) may be from a similar operation. One tuyere fragment from Agucha (Fig. 6.74) was found set in clay which had a flat side. Tylecote (1981) suggested that such tuyeres with square external sections were made to sit on the ground. The Agucha tuyere is probably to be interpreted as a rather long tuyere support still containing a fragment of a thin round tuyere, rather than a square-sectioned tuyere. The tuyere was almost horizontal but may have had an angled tip such as were found on some of the square-sectioned tuyeres from the Lydian gold refinery at Sardis (Ramage and Craddock 2000: 92–4, 223–5). It is suggested that the tuyere from Agucha may have sat on the clay support, supplying air to one of the cupellation hearths also set in a hollow in the ground and surmounted by the distinctive heavily lead-glazed ceramic hoods (see Chapter 6, Fig. 6.70 and Chapter 10).

The smelting units, hearths and furnaces

No remains of the installations for the smelting of copper or silver/lead were encountered *in situ* on any of the sites investigated in this project. However, many fragments of refractories of various kinds were found in the slag heaps. They are of clay often tempered with crushed stone, animal dung, straw and other chopped plant remains (Chapter 7). It is common experience that all South Asian metallurgical refractory ceramics, both past and present, contain large amounts of plant remains. Dung is also mentioned as a filler in many accounts of simple metallurgy in the 18th–20th centuries, and it was cow rather than horse dung that was usually specified. Donkey dung was used as a temper by Rajasthani potters in the recent past (Kramer 1997: 56). The other common temper was silica (SiO_2) as quartz, usually in the form of sand, although crushed gangue material could be used. Research has shown that both plant remains and quartz improve the resistance to thermal shock although they reduce the overall strength (Tite *et al.* 2001) (Chapter 7).

From the excavated fragments it has proved possible to deduce something of the size and shape of the various installations and their ancillary fittings, and rather more about their construction and function. The most common finds were curved clay fragments that are slagged and vitrified on their concave faces. From these it is reasonably certain that the majority come from smelting units that were of circular, or possibly oval, plan. What is less certain is the form of these units. On the early Dariba Sites 7 and 9 the only refractory remains are small fragments of ceramic lining where the curves seem to be in all directions. Without additional evidence these could equally well have come from the lining of a hemispherical smelting hearth or sides of a bowl furnace. They are associated

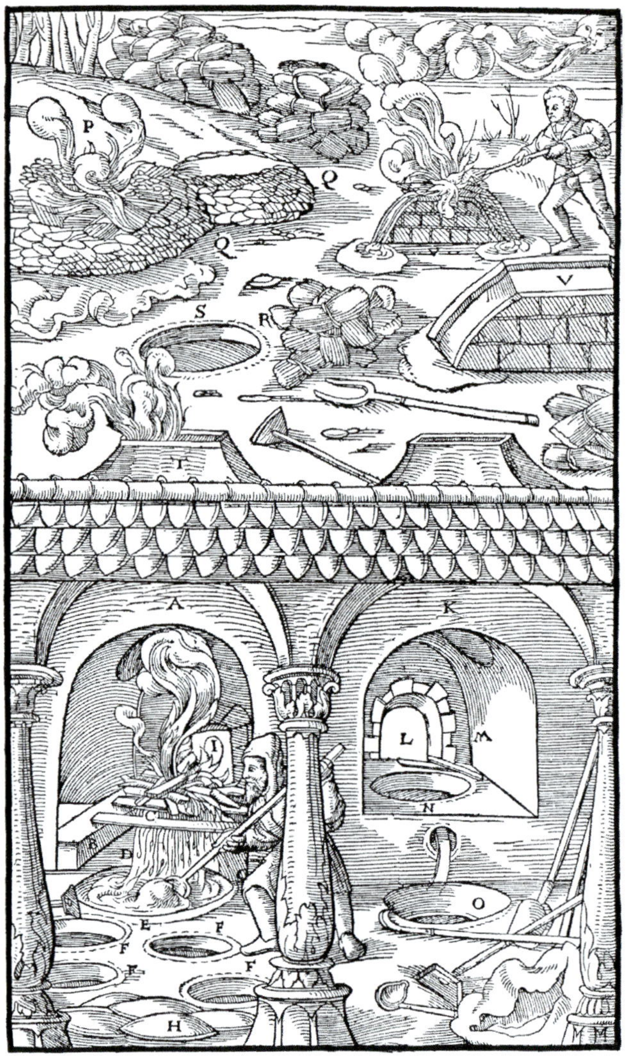

Fig. 9.5 Mid-16th-century European hearths for smelting lead. The simple raised hearth marked V (shown in operation above) was known as a Polish hearth. The lower two are simple semicircular raised hearths set in furnace chambers, blown from behind (A was known as a *Carni* (Slovenian) hearth, K was known as a Saxon hearth). The *Carni* hearth is shown in operation with the smelted lead running into the crucible below (from *De re metallica*, Hoover and Hoover 1912: 393).

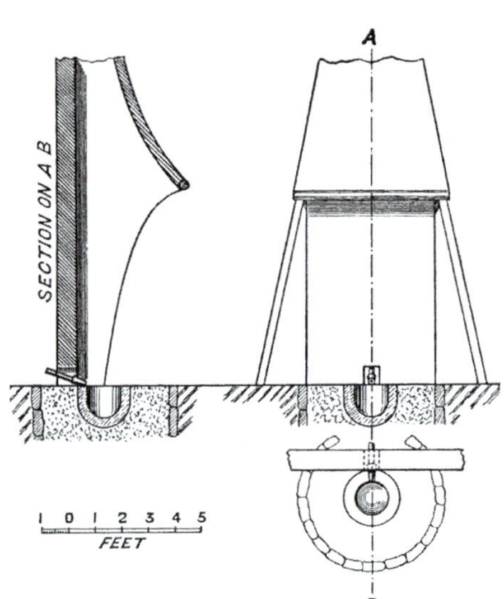

Fig. 9.6 Elements of a late 19th-century Japanese hearth for smelting copper (from Gowland 1899, fig. 5).

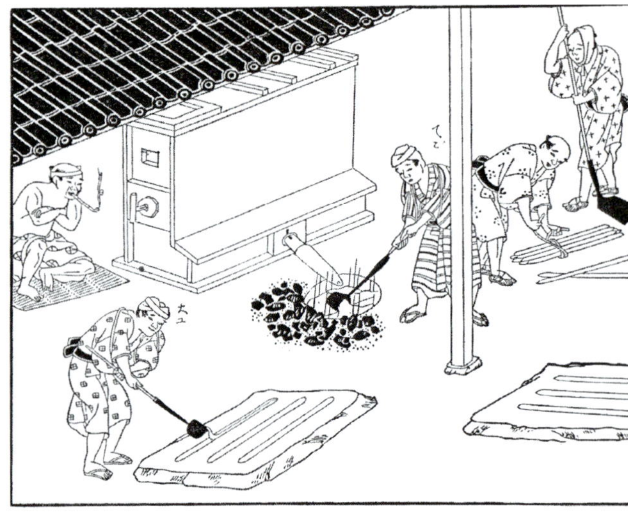

Fig. 9.7 Late 19th-century Japanese hearth for smelting tin (from Gowland 1899, fig. 15).

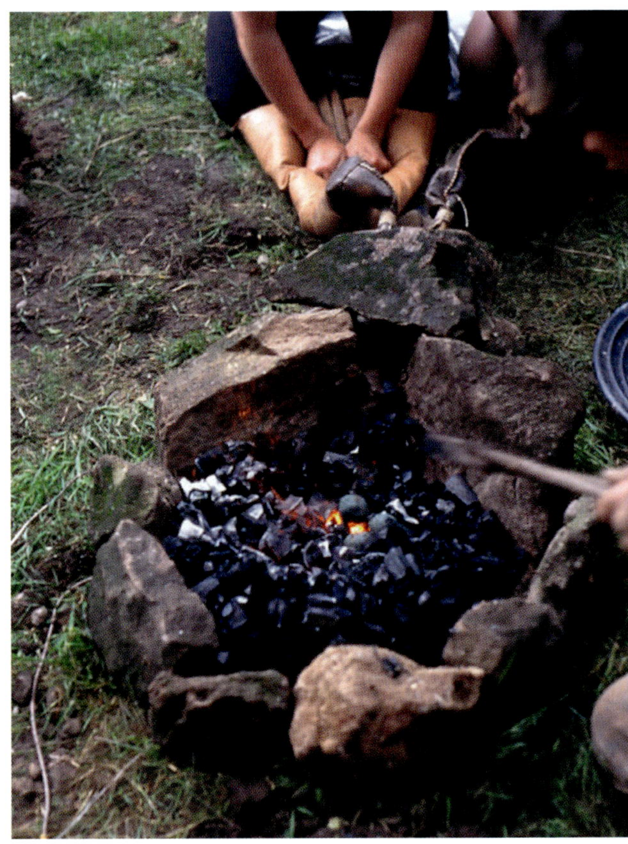

Plate 9.4 EMRG hearth smelting with partially smelted dung balls on the charcoal.

with the silver/lead smelting and are discussed further in Chapter 10.

In contrast the furnace fragments from the Zawar sites of all periods and the later Dariba, Sites 13 and 14, have either the imprints of stone walls or evidence of external plastering and were clearly freestanding. These freestanding fragments seem to curve in one direction only, suggesting that they were from cylindrical structures, the classic shaft furnace. However, the true internal profile of these furnaces is less clear: were they simple cylinders or were they more complex, such as the so called 'village' iron smelting furnaces of the 19th and early 20th century, which still stand intact in parts of Sri Lanka (Craddock 1995: figs 7.1–7.4)? It could be said that the 19th-and 20th-century descriptions are all of quite simple shapes, normally cylindrical. However, in the case of the Sri Lankan furnaces mentioned above, where the contemporary description can be checked against reality, the furnace was found to be much more complex than previously described or illustrated. In this context the comments of Major James Franklin on the very exact interior proportions of the apparently crude iron smelting furnaces he observed in central India in the 1820s are suggestive (Dharampal 1971: 229–30).[11]

Smelting hearths

Smelting in hearths was widely practised in the past worldwide. Good descriptions of traditional European post-medieval practice are given in Agricola for the 16th century (Fig. 9.5) and in Percy (1870: 279–93) for the 19th century. Hearth smelting is also described for 19th-century Japan before the Meiji reforms by Gowland (1899, 1901, 1915) and Masuda Tsuna (1983) (Figs 9.6 and 9.7). Hearth smelting of lead ores primarily to recover the silver are also well attested, for example in medieval England (Blanchard 1992; Claughton 1992), in post-medieval central Europe (Hoover and Hoover 1912: 390) and in Japan (Gowland 1901). Hearth smelting was usually practised where the smelting conditions were not required to be highly reducing, such as for the smelting of lead and for the matte smelting of copper or oxidising processes such as cupellation. Some hearth arrangements were rather more elaborate, being set on a step as exemplified by the Sikkim and Nepalese hearths (Fig. 9.9 and Plate 9.7), and others were set against a wall that could contain a chimney or hood to remove the majority of the fumes. This was done partly to protect the workers from the harmful fumes but also to catch the volatile metal oxides, notably the lead and zinc oxides that could be periodically recovered.[12] The air was invariably supplied by bellows and the fuel could be carefully arranged over the ore to create the desired temperature and oxidising/reducing conditions. A low wall may have been built to retain the fuel (Plate 9.4). As the hearth was open the ore and slag could be manipulated with iron rods during the smelting. Further control could be achieved by placing turves over the furnace charge, effectively turning the hearth into a bowl furnace, but this practice would leave no archaeological evidence.

Furnaces

Verrier Elwin devoted a whole chapter of his book on the Agaria to the technology of the iron smelting process, including a description of the construction of a new shaft furnace (1942: 191):

> The mud cone is first prepared. A red clay, called in Bilaspur (now part of Madhya Pradesh) *lāl-pila* earth, is mixed with water and the chaff of *kuāri kodon*, or *kodon* (*Paspalum scrobiculatum*, commonly known as ditch millet) ground in a new mill. This is plastered by the women on a grass frame which is afterwards burnt. The Chokh of Bilaspur build up the cone without a frame, doing about a foot (30 cm) every second day: when the first layer is dry and hardened they add another story (*sic*). When it is ready, the cone is left to dry and often remains for months before being used.

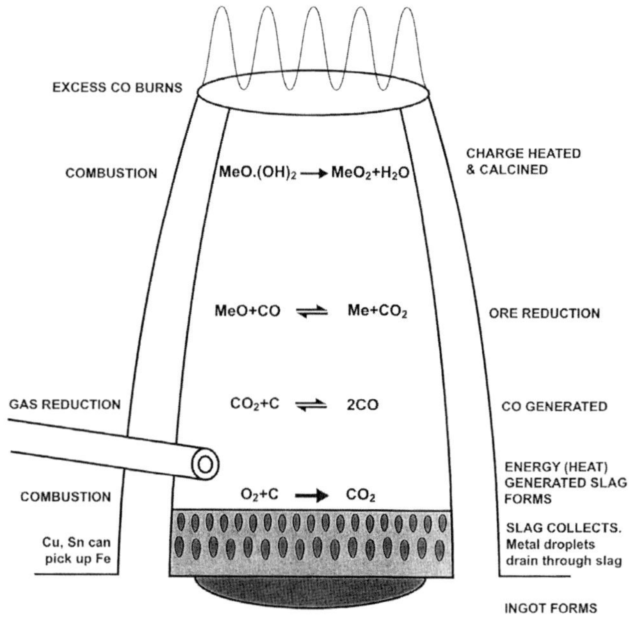

Fig. 9.8 Schematic section through a shaft furnace showing the principal reactions taking place in the reduction of the ore of a metal (Me) that was molten on smelting (drawing: A. Simpson).

The description goes on to make it clear that the furnace cone was not made *in situ* but was rolled into position when required.

The pieces from freestanding furnaces recovered from Zawar and Dariba appear to be from simple vertical cylinders. The internal diameters of the earlier furnaces seem to be somewhat greater than for the later furnaces. However, as some at least of the furnaces are likely to have been fustrums, tapering towards the top, rather than true cylinders, it is difficult to make valid comparisons except in the very few cases where fragments of the furnace rim can be recognised.

The internal diameters of the early furnaces were estimated to be between about 30 and 40 cm while the later furnaces are only between about 20 and 30 cm. The internal diameter and thus the volume of the reaction zone of the furnace are of some significance (Rehder 2000). The ratio of the surface area to the volume increases as the volume decreases, therefore the smaller the diameter of the furnace, the greater the heat loss by conduction through the walls. Rehder (2000: 15–16, fig. 1) has shown that for a typical small shaft furnace smelting operation of about five hours duration with charcoal as fuel, maintaining an internal wall temperature of 1200°C becomes increasingly difficult as the internal diameter decreases. The rate of heat loss rises dramatically as the internal diameter in the reaction zone falls below about 40 cm, and it becomes virtually impossible to maintain the necessary temperatures below 20 cm in a shaft furnace, placing a practical minimum internal diameter on furnaces.[13]

The heights of the shaft furnaces recorded in the 19th century varied between 40 and 100 cm. The height is influenced to some degree by the size of the charcoal used (Rehder 2000: 92–3). In a simple shaft furnace, the height necessary to ensure that the top of the furnace charge covers the region in which the carbon monoxide gas has been generated can be calculated to have taken place at about 10 to 20 lump size above the tuyere level, usually the higher number. For maximum efficiency the charge should continue for some way above this, say a further 20 cm. Thus for the earlier furnaces with charcoal sizes estimated to be about 3–5 cm, averaging 4 cm, the minimum height can be calculated at about 100 cm above the tuyere level (4 × 20 + 20 cm). In the smaller, later furnaces with typical charcoal lump sizes of about 2.5 cm, the minimum height should be about 70 cm above the tuyere level (2.5 × 20 + 20 cm). Overall, the constraints on furnace size using manually operated bellows and charcoal fuel limit the effective diameter to between 20 and 60 cm with a height of about 100 cm.

The fragments from the early furnaces at Zawar were of burnt and partially vitrified straw-tempered clay.[14] They have flat impressions on the cool face and a curved profile in one plane on the vitrified face. These can be interpreted as coming from basic freestanding stone structures lined with clay to form a cylinder internally. If the furnace structure had been of brick, then some brick fragments should have been present in the slag heaps, such as were found in quantity in the heaps associated with the later brick-built zinc smelting furnaces at Zawar.

There is a general similarity in the post-medieval furnace fragments from Zawar and Dariba. The furnaces from the 7th century AD on (as evidenced by Zawar, Site 7) at both Zawar and Dariba were built of clay coils that were then lined with clay on the inside, and sometimes on the outside as well. The coil walls are typically about 2–3 cm thick and where the internal and external lining survives about 4–6 cm thick (for example, Zawar Site 5, Figs 6.42 and 6.43). These walls would have been freestanding, but it is not known whether the whole of the cylinder of the furnace would have been freestanding, just sitting upon the ground or if they were partially buried with soil to provide additional support and heat insulation at the lower levels with just one side (to allow slag tapping) and the tops standing free. The furnace walls did not need to be physically very strong because although a great mass of material might pass through the furnace during the course of a smelt, the volume of the charge at any moment would have been made up of about 90% charcoal and therefore quite light. There would only have been a risk of physical collapse had a large volume of molten slag been allowed to accumulate in the base of the furnace where a combination of the weight and thermal shock could bring about disintegration of the furnace.

The degree of heat insulation provided by the walls and by burial does have implications for the operation. There are two factors to consider here: the thicker the walls the less heat loss once the walls have reached temperature equilibrium. However, the walls absorb heat so the thicker the wall the more heat is actually stored within it, ultimately to be lost. Rehder found that for furnaces with thin walls, radiation from them was a significant source of heat loss once they had warmed up, resulting in a considerable overall lowering of temperature. This was only after some hours so other factors being equal, a thick-walled or partially buried furnace is likely to be more thermally efficient only where the process is of long duration.

The walls of furnaces for smelting non-ferrous metals in the 19th century seem to have been freestanding. Brooke (1864) for example, shows that the Singhana copper smelting furnaces were just set upon the ground (see Chapter 2, Fig. 2.3). The duration of the copper smelt in Sikkim was about six hours, but elsewhere was usually longer. For example, at Ajmer the lead smelt was recorded as lasting about nine hours, and at Singhana it was apparently much the same at nine to ten hours and thus would have benefited from burial. However, for furnaces with multiple tuyeres set around the furnace at 90° and a tapping hole in front, such as those at Singhana and Ajmer, burial in soil was probably impractical. For short duration processes such as the zinc distillation and most of the smelting operations carried on at the sites reported here, to judge from the study of the vitrification into the bodies of the ceramics (see Chapter 7), a thin furnace wall of low thermal capacity would have been the most efficient.

The reactions in the upper regions of the furnace are endothermic (Fig. 9.8) and this coupled with the heat losses due to conduction and radiation rapidly reduces the temperature. It is therefore highly advantageous for the ore to have previously been rendered friable by roasting in a separate operation. The residual carbon monoxide will burn off at the top of the furnace where the characteristic near-transparent blue flame provides the evidence that there is the necessary over-pressure of carbon monoxide through the system. At the end of the smelt no more fuel is charged and the only reaction taking place is the initial burning of the fuel to form carbon dioxide without further reduction. The blue flame is replaced with an orange flame signalling the end of the reduction process.

It is very likely that most of the smelting units would have been inside some form of structure, as recorded in many of the 19th-century accounts (see Chapter 2, Fig. 2.3) in order to keep the furnaces dry, especially when not in use. A wet furnace would require slow and careful heating to dry it before use, and attempting to smelt in a furnace that was still damp would risk disintegration due to steam formation. At Zawar the large retort furnaces were contained within walled structures with tiled roofs (see Chapter 6, Fig. 6.18) and it is very likely that the other furnaces would have been positioned under a shingled or lightly thatched roof. All the smelting processes produced noxious gases, especially carbon monoxide. Elwin (1942: 192) recorded that the Agaria, just after lighting their furnaces at the base, were careful to ignite the invisible gases emanating from the top of the furnace 'otherwise they get drunk with it'. The consequences could be more serious so the structures would be open. Thatch was especially convenient as it could be partially removed during the smelting process to allow the fumes to escape as depicted in some of the illustrations.

The smelt

Smelting usually took place at night. The Agaria actually had a taboo against their furnaces being exposed to the sun (Elwin 1942: 97–8) for several practical reasons: it was much easier to observe the flame colours in the dark, especially the near-transparent flame of the burning carbon monoxide and colours of the tapped slags, as well as to judge the temperature of the interior of the furnace, inspected through the tuyere hole etc. It was also much more pleasant to work at such an arduous task in the cool of the night.

Smelting furnace operation

The furnace would have been preheated, probably by burning brushwood and dung initially, followed by charcoal, to bring it up to temperature in an operation that would in itself typically take several hours. The Agaria often filled the cold furnace with charcoal which was then ignited from the tuyere hole (Elwin

1942: 189, 213). When judged fully heated, more charcoal and the first ore and any fluxes could be charged, creating in effect a column of fuel and ore through and in which the reduction and associated reactions would take place (Fig. 9.8). Although many descriptions of traditional processes describe the apparent careful alternate layering of the ore and fuel in the furnaces there should not have been too much need for this as the ore and fuel would become intimately mixed as they moved down the shaft, as demonstrated by the EMRG experimental smelts. This is because fuel was consumed in the region of the tuyeres towards the base of the centre of the column, causing the material in the centre to drop more rapidly; this was replaced in part by more material tumbling in from the sides.

The reactions therefore took place in the column above the tuyere zone exposed to the hot rising gases during the time that it took for the materials to move down to the tuyere zone. This time, known as the residence time, depends, as outlined above, on the height of the column as well as on the relative quantity, size and nature of the charcoal, and the rate of combustion, which is directly related to the quantity of air delivered. During normal furnace operation the combustion of the charcoal would take place in the tuyere zone, producing mainly carbon dioxide:

$$C + O_2 \rightarrow CO_2 + 393\ KJ$$

Although there is no volume change in the combustion reaction, it is exothermic, generating tremendous heat and thereby bringing about a large expansion of the gases due to the temperature rise of about 1200°C. The pressure of the air entering through the tuyeres will therefore be greatly augmented by this heat expansion driving the gases up the shaft. The carbon dioxide will rise through the column encountering more hot charcoal where the bulk of it will be rapidly converted to carbon monoxide:

$$C + CO_2 \rightarrow 2CO - 174KJ$$

bringing about further volume expansion, although reduced a little as the temperature drops, the reaction being endothermic. The hot friable metal oxides of the ore become exposed to the still extremely hot and powerful reducing agent, carbon monoxide, which rapidly reduces them to metal. The oxides on the surfaces of the exposed ore particles are reduced first, forming a shell of metal, but the interior regions of the individual particles are also rapidly reduced.

The specific reactions taking place inside one of the briquettes of ore and animal dung during smelting have been studied in some detail in the copper smelting experiments carried out by the EMRG (Timberlake 2005; Craddock *et al.* 2005) (Plate 9.4). The initial experiments were not particularly successful in producing liquid copper, but examination of the briquettes afterwards showed the smelting process effectively frozen in the formation of droplets of copper. Most, if not all, of the mineral (in this case the copper hydroxy carbonate mineral, azurite ($Cu_3(CO_3)_2(OH)_2$) smelted without pre-roasting) had been reduced to metallic copper dispersed throughout the briquette. What appeared to be a solid droplet of metal (Plate 9.5) in the briquette transpired to be just a thin but coherent skin of metal surrounding the individual particles of ore reduced to metal. Immediately outside the skin there was no metal remaining; inside the skin the tiny particles of metal had not coalesced (Plate 9.6). Clearly the skin of molten copper, once formed, began to shrink inwards due to surface tension and as it did it incorporated the metal it encountered. In our experiments this stage was slow in forming and was preserved quite simply because the recorded furnace temperatures of between 1000 and 1100°C were too low for a successful smelt, being sufficient to reduce the metal oxides to metal but insufficient to completely melt the metal. In reality the smelting temperature would need to be well over the melting temperature of the metal when it would then rapidly form into a molten droplet which, being much more dense than its surroundings, would itself quickly fall out of the reaction zone down into the forming slag layer below. This is of some significance in understanding the mechanism of the smelting process.

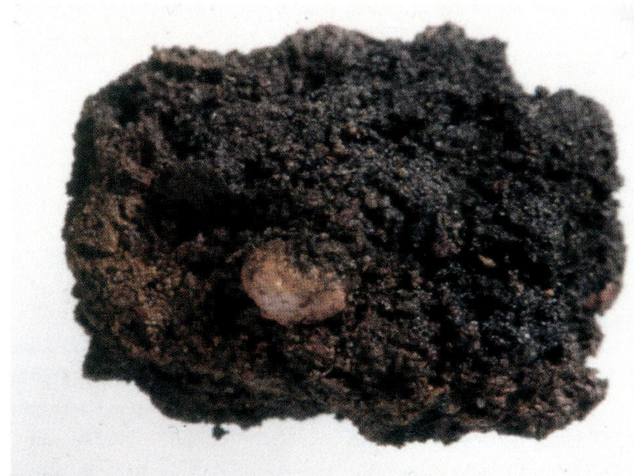

Plate 9.5 Ball of crushed azurite copper ore mixed with dung, removed from the furnace after smelting. Part of the ball has broken away revealing a forming copper prill inside.

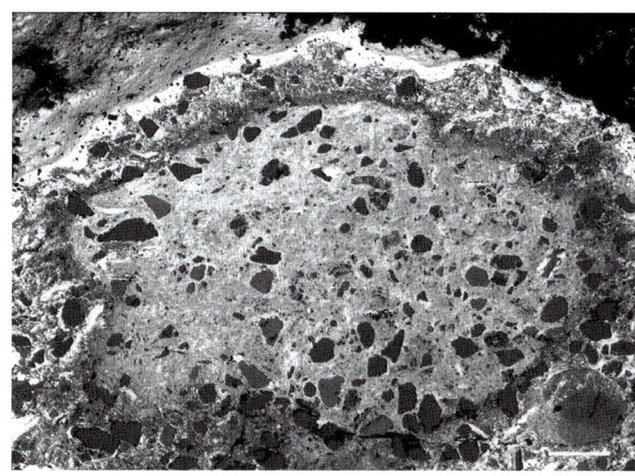

Plate 9.6 SEM micrograph of a section through the forming copper prill taken from the smelted dung ball (cf. Plate 9.5). The light areas are smelted copper – note how the copper has formed a coherent skin which is coalescing into a droplet and absorbing the solid smelted copper inside (scale bar: 200 μm) (N. Meeks/British Museum).

Where the metals formed as a liquid, such as copper and lead, the system never attained equilibrium but was dynamic, as the basic reduction reaction

$$MeO + CO \rightarrow Me \downarrow + CO_2$$

was constantly being driven to the right by the removal of the metal from the system. By contrast, in the solid state smelting of iron, the forming bloom remains in the reaction zone therefore it is necessary to maintain equilibrium conditions:

$$FeO + CO \leftrightarrow Fe + CO_2$$

Slag formation

In most shaft furnace operations the gangue material was separated from the forming metal and removed from the system by slagging, that is, the gangue was melted and could be run out of the furnace (Bachmann 1982, 1993b). This slag was formed by the reaction between the metal oxides and the silica to form a series of minerals that melt at temperatures usually between about 1100 and 1300°C and are reasonably mobile when molten. For a successful smelt it is important to have a free-flowing slag that can be easily tapped periodically from a furnace or ladled from a hearth, and through which the denser metal droplets can drain freely to form the ingot in the base of the smelter (Fig. 9.8). The formation of a free-running slag was also an important indicator to those carrying out the smelt that the process was taking place satisfactorily, as recounted by Franklin for the iron smelters of central India in the early 19th century (Dharampal 1971: 241). The slag also performs a useful insulating function, protecting the metal beneath from reoxidation and solidifying before forming a coherent ingot. The slags could be periodically drained: when the molten surface reached tuyere level, a very distinctive catarrhic sound would be heard, warning that tapping must not be delayed or the tuyeres would block. In short smelts the slag could be allowed to accumulate in the furnace. The molten slag pool was usually so reducing that the metal droplets, especially copper, draining through it could reduce and dissolve appreciable quantities of iron (Fig. 9.8) (Craddock and Meeks 1987; Craddock 1995: 137–44). This is the origin of the iron in freshly smelted copper which, together with any residual copper sulphides, had to be removed in the refining stage.

In the mines studied here, the main gangue minerals in the ore charge were silica, principally in the form of quartz, pyrites and dolomite ($CaMg(CO_3)_2$) at Zawar and quartz, pyrite and alumina (Al_2O_3) at Dariba and at Agucha (see Chapter 10 for more information on the Dariba and Agucha slags). As the lead and zinc ores from all three mines are intimately associated with iron sulphides, some of which would still be present after beneficiation and roasting as iron oxides, separate additions of flux may not have been necessary. Also the zinc and some of the lead in the ore was available to combine with the silica and form silicates, therefore at Zawar the early silver/lead slags have quite low but variable amounts of iron, suggesting that the material was smelted without additional flux whereas the later silver/lead slags have high iron contents indicating a deliberate addition of iron oxide flux to an ore which contained very little other than lead (see below and Chapter 7). Several 19th-century accounts state that old iron smelting slags, which would have been rich in iron oxide (FeO), were added as a flux. It is possible that iron metal or old iron smelting slags were added to reduce the galena, which had not been roasted prior to smelting, directly to lead (see below).

Copper smelting[15]

There were potentially three processes of smelting copper sulphidic ores: dead roasting followed by a smelt, matte smelting and iron or iron slag reduction.[16]

Dead roasting and smelt

The copper ore would have been beneficiated in the ways described in Chapter 8 prior to roasting. The 19th-century descriptions of Indian processes concur that the sulphidic copper ores were almost invariably roasted. At the copper mines of Kumaun in the southern Himalayas, for example, Stephens (1901–2: 407) noted that only the sulphidic copper ores were roasted prior to smelting, not the carbonate ores. Details of the roasting operations are often included in the descriptions, notably in Rajasthan at Daribo (by Cadwell) and at Singhana (by Boileau and Dixon) (see Chapter 2). As exemplified at both Kumaun and Singhana, the finely ground copper ores were usually mixed with cow dung, made into cakes and left to burn in heaps with additional cow dung as fuel for many hours to burn out the majority of the sulphur. The roasted ore was then mixed with more dung to make small cakes ready for the main smelting process in small shaft furnaces as described in Chapter 2.

The following account relates to the copper smelting operations at Kumaun and Chamoli (formerly Garhwal) Districts in the foothills of the Himalayas in Uttarakhand, compiled by Stephens (1901–2: 407), and reproduced in Chakrabarti and Lahiri (1996: 161–2):

> Native smelting arrangements are ingenious, but clumsy and of most meagre out-turn. The furnace is built of common stone and clay, luted with a compost of clay and cow-dung beaten up together. It is about 4 in by 3½ in (9 to 10 cm) broad (wall thickness?), with a place for ashes in front. The operation of smelting takes place about 6 hours, during which time a continuous blast is kept up by two men, in relays.
>
> There are two bellows of oxhides, with an iron nozzle protected by a burnt clay pipe. The bellows work alternately. The broad end has two flat pieces of wood nailed or laced on to the leather. These are grasped in one hand by the operator, who closes the hand, thus shutting the open end of the hide. He then pushes the hide forward the full length of his arm, and opening the hand, draws back his arm, repeating the operation until relieved by another operator. It is astonishing what a powerful blast can be maintained by this crude process, with two hides working.
>
> The smelter sits in front of the furnace, and feeds the ore and the charcoal, in the proportion of one to three, into the furnace for the whole period. The ore is always powdered to a very fine point. If carbonates or oxides, it is not calcined; if pyrites, it is burned with cunda (cow-dung), stacked in alternate layers, for a period of 8 hours. The whole thing is then powdered once more, ore and cunda dust together.
>
> The charcoal preferred is either made from scrub oak or from rhododendron wood. Pine is of no use at all.
>
> At the end of the run, when the smelter judges the process to be complete, the blast is stopped, the glowing charcoal on top of the molten mass is raked off, and the slag is lifted off the metal with two long sticks. This is

repeated half a dozen times, the last two or three 'skims' being remelted in the next smelt. The furnace is then broken down in front, and the cake of molten copper is taken out. It is very pure, notwithstanding the roughness of the process (but see below, the copper content of 96% would hardly qualify for the description 'pure'). Sometimes it is thought advisable to once more melt the copper and refine it, which is done by stirring it while molten with two green oak sticks, being, in fact, 'poling'.

The Kumaon ore is very free from arsenic. When the ore is very foul, the smelter adds a few broken bits of iron [cast in the hills] from time to time. From a maund [80lb (36.3 kg)] of calcined copper pyrites is obtained 6 seers [12 lb (5.5 kg)] of metal, which is about 96% pure copper. The loss is very great about 10% or more going away in the slags. There are thousands of tons of copper slag in the hills.

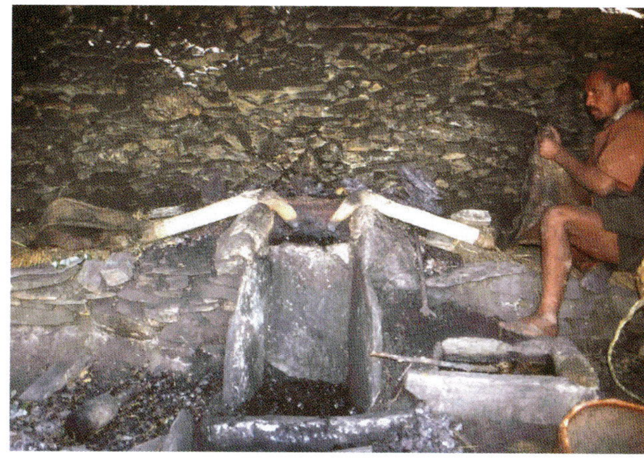

Plate 9.7 Copper smelting in late 20th-century Nepal, identical to the matte smelting process described by Boileau in the 1830s in Sikkim (cf. Fig. 9.9) (from Anfinset 2011: pl. 5).

Several points of interest emerge from this description. The first is that apparently the bellows were worked alternately to provide as continuous and uniform a blast as possible. Secondly, the addition of iron metal when the ore was 'very foul' (that is rich in sulphur) is interesting. Finally, this is one of the few accounts from South Asia to mention the process of poling (see refining, below).

The matte process

The matte smelting process is a variant of the direct reduction process designed to facilitate the smelting of poorer grade ores (Biswas and Davenport 1976: 80–98; Moore 1981: 200). Typically the ores were copper-iron sulphides such as chalcopyrite ($CuFeS_2$) or bornite (Cu_5FeS_4) often intimately intermixed with pyrite (FeS_2) such that even after beneficiation the ore still contained insufficient copper to be satisfactorily smelted by direct reduction. The aim of the matte process was, by successive stages of oxidation and slagging, to separate the bulk of the iron, silicon and sulphur, thereby concentrating the copper prior to its reduction.

The process is of considerable antiquity generally (Craddock 1995: 149–53; Craddock *et al.* 2003) and extensive remains of matte smelting operations of the Early Islamic period (9th–10th century AD) have been excavated at El Maysar in Oman for example (Hauptmann 1985). The smelting of copper in Sikkim and Bhutan by a primitive version of matte smelting was reported by H.F. Blandford and later by Mallet and is recorded in Ball (1881: 276–7) and Percy (1861: 389–90). The ore here was described as being low grade and was clearly iron-rich for the smelting of which the matte process is ideally suited. The two descriptions differ in some details: Blandford's account is given here, taken from Percy, with significant differences, supplied by Mallet, from Ball's text added in { }.

The furnace, formed of a sandy clay, was of the form and dimensions shown in the accompanying engraving (Fig. 9.9). It was built in a bank about two feet high, {built of refractory clay, or is simply a hole dug in the ground} and consists of a shallow square cavity {about a foot square at the top, and plastered with refractory clay, if it is not sufficiently so naturally}, the bottom of which was slightly concave. {Charcoal was rammed down to form a floor through which the molten metal cannot sink. There was no orifice at the base of the furnace} (or anywhere else for that matter). The back wall of the furnace – which, as a whole, may well be compared to an arm-chair – was carried up to a height of about eighteen inches, in the form of a truncated pyramid, the top being hollowed out for the purpose of receiving the slags as they were removed with pincers from the furnace. The front wall was very low, not more than about six inches above the bed, while the side walls were of intermediate height, being about one foot above the furnace-bed and six inches above the earthen platform on which rested the bellows.

These bellows were of simple construction, two in number, one being placed on each side of the furnace. They consisted of a seamless bag of goat-skin, formed of the skin of the body and fore-limbs of the animal. The bottom, formerly covering the neck of the animal, now embraced in like manner the earthen nozzle of the bellows. The mouth of the bag was gathered in, so as to leave a small opening only, and was grasped by a boy who squatted beside it and worked the bellows, alternately loosening and tightening his grasp as he raised or depressed the bag, thus producing an effectual, though intermittent, blast. The nozzle, moulded by hand, resembled in form the common mouth blowpipe, the end being bent at a right angle, so that, while the stem rested on the side wall of the furnace, the entrance aperture reached to within three or four inches of the furnace-bed, on which the blast impinged at a rather obtuse angle. Charcoal was the only fuel employed in the furnace.

The first smelting operation was a simple fusion. The furnace being heated with charcoal, a few handfuls of the washed ore, previously dried and mixed with charcoal, were thrown on, and the bellows worked by boys, as above described. More charcoal was added as required until a perfect fusion of the ore was effected. The fused 'metal' [regulus] (a 19th-century term for matte, copper and iron sulphides) then formed a small pool at the bottom of the furnace, covered with a layer of fused slag, while the burning charcoal floated on the surface. The fusion being complete, the charcoal was removed, water {a wisp of wet straw} was sprinkled onto the slag to solidify it, and it was then lifted off in successive cakes with a pair of pincers and placed to cool in the shallow basin at the back of

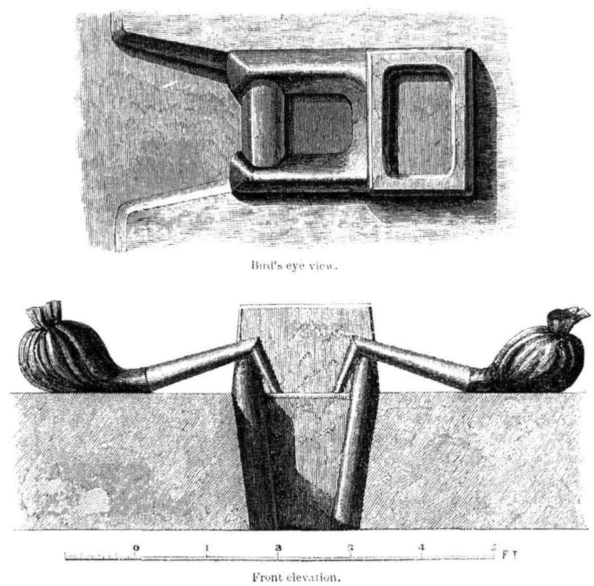

Fig. 9.9 Front elevation and plan of the 'armchair' smelting hearth used to smelt copper by the matte process at Sikkim in the 19th century (cf. Plate 9.7) (from Percy 1861: 389, 390).

the furnace. A fresh charge was then thrown on, and the same series of operations repeated, until a cake of 'crude metal' weighing eight or ten pounds had accumulated in the bottom of the furnace. This was removed when cold, pounded, and kneaded with cow-dung into small balls, which were dried in the sun and then roasted in a shallow furnace formed of a ring of slag-cakes placed on edge. The roasted 'metal' was afterwards refined in the same furnace in which the fusion of the ore had been effected, and in a precisely similar manner, the result being – 1st. refined copper, which collected in the bottom in a cake, weighing four or five pounds; and 2nd slag, which was not, so far as I could learn, subjected to any further process, the probability being that the ash yielded by the very large amount of charcoal consumed in the process is sufficient to form a highly basic slag, and thus allow the whole of the copper to be reduced to the metallic state. Three pounds of 'crude metal' were said to yield one pound of refined metal. {In this state it is sold at a rate of 10d. (pence) a pound, but it has to be still further refined before it is available for ordinary use.}{One maund picked ore = 6 or 7 seers of washed ore; 7 seers of which = 4 of regulus; 4 seers of regulus = 1¾ seers of copper, or 3¾ to 4½ per cent. of copper from the picked ore. Eight or nine seers of regulus are obtained at one operation in about six hours, and 3 or 4 seers of copper in about three hours more.}

(NB 1 maund = 40 seers, 1 seer = 2 pounds or 0.9 kg)

This is a rather rare eye-witness description of a simple matte smelting process in operation. There seems to have been no preliminary roasting stage – instead, the first smelt of the ore, which was copper-iron sulphide with some silicate waste material, got rid of much of the waste and iron mineral as the lighter slag, and the remainder which was the dense 'regulus' or matte was more or less concentrated copper sulphide. This was then broken up and roasted in air to convert the sulphides to oxides 'crude metal'. This, in turn, was again smelted, this time to reduce the copper oxides to metal and remove more of the remaining iron as slag. Although called refining by Blandford, this was in fact a smelting process, and the metal would still have contained some sulphur and iron, which could have been removed by melting in an open crucible to oxidise the impurities to form yet another slag layer, followed by stirring in green twigs, known as poling (as described above) to reduce the copper that had oxidised in the last operation back to metal. This process survived almost unchanged into the 21st century in Nepal (Anfinset 2011) (Plate 9.7).

The first stage was to smelt under oxidising conditions both to remove the majority of the sulphur as sulphur dioxide and to separate the remainder of the ore into two fractions, an iron silicate slag and the matte which would have been a mixture of copper and iron sulphides with some additional oxides. The reactions can be expressed as follows:

$$\underset{\text{pyrites}}{4FeS_2} + 11O_2 \rightarrow 2Fe_2O_3 + 8SO_2 \uparrow$$

$$\underset{\text{chalcopyrite}}{CuFeS_2} + 3O_2 \rightarrow FeO + CuO + 2SO_2 \uparrow$$

$$\underset{\text{chalcopyrite}}{4CuFeS_2} + 9O_2 \rightarrow 2Fe_2O_3 + \underset{\text{matte}}{2Cu_2S} + 6SO_2 \uparrow$$

$$\underset{\text{chalcopyrite}}{4CuFeS_2} + 7O_2 + \underset{\text{pyrite}}{2FeS_2} \rightarrow \underset{\text{matte}}{2Cu_2S} + 4FeS + 2FeO + 8SO_2 \uparrow$$

The iron oxides would have formed into a separate slag with additions of a crushed quartz flux as necessary:

$$2FeO + SiO_2 \rightarrow \underset{\text{fayalite slag}}{Fe_2SiO_4}$$

The mixed copper sulphide/iron sulphide matte formed what was described as a 'small pool' beneath the slag, probably with some unchanged chalcopyrite, which melts at only 880°C. When sufficient matte had accumulated, it was removed, crushed, mixed with cow dung and re-smelted in the same smelting hearth. This reaction can be represented as:

$$\underset{\text{matte}}{Cu_2S + 2FeS} + 5O_2 \rightarrow 2CuO + 2FeO + 3SO_2 \uparrow$$

In standard matte smelting processes the copper oxide produced is then again separated and reduced to copper metal in a reducing atmosphere, but at Sikkim Blandford stated specifically that the matte was converted to copper in one operation that must have been essentially oxidising. The process probably relied on the reactions between the mixture of gases generated by the oxides and sulphides in the matte, and between the oxides and any remaining unchanged chalcopyrite ore, expressed over-simply in the reactions:

$$3CuO + FeS \rightarrow FeO + SO_2 \uparrow + 3Cu$$

$$5CuO + \underset{\text{chalcopyrite}}{CuFeS_2} \rightarrow FeO + 2SO_2 \uparrow + 6Cu$$

Thus copper metal could be produced without the conditions ever having been very reducing. The latter reactions are based on the researches of Rostoker *et al.* (1989) into the primitive smelting of sulphidic copper ores, and seem to fit the Sikkim process better than the reaction for the standard matte processes.

Refining

The resulting copper would still contain some copper sulphides, arsenic and typically several percent of metallic iron. Traditionally, the method used to purify the copper smelted by the matte process was known as fire refining (Peters 1911: 530–77). The raw copper would be remelted in an open crucible and exposed to a blast of air from the bellows while stirring to oxidise any arsenic present in the metal, almost all of the iron and some of the remaining sulphides. The oxides being relatively light would float and clean sand or crushed quartz could be sprinkled on to form a slag which could be more easily removed by skimming the surface. Such slags, sometimes referred to as crucible slags, are quite recognisable as they are light and gassy compared to the smelting slags. They are also often inhomogeneous and contain relatively large amounts of copper – that had oxidised in the refining process – compared to smelting slags. The oxidised copper remaining in the melt could be reduced back to metal by poling. Bundles of green twigs were stirred into the molten metal. This operation, spectacular to watch when done on a large scale, instantly forms large volumes of gaseous hydrocarbons which penetrate into the violently agitated molten copper and react with the oxygen to create water vapour that is expelled harmlessly as steam.

The usual refining process in the 19th century in India seems to have been much simpler, as noted above. Where reference was made to refining, the copper was remelted in an open crucible thereby oxidising some of the sulphides, arsenic and the iron which could be skimmed from the surface of the copper, but the molten metal was apparently not usually poled. Thus the resulting copper probably still contained some dissolved iron and copper sulphides and oxides. At both Singhana and Sikkim it was specifically stated that the copper after refining was still of poor quality; only at the copper smelting operations at Kumaun and Chamoli was reference made to poling the metal with green poles after the remelting.

The only evidence for regular copper smelting at the sites investigated in this project was in post-medieval Dariba, at Sites 13, 14 and 15, which all produced slags and furnace linings; in addition Site 14 produced abundant tuyeres. The available ores were all sulphidic so would have required roasting, after which the ores were smelted, possibly with the addition of iron metal or iron smelting slags, in the small furnaces built of clay coils (described above and in Chapter 6). The furnaces were apparently cylindrical with an internal diameter of about 20–25 cm. Air was supplied through steeply angled tuyere(s) each set in a circular clay collar. In the absence of intact furnaces, the arrangements for air supply and slag-tapping at the base of the furnace are not known, but comparisons with the contemporary copper smelting furnaces at Singhana (see Chapter 2, Fig. 2.3) would suggest three tuyeres set at 90° to each other with one side reserved for tapping operations. The inside walls of the furnaces reached temperatures of about 1150°C and a smelt would typically be maintained at these temperatures for about two to three hours (see Chapter 7). The copper content of the slags from Sites 14 and 15 varies but is usually between about 0.2 and 1%, with some higher values (see Chapter 7, Table 7.15).

Lead smelting[17]

Most of the 19th- and 20th-century accounts of lead smelting in India and the surrounding region describe small shaft furnaces such as those from the post-medieval period at both Zawar and Dariba. However, Collins (1893–4) does describe a hearth process carried out at Ghorband in Afghanistan. The hearth was a small depression of about 30–40 cm in diameter and lined with clay, similar to that proposed for the early silver/lead smelting process at Dariba. Hearth smelting was widely used to smelt lead in post-medieval Britain and elsewhere until the 19th century (Gowland 1901; Tylecote 1976: 100; Willies 1991b, 1993; Percy 1870: 278–93; Hofman 1918: 96–135).

There are three potential methods by which lead was smelted in the past: dead roasting followed by smelting, matte smelting and iron reduction. The conditions necessary for smelting lead are not as rigorous as for the other metals discussed here, allowing simpler smelting arrangements (Gill 1992; Willies 1993). The smelting of the lead, therefore, could be represented as a partial oxidisation of the ore to the oxide or sulphate, followed by a reaction between the unchanged lead sulphide ore, galena, PbS and the oxidised part, presented conventionally as:

$$2PbS + 3O_2 \rightarrow 2PbO + 2SO_2 \uparrow$$

and

$$PbS + 2O_2 \rightarrow PbSO_4$$

followed by

$$2PbO + PbS \rightarrow 3Pb \downarrow + SO_2 \uparrow$$

and

$$PbSO_4 + PbS \rightarrow 2Pb \downarrow + 2SO_2 \uparrow$$

In fact reactions between reactants that are in the solid state are at best very slow and it is more likely that the second reaction should be between the lead oxide and sulphate and the carbon monoxide gas:

$$PbO + CO \rightarrow Pb + CO_2$$

and

$$PbSO_4 + 2CO \rightarrow Pb + 2CO_2 \uparrow + SO_2 \uparrow$$

Note that these reactions are all driven to the right by the removal of the molten lead metal and sulphur as sulphur dioxide from the system.

The traditional European hearth processes were two-stage: a preliminary smelt of the galena ore that produced lead and a slag containing large amounts of lead sulphide, and the second, hotter stage, in which the slag was melted and the majority of the lead sulphide reduced to metal.

Iron reduction

The traditional iron-reduction method reduced the lead sulphides to metal with powdered iron and charcoal. The process was used widely in Germany in the 19th century but has a much far-ranging and longer history. Liu *et al.* (2015) suggested that the process was used in medieval China and it was recorded in Japan by Pumpelly (quoted in Percy 1870: 384). The process is described in some detail in the *Dhātūtptti* compiled by the Indian scholar Thakkura Pherū, probably in the early 14th century (Dube 2006):

> The ore rocks obtained from lead ore mines are crushed, ground and washed with water for beneficiation. The resulting material, containing gangue fraction as well, is mixed with iron in one-sixth proportion. The mixture is charged into crucibles of capacity 1000 *palas* (approx. 4–5 kg). Thirty such crucibles are placed in a furnace and heated/fused. Lead is obtained in weight equivalent to one-fourth of the starting material.

The reaction between the powdered solid iron and liquid 'fused' lead minerals can be represented conventionally as:

$$PbS + Fe \rightarrow FeS_2 + Pb$$
and
$$PbO + Fe \rightarrow Pb + FeO$$

There is no mention of charcoal as part of the charge in this account but it should have been present in order to create the necessary carbon monoxide gas phase as mentioned above. The German and Japanese processes certainly included charcoal.

No physical evidence of Pherū's crucible process has yet been identified but as already noted above and in Chapter 2, many of the 19th-century records of the Indian processes for smelting chalcopyrite and bornite ores specify the addition of iron smelting slags and sometimes iron metal itself; later modification of the German processes, operating at higher temperatures and more reducing conditions, also substituted iron refining slags for iron metal. Both the Indian bloomery slags and the German refining slags would have contained appreciable quantities of wustite (FeO), which could be transiently reduced to iron metal and participate in the process. No direct evidence for the process was found at the Aravalli sites investigated although the presence of iron smelting slags at Dariba, Site 9 could be significant.

Refining

The usual refining process for lead was to remelt the metal in an open crucible allowing some of the unwanted elements, notably sulphur and arsenic, to oxidise and evaporate and others, such as iron, to partially oxidise and float to the surface. Other dross, slag etc, would also float and could be removed by skimming the surface.

Notes

1. Principal sources: Craddock 1995; Percy 1861, 1864, 1870; Rehder 2000; Rothenberg 1990b; Tylecote 1976, 1987.
2. Our surveys of the slag heaps etc. at Dariba, where sulphidic copper ores were smelted in the post-medieval period, as evidenced by Sites 13, 14 and 15, failed to locate any heaps of roasting debris which confirms the impression that copper production was never included in the main activities of the mine.
3. See Peters (1911: 426–30) for the briquetting of copper ores, and Almond (1998: 176–8) for the development of sintering in the zinc smelting industry during the 20th century.
4. Wood: Percy 1861: 107–42, 1861: 62–73; dung: Reddy 1998; charcoal: Smith and Gnudi 1942: 173–9).
5. Hegde (1991: 75) suggested that cow dung was likely to have been the main fuel used in the zinc smelting *koshthi* at Zawar, a claim that must be carefully considered. Neither the distinctive vitrified material of cow dung burnt at high temperature under oxidising conditions nor wood ash was encountered in any of the furnaces excavated, even though they still carried their last charge of retorts and obviously had not been cleared out except in one instance to remove the zinc that had spilt onto the floor of furnace 4. On the partially vitrified and fuel-glazed surfaces of the retorts the potassium content was enhanced such as might have been expected if either dung, wood or charcoal had been used as the fuel (Freestone *et al.* 1985a; see Chapter 7, Plate 7.11). Also, the space for the fuel in between the retorts is very restricted; typically the tips of the retorts are unvitrified but the bodies of the retorts beneath are heavily vitrified. This indicates that the main body of the fire must have been confined to a vertical dimension of no more than about 10 cm, interspersed by the retorts, but nevertheless was very hot. Cow dung does not have a sufficiently high calorific value to generate the necessary heat in such a small space – only wood charcoal would suffice. The early description of zinc smelting given in the *Rasaratna Samuchchaya* specifies that the retort is to be 'strongly heated by means of the fire of *jujube* charcoal (*Ziziphus jububa*)' (Rây 1956: 172) (see Chapter 11), a species quite common in the Zawar area. One hard and dense lump of charcoal actually found in one of the furnaces at Site 30 was confidently identified by the locals as coming from the *nin* tree, very probably the neen (*Azadirachta indica*). Fragments of charcoal were fairly common throughout the retort mound of Site 30 that was composed largely of used retorts and furnace fragments, and where other forms of debris, even sherds of pottery, were almost entirely absent, suggesting that the charcoal was associated with the smelting operations rather than domestic activity. As a final argument for charcoal smelting, in the one near contemporary description of the zinc smelting process at Zawar (Brooke 1850: 213) (see Chapter 3), the loaded retorts were smelted using charcoal as fuel.
6. Thus far from causing deforestation in countries such as Britain, without large areas of natural forest, the post-medieval smelting operations were responsible for the existence of huge areas of woodlands. It is estimated that in the region of 150–200,000 acres of forest were required to provide the charcoal requirements of the iron smelting industry alone in Britain in the mid-18th century (Rehder 2000: 138).
7. Forbes 1971:113–20; Tylecote 1981; Weisgerber and Roden 1985, 1986; Craddock 1995: 74–85; Davey 1979, 1988; Merkel 1990: 103–9; Gamage 2002; Mishra 2003: 114–16.
8. Tylecote 1981; Rothenberg 1990b: 29–39; Craddock 1995: 185–9.
9. See Boileau and Brooke's reports, Chapter 2, pp. 12 and 13.
10. Their materials are described in Chapter 7, p. 156.
11. Quoted in Chapter 12, p. 234.
12. See Chapter 11, p. 214 for descriptions of more elaborate arrangements for collecting zinc oxide fumes.
13. This is not necessarily the smallest unit that can reduce ore. For example, one of the standard ore assay methods uses a charcoal block and blowpipe in which high temperatures and intense reducing conditions are maintained in a tiny volume. However, this is not applicable to the volume reduction of metal ores.
14. No recognisable smelting hearth or furnace fragments were excavated at Agucha. See Chapter 7, pp. 139 and 153, for more information on the materials.
15. Percy 1861; Peters 1907, 1911; Rehder 1999, 2000.
16. Evidence for the latter process are the various 19th-century descriptions of iron or iron slags being added to the copper smelting charge, but the main early written descriptions of the iron reduction process refer to the reduction of sulphidic lead ores so the process is discussed in the lead smelting section below.
17. Simple lead smelting is described in Chapter 2 and the silver/lead smelting operations performed at the Aravalli mines are discussed in Chapter 10.

Chapter 10 Process silver

P.T. Craddock

Silver/lead production[1]

It is clear that at both Dariba and Agucha in antiquity the primary impetus behind the mining activities was the production of silver from argentiferous lead. At Zawar the situation is more equivocal, but some silver is likely to have been produced in the post-medieval period at least when the mines at Baroi Magra were in operation. The remains of production at Dariba and Agucha attest to the enormous scale of the operations, ultimately stimulated by the introduction of a silver coinage in northern India after the 5th century BC (Cribb 1985), and the recognition of these mines as early silver producers goes some way to providing a solution to the long-running problem of the sources of silver for the India silver punch-marked coinage (see Chapter 8). These mines were comparable in scale to the contemporary mines at Rio Tinto and Laurion in the Classical world, where the impetus behind their expansion was also the need to provide silver for coinage.

Description of silver production and assaying in the Mauryan period

The organisation of mines generally is described in the *Arthaśāstra* (see Chapter 8) which includes sections more specifically on silver production. As it is likely that the text is broadly contemporary with the operations at Agucha and some of those at Dariba, which seem to be mainly engaged in the assaying and smelting of silver ores, this account is very relevant. Both lead and silver ores are described in some detail:

> The earth which is black like a crow, grey like a pigeon or *gorchana*, marked with white lines and smells like raw meat. This is the place where lead ores can be found (2.12.13).
>
> These ores have the colour of a conch shell, camphor, butter, a pigeon, turtle-dove, a red-white gemstone (*vimalak*) or the neck of a peacock. They are as resplendent as blue or green (*sasyaka*) and yellow (*gomedaka*) gemstones (possibly opal and agate), cane and grain sugar and have the colours of *kovidara* flowers, lotus, *patali* (*Stereospermum suaveolens*), *kayala* (a type of *Phaseolus* bean) or of flax. These are associated with lead (*nag*) and zinc (*anjan*). They smell of raw meat and on breaking show black or white spots and lines, and which on roasting do not split but effloresce and emit smoke. These are silver ores ... In all cases the heavier the mineral the higher the metal content (2.12.6–7).

The black ore, grey colours and mention of iridescence probably refer to galena, lead sulphide or possibly the complex mineral freibergite ($(Ag, Cu, Fe)_{12}Sb_4As_4S_{13}$) mined at both Dariba and Agucha. The light colours could indicate the lead carbonate ore, cerussite ($PbCO_3$). The reference to white spots and lines could well be streaks of native silver such as occur in the deposits at both Zawar and Dariba. The description of the ores efflorescing on heating is fairly common to all sulphidic ores, but the generation of smoke suggests oxidised zinc vapour from a mixed argentiferous lead-zinc ore, which are the prevalent ores in the Aravalli deposits generally.

There is no evidence at any of the three main mines for the occurrence or treatment of the oxidised ores, such as cerussite, as has been postulated for some mines elsewhere (Meyers 2004) (see Chapter 7). The very small quantities of cerussite that were found on the excavations (notably a single piece from Agucha, BMRL 26277V) were almost certainly originally galena, or even litharge, when the sites were in operation that have oxidised subsequently. The galena ores used in the past in the Aravallis seem to have contained only relatively minor amounts of other non-ferrous metals, apart from zinc in the form of sphalerite with which the lead ores were associated, and antimony in the form of freibergite at Agucha. The latter element occurs in the slags as the iron antimonide, known as speiss (see Chapter 7, Plate 7.27). Speiss is a problem, especially in silver smelting, because not only does it tend to block the furnace but it also absorbs silver, which can be difficult to recover (Craddock *et al.* 1987).

The cupellation process is described rather laconically in the *Arthaśāstra*. The instructions are complex and there are considerable variations in the translation; that of Kangle (1965–72) is probably the most authorative. In Book 2, chapter 13 (Kangle 1965–72, II: 111), it states:

> 10 That originating in Tuttha, that from Gauda, that from Kambu and that from Cakravāla, – these are [types of] silver. 11 [Silver which is] white, smooth and soft is best. 12 The opposite kind and that which tends to burst is bad. 13. He (the Superintendent of Gold) should purify that with one-fourth part of lead. 14. That in which a crest

has appeared at the top, which is clear, lustrous and of the colour of curds is pure.

The impure silver was to be purified with lead. An important point to note is that here Tuttha, about which there is much debate (see below), is identified as a place, apparently a mountain. In Book 2, chapter 13 at 49 there is a more detailed description of cupellation (Kangle 1965–1972, II: 114):

> Or, silver, purified, four times in a crucible of bones, four times in one with an equal amount of lead, four times in a dry crucible, three times in a pot-sherd [and] twice in cow-dung, thus passed through seventeen crucibles, [and] made bright with Indus-earth (saltpetre (KNO_3)) [is a base for colours].

Shamasastri (1915), quoted in Rây (1956: 54), translates this passage very differently:

> Pure or impure silver may be heated four times with copper sulphate (*asthituttha*) mixed with powdered bone again four times with an equal quantity of lead, again four times with dry copper sulphate (*sushkatuttha*), again three times in skull and lastly twice in cow dung.

There are a number of points of interest in these accounts. Bone is mentioned as a material in which the refining was to take place – this is the earliest reference anywhere to the use of bone in cupellation, although no evidence for the use of bone was found in any of the cupels or in the litharge from Agucha. The instruction that the silver was to be refined in potsherds at one stage is very interesting given the quantities of heavily lead-glazed potsherds at Dariba, Site 9, especially 27294. The multi-stage process could be a reference to a preliminary scorification stage; this would accord well with the 'dry crucible' stage, although this should have preceded the main cupellation.

A real problem occurs with the identification of the word *tuttha*. Shamasastri (1915) and Meyer (1925–6) translate *asthituttha* as blue vitriol, that is copper sulphate, thus *sushkatuttha* is dry or anhydrous copper sulphate. The justification for this is that there is a much later medieval reference to *tuttha* being formed by the action of 'burning water' on copper. Presumably the burning water is acid, which is actually more likely to have been nitric acid – made from potassium and sodium nitrate, saltpetre, the Indus earth of the alchemists – than the sulphuric acid necessary to make copper sulphate, blue vitriol, but the reference would still suggest that a copper salt formed. Some of the medieval Islamic sources state that *tūtīyā*, which in the context in which it appears is definitely zinc oxide, was the vapour from copper. It is possible that the Islamic sources were referring to the heating of brass to cause the zinc to evaporate as zinc oxide, as described in Dioscorides (see Chapter 11). However, there would seem no point in adding copper sulphate or nitrate, the mineral of a base metal, to the silver to be refined when the whole object of cupellation was to remove base metals such as copper. None of the cupellation debris from Agucha, for example, contains more than small amounts of copper.

Kangle discusses these and the suggestions of other commentators who believed that *tuttha* should be translated as the basic material or earth from which the crucible or cupel was made, earth and bones, earth and lead etc, or that *tuttha* refers to materials added to the silver with the lead in the process. Kangle points out that in 10 *tuttha* seems to be the name of a locality, but at 49 he translates *tuttha* just as being the crucible or cupel. Falk (1991) has suggested that *tuttha* should be translated as zinc ores, being related to the Sanskrit and Old Persian word *tūtīyā*, which literally means smoke, plausibly derived from the clouds of zinc oxide sublimate produced when zinc ores are smelted in conventional furnaces, and certainly in Islamic times *tūtīyā* was the usual term for zinc oxide (Allan 1979: 39–45; Craddock *et al.* 1998c) (see Chapter 11). *Tūtīyā* was the material mined at Zawar according to the *Āīn-I Akbarī*. Given the association of the argentiferous lead ores with zinc, this suggestion requires careful consideration.

Translating *tuttha* as zinc ore (the first mention at 10), Falk would interpret this as meaning silver extracted from *tuttha*, that is, from zinc ores. If the lead residuum from the processes to produce zinc oxide at Zawar was further treated to recover any silver it may have contained (as described in Chapter 11), then this could indeed be a reference to silver produced from zinc ore. Alternatively it could be taken to mean silver from the Tuttha Mountain, which if zinc oxide was already being produced at Zawar, could in turn be a reference to the contemporary name of the Aravallis as the Zinc Mountains.

The reference at 49 is more detailed and suggests that the impure silver was to be treated four times with *tuttha*, four times with lead and then four times with the 'dried' *tuttha* and three times with bone and twice with cow dung. Heating the impure silver or argentiferous lead with zinc oxide makes little sense, but if the *tuttha* was to be understood as zinc metal then all would become very much clearer and more plausible.

One of the standard methods of the separation of silver from argentiferous lead introduced in Europe in the 19th century was the Parkes process (Percy 1870: 148–60; Schnabel 1905: 641–93; Hofman 1918: 494–591) which is still widely used. For example, HZL employed the process to separate silver from the lead at their Tundoo smelter in Bihar (Mukherjee *et al.* 1991). In this process a small quantity of zinc metal is stirred into the molten argentiferous lead and the silver goes preferentially into the zinc. As lead and zinc are immiscible when the stirring stops the lighter zinc, now containing the silver, floats on the desilvered lead. The zinc can be physically separated either by ladling it out in the molten state, or as at Tundoo, by cooling to between 450 and 500°C, sufficient for the zinc to solidify but the lead to remain liquid. Inevitably while this happened much of the zinc oxidised and the resulting product was a crusty mass containing about 4–5% silver, 15–18% zinc and zinc oxide, and other metal oxides; the balance was lead and lead oxide. The metal was smelted to remove most of the zinc as vapour, but some zinc oxide crusts remained. Finally, the metal was cupelled in the usual manner at about 900°C to recover the silver which after the refining stage was about 99.6% pure silver.

If *tuttha* could be interpreted as zinc metal then the first stage would be the Parkes process, followed by the treatment with lead, resulting in a mass still rich in zinc oxide, the so-called dry *tuttha* which would then be cupelled on a bone cupel to absorb the litharge and other dross. However the investigations at the sites in the Aravallis leave little doubt that metallic zinc was not produced beyond possibly the occasional droplets in the furnace flues at Zawar (see Chapter 11), and it seems inevitable that if *tuttha* was zinciferous then it must have been a mineral, most likely the oxide, which may well have been produced in quantity at Zawar. Mixing zinc oxide with argentiferous lead would produce little transference unless the conditions became sufficiently hot and reducing enough to transiently reduce some of the zinc oxide to zinc metal in the lead long enough for it to dissolve the silver before reoxidising but retaining the silver. Chemically this can be done, but the material evidence from Agucha can be fitted into a standard simple cupellation process, described below, without having to invoke a much more complicated ancient version of a putative Parkes process.

Ore smelting in the Aravalli mines

Mauryan smelting

The copious heaps of beneficiation debris at both Dariba and Agucha indicate that the argentiferous galena ore was crushed and sorted in centralised beneficiation plants. There would have been no need to roast the ore prior to smelting as evidenced by the early slags from Dariba which contain many sulphide inclusions (Plate 7.13), suggesting that the ores had not been roasted before smelting. The process would have been two-stage: the smelting of the ore to produce argentiferous lead, and the separation of the silver from the lead by selective oxidisation in the cupellation process. Remains of the latter process were found at Dariba and Agucha, although not at Zawar.

Metal smelting in hearths in the past is well attested (Chapter 9), and there is strong evidence that they were the usual smelting units employed at Dariba in antiquity where the early smelting refractories are continuously curved on both faces, suggesting that they are either from a smelting hearth set in the ground or a freestanding bowl furnace (see Chapter 6). One of these, 27298, is a rim sherd on which it is clear that the glaze is running away from the rim. Thus during use the rim was above the reaction, indicating that the ceramic was a crucible or hearth lining rather than a bowl furnace where it would have been inverted. The absence of holes where tuyeres could have penetrated the walls on any of these fragments, such as are found with the later freestanding furnaces, also suggests hearths are more likely than bowl furnaces. The hearths were typically just small clay-lined depressions in the ground, approximately hemispherical, and generally between 20 and 50 cm in diameter.

As already reported in Chapter 7, the silver/lead ores mined at both Dariba and Agucha were concentrated in graphite schists with a complex alumina-rich mineralisation therefore the resulting slags at both mines had high aluminium contents, especially at Agucha; those from Dariba also had high barium contents. Their elevated aluminium content gave the slags a high melting point (see Chapter 7, Fig. 7.6), causing them to be very viscous at the estimated smelting temperatures of between 1000 and 1300°C. Eissler, in his authorative work on lead smelting (1891: 36), stated that 'if lime and alumina are the prevalent bases, the heat required for their formation in their (the slags) formation is much higher. ... Such slags are generally pasty, run short, and form incoherent lumps. An earthy singulo-silicate is the least desirable slag for a lead smelter.' The smelters at Dariba and Agucha would doubtless have agreed!

There are other factors contributing to their inhomogeneity. Quite recognisable fragments of ore and quartz are present in the slags which, from their condition, suggests they had not been exposed to high temperatures for long periods (see Chapter 7, Plates 7.14 and 7.15).[2] This observation is supported by the associated slagged furnace or hearth fragments that only seem to have been exposed to high temperatures for rather short durations, which is the more likely explanation for the inhomogeneity.

The slags at Dariba and Agucha would have been removed periodically from the hearth, perhaps by inserting an iron rod and pulling them out as their postulated viscosity was too great for them to have been ladled out, as was practised in Japan. The molten argentiferous lead had to drain through the slag so the postulated viscosity and short duration should have resulted not only in the observed inhomogeneity but also the retention of large quantities of metal trapped in the solidifying slag. In practice the slags from both mines have only a very low metal content, represented by numbers of tiny blebs distributed throughout the slag and concentrated in the antimonide ore fragments. Therefore, despite the conditions, the smelters achieved an excellent metal/slag separation. It is possible that the semi-molten pasty slags were mechanically worked with iron tools as was the practice at some 19th-century hearth smelting processes in Europe, although the resulting slag compositions are different (see Chapter 9). Whatever method was used, the separation achieved in antiquity is impressive.

Post-medieval smelting

Both documentary (see Chapter 3) and archaeological evidence suggest that argentiferous lead was being smelted to produce silver in the post-medieval period at Zawar. The remains at Site 5 (see Chapters 6 and 7) confirmed that the smelting process was very different from that carried out at Dariba or Agucha. The beneficiated but unroasted galena was smelted directly in small bellows-blown shaft furnaces. The gangue material was mainly silica and the ore was smelted with the addition of an iron oxide flux. In total contrast to the earlier mines, this produced a free-flowing fayalite slag which must have been much easier to handle than the aluminium- and calcium-rich earlier slags (Plate 7.13). Spills of metal together with fragments of unreduced ore in the slags suggest that the silver of the content lead was of the order of 1000 ppm silver. This ore probably came from the mines at Baroi Magna, which are believed to be of post-medieval date. The limited surveys and excavations at Zawar failed to locate any evidence of the second, cupellation, stage of the operation, so well represented at Agucha.

Scorification and cupellation

The next stage was to remove the lead by selective oxidisation: cupellation. This has a very long history in the Middle East dating back to the fourth millennium BC (Hess *et al.* 1999; Pernicka *et al.* 1998). In medieval and post-medieval Europe, as a preliminary process, the products of the smelting, metal and any unchanged ore could be heated in air on a ceramic tray or tile, in a process known variously as scorification, scourification or raffination (Sisco and Smith 1951: 37–44; Hoover and Hoover 1912: 228–30; Percy 1870: 209–11, 1880: 242–3; Potts 1987: 489). Scorification caused much of the lead to oxidise, which in turn reacted with the silica in the ceramics to form lead silicate, thereby concentrating the silver in the remaining lead which could then be treated by full cupellation. Many early silver smelting sites of the first millennium in Spain produced large numbers of heavily glazed potsherds that are usually interpreted as forming part of scorification

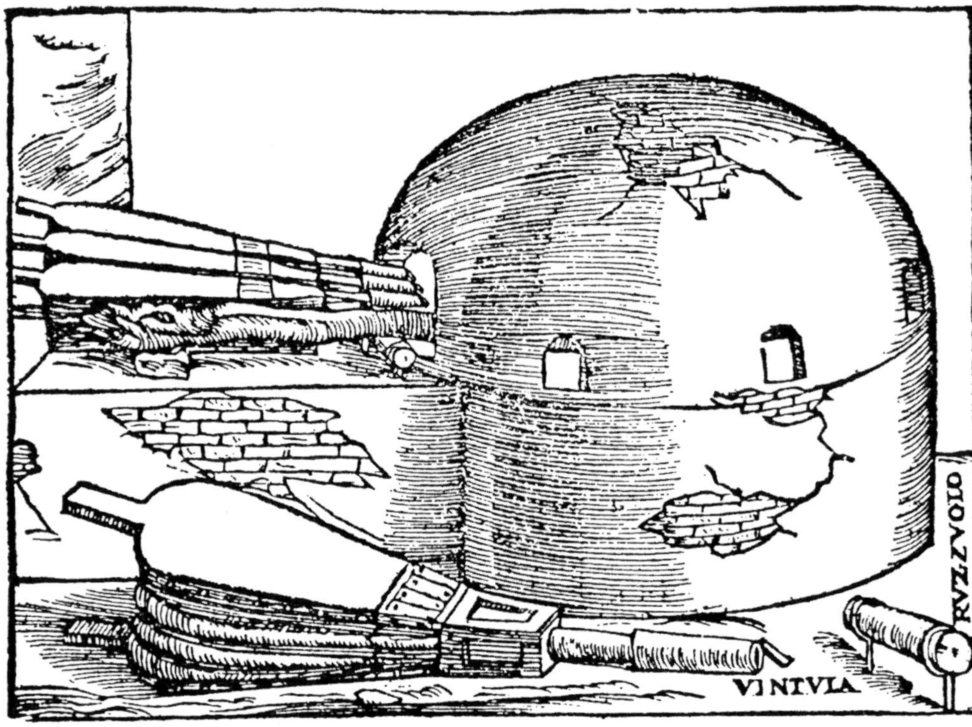

Fig. 10.1 Sixteenth-century Italian cupellation hearth with brick dome. The double manually operated hinged concertina bellows have the iron tuyere attached with a hinged cap at the end to prevent suck-back of burning gases from the hearth. Note the small ceramic roll support depicted to the far right and in place beneath the tuyere at the hearth (from Biringuccio's *Pirotechnia*, Smith and Gnudi 1942: 166).

units (Hunt Ortiz 2003: 366). It is possible that some of the multi-stage processes referred to in the *Arthaśāstra* included a scorification stage.[3]

In the cupellation process the lead was heated typically to over 1000°C while exposed to a strong blast of air from bellows that oxidised the lead and any other base metals to their oxides which entered the litharge phase, but the silver, together with any gold present, remained unchanged as a separate phase of molten metal. Traditionally the crucibles in which the operation was performed, cupels, were of bone ash with a clay binder to absorb the litharge (Craddock 1995: 228–9); more recently they were made from pressed magnesia, magnesium oxide.

The molten silver, together with any gold that might be present, would be immiscible in the molten litharge and would tend to float on the surface 'like oil on water' in the apt phrase of the Roman author, Pliny (*Natural History* 33.95) (Rackham 1952: 72) in his description of Roman cupellation practice. Surviving fragments of cupellation litharge quite often still carry the impression of the button of silver where it had been allowed to set in the litharge before removal. The fragment of the small cake of litharge recovered from Agucha did not carry any such impression – either the particular batch of lead being processed was poor in lead, or more likely, the silver was removed periodically from the litharge by ladling in what was very probably a continuous process.[4]

Mauryan cupellation

Evidence for cupellation both for production and for assay are represented at Agucha, as excavated from the pit or ditch fills in the side of the well section (see Chapter 6). The evidence for fire assay are the many small cupels, some with a tuyere still attached and the small tuyere stands, and the evidence for cupellation as a production process are the many heavily lead-glazed or vitrified fragments of curved ceramic furnace wall and the litharge. Scientific examination of thin sections from the refractories indicate that whereas the cupels had been held at temperatures approaching 1000°C, but for only a very short time, the furnace walls had been maintained at temperatures of around 1100°C for many hours, suggesting that two operations are represented here (as discussed in Chapter 7).

For silver production

The droplets of lead from Agucha contain between 800 and 1800 ppm silver, a typical product of smelting a rich argentiferous lead ore. The silver would have been separated by cupellation.

The material from the well section at Agucha included many large fragments of hemispherically curved refractory with very distinctive and quite massive lead silicate glazing on the concave surfaces. These surfaces had clearly been exposed to lead oxide fumes at high temperatures under oxidising conditions for a long duration. This was confirmed by the scientific examination and refiring of test pieces (see Chapter 7) which showed incipient vitrification throughout the thickness of the ceramic body, suggestive of being held at temperatures around 1100°C for what probably amounted to tens of hours. The ceramic is different from that used for the tuyeres and cupels etc., being much more coarse and containing many very poorly sorted inclusions. Many of these are particles of quartz, thus raising the overall silica content. This is probably no more than an unintentional consequence of the rather coarse sand temper used, resulting in more silica being available to form lead silicate with the lead oxide fumes. Also it is likely that the

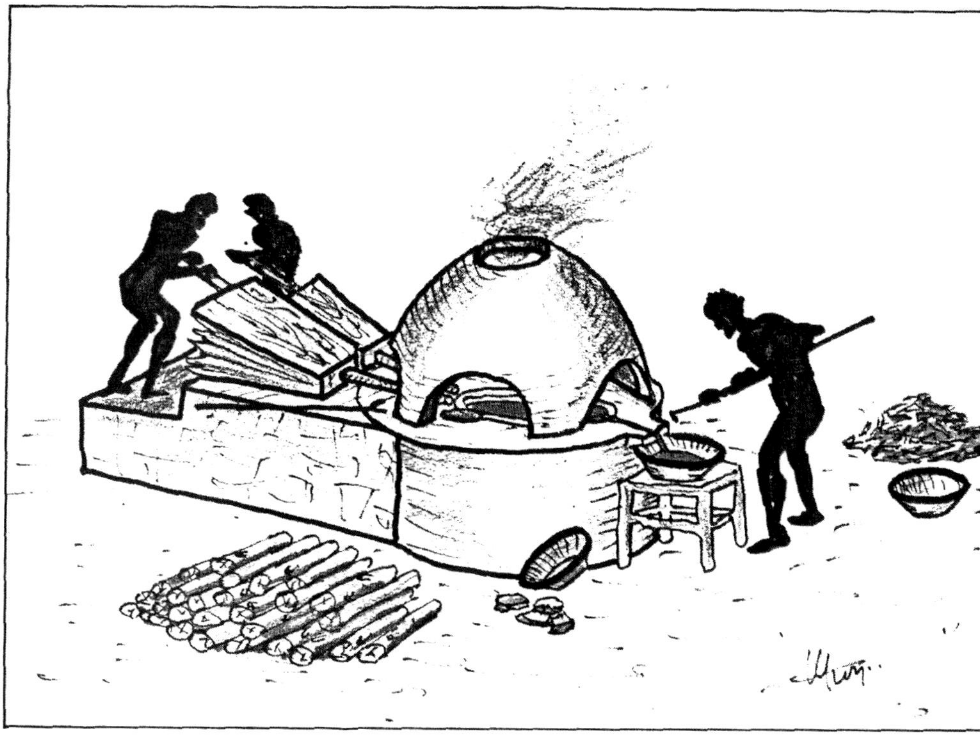

Fig. 10.2 Reconstruction of how the cupellation hearths may have appeared and been operated at Laurion, near Athens in Greece, in the 4th and 5th centuries BC (from Conophagos 1980: 313, fig.12.1, based on Renaissance cupellation hearths from Europe) (cf. Fig. 10.1).

inner surfaces of the wall were dimpled thereby encouraging the forming lead silicate to drip back down into the mass of material being cupelled rather than run down the walls to the sides. The litharge in the hearth below would be largely transformed into a much more viscous lead silicate-oxide mix (as suggested by the composition of the excavated fragment) from which the more mobile silver would separate easily. The well-formed drips of lead silicate enabled the orientation of the furnace fragments to be determined with some precision (see Chapter 6, Fig. 6.75). From this it is clear that some pieces were almost horizontal over the source of heat whereas other pieces were closer to vertical. The pieces appear to have come from an almost hemispherical bowl furnace. Some of the pieces have a flat lower edge suggesting that they formed the base of the furnace where it sat on the ground, but others have flat edges that were vertical. This suggests that the hemisphere was not complete but had segments removed to allow the tuyeres access and the fumes to escape – indeed they may have only been half hemispheres (as depicted in Chapter 6, Fig. 6.70). The dimensions of the furnaces, as calculated from the curves of the fragments, indicate that they had an external diameter of about 40–50 cm, an internal diameter of around 30–40 cm and a rather flattened top. The thickness of the ceramic varies from about 2–3 cm at the top to 5 cm at the base. Conophagos attempted to reconstruct the ancient Greek cupellation installations at Laurion based on the 16th-century illustrations of much larger cupellation installations in central Europe (Fig. 10.1), as he had few, if any, recognisable fragments of original furnace wall with which to work (Figs 10.2 and 10.3). His reconstruction differs from that postulated at Agucha in several ways. In particular, no original curved edges were found on the Agucha pieces suggesting a different arrangement for admitting the tuyeres, and more importantly, implying that there was

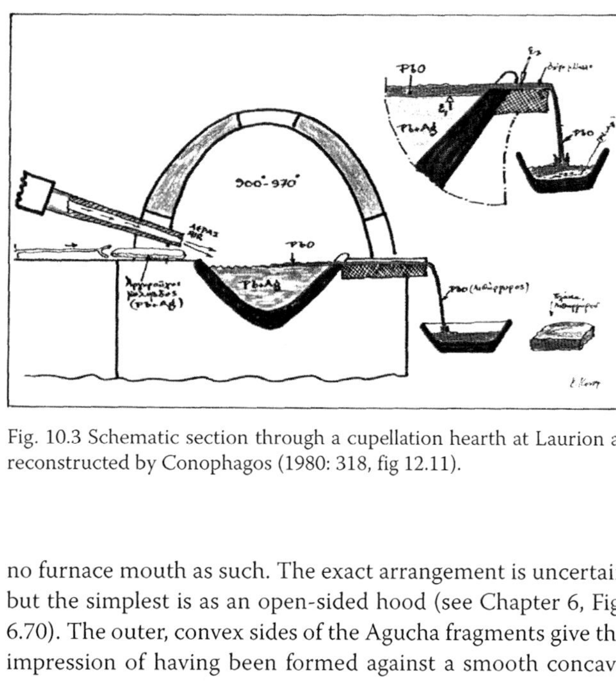

Fig. 10.3 Schematic section through a cupellation hearth at Laurion as reconstructed by Conophagos (1980: 318, fig 12.11).

no furnace mouth as such. The exact arrangement is uncertain but the simplest is as an open-sided hood (see Chapter 6, Fig. 6.70). The outer, convex sides of the Agucha fragments give the impression of having been formed against a smooth concave surface – the ceramic hood may have been moulded against a wooden former that was then removed leaving the hood free-standing. Alternatively, the refractory could have been a lining on the inside of a freestanding structure of brick or stone. However, the smooth surfaces have no impressions of bricks or stones as were clear on the contemporary furnace structures at Zawar, Site 14, for example. It has been shown that the refractory was extremely hot throughout its thickness – if the very hot outer face had been against a stone or brick surface some interaction could have been expected, but the ceramic faces were clean and no fragments of brick or of burnt stone were found. Conversely, although there is no evidence for an outer structure, the very fact that the temperature was so high and with so little evidence of a gradient suggests that the outer surface was not exposed; the heat loss by radiation would have been enormous. Based on later practice in India it seems likely that much of the furnace would have been buried under the local sandy soil to conserve heat.

It is likely that at Agucha the ceramic hoods were built on a flat surface over a shallow hollow in the ground that would probably have been lined with clay and ashes as described above and served as the cupel. As already noted, the litharge piece and the glazing on the cupels contained no more than traces of phosphate showing that bone or bone ash linings were not used at this time. The segment of litharge found with the ceramic hood fragments came from a cake that must have originally been about 6 cm in diameter with a depth of about 3 cm at the centre, giving some indication of the minimum dimensions of the cupellation hollow and according well with the dimensions suggested by the ceramic hood fragments which would have surrounded it. Clearly the process operated for many hours and periodically the silver must have been ladled or run from the furnace. Similarly, the viscous litharge must have been periodically scooped or run from the hearth into an adjacent hollow. Both the post-medieval European illustrations and Conophagos' Laurion reconstruction based on them show the hearth mounted on a plinth, thereby making the tapping a little easier. However there is no evidence for such plinths in antiquity. The cupellation operations at Sardis, for example, were all performed in small hollows in the ground (Ramage and Craddock 2000: figs 4.17–23). One tuyere support was of square section (see Chapter 6, Fig. 6.74) that would have sat upon a flat surface adjacent to the furnace (see Chapter 9), suggesting that the cupellation hollows at Agucha were set in the ground.

Some estimate of the efficiency of the removal of silver from the lead can be judged by the composition of the litharge. The litharge cake from Agucha contains about 70% lead and 120 ppm silver, that is about 170 ppm in the lead, which is similar to the 200 ppm found in Roman litharge from Rio Tinto (Palmer 1927).

Avadhanulu (1979) and Subbaiah (1984) analysed lead coins and other artefacts of the Satavahana period (approximately 2000 years ago) from Andhra Pradesh in central India. They found that the silver content varied fairly consistently between about 30 and 170 ppm and the zinc content between 20 and 50 ppm, suggesting that the lead had been cupelled to remove the silver, an indication of the efficiency of the process.

For assay

The small distinctive ceramic cupels from Agucha are sub-conical with a capacity of between about 20–30 ml. Their lips are drawn out on one side to accommodate the tuyere, which often became attached during the process (see Chapter 6, Plates 6.68 and 6.72, Fig. 6.72). They have far too small a capacity for industrial use but would be an ideal size for carrying out assays. The cupels could contain about 20 ml or about 230 g argentiferous lead. If the figures of around 0.1–0.2% of silver are typical and the cupellation process was 100% efficient, then the amount of silver separated would have been in the region of 0.35 g. As the cupels are rather deep there could have been a problem if most of the lead formed liquid litharge through which the silver sank. However, the lead glaze extends all over the interior surfaces of the cupels suggesting that the lead was almost completely vaporised or absorbed as lead silicates and little or no liquid litharge formed. The silver would have coalesced as a small bead at the base of the cupel which should have been easy to remove and weigh with some accuracy using the balances then available. Thus an assessment of the quality of the ore could have been made.

The cupel bases are drawn down to two small points upon which it rested, and this would have allowed the fire in which they would have sat to penetrate beneath. The limits of the lead glazing on the outside suggest that the cupels sat in the fire, presumably of charcoal, with only their upper half exposed to the air and with the tuyeres at an angle of about 30° to the horizontal, as also implied by the angle of the channel in the tops of the tuyere supports. This would have increased the surface area of molten argentiferous lead exposed to the blast. The estimated firing temperatures of below about 950°C suggest that the cupels had not been exposed to high temperature for long enough for significant vitrification to have commenced, but they must have been held at about 1000°C at least for a short time in order for the silver to become molten (MP 961°C) and coalesce. The cupel must have been placed in the hearth and the tuyere inserted and held in place on a support of clay. Their very wide range of size and shape suggest that the supports were just lumps of unfired clay moulded to suit, which then became fired during the operation and so must have been situated immediately behind the cupel as shown in the reconstruction (see Chapter 6, Fig. 6.69), although not close enough to become lead glazed.

All silver mining operations must have regularly tested the silver content of each batch of smelted lead by cupellation to ascertain if it was worth extracting the silver and if so, then to estimate the likely yield. However, the material excavated at Agucha seems to be the only direct evidence for a mine laboratory from antiquity to have been recognised so far. In Classical antiquity it is estimated that the silver could be recovered economically from silver ores down to about 100 ppm (Craddock *et al.* 1985b; Pernicka 1987).

Later Indian descriptions of cupellation for purifying silver metal

Medieval descriptions of cupellation can be found in some of the Indian alchemic and iatrochemical treatises. The treatise known as the *Rasaratnakara*, ascribed to Nagarjuna, but probably compiled in its present form in the 6th and 7th centuries AD, notes 'Silver alloyed with lead and fused with ashes becomes purified' (Rây 1956: 130). The *Rasahridaya* compiled by Bhiksu Govinda in the 11th century AD, in the Himalayas, in *patala* 11, describes cupels made from the ashes of goat bones (Rây 1956: 148). The purification of silver by cupellation is described in Book 5 of the *Rasaratna Samuchchaya* (Rây 1956: 181), an iatrochemical treatise, which was compiled in the medieval period as follows:

> Silver melted with lead and borax undergoes purification ... Arrange on an earthen dish a mixture of lime and ashes in circular row and place in it silver with its equal weight of lead. Now roast it over fire until the lead is

consumed. Silver thus purified is to be used for medicinal purposes.

The most detailed early description of cupellation is given in the *Ā-īn-I Akbarī* (Phillott 1927, I: 23–4), compiled in the late 16th century. Under the heading of 'The method of refining silver' it states:

They dig a hole, and having sprinkled into it a small quantity of wild (this probably just means untethered) cow dung, they fill it with the ashes of *mughīlān* wood (babul, *Acacia farnesiana*); then they moisten it, and work it up into the shape of a dish; into this they put the adulterated silver, together with a proportionate quantity of lead. First, they put a fourth part of the lead on top of the silver, and having surrounded the whole thing with coals, blow the fire with a pair of bellows, till the metals are melted, which operation is generally repeated four times. The proofs of the metal being pure are a lightning-like brightness, and its beginning to harden at the sides. As soon as it is hardened in the middle, they sprinkle it with water, when flames resembling in shape the horns of wild goats, issue from it. It then forms itself into a disc and is perfectly refined.

O'Shaughnessy (1847: 559) introduced European cupellation practice at the Calcutta mint, but first described the existing traditional procedures of 'refinage':

The object of refinage for the mint is therefore usually to remove the lead, tin, zinc or sulphur and to leave the silver and copper, or occasionally, when pure or rich silver is not available, to bring up inferior alloys to standard or even superior fineness.

In the previous section O'Shaughnessy had stated that silver containing lead, tin, zinc or sulphur was liable to be brittle, whereas if it only contained copper it would still be malleable. It is difficult to understand how cupellation could remove everything except the copper. Possibly just melting under oxidising conditions in a scorification process could expel the zinc and sulphur, and cause much of the tin and lead to oxidise while leaving the copper largely unaffected. O'Shaughnessy continued:

The process followed by the native refiners in the bazaar is that of *cupellation*, and is performed by them with great success and economy. They use for the cupel a mixture of one part by weight of recently burned lime, sifted but unslaked, and two parts by weight of chaff ashes. With this they make a basin like mass, usually 18 inches (45 cm) in diameter below and 4 to 6 inches (10 to 15 cm) deep. This they moisten well with water and beat with the hands into a firm consistence. Pieces of brick are placed around the sloping sides to give support, and two pairs of bellows are arranged so that by their alternate use a constant blast of air is kept up during the process.

While still wet the basin is charged with charcoal and an active fire kindled, the silver is then introduced and the lead added till all is melted and red hot. Two large logs of firewood are then placed over the charcoal so as to form a dome to the heap, and at the interstices torch-like pieces of wood are continually introduced, so that a powerful flame is reverberated from the blazing dome above.

By this manipulation the lead is oxidized, and the oxyde of lead [litharge] formed is absorbed with oxides of copper, and other base metals usually present, by the porous mixture of lime and ashes. None of the litharge is removed by skimming. In refining 2500 tola wt. (2.91 kg) [The tola is 180 Troy grains (= 11.65 gm). 32 tolas = one Troy pound] (= 373 gm) from 16[5] 'worse' they use 1200 tolas (1.4 kg) of lead, and the operation is completed in less than 3 hours, yielding a cake of silver 16 to 17 dwt (67–70 gm) 'better' than standard (standard was 220 dwt of silver per 240 dwt of alloy). The bazaar refiners contract to return all the silver according to official assay; and finding all materials, they receive for their labour 8 annas, or ½ per 100 on the value of the metal. The cake of litharge when cold is ground and sifted and yields granules of silver. The sifted powder is made into a paste with cow-dung, and the lead recovered in a furnace of particularly ingenious and effective construction.

The refractory of the cupels was apparently just burnt lime and chaff. Percy (1880: 239) mentions the use of marl, a mixture of lime and clay, or burnt lime with clay, both with wood ash as a substitute for bone ash, but warns that if too much lime is present this makes the cupel very loose with the attendant danger of the loss of silver. In the cupellation process the molten silver normally coalesces into a single button of silver on or in the molten, but often viscous, litharge. The segment of a cake of litharge found at Agucha, had only traces of silver suggesting that the separation there had been almost complete.[6]

O'Shaughnessy continued with a description of the recovery of silver from the litharge and sweepings, one of the few accounts of this process.[7] Another description of how metal was to be recovered afterwards from the refractory debris was given by Viccajee (1908: 38–9)[8] in his description of traditional cupellation processes.

Later Indian descriptions of cupellation for assay[9]

As well as being used as the main production technique of separating the silver from lead, cupellation has also traditionally been the most reliable method of determining the purity of silver artefacts, or the silver content of base metals and of potential silver ores, in the technique known as fire assay. There are several early Indian descriptions of the assay of silver artefacts by cupellation, all associated with mints, and there are some good later European descriptions of continuing traditional assay by cupellation at mints in India. The most detailed early description is that given in the *Ā-īn-I Akbarī* (Phillott 1927, I: 24):

In the case of silver, he takes one tola (approx. 12 g) with a like quantity of lead, which he puts into a bone crucible, and keeps it on the fire till the lead is all burnt. Having then sprinkled the silver with water, he hammers it till it has lost all smell of the lead; and having melted it in a new crucible, he weighs it and if it has lost in weight three *birinj* [rice grains] it is sufficiently pure; otherwise he melts it again, till it comes to that degree.

Note that to test the silver a bone crucible (presumably of bone ash) was specified, rather than a dish formed out of ashes which was specified for the main production process.

Another detailed later description of the Indian processes was made by F.K. Viccajee, Assistant Superintendent at the mint of the Nizam of Hyderabad, who trained in metallurgy at the Royal School of Mines in London. His book describes in detail the traditional Indian procedures still in use at what was the last 'native' mint to remain in operation, the others having been taken over by the British. It includes a description of cupellation specifically for assay (1908: 33–40):

(a) Silver Assay

This operation is termed by the natives *'Chashni'* which literally translated means in Persian *'taste or trial'.* In the presence of a responsible officer of the Mint, called the Sadar [Head] Daroga, the blower accurately weighs on a pair of hand-made fine scales one tola [172.5 grains (11.2 g)], of the silver to be tested. He then prepares his fire for cupellation, by filling wood ashes almost to an inch of its top, in an earthen ware tumbler shaped vessel, about 8 to 10 inches (20 to 25 cm) in diameter at the top 4 to 6 inches (10 to 15 cm) at the bottom, and about 9 inches (23 cm) high. The surface layer of the ashes is smoothed and levelled by slight pressure of the fingers, and then the cupel is placed in the centre of the ash-bed, and to prevent the edges of the cupel being broken by the weight of charcoal around it, it is buried in ashes; the surface is sloped towards the inner side of the furnace.

The cupels are made of calcined bones reduced to a coarse powder, mixed with cow-dung, shaped like little saucers about 1" or 1¼" (3 to 4 cm) in diameter. The weight of each of these cupels is from 300 to 350 grains (20 to 23 g). The cow-dung is used as a binding material with just sufficient water to allow the dough to be modelled by hand into shallow saucers. The cupels are air then (air?) dried.

In this cupel the weighed metal is placed with 1 to 1½ tolas (11 to 17 g) of pig-lead, and large pieces of charcoal are placed around it, so as to completely cover the cupel on all sides, leaving just sufficient opening in front of the operator to allow his blow-pipe [a broken piece of gun-barrel from 18 to 24 inches long (45 to 60 cm), of a ½ to ¾ inch (1.5 to 2 cm) bore or a gas pipe of the same dimension] to be inserted, so that the flame can be made to play directly on the cupel. Even this opening is closed round the blow-pipe by smaller pieces of charcoal, so as to prevent any draft of air playing on the cupel. Now a few pieces of ignited charcoal are placed about the pile and the operator commences to blow. No further precautions are taken to heat the cupels before charging, so as to prevent the spurting of the molten mass by the escape of steam, from insufficiently dried cupels, or escape of CO_2 from the unburnt carbon of the bone-ash used, by contact with PbO.

In one charge of the furnace, an expert blower can blow two assays at a time. As the charcoal gets burnt the operator piles fresh charcoal on the top. Should ashes fall into the molten mass on the cupel it is blown off by a stronger blow, and pieces of charcoal too heavy to blow off are removed carefully by a pair of tongs.

This process of cupellation continues for about 20 to 30 minutes. When the button 'brightens up', the operator immediately extinguishes the fire by sprinkling water over his refined button allowing very little time for 'spurting'. An expert can manage this very well, but I have seen many a time that 'spurting' does take place.

'Spurting' or 'spitting' occurs when the molten silver has dissolved oxygen which is then quite violently expelled with a potential loss of droplets of metal as the metal cools and the oxygen comes out of solution (Percy 1880: 14–16). One method of preventing this was to sprinkle the molten silver with water (Percy 1870: 211). Viccajee continued:

When perfectly cooled by water, the cupel with its adhering ashes is taken out of the furnace, and the bottom is removed by breaking the cupel. It is then scratched with the nails or a sharp pin of iron to remove the adhering bone ash and washed. The appearance of the button on the bottom indicates the adherence of the oxides of the base metals. These are removed by hammering on an anvil and washing, and again hammering wet. Thus a greater part of the oxides is removed, but a part of it in fine powder is imbeded (*sic*) in the button which under a magnifier distinctly shows itself, as a black or brownish stain on the dead shining surface.

When the button is taken off the cupel, it often presents a globular or even more sharply rounded on its under (side?) than on its upper surface, showing the presence of lead. The proportion of lead used for cupellation being very small (i.e. compared to contemporary European practice), the button is always impure and from several analyses I have made of such buttons, I found that the silver which was taken as 1000, fine is invariably between 997 and 998 per mille. This button is then weighed against the weight taken for assay and the excess or the deficiency per 9 mashas (1 masha =15 grains troy = 0.97 g) of pure silver is calculated in Ratties (1 ratti = 1.67 grains troy = 0.1 g) and its fractional parts. The whole of the metal tendered for minting is then calculated for alligation (alloying). If the silver is found to be more than 9 mashas in the tola (i.e. 3 parts in 4) of alloy taken for assay, copper is added, and if less, silver, to bring the metal to the standard fineness.[10]

From the above it will be observed that no allowances are made for the presence of silver in the lead used, nor for the cupellation losses. No 'Checks' are used. Invariably the cupelled button is cut in two and its fracture and colour noted. If the button shows larger grains and a white colour, the assay is said to be 'incomplete', and a fresh assay is taken, if it is of a fine grain and bluish gray colour it is said to be 'properly finished'.

The whole of the cupellation work is done by the operator in a sitting position.

Note that in these cupellation processes bone cupels only seem to be used for assay, not for primary production or even for the assay of ores (see Chapter 7). This seems to have been general practice in antiquity (Craddock 1995: 228–9). Viccajee (1908: 45–50) also described in detail the procedures adopted for the melting and casting the silver into ingots.[11]

Notes

1. Principal sources. General: Bachmann 1991, 1993c; Pernicka 1987; Craddock 1995: 221–31; Hoover and Hoover 1912: 464–83; Conophagos 1980, 1982, 1989; Percy 1870: 121–212; Schnabel 1905: 380–918; Eissler 1891; Hofman 1918. Traditional cupellation processes: Percy 1870: 178–208, 1880; Schnabel 1905: 693–731), and see also Smith and Gnudi 1942: 136–41, 161–9 (Biringuccio); Hoover and Hoover 1912: 464–83 (Agricola); Sisco and Smith 1951: 57–64, 71–8 (Ercker) and Gale and Stos-Gale 1981.
2. It is possible that the quartz was only added at a late stage in the smelt as exemplified by the free silica slags from the Phoenician and Tartessian silver smelting sites in southern Iberia (Craddock *et al.* 2013). Also some 19th-century lead smelting processes added lime at a late stage in the smelt (Percy 1870: 232–5).
3. The glazed sherds from Dariba, Sites 7 and 9, were initially suspected as being from a scorification process, but the presence of appreciable quantities of zinc show that they must have been involved in the primary smelting of the ore. Ball (1881: 234) stated that at Kyouktat, in Upper Burma, the silver was separated from the lead 'without the aid of cupellation', using a reverbatory furnace in a process that sounds like scorification. However, the metallurgical operations carried out there seem to have been closer to Chinese than to Indian practice.
4. Note, the silver, being denser, should not have floated on the litharge. However, in the ancient processes using ceramic cupels some of the molten litharge would have attacked the clay lining of the cupellation hearth to form lead silicates, so the material would actually have been a stiff, pasty mass of lead oxides and lead silicates on which the molten silver sat (Bayley 1992). The Agucha 'litharge' segment, for example, actually contained approximately 50% lead silicate.
5. Almost certainly dwt should have followed 16; that is, the silver was 16 pennyweights per pound or 67 g per kg below standard.
6. In contrast, for example, one of the six litharge fragments analysed from the silver recovery unit at the Lydian gold refinery at Sardis had 13% of silver (Ramage and Craddock 2000: 160, table 6.2), showing that the separation was not always complete, presumably because the litharge was too viscous for the silver to coalesce.
7. 'A barrel-shaped clay cylinder is made, open at both ends, nine inches (22 cm) diameter below by 12 to 15 (30 to 37 cm) above, and usually 24 inches (60 cm) high. A bellows pipe of refractory clay 3 inches (7 cm) in diameter enters at the side about 4 inches (10 cm) from the top, and is led down the cylinder so that the nozzle of the pipe is within six inches (15 cm) of the bottom. The cylinder stands over a cup-shaped hollow made in the ground and sifted over with a little wood ashes. To use this furnace it is first half filled with charcoal and the fire kindled. The mixture of litharge and cow-dung is then introduced in balls the size of an orange, with layers of charcoal and fire urged. The litharge is quickly reduced to the metallic state and the lead containing any silver present in solution, collects in the cup-shaped hollow – 100 lbs (37 kg) of litharge can be thus worked off in about 4 hours. This process is applied with remarkable success to the treatment of sweepings and other rubbish containing not more than 1 per 100 or 2 per 100 of silver, but in this case a small and variable quantity of borax is added to the mass of litharge, sweepings and cow-dung. ... The basin below the cylinder is open at one side, but during the process is kept partially closed by a heap of charcoal and a brick. This being removed occasionally, the surface of the melted lead is raked free from earthy slags by an iron rod, and the firing continued till the balls are all consumed. The cylinder is then removed, water thrown on the lead – and this containing silver, is used for the next cupellation refinage. ... The skill exhibited by the native refiners in conducting these processes is beyond all praise, and for the scale on which they have to operate, it would, I conceive, be scarcely practicable to effect any improvement on their system. But it has serious inconveniences when we attempt to follow it in large operations. Each operation is limited to about 2500 tola wt. (2.9 kg) and this may be repeated, so as to give 5000 (tolas) as the day's work of six men. The heat is almost intolerable, the lead fumes most deleterious. These objections might be obviated by the erection of suitable screens and hoods, but the refinage never proceeds so successfully as when the native operator is left to his own fashion. Superintendence and the prevention of pilfering become exceedingly difficult also when a large quantity of bullion has to be operated on, from the great number of people employed, the large space occupied by each gang, and the dense smoke and fumes which fill the refinery' (O'Shaughnessy 1847: 560). O'Shaughnessy's comment on the viability of a small-scale operation is interesting.
8. All the pieces of broken cupels are collected, and, when in sufficiently large quantities, silver and copper are recovered from them by the following processes. ... This is a process of liquation on the primitive style. ... The bone ash, with its adhering litharge and ash, is ground on a slab of stone with a cylindrical stone mixed with fresh litharge in the following proportions, made into balls, about the size of an egg, air dried and liquated.

   Cupel with its litharge called *"Kharal,"* -- 2 parts,
   Fresh Litharge called *"Ada,"* --------------- 1 part,
   Cow-dung and clay --------------sufficient to bind.

   These balls are heaped in a conical form in an earthenware bowl shaped-furnace with a spout to let the liquated argentiferous lead run out, charcoal is placed over them and blown with a pair of belows (*sic*). As the scum forms on the surface it is cleared to one side with a piece of bamboo stick. This scum, with the slag contains the copper. ... When it is found that no more lead liquates, the scum with the slag is removed, cooled, ground, washed clear of wood ashes, melted with borax, and cast into finger bars. ... The cost of liquating 50 lbs. (18.7 kg) of cupel dust is.

   Charcoal 20 lb (7.5 kg) 6 annas or 5*d* (2p)
   Litharge 25 lbs (9.8 kg) Rs. 4 or 3*s* (15p)

   A man employed at a cost of 8 annas [6*d*] (2p) a day can work 50 lbs (18.2 kg). ... When a supply of litharge is not at hand, lead is melted and granulated in wood ashes and well ground while hot with a wooden muller and used instead. The lead and ashes are taken in equal quantities.

   It is possible that the refractory debris of used cupels, tuyeres, furnace fragments and litharge, found in the pit or ditch excavated at Agucha, was waiting reprocessing in the manner described above, although their overall silver content was very low.
9. Percy 1880: 279–82; Smith 1923–4, 1947; Potts 1987: 486–90.
10. That is 3 parts silver to 1 part copper, the same as for the Mauryan punch-marked coins (Rao *et al.* 1990; Agrawal *et al.* 1990).
11. **The process of melting called 'Gudakht'**

   This operation is termed the *'Gudakht'* which means in Persian "melting". Before describing this process it would be better to know how the melting pots, moulds, furnaces etc., are made.

   (a) *The melting pots*
   A special kind of red earth [clay with ferric oxide] 20 parts
   Charcoal 2½ parts

> Rags or Bengal gunny bags 4 parts
> [a kind of coarse hemp cloth]
> Water sufficient to bring the whole stuff to the contingency (consistency?) of a plastic paste for moulding by hand.
> All these are pounded to such an extent that the fibres of the canvas are no more visible, and the whole forms a plastic mass, capable of being moulded by hand.

This use of coarse cloth jute fibres as a filler is interesting: cotton was a more usual addition to crucibles and moulds (Craddock 2015).

> Oval crucibles are moulded by hand from this mixture to melt from ½ a tola [86.25 g] (that is, grains, 5.6 gms) to 5 to 6 thousand tolas [125 to 150 lbs] (48 to 56 kg) of metal. ... The outward dimensions of one of the largest crucibles are 18" × 12" × 2¾" (45 × 30 × 7.7 cm) the bottom being from two to two and a quarter inches thick (5.1 to 5.7 cm), these crucibles are air dried, and when found to be perfectly dry, a coating of half an inch in thickness (1.2 cm) is made inside and outside with a mixture of 5 parts of the prepared clay, 3 parts fine sifted sand and glazed with saturated borax water and air-dried. ... The upper edge of the crucibles to about 2½ to 3 inches (6 to 7.5 cm), is made from the same mixture as the coating. When the coating and the edge have perfectly dried, they are used for melting. The object of this coating is the same as that of the coating of borax given to Morgan's Plumbago crucibles. (The borax rendered the crucible impervious: Percy 1880: 412.) ... The crucibles used by the native jewellers in India are made in the same way, but those used for melting gold are glazed inside by melting borax in them before use.

Viccajee then described the production of the ingot moulds which were made in the same manner as the crucibles, concluding that:

> When these moulds are perfectly air-dried they are well smeared with castor oil and burnt on a charcoal fire. These moulds are oiled with castor oil when being used for work.

He then continued, describing the melting hearth and its operation:

> *The Furnaces*
> A circular pit about 2 feet (60 cm) in diameter, and of the same depth is dug and built up with red earth bricks to about a foot (30 cm) on the surface of the ground. A slant hole about 3" (7.5 cm) in diameter is left from the surface of the ground to the middle of the furnace for a pair of hand bellows made from raw cow hides; a wall about 2 feet (60 cm) high by 9 inches (23 cm) thick is built on one side to protect the bellows-man from the radiation of heat, and to prevent the bellows being burnt by sparks of fire. ... The crucible is placed on four crucible stands, made of the same material as the crucibles, representing a fruston (frustrum?) of a cone so as to allow the draft to play all around the crucible.
>
> (d) *Fuel and the melting process*
> Wood charcoal is used for melting purposes. In order to expedite the melting a pair of bellows is used on the top of the fire with a long hollow cylindrical nozzle like a test tube, made of moulded clay, having a side hole about ½' (presumably ', i.e feet, is wrong and ", i.e. inches or 1.5 cm, was intended) in diameter, about 2 to 2½ inches away from the closed end. ... When the crucible becomes red hot the charcoal fire from it is scooped off with an iron ladle and the silver placed in it, covered up with fire and fresh charcoal added. The blowing is continued until the metal is melted. For a pot of five to six thousand tolas [120 to 150 lb] (98 to 56 kg) it takes 3 hours for the 1st charge and 2½ hours for the subsequent charges, three charges in each furnace in 8 hours are taken. ... When the mass is in a perfect liquid state it is well stirred with an iron rod, and the alloyed metal ladled out with a small crucible, into the open oiled moulds placed on the floor. This casting continues till the whole or nearly all the metal in the large crucible is finished. When fresh metal is added and melting continued. During a working day of 8 hours with one man for blowing and one man for ladling and casting, and one man to supervise, 18,000 tolas or [about 443.5 lb (165 kg)] of metal can be turned out for coinage at a cost of 1 Rupee or 1*s.* (5p) per day for wages. It takes [one maund] or 80 lbs. (30 kg) of charcoal, and ¼ lb. oil, per 1000 tolas [25 lb (9.2 kg)] of metal cast.... After the requisite amount of work is turned out, the furnaces are cooled down with the crucible in it by throwing water on them, and when found perfectly cool the droppings of the metal and the silver particles from the crucible are recovered next day by pulverizing and washing the pounded stuff in water.

Very similar operations must have taken place at the Aravalli mines.

Chapter 11 The production of brass and zinc

P.T. Craddock

From the combined archaeological and textural evidence, it does seem that brass was already in limited, albeit somewhat erratic, use in the later first millennium BC, and there is some evidence that metallic zinc was also known in the north of India.[1] This mirrors the contemporary situation in the Near and Middle East (Craddock 1998a). From then on the evidence from India is markedly different from elsewhere. In the first millennium AD there are references both to the production of brass by the direct process of reacting zinc minerals with molten copper or by cementation and to the distillation of metallic zinc. The last reference to the cementation process occurs in the 12th century AD, and in the medieval period the descriptions of the zinc distillation process become increasingly detailed and would seem to be the laboratory precursors of the industrial processes that were taking place at Zawar. The descriptions of the production of calcined zinc ore and of zinc metal given below necessarily concentrate on the evidence from Zawar.

One of the enigmas of the investigations at the three mines in the Aravallis is that even though the orebodies are in general quite similar, the history of metal exploitation at Zawar is quite different from the other two mines. It is necessary to appreciate this difference in order to interpret the evidence and reconstruct the processes. Although the orebodies at all three are predominantly made up of the sulphides of iron, zinc and lead, there are significant differences. The ores at Zawar exploited in the early period contained relatively little silver and the minor quantities of lead mineral were macroscopically separable from the zinc sulphide which predominated. After the heady days of the Mauryan Empire during which Agucha and Dariba produced silver and lead on a prodigious scale, Agucha ceased production altogether and Dariba became a minor producer of lead with possibly a little silver, and copper. At Zawar the story is quite different. In the earlier periods, contemporary with the periods of maximum activity at Agucha and Dariba, there was significant mining activity but relatively little evidence on the surface for metal production. That which there is seems quite different from that at Agucha and Dariba so it was not immediately apparent whether the metals sought were primarily silver, lead or zinc. In the medieval and post-medieval periods the mine produced large quantities of metallic zinc together with some lead and silver.

In the quotations given below from the contemporary early scientific literature on brass and zinc production, some of the translations given here from the Sanskrit texts were made by L.K. Gurjar (see also Paliwal *et al.* 1989), and other translations are those given in Tripathi (1978), Dash and Kashyap (1994) and Rây (1956). For the identification of the plant names Gamble (1902) and Tewari (1992) were used.[2] Through the centuries many different terms were employed for the various materials and processes, but the high volatility of zinc leading to special methods of preparing the metal and its alloys means that it is usually possible to identify it unambiguously in the texts. The main problems concern the dating of the original texts: none of the extant documents are more than a few centuries old and are therefore far removed from the supposed dates of their original composition. This creates potential problems of interpreting the original meaning due to possible mistakes and misunderstandings by the copyist together with later interpolations and omissions. In addition, the texts rarely mention contemporary events or well-known personages that might allow them to be dated. In this book the dating given by the translators and editors of the various texts has been followed.

Brass[3]

The early word for brass is usually taken to be *ārakūta* which is found in a number of Vedic texts dating from the late first millennium BC (Chakrabarti and Lahiri 1996: 137–50), and in the *Arthaśāstra* of Kautīlya (2.12.23, 2.17.14, 4.1.35) (Kangle 1965–1972, II) of similar date. In the latter it is regarded as a loan word from the Greek *oreichalkos*, occurring at much the same time as the word for tin, *yrapu*, changed to *kastīra*, derived from the Greek *kassiteros* (Biswas and Biswas 1996, I: 361 and II: 51), suggesting that at that time brass itself may have been a relatively recent introduction from the West introduced by Alexander's armies. The medieval Sanskrit term for brass was *pittala*.

The first brasses in South Asia

As yet little metalwork from the archaeological excavations carried out at the large urban sites of the first millennium BC has been analysed. Kharakwal (2011: 83–115) has brought together examples of early brasses from India, and from these there is growing evidence for copper alloys containing zinc.[4] Brass seems more prevalent in the far north west of the region. Analysis of 39 copper alloy artefacts from the excavations carried out at Taxila, now in northern Pakistan near to Rawalpindi (Ullah and Marshall 1951: 567–9; Prakash and Rawat 1965: 50–51, table 5.3), dating from the latter part of the first millennium BC and into the first centuries AD, revealed a wide range of alloys including high tin bronze, one cupronickel item and four of brass. The brasses had zinc contents of 13.07%, 12.88%, 19.7% and 34.4% with small amounts of tin and lead. The brass with 34.4% zinc came from the neck of a metal vessel dated to around 300 BC and also contained 4.25% tin, 3.08% lead together with 0.26% antimony, 0.4% nickel and 1.77% iron. Some small copper alloy rods and bangles from the site of Senuwar in the Kaimur region of the Ganges Valley dated to the early centuries AD are mainly of copper with a few percent of zinc, but two have 35–36% zinc with only traces of other metals (Singh 2004). The high zinc content of these pieces suggests that the alloy was made by mixing copper and zinc metals, the significance of which is discussed below.[5]

Early descriptions of brass production[6]

Ores

The *Rasarnavam Rastantram*, normally referred to as the *Rasarnava*, lists three kinds of zinc ore before going on to describe how they may be used to make brass by the cementation process (see below). The date of the work is very uncertain: Tripathi (1978: 6) dated it to the second part of the first millennium BC, but this would appear too early: many of the processes involve mercury and would seem to be Tantric in origin, therefore are unlikely to date before the mid-first millennium AD. Rây (1956: 118) dated it to the 12th century AD, which is more likely. The description of brass production is also included in the compilation made by Todarananda in the 16th century (Dash and Kashyap 1994: 328–31).

The three types of zinc ores (*rasak*) are described in the *Rasarnava* (VII 31–4) (Rây 1956: 138) as follows:

- *Mraica rasak*, which was described as the best quality and like yellow-coloured soil. This is almost certainly smithsonite (calamine), zinc carbonate ($ZnCO_3$).
- *Gud (jaggary) rasak*, which was described as being of medium quality and was the colour of treacle. This is probably sphalerite (blende), zinc sulphide (ZnS).
- *Pushan rasak*, described as being the poorest quality and hard like stone. This could be either hemimorphite, the hydroxy silicate ($Zn_4Si_2O_7{\cdot}(OH)_2{\cdot}H_2O$) or willemite ($ZnSiO_4$), both of which form in the upper, oxidised parts of the orebody, although apparently not at Zawar.

This is one of the few descriptions of the main zinc ores; elsewhere the general term *raska* for zinc ore is used without further elaboration, and which most modern translators assume to be calamine (smithsonite), zinc carbonate, whereas in fact smithsonite is rare in the Indian deposits: the usual ore is sphalerite, the sulphide.

Brass production by the cementation process

There is a brief reference to the production of brass by the cementation process in the *Rasaratnakara*. This work is believed to have been compiled by the greatest of the early Indian scientists, Nagaruna, who probably lived at some time in the 2nd–4th century AD. In its present form it probably belongs to the 7th or 8th century AD and also includes a description of the preparation of metallic zinc (see below): 'What a wonder is it that zinc ore ... roasted three times with copper converts the latter to gold?'(3) (Rây 1956: 129). Brassmaking is also described in the *Rasarnava*, but the description is clearly directly derived from the *Rasaratnakara*: 'What a wonder is it that *rasaka* mixed with certain organic matter and roasted three times with copper becomes gold?' (VII 37–8) (Rây 1956: 138).

In the *Rasarnavakalpa*, which is believed to have been compiled between the 10th and the 12th century AD, a more explicit description of the production of brass by the cementation process is given:

> One *māsa* of the powder of blue vitriol (*tuttha*)[7] and an equal weight of copper leaves are to be placed in order on the above substances (killed sulphur and the four alkaline substances).
>
> One *māsa* of *rasaka* (translated as calamine, but roasted sphalerite seems more likely) is to be arranged on it (i.e. the copper leaves). Powdered *kāca* (salt) again arranged all over the substances. The mouth of the crucible is to be sealed (370–71).
>
> The whole substance thus contained in the crucible is to be roasted over a fire of cow dung, until the copper becomes liquid. This copper will, no doubt, assume the form of gold, displaying eight shines (372) (Roy and Subbarayappa 1976).

Although these are very short descriptions of brassmaking, they are nonetheless of great interest. The medieval European metalworker, Theophilus (Hawthorne and Smith 1963: 143) and the Islamic metallurgists, al-Hamdānī and al-Kāshānī (Toll 1968; Allan 1979: 42; Craddock *et al.* 1998c), all describe brassmaking by adding the zinc ore to molten copper which is not a cementation process. The Indian processes described above where zinc minerals are roasted with copper leaves in the presence of other materials are clearly genuine cementation processes in which the copper remains in a solid state throughout while being converted to brass by exposure to zinc vapour. True cementation processes were not described in Europe until the 16th century (Craddock and Eckstein 2003; Martinón-Torres and Rehren 2002), although there are some early descriptions from China (Zhou 2001). These would seem to be the latest references in the Indian literature to the production of brass by treating copper with zinc ore. As metallic zinc was already being prepared as an industrial commodity, with descriptions of laboratory preparation given in both the *Rasaratnakara* and the *Rasanarva* (see below), henceforth brass was presumably made by mixing copper and zinc metals (speltering).

The earliest surviving description of brass as an alloy made by mixing copper and zinc metals occurs in the *Bhāvaprakāśa Pūrvakha* compiled in the 15th or 16th century AD by Bhāvamiśra (Srikantha Murthy 2004; Bhowmik and Jani 1998), where in the first line of couplet 70 it states that brass is a semi-metal made of copper and zinc metals.

In the *Rasaratna Samuchchaya* (Book 8, 24–25), which is likely to have been compiled sometime in the 14th–16th centuries, there is an extraordinary description of the direct production of brass by leading zinc vapour, presumably from the retort in which it had been smelted, through a tube into molten copper (Biswas 1987). Such a process should work in theory, and might be thought to be a very efficient method of collecting and utilising the zinc, most of which was destined for brass production anyway. The problem would be the volatility of the zinc in the molten copper (MP 1061°C): it would be very difficult to maintain the zinc content, and even if one was to drip molten zinc, it would still volatise in the molten copper.[8]

Process evidence[9]

Although the evidence for the deliberate production of brass in antiquity by the interaction of calcined zinc ore and copper metal is strong both in surviving artefacts of brass and in accounts of the process, the physical evidence for the preparation and collection of the calcined ore and production of brass has remained largely elusive. The principal mineral used in the West in Classical antiquity was known as *cadmea*, and it is clear from a variety of sources that *cadmea* must refer to either zinc carbonate ore, smithsonite or the zinc oxide deposited on the walls of furnaces smelting copper or silver/lead ores containing some zinc. Different sources and varieties of *cadmea* are discussed at length in the Classical literature, particularly the accounts of Pliny, Dioscorides and Galen, who were almost contemporaries of the early operations at Zawar. It must be made clear at the outset that the primary concern of these authors was the pharmaceutical uses of *cadmea* – its use in the production of brass is very secondary to the extent that Dioscorides cites brass as a source for making high quality *cadmea*. Pliny describes three basic sources of *cadmea* (*Natural History* 34: 100–103) (Rackham 1952: 200–205): first, the zinc mineral obtained from the ground, which was used to make brass but which was of no use pharmaceutically; secondly, the variety collected during the smelting of copper ores containing some zinc, which was certainly practised on Cyprus as recorded both by Pliny (34:103) and Galen (Walsh 1929); and thirdly, that collected from the furnace walls and chamber roofs where silver ores were being smelted, that is, from argentiferous lead/zinc ores similar to those found in the Aravalli mines. For pharmaceutical uses, *cadmea* produced from copper (either brass or copper ores) was adjudged the best. This is understandable as smithsonite ore, even after roasting, would contain much gangue material as well as lead and some cadmium; and even the *cadmea* resulting from the sublimation of zinc oxide in the silver smelting processes would still contain much lead, as exemplified by the samples from Zawar (see Chapter 7). That from copper would be substantially free of lead.[10]

Here we are concerned with production of brass and for this the *cadmea*, or *tūtiyāh* as it later became known in the Islamic world and as *rasak* in India, would have come from ore deposits. Lead and zinc minerals are almost always associated, and may or may not contain sufficient silver to warrant its recovery. Deposits containing zinc in the form of the carbonate ($ZnCO_3$) smithsonite (but still popularly called calamine), which can be used to make brass with little pre-treatment, are common in western Europe and in China, but are quite rare across the Mediterranean lands, the Middle East and South Asia where the sulphide mineral sphalerite (ZnS) predominates (Coggin Brown and Dey 1955: 163–7; Khrishnaswamy 1988: 454–9). If the primary aim was to recover the lead and any silver that it contained, then the sulphidic ore could be smelted directly under moderately reducing conditions (as at Agucha, Dariba and at Site 5 at Zawar, described in Chapter 10), and the majority of the zinc would have gone into the slag. If, as at Zawar, and also, for example, at medieval Kerman in Iran, the sphalerite contains little lead or that which it does contain can be manually separated before smelting, then it is possible to produce and collect zinc oxide. To do this it is necessary to first roast the ore containing the zinc and any residual lead sulphides to convert them to oxides, and then to smelt these under strongly reducing conditions. This would produce zinc as a vapour which would then promptly oxidise in the upper parts of the furnace. Once the pharmaceutical and metallurgical value of the zinc oxide so produced was realised then a variety of methods was introduced to collect it. These methods already seem to have been fully operational in Late Hellenistic times, contemporary with the Aravalli mines.

The most detailed early descriptions of the production of zinc oxide from its ores are those given by Dioscorides in his *Materia medica* (Book V, chapter 84) (Gunther 1934: 623–5; Riddle 1985). The translations that follow were originally made by John Goodyer in 1655 (hence the archaic English) and republished by Gunther:

> But Pompholyx [calcined smithsonite, zinc oxide] is also made from Cadmia purposely blown with ye bellows for ye making of it. And it is made thus. In an house having a double roof a chimney is built, and by it towards ye loft, a suitable window, and open at ye parts above. But ye wall of ye house next to ye chimney, is beaten through with a small hole, into the very furnace for ye receiving of the bellows. It hath also a proportionable door made for ye going in, and coming out of the workman. There is joined unto this room another room, wherein ye bellows and ye bellows-blower work. Then the coals are put into ye furnace and kindled, afterward ye workman standing by doth sprinkle on from ye places over the head of the furnace the Cadmia being beaten small. And ye servant that is under withall doth do ye same, and casteth on more coals, until all ye Cadmia that was laid on be consumed, so that by the burning, the thin and light part is carried into the upper room, and sticks to the walls and to the roof thereof. But the body that is made of the those things carried up, at ye beginning indeed is like to ye bubbles standing upon ye waters, but at last more increase coming, it is like to the fleeces of wool. But that which is heavier goeth into ye places under foot and is dispersed about, some to ye furnace, and some to ye floor of ye house. This is thought to be worse than that of thin parts, because it is earthy and full of filth by ye gathering of it.

Note that there is a clear distinction between the purity of the sublimed material collected from above the furnace which should equate with the flat plates of sublimate from Zawar (illustrated in Chapter 7, Plates 7.3 and 7.4) and the material remaining in and around the furnace. Dioscorides also described a very different arrangement for the production of calcined zinc oxide:

> But Cadmia also comes from brass red hot in a furnace, the soot sticking to the sides, and to ye top of the furnaces.

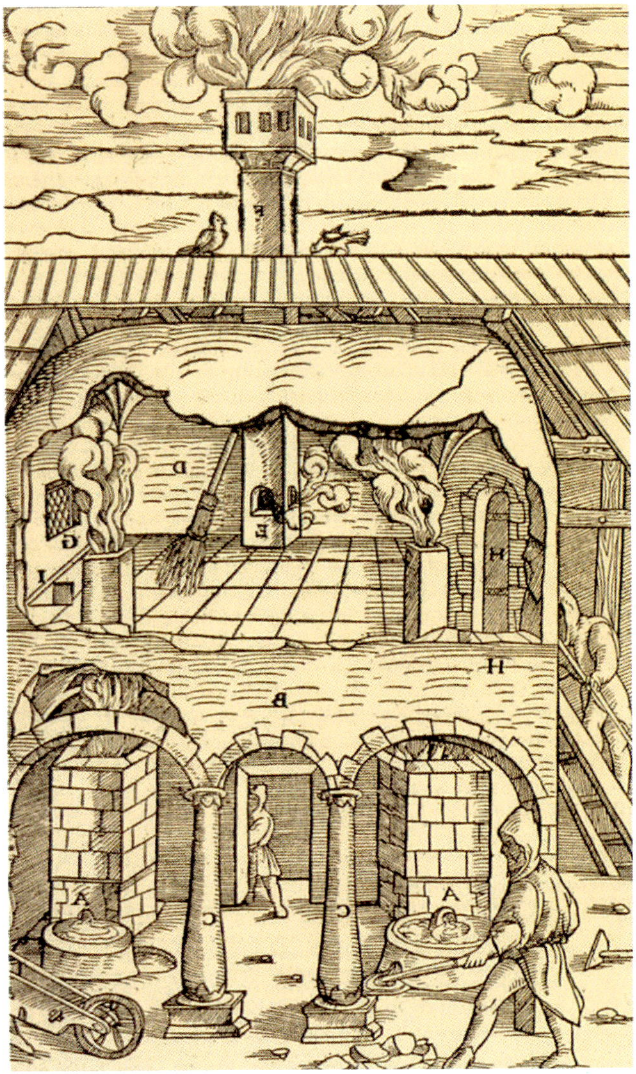

Fig. 11.1 Furnace for the smelting of zinc ores. The zinc vapour reoxidised and was collected as a sublimate on the walls and ceiling of the upper vaulted chamber (from *De re metallica*, Hoover and Hoover 1912: 394).

> Now [rods] of iron, of a great bigness, called by ye metal men acestides are joined together at ye top to ye end that ye bodies that are carried up from the brass may be held and settled there, which also sticking ye more, grow into a body, and sometimes one kind of it, sometimes two, or all are made. It is made also of a burned stone called Purites digged out of an hill that lies over Solis [Cyprus].

In this instance the source of the zinc vapour was from the molten brass, but the process would work equally well if an oxidised zinc ore was smelted. In the latter part of the description clearly a mixed sulphidic mineral including sphalerite must be intended by 'purites', rather than just iron sulphide, and elsewhere (Book V, 120) Dioscorides gives more details of how the sphalerite mineral was to be treated:

> They which are over these works putting ye stone called Pyrites into a furnace, they burn it as chalk for many days. But when the colour becomes Rubrica-like they take it out and set it up. But some say that ye aforesaid is made of ye only matter that doth perfect brass stone, when being roasted in ye said ranges it is carried thence into the ditches (flues?), and is there burnt, for it takes up ye place round ye ditches, and it is found nevertheless after ye taking away of ye stones.

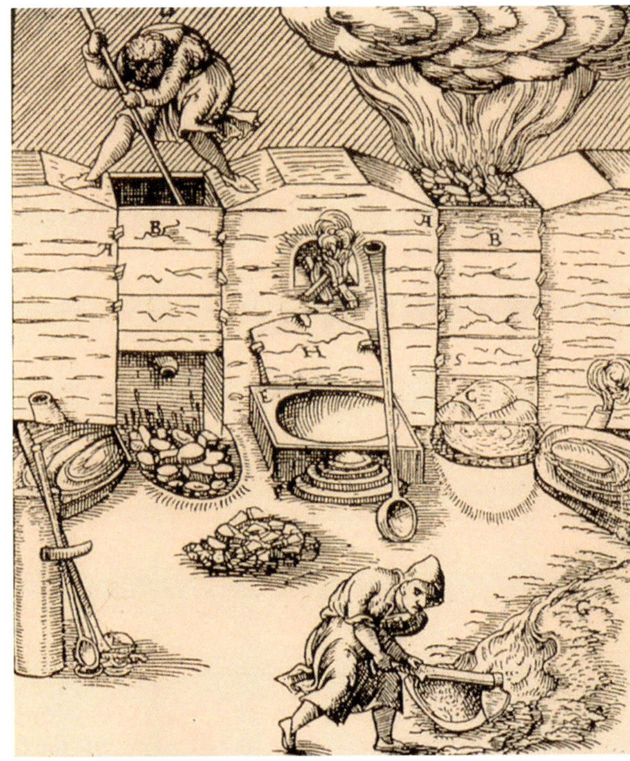

Fig. 11.2 Scraping zinc oxide from the furnace walls at the conclusion of smelting lead/zinc ores at Rammelsberg, Goslar (from *Pirotechnia*, Sisco and Smith 1951: fig. 36).

This again sounds like a conflated description of a roasting followed by a smelting of the ore to produce a sublimate of zinc oxide.

A somewhat similar arrangement was described and illustrated by Agricola (Hoover and Hoover 1912: 394–5) in central Europe 1500 years later (Fig. 11.1). Ercker (Sisco and Smith 1951: 271) described and illustrated (Fig. 11.2) how zinc oxide was scraped off the walls of the furnace at the conclusion of smelting the roasted mixed lead/zinc sulphide ore. Very similar installations were described in Iran by al Birūnī in the 11th century AD, by al Kāshānī in the 13th century AD (Allan 1979: 40; Craddock *et al.* 1998c), and by Marco Polo also in the 13th century (Latham 1972: 39).

It might be argued that the preliminary roasting would yield a product sufficiently pure for brassmaking, if not for medical use. However, many of the sphalerite ores as well as some lead ores also contain a great deal of iron, which would have become incorporated in the forming brass and could only then have been eliminated by remelting and oxidisation, which would also have removed much of the zinc.[11] Also, the mineral added to the copper has to be substantially free of sulphur, something that a simple heap-roasting operation could not achieve.

The sublimation of zinc minerals to produce pure zinc oxide for medical purposes was also described by Galen in the 3rd century AD on Cyprus, where it was prepared for him by throwing the zinc mineral onto the fire and collecting the zinc oxide from above (Walsh 1929). In all of these descriptions

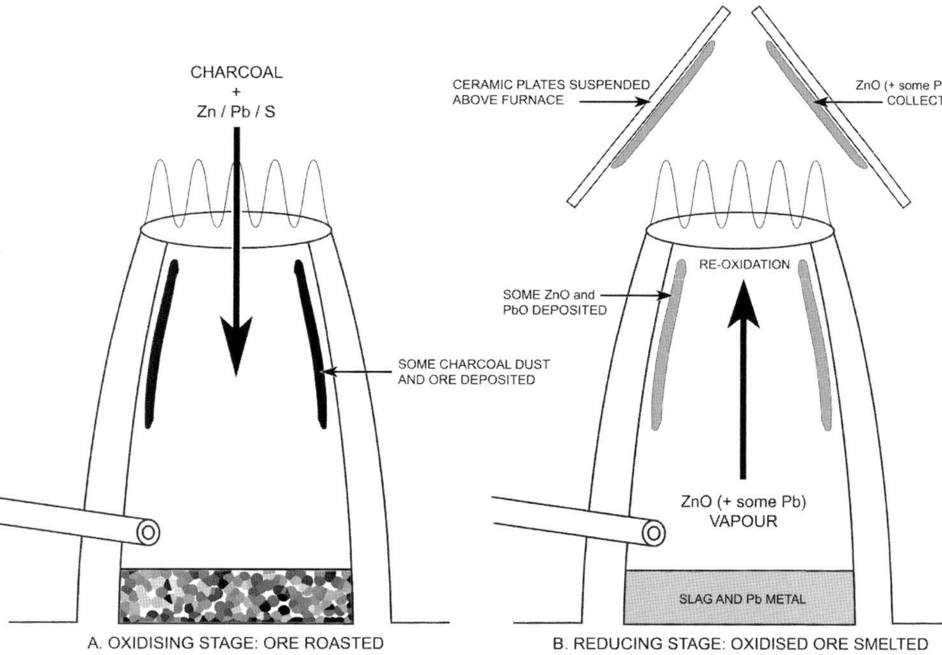

Fig. 11.3 Flow diagram of the proposed production of ZnO at Zawar in antiquity (diagram: A. Simpson).

the zinc minerals would have been reduced to the volatile zinc metal, the vapour of which reoxidised above the fire and was collected.

The product of the sublimation process was not uniform as all the ancient sources make clear (*Natural History* 34: 100–105). That which settled farthest away – on the roof or outside of the furnaces – was called smokey *cadmea* and seems to have been too contaminated with smoke particles (and just possibly with cadmium); that which collected in the 'vaults of the roof chamber' (presumably the deliberately placed collecting chamber) was adjudged the best because it would be likely to contain less of the heavier, less volatile, vapour of lead oxide. Pliny also described a third sort:

> A third sort is deposited on the sides of the furnaces, not having been able to reach the vaults because of its weight; and this is called in Greek 'placitis,' 'caked residue,' in this case by reason of its flatness, as it is more of a crust than pumice, and is mottled inside; it is more useful for itch-scabs and for making wounds draw together into a scar. Of this kind are formed two other varieties, onychitis which is almost blue outside but inside like the spots on onyx or layered quartz.

The zinc-rich crust from Site 7 at Zawar (see Chapter 7, Plate 7.3) would seem to be a good candidate to be classed as an example of *onychitis*.

Study of the enigmatic early smelting remains from Zawar, interpreted in the light of some of these contemporary western descriptions of the production, suggests that zinc oxide was being prepared. The process evidence at Zawar consists of a number of relatively small slag heaps and scatters of early date (especially Sites 2 and 14, both dated to about 2000 years ago and Site 7 dated to around 1300 years ago) and which are seemingly associated with the production of lead (see Chapter 6). The materials excavated from these heaps have been studied in detail (see Chapter 7) and found to contain evidence for the treatment of sulphidic minerals as well as the smelting of oxidised material. It is proposed to interpret this evidence as the remains of two quite separate processes mixed together, namely the roasting of the ore followed by the smelting of this to produce zinc vapour which would be collected predominantly as the oxide, as well as the minor quantities of lead contained in the ore (Fig. 11.3).

To recap the evidence from the early slag heaps at Zawar, the principal materials are the slags themselves. These are glassy tap and furnace slags that were clearly fully mobile while molten. The main components are pyroxenes and olivines, and they have rather complex compositions, containing mainly silica with some iron, aluminium, calcium, magnesium, lead and zinc. The source of the alkaline earths could have been the dolomite vein rock or fuel ash. In contrast the later Zawar slags contain much less zinc, calcium and magnesium but correspondingly more iron (see Chapter 7, Table 7.1). These slags are located near to Old Zawar and probably relate in part to the exploitation of the nearby mines on Baroi Magra (see Chapter 6, Fig. 6.2) where, uniquely in the Zawar mineralisation, lead predominates in the ore over the zinc. These deposits were exploited in the post-medieval period, probably primarily to recover the silver they contained (lead samples from Site 5 contained 2300 and 490 ppm silver).

The early slags are free of inclusions with little or no galena, suggesting that the material smelted was in an oxidised state in contrast to the contemporary slags from Agucha and Dariba and the later slags from Zawar itself in which sulphides are abundant. It must be considered that there is no evidence for any oxidised ores from Zawar and it is difficult to see why the sulphidic ores from Zawar should have been oxidised in order to smelt the lead; as outlined in Chapter 10, it was not necessary. The implication must be that the oxidation of the ore was being done to recover something else, the zinc being the obvious candidate from an orebody predominantly of sphalerite. The zinc content of the slags does vary considerably from one site to another which is somewhat surprising if zinc was the product. However, only three of the sites have carbon dates, none of which are contemporary with the earliest phases of the mining therefore it is possible that the varying zinc contents reflect changes in the extraction technology over time.

In among the slags are fragments of furnace lining including rim fragments with crusts rich in silica and the oxides of zinc and lead (see Chapter 7, Plate 7.2). These crusts also contain some lead and zinc sulphides. There seems to be a general ordering of deposition with the sulphides being directly against the ceramic and the lead becoming less prevalent away from the ceramic. Such crusts have only been found on the early Zawar smelting sites (Sites 2, 14 and 7). The presence of the sulphides confirms that the ores used were originally sulphidic. It would seem most likely that the crusts had formed on the upper surfaces of a furnace where initially sulphidic ores were being charged to the furnace and roasted to convert them to oxides. This was followed by a more reducing process during which the zinc and lead vapours formed and some deposited on top of the unchanged sulphide ore dust on the upper sides of the furnace.

Somewhat similar crusts have been found at Kerman in central Iran, dated to the 9th century AD (Hezarkhani-Zolgharnian *et al.* 1994); these were interpreted as resulting from the smelting of zinc-rich copper ores. This is possible, but the absence of copper and the presence of lead and zinc sulphides at Zawar suggests that it was the latter two metals that were being processed there, and of course medieval Kerman was famed for the production of zinc oxide.

The ceramic linings from Site 2 had vitrified suggesting temperatures in the region of 1150°C, at which temperature the zinc oxides should have reacted with the hot silica to form zinc silicates, but these are not present. This infers that during the pre-heating phase prior to the roasting operation the burning charcoal must have nearly filled the furnace, resulting in high temperatures and the observed vitrification near to the top, which had cooled down significantly before the zinc ore was charged. The fragments of lining from Site 7 were also partially vitrified but in this case zinc silicates had formed, showing that the lining had been very hot while the lead and zinc were being deposited.

A second group of material from Sites 2 and 14 comprises the thin flat plates of high density made up of layers of zinc and lead oxides (see Chapter 7, Plates 7.3 and 7.4). The layers are quite distinct: the lead-rich layers are yellow-white in colour whereas the zinc-rich layers are green, recalling the ancient Sanskrit and Greek descriptions of layered zinc oxide, suggesting a progressive build-up. The plate from Site 14 preserved a thin layer of ceramic against which there was a very little sulphidic material, but otherwise no sulphidic material was found in these plates. The lead-rich layers were adjacent to the ceramic, with the zinc-rich layers being outermost as was the case with the linings. The contact surfaces of the plates are grooved or indented suggesting that the surface against which the plates had formed was deliberately keyed, presumably to stop them falling off prematurely. In a few places traces of ceramic still adhere, probably the last remains of the original surface on which the condensation took place. The presence of charcoal fragments on and in a sample from Site 2 show that the temperatures cannot have been very high during deposition of the oxides otherwise the charcoal would have oxidised. Similarly, grains of quartz which had not reacted with the surrounding zinc oxide in the plate from Site 14 also indicate a relatively low temperature. Such plates have not been recognised at any other smelting sites within India but have been reported from contemporary Cyprus (Zwicker 1992).

The most tenable explanation for the finds from the early slag heaps is that they represent the debris from a multi-stage process to produce zinc oxide, which may be reconstructed as follows (Fig. 11.3). The beneficiated sulphidic ore, which would have been predominantly of zinc sulphide but with some lead and iron sulphide present, would have been roasted in the pre-heated small shaft furnaces to convert the bulk of the zinc sulphide to the oxide, while at the same time oxidising the lead and iron sulphides. This stage is evidenced by the crusted linings, predominantly of lead and zinc oxides but in which small amounts of unchanged mineral persisted, presumably having stuck to the sides as the ore was charged and failed to oxidise completely, or having volatilised without oxidising. When it was judged that the oxidation was complete the conditions would have been made much more reducing and the zinc oxide smelted to produce clouds of zinc vapour that would have promptly oxidised in the flue and then been deposited as a sublimate against flat ceramic baffles that were positioned above the furnace. The presence of a little sulphidic material on the surface of the ceramic adhering to the piece from Site 14 suggests that baffles were in place throughout the process. Apart from that, the almost complete elimination of sulphides from the zinc, necessary for brass production, would have been achieved. The lead would have been reduced to metal, some of which also volatilised and deposited together with zinc on the furnace sides and the baffles, but the majority would have drained through the slag to form an ingot at the base of the furnace. The iron oxides, together with some of the zinc and lead, would have reacted with the silica to form the slag. It is possible that the lead could have been further treated to recover any silver it may have contained, but unfortunately the blebs of metal in the slags were too small to allow an accurate estimate of their silver content to be made. No evidence of cupellation was found at any of the Zawar sites investigated, including those of the later, post-medieval period when the surviving records suggest that silver was being extracted from the lead. If our interpretation is correct and the plates are the zinc oxide product of the operation, then the lead content is very high. It is possible that these pieces had been discarded on the slag heap because they were perceived to be too impure for use.

Zinc[12]

As already noted, it is not difficult to reduce zinc oxide to zinc – the problem is collecting it as a metal. At the temperatures at which the reduction takes place the zinc is produced as a highly reactive gas. Thus in a traditional metal smelting hearth or furnace, instead of a dense liquid metal collecting at the base of the furnace under the slag, the zinc vapour would rise and rapidly reoxidise, as described in the previous section. Clearly the zinc vapour had to be condensed out of contact with air so some sort of condenser system was necessary. The solution to this problem and the adaptation of the scientific method to form an industrial process is one of the great achievements of early metallurgy.

Early references to zinc metal

The earliest reference in the Sanskrit sources to a metal that could be zinc is found in the *Charaka Samhita*, a medical treatise that was compiled in the early Buddhist period and which in its original form should date to around the mid-first millennium BC (Rây 1956: 60). In chapter 26, it describes how a salve for the treatment of infected eyes and open wounds, known as *pushpanjan* (the *-anjan* element denotes a salve), was to be made by burning a metal in air. The salve is surely zinc oxide, the familiar zinc ointment, therefore the metal to be burnt must have been zinc. *Pushpanjan* is also mentioned in the slightly later *Susruta Samhita*, and in the medieval *Rasaratna Samuchchaya* (109–112) (Rây 1956: 175). The mysterious word *tuttha*, which could mean a copper salt or zinc oxide, occurs in the *Arthaśāstra* and is discussed in Chapter 10. The medieval *Dhautukriya* or *Dhatumanjari* treatise describes the amalgamation of zinc with mercury and also a red gold made of 4 parts zinc to 1 part gold (Rây 1956: 197, 198).

Commercial zinc production

The quite detailed early descriptions of the ore charge for the zinc-smelting retorts (see below) are of some interest. It is important to realise that these descriptions were of laboratory operations that were either spagyrical or iatrochemical in their objectives. Therefore some of the more exotic and expensive reagents, such as lac or turmeric, which also feature in many of the recipes given for other processes, were almost certainly not used in the industrial processes carried on at Zawar. Similarly, the apparatus outlined in the early manuals have only one crucible or retort. There is nothing in the contemporary literature to suggest that the 252 retorts fired simultaneously in each *koshthi* bank at Zawar, Site 30 (frontispiece). The industrial processes which took place at Zawar represent the adaptation of the small-scale laboratory processes to form industrial operations; this was chemical engineering, not alchemy. However, before the excavations there was no inkling of their existence – their record is purely archaeological and contradict Forbes' (1971: 281) pejorative comments that:

> Though the value of the old Indian alchemists and their modern commentators is very doubtful it seems that zinc was prepared by Indian chemists since the 12th century, but that this remained a laboratory experiment and was never applied to industrial production.

The earliest dated reference to Indian zinc from outside India was made by the Iranian writer Abū Dulaf in his *al-risālat al-thāniya*, compiled around AD 950 (Allan 1979: 44) in which he noted many kinds of *tūtīyā* that he claimed came from the vapour of copper, except for the Indian *tūtīyā* which came from the vapour of tin. This statement has puzzled many, including Allan, as zinc oxide does not come from tin. One of the methods of preparing zinc oxide for medical usage in the Greek world as described by Dioscorides was by heating brass and collecting the vapour, and this would be the only indirect source available elsewhere, except of course in India, where a pure form, free of lead and other harmful elements, could be obtained by burning zinc metal.[13] Abū Dulaf's comment is interesting as it demonstrates that by the 10th century AD metallic zinc was already known, if only partially understood outside India itself, and that Indian *tūtīyā* made by burning zinc metal was by then an item of international trade. Similarly, the Persian scholar al-Khwarizmi, writing in the 9th century AD, noted various varieties of *tutty*, including the Indian described as being artificial (Ryding 1994) and which was white and precious, as well as a kind in the form of thin plates like palm leaves, recalling the plates of zinc oxide from Zawar.[14]

Other Iranian writers such as Ibn al-Faqīh (c.AD 900) and al-Tha'ālibī (c.AD 1000), also mention Indian *tūtīyā*, but without giving any details of how it might be different. Al-Khwarizmi writing in the 9th century, and Al-Kāshānī, in the 13th century AD, believed that Indian *tūtīyā* came from the sea as a sort of surf.[15]

The evidences of the brasses

The last practitioners of the cementation process in Europe gave various upper limits of the amount of zinc that could be absorbed by the copper (between 29% and 33%). Among the early brasses from India only a few have zinc contents exceeding these figures; these have been cited as evidence for the existence of metallic zinc. Perhaps the most frequently cited is the piece from Taxila (mentioned above) with 34.3% and small amounts of tin and lead (Ullah and Marshall 1951: 567–9; Prakash and Rawat 1965: 50–51, table 5.3), more recently joined by the two bangles from Senuwar, dated to the early centuries AD which were reported to contain 35.52 and 35.93% zinc with only traces of other metals (Singh 2004).

It is clear that these are very isolated examples: there seems to be no regular tradition of making particular classes of metalwork in brass at this period, much less of making them of high-zinc brass. It is also unfortunate that the zinc figures reported are only a few percentage points above the maximum level found in cementation brass. Such small numbers of examples and such a small percentage difference inevitably leads to doubts both about the analytical accuracy and the certainty that the 33% level was never exceeded; it was after all a practical, not theoretical limit. Experimental replication of the cementation process conducted by several groups including Welter (2003) and Bourgarit and Bauchau (2010) obtained brasses with zinc contents up to 40%. In addition it is known that the smelting of mixed minerals of lead and zinc sulphides can result in small amounts of metallic zinc solidifying before they have time to oxidise (see Chapter 13). Even more significant are the reports from silver mines in Anatolia recorded in Strabo's *Geography* (XIII.56) (Jones 1929: 115), where such ores were treated contemporary with the brasses from Taxila, that the minor quantities of metallic zinc produced as a by-product and known as 'mock silver' could be added to copper to make brass (Craddock 1998a). If the analyses are correct this is the most likely source of the zinc found in the high zinc brasses.

Early descriptions of the production of zinc

The earliest reference to the production of zinc by distillation is given in the above-mentioned *Rasaratnakara*:

> *Rasak* (zinc ore) should be treated with *dhanyamia* (a fermented spirit similar to present day *paddy*) alkalis, and ghee (clarified butter) and then mixed with wool, lac, *harad* (or *hara*) (*Terminalia chebula*) *kenchua*

(earthworms) and *suhaga* (borax). Upon heating it yields an extract which is white, shining like tin; of this there is no doubt (31–32) (see also Rây 1956: 130).

The somewhat later *Rasarnava* commenced its account with a description of the preparation of the ore:

> The *rasak* powdered by the wise *vaidhya* and put into a cloth bag is dipped into a container of women's urine for seven nights. The soaked powder is then treated with juices of yellow and red flowers (saffron, acacia, catechu, lac tree, *Terminalia chebula* and *dhak* (*Butea frondosa*)), and then treated repeatedly with alkaline, neutral and acidic solutions respectively. Then wool, lac, turmeric (*Curcuma roxob*), *harra* (*Terminalia chebula*), *kenchua* (earthworms), *suhaga* (borax) and *grahdhoom* (soot?) are mixed together and added to the treated ore. The product is then placed in the *mook musha* (closed crucible, or more likely a retort) and heated strongly. The extract released on heating has a diamond-like shine which is undoubtedly zinc (VII: 33–38) (Tripathi 1978).

Rây (1956:138) translates this rather differently:

> *Extraction of Zinc from Calamine*: Rasaka mixed with wool, lac, *Terminalia chebula* and borax and roasted in a covered crucible (retort) yields an essence of the appearance of tin; of this there is no doubt (VII: 37–38).

The *Rasakalpa*, probably compiled in the 12th or 13th century AD, also gives a detailed description of how the ore is to be treated before smelting:

> *Extraction of Zinc from Calamine*: Calamine, enclosed in fourfold pieces of cloth is heated in steam in a vessel for five months. Then it is powdered in a mortar and then mixed with treacle, soot etc. The mixture is then made into balls. These are enclosed in a crucible and strongly heated. The essence possessing the lustre of tin is thereby obtained (Rây 1956: 157).

Steaming seems a most unlikely treatment for a zinc ore. If it was the sulphide then the sulphate would most likely be produced, which would be very difficult to treat; if it was the silicate the treatment would ineffectual; and if it was the hydroxy-carbonate, the treatment would be unnecessary.

The most detailed accounts of zinc smelting are contained in the *Rasaratna Samuchchaya*, which was probably compiled in the late medieval period, between the 14th and 16th century AD. Rây believed it was compiled immediately prior to the arrival of the Portuguese. The work contains much earlier material, especially from the *Rasarnava* and the *Rasaprakasasudhakara*, the latter ascribed to Yasodhara, who is believed to have worked in the 12th or 13th century.

The refractories

Book 10 contains a description of the materials used for the refractories:

> Earth which is heavy and of a pale colour, sugar or earth from an ant-hill, or earth which has been mixed with the burnt husks of paddy, fibres of the hemp plant, charcoal and horse-dung pounded in an iron mortar, and also rust of iron, are to be recommended for crucible-making (5–6) (Rây 1956: 191).

Satpute's (2003) more complete translation also mentions clay from the bottom of lakes as a suitable clay for refractories, supporting the suggestion that a source of the finer clays needed for the refractories at Zawar would have been the silt deposited behind the dam (see Chapter 6). Various refractories to make the retorts, known as *vajramūṣā* (literally thunderbolt crucible), for the distillation of mercury, are also described in Todarananda's *Ayurveda* 16th-century compilation (Dash and Kashyap 1994: 166–7):

> For the preparation of this crucible, ash of paddy husk [2 parts], mud from the ant-hill [one part], *maṇḍūra* or iron-rust [one part], white clay [one part] and ash of human hair [one part] should be strongly triturated by adding a small quantity of goat's milk for two *yāma-s* (approximately six hours). Out of this paste the crucible of the above mentioned shape should be prepared and burnt.
>
> Ash of paddy husk [one part], ash of cloth [one part], mud [one fourth part] and stone dust [one fourth part] should be triturated for one day by adding the juice of *vajra-vallī* (*Cissus quadrangularis*). Out of this clay *vajramūṣā* should be prepared.

Another recipe is given this time for the furnace walls and 'bent pipe', which was probably the condenser or tuyere:

> *Gaurā* [chalk], paddy-husk, mud of ant-hill and stool of elephant and horse should be burnt till they become black. This ash and the leaves of *Pāṣāṇabheda* (*Bergania ligulata*) should be taken in equal quantities and strongly triturated by adding the juice of *vajra-vallī* (*Cissus quadrangularis*) for one day. With this clay *Koṣṭhī* [oven], *vakra-nālī* [bent pipe] and *vajra-mūṣā* should be prepared.

There are also descriptions of the materials used to lute the various vessels:

> For sealing the joints of crucible, etc., pieces of cloth should be smeared with the clay of *Khaṭikā* [chalk], salt, paste of *māṣa*, (*Phaseolus radiatus*) jaggery (raw sugar), *guggulu* (*Commiphora mukul*), *atasī* (*Linum usitatissimum*) and the powder of stone.
>
> Mud of ant-hill, charcoal, *maṇḍūra* [rust of iron] and white clay, all these taken in equal quantities should be made into a powder. To this, equal quantity of the ash of paddy husk should be added. By adding water, these ingredients should be triturated and made to a paste. This paste should be used for sealing the joints of crucibles etc.
>
> Two parts of paddy-husk, one part of cloth-pieces and three parts of mud should be added with water and pounded. To this, some human hair should be added and further pounded till the paste becomes sticky. Effort should be made to keep this paste wet so that it should not get dried up. After keeping it wet for seven day, the paste should be used for the purpose of wrapping around

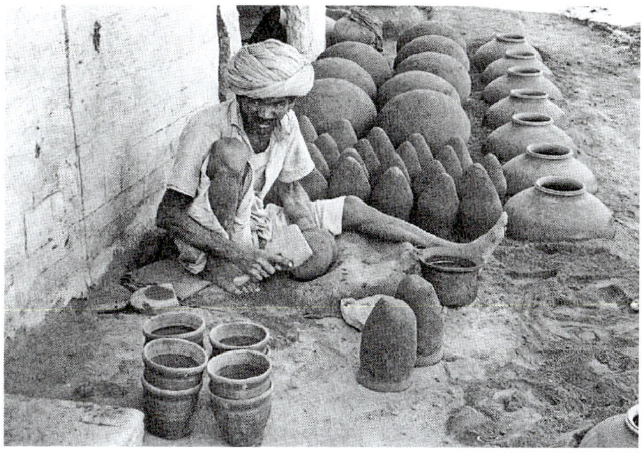

Plate 11.1 Making the bucket chain *noria* pots in Udaipur. Production of the retorts must have been very similar (from Kramer 1997).

glass bottles and sealing the joints of crucibles, etc. (Dash and Kashyap 1994: 170).

All the refractories from the excavations were heavily tempered with straw and chaff etc. either added directly or in dung (see Chapter 7), and the reference to paddy husks recalls the instructions in the *Arthaśāstra* for the Superintendent of the Magazine to deliver chaff to the metalworkers (see Chapter 9). O'Shaughnessy (1847) (see Chapter 10) stated that the cupels used by the silver assayers in Calcutta contained chaff.

In Book Two of the *Rasaratna Samuchchaya* there is a section devoted to *raska* (zinc ore). Rây translates this as calamine, zinc carbonate, now usually known as smithsonite, without comment, but the ore with which the Indian chemists would have been most familiar was sphalerite, although as mentioned above it was possibly already in a roasted state:

> *Rasaka* (calamine) is of two kinds: the one of laminated structure is known as *dardura*; the other non-laminated is called *karavellaka* (149).

> Calamine is to be heated and plunged seven times into the juice expressed from the seeds of lemon, or immersed in the urine of man or of horse, or in sour gruel or sour milk, and thus purified (154–55) (Rây 1956: 171).

These two paragraphs, or *slokas*, are taken directly from the *Rasaprakasasudhakara* of Yosadhara. The laminated structure of one form of *rasaka* recall the similar description given to some *cadmea* by Pliny and could be a reference to the layers of zinc oxide which condensed on the plates suspended over the smelting furnaces as described by Dioscorides (Fig. 11.1) and possibly found at Zawar, Sites 2 and 14.

The various classes of refractory were made using a number of different processes. It is likely that the retorts were thrown as a pot with a flat base which was then inverted, paddled and hand-fettled to make the pointed base (see Chapter 6, Fig 6.28), very much in the manner that the *gharalis* (the pointed pots used on the ubiquitous bucket chain *noria* water wheels) were made in Rajasthan until the recent past (Plate 11.1) (Kramer 1997: 26–7). Some of the early condensers and plates at Zawar are likely to have been handmade almost after the manner of thumb pots, but later were moulded. The actual condenser tubes were clearly

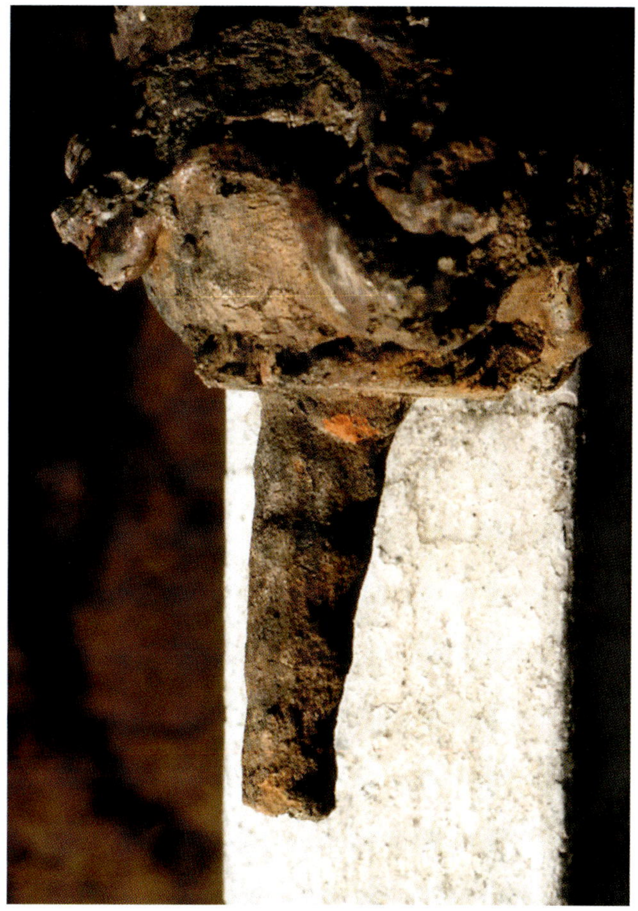

Plate 11.2 Broken end of a retort and condenser still with a length of condenser tube attached (cf. Fig. 6.16). Note the fingermarks where the clay was hand-moulded around a cylindrical former.

hand-moulded around a cylindrical mould, the fingermarks still being clear on some condensers (Plate 11.2).

The smelting operation

The descriptions of the smelting of the zinc ores are as follows:

> Extraction of Zinc: Rub calamine with turmeric, the chebulic myrobalans, resin, the salts, soot, borax, and one-fourth its weight of *Semicarpus anacardium*, and the acid juices. Smear the inside of a tubular crucible with the above mixture and dry it in the sun and close its mouth with another inverted over it, and apply heat; when the flame issuing from the molten calamine changes from blue to white, the crucible is caught hold of by means of a pair of tongs and its mouth held downwards and it is thrown on the ground, care being taken not to break its tubular end. The essence possessing the lustre of tin, which is dropped, is collected for use (157–61) (Rây 1956: 171).

This was taken from the *Rasaprakasasudhakara* of Yosadhara. Another contemporary account in the *Rasachintamani* of Madanantadeva is similar but with the added information that the process was judged to be complete as soon as the white fumes appeared (Rây 1956: 155). The blue flame would be made up in part from zinc vapour burning with a blue-green flame and carbon monoxide also burning with a blue flame. When this turned white or white fumes appeared this would suggest that the conditions were no longer reducing and zinc was being turned into zinc oxide so the process should be terminated. Observing the flame colour and material issuing from the condenser end was the only way of determining what might be going on inside the retort, and similar comments that the process was judged complete when the flame or issue from the condenser turned from blue to white can be found in European manuals on retort smelting (Percy 1861: 570; Lones n.d.: 101) and in descriptions of the traditional Chinese process (Craddock and Zhou 2003).[16]

From the instructions to pour the molten zinc from the vessel it would seem that the zinc metal collected inside the retort, which must have been heated with its neck pointing up. Other references in the *Rasaratna Samuchchaya* describe the neck pointing down and the zinc leaving the retort during the process as was the practice at Zawar:

> *Vrintaka Crucible* (or retort): Prepare a crucible in the shape of a *brinjal* (aubergine) and put onto it a tube eight or twelve anguls (inches) long, which opens out like a thorn apple flower (*Datura stramonium*), at one end. The tube should be hollow and have a circumference at the expanded end of eight *anguls*. Such crucibles are known as *vrantak moosha* and are used for the distillation of *kharpar* (zinc ore) and other minerals (23–24) (Rây 1956: 191).

The actual process of *tiryakpatana yantram*, or distillation by descending, is described in Book 2 as follows:

> Take equal parts of lac, the bark of the papal tree, *hara*, myrobalans (*Terminalia chabula*), turmeric *haldi* (*Curcuma longa*), treacle, resin, rock salt, *suhaga* (borax), and *kharpar* (zinc ore). Bind them together and bake with cow's milk and *ghee* (clarified butter). Then make into balls and put into a *vrantak moosha* and heat strongly. The contents are then poured onto a slab of stone – the essence of *kharpar* of the beautiful appearance of tin is to be used (163–64) (Rây 1956: 172).

Alternatively, having prepared the retort and its charge, as described above and fitted the condenser:

> Dig a hole in the bottom of a *koshthi* [the furnace] and place inside a water-filled vessel with a perforated plate or saucer over its mouth. Then fix the *vrantak moosha* in an inverted position over the perforated plate or saucer. Put charcoal of the *jubube* tree (*Ziziphus jububa*) over and around the *vrantak moosha* and heat strongly with the help of bellows. On heating the zinc extract descends and is collected in the vessel bottom (165–68) (Rây 1956: 172).

These descriptions become more detailed through the centuries and clearly describe processes very similar in principle to those being carried out at Zawar, although only for a single retort.

The frequent mention in the recipes quoted above of a variety of plant material, bark, leaves etc, which were apparently not treated in any way or turned into charcoal but used in their natural state, may initially seem very unlikely, but it should be considered that the descriptions of the charges used in the crucible steel processes often mentioned similar plant material (Bronson 1986: 37), including descriptions dating back to Classical antiquity (Craddock 1998b). These could also act as initiators at the commencement of the process generating CO_2 which in turn was reduced to CO. There is no description of the zinc smelting process in any of the Mughal sources compiled in Persian, but the *Āīn-I Akbarī* of Abū `L-Faẓl Allāmī (Phillott 1927: 40–42) does make a specific reference to Zawar (see Chapter 3):

> Jast [zinc], which according to the opinion of some, is *Rūḥ-i-tūtiyā* (that is, made from or originating from *tūtiyā*, zinc oxide), and resembles lead, is nowhere mentioned in the philosophical books, but there is a mine of it in Hindustan, in the territory of *Jālor*, which is a dependency of the Ṣūba of Ājmīr.

Therefore Abū `L-Faẓl Allamī in India clearly understood that the zinc was made from *tūtiyā*, zinc oxide,[17] although he then goes on to state that the Hindus believe that the metal called *riṣāṣ* (presumably derived from the Hindi *rasaka*) was silver in the state of leprosy, seemingly not realising that they were one and the same. Most of what Abū L-Faẓl had to say on metals was fairly standard Islamic scientific/alchemic theory which was itself rooted ultimately in the Classical tradition and largely ignored the Hindu philosophical works, hence his comment that *Rūḥ-i-tūtiyā* is nowhere mentioned in the philosophical books. He also describes three alloys of brass made by mixing the *Rūḥ-i-tūtiyā* with copper (Phillott 1927: 42):

> *Brass*, which the Hindūs call *pītal*, is made in three ways: *first*, 2½ sers of copper to 1 ser of *Rūḥ-i-tūtiyā*, (that is 28.5 % of zinc) which is malleable, when cold; *secondly*, 2 sers of copper to 1 ser of *Rūḥ-i-tūtiyā*, (that is 33% zinc) which is malleable, when heated; *thirdly*, 2 sers of copper to 1 ser of *Rūḥ-i-tūtiyā*, not worked with the hammer, but by casting.[18]

This must be the first metallurgical description of the working properties of the brasses. It includes the familiar α brasses containing up to about 30% zinc, which are suitable for cold working, and the β brasses containing between 30 and 40% zinc which are hot worked or cast, and probably the high zinc brasses that are too brittle to be shaped by hammering and can only be cast.

Process evidence from Zawar

The process and apparatus for the production of zinc remained remarkably constant throughout the many centuries of production at Zawar. Basically the process was one of *tiryakpatana yantram*, that is, distillation by descending, and apart from some changes to the retort charges, the main changes were in the dimensions of the furnace (*koshthi*) and of the retorts (*vrantak moosha*).

Beneficiation and roasting

The zinc ore from the Aravallis, then as now, was almost exclusively the common zinc sulphide mineral, sphalerite (ZnS). It is estimated that the crushed ore was beneficiated to about 50%

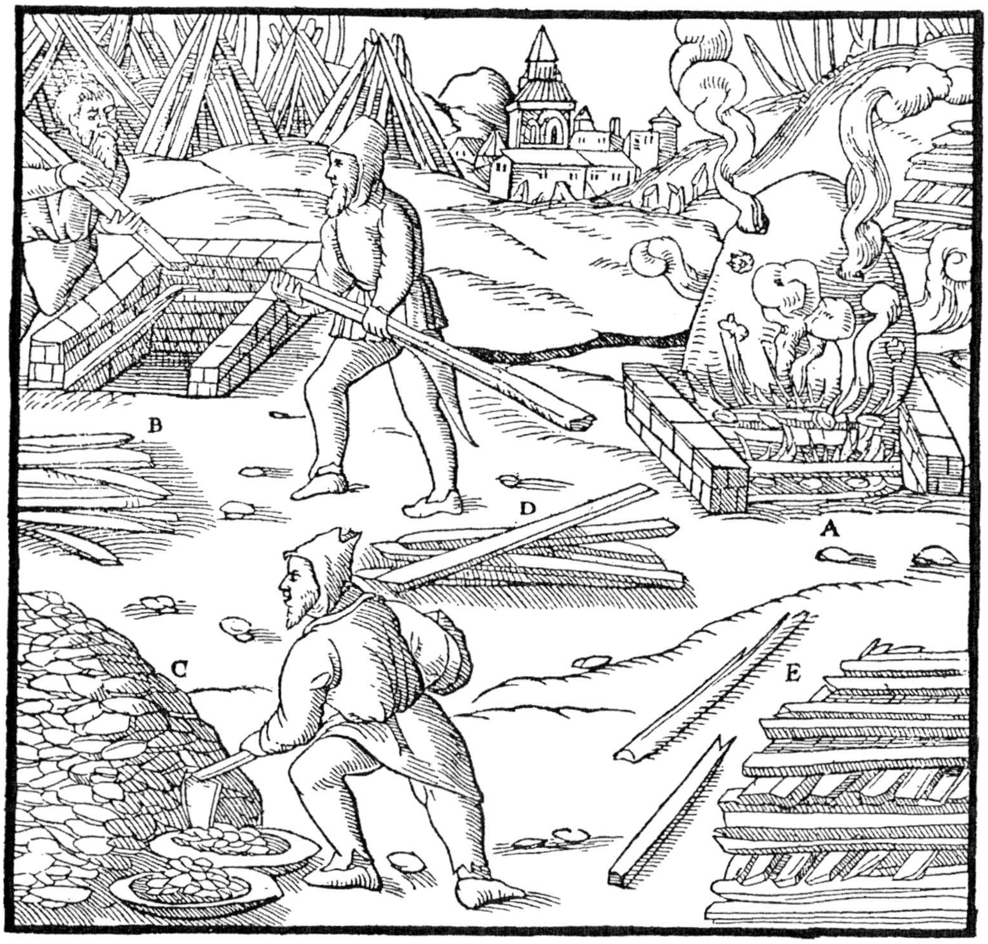

Fig. 11.4 Roasting operations in small conical mounds of ore set on timber and within low-walled stalls: C: the ore; D and E: the wood; B: a heap under construction; A: a burning heap. Note that in the text (273) Agricola describes the heaps as a gently sloping cone about a metre in height, more similar to the white heaps at Zawar. However the figures in *De re metallica* were not prepared until after Agricola's death (from Hoover and Hoover 1912: 275).

sphalerite by hand picking, removing as much of the associated galena and pyrite as possible (see Chapter 7). It was essential to remove the bulk of the galena otherwise during the roasting stage this would have converted into lead sulphate and lead oxide which would have reacted with any silica in the ore or hearth lining, forming molten lead silicates that would have coated the zinc ore, rendering it incapable of further treatment. Excess pyrite would also have been inimical to the successful production of zinc.

The sulphides also have to be converted to oxides before reduction can take place. In the later European retort process the sulphur content of the roasted ore did not exceed 1% and was often less (McBean 1959; Smith 1918: 82). Any remaining zinc sulphide was incapable of reduction to zinc and was thus a direct loss, and it was believed that due to a variety of reactions, its presence reduced the efficiency and yield generally. It was also found that any sulphides present could react with iron oxides in the ore charge to form iron sulphides that were very corrosive to the walls of the retorts and which were otherwise reusable. The latter consideration was obviously not a factor at Zawar where the retorts were used but once so there was no overriding reason for almost total elimination of the sulphides. In the account of the process at Zawar made by Straczek and Srikantan (1967: 50), they suggested that the rather high zinc content of the sinter in the retorts may have been caused by a high sulphur content, indicative of imperfect roasting. In fact, analysis of the residues from the large retorts suggests that the sulphur content was of the order of about 0.5%, and that the roasting operations must have been fairly efficient.

The conversion of zinc sulphide to zinc oxide, zincite, is not difficult. Temperatures in excess of 700°C must be maintained but this is fairly easy as the reaction is exothermic (Morgan 1985: 40):

$$2ZnS + 3O_2 \rightarrow 2ZnO + 2SO_2\uparrow + 832\text{ KJ at }1200°C$$

In post-medieval Europe it was difficult to concentrate lead/zinc ores with a high pyrite content except by roasting (see below). The presence of pyrites in the ore to be roasted was a mixed blessing – the iron sulphides oxidised preferentially to the zinc sulphides and there was also the danger of forming zinc ferrite compounds such as $ZnO{\cdot}Fe_2O_3$. Smith (1918: 82), for example, described highly ferruginous sphalerite ores, such as those from Zawar, as being 'probably the most difficult to roast'. However, the pyrite burns freely from quite low temperatures, generating a great deal of heat, sufficient to keep the temperature of the whole mass above that necessary for the oxidation of the sphalerite, which is itself an exothermic reaction, with little additional

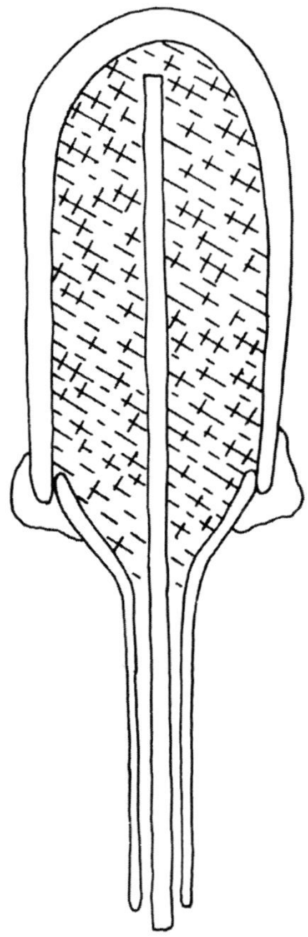

Fig. 11.5 Schematic section through a charged retort with its condenser luted on with clay, ready for firing in the *koshthi* furnace (cf. Figs 6.9 and 6.10). Note the small stick inserted to form the central channel and to hold the ore balls in place when the retort was inverted into the furnace (cf. Plate 11.6).

fuel. Even so, roasting ores in open stalls required some fuel, usually brushwood; the offcuts from charcoal burning could also be usefully employed. In post-medieval Europe the smithsonite ore was usually calcined in heaps or open stalls. The roasting of the mixed sphalerite-galena-pyrites ores, quite similar to those mined in the Aravallis, at Rammelsberg in 16th-century Germany was described in detail by Ercker (Sisco and Smith 1951: 266–7). He reported the heaps just as 'very large piles', but Agricola (Hoover and Hoover 1912: 274–6) described gently sloping conical mounds about the size of a man. The base was of timber and brushwood upon which the ore was stacked, the largest pieces at the bottom and the finer ore at the top, covered with a coat of very fine ore to control the combustion and all contained within low walls (Fig. 11.4). Simple heaps were still used for calcining smithsonite ore in the 18th century at Namur in Flanders according to Galon (1764). He described the construction of conical heaps made up of alternate layers of charcoal and wood, with the ore sitting on a base of more substantial logs. The heaps were from 3 to 4 m in diameter and about 3 m tall, and the openings and central chimney were initially filled with wood. The heaps burned for 8–10 hours. About 7 tonnes of ore was burnt in each heap and afterwards the calcined ore was sorted and the stones and other waste removed.

Any calcite in the ore tended to form a mixture of calcium oxide and calcium sulphate, which did not interfere with the following reduction process in the retort. Where dolomite was present, most of the magnesium sulphate that initially formed would have decomposed to magnesium oxide while still in the roasting stage. However, any magnesium sulphate that survived to be included in the charge would have decomposed in the reduction stage, releasing sulphur which could have converted some zinc back to zinc sulphide (Ingalls 1906: 33).

The open roasting of zinc ores did not reduce the sulphur contents sufficiently for the 19th-century European processes so some form of secondary furnace roasting was also required. The problem was that either the ore lumps were sufficiently large to allow the necessary passage of gases, in which case their interiors were unreduced, or the ore was finely ground and exposed but the air could not penetrate the powder and the fire could not be maintained. This problem was resolved in India by the traditional roasting methods such as those recorded for copper smelting in the 19th century (see Chapters 2 and 9). The beneficiated and ground ore could be mixed with cow dung to make small cakes or briquettes which could then be roasted in heaps with brushwood.

The obvious remains of the ore roasting operations at Zawar are the white heaps that ring the main retort heaps at Old Zawar. They are described in more detail in Chapter 6 (Plates 6.33 and 6.34, Fig. 6.33) and are made up of successive layers of fine white calcine debris and thin dark layers containing many charred twigs, together with some vitrified material. The calcined material is comprised principally of calcite and magnesite, presumably from the dolomite vein-stuff associated with the zinc and iron ore minerals, together with some phylite and occasional crumbs of ceramic and slag. Analysis showed that the white material contains relatively low zinc, lead, silver and iron but higher cadmium contents compared to the material in the retorts (see Chapter 7, Table 7.6). These differences must be the result of careful selection of the roasted mineral to include most of the metal oxides while leaving the majority of the dolomite and phylite behind. During the roasting process the cadmium sulphide would have oxidised along with the other sulphides present, but cadmium oxide is volatile at the roasting temperatures and some would have precipitated on the calcine debris. It was usual practice for most of the cadmium to leave the ore in the roasting stage (Ingalls 1906: 585).

It was believed originally that the white heaps were the remains of roasting operations (Craddock *et al.* 1985a), but then it was suggested that they were possibly ash collected from inside the *koshthi* furnaces (Craddock *et al.* 1998a). Further considerations have confirmed the original view. The layers in the heaps are typically 10–20 cm thick and continuous all over the heap with little intermixing between the fuel, represented now by the charred material and the charge, represented by the white material. Furthermore it is obvious from the stratigraphy of the white heaps that there are no tip lines, etc. such as are clear in the main retort dumps, showing that the material was not brought from elsewhere but is *in situ* from whatever process is represented by the debris, and also that considerable quantities amounting to several tons were treated at a time. There are no remains of the foundations of stalls etc. or any other installation elsewhere that could have been associated with the roasting process.

Realistically, the heaps can only be the result of *in situ* roasting operations. The heaps built up over a considerable number of operations. In each one a bed of brushwood etc. fuel was set on the surface of the heap and the beneficiated ore, possibly briquetted with dung laid on top, and fired. The excavated heap had been levelled at least once in order to create a level working surface. The operation could have taken several weeks even if the ore was briquetted. The ore probably required periodic raking over and venting to ensure that all the sulphidic material was exposed to oxidation. The roasting operation produced large quantities of sulphur dioxide together with some fumes of cadmium oxide so it made sense to concentrate these highly toxic activities at a limited number of sites well away from the main smelting and other activities. Even so the soil in the vicinity of Old Zawar is still heavily impregnated with cadmium, as reflected in the cadmium content of the vegetation growing upon it, especially common balsam (*Impatiens balsamina*) (*gulmendi* in Hindi) as reported by Aery and Tiagi (1988). The material that remained behind was very rich in calcium oxide (quicklime) and magnesium oxide which, together with some additional silica, was used to make the mortar identified on some of the walls built out of old large and intermediate retorts (Freestone *et al.* 1985a).

Smelting and condensation

The next stage in the process was the reduction of zinc oxide to zinc metal, the condensation of the zinc vapour and its collection. As this reaction can only be carried out at temperatures above the boiling point of zinc (see Chapter 7, Fig. 7.4), the reduction must take place in a space where air is excluded and where the reactive vapour can condense to a liquid, prior to collection. At Zawar the roasted ore was reduced in the retorts and condensed in the attached condensing tubes (Fig. 11.5 and Plate 11.2) (see also Chapter 6, Figs 6.9, 6.10 and 6.16).

The charge

The charge for the retorts, at least in the later large and intermediate retort processes, was of some physical and chemical sophistication, as indicated both by the contemporary accounts, outlined above, and the scientific examination of the spent residues removed from the retorts (Kumar and Biswas 1986; Gangopadhyay *et al.* 1984; Friedlaender and Choo 1970) (see Chapter 7). The physical form of the charge was crucial to the operation. The charge in the retort had to maintain an open structure throughout the operation for two reasons: (1) the reactions taking place in the retorts generated large amounts of reactive gases, both carbon monoxide and zinc vapour – it was vital that these could move easily through the system, the carbon monoxide to carry out the reduction and the zinc vapour to escape to the condenser; and (2) the reaction is endothermic, thus all the energy needed to carry out the reaction had to be supplied to the retort contents from outside. If the charge was in the form of a powder energy transfer would have to be by conduction through the charge. This is relatively slow and would mean in practical terms that the charge adjacent to the retort walls would be reduced much more quickly than the charge towards the centre. If the charge has an open structure, then energy transfer is primarily by radiation which is instantaneous therefore the reactions proceed uniformly throughout the charge.

The material in the small retorts is generally loose, but in the large and intermediate retorts the spent charge is often firmly sintered together and to the retort walls (Plate 11.3). Although the intense heat of the process has meant that little of the original structure survives, it does support the hypothesis that the original charge was in the form of small balls approximately

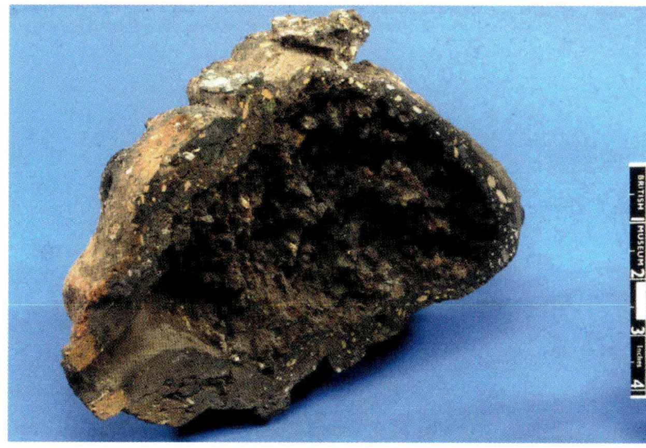

Plate 11.3 Half-sectioned used large retort (5048-49). Note the distortion of the walls of the retort walls, with relatively large fragments of quartz filler, and the sintered mass of the spent charge still apparently retaining something of the original open structure (A. Milton/British Museum).

1 cm in diameter. In the European horizontal retort process in the 20th century the ore was also normally charged in the form of briquetted balls or pellets (McBean 1959) therefore the descriptions given in the early literature of forming the ore and other reagents into balls with sticky organic materials before charging to the retorts are likely to be broadly accurate.

Carbon in whatever form was essential to provide the carbon monoxide necessary to carry out the reduction. For reasons discussed below it was essential that there was an over-pressure of carbon monoxide and that carbon dioxide was effectively transient, which could only be achieved by having an excess of fuel over that required theoretically. For example, in the European horizontal retort process it was usual practice to mix the calcined ore with about 40% of its weight of coke etc., which represented an excess of about ×2.5 (Smith 1918: 99), and in the old English vertical retort process about 60% of its weight was deemed necessary (Lones n.d.: 77). At Zawar a few flecks of charcoal survive in the furnace charges but it is impossible to identify the original organic materials from which they came; the exotic materials of the written descriptions seem very unlikely, but some more common plant material may well have been included, together with the ubiquitous dung.

Even though inorganic materials have survived better, there are still problems of interpretation (see Chapter 7, Tables 7.7 and 7.10). The residues have appreciable lime and magnesia contents, which must have originated ultimately in the dolomite associated with the ore, but it is not certain whether this was just the residuum of gangue mineral that still adhered to the zinc minerals after beneficiation and which went through the roasting process with them, or whether there was a deliberate addition of extra uncalcined dolomite. To deliberately add some of the gangue material that the beneficiation processes had been trying to reduce to a minimum might seem unlikely, however, the uncalcined dolomite would have provided additional material for the sintered structure, and more importantly the carbonates could have acted as initiators. In the early stages of the process in the very reducing conditions of the retort the carbonates are broken down to oxides and carbon dioxide, which in turn reacts with the by now elemental carbon of the organic materials to produce the essential carbon monoxide (see chemical formulae below). Also the slow, steady production of carbon dioxide continuing

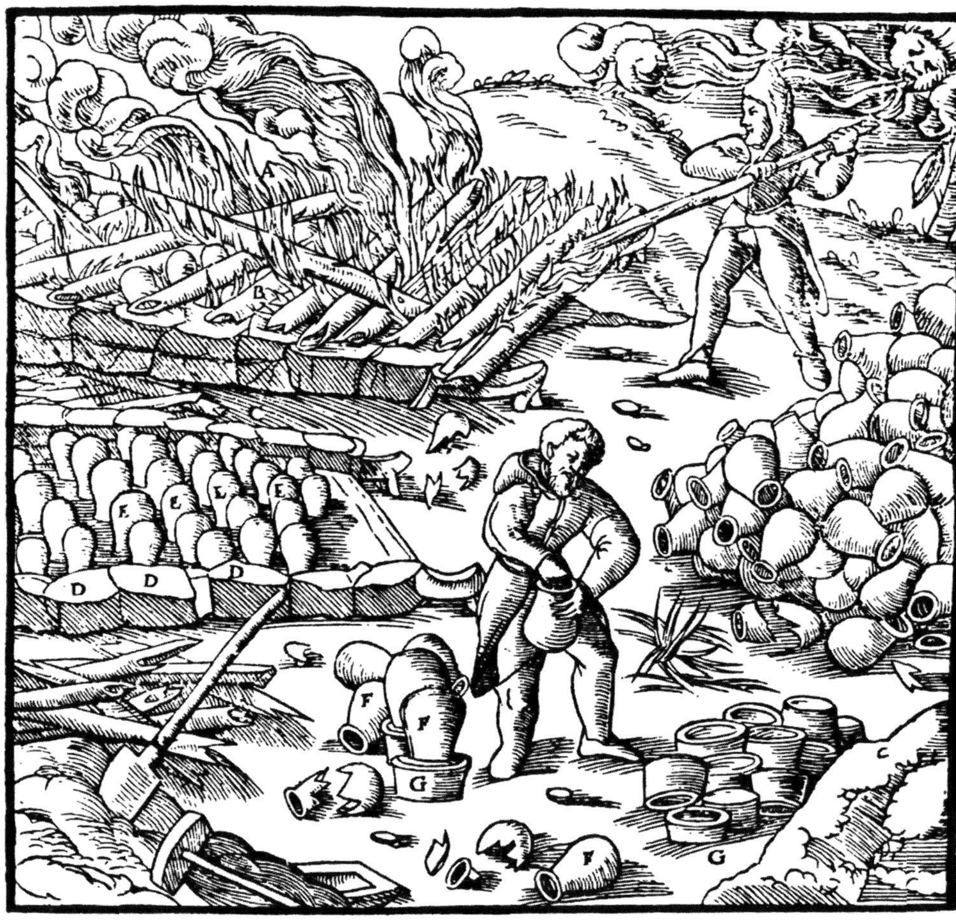

A—HEARTH. B—POLES. C—HEARTH WITHOUT FIRE IN WHICH THE POTS ARE PLACED. D—ROCKS. E—ROWS OF POTS. F—UPPER POTS. G—LOWER POTS.

Fig. 11.6 Mercury production in 16th-century central Europe. The principle of the method, including the arrangement of the retorts, is very similar to that used at Zawar and is probably itself ancestral to the Carinthian process used in the late 18th century to smelt zinc (cf. Fig. 13.4) (Almond 1998) (from *De Re metallica*, Hover and Hoover 1912: 427).

through the process would ensure a continuous over-pressure of carbon monoxide and dioxide which would protect the zinc vapour from reoxidation in the condenser tube while it condensed to liquid. Small quantities of limestone initiators are often added in other processes, thus for example limestone was usually added to the carbonaceous materials used in the case hardening of iron, and for the same reason, to promote the formation of carbon monoxide. Wagner (2008: 66) states that barium or calcium carbonates (the latter sometimes in the form of cow bones which contain both calcium phosphate and carbonate) were added as initiators by the Chinese in their crucible ironmaking processes. None of the early zinc smelting recipes mentioned any such separate addition, but they were not describing industrial processes. To an extent the plant material added as the source of the carbon instead of charcoal would also have acted as initiators, burning to produce carbon dioxide which would then react with the charred carbon to produce carbon monoxide.

The alumina content of the residues strongly suggests that clay was present. Small quantities of clay, probably mixed with the dung as a binding agent, could have been added to make the balls. The necks of the condensers were almost certainly recycled to recover the zinc that they contained and this would have introduced approximately 50 g clay into the charge (see Chapter 7).

The other two mineral additions mentioned in the early texts are borax and common salt. Samples of the retort residues were analysed by emission spectrography specifically for borax, which was detected but only in the trace amounts commonly found in clays; there was no evidence of deliberate addition. Both sodium and potassium together with chlorides are present in the spent charges, and were essential to bring about the slight vitrification of the silica in the charge to produce the sintering. These glass phases in the charge contain between 1 and 2% sodium and potassium, although they need not necessarily have originated from common salt, sodium chloride (NaCl) alone as there is relatively little chloride present. The approximately equal amounts of potassium and sodium suggest that saltpetre, the Indus earth of the early scientific writers, may also have been added. The traditional Indian saltpetre in the early 20th century comprised mainly potassium nitrate (KNO_3) together with some sodium nitrate ($NaNO_3$) as well as some sodium chloride (Leather and Mukerji 1911). If the balls were made up with dung then this too would be a source of sodium and potassium salts. Soaking the various components of the ore in urine – a good source of

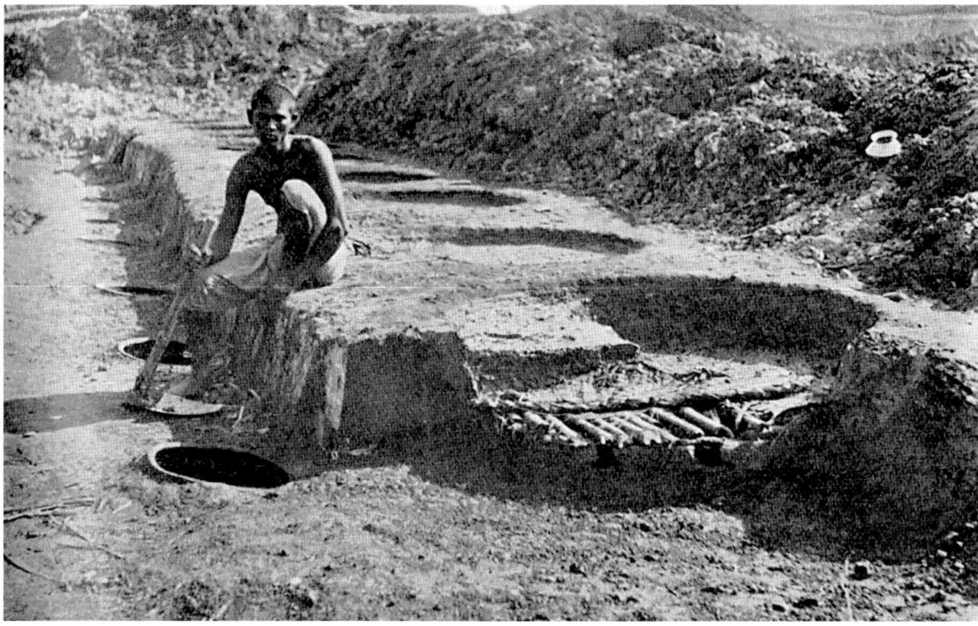

Plate 11.4 Bank of *kothi* percolators for extracting nitre from the soil (from Marshall 1915: 47).

both sodium nitrate and chloride – was frequently mentioned in the texts. A mixture of saltpetre and common salt seems the most likely flux but the quantities would have to be carefully controlled: too little and no sintering would take place, too much and excessive vitrification would seal the zinc minerals in glass.

Salt was certainly added to the retort charges in the European horizontal retort process – indeed the earliest specification by Dony, the inventor of the process, included salt in the retort charge. Additions of about 1 or 2% increased the yield of zinc. It was believed that it did this by slightly vitrifying the walls of the retorts and the glazing so produced stopped the absorption of the zinc vapour (Gatelier 1862, who recommended swabbing down new retorts with seawater and gum arabic). New retorts at Seaton Carew, Durham, in northern England were soaked in seawater before their first use.[19] The addition of small quantities of salt was also supposed to suppress the formation of the pernicious blue powder, which was the bane of more recent zinc producers. This was made up of tiny droplets of partially reoxidised zinc that blocked condensers and were difficult to treat (Lones n.d.: 101–2; Ingalls 1936: 347; Smith 1918: 101). The problem was especially serious at the outset of the process when the conditions in the condenser and collecting prolong were rather cool and oxidising and the first few droplets of zinc to condense would oxidise on the surface, preventing them from coalescing. The salt volatilised relatively early on and formed either volatile zinc chloride which dissolved the zinc oxide (Foster 1970: 485) or sodium oxide vapour which reduced the zinc oxide layer (Dean 1919). Either way the zinc oxide was removed from the surfaces allowing the droplets to coalesce. This would be essential in the Zawar process where there was a danger of the droplets condensing early on in the condenser and blocking it. At the end of the process when the charge was almost exhausted the last droplets would again be liable to oxidise, but the problem was less serious and the tubes could be broken off, ground up and included in the next ore charge. Furnace 1 of the Site 30 small retort furnace

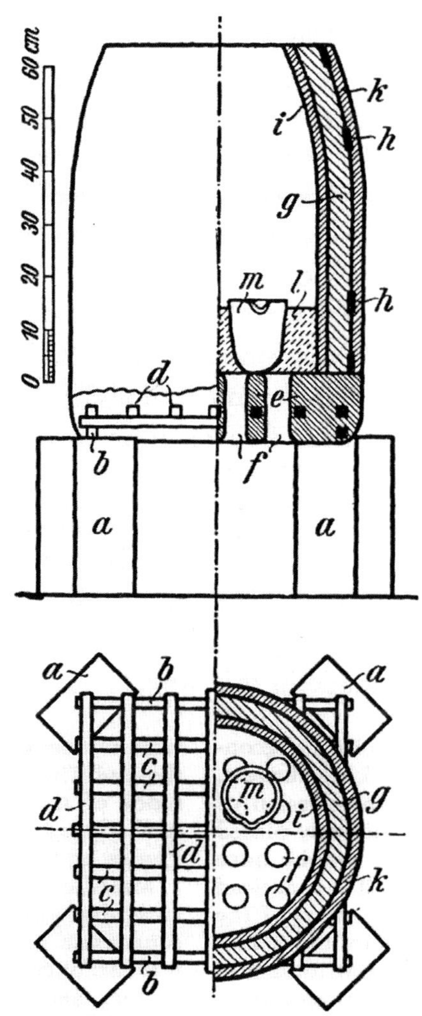

Fig. 11.7 Reconstruction of a medieval European furnace for the production of brass based on the description given by Theophilus: *a*: stone supports; *b*, *c* and *d*: iron grid; *e*: clay and dung surround, leaving circular holes; *f* and *g*: furnace wall of clay and dung tempered with pebbles; *h*: iron bands; *i* and *k*: clay linings; *l*: charcoal; *m*: the crucibles (from Hawthorne and Smith 1963: 142, fig. 16).

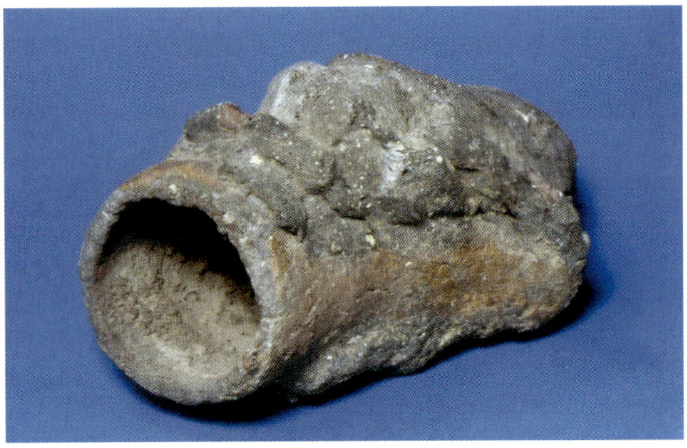

Plate 11.5 Used large retort (5048-50) which is empty, relatively undistorted, and from which the condenser has come away cleanly without breaking the retort's rim, showing how the original unused retorts must have appeared (A. Milton/British Museum).

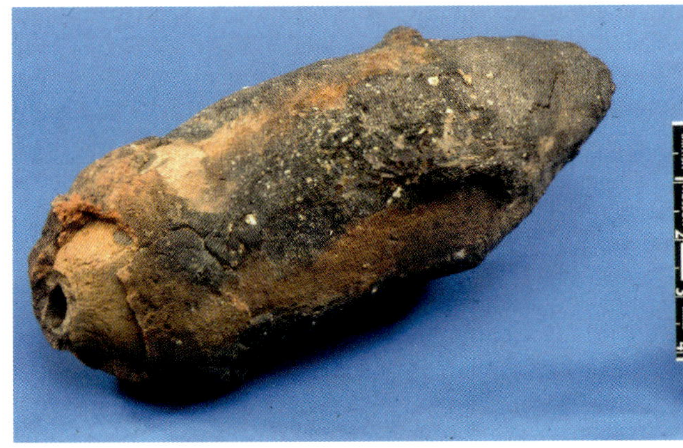

Plate 11.7 Intermediate retort with the condenser head still firmly luted in place (5048-7) (cf. Fig. 6.28). Note as usual the zinc-laden condenser end is missing, presumably recycled to recover the zinc.

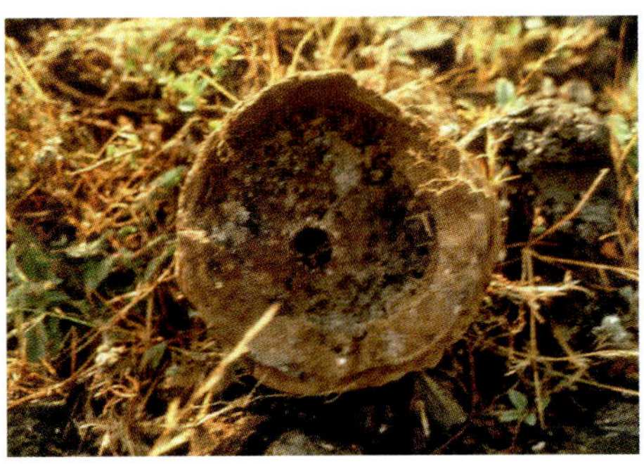

Plate 11.6 Small retort broken approximately in half exposing a section of the spent charge with the central channel in the middle (cf. Fig. 11.5).

block, the north furnace of the Site 37 large retort furnace block and the excavated furnace of the Site 33 intermediate retort furnace block all still had their last charge of retorts in place but all were missing their condenser ends (see Chapter 6, Plate 6.23), suggesting that they were removed when the putative collecting vessels were taken out rather than when the retorts themselves were removed.

Principle of the process

The principle of *tiryakpatana yantram* – literally distillation by descending – was probably already well known in India long before the distillation of zinc began, and a variety of systems is described in the early literature (Subbarayappa 1999; Rây 1956: 189–91). The descriptions of the production of mercury, which also necessitates the use of some distillation technology in order to condense and collect the vapour, are especially relevant here, particularly as *rasak* (the medieval Hindu term for zinc ore) derives from *rasa* (the Tantric word for mercury). Mercury-gilded metalwork has been found in India dating from the last centuries of the first millennium BC, and there are frequent references to the metal from the time of the *Arthaśāstra* (2.13.44) (Kangle 1965–72, II: 114). Early descriptions of the production of the metal in Europe suggest that the process was that of distillation by descending, with the mercury ore in the upper vessel surrounded by the fire, dripping down mercury into the lower cool vessel as described by Agricola (Hoover and Hoover 1912: 426) (Fig. 11.6).

Apparatus for downward distillation is also described in the *Rasendrachudamani*:

> *Adhaspatna Yantram*: a modification of the above apparatus in which the bottom of the upper vessel is smeared with the substance, the vapour or essence thereof condensing into the water of the lower one. Heat is applied on the top of the upper vessel by means of the fire of dried cow-dung cakes (Rây 1956: 189–90).

The *Rasaratnakara* describes the process for the smelting and collection of mercury:

> *Darada* [cinnabar], when distilled in a *patina yantram* [distillation or sublimation apparatus] into a vessel containing water, yields an essence identical with mercury. There is no doubt about it (Rây 1956: 131).

This seems very similar to the process described for producing zinc, although *tiryakpatana yantram* is not actually mentioned. However, other medieval Indian descriptions of the apparatus used specifically for mercury production are rather different. In the *Rasendrachudamani* of Somadeva, compiled in the 12th or 13th century AD, and quoted in the *Rasaratna Samuchchaya*, it is stated that:

> *Vidyahara Yantrum*: Vidyahara Yantrum is for the extraction of mercury from cinnabar. Two earthen pots are arranged one above the other. Place cold water in the upper vessel and heat the lower one by burning fire below. Cinnabar is placed in the lower vessel, and mercury condenses at the bottom of the upper vessel (Rây 1956: 190–91).

Although possibly not used for mercury making in India, the principle of *tiryakpatana yantram* was already known and used. The adoption of the local distillation technology to the condensation of zinc is paralleled around the world (Craddock 1987, 2009c).

Origins and parallels for the *koshthi* furnace

The *koshthi* in which the zinc smelting took place have been fully described in Chapter 6 and only varied in size during the centuries that zinc was smelted at Zawar. The form of the *koshthi*, a furnace divided horizontally by a perforated floor, is paralleled in a number of other installations. The standard pottery kiln is the most obvious analogy with the pots stacked in the upper chamber but with the fire beneath, separated by a perforated ceramic floor. Also analogous were the *kothi* installations for draining and condensing the salt-laden liquors in the traditional production of saltpetre in north India (Plate 11.4). If the distillation of zinc was developed by the brassmakers, as was certainly the case in Europe and probably also in China (see Chapter 13), then the existing furnaces for heating the crucibles containing the copper and zinc ore would possibly have been a more direct parallel. No brassmaking furnaces have been found in India and there are no descriptions, but the furnace described in great detail by Theophilus (Hawthorne and Smith 1963: 142, fig. 16) (Fig. 11.7) in 12th-century Europe for brassmaking is very close to the form of the Indian *koshthi*. The main difference is that in both the pottery kilns and brassmaking furnaces the fire was located in the lower chamber, whereas in the *koshthi* the fire was positioned in the upper chamber. Accordingly, a combination of the principle of *tiryakpatana yantram*, taken from existing distillation technology, combined with the form of the potter's kiln or brassmaker's furnace produced the *koshthi* in which zinc could be produced on an industrial scale.

The smelting operation

The open retort (Plate 11.5) would be filled with the sticky ball, and a stick of about 1 cm diameter pushed into the charge. This small practical detail, not mentioned in any of the written descriptions, is evidenced at Zawar by the central hole running through the centre of the retort filling in some retorts (Plate 11.6). Occasionally fragments of the charred sticks were found in the channel, evidence that they were not removed prior to firing. The stick fulfilled the dual function of creating the central channel down which the zinc vapour was to pass, as well as preventing the charge from falling out once the sticky organics had charred but before sintering had taken place to hold the charge. Finally, the condenser would be placed over the retort end and securely luted in place with clay to hold it in position and to prevent any air entering the retort (Plate 11.7), ready for loading into the *koshthi* (frontispiece).

Consideration of the free energy diagram (see Chapter 7, Fig. 7.4) shows that the reaction can only proceed at temperatures above 900°C. The principal reactions taking place inside the retorts may be expressed as follows:

$$\text{Ca Mg(CO}_3)_2 \rightarrow \text{CaO} + \text{MgO} + 2\text{CO}_2$$
$$\text{Organic material} \rightarrow \text{C} + \text{H}_2\text{O} + \text{CO}_2$$

The main reactions in the retort are those between the charred organics and CO_2 and between CO and ZnO, as represented by the two equations below. As can be seen they are both strongly endothermic therefore all the energy necessary has to be supplied from outside the retort:

$$\text{C} + \text{CO}_2 \leftrightarrow 2\text{CO} - 174\ \text{T J mol}^{-1}$$
$$\text{ZnO} + \text{CO} \leftrightarrow \text{Zn} + \text{CO}_2 - 115.6\ \text{T J mol}^{-1}$$

The reaction between the zinc oxide and carbon monoxide at high temperature lies very much to the left, that is, the carbon dioxide will oxidise the zinc unless there is excess carbon present, which will preferentially react with the forming carbon dioxide to convert it back to carbon monoxide. In the European horizontal retort process even as little as 0.25% of carbon dioxide in the vapours would cause significant losses (Smith 1918: 93).

In the only near-contemporary description of the traditional zinc smelting process at Zawar (Brooke 1850) (see Chapter 3) it was stated that the process took three or four hours and this is supported by the scientific examination of the ceramics of the furnace walls, where the depth of penetration of vitrification (see Chapter 7, Table 7.9) also suggests that the furnaces were held at maximum temperature for approximately four hours.

Studies on the degree of vitrification of the retort walls and furnace linings suggest that the maximum operating temperatures of both the small and large retort furnaces were in the region of 1200°C, and the usual operating temperatures around 1100°C (see Chapter 7, Fig. 7.3). These temperature estimates are supported independently by the study of the mineralogy of the large retort residues carried out by Friedlaender and Choo (1970), who noted that the presence of 'gahnite and willemite, and of clinohypersthene and hypersthene place the temperature of the smelting process around 1100°C'. This is relatively low compared to the usual operating temperatures of the European horizontal retort process where temperatures of 1400°C were usual, and the reaction only proceeded very slowly at temperatures below 1225°C (Smith 1918: 93). However, it was noted that if soft coke or charcoal was used in the retort, then zinc could be produced at temperatures as low as 1000°C. For temperatures in this range a pO_2 of at least10^{-16} is necessary, and in fact the pO_2 was typically around 10^{-20} (see Chapter 7, Fig. 7.4). If the temperature had reached much higher the relatively iron-rich charge in the Zawar retorts would have formed a molten slag with the silica in the charge and retort walls, thereby terminating the process. Overall, the temperature in the retorts can be seen to have been contained within tight limits: a few degrees too low and the reaction falls away to almost nothing, a few degrees too high and the charge begins to melt.

The temperature and reducing conditions within the condenser tube were, if anything, even more critical. There must have been a very considerable temperature drop down the approximately 10 cm length of the condenser tube. Zinc vapour must have entered at temperatures in the region of 1000°C but have condensed to a liquid at around 500°C at the bottom. Although zinc boils at 907°C, in reality the vapour only condenses very slowly unless the temperature is below 500°C, but the temperature cannot drop much below 419°C (the solidification temperature) otherwise the zinc could freeze and block the tube. In fact there was probably a greater danger from the formation of blue powder blocking the tube, as observed in several tubes. The formation is temperature dependent and in the traditional zinc smelting industry in Europe the prolongs of the condensers had to be kept at about 500°C to minimise the formation. In practice the range of permissible temperature for the condensation in the European horizontal retort process was between 415 and 550°C (Smith 1918: 93). Thus at Zawar the temperature of the zinc had to fall by approximately 500°C to within a temperature range of about 50°C in a very short distance and, presumably, time. The material issuing out of the condensers could be observed either directly or by the colour of the flame as the necessary excess carbon monoxide burned. The conditions in the condenser could only have been controlled indirectly by the fire around the retort above: the hotter the retort the faster the production of CO and zinc gases and therefore the faster they would have been driven down the condenser. It was also essential to maintain the reducing conditions, especially within the condenser, by maintaining an over-pressure of carbon monoxide until the zinc vapour had condensed.

It was necessary to maintain this precise temperature regime rigorously for each retort throughout the whole furnace block of seven furnaces each with its load of 36 retorts. The obvious way to achieve this was by the judicious supply of fuel. As the spaces between the retorts are very restricted it must have been necessary to frequently add more charcoal. Through careful selection of the appropriate size of lump and quantity it was probably possible to achieve considerable local control. The overall rate of combustion was apparently achieved by controlling the flow of air through the entrance to the lower chamber, some of which still contained their blocking stones.

Great care was obviously taken to try to ensure that the retorts were evenly spaced and the spacing between them was regular and open. This was to maintain an even passage of air through the furnace thereby promoting uniform combustion, and thus heating, around each retort. In some of the earlier small retort furnaces, occasional retorts were found lying across and fused to the vertical retorts which were clearly wrongly oriented to have been part of the original load in the furnace, suggesting that old discarded retorts from the surrounding heaps were being used as convenient spacers. On the later large and intermediate retorts there were regular deliberately made wedges that fulfilled the same function (see Chapter 6, Figs 6.22 and 6.23). In the latter part of the process slumping of the ceramic bodies of the retorts meant that many had actually slumped so much that they fused together, effectively shutting off the air supply which must have severely affected the combustion locally. As this would have taken place towards the end of a relatively short process, it probably did not matter too much.

It has been suggested that *koshthi* would have needed to have some sort of superstructure in order to achieve the updraught necessary for the high temperatures required, effectively turning them into a bowl or even a sub-conical shaft furnace (Hegde *et al.* 1986: 254, fig. 24.6; Hegde 1991: 82, fig. 35). There is absolutely no evidence for any such putative upper parts, either fragmentary in the retort dumps or on the furnaces themselves (as discussed in more detail in Chapter 6).

The one element of the process for which the evidence was totally absent from our investigations was the collection of the zinc in the cool chamber (hence no collecting vessels are shown in the reconstruction diagrams). There are two possibilities: either the molten zinc dripped down from the condenser neck directly onto the chamber floor or there was a separate collecting vessel or crucible set beneath each neck. Morgan (1985: 14) favoured the former method, citing the Overpelt process as a parallel. The evidence for this at Zawar was the small amount of metallic zinc recovered from the floor of one of the Site 30 cool chambers (although unusually here the retorts and plates had been removed suggesting there may have been an atypical spillage). The absence of recognisable crucible fragments from the heaps of retort debris could also be taken as evidence that they had never existed. The evidence for collecting vessels is all rather negative. There is nothing to indicate how the putative zinc, once having settled on the floor, could have been managed or removed. Also, more seriously, the burning ash which certainly exited through the small holes in the plates would have fallen onto the forming sheets, not only contaminating them but almost certainly causing appreciable oxidisation of the zinc. It is most unlikely that the chamber floor was sufficiently hot for the zinc to remain molten and flow out of the entrance. Furthermore there are no channels and the floors do not slope, so it would have been necessary to dismantle the floor of perforated plates after each smelt and pull out the thin sheet of metal that would have formed. This is not impossible; the perforated bricks do seem to have been deliberately made to be easily removed without damaging the rest of the furnace. However it is more likely that this was done so that they could be individually replaced when broken. Furnaces 1 and 2 at Site 30 still had their retorts in place but there was no zinc on the floors of their cool chambers. Another serious objection is that the floors were quite badly cracked and there was no evidence of metal having penetrated into the cracks, or that any attempt had been made to seal them, even though they could potentially have led to a serious loss of metal. The tapered central support for the four plates shows that it was paramount to keep the lower space uncluttered and accessible, suggesting that there were some individual vessels that might have needed attention while the process was in operation and could be removed afterwards without having to disturb the plates at all. Another piece of indirect evidence for individual vessels are the missing condenser necks from the retorts that were still place from the last operation of the furnaces, as mentioned above (see Chapter 6). If the necks were protruding into collecting vessels, then removing the latter would break off the former otherwise it is difficult to explain why the necks are missing from retorts which were still in place. These putative collecting vessels could have been of ceramic that, like the condenser ends, would have become soaked in zinc oxide and blue powder and were therefore crushed and added to the next retort charge. It is also possible that they were of iron.

The contemporary descriptions of the distillation process as carried on in a laboratory, while similar in principle, differ in details. Some of these describe the zinc as being collected in vessels under water. It is difficult to see how water in the catchment vessels would survive for approximately four hours at the quite high temperatures prevalent in the lower chamber surrounded by burning coals and ash, collecting the molten zinc itself at temperatures in excess of 400°C.

Production estimates

The amount of zinc produced in the retorts can be estimated with some accuracy from their dimensions and the study of their contents (see Chapter 7, Tables 7.7 and 7.10). The concentration of the involatile metals such as silver and iron left in the retort residues gives an indication of the amounts of zinc in the original retort charges, and if the amounts of these metals were the same as in the concentrates currently produced, which seems reasonable, then approximately half the charge would have been of zinc. Each small retort held about 300 g of residues, which equates to about 150 g zinc in the original charge. Of these 150 g, not all, would have been recovered as metallic zinc – some remained in the residues or was absorbed in the retort walls, and more would have been converted to zinc oxide in the condenser or lost to the atmosphere. In the European horizontal retort process losses were typically in the region of 15–20% (Smith 1918: 93) so figures of between 25 and 33% seem not unreasonable for Zawar. Consequently, each small retort could have produced about 100–125 g zinc or about 25–30 kg per furnace block of 252 retorts per day, assuming a daily cycle.

The large retorts contain around 1000 g of debris so would originally have contained about 650 g zinc, which on the estimates made above should have produced about 500 g per retort or approximately 50–60 kg per furnace block of 108 retorts per day.[20]

Refining

It could be thought that zinc, the product of distillation, would be relatively pure. However, the product of the European downward distillation process (the so-called English process) was highly impure, containing considerable quantities of zinc oxide and the inevitable blue powder, with other debris from the retort above (Lones n.d.: 78), and the metal produced at Zawar is likely to have been similarly impure. The common elemental impurities in zinc produced by distillation are lead with a little cadmium, germanium and iron. The tiny spill of metallic zinc found on the floor of furnace 3 at Site 30 contained about 2% lead (see Chapter 7).

In 19th-century Europe the zinc was usually remelted in cast iron pots and the zinc oxides and other dross skimmed off before ladling into the ingot moulds. In the present-day traditional Chinese process the usual practice is very similar, that is to melt the freshly distilled zinc in an open iron vessel under burning rubber and plastics to maintain a reducing atmosphere and skim the surface of solid impurities, mainly blue powder particles, before ladling into the ingot moulds (Craddock and Zhou 2003).

Notes

1. General sources: Ingalls 1936; Craddock 1998c. India: Craddock *et al.* 1998a; Darling 1990; Kharakwal 2011. China: Zhou *et al.* 2012. Europe: Percy 1861; Day 1973, 1991, 1998.
2. We are very grateful to Caroline Cartwright for guidance in these matters.
3. General sources: Craddock 1978, 1995: 292–302; Craddock and Eckstein 2003; Eckstein *et al.* 1996. Middle East: Allan 1979; Craddock *et al.* 1998c. Europe: Theophilus, Hawthorne and Smith 1963: 139–45; Smith and Gnudi 1942: 70–76ff. (Biringuccio); Martinón-Torres and Rehren 2002 (Agricola); Sisco and Smith 1951: 254–8 (Ercker); Woolrich and den Ouden n.d. (Galon); Percy 1861: 606–18; Day 1973, 1998.
4. Four analysed copper alloy artefacts from Atraṇjikhera, in Uttar Pradesh, dated to between 1200 and 600 BC, all contained tin, but one was reported to contain 6.28% zinc and another 16.2% (Gaur 1983: 483–90). Ten copper alloy artefacts from Prakash, an important settlement site in the Shahad Taluk of the Dhule District of Maharashtra situated at the confluence of the Tapti and Gomai rivers were analysed and two, from the NBP phase dated to between 600 and 100 BC, were found to be of brass (Thapar 1964–5). One was a bracelet with 25.9% zinc and 8% lead, the other was a coiled ring with 17.8% zinc, and neither contained any detectable tin, which was the main alloying element in the other pieces (Athavale 1967: 136, table IV, also reported in Bhardwaj 2000: 119, table 44). Conversely, 35 copper alloy artefacts from Kausambi in Uttar Pradesh, dating from the second half of the first millennium BC (Sharma 1960), were found to be mainly of bronze, including some pieces with over 20% tin and a number of leaded coppers, but with no significant amounts of zinc (Prakash and Rawat 1965: 53–4, table 5.5). Some copper alloy artefacts from early historic period Asura sites in Chhotanagpur were found to contain between 19 and 25% zinc (Caldwell 1920).
5. The brasses from Taxila are paralleled elsewhere in the region. L.P.H. Hurtel and F. Tissot at the *Crossroads of Asia* conference (Madingley Hall, 6 October 1992) reported analyses carried out on a series of cast copper alloy vessels from Begram in the Kabul Valley of Afghanistan, and dating from the second century BC. They were all found to be of leaded brass. This near total reliance on brass at the expense of bronze at such an early date is presently unparalleled and was not to take place in the West for at least a century. Furthermore, the Begram brasses had low iron contents, suggesting either the use of metallic zinc or more likely a sublimated ore.
6. Rahman *et al.* 1980; Rây 1956; Tripathi 1978; Dash and Kashyap 1994; Roy and Subbarayappa 1976 (*Rasarnavakalpa*); Satpute 2003 (*Rasaratna Samuchchaya*).
7. The problems of the translation of *tuttha* have been discussed in Chapter 10 (p. 202), but in this context it must be some form of zinc oxide.
8. This recalls a somewhat similar process for using the zinc vapour directly that was supposedly patented by John Champion, the brother of William Champion, who pioneered zinc production in 18th-century Britain (Chapter 13, p. 247). In this process the zinc vapour was to be absorbed directly into hot but solid copper plates (Watson 1796: 65). It is unclear how the process would have achieved more than a brass plating, and the process seems unlikely to have met with any commercial success. Staunton (1797) who accompanied Lord Macartney's delegation to China in 1793–4, also reported a process for the production of brass by exposing copper sheets to zinc vapour (1797: 541) as well as an equally unlikely process for the production of zinc by a European-style distillation unit (1797: 540), totally different from that actually used in China.
9. Craddock and Eckstein 2003; Eckstein *et al.* 1996.
10. There is some evidence that in contemporary India the zinc oxide for pharmaceutical use was prepared from metallic zinc, which would be the purest of all (see p. 217).
11. One of the uses claimed for zinc in 18th-century Europe was to produce a superior brass, free of the iron found in ordinary cementation brass (Day 1973: 123–5).
12. General sources: Ingalls 1902, 1906; Morgan 1985; Lones n.d.; Smith 1918; Percy 1861: 518–605). Historical: Day 1991, 1998; Craddock 1995: 309–17, 1998a, c. Europe: Ingalls 1936; Almond 1998. India: Ray 1902; Rây 1956; Freestone *et al.* 1985a and b; Craddock *et al.* 1985a; 1998a and b; Kharakwal 2011. China: Zhou 1993, 1996; Craddock and Zhou 2003; Liu *et al.* 2007; Zhou *et al.* 2012.
13. Moreover, before the 19th century zinc was frequently confused with tin, and lacking any other name, the metal was often referred to as Indian tin; as late as the 1820s no less an authority than Tod (1978, I: 117, 223ff.) (Chapter 3, p. 20) described the workings at Zawar as tin mines.

14. Recent Āyurvedic descriptions of the preparation of zinc oxide from metallic zinc have been published by Dash (1986: 127–8) as follows: 'Zinc should be melted in an iron pan. To this, equal quntity (*sic*) of the powder of *bhaṅgā* (*Cannabis sativa* Linn.) and poppy pod should be poured in small quantities. While pouring this powder, zinc should be rubbed with the help of a strong iron spoon. When all the zinc gets reduced to ash, another iron plate should be kept over the pan and strong heat should be applied below till the iron pan becomes red hot. Thereafter, it should be allowed to cool of its own. It should then be washed for separating the ash of the powder added to it and dried. Then it should be triturated by adding the juice of *kumārī* (*Aloe barbadensis* Mill.). Small cakes should be made out of the paste, dried in sun and kept in *śarāva samuta* and cooked in *gaja puṭa*. This process should be repeated for seven times.'
15. Yet at the time of Marco Polo's visit in the late 13th century *tutty* (zinc oxide) was being imported into India (Chapter 13, p. 245).
16. Note that in the early stages of the process the flame would have been orange if the charge contained small quantities of common salt as often specified.
17. European curiosity was aroused by this strange metal *caleum* or *calim* encountered in the East. For example De Laval in 1679 noted that *calim* was whiter than tin but more beautiful, and that the metal was highly prized throughout India (quoted in Partington 1961: 109). Europe's first encounter with zinc were the strange droplets of metal that appeared in the flues of certain lead/silver smelters and the nature of this *counterfrei* aroused considerable curiosity. Brass was produced by cementation of calcined zinc carbonate smithsonite ore and copper, but even though *counterfrei* was sometimes added, the connection between the two seems not to have been made. Not until 1721 did Henckel establish that smelting and condensing smithsonite mineral produced zinc. Even so there was considerable uncertainty over the nature of zinc, probably due in part to its brittleness and flammability and up to the beginning of the 19th century it was often designated as a half metal.
18. The second and third alloys are identical as given in the 1989 print of the Phillott edition, possibly 1:1 copper:zinc was intended, which would be a hard brittle brass.
19. J.K. Almond, pers. comm., 1985.
20. This compares to about 300 kg per day for the six large retorts in the Old English vertical retort process as practised in the mid-19th century (Percy 1861: 557), 20–25 kg per day for the 36 retorts in the recent Chinese traditional trough furnace process, and about 100 kg per day in the late 20th century in furnaces with between 60 to 90 retorts (Craddock and Zhou 2003). A 19th-century European horizontal retort furnace typically held about 800 retorts, each producing about 35 kg of zinc per day (Morgan 1985: 61).

Chapter 12 Extinction and revival: Indian metallurgical industry from the 17th to the 21st century

P.T. Craddock

A major consequence of the development of international maritime trade in the post-medieval period was the exposure of local economies around the world to more sophisticated systems. The results were often calamitous, especially for indigenous production when faced with cheaper imports. Yet from this nadir new discoveries and new industries ultimately developed. This worldwide phenomenon is exemplified by the experience of the various mining and metallurgical industries in India. Mines and smelters such as those in the Aravallis have revived and continue in production.[1] A related issue is the misconceptions of 19th-century Europeans of the Indian mining and smelting activity that they observed. A consideration of the broader context of the overall Indian experience enables a better understanding of the fundamental causes of the decline and rejuvenation generally.

Traditional metal trade

India traditionally produced metals, including gold, base metals and iron and steel in quantity for domestic use. The overall trends in India's external trade through the centuries was the import of precious and base metals and the export of iron and steel (Craddock 2013). This was to change at the end of the medieval period when European vessels made contact and true global trade began. European purchases were made in silver from American sources, and soon other metals, notably copper and lead from Europe and zinc from China, found ready markets in India and began to supplant local production. Large quantities of copper were being produced in Rajasthan at Khetri etc. up to the end of the 16th century but declined sharply thereafter (see Chapter 2), the same situation as was taking place at Zawar. Throughout the post-medieval period India's requirements of copper together with tin, lead, zinc and silver were increasingly met by imports (Khan 1986).

Initially iron and steel production was less affected. With huge forests and vast iron ore deposits, wrought iron and crucible steel had been a major Indian export commodity for millennia to Southeast Asia and across the Arabian Sea to the Middle East (Goitein and Friedmann 2008). Indeed iron and steel could be regarded as forest products so reliant was the production on abundant charcoal (Craddock 2009b). This initially continued in the early days of the international maritime trade with European vessels engaged in country trade around the Arabian Sea and Indian Ocean and into Southeast Asia, regularly carrying cargoes of Indian iron and steel, although apparently not to Europe (Bronson 1986, 1992). The export of Indian iron to the Middle East probably ceased in the 18th century, replaced by European iron, and by the end of the 18th century the Indian domestic market itself was being increasingly supplied by imports from Britain although some indigenous iron and steel production persisted and from the mid-19th century the introduction of European blast furnace technology to India had already commenced although with little initial success (see below).

The demise of indigenous metal production

Many have speculated on the reasons for the decline of the indigenous Indian metallurgical industry, for example Sundaram *et al.* (1999: 51) enquired:

> if we could achieve such heights in the past with respect to metals-technology, where and how did we lose the benefits of all our experience, our heuristics and our methodologies? Was this because of the absence of objective documentation? Or, was it due to foreign invasions? Whatever may be the reasons, the paramount task before us is to critically study these ancient technologies and achieve new heights of excellence in metals and materials engineering, relevant to the needs of the present and the future.

A number of factors have been suggested including the exhaustion of the most accessible and richest ores at many mines, the unsettled political conditions following the collapse of the Mughal Empire, competition from imports coupled with a general pessimism on the abilities of the indigenous workforce and infrastrucure. These will be examined below.

Exhaustion of ores

India had a long tradition of mineral prospection and working, leaving few virgin deposits to be discovered as was the case in some other parts of the world. As the geologist and gold prospector Malcolm Maclaren (1908) gloomily acknowledged:

> In India the prospector is offered an uninviting field, the whole auriferous zones have been exploited anciently. The spoil heaps only mark the valueless fields, the richer having been carted away to the rivers for crushing and washing.

There is some truth in this; the major indigenous mining operations of the past had indeed removed the richest ores often to very considerable depths as our surveys in the Aravallis have demonstrated. Therefore new operations would require considerable experience and investment to succeed. The freshly discovered virgin deposits of the Americas, Australasia and sub-Saharan Africa were much easier and cheaper to work. Also, although an ever-increasing proportion of Europe's non-ferrous metals came from overseas, it was usually in the form of the beneficiated ore to be smelted at centres such as Swansea in South Wales rather than as metal (Craddock and Hook 2012).

Politics

By the time the first serious geological, topographical and economic surveys were being undertaken in the 19th century much of India had faced economic and political disruption for well over a century. The collapse of the Mughal Empire following the death of Aurangzeb in 1707 created long-term instability such as the more immediate disruption brought about by the Mahratta wars. These ranged over parts of the subcontinent in the late 18th and early 19th century, and were especially destructive in Rajasthan for at least a generation (Tod 1978, I: 363–4) (Chapter 3). This was the long-term confused state of affairs that the European powers encountered and believed to be endemic. In Rajasthan the British certainly exploited the situation enabling them to extend their influence in the 19th century. However this did not necessarily result in the stability and peace that Tod claimed. In Mewar, in the vicinity of Zawar, for example, unsettled conditions involving the local tribal Pindars and Bhils actually continued throughout much of the 19th century, once again undermining any serious attempts to restart mining (see Chapter 3).

Faced with these conditions large-scale, well-organised operations came under great pressure – not only from direct local disruption and destruction brought about by war, but more generally the long-term absence of security meant that established skills were lost and the necessary training and apprenticeships not taken up. As most major metal production operations were state owned, in the absence of stable governance over much of the country effective control ceased and such operations that continued were for short-term profit only. With no proper investment in infrastructure the operations inevitably declined and many mines became unworkable as exemplified by the lamentable conditions prevailing in the early 19th century at the Singhana and Khetri copper mines that had formerly been major producers (see Chapter 2).

By the time of the first serious European reports, the large-scale technically proficient metal industries had gone, and rather surprisingly were already largely forgotten. For example James Tod, who took a serious interest in all matters concerning Mewar where he was based for a number of years, was completely ignorant of the zinc distillation processes that had taken place at Zawar only a few years before his arrival, believing instead that the mines produced tin (Tod 1978, I: 117).[2]

Competition

Metal production was severely curtailed by competition from imports. This was widely acknowledged at the time, but after the introduction of free trade with unprotected markets throughout the British Empire in the 19th century there was little prospect of any protection by import tariffs. This threatened the survival of the traditional metallurgical Indian industries and discouraged the creation of new ones. Laissez-faire policies were completely alien to the traditional Indian paternalistic system with assets such as mines being administered by the ruling authority in a market protected both by legislation as well as by poor communications. This exposure has sometimes been represented as a deliberate British policy of curtailing Indian industries (Gupta 1960). Rahman *et al.* (1980: ix) for example, stated that: 'The British carried out their objective of de-developing India through the destruction of her industries, which must have given them great satisfaction.' De-development would not have been a sensible way to make India economically viable or, more practicably, profitable. It was the principle of free trade rather than any deliberate policy of suppression that prevented the resurgence of Indian metal production in the face of imports from Britain and elsewhere. The same principle just as surely allowed the British non-ferrous metal mining industry itself to die unprotected in the late 19th century in the face of competition from the vast easily won deposits of the Americas and Australasia.

The situation at Zawar through the centuries provides a good example of the problems faced by economies seeking to recover from local misfortunes undermined by outside competition. Zinc production at Zawar was almost certainly curtailed by military disruption on several occasions in the 13th, 16th and late 18th centuries (Chapter 3). From the first Zawar could recover but by the time of the second disruption in the 16th century, imports of Chinese zinc in European vessels permanently captured much of the trade. After the Mahratta incursions at the end of the 18th century there was no possibility of the traditional process recommencing in a market now completely dominated by imports from Europe.[3]

The damage to local production caused by imports was widely understood. The eminent geologist, Valentine Ball – whose *Manual of the Geology of India* (1881) is a major survey of the mineral sources and the state of production of 19th-century India – was in no doubt as to the ultimate cause of the decline. For example, the manufacture of iron 'has in many parts of India been wholly crushed out of existence by competition with English iron, while in others it is steadily decreasing, and it seems destined ultimately to become extinct' (Ball 1881: 338). Similarly for copper production he noted (1881: 268) that English copper had been undercutting locally produced metal at Almora in Uttarakhand, hundreds of kilometres from the coast, back in 1838, even before the advent of the railways. Abraham Locket recorded a similar situation at the copper mines at Singhana in the Aravallis where by 1831 imported English copper had already undercut the better quality local copper, leaving the local industry no choice but to produce inferior copper that could compete in price (see Chapter 2).

Development of the inland transport infrastructure, notably the railways but also the road system, together with increased security and relief from local customs levies and dues, had the effect of exposing inland markets previously protected by poor communications as well as local patronage. By the turn of the century, the economic geologist Sir Thomas Holland (1908: 1–5) could state that:

> Today manufacture of iron by primitive *lohars* is restricted to the areas far removed from the railways which distribute the imported foreign articles. Copper and brass vessels are made exclusively from imported materials and every attempt to work the known copper-deposits has come to grief. No lead mining now exists in India, although the mineral galena is known in many localities.

In fairness he continued in a more upbeat manner, noting that:

> There is now a proposal of introducing the production in India of iron and steel. Copper, lead and some of the inorganic chemicals are also attracting the attention of investors, and active prospecting operations are now being undertaken. ... since 1899 the number of prospecting concessions has gradually, and more recently rapidly increased.

Indigenous production declined steadily through the 19th century, as revealed when reports were compiled on operations over a number of years, exemplified by the copper workings at Khetri-Singhana and the lead workings at Ajmer (see Chapter 2) in the Aravallis. Similar declines were observed in the Himalayas where Stephens (1901–2: 393–414) recorded the fortunes of the mining industry at the Dhanpur, Dhobi and Rai mines in the Garhwal District of Uttarakhand through the 19th century. In 1830 there had been 30 copper and iron mines working, which had increased to 76 in 1865, but by 1873 the number had fallen to six, partly due to the low price of copper on the market and partly to new forest laws limiting the cutting of timber for charcoal production. At the time of Stephens' report the industry was effectively moribund. The government of India had sought to increase production, instigating at least six reports between 1820 and 1870, and Stephens' report was yet another. Stephens believed that the few attempts to reopen the mines had failed through undercapitalisation. Traditional Indian industries failed to attract the capital necessary for them to develop and evolve while remaining in production in the economic climate of unrestricted imports. It was extinction by neglect, rather than deliberate suppression.

European support and investment in indigenous Indian industry now appears somewhat contradictory (McPherson 1993). There were numerous attempts to reopen old mines with new processing and smelting facilities but these were almost always undercapitalised and lacked sound planning and management so were largely unsuccessful, as recorded in Ball (1881). In the meantime the traditional indigenous extractive metallurgical industries, including the reasonably sophisticated technologies producing zinc and crucible steel, were allowed to dwindle away to nothing. Meanwhile the new industries of the European industrial revolution were vigorously developed in India resulting in considerable investment in the railways and the coal industry throughout the 19th century. It was not just foreign investment that was missing, Khan (1986) believed that the local aristocracy was not interested in investing in industry generally. It is interesting to compare this situation with that in contemporary Japan where indigenous capital does seem to have been able to support the modernisation of the existing industries, including mines and smelters, in the Meiji period (Craddock and Zhou 1998).

In these circumstances the only operations that were feasible and could compete were those that required minimal capital investment or skill and could be set up quickly, often on a seasonal basis such as that of the non-ferrous metallurgical metal production that operated in the early 19th century. It had probably sprung up relatively recently in an opportunist manner with little or no continuity with the past, local farmers working the old mines in the best way they could in the drier slack seasons of the year and the mine owners glad to get what they could from them, but unwilling to invest. This suppressed labour costs, which itself was a further disincentive to more sophisticated methods of working. The products were of low quality, could only compete on price, and were especially suited to areas with poor communications which became ever more limited with the spread of the railways.

Perceptions

As noted above, the collapse of the Mughal Empire brought about long-term disruption such that by the beginning of the 19th century the confused state of affairs in much of native India had come to be regarded as endemic. These attitudes had their roots in the common European perception that the economic and technical conditions encountered elsewhere had always been like that and were incapable of indigenous change or development.[4] There was also the general misunderstanding by Europeans of the conditions within India and the circumstances in which they had arisen. As William Dalrymple stated in *White Mughals* (2002: 284) the British thought that they were observing timeless rituals, when in fact everything was in a state of transition. This, together with a growing prejudice against the indigenous population and its institutions from the early 19th century, was inimical to investment in the regeneration of local industry along modern lines.

More specifically to the metal mining and production industries, the prevailing economic conditions noted above resulted in a negative perception among 19th-century European geologists and surveyors that the present much-reduced state of metal production that they encountered was all that native Indians had ever been capable of, and that the miners and smelters were the most backward and lowest class of workers. This attitude, repeated in many reports, was again not conducive to encourage investment. These operations when encountered by the European surveyors were generally mistakenly described, usually disparagingly, as being the survivors of India's metallurgical industry in its original condition. Percy (1870: 293) for example, introduced Captain Dixon's description of lead smelting in Ajmer (see Chapter 2), by commenting that 'the process of lead-smelting as conducted by the natives of India is, as might be anticipated, simple, primitive, and on a very small scale, *representing the infancy of the art*' (our italics). The

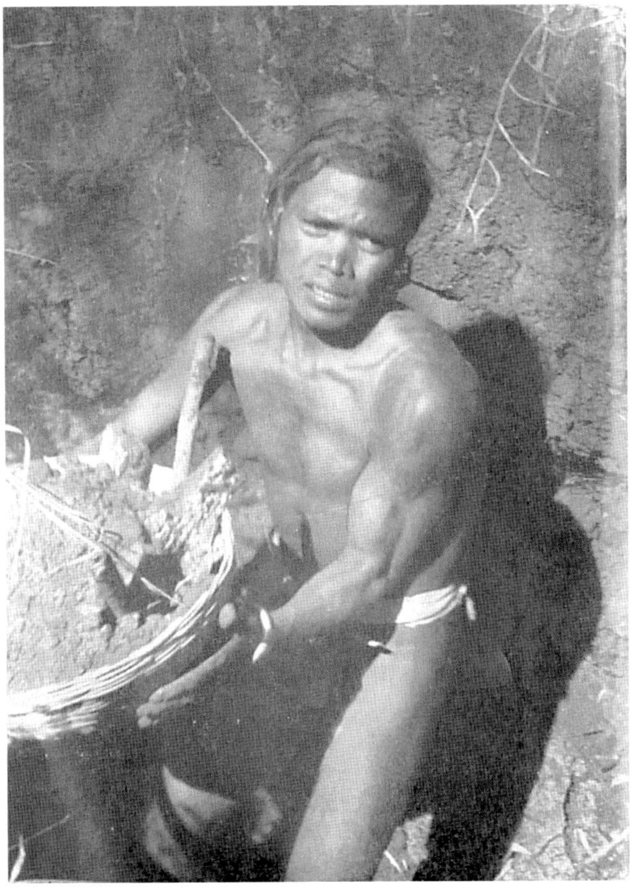

Plate 12.1 Patharia Agaria bringing iron ore out of the mine. Images such as these helped to create the impression that this was all that Indian mining had ever aspired to (frontispiece from Elwin 1942).

following story told by European mining engineers is illustrative. Their attempts to modernise a copper mine at Pokhri in the Kumaon District of the North West Provinces (now Uttarakhand), recorded in Ball (1881: 268), by the installation of proper timbering were resisted by the inherent conservatism of the native miners until they discovered ancient timbering in an old working installed by their ancestors. The point missed by the mining engineers was that the working practices had clearly regressed in recent centuries. The venture was abandoned because the copper produced could not compete in price with that imported from England.

Again, introducing the geologist H.F. Blandford's description of copper smelting in Sikkim, Percy (1861: 387), referring to India, stated that 'in some countries no progress has been made, and smelting processes are still carried on just as they appear to have been at their commencement'. Blandford himself (quoted in Percy 1864: 261) refers to the iron smelters of Orissa as the 'degenerate relics' (Plate 12.1). Pejorative descriptions of current mine workings occur frequently in the geological reports. Mallet (1875), recorded in Ball (1881: 276–7), compared the copper workings in Sikkim in the 1850s to 'magnified rabbit holes'. Stephens (1901–2: 393–414) gave detailed descriptions of the operations he observed at the Dhanpur, Dhobi and Rai mines in the Garhwal District of Uttarakhand:

> The ancient underground works are of the rudest description. A gallery was cut rarely over 4 ft (1.2 m) high, and almost always inclining inwards, drainage of the mine being ignored. Every little bunch was followed up and fossicked out, with no regard to safety or to regularity. The workmen did not understand attacking a deposit by shafts or levels, and their natural desire was to burrow. To this day it is necessary to carefully watch their descendants, as the native's instinct is to go after each little string or bunch he meets with, and levels or shafts soon get out of line. On one occasion the writer was absent for a month, and on his return he found it necessary to fill in the floor, and reopen a level, which his Paharris had turned into an incline during his absence. ... Such operations cannot be called mining. Probably these Paharris are about the most primitive of all underground workers. Content to live on about 2 or even 1½ annas a day [2d to 1½d], in a country yielding two crops a year on the merest scratching of the soil, necessity has never sharpened their inventive faculties.

These reports merely served to reinforce the European belittlement of Indian technology generally and totally belied earlier achievements.

However, some observers realised that things had once been very different. The eminent geologist Valentine Ball understood this. Perhaps his long and direct experience of the ancient mining and smelting remains in his geological pursuits in the subcontinent had made him more aware of the true situation and how it had arisen. He mused (1881: 338) in a rather convoluted comparison with nature that:

> As in the animal world, the process of degeneration has produced forms which are but dwarfed representatives of their earliest progenitors, so it is with the rude smelting furnaces of the natives, which, though they may not now in some cases be much superior to those which the Celts erected on hill-tops to catch the passing breezes, are probably to a great extent the lineal descendants of a system of iron manufacture, which in the earliest times of which we have any record must have been on a scale of considerable magnitude.

Ball (1881: 339) went on to cite the famous Delhi Pillar of iron as an example of past achievements, continuing the theme of comparison of the animal world to the demise of technically sophisticated industry:

> as we find in certain animals aborted or partially extinguished organs which point to certain functional powers having been possessed by their progenitors, similarly if we take a survey of the systems of iron manufacture, as practised by the natives of India, we meet here and there traces of what may be the remnants of higher systems of working than those now existing. ... First in importance is the manufacture of cast-steel, in crucibles, which attracted so much notice many years ago.

In a similar vein Major James Franklin in his description of the surviving iron smelting industries of eastern India in 1829 made the following observation (quoted in Dharampal 1971: 236), with its suggestion of regression:

> the original plan of this singular furnace must have been the work of advanced intelligence, and that its

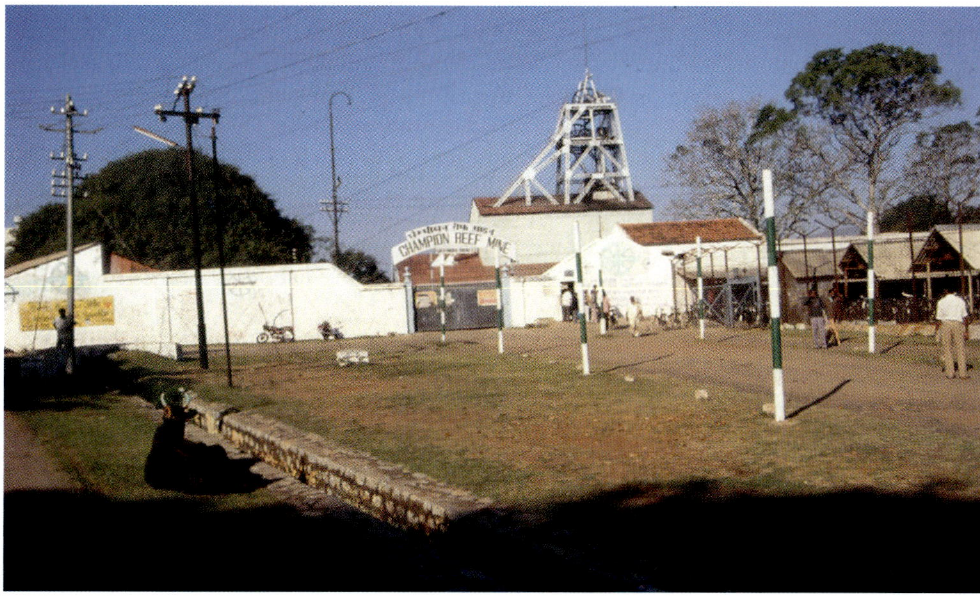

Plate 12.2 Champion Reef Mine at the Kolar Gold Field in 1990, then still much as built by John Taylor nearly a century previously.

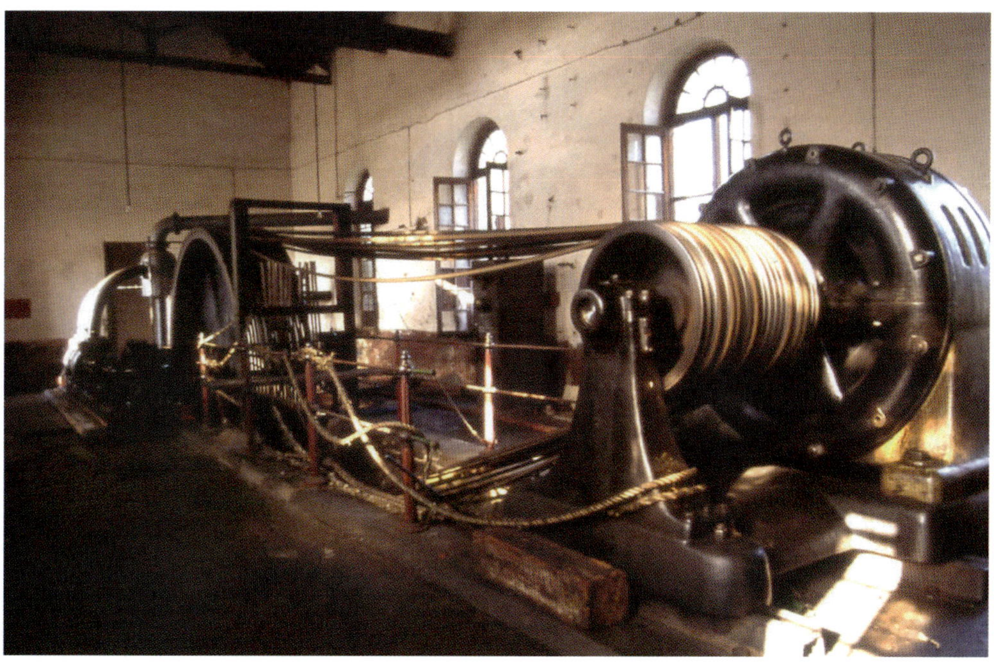

Plate 12.3 Dynamo (right) rope-driving compressor (left) for blowing cool air through the underground workings, which at the time were the deepest in the world (McTaggart's Shaft). John Taylor's workings at Kolar were very advanced being all electrically powered in the 1890s (Kolar Gold Mines 1990).

geometrical proportions have been preserved by simple measures; hence though its original form may be changed by caprice or ignorance, its principle never can be lost so long as hands and fingers remain.

A few other Europeans, usually those with extensive direct contacts with the country, realised that the present disarray had not always been the case. James Tod, in his *Annals and Antiquities of Rajputana* (1978, I: 108), warned against judging the chaotic conditions and absence of the proper forms of government that until very recently had held sway in Mewar as being endemic, instead of being seen as a more sophisticated administration brought low by 'Mahratta cunning, engrafted on Mahomedan intolerance'. He also observed that 'we cannot march over 50 miles of country without observing traces of the genius, talent and wealth of past days' (1978, I: 516).

The introduction of European metal mining and metallurgical industry into South Asia

Non-ferrous mining and extractive industries

European prospection commenced early in the 19th century, soon to be followed by comprehensive geological surveys, concentrating on minerals of potential commercial significance, as exemplified by the magisterial work of Ball on the economic geology of South Asia published in 1881. However, all too often the reports speak of old workings picked clean, and the pessimistic comments of Maclaren quoted above on the prospects for the discovery of viable gold deposits were applicable generally. The accounts of Ball and his colleagues in the publications of the GSI abound with European attempts to reopen mines, but almost without exception they were unsuccessful. The usual reasons given were that they were underfunded coupled with a lack of competence and experience and, as always, competition from imports.

Captain C.G. Dixon was prospecting in the Aravallis for copper and carrying out assays at Ajmer and Singhanna (see Chapter 2, Fig. 2.1) in the 1830s, for example (Prinsep 1835), but nothing seems to have come of them. Some copper mining ventures were established in the Singhbhum District of Bihar, but insufficient financing seems to have been the root cause for their early demise. By the early 19th century indigenous production had long since ceased there but by 1847 Captain J.C. Haughton (1854) had already recognised the potential of the copper deposits, and in 1857 a company was floated to reopen the copper mines. However, despite producing about 500 tons of ore per month, the mines were soon closed. Ball (1881: 248) attributed this to the heavy royalty payable to the local rajahs, 'an expensive establishment of Saxon miners and English smelters, the erection of a costly foundry and engine, and other premature and excessive expenditure'.

Medlicott and Blandford (1893: 63), writing of the geology of the Singhbhum District, noted that the rocks carried gold, silver, copper, lead etc, but that as in southern India, 'so far none have proved remunerative – except to promoters of joint stock companies and a limited number of speculators in mining scrip.': a rather cynical recognition of failure, but at least showing that attempts had been made. By the end of the 19th century more serious attempts to establish the production of non-ferrous and precious metals in India were at last taking place.

The first truly successful development of a major modern metal mining industry in India was in the goldfields of the south, principally at Kolar (Hatch 1901; Gupta 1986; Radhnakrishna and Curtis 1999: 63–6) and at Hutti, both now in Karnataka (Anon. 1963; Curtis *et al.* 1990; Radhnakrishna and Curtis 1999: 66–72). A great deal of detailed, if uncoordinated, exploration of potential gold sources had already been carried out, and many poorly planned and underfunded mining ventures had sprung up from the mid-19th century, but these enterprises failed, not least because all the easily mined surface deposits had long since been removed and the ventures lacked the capital to sink deep mines. This was the position at the Kolar Gold Field where Ball (1881: 186) presciently predicted that 'a proper system of working would prove the existence of a considerable quantity of gold'. A group of small independent mines had already become established in the 1870s, and again Ball predicted 'the half a dozen companies that have recently started in Mysore, whose management is good and whose preliminary expenditure has not been too heavy, have a profitable career before them'. In this assessment he was wrong however; they were soon collectively on the verge of financial collapse and the mines were taken over by John Taylor, the British-based international mining group. It had both the expertise and resources to effectively develop deep mines and almost immediately encountered rich gold deposits at depth, so that Medlicott and Blandford (1893: 49) were able to state, 'that gold mining has, within late years, passed through the phase of fierce speculation into a well established and remunerative industry' (Plate 12.2). The initial success was wisely exploited by further massive investment and by the end of the century the Kolar Gold Mines were among the most advanced in the world with features such as all-electric haulage and pumping supplied from their own power station (Plate 12.3). The company continued to produce considerable quantities of gold and remained very profitable throughout the first half of the 20th century, with over 400 tonnes of gold produced between 1901 and 1930, showing what could be achieved at the right location with proper investment, an object lesson not lost on the government of independent India.[5]

The Kolar deposits eventually became exhausted.[6] The much smaller operations in the Hutti Gold Field, in what was then the state of Hyderabad but is now part of the Raichur District in the north of the state of Karnataka, started a modern operation under private control in about 1887 and worked until 1920. The state government of Hyderabad reopened the Hutti mines in 1938 and, apart from the war years, they have continued in production ever since under state or national control as the Hutti Gold Mines Co. Ltd.

As Holland had stated (see above), at the beginning of the 20th century serious developments in other metals were being undertaken. A systematic drilling programme was carried out by the GSI in 1906–8 in the Singhbhum copper deposits, proving the existence of substantial reserves (Coggin Brown and Dey 1955: 148–9). Following directly from this the Cape Copper Company developed the Rakha mine in the Singhbhum District, together with a smelter between 1908 and 1914, commencing production in 1918. This proved unsuccessful – a victim of the economic uncertainties of the 1920s, the operation closed in 1924, but not before 3550 tons of copper had been produced. Production recommenced in Singhbhum with the opening of the Mosaboni mine, a few kilometres south of Rakha, by the Cordoba Copper Company, managed by John Taylor, the international mining group, in 1928. This was reorganised in the 1930s as the Indian Copper Corporation and through the years other deposits in the Singhbhum Thrust belt were opened and copper production increased considerably to about 10,000 tonnes per annum. New smelting capacity raising capacity to 16,500 tonnes came on stream in 1972 (Khrisnaswamy 1988: 172), just prior to nationalisation (see below).

Before the Second World War almost all of the indigenous lead, silver and zinc production in the British Asia possessions came from the Bawdwin mine in the Northern Shan State of what is now Myanmar (Coggin Brown and Dey 1955: 156–61). The Burma Corporation was established in 1918 to work the mines following the discovery of the Chinaman Lode beneath the existing abandoned workings and by the 1930s the mine was producing approximately 100 tons of silver and 75,000 tons of lead annually at its smelter at Nam Tu, and about the same quantity of zinc concentrates that were sent to Belgium for smelting. The loss of this mine first to the Japanese and then to an independent Burma, was one of the prime catalysts in the reopening of the mines at Zawar. The story of lead and zinc production in India itself in the second half of the 20th century is the story of the Mining Company of India, latterly after nationalisation, Hindustan Zinc Ltd.[7]

Iron and steel

The story of iron and steel production in India is rather different and more complex than for the non-ferrous metals. As stated above, India was a major producer and exporter of both iron and steel and to some degree this survived through the recent centuries, although there was no direct continuity between traditional indigenous production and modern industry. The iron was charcoal smelted and the wrought iron was made by the direct solid state process, known in English as the bloomery process. The crucible steel was made either by carburisation of wrought iron or by a co-fusion of wrought and cast iron. This cast iron was indigenous, which shows that there must have been some local cast iron production by a blast furnace process (Craddock 2007).

The arrival of European-style ironworks in India further exacerbated the problems already caused by imports to the indigenous traditional industry and led to the cessation of the larger, more sophisticated producers such as those recorded at Birbhum (Elwin 1942), and ultimately the other major traditional iron smelting centres, such as those at Jahazpur, Gawgrar and Bar Sidri in Rajasthan, all ceased production in the 19th century (Hooja 2006: 816). Elsewhere the decline was slower with some traditional iron smelting still taking place well into the 20th century.[8] To some extent the pattern of closure was local, the establishment of an ironworks bringing about the demise of traditional smelting in the surrounding region. For example, the iron smelters at Bukkasagara (Buchanan 1807) in the Chitradurga District of Karnataka, who had supplied the iron for the crucible steelmakers at Ghattihosahalli (Sambasiva Iyer 1898–99; Anantharamu *et al.* 1999), ceased production when the Visveswayara Iron Company was established during the First World War at Bhadravati, in the Shimoga District, now part of Karnataka. This was a charcoal-smelting operation, established by the government of Mysore. The Bukkasagara smelters were aware that this development would take away their markets; to preserve their memory and go out in style, they used the last of their iron to forge a huge bow and arrow, weighing several hundred kilograms for the local temple.

Attempts to preserve or restart iron production in India took a number of forms. Very basic rural iron production and crucible steel production struggled on into the 20th century, usually located in remote areas, with attempts at adaptation or preservation. There was also some European involvement in iron production almost from the establishment of the first trading factories. Attempts made to establish European-style iron production from the 18th century, were finally successful in establishing major modern integrated iron- and steelworks by the beginning of the 20th century. Each aspect will be considered.

Small-scale indigenous smelting

Small bloomeries, operated by the semi-itinerant Agaria, continued to produce iron in the more remote parts of eastern India, such as in the tribal areas of Orissa and Bihar, until at least the 1960s (Ahmed 1963; Elwin 1942; Leuva 1963) and apparently may still continue (Mishra 2003; Pande and Geijerstram 2002). In one instance the local direct-process charcoal-fuelled iron smelting industry was investigated to ascertain whether it could be organised and operated to enable it to compete in price with imported iron, as it certainly did in quality (Ball 1881: 341–4). Captain J. Campbell (1843), Assistant Surveyor General to the Madras establishment, studied the commercial viability of producing wrought iron by the traditional solid state bloomery process in direct competition with British imports (Dharampal 1971: 249–63; Percy 1864: 266–7). In this very interesting experiment, teams of four workers operated the small furnaces which produced three blooms of iron, each weighing about 11 lb (5 kg), per working day. After bloom smithing and forging the blooms were reduced to between 3 and 6 lb (1.4–2.7 kg) of bar iron, which sold for about £9 per ton. Percy, used to the furnaces and production of mid-19th-century Britain, was astonished by the tiny output per furnace and per man. When Campbell tried using small blast furnaces, which presumably produced liquid cast iron to be fined to produce wrought iron, teams of two workers each produced about 40 lb (18 kg) of wrought iron per day, with a commensurate saving in fuel and time. In fact this iron was competitive in price with imported iron, but the quality may not have been as good as that produced by the old bloomery process, which was generally regarded as excellent. Campbell concluded that traditional iron smelting was not viable in competition with imported bar iron near the coast, but that inland with abundant high quality ore, cheap charcoal and poor communications it was viable. These experiments were carried out just before the great railway boom began in South Asia and for whatever reason the project was discontinued.

In the 1930s Elwin (1942) decried the demise of the traditional Agaria iron smelters and suggested that their industry should be revived. This was to be done partly by tax concessions and partly by improving the smelting methods. The main technical improvement was to be the addition of a limestone flux (1942: 194) which would have had the effect of reducing the amount of iron lost to the slag. However Elwin seems not to have appreciated that the slag so formed would have required a much higher temperature to melt and the overall result would have been the production of liquid cast iron, rather than the bloomery iron that was required.

Some attempts continue to be made to encourage traditional smelting industries. These include efforts to re-establish the bloomery iron smelters of Orissa and elsewhere (Srinivasan 1998) and to regularise the tin smelters in the forests of Madhya Pradesh and Orissa by making them a cottage industry (Babu 1994: 69–99, 2003). In reality even in the past these operations were probably always peripheral to the main industrial effort and if they have any place at all in the modern Indian economy it is surely as heritage enterprises in the tourist sphere, similar to the Japanese *Tatara* process (Craddock and Zhou 1998).

European involvement in establishing ironworks

The first attempts to establish European-style metal industries within South Asia were centred on iron (Pande and Geijerstram 2002). This was quite logical since iron is a cheap metal where transportation charges are likely to be a significant factor in overall costs. Also, abundant resources of iron ore and fuel, both charcoal and latterly coking coal, were available within South Asia, and there was a large and growing market.

By the 17th century the Dutch East India Company had several large ship building and repair enterprises in South and Southeast Asia which required very considerable quantities of iron. It is estimated that a typical merchant ship of the type used by the company would contain about 80 tonnes of iron (Lowe 2002), of which nails would account for a considerable proportion. The Dutch established a number of quite sizeable nail-making factories in the iron-producing areas on the Coromandel Coast of central India at centres such as Narsapur on the Godavary estuary (Alam 2002; Lowe 2002). These were set up along European lines employing large numbers of local smiths who were taught the European fabrication methods by technical experts from Sweden and the Walloon region of Flanders. The factories used locally produced iron in considerable quantities (for example 40,000 lb [Dutch] of ships' nails were produced in 1664). Thus in the 17th century the Dutch were content to use the iron smelted by the local technology and establish factories using trained local craftsmen to work it, although as Alam claims, this seems neither to have stimulated the indigenous smelting technology to change nor to have produced lasting changes to the methods of blacksmithing in South Asia. It was, however, in marked contrast to the European, largely British, efforts of two centuries later to establish local metals industries where the local industries and craftsmen were ignored. This partly reflects the great strides made by European industry in those two centuries, but also more subtly the change in attitude of Europeans towards South Asia.

Captain B. Hamilton in his unpublished *A New Count of the East Indies* (vol. 1, p. 302) refers to the production of anchors in 1708 at Balasor (modern Bāleshwar), a port on the coast of Orissa, where the factories of several European nations were located. The significance of this has been noted by Ball (1881: 361), Goitein (1966: 339) and Qaisar (1982: 28) as probably the earliest reference to iron production with European technology in South Asia, but nothing more is known of this venture. The reference is a little puzzling because the account states that the anchors were cast, whereas anchors were traditionally made by forging iron, not by casting. If they were forged then the traditional Indian skills in the working of large masses of hot iron as exemplified by the large wrought iron bombards of the Deccan, and iron beams in Orissa, could well have been utilised in the production of anchors.

First attempts at establishing modern ironworks

Several ironworks were projected in various parts of South Asia throughout the 18th and 19th centuries (e.g. Watson 1834), but all met with great difficulties (Ranade 1906: ch. 6; Pillai 1925: 213–33). The first were probably those of Motte and Farquhar, who were granted exclusive rights to manufacture iron in the East India Company's territories at the Birbhum District of West Bengal in 1778, in an area where iron smelting on some scale was already practised. Little is known of this enterprise, even whether or not it actually came to fruition, and the grant was relinquished in 1795 (Ball 1881: 362–3; Coggin Brown and Dey 1955: 176–7). In 1845 the Birbhum ironworks were described as being run by the natives and producing iron by traditional methods (Ball 1881: 363). In about 1855 the Birbhum Iron Works was established at Mahomed Bazaar in Bengal by Mackey, utilising the local iron ores and limestone while still relying on charcoal for much of the fuel, but the enterprise ceased around 1875.

With the coming of the railways to South Asia there was the potentially huge market for both coal and iron. As Karl Marx presciently observed in 1857, such a major system would inevitably bring into being a locally based engineering and metallurgical industry to service its requirements (Marx and Engels 1959). Many companies were established with both coal and iron production in their prospectuses and even in their names, but few if any produced anything other than coal.

The first European-style blast furnaces were founded in 1833 by that pioneer of industrial iron production in South Asia, Josiah Marshall Heath, at Porto Novo (modern Parangipetta), South Arcot District, on the coast of Tamil Nadu (Ball 1881: 34–50; Heath 1832, 1839). The works were in production for a number of years, initially producing good quality charcoal-smelted iron and steel, fluxed with seashells as the source of lime. The company briefly supplied wrought iron to Sheffield in Britain. It was hoped that Indian iron could be a viable alternative to Swedish iron as a source of charcoal-smelted bar iron for conversion into steel – the company even supplied wrought iron for the Menai and Britannia Tubular bridges in North Wales. But in spite of, or perhaps because, of this brave attempt to take on the British iron industry on its own ground, the venture ran into difficulties and the company was taken over by the East India Company with more works being established at Tiruvannamalai and at Palampatti in the North Arcot District of Tamil Nadu, and at Beypore near Calicut on the coast of Kerala. The latter works was also in difficulties by 1859, and the comments of the German workmen are instructive. They had returned home because they found it too hot to work, and it 'was impossible to get naked savages to do such work as puddling; if they themselves left for a few moments all went wrong and the natives would fall asleep' (reported in Ball 1881: 352). Apparently the admittedly arduous, traditional European puddling process was superseded shortly afterwards by the installation of a Bessemer converter. What the 'naked savages' thought of a Bessemer converter in full operation is not recorded, but shortly afterwards the operations ceased. Later attempts were made to establish charcoal smelting in the Chota Nagpur region, now in Jharkhand, but before the century was out Thomas Holland (1893) noted bleakly that 'their ruins stand as monuments to misguided efforts'.

Other charcoal iron smelting works were established in the latter 19th century at locations all over South Asia, at Indore and Chanda in Madyha Pradesh (Ball 1881: 389–93), for example, and at Kaladhungi, in Uttarakhand, where substantial remains of the blast furnaces still survive (Pande 2002), but none met with permanent success and all soon ceased trading.

Government encouragement

There was some, admittedly sporadic, enquiry into how industry could be encouraged by the European authorities. For example, a list was prepared of the suitable iron ores, charcoal-producing areas and existing iron smelting centres, and the costs of production and distribution of the iron to establish whether a viable indigenous industry could be supported (Balfour 1855). The East India Company, which still administered much of northern India, experimented with the establishment of major modern ironworks and thereby kick-started a modern iron industry (albeit still charcoal-fuelled). These were the ill-fated Burwai ironworks, in West Nimar, Madyha Pradesh, and the Kumaon ironworks in Uttarakhand (Ball 1881: 406–11; Pande 2002; Geijerstam 2002, 2004). Following the demise of the East

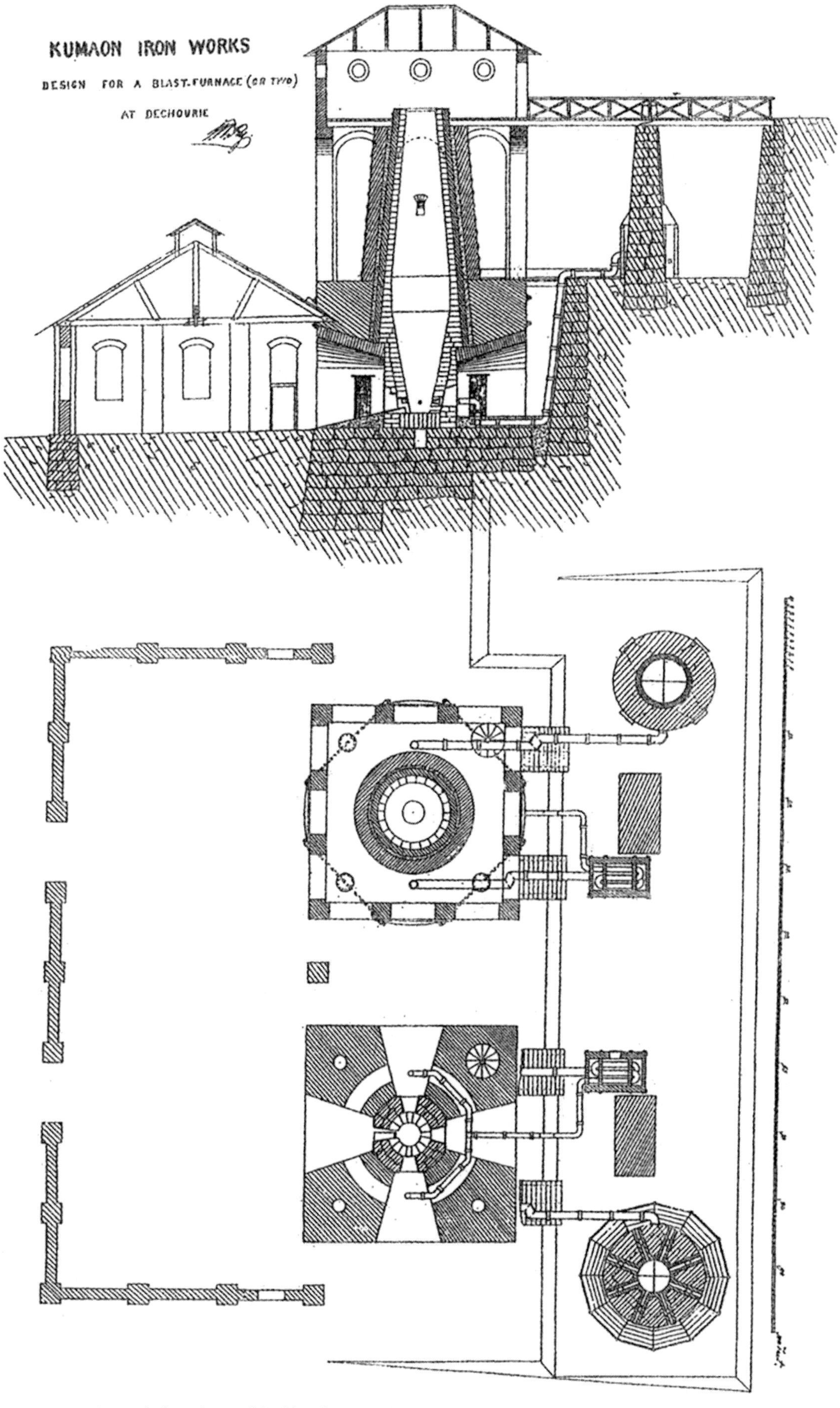

Fig. 12.1 Contemporary section and plan of two of the blast furnaces that were at least partially constructed at the Kumaon ironworks at Dechauri in the 1860s (from Geijerstram 2002).

India Company, these became imperial government initiatives, undertaken specifically to show that India could successfully compete with imported iron and as model enterprises to encourage private investment. However the Burwai ironworks closed after the failure of its first and only smelt in 1864, but the Kumaon works floundered on for 20 years from the late 1850s until its final closure in 1879 (Fig. 12.1). The well-documented reasons for the failure of such a major project give considerable insight into the problems faced when trying to establish a large undertaking, well away from the necessary infrastructure, including access to a railway.

The project seems to have been too ambitious for the finance and expertise available to set it up and run it. Also there was continuing uncertainty as to whether the price of the locally produced iron could ever compete with that of the imported iron and steel in an unprotected market. Overall the attitudes of the East India Company and its colonial successor were positive, with one top government official stating that 'it should be thought right to encourage the production of native metal, and make an Indian Railway of Indian iron and Indian wood'. However, as with government-sponsored or assisted projects the world over, there were problems with efficiency, interminable reports resulting in delays with decisions taking years to make and longer to implement, if at all.

Coke-smelted iron into the modern world

The previous attempts to establish viable ironmaking had been based on charcoal, the traditional fuel of India. But throughout the 19th century India's huge reserves of coal began to be exploited for ship bunkering and to supply the ever-expanding railway network. In 1874 the crucial step for Indian extractive metallurgy was taken with the establishment of coke-fuelled iron smelting blast furnaces at Kulti on the Bihar–West Bengal border. This was situated on the Raniganj coalfield and used ore collected from local surface deposits. The venture was not particularly successful and went bankrupt, but was soon re-established as the Bengal Iron Company. Despite official encouragement in the form of sizeable government contracts for castings, it too had failed by 1879, but not before it had produced about 12,700 tons of pig iron (Ball 1881: 363). Ball attributed the failure entirely to initial underfunding. However, iron production recommenced at Kulti, producing about 35,000 tons of iron per annum in the early 1900s, through to the mid-20th century, latterly supplying iron to the Steel Corporation of Bengal (Nanavati 1960; Ghandy 1960). From the 1880s other coke-fuelled blast furnaces were established and by 1900 approximately 40,000 tons of coke-smelted iron was being produced annually in India. This represented a considerable advance, but was tiny by world standards, and much less than India's requirements, most of which still had to be met from imports.

Technology was also moving on, and the era of cheap steel had arrived. Since the 1880s world production of wrought iron had remained static, but bulk steel production had soared. The establishment of a viable integrated steel plant making bulk iron and steel by Jamshejki Tata was a major milestone in India's economic resurgence, not least because it was initiated and financed from indigenous sources (Elwin 1958; Dutt *et al.* 1960). In contrast to the Kumaon ironworks, this was a properly researched, financed and managed project. After Tata had spent many years prospecting for suitable iron ore sources in conjunction with the geologist, P.N. Bose, in the huge iron ore belt that straddles Bihar and Orrisa, Dhalli-Rajhara, in the Dhalbhum region of Bihar was chosen as the source. The works and settlement, named Jamshedpur, were built in Bihar on the confluence of two rivers with good rail access to the iron ore mines, to the Jharia coalfield and to Calcutta. The works were completed in 1907, the first pig iron was smelted in 1911 and the first steel rolled in 1912, and through the war years production grew, outstripping the original target of 125,000 tonnes per annum to reach 420,000 tonnes in 1920. Other privately financed steel firms were also now being established, and further expansion was planned in the immediate post-war boom. However, this boom soon came to an end leaving the Indian companies dangerously exposed, especially as the colonial Indian government of the day still espoused free trade, and steel was being dumped in India at prices well below the cost of production. The future for indigenous steel production looked bleak in the mid-1920s and its plight was to no small degree instrumental in bringing about a fundamental change in policy with a limited degree of protection for the developing home industries against imports, championed by Nehru and the Congress Party. This proved very successful and the major Indian steel producers survived the world trade depression of the 1930s. By 1940 Tata was producing about 800,000 tonnes of steel per annum, while other companies, notably the Indian Iron and Steel Company, were outputting a further 400,000 tonnes. The success of this policy must have encouraged the government of independent India to pursue a policy of self-sufficiency.

However, up to the Second World War India still continued to import the great majority of the metals it needed. The shock of that conflict – and especially the loss of Southeast Asia and Burma – highlighted India's vulnerability and the need to exploit indigenous sources. The excellent surveys of the GSI had already laid much of the geological groundwork and known sources such as Zawar were systematically surveyed and assessed as potentially viable producers.[9] After independence in 1947, socialist policies of a central planned economy were followed, reminding some commentators of the centrally controlled economy of the Mauryan Empire as described in the *Arthaśāstra* by Kautilya, some two thousand years previously (Kangle 1965–72, III: 191). The Congress Party, with its economic directives based on Gandhi's urging for home production and Nehru's more pragmatic plans for import substitution, continued the existing policy of encouraging self-sufficiency in metals wherever possible. This was to be achieved by investment and the creation of protected markets.

Lacking sufficient smelting capacity, the export of metal ores was initially encouraged which was, for example, the main activity at Zawar until the commissioning of the zinc smelter at Udaipur in 1968. Latterly only iron ore was exported in significant quantities. After independence large-scale expansion programmes were initiated under two 5-year plans in the 1950s and 1960s, in particular for the iron and steel industry which was seen as central to India's industrial development (Chaudhuri 1975). The plans acknowledged the place of private industry, and called for expansion in both the private and state sectors, stating that 'The iron and steel industry was included among those fields in which the State will be exclusively responsible for the establishment of new undertakings, except where in the national interests, the State will find it necessary to secure the co-operation of private enterprise, subject to such control and regulations as the central Government may proscribe' (Pande 1960: 4). Thus the government of India, with foreign technology and aid programmes, instigated the construction of large integrated steelworks situated at Durgapur in the Birbhum District of West Bengal with a British consortium (Harley 1960), at Bhilai in the Durg District of Madhya Pradesh, with Russian assistance, and at Rourkela in the Sundargarh District of Orissa aided by West Germany (Dutt *et al.* 1962). These came on stream in the 1950s and 1960s, each initially producing about 1,000,000 tons per annum and operating as Hindustan Steel Limited. At the same time the two largest private firms, Tata and the Indian Iron and Steel Corporation, were authorised to increase annual production to 2,000,000 and 1,000,000 tonnes respectively, thereby increasing India's capacity to about 6,000,000 tonnes of pig iron, from which about 4,500,000 tonnes of steel could be produced. In addition some small-scale works were established in the south of India.

Recent developments

The extractive industries in the private sector were brought increasingly into a variety of state or national owned enterprises, with the Metal Corporation of India, who ran the Zawar mines, nationalised to become Hindustan Zinc Ltd. in 1966 (Bansal and Ahluwalia 1990) (see Chapter 3). The copper mines were also nationalised, subsequently becoming the Indian Copper Complex of Hindustan Copper Limited (Anon. 1983c), and mining recommenced at a number of old mines throughout South Asia including those at Agnigundala, in the Guntur District of Andra Pradesh, Chitradurga, in Karnataka, Malanjkhand, in the Balaghat District of Madyha Pradesh (Anon. 1983b), and at the Khetri copper belt in the Jhunjhunu and Sikar Districts of Rajasthan. At the latter a smelter with a 31,000 tonne capacity was opened at Singhana in 1974 and further expanded to 45,000 tonnes in the late 1980s.

Although state control enabled properly designed and well-funded expansion projects to go ahead, and ensured that the working and living conditions of the workers were protected, it also led to inefficiencies and some lack of economic reality in the protected home markets. Some of the problems inherent in the nationalised industries began to be addressed through the 1990s and there has been a considerable liberalisation and loosening of state control, leading to privatisation and serious involvement with the international market.

Summary

To summarise, free trade policies of successive colonial governments in the 19th century exposed what was left of the indigenous extractive metal industries to imports with which they could not compete. This created an economic climate that was unfavourable for investment in modernising old industries or creating new ones although, reading through the geological reports of the period, one is struck by the frequent attempts that were made to work existing or new deposits. Initial efforts at establishing mining and smelting operations throughout the 19th century were generally unsuccessful, usually due to a lack of proper planning and inadequate finances. The first wholly successful mining operations, those at the Kolar goldfields, succeeded where others had failed because the John Taylor Group had the necessary expertise and was prepared to invest very heavily in the project. This example of foreign investment was soon matched by Jamshejti Tata's vision of an Indian-financed steel industry.

This initiative was followed by other firms in the early 20th century, but these ran into problems as the world economy worsened in the 1920s, necessitating some form of protection which enabled Indian metal production to continue expansion through the 1930s. The disruption of the Second World War, when access to metals was severely curtailed, showed that India needed to expand and broaden its mineral and metal production capability. Following independence there was a move towards self-sufficiency in an economy dominated by central planning. The expansion in metals production was led by large-scale investment in exploratory and development work, and nationalisation of many existing facilities, protected by tariff legislation. From the 1990s this began to change, with Indian companies able to collaborate with international companies both within India and abroad. After an absence of some centuries, the metal industries of India are once again participating in the global economy.

Notes

1. Mines and smelters such as those in the Aravallis have revived and continue in production; the specific experience of the Zawar and Dariba mines has already been described in some detail in Chapter 3.
2. This apparent lack of European interest in the more sophisticated traditional Indian industries, as exemplified by zinc at Zawar and also crucible steel in the south, while they were still in production and manufacturing commodities that were extensively traded on the international market is surprising. It would be thought that in the early 18th century when metallic zinc and crucible steel were not yet produced in Europe but had to be imported that there would have been an interest in discovering the secrets of the processes. The process of downward distillation adopted by William Champion in Bristol was similar to that used in India, but a direct link is difficult to establish, the first actual description of the Indian process only being made in the mid-19th century (Chapter 13, p. 247). There are no descriptions of the Indian process for making crucible steel before the end of the 18th century and the first European crucible steel process is very different from the Indian processes, yet 'hindostan' steel was widely appreciated in Europe. The Indian processes of *in situ* carburisation and co-fusion of cast and wrought irons may have influenced European practice much later in the 19th century, but again if so then only very indirectly (Craddock *et al.* forthcoming). The first really detailed descriptions of the operations only appeared in the latter part of the 19th century long after crucible steel manufacture had become established in Europe. They only appeared when the Indian steel industries were in terminal decline and seem to have been academic exercises rather than for serious instruction (Bronson 1986).
3. See Chapter 13, note 8.
4. As trenchantly expressed by Jones (2003: 45) who, musing why the industrial revolution had only taken place in Europe, opined that 'the fires of modernisation and industrialisation, once lighted in Britain and Belgium and the Rhineland, burned quickly to the fringes of this European system. But at the asbestos edge of the Muslim sphere the fires abruptly died. They never took light

over most of the non-European world.' In reality the fires had burned centuries earlier over many parts of the world before being extinguished by the European competition. The universality of this perception, together with its survival, is exemplified by European attitudes to China. Joseph Needham, the great historian of Chinese technology and scientific thought, could on occasions use evidence quite uncritically in order to show that particular technologies had origins (always Chinese) much earlier than hitherto accepted. For example, Needham cited secondary sources using a particular word that latterly came to mean zinc, to claim that the metal was produced in China over two thousand years ago (Needham 1974: 212–19). Pushing the dates of major technical advances back into the distant past had the effect of creating an apparent innovation vacuum for the last thousand years; again the statis that so many Europeans believed was the situation over most of the world outside of Europe. In this context it is significant that his historical surveys terminated at the end of the Ming period. Clearly, at least subconsciously, for him technical and scientific advancement in China had ceased.

5. The Kolar Gold Mines were taken over amicably by the state of Mysore in 1946, transferred to the national government in 1962, and in 1972 became the Bharat Gold Mines Ltd., but suffered ever increasing losses as the gold deposits became exhausted (Radhnakrishna and Curtis 1999: 63; Gupta 1986). The government finally closed the mine in 2001 and its future remains (2015) uncertain.
6. See note 5 above.
7. These activities are described in Chapter 3, p. 29.
8. The exposure of China to European imports of iron came rather later than in India, but was equally devastating to local production (Wagner 1997). Following a series of trade agreements in the mid-19th century, forcing China to remove tariffs on imports, the import of European iron and steel expanded rapidly through the second half of the 19th century. Wagner believed that as late as the 17th century China 'probably had the world's largest and most efficient iron industry' (1997: 5). As in India exposure to cheap imports led to the rapid demise of the larger establishments 'where the most technically sophisticated and capital intensive techniques were in use' (1997: 9). The operations that survived tended to be small scale in remote areas serving just local needs and were labour intensive using simple low capital methods.
9. See Chapter 3 for the story of the re-establishment of zinc production at Zawar from the 1940s.

Chapter 13 Summary and overview: beyond the Aravallis and beyond India

P.T. Craddock

The Aravalli mines and mining

This project in the Aravalli Hills concerned the survey and excavation of three major mine enterprises at Zawar, Dariba and Agucha, where the main exploitation overall began in the Mauryan and immediate post-Mauryan periods in the later first millennium BC. Although all three mines worked the same geological formation, running through the Aravallis, there are very considerable differences in the local topography, geology and mineralogy: the mines at Zawar are in the hills; those at Dariba and Agucha are situated in the plains. The ores of Zawar are in dolomite; at Dariba and Agucha the dolomite has mainly weathered to phylite/graphite schists. The ore minerals were sphalerite and galena at all three mines together with lesser amounts of freibergite at Dariba and Agucha. There is no evidence that oxidised ores such as smithsonite, willemite or cerussite were ever present in significant amounts. At Zawar, sphalerite predominated and could be separated macroscopically from the bulk of the galena. Furthermore the galena in the deposits worked at Zawar in the Mauryan period was not especially argentiferous. At Dariba and Agucha the galena, together with the freibergite, was argentiferous and in antiquity this was the main focus of the mining operations with the sphalerite apparently ignored.

At all three mines the deposits are at a steep angle and the mining strategy adopted was first to follow the ore down from where it outcropped by underhand stoping, coupled with some relatively minor opencast work. The mining operations were well planned working down through the richest ore before retreat mining to take out the lower grade ores. In antiquity at Dariba and Agucha, great opencasts were already being dug above the existing deep mines, that at Dariba having one side secured by a huge timber revetment, arguably the largest wooden structure to survive from antiquity anywhere. In the medieval/post-medieval periods at both Zawar and Dariba it is likely that many of the mining operations were also carried out by opencasting.

The ancient underground workings are on a grand scale, with wide stairways indicating the presence of a large workforce. These extensive workings required careful planning to ensure adequate ventilation and drainage. At Zawar where the workings are up in the hills, adits could have been driven in from the sides to facilitate access and drainage, but this was not done. In fact the vast majority of the workings are in ore. Drainage must have been a major problem at Dariba and Agucha where the workings descend from the floor of a plain. A bund was built at Agucha to hold back floodwater during the monsoon. However, at both mines the workings penetrate over 100 m below the water table in permeable rock; at Agucha in particular there is evidence of very careful water management. At Dariba remains of water bailing systems can be found, indicating that the necessity for the removal of very considerable quantities of water must have been an enduring problem.

There is no evidence of pumps or of mechanical raising devices in any early mines in South Asia. The ore-crushing operations also seem to have been entirely manual. In contrast with contemporary European mines (Craddock 2008), the absence of mechanical devices and of the use of water power is striking – even animal power seems absent which is a little surprising considering that from the medieval period on clay-pot chain *noria*, turned by oxen, were widely used to raise water for irrigation. Instead it seems that the mines relied on a large, well-organised and directed human labour force.

The workings demonstrate a determination to overcome major mining problems, but there were also other equally daunting tasks if the major operations were to succeed in what must have been remote and potentially hostile environments. There may have been some small-scale workings before the mid-first millennium BC but the expansion of operations were effectively new enterprises in what is regarded as being very underdeveloped regions in the Iron Age (Possehl and Gullapalli 1999). The mining and smelting operations were on a scale that must have consumed enormous quantities of timber on a regular basis that realistically could only have been supplied from managed woodland. The considerable labour force required feeding and the evidence of the refractory ceramics tempered with the chaff from a variety of cereal crops demonstrates that arable farming must have been carried on locally on a considerable scale. This all indicates thoroughly planned enterprises – not just in the mines themselves but in the infrastructure, requiring very considerable investment. Realistically this can only have come from the Mauryan state.[1]

At all three mines the massive scale of the ancient operations are immediately obvious, betokening a highly organised industry, vastly superior to the few mines that were operational in the 19th century pejoratively recorded by Europeans. Very clearly the industry had regressed.

Indian silver production

In global assessments of silver production in antiquity the potential contribution of South Asia has been paid scant attention. For example, Patterson (1972) in her major study of world silver stocks in antiquity, took a not uncommon Helleno-Roman viewpoint and failed to even mention South Asia, although the existence of an Indian silver coinage on a scale commensurate with that of the ancient Greeks has long been recognised.

In India itself little evidence of ancient silver production had been recognised before this project uncovered extensive evidence in the Aravallis. Ball (1881: 231) for example, stated that 'there appears to be no very direct proof forthcoming that silver was ever produced to any large extent in the peninsula'. An additional problem lay in the difficulty of establishing whether some argentiferous deposits with evidence of early working were mined primarily for lead, silver or even zinc in the past (Bhardwaj 1979: 133). In general, the consensus of opinion within India seems to have been that no major sources of silver were worked in ancient times, concurring with Prasad and Ahmad's (1998: 189) opinion that 'the silver used in our country was mainly imported from Burma, Afghanistan and Iran'. The scale of the workings at Dariba and at Agucha, would suggest otherwise, at least during the Mauryan period. As noted in Chapter 2, there is no evidence of silver production in the Bronze Age in the Aravallis; the exploitation of silver on a massive scale coincides with the introduction of silver coinage after the mid-first millennium BC.

The apparently contradictory Classical references to metal availability have been made clear by the dates of the mines. The comments of Strabo and of Pliny, that metals were available in India, were based on the earlier work of Megasthenes, who had accompanied Alexander's expedition in the 3rd century BC when the Aravalli mines were in full production. Trade manuals such as the *Periplus*, which stated that metals were imported into north western India, were compiled in the first centuries AD when production at the mines is likely to have been in serious decline.

In the post-Gupta period (c.AD 700–1200) there was a shortage of silver throughout northern South Asia as reflected in the coinage (Deyell 1990: 1–8). Following the Islamic incursions silver would have become more available, although the Islamic world itself generally suffered a severe shortage of silver around the 10th–11th centuries AD (Allan 1977; Watson 1967). After the establishment of the Delhi Sultanate in the 12th century AD, silver came to South Asia from the mines of Afghanistan and Central Asia, which were probably the world's major producers in these times (Blanchard 2001), from sources such as the Panjhīr, Jārbāya, Andarābah, Zebak and Wakman mines in Afghanistan (Deyell 1990: map 11; Allan 1979: 14–16; Cowell and Lowick 1988) and beyond in Central Asia until disrupted by the Mongol invasions in the early 13th century. In the medieval and Mughal periods some silver may have been produced from within South Asia. Habib (1982: maps 1B, 2B, 3B, 4B, 8B, 11B) believed silver was mined at Ghorband, Jaitaran, Sojat, Kumaon, Tiparah and Guge, but the majority was certainly imported (Deyell 1990: Hasan 1969).[2] Silver is likely to have been produced at Zawar in the post-medieval period.

The entry of silver from the Americas in the 16th century into world markets had an immediate and profound effect on silver supply globally, not least in India. Certainly the purchases of the European trading companies were made almost exclusively in silver (Craddock 2013).

Indian zinc production

Zinc minerals

It is clear that at Dariba and at Agucha silver was the main metal sought but the situation at Zawar in antiquity is much less certain. Lead, silver and zinc are all present, but as noted above, the silver content of the ores in the mines worked in antiquity was relatively poor. The mineralisation was primarily of sphalerite but there is no evidence that it was discarded or avoided (unlike Agucha, where the zinc-rich mineralisation had been ignored). Above ground, the problems of interpretation are compounded by the absence of any smelting evidence directly contemporary with the mines. These date through the latter half of the first millennium BC, whereas the slag heaps date from the turn of the millennium, towards the later end of the date range for the mines. These slags and the associated debris clearly do not belong to a simple smelting of galena, but could be associated with a process for producing zinc oxide in a method reminiscent of those described by the broadly contemporary Greek authors (Chapter 11). This raises the question of whether these major mines at Zawar were worked only for zinc oxide and a little lead or if silver was also being exploited in antiquity, but the distinctive process evidence such as cupellation debris was not encountered in our surveys. In the post-medieval period argentiferous lead was mined and the smelting slags found, which indicated that the galena was being smelted directly as would be expected. It is perhaps significant that the earlier heaps containing the oxidised material are all adjacent to the mine systems with early dates.

It is also necessary to consider the uses for which the zinc oxide was intended. The references to zinc oxide in both the Classical and Indian sources are mainly pharmaceutical,[3] and there is also some evidence that it may have been used as a pigment in India (Stós-Fertner 1980; Lehmann 2000), the familiar zinc white of more recent times, but its main use would surely have been in the production of brass by the cementation process. A few brasses have been found in South Asia dating from the latter part of the first millennium BC, but these tend to be single items rather than regular series of items, such as some contemporary Roman brooches (Istenič and Šmit 2007) or Late Hellenistic and Roman coins (Craddock *et al.* 1980) which were always of brass. More locally, the few contemporary copper alloy fragments found at Dariba were of bronze and the major foundry excavated there also seems to have been working with bronze rather than brass. However, from the beginning of the first millennium AD the use of brass did become more widespread in north India and the Himalayas, as exemplified by the numerous small-cast copper alloy figures, so it is very possible that Zawar did supply at least some of the zinc oxide for their brasses in the first centuries AD. In the absence of major tin supplies, such as existed in other parts of the world, brass was likely to have been an attractive alternative to bronze in South Asia.

Zinc metal

By the latter part of the first millennium BC, some Indian references to a metal with properties commensurate with zinc can already be found, together with a few instances of early brasses with zinc contents that are apparently too high to have been made by the cementation process and could only have been produced by mixing copper and zinc metals (speltering). These

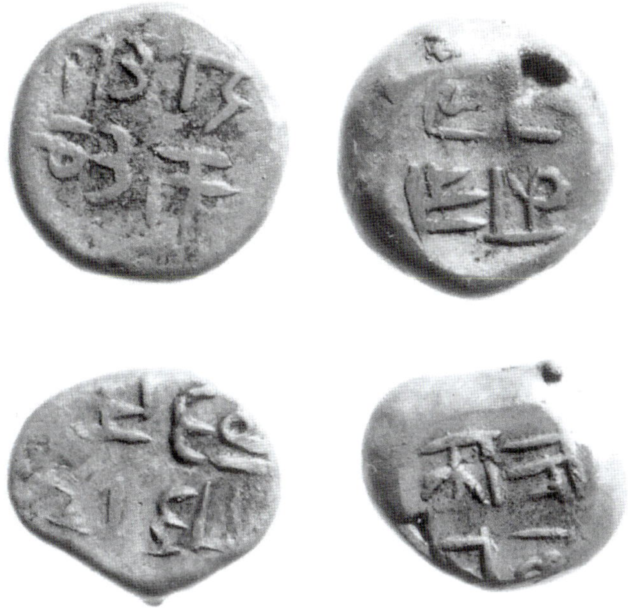

Plate 13.1 Two *jittal* coins of zinc from northwest India, Series B, obverse and reverse (C & M Reg. 1892, 0207. 46-7) (A. Milton/British Museum).

minor amounts of zinc may have been a by-product from the production of silver from pyritic lead and zinc ores such as those found in the Aravallis. A similar phenomenon in the Classical world was noted by Strabo, and tiny amounts of zinc metal have actually been found on some Roman lead smelting sites (Craddock 1998a) (Chapter 11). This was to be the only local source of zinc in post-medieval Europe (see below).

It is difficult to assign precise dates to early Indian scientific literature or to ascribe certain authorship as many works seem to have been progressively copied and rewritten through the centuries. However, it does seem likely that some early scientific texts do contain elements dating from the first millennium AD that describe the production of zinc by distillation. In the 10th century AD, the Persian author, Abū Dulaf described Indian zinc oxide as being superior because it was made by burning the metal (Chapter 11). The last Indian description of the production of brass by cementation was contained in the *Rasarnavakalpa* which is believed to have been compiled between the 10th and 12th century AD, and through the medieval period there are increasingly detailed descriptions of the production of metallic zinc by *tiryakpatana yantram*, which could then be alloyed with copper to form brass. These are laboratory descriptions of the processes carried on at Zawar on an industrial scale.

Zinc production probably originated at Zawar as a commercial process about a thousand years ago, developing out of the existing zinc oxide industry. The industrial process was based on the principles of the existing laboratory techniques, an early and important example of the establishment of a chemical industry. This would have required considerable developmental stages involving both the scientific and business community under the control and encouragement of an enlightened administration. Our surveys encountered little evidence for this stage of the industry at Zawar. When Marco Polo visited Cambay, in present-day Gujarat, the port nearest to Zawar, in the late 13th century, he reported that zinc oxide was an import commodity. This suggests that there was a requirement for zinc oxide

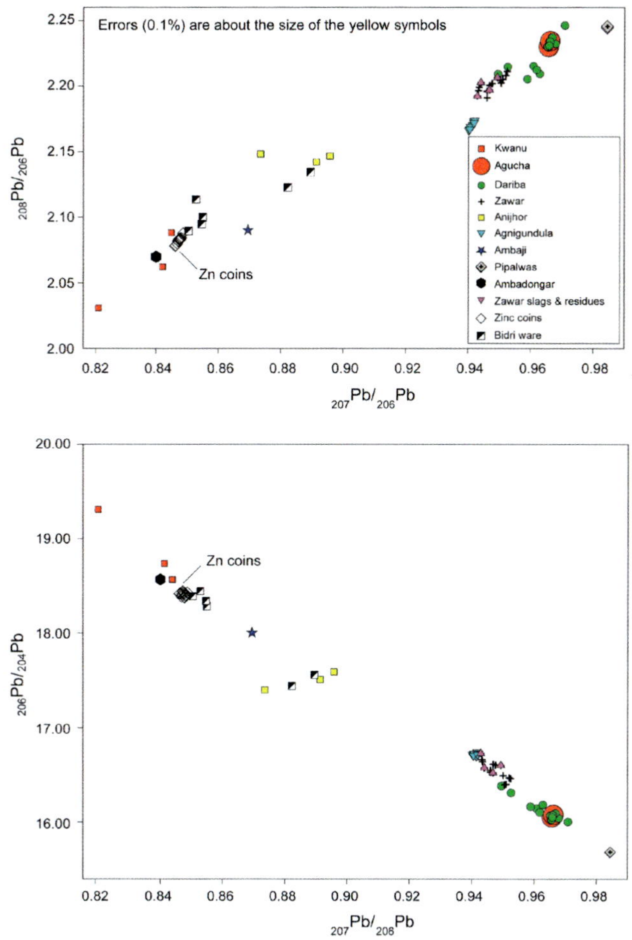

Fig. 13.1 Lead isotope plots of some of the principal sources of zinc, copper, lead and silver worked in the past, together with those of a selection of the zinc *jittal* coins and zinc-based *bidri* ware. Note the Aravalli sources on the right are very different from those of the zinc coins, but some mineral samples from Kwanu in Uttarakhand are much closer, but not coincident (from Craddock *et al.* 2015).

Table 13.1 Lead isotope values of the selected *jittal* zinc coins.

BM CM Reg.	208Pb/206Pb	207Pb/206Pb	206Pb/204Pb
1877, 0707. 2	2.08174	0.84710	18.446
1877, 0707. 5	2.08311	0.84722	18.443
1877, 0707. 6	2.07993	0.84672	18.389
1877, 0707. 8	2.08191	0.84698	18.418
1889, 01203. 49	2.07870	0.84620	18.423
1889, 01203. 51	2.08429	0.84791	18.401
1889, 01203. 55	2.08365	0.84773	18.423
1889, 01203. 56	2.08319	0.84757	18.385
1889, 01203. 57	2.07809	0.84607	18.419
1892, 0207. 51	2.08360	0.84735	18.428
1900, 0805. 4	2.08312	0.84767	18.379
Seeley coin 4th–5th century AD	2.08775	0.84857	18.431

that could not be met locally at that time. Presumably the local brassmakers had previously relied on Zawar for supplies of zinc metal or zinc oxide and when that supply was terminated, at least temporarily, alternative supplies had to be found. As at that time there were no other sources of metallic zinc they

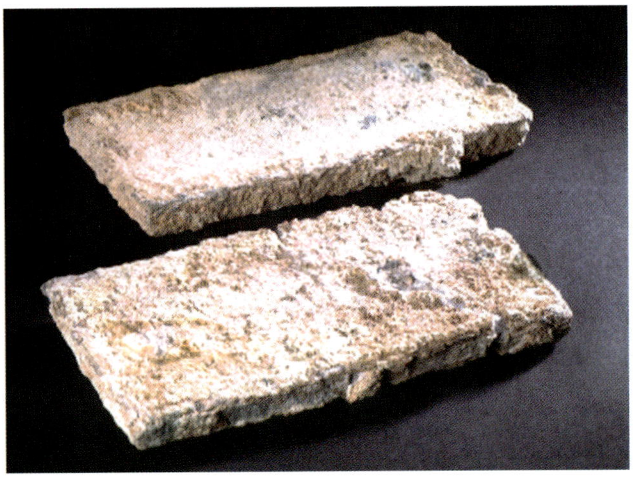

Plate 13.2 In the post-medieval period most zinc entering international trade came from China. Zinc ingots, part of a cargo on the EIC *Diana*, wrecked in the Malacca Straits in 1812 en route from China to India (BM Reg. PE 1997, 0202.1 and 2) (A. Milton/British Museum).

had to fall back on zinc oxide, most probably from Iran. This imported zinc oxide could either have been distilled locally to make zinc metal or used directly in a revival of the cementation process. The situation at Zawar becomes much clearer in the 14th century when Maharana Lakha Singh (AD 1382–1421) re-established the industry at Zawar. The historical record is matched by the abundant field evidence of zinc production by the small retort process and also by the foundation of many temples (Kumar forthcoming). This period from the 14th to the 16th century until the Mughal invasions was to prove the most prosperous at Zawar, both architecturally as well as technically, with the small retort process developing into the large retort process. The small retort process was widely used all over Zawar, including Site 30 up on Zawar Mala and at outliers such as Kaya a few kilometres distant. The large retort followed by the intermediate retort processes were restricted to the area around the settlement known as Old Zawar.

There is also evidence for zinc production elsewhere in India such as for the zinc *jittal* coins minted by the Kangra kings in the Himachal Pradesh–Punjab region, probably during the 15th–16th centuries (Plate 13.1) (Cribb 2014). Lead isotope analysis revealed that the zinc of the coins was from a single source and that moreover this source cannot have been in the Aravallis (Craddock *et al.* 2015). In addition it is reported that retorts very similar to the Zawar large retorts have been found at Kwanu, a small settlement in the Tons Valley that forms the present boundary between Himachal Pradesh and Uttarakhand (Dey 2008).[4]

The fate of this industry is unknown, but Zawar together with the rest of Mewar, underwent major disruption for decades following the Mughal invasion and occupation in the late 16th century. The hoard of freshly minted silver coins dated 1593 (see Chapter 3, Plates 3.12 and 3.13) buried in an abandoned large retort furnace at Old Zawar and never reclaimed succinctly encapsulates Zawar's misfortunes at this time.

Following eventual peace in the early 17th century, zinc production, now using the intermediate retort process, resumed but on a much reduced scale. During the disruption much had changed: the markets for Zawar's zinc, both international and internal, had been lost to imports brought in European vessels from China – the age of international maritime trade had arrived (Prakash 1984; 2007). Unfortunately documentation of the European maritime trade in the Indian Ocean during the 16th century, when it was dominated by the Portuguese, is sparse and there are only indirect hints at a trade in zinc at this time.[5] The German chemist, Andreas Libavius, reported in the 1590s that the Dutch had captured a Portuguese vessel from Malacca, now in Malaysia, with a cargo of zinc (Partington 1961: 109). This is very likely to have come from China. By the early 17th century the Dutch and English trading companies dominated the trade and their documentation through the next two centuries is abundant (Craddock 2013). From this there can be no doubt that there was a considerable trade in zinc from China to India (Plate 13.2), including to Surat and Diu, almost the closest ports to Zawar itself.[6] Zinc production was again disrupted at Zawar by the Maharatta wars, and production by the traditional process ceased completely at the beginning of the 19th century.

Chinese zinc

Brass was probably introduced into China by Buddhists from the north of India in the Han period, approximately two thousand years ago, and for many centuries brass seems to have been imported and used only on a small scale (Zhou 2001). It is believed that local production of cementation brass probably began about a thousand years ago and from this evolved the production of zinc metal.

The method developed by the Chinese is completely different from the Indian process (Craddock and Zhou 2003). In this distinctive process the condensation and collection all take place within the retort itself (Fig. 13.2). This is the traditional Chinese method used for aqueous distillation, sometimes referred to as the Mongolian still, therefore it seems most likely that the Chinese method for zinc production was an indigenous development. The dates for the start of zinc production within China are uncertain but recent excavations of major industrial production units in Chongqing, commensurate in scale to those at Zawar, are dated to the 16th century (Liu *et al.* 2007; Zhou *et al.* 2012). These belong to the developed industrial phase suggesting there are likely to be antecedents. However brass did not become popular in China until the 16th century and then within a short space of time spelter brass replaced bronze as the usual copper alloy throughout China as evidenced by the *cash* coins (Wang *et al.* 2005). In addition, by the very beginning of the 17th century at the latest, zinc metal was already a major export commodity in the international maritime trade (Bonnin 1924; Souza 1986, 1991). As has been noted, India was an important market for this zinc – it was one of the few places that was already familiar with zinc and was using it extensively to produce brass so there was a ready market when internal supplies were disrupted. Conversely, only relatively small amounts seem to have been sent to Europe, where it was an expensive and exotic commodity, arousing considerable academic interest as to its nature, and what to do with it.[7] Matters were different in Asia. Ships sailed to China with lead as the cheap ballast cargo, where it was unloaded and replaced with zinc as the cheap durable cargo for the voyage to India (Souza 1991). It would no doubt have surprised the European savants to learn that this technically sophisticated material was already affordable and commonplace in the East.

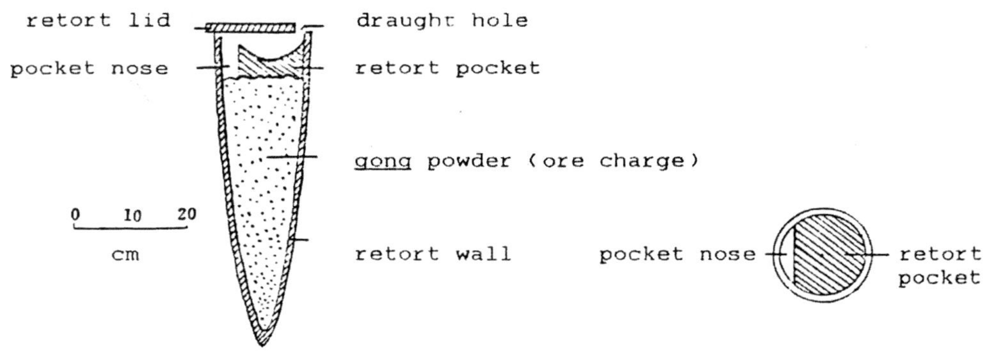

Fig. 13.2 Principle of the Chinese method of zinc production using an internal condenser, very different from either the Indian or European processes (from Xu 1998: fig. 4).

Exports of zinc to Europe, mainly in Dutch vessels, ceased in the late 18th century, but exports to India continued until the early 19th century.[8]

In terms of costs the Chinese production had three distinct advantages. First, the ore used was the carbonate, smithsonite, which only required moderate calcining. This was normally carried out in the open retorts as the first stage of the process before reduction, rather than the full separate roasting stage to treat the sulphide ores, thereby minimising the zinc losses. Secondly the Chinese were able to use mineral coal – not only ordinary coal to heat the retorts from outside, but also special low sulphur coal within the retorts acting as the reductant. Finally the traditional Chinese retorts were reusable (typically seven times in the late 20th century: Craddock and Zhou 2003). The sizes of the 16th-century Indian and Chinese retorts are similar therefore the yield per retort is likely to have been broadly the same. Both the Indian and Chinese fired multiple retorts together so the overall yield per furnace block is also likely to have been mainly similar.

Zinc in Europe

The inspirations and stages by which zinc production in Europe developed are rather unexpected. In some central European silver mines there was already a very limited production of zinc. During the initial smelting of the silver-rich mixed galena and sphalerite ores some zinc oxide was collected on the flue walls as described in Classical antiquity and as practised at Zawar (see Chapter 11, Figs 11.1 and 11.2). In addition to the zinc oxide, small droplets of metallic zinc also condensed on the flue walls, and from the post-medieval period the zinc was encouraged to condense in special cool chambers inserted into the walls of furnaces such as those at Rammelsberg, Goslar (Craddock 1998a). Mercury was also smelted in central Europe as a vapour that was collected in a downwards distillation arrangement reminiscent in principle of the Indian *tiryakpatana yantram* distillation method (see Chapter 11, Fig. 11.6). Thus it might have been thought likely that the distillation of zinc would have developed in central Europe by applying the principle of the local mercury production to the zinc ores, but this was not immediately to be the case (Craddock 2009c). Instead it seems that it was the zinc imported from Asia that inspired interest in developing a viable production method.[9]

Once again the actual course of events turned out to be different from what might have been expected. Given that Chinese zinc dominated the market it would be reasonable to assume that the Chinese method, with an internal condenser, would be the inspiration behind the method adopted. However, Watson (1796: 28) made it clear that he regarded zinc production and importation into Europe as very much Indian concerns:

> The eastern Indians have long since been in possession of the method of extracting pure zinc from the ore; at least in the course of the last century this metal has been brought from thence to Europe. Jungius mentions the importation of zinc from India in 1647 (in his *De Mineralibus*); a metal of this kind under the name of Tutenag is still brought from thence.

In fact the first large-scale production of zinc in Europe used a process of downward distillation very similar in principle to that used at Zawar (Fig. 13.3). This process was the creation of William Champion who ran an innovative brassmaking business in Bristol, England, and patented zinc production in 1738 (Day 1973, 1988). The exact nature of his furnaces and arrangements of the retorts were not revealed, but it is likely that a few very large retorts were inverted in furnaces based on those of the local glassmaking industry. The zinc vapour condensed and ran down iron pipes into jars of water in a cool chamber beneath. There was no industrial plant or process remotely like this already operating in Britain for Champion to have copied, and local legend and history strongly hints at the process having been learnt from the East. It is claimed that various entrepreneurs and adventurers travelled either to Holland or all the way to China to learn the process: this is a feasible suggestion as such zinc as arrived in Europe came mainly from China in Dutch ships (Bonnin 1924). However, if that really had been the case, then it would have been the internal condenser process that was encountered. In practice it is possible that Champion heard indirectly of the principle of the Indian process, but without any direct or detailed descriptions or knowledge of similar processes such as those for distilling mercury he had to set up his own plant from scratch. A consequence of this inexperience was that his process had serious flaws. The zinc reduction process is endothermic and all the thermal energy necessary has to come into the retort from outside. Therefore for a rapid and efficient process it is essential to have a relatively large surface area of the retort exposed to the heat. Also any heat generated which does not penetrate the retorts is effectively lost so it is desirable to keep the burning fuel close up against the retorts. At Zawar the large number of

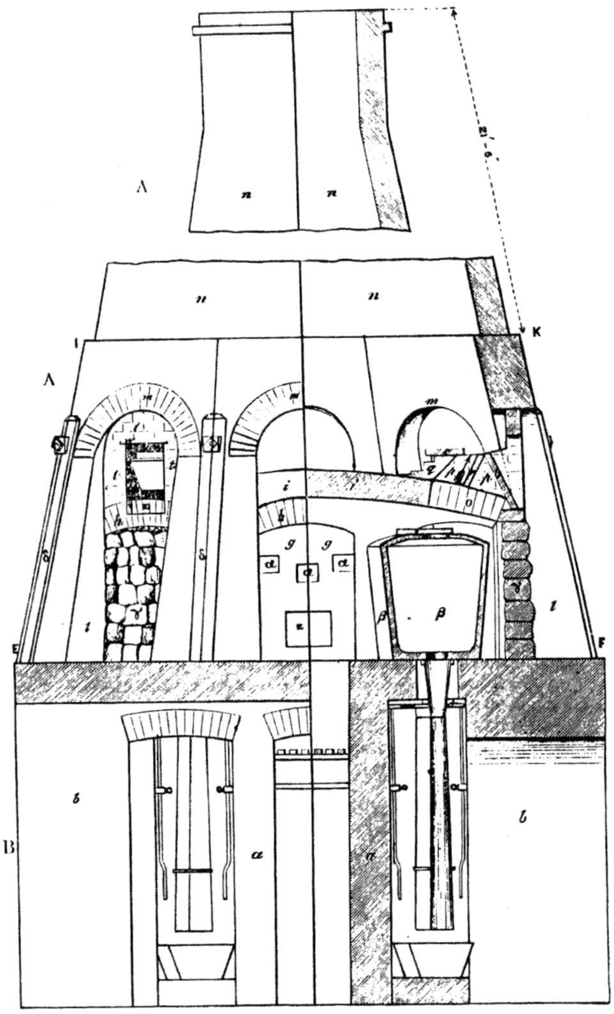

Fig. 13.3 The so-called 'English process' based on Champion's furnace arrangement of distillation by descending. The width at the base was 16 feet (4.5 m) (from Percy 1861: fig. 125).

small retorts packed into as small a furnace space as possible achieved this, as did the central European mercury distillation process. In contrast, Champion's process with just a few very large capacity retorts must inevitably have been much slower and less efficient with much of the heat generated missing the retorts altogether. Perhaps it is not too surprising that after only a few years Champion became bankrupt and his furnaces closed, but not before the process attracted attention from Europe. Industrial spies from central Europe visited Bristol and reported home, confirming that Champion had indeed set up an industrial process to produce zinc in quantity, but that his method as far they could ascertain, was impractical (Day 1984). The Champion process (which became known as the English process) struggled on for a few more years but in Europe existing techniques of industrial distillation were adapted and by the early 19th century a variety of methods was in use (Craddock 1987, 2009c; Almond 1998). These were often based on the local mercury production technology and included the Carinthian process (Dumas 1847: 16–17, pl. 40) (Fig. 13.4). This incorporated a large number of small retorts, set vertically and which dripped down their zinc. This was remarkably similar to the Zawar process, although almost certainly based on the local mercury production methods. These early 19th-century techniques formed the basis for all subsequent retort processes for the production of zinc, motivated by Champion's process that itself was almost certainly inspired by Zawar.

Perceptions and regeneration

The technically sophisticated processes, exemplified by the zinc and steel industries in India and the zinc industry in China, challenge the usual perception of the origins and scope of the industrial revolution. This is usually seen as an essentially European phenomenon with its origins in the post-medieval period, but only really gaining momentum from the mid-18th century (Day and Tylecote 1991). It has been acknowledged that European voyages of discovery and trade, often culminating in administrative control, frequently encountered a variety of novel materials, often produced by distinctive and complex technologies. Yet to a degree they were, and to a large extent still are, regarded as exotic – in a pejorative sense – in that they had no relevance to the modern developments in science or industrial technology. Also they were, and are, regarded as essentially static or even in decline (Chapter 12). These industries were generally considered in some sense as being a local peculiarity of whatever region in which they were found, with no sense of change and development much less capable of innovation and influence.[10] As noted previously, this fitted within the general European perception of statis for the whole of society outside Europe. For India this was especially unfortunate as by the time that serious recording and investigation of the country, including its mines and metallurgical industries, began in the 19th century, a century and more of upheaval and attendant decay had largely terminated advanced industry, and even in some cases, its memory.

The programmes of investigation of the early zinc industry in India and in China have both revealed the application of scientific principles and techniques being developed into industrial processes through the medieval period, and evolving through that period up to the time of the first European contact, and beyond in the case of the Chinese zinc industry (Craddock and Zhou 2003). The influence of these established industries on European developments is difficult to gauge. For example, the first European crucible steel process, the Huntsman process in which cementation steel was homogenised by melting, is very different from the Asian processes and was almost certainly an indigenous development. The later 19th-century European crucible steel processes did embrace both of the Asian processes of *in situ* carburisation and co-fusion, although without any acknowledgement (Craddock *et al.* forthcoming). The zinc production process patented by William Champion in 1738 certainly adopted the Indian method, again without acknowledgement.[11] However, it is clear that these new metals, imported from the East, were at last generating interest as well as being potentially commercial viable products, for which uses could be found.[12] It is also clear, especially in the case of zinc production, that it was understood that information on the new products should be sought in their countries of origin. Thus the imported materials were not only a direct inspiration in their own right but also encouraged investigation into the technologies by which they had been made.

Increased knowledge of the history and technical development of these advanced industries outside Europe are leading to a fuller understanding of the whole process of industrial development on a global scale of which the European industrial revolution formed but a part.

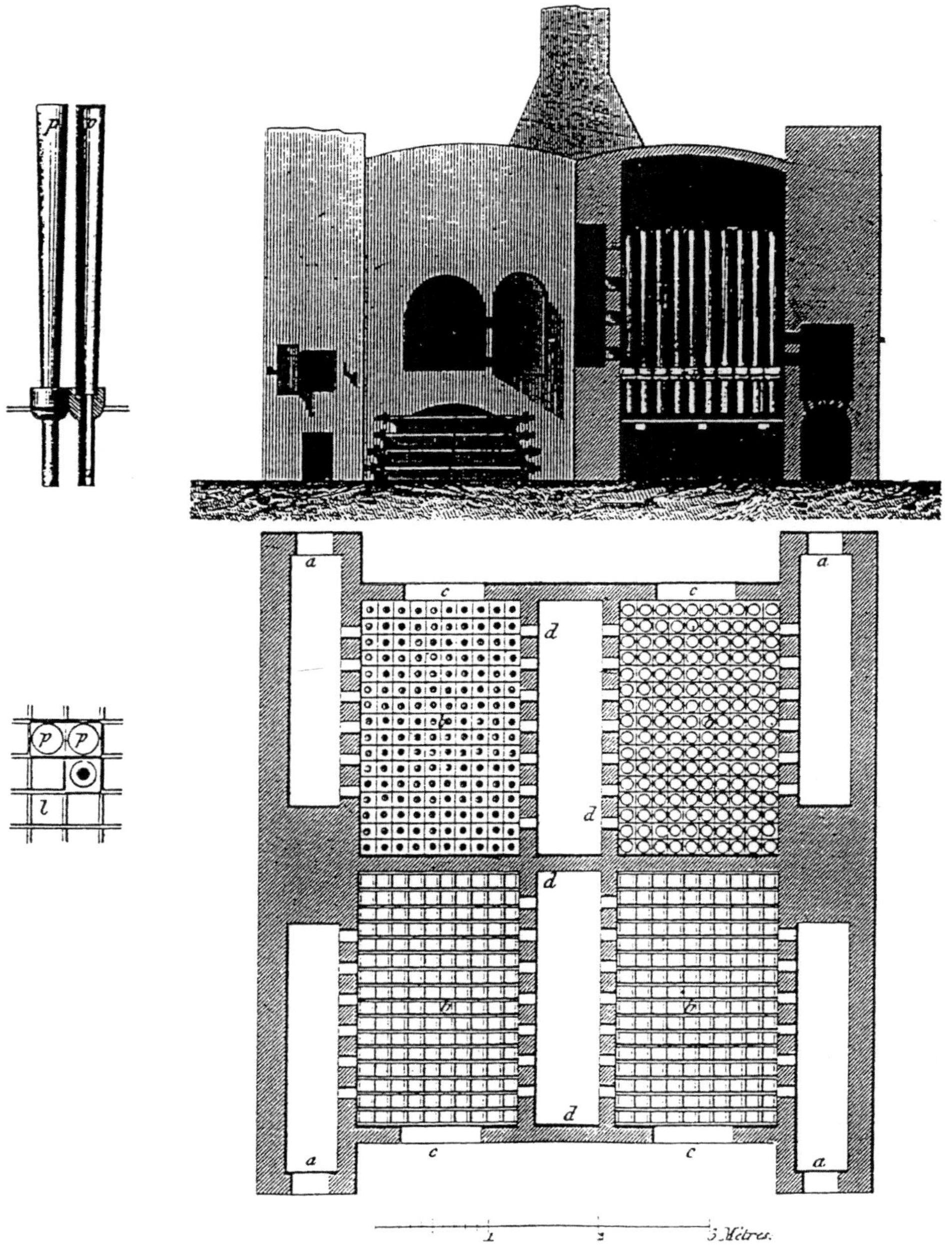

Fig. 13.4 Carinthian process furnace. The large number of small vertical retorts in the hot chamber drip down their zinc into the cool chamber beneath (cf. Figs 6.9 and 11.5) (from Dumas 1847: pl. 40).

Notes

1. As mentioned in Chapter 8 (note 3) and Chapter 9, Kautīlya's *Arthaśāstra* describes in some detail the establishment of mines in remote areas, together with their defence and management. One of the main arguments previously advanced for dating the *Arthaśāstra* after the Mauryans (discussed and refuted by Kangle 1965–72, III and Mital 2000, esp. p. 32) was that they had no experience of mining. It can now be appreciated that these mines in the Aravalli Hills were Mauryan enterprises, set up far from their main centres, and it is not impossible that a very real Kautīlya had these operations in mind when he compiled his treatise on governance in the 4th century BC.
2. Deyell (1990: 251) made much of the absence of silver sources in South Asia, stating that: 'It can be asserted with tolerable confidence that the medieval silver coinage of North India was not struck from locally mined silver.' This may well be correct – the Aravalli mines in our project do not seem to have been producing silver at this time – but unfortunately Deyell sought to support this argument on fallacious technical grounds, stating that medieval technology was incapable of extracting silver from argentiferous lead containing between 0.05 and 0.18% of silver, whereas in fact the lower limit in both South Asia and the Mediterranean world was of the order of about 0.02% (Chapter 10, p. 206). More specifically Deyell denied that Zawar could ever have been a significant producer. This is probably correct, but Deyell based his supposition on Hegde's (1967) somewhat dubious claims based on trace element analysis (Chapter 2, note 5). Deyell (1990: 221–6) also contrasted the relatively abundant pure silver coinage of the Delhi Sultanate up to the 13th century, which already had access to the silver sources to the north, with the debased coins of the Rajput kingdoms of Rajasthan and Gujarat, further suggesting to him that no local supplies in the Aravallis were being worked at this time.
3. A sealed tin container found with other vials and physicians' tools on a Roman wreck contained tablets made up of a variety of herbal extracts in a medium of zinc oxide and carbonate, the familiar calamine zinc ointment (Giachi *et al.* 2013).

4. The Dutch traveller, Abraham Rogerius, who spent many years travelling in South Asia in the 1630s, and who produced the standard work on Hinduism (Caland 1915), stated that taxes were paid to the temples in the Mughal Empire in part in *spiauter* (zinc), suggesting that zinc coins were quite prevalent in north India. There were also some low denomination coins of zinc minted by European traders in 18th-century India, known locally as *bazarucos*, but the zinc probably came from China (Yih and de Kreek 1993).
5. For example, there is the intriguing letter dated 12 October 1513 from the Italian diplomat, Antonio Bavarin, in London, claiming that he had received reports from Lisbon stating that three vessels had arrived there with cargoes including 400 pieces of tin, which would be very detrimental to the English (*Calendar of State Papers* n.d.). This could be tin as the Portuguese had reached Southeast Asia in 1512, but if so then these must have been the very first return vessels. It is more probable that the vessels had returned from India where the Portuguese were already well established and the 'tin' was 'Indian tin', namely zinc.
6. Lead isotope studies on a selection of Indian *bidri* items composed mainly of zinc failed to reveal any with zinc that matched any Indian source (Craddock *et al.* 2015).
7. European curiosity was aroused by this strange metal *caleum* or *calim* encountered in the East. For example De Laval in 1679 noted that *calim* was whiter than tin but more beautiful, and that the metal was highly prized throughout India (quoted in Partington 1961: 109). Europe's first encounter with zinc had been the strange droplets of metal that appeared in the flues of certain lead/silver smelters (see below) and the nature of this *counterfrei* aroused considerable curiosity. Brass was produced by cementation of calcined zinc carbonate smithsonite ore and copper, but even though *counterfrei* was sometimes added, the connection between the two seems not to have been made until 1721, when Henckel established that smelting and condensing smithsonite mineral produced zinc. Even so there was considerable uncertainty over the nature of zinc, probably due in part to its brittleness and flammability, and up to the beginning of the 19th century it was often designated as a half metal.
8. East India Company records of zinc imported into England from the 1760s show that an average of about 40 tons per annum of zinc came from East India (i.e. China), declining until in 1779 only about 3 tons were imported. Although zinc was no longer then being sent to Europe, there was still a considerable country trade to India, continuing well into the 19th century (Cranmer-Byng 1962), as exemplified by the EIC *Diana* which sank on her way to Calcutta laden with Chinese zinc in 1812 (Plate 13.2) (Ball 1995). Sir Thomas Dick Lauder stated in 1820 that 'Tutanag is an article of very extensive commerce between China and India ... It is employed by the natives of India as an alloy for copper to make brass for their domestic utensils' (Holland 1834). However, only a short time later, John Holland (1834) claimed that in 1826 the importation of Chinese *tutanag* into Calcutta had ceased and was replaced by imports of German *spelter*.
9. Malachy Postelthwayt in *The Universal Dictionary of Trade and Commerce* (1751) observed that: 'Our artificiers have long been acquainted with zink under the name spelter ... We have much brought also from the *East Indies* under the name of tutenage, yet nobody ever knew from what or how it was produced there ... There is great reason to believe that all the zink or tutenage brought from the *East Indies* is produced from Calamine: and we have now on foot at home a work established by the great discoverer of this ore, which will probably very soon make it unnecessary to bring any more zink from elsewhere into England.'
10. See Chapter 12, notes 2 and 4.
11. To a certain degree the failure is understandable, as to acknowledge that the process was inspired by a similar process elsewhere would have invalidated any patent application.
12. Zinc production grew slowly before improvements in the European zinc production methods at the end of the 18th century enabled brass to be made more cheaply by speltering than by cementation. Also in the early 19th century, the production of rolled zinc sheet was perfected and the use of the familiar zinc white pigment introduced in Europe.

Bibliography

Aery, N.C. and Tiagi, Y.D. 1988. 'Accumulation of cadmium by plants of Zawar Mines, Rajasthan, India', *Acta Biologica Hungary* 39(1): 87–98.

Aggarwal, S.C. 1956. *The Salt Industry in India*, 2nd edn. Delhi: Manager of Publications, Government of India.

Agrawal, B.L. 1999. 'Expansion of Rampura-Agucha: a step in HZL growth', *HindZinc Tech* January 1999: 11–19.

Agrawal, D.P. 1984. 'Metal technology of the Harappans', in *Frontiers of the Indus Valley Civilisation*, B.B. Lal and S.P. Gupta (eds), 163–7. New Delhi: Indian Archaeological Society.

Agrawal, D.P. 2000. *Ancient Metal Technology and Archaeology of South Asia*. New Delhi: Aryan Books.

Agrawal, D.P. and Kharakwal, J.S. 2003. *Bronze and Iron Ages in South Asia*. New Delhi: Aryan Books.

Agrawal, D.P. and Seshadri, R. 1998. 'The metallurgical tradition of the Harappans', in *Archaeometallurgy in India*, V. Tripathi (ed.), 10–16. Delhi: Sharada Publishing House.

Agrawal, D.P., Margabandhu, C. and Shekar, N.C. 1976. 'Ancient copper workings: some new C^{14} dates', *Indian Journal for History of Science* 2(2): 133–6.

Agrawal, O.P., Narain, H. and Prakash, J. 1990. 'Technical studies of silver punch mark coins from Narhan (Gorakhpur U.P.), India – 600 BC–200 BC', *Journal of the Historical Metallurgy Society* 24(2): 81–8.

Agrawala, R.C. 1958. 'An inscription from Javar, Rajasthan', *Journal of Indian Historical Quarterly* 34(3–4): 15–25.

Agrawala, R.C. 1984. 'Aravalli, the major source of copper for the Indus civilisation and Indus-related cultures', in *Frontiers of the Indus Valley Civilisation*, B.B. Lal and S.P. Gupta (eds), 157–62. New Delhi: Indian Archaeological Society.

Ahmed, M. (ed.) 1963. 'Indigenous smelting of iron', in *Census of India 1961* (12) (7A) *Survey of Traditional Crafts of Orissa. Book 1*. Delhi: Government of India.

al Hassan, A.Y. and Hill, D.R. 1986. *Islamic Technology*. Cambridge and Paris: Cambridge University Press and UNESCO.

Alam, I. 2002. 'Iron manufacturers in Golconda in the 17th century', in *Tradition and Innovation in the History of Iron Making: An Indo European Perspective*, G. Pande and J. af Geijerstam (eds), 98–111. Nainital, Uttarakhand: PAHAR.

Allan, J.W. 1977. 'Silver: the key to bronze in Early Islamic Iran', *Kunst der Orients* 11: 5–21.

Allan, J.W. 1979. *Persian Metal Technology, 700–1300 AD*. London: Ithaca Press.

Allchin, F.R. 1962. 'Upon the antiquity and methods of gold mining in ancient India', *Journal of the Economic and Social History of the Orient* 5(2): 195–211.

Allchin, F.R. 1963. *Neolithic Cattle-Keepers of Southern India*. Cambridge: Cambridge University Press.

Allchin, F.R. (ed.) 1995. *The Archaeology of Early Historic South Asia*. Cambridge: Cambridge University Press.

Allchin, B. and Allchin, F.R. 1982 *The Rise of Civilisation in India and Pakistan*. Cambridge: Cambridge University Press.

Almond, J.K. 1998. 'Zinc-production technology 1801–1950: a review', in *2000 Years of Zinc and Brass*, P.T. Craddock (ed.), 2nd rev. edn, 159–228. London: British Museum Press.

Anantharamu, T.T., Craddock, P.T., Nagesh Rao, K., Murthy, S.R.N. and Wayman, M.L. 1999. 'Crucible steel of Ghattihosahalli, Chitradurga District, Karnataka, southern India', *Journal of the Historical Metallurgy Society* 33(1): 13–25.

Anfinset, N. 2011. *Social and Technological Aspects of Mining, Smelting and Casting Copper*. Bochum: Deutschen Bergbau-Museum.

Anon. 1872. 'The mines of Mewar', *The Indian Antiquary* February 1872: 63–4.

Anon. 1930. 'General report for 1929', *Records of the GSI* 72: 79.

Anon. 1963. 'The Hutti Gold Mines Company Limited', in *Gold Mining Industry in India*, Committee (ed.), 126–3. Bangalore: Geological Society of India.

Anon. 1983a. 'Malanjkhand copper project', *Mining Magazine* 149(5): 334–41.

Anon. 1983b. 'The Zawar mines', *Mining Magazine* 149(5): 323–33.

Anon. 1983c. 'HCL's Indian copper complex', *Mining Magazine* 149(5): 343–5.

Anon. 1989. 'Zinc: a heritage', *Hindzinc Tech* 1(1): 3–12.

Aralere, T.S. 2016. 'The crossroads of Hellenic and Sanskrit science', in *Companion to Science, Technology and Medicine in Ancient Greece and Rome*, G.L. Irby (ed.), 943–57. Chichester: Wiley Blackwell.

Armstrong, L. 1978. *Woodcolliers and Charcoal Burning*. Horsham: Coach Publishing House.

Athavale, V.T. 1967. 'Chemical analysis and metallographic examination of metal objects (Prakash 1955)', in 'Prakash 1955', B.K. Thapar, *Ancient India* 20–21: 5–167 (pp. 135–9, appendix, Table IV).

Atkinson, E.T. 1882. *The Himalayan Districts of the North Western Provinces of India I (1)*. Allahabad: North West Province and Oudh Government Press [reprinted 1973 Delhi: Cosmo].

Avadhanulu, A.B. 1979. 'Analyses of lead coins of the Satavahana period', *Birla Archaeological and Cultural Research Institute Research Bulletin* 1: 13–26.

Babu, T.M. 1994. *Tin in India*. Bangalore: Geological Society of India.

Babu, T.M. 2003. 'Advent of the Bronze Age in the Indian subcontinent', in *Mining and Metal Production through the Ages*, P.T. Craddock and J. Lang (eds), 174–80. London: British Museum Press.

Bachmann, H-G. 1982. *The Identification of Slags from Archaeological Sites*. London: Institute of Archaeology.

Bachmann, H-G. 1991. 'Archäometallurgie des Silbers', *Die Geowissenschaften* 9(1): 12–16.

Bachmann, H-G. 1993a. 'Zur metallurgie der römischen Goldgewinnung in Três Minas und Campo de Jales in Nordportugal', in *Montanarchäologie in Europa*, H. Steuer and U. Zimmermann (eds), 153–60. Sigmaringen: Jan Thorbecke Verlag.

Bachmann, H-G. 1993b. 'The identification of slags from archaeological sites', in *Archeologia delle attivitá estrattive e metallurgiche*, R. Francovich (ed.), 563–92. Florence: All'Insignia del Giglio.

Bachmann, H-G. 1993c. 'The archaeometallurgy of silver', in *Archeologia delle attivitá estrattive e metallurgiche*, R. Francovich (ed.), 487–95. Florence: All'Insignia del Giglio.

Balfour, E. 1855. *Manufacture of Iron and Steel and the Coals of the Madras Presidency*. Madras: Report of the Government Central Museum.

Ball, D. 1995. *The Diana Adventure*, Kuala Lumpur and Kemper, Gronigen: Malaysian Historical Salvors.

Ball, V. 1881. *A Manual of the Geology of India Pt. III: Economic Geology*, Calcutta: Geological Survey of India.

Bansal, N.C. and Ahluwalia, M.S. 1990. 'Zinc-lead mining in India: changing technology', *Hindzinc Tech* 2(1): 3–17.

Bayley, J. 1992. 'Metalworking ceramics', *Medieval Ceramics* 16: 3–10.

Bender, F. (ed.) 1983. *Geology of Burma*. Berlin and Stuttgart: Gebrüder Borntraeger.

Bender, F.K. and Raza, H.A. (eds) 1995. *Geology of Pakistan*. Berlin and Stuttgart: Gebrüder Borntraeger.

Bhardwaj, H.C. 1979. *Aspects of Ancient Indian Technology*. Delhi and Varanasi: Motilal Banarsidass.

Bhardwaj, H.C. 2000. *Metallurgy in Indian Archaeology*. Delhi: Tata Book Agency.

Bhat, M.R. (ed. and tr.) 1981.*Varahamihira's Brhat Samhita I*. Delhi: Motilal Banarsidass.

Bhati, H.S. (ed.) 1995a. *Maharana Rajsingh Patta Bahi Pattedaran ri Vigat*. Udaipur: Himanshu Publication.

Bhati, H.S. (ed.) 1995b. *Maharana Rajsingh Paragana Bahi*. Udaipur: Himanshu Publication.

Bhatnagar, R.P. 1979–80. *Sahitya Majhamika*. Udaipur: Pratap Sodha Sanstan.

Bhatnagar, S.N. and Mathur, S.B 1989. 'Geology of lead-zinc resources', *Hindzinc Tech* 1(2): 15–28.

Bhowmik, S.K. 1964. 'Technical study and conservation of a Nepalese metal image from the Varodara Museum', *Journal of the Oriental Institute of the M.S. University of Baroda* 13(4): 393–8.

Bhowmik, S.K. and Jani, M. 1998. 'Literary references on metals, metallic objects of art and metal technology', in *Archaeometallurgy in India*, V. Tripathi (ed.), 319–38. Delhi: Sharada Publishing House.

Biek, L. 1970. '"Vitrified products"'– a comment', *Bulletin of the Historical Metallurgy Group* 4(2): 81–2.

Biswas, A.K. 1987. '*Rasa-ratna-samuccya* and mineral processing: state-of-art in the 13th century A.D. India', *Indian Journal of History of Science* 22(1): 29–46.

Biswas, A.K. and Biswas, S. 1996. *Minerals and Metals in Ancient India*, 2 vols. New Delhi: D.K. Printworld.

Biswas, A.K. and Davenport, W.G. 1976. *Extractive Metallurgy of Copper*. Oxford: Pergamon.

Blanchard, I. 1992. 'Technical implications of the transition from silver to lead smelting in twelfth century England', in *Boles and Smeltmills*, L. Willies and D. Cranstone (eds), 9–11. London: Historical Metallurgy Society.

Blanchard, I. 2001. *Mining, Metallurgy and Minting in the Middle Ages: I Asiatic Supremacy, II Afro-European Supremacy*. Stuttgart: Franz Steiner Verlag.

Boileau, A.H.E. 1831. 'On the copper works at Singhana near Khetri in the Sekhawati country', *Gleanings in Science* 3: 380–84.

Boileau, A.H.E. 1845. *Miscellaneous Writings in Prose and Verse*. Calcutta: Thacker and Ostell & Lepage.

Bonnin, A. 1924. *Tutenag and Paktong; With Notes on Other Alloys in Domestic Use during the Eighteenth Century*. Oxford: Oxford University Press.

Bourgarit, D. and Bauchau, F. 2010. 'The ancient brass cementation processes revisited by extensive experimental simulation', *Journal of Metals* 62(3): 51–7.

Brill, R.H. and Shields, W.R. 1972. 'Lead isotopes in ancient coins', in *Methods of Chemical and Metallurgical Investigation of Ancient Coinage*, E.T. Hall and D.M. Metcalf (eds), 279–303. London: Royal Numismatic Society.

Bronk Ramsey, C. 1995. 'Radiocarbon and the analysis of stratigraphy: the OxCal program', *Radiocarbon* 37(2): 425–30.

Bronson, B. 1986. 'The making and selling of wootz: a crucible steel of India', *Archeomaterials* 1(1): 13–51.

Bronson, B. 1992 'Patterns in the early Southeast Asian metals trade', in *Early Metallurgy, Trade and Urban Centres in Thailand and Southeast Asia*, I. Glover, P. Suchitta and J. Villiers (eds), 63–114. Bangkok: White Lotus.

Brooke, J.C. 1850. 'Notes on the zinc mines of Jáwar', *Journal of the Asiatic Society of Bengal* 19: 212–15.

Brooke, J.C. 1859. *History of Meywar*. Calcutta: Baptist Mission Press.

Brooke, J.C. 1864. 'The mines of Khetree in Rajpootana', *Journal of the Asiatic Society of Bengal* 33: 519–629.

Buchanan, F. 1807. *A Journey from Madras through the Countries of Mysore, Canara and Malabar 23rd April 1800 to 6th July 1801*. London: T. Cadel and W. Davies and Black, Perry and Kingsbury [reprinted 1999 Delhi: Asian Educational Services].

Buchanan, P.T. 1992. 'Metaliferous plant communities: the flora of lead smelting in the Upper Nent Valley', in *Boles and Smeltmills*, L. Willies and D. Cranstone (eds), 58–61. London: Historical Metallurgy Society.

Burnham, B. and Burnham, H. 2004. *Dolaucothi-Pumsaint: Survey and Excavations at a Roman Gold-mining Complex 1987–1999*. Oxford: Oxbow Books.

Caland, W. (ed.) 1915. *Abraham Rogerius: De open-deure tot het verbogen Heyendom*. Gravenhage: M. Nijhoff.

Calendar of State Papers n.d. Researches in Foreign Archives, Italy 1509–1519. London: Public Record Office.

Caldwell, K.S. 1920. 'The result of analyses of certain ornaments found in Asura sites', *Journal of the Bihar and Orissa Research Society* 6: 390–423.

Campbell, J. 1843. 'Manufacture of bar iron in southern India', *Calcutta Journal of Natural History* 3: 386–400.

Cannon, H.L. 1960. 'Botanical prospecting for ore deposits', *Science* 132: 591–8.

Casson, L. 1989. *The Periplus Maris Erythraei*. Princeton, NJ: Princeton University Press.

Chakrabarti, D.K. 1979. 'The problem of tin in India: a preliminary survey', *Man and Environment* 3: 61–74.

Chakrabarti, D.K. 1985/6. 'The pre-industrial mines of India', *Puratattva* 16: 65–71.

Chakrabarti, D.K. and Lahiri, N. 1996. *Copper and its Alloys in Ancient India*. New Delhi: Munshiram Manoharlal.

Chakravarti, N.P. 1934/5. *Annual Report of the Archaeological Survey of India*. Poona.

Chattopadhyay, P.K. 2004. *Archaeometallurgy in India*. Patna: Kashi Prasad Jayaswal Research Institute.

Chaudhuri, M.R. 1975. *The Iron and Steel Industry of India: an Economic–Geographic Appraisal*. Calcutta: Oxford University Press and IBH.

Claughton, P. 1992. 'Medieval silver-lead smelting in Devon', in *Boles and Smeltmills*, L. Willies and D. Cranstone (eds), 12–14. London: Historical Metallurgy Society.

Coggin Brown, J. and Dey, A.K. 1955. *India's Mineral Wealth*, 3rd edn. Oxford: Oxford University Press.

Collins, A.L. 1892–3. 'Firesetting: the art of mining by fire', *Transactions of the Federated Institute of Mining Engineers* (5)1: 82–92.

Collins, A.L. 1893–4. 'The Ghorband lead mines, Afghanistan', *Transactions of the Federated Institution of Mining Engineers* 6(3): 449–56.

Conophagos, C.E. 1980. *Le Laurion antique et la technique grecque de la production del'argent*. Athens: Ekdotike Hellados.

Conophagos, C. 1982. 'Smelting practice at Laurion', in *Early Pyrotechnology*, T.A. and S.F. Wertime (eds), 181–91.Washington, DC: Smithsonian Institution Press.

Conophagos, C.E. 1989. 'La technique de la coupellation des Grecs anciens au Laurium', in *Archaeometry*, Y. Maniatis (ed.), 271–89. Amsterdam: Elsevier.

Conrad, H-G. and Rothenberg, B. (eds) 1980. *Antikes Kupfer im Timna-Tal*. Bochum: *Der Anschnitt* Beiheft 1.

Constantine, A. 1975. *The Manufacture of Charcoal*. Perth, Western Australia: Iron and Steel Institute.

Coomaraswamy, A.K. 1908. *Medieval Sinhalese Art*, 2nd edn. Broad Camden: Essex House Press [reprinted 1956 New York: Pantheon Books].

Cooper, I. 2008. *The Guide to Painted Towns of Shekhawati*. New Delhi: Prakash.

Cowell, M.R. and Lowick, N.M. 1988. 'Silver from the Panjhīr mines', in *Metallurgy in Numismatics* 2, W.A. Oddy (ed.), 65–74 London: Royal Numismatic Society.

Cracroft, W. 1832. 'The smelting of iron in the Khasya Hills', *Journal of the Asiatic Society of Bengal* 1(4): 150–55.

Craddock, P.T. 1978. 'The composition of the copper alloys used by the Greek, Etruscan and Roman civilisations 3: The origins and early use of brass', *Journal of Archaeological Science* 5(1): 1–16.

Craddock, P.T. 1987a. 'A history of the distillation of metals', *Bulletin of the Metals Museum of the Japan Institute of Metals* 10: 3–25.

Craddock, P.T. 1987b. 'The early history of zinc', *Endeavour* 11(4): 183–91.

Craddock, P.T. 1989. 'The scientific investigation of early mining and metallurgy', in *Scientific Analysis in Archaeology*, J. Henderson (ed.), 178–212, Oxford: UCLA.

Craddock, P.T. 1992. 'A short history of firesetting', *Endeavour* 16(3): 145–50.

Craddock, P.T. 1995. *Early Metal Mining and Production*. Edinburgh: Edinburgh University Press.

Craddock, P.T. 1996. 'The use of firesetting in the granite quarries of south India', *Mining History* 13(1): 7–11.

Craddock, P.T. 1997. 'Low technology zinc mining in the south west of China', *Mining History: The Bulletin of the Peak District Mines Historical Society* 13(3): 41–51.

Craddock, P.T. 1998a. 'Zinc in Classical antiquity', in *2000 Years of Zinc and Brass*, P.T. Craddock (ed.), 2nd rev. edn, 1–6. London: British Museum Press.

Craddock, P.T. 1998b. 'New light on the production of crucible steel in Asia', *Bulletin of the Metals Museum of the Japan Institute of Metals* 29: 41–66.

Craddock, P.T. (ed.) 1998c. *2000 Years of Zinc and Brass*, 2nd rev. edn. London: British Museum Press.

Craddock, P.T. 2002. 'Giacimenti, miniere, methodi di riduzione', in *Le Arti di Efesto*, A. Gimlia-Mair and M. Rubinich (eds), 1–15. Trieste Silvana Editoriale.

Craddock, P.T. 2003. 'Cast iron, fined iron, crucible steel: liquid iron in the ancient world', in *Mining and Metal Production through the Ages*, P.T. Craddock and J. Lang (eds), 231–57. London: British Museum Press.

Craddock, P.T. 2007. 'Cast iron: the elusive feedstock of crucible steel', *Indian Journal of History of Science* 42(4): 593–607.

Craddock, P.T. 2008. 'Mining and metallurgy', in *Engineering and Technology in the Classical World*, J.P. Oleson (ed.), 94–120. Oxford: Oxford University Press.

Craddock, P.T. 2009a. 'Metals, minerals and medicine', *Indian Journal of History of Science* 44(2): 209–30.

Craddock, P.T. 2009b. 'Perceptions and reality: the fall and rise of the Indian mining and metal industry', in *Metals and Societies*, T.L. Kienlin and B.W. Roberts (eds), 453–64. Bonn: Rudolf Habelt GMBH.

Craddock, P.T. 2009c. 'The origins and inspiration of zinc smelting', *Journal of Material Science* 44: 2181–90.

Craddock, P.T. 2013. 'Two millennia of the sea-borne metals trade with India', *Indian Journal of History of Science* 48(1): 1–37.

Craddock, P.T. 2015. 'The metal casting traditions of South Asia: continuity and innovation', *Indian Journal of History of Science* 50(1): 55–82.

Craddock, P.T. and Eckstein, K. 2003. 'Production of brass in antiquity by direct reduction', in *Mining and Metal Production through the Ages*, P.T. Craddock and J. Lang (eds), 216–30. London: British Museum Publications.

Craddock, P.T. and Gurjar, L.K. 1991. 'Traditional zinc distillation in India and China', in *Role of Chemistry in Archaeology*, M.C. Ganorkar and N. Rama Rao (eds), 112–20. Hyderabad: Birla Institute of Scientific Research.

Craddock, P.T. and Hook, D. 2012. 'An economic history of the post-medieval world in 50 ingots: the British Museum collection of ingots from dated wrecks', *British Museum Technical Bulletin* 6: 55–68.

Craddock, P.T. and Hughes, M.J. (eds) 1985. *Furnaces and Smelting Technology in Antiquity*. London: British Museum Occasional Paper 48.

Craddock, P.T. and Lang, J. (eds) 2003. *Mining and Metal Production through the Ages*. London: British Museum Press.

Craddock, P.T. and Meeks, N.D. 1987. 'Iron in ancient copper', *Archaeometry* 29(2): 187–204.

Craddock, P.T. and Zhou, W. 1998. 'The survival of traditional zinc production in China', in *The 6th International Conference on the Beginning of the Use of Metals and Alloys (BUMA IV)*, Organizing Committee (ed.), 83–96. Sendai: Japan Institute of Metals.

Craddock, P.T. and Zhou, W. 2003. 'Traditional zinc production in modern China: survival and evolution', in *Mining and Metal Production through the Ages*, P.T. Craddock and J. Lang (eds), 267–92. London: British Museum Publications.

Craddock, P.T., Burnett, A.M. and Preston, K. 1980. 'Hellenistic copper-base coinage and the origins of brass', in *Scientific Studies in Numismatics*, W.A. Oddy (ed.), 53–64. London: British Museum Occasional Paper 18.

Craddock, P.T., Gurjar, L.K. and Hegde, K.T.M. 1983. 'Zinc production in medieval India', *World Archaeology* 15(2): 211–17.

Craddock, P.T., Freestone, I.C., Gurjar, L.K., Hegde, K.T.M. and Sonawane, V.H. 1985a. 'Early zinc production in India', *Mining Magazine* January 1985: 45–51.

Craddock, P.T., Freestone, I.C., Gale, N.H., Meeks, N.D., Rothenberg, B. and Tite, M.S. 1985b. 'The investigation of a small heap of silver smelting debris from Rio Tinto, Huelva, Spain', in *Furnaces and Smelting Technology in Antiquity*, P.T. Craddock and M.J. Hughes (eds), 199–218. London: British Museum Occasional Paper 48.

Craddock, P.T., Freestone, I.C. and Hunt Ortiz, M. 1987. 'Recovery of silver from speiss at Rio Tinto (SW Spain)', *IAMS Bulletin* 10(1): 8–11.

Craddock, P.T., Freestone, I.C., Gurjar, L.K., Middleton, A. and Willies, L. 1989. 'The production of lead, silver and zinc in ancient India', in *Old World Archaeometallurgy*, A. Hauptmann, E. Pernicka and G.A. Wagner (eds), 51–70. Bochum: *Der Anschnitt* Beiheft 7.

Craddock, P.T., Ambers, J., Hook, D.R., Farquhar, R.M., Ckikwendu, V.E., Umeji, A.C. and Shaw, T. 1997. 'Metal sources and the bronzes from Igbo-Ukwu, Nigeria', *Journal of Field Archaeology* 24: 405–29.

Craddock, P.T., Freestone, I.C., Middleton A., Gurjar, L.K. and Willies, L. 1998a. 'Zinc in India', in *2000 Years of Zinc and Brass*, P.T. Craddock (ed.), 2nd rev. edn, 27–72. London: British Museum Press.

Craddock, P.T., Freestone, I.C., Willies, L., Paliwal, H.V., Gurjar, L.K., and Hegde, K.T.M. 1998b. 'Ancient industry of lead, silver and zinc in Rajasthan' in *Archaeometallurgy in India*, V. Tripathi (ed.), 108–33. Delhi: Sharada Publishing House.

Craddock, P.T., La Niece, S.C. and Hook, D.R. 1998c. 'Brass in the medieval Islamic world', in *2000 Years of Zinc and Brass*, P.T. Craddock (ed.), 2nd rev. edn, 73–114. London: British Museum Press.

Craddock, P.T., La Niece, S. and Hook, D.R. 2003. 'Evidences for the production, trading and refining of copper in the Gulf of Oman during the third millennium BC', in *Men and Mining- Mensch und Bergbau*, Th. Stöllner, G. Körin, G. Steffens and J. Cierny (eds), 103–12. Bochum: *Der Anschnitt* Monograph 16.

Craddock, P.T., Cowell, M.R. and Meeks, N.D. 2005. 'Appendix 1: Report on the scientific examination of materials from experimental copper smelting at Alderley Edge', in *The Archaeology of Alderley Edge: Survey, Excavation and Experiment in an Ancient Mining Landscape*, S. Timberlake and A.J.N.W. Prag (eds), 211–16. Oxford: BAR British Series 396.

Craddock, P., Cartwright, C., Eckstein, K., Freestone, I., Gurjar, L., Hook, D., Middleton, A. and Willies, L. 2013. 'Simple sophistication: Mauryan silver production in north west India', *British Museum Technical Bulletin* 7: 79–93.

Craddock, P.T., Cribb, J., Gale, N.H. and Gurjar, L.K. 2015. 'Sources of zinc in early India: the evidence of numismatics trade and lead isotope analysis', in *Metals and Civilisation*, S. Srinivasan, S. Ranganthan and A. Giumlia-Mair (eds), 174–84. Bangalore: National Institute of Advanced Studies.

Craddock, P.T., Andrews, P. and Spataro, M. forthcoming. 'Not even if we offered him £50': the composition of the 18th century Huntsman crucibles revealed', *Journal of the Historical Metallurgy Society*.

Cranmer-Byng, J.L. (ed.) 1962. *An Embassy to China: Being the Journal kept by Lord Macartney 1793–4*. London: Longmans Green.

Cribb, J. 1985. 'Dating India's earliest coins', in *South Asian Archaeology 1983*, J. Schotsmans and M. Taddei (eds), 535–54. Naples Istituto Universitario Orientale Series Minor 23.

Cribb, J. 2003. 'The origins of the Indian coinage tradition', *South Asian Studies* 19: 1–19.

Cribb, J. 2014. 'Indian zinc coins', *Journal of the Asiatic Numismatic Society* 219: 321–31.

Crookshank, H. 1947. 'Zawar silver-lead-zinc mines', *Indian Minerals* 2: 22–7.

Curle, J.H. 1899. *The Gold Mines of the World*. London: Waterlow.

Curtis L.C., Radhakrishna, B.P. and Nadu, G.R.1990. *Hutti Gold Mine with a Future*. Bangalore: Geological Society of India.

Dalrymple, W. 2002. *White Mughals*. London: Harper Collins.

Darling, A.S. 1990. 'Brass and zinc', in *An Encyclopaedia of the History of Technology*, I. McNeil (ed.), 72–96. London: Routledge.

Dash, B. 1986. *Alchemy and Metallic Medicines in Āyurveda*. New Delhi: Concept Publishing Co.

Dash, B. and Kashyap, L. (eds and trs) 1994. *Iatro-Chemistry of Āyurveda (Rasa Śāstra)*. New Delhi: Concept Publishing Co.

Davey, C.J. 1979. 'Some ancient Near-Eastern pot bellows', *Levant* 11: 101–11.

Davey, C.J. 1988. 'Tell Edh-Dhiba'I and the southern Near Eastern metalworking tradition', in *The Beginning of the Use of Metals and Alloys*, R. Maddin (ed.), 63–8: Cambridge, MA: MIT Press.

Day, J. 1973. *Bristol Brass*. Newton Abbot: David and Charles.

Day, J. 1984. 'The continental origins of Bristol brass', *Industrial Archaeology Review* (7)1: 32–55.

Day, J.M. 1988. 'The Bristol brass industry: furnace structures and their associated remains', *Journal of the Historical Metallurgy Society* 22(1) 24–41.

Day, J. 1991. 'Copper, zinc and brass production', in *The Industrial Revolution in Metals*, J. Day and R.F. Tylecote (eds), 131–99. London: Institute of Metals.

Day, J. 1998. 'Brass and zinc in Europe from the Middle Ages until the nineteenth century', in *2000 Years of Zinc and Brass*, P.T. Craddock (ed.), 2nd rev. edn, 133–58. London: British Museum Press.

Day, J. and Tylecote, R.F. (eds) 1991. *The Industrial Revolution in Metals*. London: Institute of Metals.

Dean, R.S. 1919. 'Effect of sodium chloride on blue-powder formation', *Chemical and Metallurgical Engineering* 20(12): 628.

Deb, M., Thorpe, R.I., Cumming, G.L. and Wagner, P.A. 1989. 'Age, source and stratigraphic implication of Pb isotope data for conformable, sediment-hosted, base metal deposits in the Aravalli-Delhi Orogenic Belt, northwestern India', *Precambrian Research* 43: 1–22 .

Dey, R.C. 2008. 'Ancient smelting devices (retorts) from different parts of India', *Journal of the Geological Survey of India* 72: 527–32.

Deyell, J.S. 1999. *Living without Silver: The Monetary History of Early Medieval North India*. New Delhi: Oxford University Press.

Dharampal 1971. *Indian Science in the Eighteenth Century: Some Contemporary European Accounts*. Delhi: Impex India.

Di Florio, M.R., Lorenzoni, S., Olivieri, L.M. and Zanettin Lorenzoni, E. 1993. 'Evidence of ancient quarrying in Lower Swat, N.W.F.P., Pakistan: a geo-archaeological study', *Science and Technology for Cultural Heritage* 2: 63–74.

Dixon, C.G. 1831. 'Some account of the lead mines of Ajmer', *Gleanings in Science* 3: 111–15.

Drummond, H. 1838. 'Report on the copper mines of Kumaon', *Journal of the Asiatic Society of Bengal* 7(3): 934–41.

Drummond, H. 1841. 'On the mines and mineral resources of north Afghanistan', *Journal of the Asiatic Society of Bengal* 10(109): 74–93.

Dube, R.K. 2006. 'The extraction of lead from its ores by the iron-reduction process: a historical perspective', *Journal of Metals* October 2006: 18–22.

Dubois, C. 1996. 'L'ouverture par le feu dans les mines', *Revue d'Archéométrie* 20: 33–46.

Dumas, M. 1847.*Traité de chimie appliqué aux arts partie inorganique*. Liège: Felix Oudart.

Dutt, M.N. (tr. and ed.) 1896. *Kamandakya Nitisara*. Calcutta: Dutt.

Dutt, R.N., Bagri, P.R, Moondra, D.R., Khanelwal, R., Ramaswamy, C.R. and Gupta, L.H. (eds) 1960. *Tata T.M.P. (Two Million Tonne Plan), Iron and Steel Review (Tata T.M.P number)*.

Dutt, R.N., Bagri, P.R, Moondra, D.R., Ramprasad Khanelwal, Ramaswamy, C.R. and Gupta, L.H. (eds) 1962. *Rourkela, Iron and Steel Review (Rourkela number)*.

Eckstein, K., Craddock, P.T. and Freestone, I.C. 1996. 'Frühe Bergbau- und Verhüttungsaktivitäten in Zawar, Dariba und Agucha (Nordwestindien)', *Metalla* 3(1): 11–26.

Eissler, M. 1891. *The Metallurgy of Argentiferous Lead*. London: Crosby Lockwood and Son.

Elwin, V. 1942. *The Agaria*. Oxford: Oxford University Press.

Elwin, V. 1958. *The Story of Tata Steel*. Bombay: Tata Steel.

Erskine, K.D. 1908. *Rajputana Gazetteers: II The Mewar Residency*. Ajmer.

Evans, L.T. and Tylecote, R.F. 1967. 'Some vitrified products of non-metallurgical significance', *Bulletin of the Historical Metallurgy Group* 1(9): 22–3.

Fairservis, W.A. Jr. 1971. *The Roots of Ancient India*. London: George Allen & Unwin.

Falk, H. 1991. 'Silver, lead and zinc in early Indian literature', *South Asian Studies* 7: 111–17.

Fell, A. 1908. *The Early Iron Industry of Furness and District*. Ulverstone: Hume Kitchin [reprinted 1968, London: Cass].

Fleming, J., Robottom, H.H. and Narayanan, P.I.A. 1949. 'The beneficiation of lead-zinc ore from Zawar', *Transactions of the Mining, Geological and Metallurgical Institute of India* 45(1): 21–78.

Forbes, R.J. 1971. *Studies in Ancient Technology VIII*, 2nd edn. Leiden: E.J. Brill.

Foster, M.K. 1970. 'Horizontal retort and acid plant Ascarco Mexicana, S.A. Nueva Rosita plant', in *Lead and Zinc II: Extractive Metallurgy of Lead and Zinc*, C.H. Cotterill and J.M. Cigan (eds), 463–511. New York: AIME.

Francovich, R. (ed.) 1993. *Archeologia delle attivitá estrattive e metallurgiche*. Florence: All'Insignia del Giglio.

Freestone, I.C. 1989. 'Refractory materials and their procurement', in *Old World Archaeometallurgy*, A. Hauptmann, E. Pernicka and G.A. Wagner (eds), 155–62. Bochum: *Der Anschnitt* Beiheft 7.

Freestone, I.C. and Tite, M.S. 1986. 'Refractories in the ancient and preindustrial world', in *High-Technology Ceramics Past, Present and Future. Ceramics and Civilisation III*, W.D. Kingery (ed.), 35–64. Westerville, OH: American Ceramic Society.

Freestone, I.C., Craddock, P.T., Gurjar, L.K., Hegde, K.T.M. and Paliwal, H.V. 1985a. 'Analytical approaches to the interpretation of medieval-zinc smelting from Zawar, Rajasthan', *Journal of Archaeological Chemistry* 3. Hyderabad: Birla Institute, 1–12.

Freestone, I.C., Craddock, P.T., Hegde, K.T.M., Hughes, M.J. and Paliwal, H.V. 1985b. 'Zinc production at Zawar, Rajasthan', in *Furnaces and Smelting Technology in Antiquity*, P.T. Craddock and M.J. Hughes (eds), 229–44. London: British Museum Occasional Paper 48.

Freestone, I.C., Middleton, A.P., Craddock, P.T. and Hook, D.R. 1991. 'Role of materials analysis in the reconstruction of early metal extraction technology: zinc, and silver-lead smelting at Zawar, Rajasthan', in *Materials Issues in Art and Archaeology* II, P.B. Vandiver, J. Druzik and G.S. Wheeler (eds), 617–26. Pittsburgh: Materials Research Society.

Friedlaender, C.G.I. and Choo, K. 1970. 'Mineralogical observations on slags from Zawar Mines, Rajasthan', *Indian Mineralogist* 1: 1–10.

Fulford, M.G. and Allen, J.R.L. 1992. 'Iron-making at the Chesters Villa, Woolaston, Gloucestershire: survey and excavations 1987–91', *Britannia* 23: 159–215 [appendix by I. Figueiral: 188–91].

Gaimster, D. 1986. 'Dung-tempering? A late Norse case study from Caithness', *Medieval Ceramics* 10: 43–7.

Gale, N.H. and Stos-Gale, Z.A. 1981. 'Cycladic lead and silver metallurgy', *Annual of the British School of Archaeology at Athens* 76: 169–224.

Galon, M. 1764. *L'art de convertir le cuivre rouge ou cuivre de rosette en laiton ou cuivre jaune au moyen de la Pierre Calamine*. Paris: Descriptions des Arts et Métiers, Academie Royale des Sciences.

Gamage, D.H. 2002. 'The production and use of bellows in traditional metallurgical operations', in *The Beginnings and Use of Metals and Alloys V*, Gyu-Ho Kim, Kyung-Woo Yi and Hyung-Tai Kang (eds), 107–11: Seoul: The Korean Institute of Metals and Materials.

Gamble, J.S. 1902. *A Manual of Indian Trees*. London: Sampson Low, Marston & Co.

Gandhi, S.M. 1983. 'Rampura-Agucha lead/zinc deposit', *Mining Magazine* 149(5): 315–21.

Gandhi, S.M. 2000. 'Ancient mining and metallurgy in Rajasthan', in *Crustal Evolution and Metallogeny in the Northwestern Indian Shield*, M. Deb (ed.), 29–48. New Delhi: Narosa Publishing House.

Gandhi, S.M., Rampura, T.C. and Sharma, N.K. 1984. *Old Workings and Ancient History of the Area around Rampura-Agucha Zinc/Lead Deposit, Bhilwara Distt., Rajasthan*. Agucha Mine: Hindustan Zinc Ltd.

Gangopadhayay, A., Ramachandran, T.R. and Biswas, A.K. 1984. 'Phase studies on zinc residues of ancient Indian origin', *Transactions of the Indian Institute of Metals* 37(3): 234–41.

Gatelier, M.E. 1862. *Annales des Mines* 2(4): 145.

Gaur, R.C. 1983. *Excavations at Atranjikhera: Early Civilisations of the Upper Ganga Basin*, New Delhi: Aligarh Muslim University /Motil Banarsidass.

Geijerstam, J. af 2002. 'The Kumaon Iron Works: a colonial project', in *Tradition and Innovation in the History of Iron Making: An Indo European Perspective*, G. Pande and J. af Geijerstam (eds), 157–201. Nainital, Uttarakhand: PAHAR.

Geijerstam, J. af 2004. *Landscapes of Technology Transfer*. Stockholm: Jernkontorets Berghistoriska.

Ghandy, J.J. 1960. 'The place of T.I.S.C.O. in India's steel industry', in *Tata T.M.P. (Two Million Tonne Plan), Iron and Steel Review (Tata T.M.P number)*, R.N. Dutt, P.R. Bagri, D.R. Moondra, R. Khanelwal, C.R. Ramaswamy and L.H. Gupta (eds), 41–9.

Gill, M.C. 1992. 'An outline of the chemistry of lead smelting', in *Boles and Smeltmills*, L. Willies and D. Cranstone (eds), 3–4. London: Historical Metallurgy Society.

Giumlia-Mair, A., Jeandin, M. and Ota, K. 2009. 'Metal trade between Europe and Asia in classical antiquity', in *Metallurgy and Civilisation: Eurasia and Beyond*, J. Mei and Th. Rehren (eds), 35–43. London: Archetype Publications.

Goitein, S.D. 1966. *Studies in Islamic History and Institutions*. Leiden: Brill.

Goitein, S.D. and Friedman, M.A. 2008. *India Traders of the Middle Ages: Documents from the Cairo Geniza (India Book.), Études sur le Judaïsme Médiéval* 31. Jerusalem: Ben-Zvi Institute/Leiden: E. J. Brill.

Golas, P. 2000. *Science and Civilisation in China V: Chemical and Chemical Technology: Pt. 13 Mining*. Cambridge: Cambridge University Press.

Gowland, W. 1899a. 'The early metallurgy of copper, tin and iron in Europe, as illustrated by ancient remains, and the primitive processes surviving in Japan', *Archaeologia* 56(2): 267–322.

Gowland, W. 1899b. 'The early metallurgy of silver and lead, Pt. 1: Lead', *Archaeologia* 57(2): 359–422.

Gowland, W. 1915. 'Metals and metal-working in Old Japan', *Transactions of the Japan Society of London* 13: 19–100.

Grover, A.K. 2005. 'Observations on the nature and type of ancient metallurgical slags present in different basement belts in parts of Rajasthan and Gujarat', *GSI Special Publication 72*: 267–94.

Gunther, R.T. (ed.) 1934. *The Greek Herbal of Dioscorides: John Goodyer's Translation of 1655*. Oxford: Oxford University Press.

Gupta, B.C. 1934. *Geology of Central Mewar*. Calcutta: GSI Memoir 65(2).

Gupta, C.D. 2000. 'Stabilization of Chanderiya lead zinc smelter', *Hindzinc Tech* January: 34–43.

Gupta, H.L. 1960. 'The economic impact of the West on Indian industries', *Journal of Indian History* 38(1): 77–100.

Gupta, P.D. 1986. 'The past 100 years and future of the Kolar Gold Field', in *Hundred Years Heritage*, Mining Committee (ed.), 1–6. Bangalore: Bharat Gold Mines Ltd.

Gupta, P.L. 1969. *Coins*. New Delhi: National Book Trust.

Gupta, P.L. and Hardaker, T.R. 1985. *Ancient Indian Silver Punchmarked Coins of the Magadha-Maurya Karshapana Series*. Nasik : Indian Institute of Numismatic Studies.

Gupta, S.N., Arora, Y.K., Mathur, R.K., Iqbaluddin, Prasad, B., Shahi, T.N. and Sharma, S.B. 1980. *The Precambrian Geology of the Aravalli Region, Southern Rajasthan and Gujarat: A Lithostatigraphic Map of the Area on 1:1 Million Scale and Explanatory Brochure*. Calcutta: Geological Survey of India.

Gupta, S.N., Arora, Y.K., Mathur, R.K., Iqbaluddin, Prasad, B., Shahi, T.N. and Sharma, S.B. 1997. *The Precambrian Geology of the Aravalli Region: Southern Rajasthan and North-eastern Gujarat*. Calutta: Memoirs of the GSI 123.

Gurjar, L.K., Craddock, P.T., Willies, L. and Paliwal, H.V. 2001. 'Zinc', in *History of Technology in India III*, K.V. Mital (ed.), 621–38. New Delhi: Indian National Science Academy.

Habib, I. 1982. *An Atlas of the Mughal Empire*. Delhi: Oxford University Press.

Habib, I. 1986. 'Capacity of technological change in Mughal India', in *Technology in Ancient and Medieval India*, A. Roy and S.K. Bagchi (eds), 1–13. Delhi: Sundeep Prakashan.

Hacket, C.A. 1880. 'Useful minerals of the Aravalli region', *Records of the GSI* 13: 243–50.

Halder, R.R. 1929–30. 'Samoli inscription of the time of Siladitya (V S 703)', *Epigraphica Indica* 20: 97–9.

Hamilton, W. 1820. *A Geographical, Statistical and Historical Description of Hindostan and Adjacent Countries*. London: John Murray.

Hammersley, G. 1973. 'The charcoal iron industry and its fuel 1540–1750', *Economic History Review* 26(3): 593–613.

Hardaker, T.R. 1992. 'Punch-marked coinage of Kosala: towards a classification', in *Indian Numismatics, History Art and Culture: Essays in Honour of Parmeshwari Lal Gupta*, D.W. Macdowall, S. Sharma and S.Garg (eds), 3–27. Delhi: AgamKala Prakashan.

Hardie, J. 1829. 'Sketch of the geology of central India', *Journal of the Asiatic Society of Bengal*: Physical Class III: 28–92.

Harley, F.H. (ed.) 1960. *A Technical Survey of the Durgapur Steelworks*. London: Iron and Coal Trades Review/ISCON.

Hasan, A. 1969. 'Silver currency output of the Mughal Empire and the affect on prices in India during the 16th and 17th centuries', *Indian Economic and Social History Review* 6(1): 85–116.

Hatch, F.H. 1901. 'The Kolar Gold Field', *GSI Memoir* 33(1): 1–81.

Haughton, J.C. 1854. 'Memorandum on the geological structure and mineral resources of the Sighbhoomm Division, South-West Frontier Agency', *Journal of the Asiatic Society of Bengal* 23: 103–22.

Hauptmann, A. 1985. *5000 Jahre Kupfer in Oman*. Bochum: *Der Anschnitt* Beiheft 4.

Hawthorne, J.G. and Smith, C.S. (eds) 1963. *On Divers Arts: The Treatise of Theophilus*. Chicago: University of Chicago Press.

Head, R. 1985. 'Recorder of the stupendous: Col. James Tod (1782–1835) and Rajasthan', *Country Life* 178(4608): 1924–6.

Healy, J.F. 1978. *Mining and Metallurgy in the Greek and Roman World*. London: Thames & Hudson.

Heath, J.M. 1832. 'Salem iron works', *Journal of the Asiatic Society of Bengal* 1: 253–7.

Heath, J.M. 1839. 'On Indian iron and steel', *Journal of the Royal Asiatic Society* 5: 390–97.

Hegde, K.T.M. 1967. 'Chemical and spectrometric studies in the Kṣhatrapa silver coins', *Journal of the Numismatic Society of India* 29(1): 63–6.

Hegde, K.T.M. 1968. 'Source of the metals in lead coins of Kshatrapa periods', *Current Science* 37(18): 518–20.

Hegde, K.T.M. 1986. 'How zinc was made in ancient India', *Science Today* February: 42–4.

Hegde, K.T.M. 1989. 'Zinc and brass production in ancient India', *Interdisciplinary Science Reviews* 14(1): 86–96.

Hegde, K.T.M. 1991. *An Introduction to Ancient Indian Metallurgy*. Bangalore: Geological Survey of India.

Hegde, K.T.M. and Ericson, H.E. 1985. 'Ancient Indian copper smelting furnaces', in *Furnaces and Smelting Technology in Antiquity*, P.T. Craddock and M.J. Hughes (eds), 59–70. London: BMP.

Hegde, K.T.M., Craddock, P.T. and Sonavane, V.H. 1986. 'Zinc distillation in ancient India', in *Proceedings of the 24th International Archaeometry Symposium*, J.S. Olin and M.J. Blackman (eds), 249–58. Washington, DC: Smithsonian Institution Press.

Heron, A.M. 1917. 'The geology of Rajputana and adjacent districts', *Memoir of the GSI* 45(1). New Delhi: Geological Survey of India.

Heron, A.M. 1953. *The Geology of Central Rajputana. Memoir of the GSI* 79. New Delhi: Geological Survey of India.

Heron, A.M. 1954. *Directory of the Economic Minerals of Pakistan*. Quetta: Records of the Geological Survey of Pakistan 7(2).

Hess, K., Hauptmann, A., Wright, H. and Whallon, R. 1999. 'Evidence of fourth millennium BC silver production at Fatmali-Kalecik, East Anatolia', in *Metallurgica Antiqua*, Th. Rehren, A. Hauptmann and J.D. Muhly (eds), 57–67. Bochum: *Der Anschnitt* Beiheft 8.

Hezarkhani-Zolgharnian, Z., Keesmann, I., Kronz, A. and Maier, P. 1994. 'Zink in alten Schlacken und technischen Produkten', *Berichte der Deutschen Mineralogischen Gesellchaft Beihefte zum European Journal of Mineralogy* 6(1): 107.

Hill, D.R. 1993. *Islamic Science and Engineering*. Edinburgh: Edinburgh University Press.

Hodges, H. 1964. *Artifacts*. London: John Baker.

Hofman, H.O. 1918. *Metallurgy of Lead*. New York: McGraw-Hill.

Holland, J. 1834. *Treatise on the Progressive Improvement and Present State of the Manufacturers in Metal*. London: Longman.

Holland, T.H. 1893. *The Iron Resources and Iron Industries of the Southern Districts of Madras Presidency*. Calcutta: Imperial Institute Handbook of Commercial Products 8.

Holland, T.H. 1908. *Sketches of the Mineral Resources of India*. Calcutta: Geological Survey of India.

Höller, W. and Gandhi, S.M. 1995. 'Silver-bearing sulfosalts from the metamorphosed Rampura Agucha Zn-Pb-(Ag) deposit, Rajasthan, India', *The Canadian Mineralogist* 33: 1047–57.

Hooja, R. 2006. *A History of Rajasthan*. New Delhi: Rupa.

Hooker, J.D. 1832. 'Smelting of iron in the Kasya Hills', *Journal of the Asiatic Society of Bengal* 1: 150.

Hooker, J.D. 1854. *Himalyan Journals*. London: John Murray.

Hoover, H.C. and Hoover, L.H. (trs and eds) 1912. *Agricola: De re metallica*. London: The Mining Magazine [reprinted 1950 New York: Dover].

Horne, L. 1982. 'Fuel for the metal worker', *Expedition* 25(1): 6–13.

Houston, D.F. 1972. 'Rice hulls', in *Rice Chemistry and Technology*, D.F. Houston (ed.), 301–52. St Paul, MN: American Association of Cereal Chemists.

Hubert, F. 1974. 'Minères néolithiques á Jandrain-Jandrenouille en Brabant', *Archeologica Belgica* 167: 8–45.

Hunt Ortiz, M.A. 2003. *Prehistoric Mining and Metallurgy in South West Iberian Peninsula*. Oxford: BAR Internation Series S1188.

Huntingford, G.W.B. (tr. and ed.) 1980. *The Periplus of the Erythraean Sea*. London: Hakluyt Soc.

ICDD n.d. *Powder Diffraction File*. Philadelphia: International Centre for Diffraction Data.

Ingalls, W.R. 1902. *Production and Properties of Zinc*. New York: The Engineering and Mining Journal.

Ingalls, W.R. 1906. *The Metallurgy of Zinc and Cadmium*. New York: The Engineering and Mining Journal.

Ingalls, W.R. 1936. 'History of the metallurgy of cinc', Chapter I of *Metallurgy of Lead and Zinc, Transactions of the American Institute of Mining and Metallurgical Engineers* 121: 339–73.

Istenič, J. and Šmit, Ž. 2007. 'The beginning of the use of brass in Europe with particular reference to the southeastern Alpine region', in *Metals and Mines: Studies in Archaeometallurgy*, S. La Niece, D. Hook and P. Craddock (eds), 140–47. London: Archetype Publications.

Jaikishan, S. 2009. *Iron and Wootz Steel Industry in Northern Telangana*. Hyderabad: Telangana Jagrithi.

Jain, N.C. and Rakesh, K.C. 1984. 'Rajpura Dariba Lead-Zinc Mine', *Mining Magazine* 150: 22–31.

Jakhu, M.R. 1990. 'Optimising concentrator returns', *Hindzinc Tech* 2(2): 37–44.

Jakhu, M.R., Subbarao, S.V.B. and Lodha, S.S. 2001. 'Performance improvement and capacity expansion of Debari and Vizag zinc smelters', *Hindzinc Tech* January: 13–25.

Jarrige, J-F. 1985. 'Continuity and change in the North Kachi Plain (Baluchistan, Pakistan) at the beginning of the second millennium BC', in *South Asian Archaeology 1983*, J. Shotsmans and M. Taddei (eds), 35–68. Naples: Istituto Universitario Orientale.

Jones, E. 2003. *The European Miracle*, 3rd edn. Cambridge: Cambridge University Press.

Jones, H.L. (tr.) 1928. *Strabo: Geography*, Vol. III. Cambridge, MA: Harvard University Press.

Jones, H.L. (tr.) 1929. *Strabo: Geography*, Vol. VI. Cambridge, MA: Harvard University Press.

Jones, H.L. (tr.) 1930. *Strabo: Geography*, Vol. VII. Cambridge, MA.: Harvard University Press.

Jones, J.E. 1984. 'Ancient Athenian silver mines, dressing floors and smelting sites', *Journal of the Historical Metallurgy Society* 18(2): 65–81.

Jones, J.E. 1988. 'The Athenian silver mines of Laurion and the British School at Athens excavations at Agrileza', in *Aspects of Ancient Mining and Metallurgy: Acta of a British School at Athens Centenary Conference, 1986*, J.E. Jones (ed.), 11–22. Bangor: Department of Classics, University College of North Wales.

Jones, W.H.S. (tr.) 1963. *Pliny: The Natural History* 8. Cambridge, MA: Harvard University Press.

Juleff, G. 1996. 'An ancient wind-powered iron smelting technology in Sri Lanka', *Nature* 379(6560): 60–63.

Juleff, G. 1998. *Early Iron and Steel in Sri Lanka*, AVA-Materialien 54. Mainz: Philipp von Zabern.

Kakoti, B. 1971. 'Old workings as a guide to the exploration of base metal deposits', in *Base Metals, Part* 1, 160–68. Delhi: Geological Society of India Miscellaneous Publication 16.

Kangle, R.P. (tr. and ed.) 1965–72. *The Kauṭilīya Arthaśāstra*, 3 vols. Bombay: Bombay University Press.

Kasaliwali, K. 1982. *Archarya Somakirti evam Brahma Yashodhara*. Jaipur: Shri Mahaveer Granth Academy.

Kelley, D.W. 1986. *Charcoal and Charcoal Burning*. Aylesbury: Shire.

Kemp, J.F. 1893. *Ore Deposits of the United States*. New York: Scientific Publishing Co.

Kenoyer, J.M. and Miller, H.M.-L. 1999. 'Metal technologies of the Indus Valley tradition in Pakistan and western India', in *The Archaeometallurgy of the Asian Old World*, V.C. Pigott (ed.), 107–51. Philadelphia: University of Pennsylvania Museum.

Kerr-Cross, D. 1950. 'Ancient mines and miners of India', *Indian Minerals* 4(1): 5–10.

Khan, I.G. 1986. 'Metallurgy in medieval India 16th to 18th centuries', in *Technology in Ancient and Medieval India*, A. Roy and S.K. Bagchi (eds), 71–91. Delhi: Sundeep Prakashan.

Kharakwal, J.S. 2011. *Zinc Production in India*. New Delhi: Pentagon Press. Infinity Foundation.

Kramer, C. 1997. *Pottery in Rajasthan*. Washington, DC: Smithsonian Institution Press.

Krishnaswamy, S. 1988. *India's Mineral Resources*, 3rd edn rev. by R.K. Sinha. New Delhi: Oxford & IBH Publishing.

Kumar, A. forthcoming. *History of Zawar*. Udaipur: Hamanshu edu.

Kumar, R. and Biswas, A.K. 1986. 'Zinc recovery from Zawar ancient siliceous slag', *Hydrometallurgy* 15: 267–80.

Laidler, C. and Sharp, W.E. 2000. 'An example of fire setting in West Africa', *Mining History: The Bulletin of the Peak District Mines Historical Society* 14(4): 56–8.

Lal, B.B. 1985. 'Report on the chemical analysis and examination of metallic and other objects from Lothal', in *Lothal: A Harappan Port Town (1955–62)*, S.R. Rao, 651–66. New Delhi: ASI Memoir 78.

Latham, R.E. (tr.) 1972. *The Travels of Marco Polo*. London: Pelican Books.

Law, R.W. and Burton, J.H. 2008. 'Nondestructive Pb isotope sampling and analysis of archaeological silver using EDTA and ICP-MS', *American Laboratory* 40(17): 14–15.

Leather, J.W. and Mukerji, J.N. 1911. *Indian Saltpetre Industry*, Agricultural Research Institute Bulletin 24.

Lehmann, J. 2000. 'Seit wann wird mit zincweiß gemalt? Chemische Untersuchungen, mittelalterliche Quellenschriften', *Restauro* 106(5): 356–60.

Leuva, K.K. 1963. *The Asur: A Study of Primitive Iron Smelters*. New Delhi: Bharatiya Adamjati Sevak Samaj.

Lewis, M.J.T. 1997. *Millstone and Hammer: The Origins of Water Power*. Hull: University of Hull Press.

Liu, H., Chen, J., Li, Y, Bao, W., Wu, X., Han, R. and Sun, S. 2007. 'Preliminary multidisciplinary study of the Miaobeihou zinc-smelting site at Yangliusi village', in *Metals and Mines: Studies in Archaeometallurgy*, S. La Niece, D. Hook and P. Craddock (eds), 170–78. London: Archetype Publications.

Liu, S., Rehren, Th., Chen, J., Xu, C., Venunan, P., Larreina-Garcia, D. and Martinón-Torres, M. 2015. 'Bullion production in imperial China and its significance for sulphide ore smelting world-wide', *Journal of Archaeological Science* 55: 151–65.

Lockett, A. 1831. *Report of a Journey thro' the Khawati by Col. Abraham Lockett and Lieut. A.H.E. Boileau in 1831*. Unpublished manuscript preserved in the National Archive, New Delhi.

London, G. 1981. 'Dung-tempered clay', *Journal of Field Archaeology* 8: 190–95.

Lones, T.E. n.d. *Zinc and its Alloys*. London: Pitman's Common Commodities and Industries.

Lowe, T.L. 2002. 'Ironing out the differences: technical and economic basis of 17th century VOC global trade in Swedish and Indian irons', in *Tradition and Innovation in the History of Iron Making: An Indo European Perspective*, G. Pande and J. af Geijerstam (eds), 94–7. Nainital, Uttarakhand: PAHAR.

Macdougall, J.D., Gopalan, K, Lugmair, G.W. and Roy, A. B. 1983. 'The banded gniessic complex of Rajasthan, India: early crust from depleted mantle at 3.5Ga', *Transactions of American Geophysical Union* 64: 351–60.

MacDowall, D.W. 1990. 'Review of P.J. Turner, *Roman Coins from India*', *Antiquaries Journal* 70(1): 34–5.

Mackay, E.J.H. 1938. *Further Excavations at Mohenjadaro*. New Delhi: Government of India.

Maclaren, J.M. 1906. 'Notes on some auriferous tracts in southern India', *Records of the GSI* 34(2): 96–131.

Maclaren, J.M. 1908. 'The auriferous deposits of India', *Mineralogical Journal* 84: 189–90.

Mactear, J. 1894–5. 'Some notes on Persian mining and metallurgy', *Transactions of the Institution of Mining and Metallurgy* 3: 2–39.

Majumdar, G.G. and Rajaguna, S.N. 1964. 'Early mining and metallurgy in Rajasthan. Part 1: Copper around Udaipur', *Bulletin of the Deccan College Post Graduate Research Institute* 23: 31–3.

Mallet, F.R. 1875. *On the Geology of the Darjiling District and the Western Duars. Memoirs of the GSI 2*. Calcutta: Government of India.

Mani, B.R. 1998. 'The search of metals and techniques for minting earliest coins in India', in *Archaeometallurgy in India*, V. Tripathi (ed.), 350–55. Delhi: Sharada Publishing House.

Maniatis, Y. and Tite, M.S. 1981. 'Technological examination of Neolithic-Bronze Age pottery from central and southeast Europe and from the Near East', *Journal of Archaeological Science* 8(1): 59–76.

Maréchal, J. 1985. 'Methods of ore roasting and the furnaces used', in *Furnaces and Smelting Technology in Antiquity*, P.T. Craddock and M.J. Hughes (eds), 29–42. London: British Museum Occasional Paper 48.

Marshall, A. 1915. *Explosives*. London: J. & A. Churchill.

Martinón-Torres, M. and Rehren, Th. 2002. 'Agricola and Zwickau: theory and practice of Renaissance brass production in SE Germany', *Journal of the Historical Metallurgy Society* 36(2): 95–111.

Marx, K. and Engels, F. 1959. *The First Indian War of Independence 1857–1859*. Moscow: Foreign Languages Publishing House.

Masuda Tsuna 1983. *Kodō Zuroku*, Zenryu Shirakawa (tr.) Norwalk, CT: Burndy Library.

McBean, K.D. 1959. 'Roasting, sintering, calcining and briquetting', in *Zinc*, C.H. Mathewson (ed.), 136–73. New York: American Chemical Society Monograph, Rheinhold.

McCarthy, E.T. 1882. 'Mining enterprise on the Gold Coast', *The Mining Journal* 1 July 1882: 797–8.

McCrindle, J.W. 1877. *Ancient India as Described by Megasthenes and Arrian*. London: Trübner.

McPherson K. 1993. *The Indian Ocean*. Delhi: Oxford University Press.

Medlicott, H.B. and Blandford, W.T. 1893. *A Manual of the Geology of India*. Calcutta: Geological Survey of India.

Mellors, J.W. 1969. *Comprehensive Treatise on Inorganic and Theoretical Chemistry IV*. London: Longmans.

Merkel, J.1985. 'Ore beneficiation during the Late Bronze Age/Early Iron Age at Timna, Israel', *MASCA Journal* 3(5): 164–9.

Merkel, J.F. 1990. 'Experimental reconstruction of Bronze Age copper smelting based on archaeological evidence from Timna', in *The Ancient Metallurgy of Copper*, B. Rothenberg (ed.), 78–122. London: Institute for Archaeo-Metallurgical Studies, University College.

Meyer, J.J. 1925–6. *Das altindische Buch vom Welt- und Staatsleben: Das Arthāçastra, des Kauṭilīya*. Leipzig: Harrassowitz.

Meyers, P. 2004. 'Production of silver in antiquity: ore types identified based upon elemental compositions of ancient silver artifacts', in *Patterns and Process*, L. van Zelst (ed.), 271–88. Suitland, MD: Smithsonian Center for Materials Research and Education.

Middleton, A.P. 1991. 'Ceramics: materials for all reasons', in *Science and the Past*, S. Bowman (ed.), 16–36: London British Museum Press.

Mishra, A.K. 2003. *Ethno-Archaeometallurgy of Iron in India*. Delhi: Agam Kala Prakashan.

Misra, M.K., Ragland, K.W. and Baker, A.J. 1993. 'Wood ash composition as a function of furnace temperature', *Biomass and Bioenergy* 4(2): 103–16.

Mital, S.N. 2000. *Kauṭilīya Arthaśāstra Revisited*. New Delhi: Centre for Studies in Civilizations.

Mookerjee, A. 1964. 'The geology of the Zawar Lead-Zinc Mine, Rajasthan, India', *Economic Geology* 59: 657–77.

Moore, J.J. 1981. *Chemical Metallurgy*. London: Butterworths.

Morgan, S.W.K. 1985. *Zinc and its Alloys and Compounds*. Chichester: Ellis Horwood.

Morgan, S.W.K. 1998. 'Development of the zinc-lead blast-furnace as a research project', in *2000 Years of Zinc and Brass*, P.T. Craddock (ed.), 2nd rev. edn, 229–41. London: British Museum Press [reprinted from *Transactions of the Institute of Mining and Metallurgy* 92 Section C, 1983: 192–4].

Moulherat, C., Tengberg, M., Haquet, J-F. and Mille, B. 2002. 'First evidence of cotton at Neolithic Mehrgarh, Pakistan', *Journal of Archaeological Science* 29: 1393–1401.

Mughal, M.R. 1980. 'New archaeological evidence from Bahawalpur', *Man and Environment* 4: 93–8.

Mukherjee, A.P., Premkumar, G., and Dhana Sekaran, R. 1991. 'Primary lead smelting at HZL', *Hindzinc Tech* 3(1): 4–25.

Mull Rajgarhia, C. 1951. *Mining, Processing and Uses of Indian Mica*. New York: McGraw-Hill.

Munn, L. 1936. 'Observations and notes on the methods of ancient gold mining in southern India', *Transactions of the Mining and Geological Institute of India* 30(3): 103–16.

Murray, E.F.O. 1940. 'The ancient workers of Western Dhalbhum', *Transactions of the Royal Asiatic Society of Bengal Letters* 6: 79–104.

Musson, A.E. and Robinson, E. 1969. *Science and Technology in the Industrial Revolution*. Manchester: Manchester University Press.

Nagarch, B.L. and Trivedi , P.K. 1981. 'A note on Agucha, District Bhilwara, Rajasthan', *Puratattva* 10: 106.

Nanavati, S.K. 1960. 'History of the growth of iron and steel industry in India', in *Tata T.M.P. (Two Million Tonne Plan), Iron and Steel Review (Tata T.M.P number)*, R.N. Dutt, P.R. Bagri, D.R. Moondra, R. Khanelwal, C.R. Ramaswamy and L.H. Gupta (eds), 69–75.

Needham, J. 1959. *Science and Civilisation in China III: Mathematics and the Sciences of the Heavens and the Earth.* Cambridge: Cambridge University Press.

Needham, J. 1965. *Science and Civilisation in China I: Introductory Orientations.* Cambridge: Cambridge University Press.

Needham, J. 1974. *Science and Civilisation in China V: Chemistry and Chemical Technology 2: Spagyrical Discovery and Invention: Magisteries of Gold and Immortality.* Cambridge: Cambridge University Press.

Ojha, G.S.H. 1928. *Udaipur rajya ka Itihas* (History of Udaipur). Ajmer: Government Press.

Ojha, G.S.H. 1931. *Udaipur rajya ka Itihas* (History of Udaipur). Ajmer: Government Press.

Oldham, T. 1856. 'Preliminary notice on the coal and iron of Talcheer in the Tributary Mehals of Cuttack', *Memoirs of the GSI* 1(1): 1–32.

O'Shaughnessy, W.B. 1847. 'On the refinage, on a large scale, by means of nitre, of brittle or understandard silver, for coinage purposes; and on a ready mode of approximately assaying silver', *Journal of the Asiatic Society of Bengal* 16(1): 557–68.

Paliwal, H.V., Gurjar, L.K. and Craddock, P.T. 1989. 'Zinc and brass in ancient India', in *All that Glitters: Readings in Historical Metallurgy*, M.L. Wayman (ed.), 31–4. Montreal: The Metallurgical Society of the Canadian Institute of Mining and Metallurgy [also reproduced in *CIM Bulletin* 79(885): 75–9].

Palmer, R.E. 1926/7. 'Notes on some ancient mine equipments and systems', *Transactions of the Institution of Mining and Metallurgy* 36: 299–336.

Pande, G. 1960. 'Steel in the public sector', in *A Technical Survey of the Durgapur Steelworks*, F.H. Harley (ed.), 1–6. London: Iron and Coal Trades Review/ISCON.

Pande, G. 2002. 'Iron making in Kumaon: a study of Kumaon Iron Works', in *Tradition and Innovation in the History of Iron Making: An Indo European Perspective*, G. Pande and J. af Geijerstam (eds), 146–56. Nainital, Uttarakhand: PAHAR

Pande, G. and af Geijerstam, J. (eds) 2002. *Tradition and Innovation in the History of Iron Making: An Indo European Perspective.* Nainital, Uttarakhand: PAHAR.

Parvathisem, P. 1990. 'HZL's strides in zinc smelting technology', *Hindzinc Tech* 2(1): 19–34.

Partington, J.R. 1961. *A History of Chemistry 2.* London: Macmillan.

Patterson, C.C. 1972. 'Silver stocks and losses in ancient and medieval times', *Economic History Review* 25(2): 205–35.

Pearson, G.W. and Stuiver M. 1986. 'High precision calibration of the radiocarbon time-scale', *Radiocarbon* 28(2B): 805–39.

Percy, J. 1861. *Fuels, Fireclays, Copper, Zinc and Brass.* London: John Murray.

Percy, J. 1864. *Metallurgy: Iron and Steel.* London: John Murray.

Percy, J. 1870. *The Metallurgy of Lead.* London: John Murray.

Percy, J. 1880. *The Metallurgy of Silver and Gold:1 Silver.* London: John Murray.

Pernicka, E. 1987. 'Erzlagerstätten in der Ägäis und ihre Ausbeutung im Altertum: Geochemische Untersuchungen zur Herkunftsbestimmung archäologischer Metallobjekte', *Jahrbuch der Römisch-Germanischen Xentralmuseums Mainz* 34: 607–714.

Pernicka, E., Rehren, Th. and Schmitt-Strecker, S. 1998. 'Late Uruk silver production by cupellation at Habuba, Kabira, Syria', in *Metallurgica Antiqua*, Th. Rehren, A. Hauptmann and J.D. Muhly (eds), 123–34. Bochum: *Der Anschnitt* Beiheft 8.

Peters, E.D. Jr. 1907. *Modern Copper Smelting*, 7th edn. New York: Hill Publishing Co.

Peters, E.D. 1911. *The Practice of Copper Smelting.* New York: McGraw-Hill Book Co.

Phillott, D.C. (ed.) 1927. *The Ā-īn-I Akbarī of Abū 'L-Fazl Allāmī*, 3 vols, H. Blochmann (tr. 1873). Calcutta: Asiatic Society of Bengal [reprinted 1989, New Delhi: Low Price Publications].

Piggott, S. 1950. *Prehistoric India.* London: Penguin Books.

Pillai, P.P. 1925. *Economic Conditions in India.* London: Routledge.

Pinhey, A.F. 1909. *History of Mewar.* Calcutta: Government of India [reprinted 1996 Jodhpur: Books Treasure].

Possehl, G.L. and Gullapalli, P. 1999. 'The Early Iron Age in South Asia', in *The Archaeometallurgy of the Asian Old World*, V.C. Pigott (ed.), 153–76. Philadelphia: The University Museum, University of Pennsylvania.

Potts, P.J. 1987. *A Handbook of Rock Silicate Analysis.* Glasgow and London: Blackie.

Prakash, O. 1984. *The Dutch Factories in India 1617–1623.* New Delhi: Munshiram Manoharlal.

Prakash, S. and Rawat, N.S. 1965. *Chemical Study of Some Indian Archaeological Antiquities.* Bombay: Uttar Pradesh State Council of Scientific and Industrial Research, Asia Publishing House.

Prasad, B, and Ahmad, N. 1998. ' Probable sources of silver in ancient India: a historico-scientific approach', in *Archaeometallurgy in India*, V. Tripathi (ed.), 184–90. Delhi: Sharada Publishing House.

Prinsep, J. 1835. 'Analyses of copper ore from Nellore; with notes of the copper mines in Ajmír and Singhána', *Journal of the Asiatic Society of Bengal* 4: 574–84.

Qaisar, A.J. 1982. *Indian Responses to European Technology and Culture.* New Delhi: Oxford University Press.

Rackham, H. (tr.) 1940. *Pliny: The Natural History*, Vol. III. Cambridge, MA: Harvard University Press.

Rackham, H. (tr.) 1942. *Pliny: The Natural History*, Vol. II. Cambridge, Ma.: Harvard University Press.

Rackham, H. (tr.) 1952. *Pliny: The Natural History*, Vol. IX, Cambridge, MA: Harvard University Press.

Radhakrishna, B.P. and Curtis, L.C. 1999. *Gold in India.* Bangalore: Geological Society of India.

Raghunandan, K.R., Dhruva, Rao, B.K. and Singhal, M.L. 1981. *Exploration for Copper, Lead and Zinc Ores in India*, Bulletin of the GSI, Series A-Economic Geology, No. 47. Calcutta: Geological Survey of India.

Rahman, A., Alvi, M.A., Khan Ghori, S.A. and Samba Murthy, K.V. 1980 *Science and Technology in Medieval India: A Bibliography of Source Materials in Sanskrit, Arabic and Persian.* New Delhi: Indian National Science Academy.

Raja-Rao, C.S., Poddar, B.C. and Chatterjee, A.K. 1972. 'Dariba-Rajpura-Bethumni belt of zinc-lead-copper mineralization,Udaipur District, Rajasthan', *GSI Miscellaneous Publication* 16: 617–26.

Ramage, A. and Craddock, P.T. 2000. *King Croesus' Gold.* London: British Museum Press.

Ranade, M.G. 1906. *Essays on Indian Economics*, 2nd edn. Madras: G.A. Natesan.

Rao, B.S., Lal, U.S., Bhatnagar, I.K. and Nair, M.V. 1990. 'A correlation study of scanning electron microscopy-energy dispersive analysis of X-rays (SEM-EDAX) with wave-length dispersive X-ray fluorescence spectrometry method (WDXRF) on punch marked silver coins', *Indian Journal of Chemical Sciences* 5: 91–5.

Ravindranath, K. 1991. 'Beneficiation in HZL: a panorama', *Hindzinc Tech* 3(1): 26–38.

Ray, P.C. 1902. *A History of Hindu Chemistry.* Calcutta: Prithwis Chandra Rây,

Rây, P. 1956. *History of Chemistry in Ancient and Medieval India.* Calcutta: Indian Chemical Society.

Reddy, S.N. 1998. 'Fueling the hearths of India', *Paleorient* 24(2): 61–9.

Reedy, C.L. 1997. *Himalayan Bronzes: Technology, Style and Choices.* Newark: University of Delaware Press.

Reeves, R. 1962. *Cire Perdu Casting in India.* New Delhi: Crafts Museum.

Rehder, J.E. 1990. 'Pressure drop in air flow through beds of charcoal', *Archaeomaterials* 4(1): 105–9.

Rehder, J.E. 1999. 'High temperature technology in antiquity: a sourcebook on the design and operation of ancient furnaces', in *The Beginnings of Metallurgy*, A. Hauptmann, E. Pernicka, Th. Rehren and Ü. Yalçin (eds), 305–16. Bochum: *Der Anschnitt* Beiheft 9.

Rehder, J.E. 2000. *The Mastery and Uses of Fire.* Montreal: McGill-Queen's University Press.

Rice, P.M. 1987. *Pottery Analysis.* Chicago: University of Chicago Press.

Riddle, J.M. 1985. *Dioscorides on Pharmacy and Medicine.* Austin: University of Texas Press.

Rostoker, W., Pigott, V.C. and Dvorak, J.R. 1989. 'Direct reduction to copper metal by oxide-sulfide mineral interaction', *Archeomaterials* 3(1): 69–87.

Rothenberg, B. (ed.) 1990a. *The Ancient Metallurgy of Copper*. London: Institute for Archaeo-Metallurgical Studies, University College.

Rothenberg, B. 1990b. 'Copper smelting furnaces, tuyeres, slags, ingot-moulds and ingots in the Arabah: the archaeological data', in *The Ancient Metallurgy of Copper*, B. Rothenberg (ed.), 1–77. London: Institute for Archaeo-Metallurgical Studies, University College.

Roy, A.B. 1995. 'Geometry and evolution of superposed folding in Zawar lead-zinc mineralized belt, Rajasthan', *Proceedings of the Indian Academy of Sciences (Earth and Planetary Science)* 104(3): 349–71.

Roy, A. and Bagchi, S,K. (eds) 1986. *Technology in Ancient and Medieval India*. Delhi: Sundeep Prakashan.

Roy, A.B. and Paliwal, B.S. 1981. 'Evolution of the lower Proterozoic epicontinental deposits: stromatolite-bearing Aravalli rock of Udaipur, Rajasthan, India'. *Precambrian Research* 14: 49–74.

Roy, A.B., Nagori, D.K., Golani, P.R., Dhakar, S.P. and Choudhary, R. 1980. 'Structural geometry of the rock phosphate bearing Aravalli rocks around Jhamarkotada Mines area, Udaipur District, Rajasthan', *Indian Journal of Earth Science* 7: 192–202.

Roy, A.B., Kumar, S., Laul, V. and Chauhan, N.K. 2004. 'Tectonstratigraphy of the lead-zinc bearing metasedimentary rocks of the Rampura-Agucha Mine and its neighbourhood, District Bhilwara, Rajasthan: implications on metallogeny', in *Sediment-hosted Lead Zinc Sulphide Deposits*, M. Deb and W.D. Goodfellow (eds), 273–89. New Delhi: Narosa.

Roy, B.C. 1958. 'Economic geology and resources of Rajasthan', *Memoir of the GSI* 86(1). New Delhi: Geological Survey of India.

Roy, M. and Subbarayappa, B.V. (eds and trs) 1976. *Rasarṇavakalpa*. New Delhi: Indian National Science Academy.

Ryding, K.C. 1994. 'Islamic alchemy according to al-Khwarizmi', *Ambix* 41(3): 121–34.

Salkield, L.U. 1987. *A Technical History of the Rio Tinto Mines: Some Notes on Exploitation from Pre-Phoenician Times to the 1950s*. London: Institution of Mining and Metallurgy.

Sambasiva Iyer, V.S. 1898–99. 'Report on prospecting work in parts of the Chitaldroog and Tumkur districts during the field seasons of the years 1896–9', *Records of the Mysore Geological Survey* II: 102–10.

Sánchez-Palencia, E-J. 1989. 'La explotación del oro en la Hispania Romana: sus inicios y precedentes', in *Mineria y metalurgia en las antiguas civilizaciones Mediterraneas y Europeas II*, C. Domergue (ed.), 35–49. Madrid: Ministerio de Culture.

Sanderson. D.C.W. and Hunter, J.R. 1981. 'Composition variability in vegetable ash', *Science and Archaeology* 23: 27–30.

Sarma, S.S. 2001. *Kauṭilya's Arthaśāstra in the Light of Modern Science and Technology*. New Delhi: D.K. Printworld.

Satpute, A.D. (tr. and ed.) 2003. *Rasaratna Samuchchaya*. Delhi: Chaukhamba Ayurvijnan Studies, Sanskrit Pratisathan.

Schnabel, C. 1905. *Handbook of Metallurgy*, L. Louis (tr.), 2nd edn. London: Macmillan.

Sen Mamata Chaudhuri, S.N. 1985. *Ancient Glass and India*. New Delhi: Indian National Science Academy.

Shamasastri, R. (tr.) 1915. *The Kauuṭilīya Arthaśāstra*. Bangalore: Government Press.

Shareq, A., Chmyriov, V.M., Stazhilo-Alekseev, K.F., Drunov, V.I., Gannon, P.J.G, Rossovskiv, L.N., Kafarskiy, A., Kh. and Malyarov, E.P. 1977. *Mineral Resources of Afghanistan*. Kabul: United Nations Development Program.

Sharma, D. 1966. *Rajasthan through the Ages*. Bikaner: Rajasthan State Archives.

Sharma, G.N. 1972. *Rajasthan ka Itihas*. Agra: Shival.

Sharma, G.R. 1960. *The Excavations at Kausambi 1957–59*. Allahabad: Department of Archaeology, University of Allahabad.

Shekar, N.C. 1983. 'Antiquity of mining and metallurgical activity at Ambaji, Kumbaria and Deri, Gujarat and Rajasthan', *Indian Journal for History of Science* 18(2): 176–83.

Shires, G.L. 1960. 'Tuyere angle and blast direction', *British Foundryman* 53: 447–54.

Shrivastava, N.C. 1991 *Zinc News: Silver Jubilee Issue*, January 1991.

Shrivastava, R. 2006. *Mining and Metallurgy in Ancient India*. New Delhi: Munshiram Manoharlal.

Siddiqui, S.A. and Sharp, W.E. 1993. 'Lead smelting slags near Nal, Balochistan Province, Pakistan', *Geoarchaeology* 8(5): 395–411.

Sim, D. and Ridge, I. 2002. *Iron for the Eagles*. Stroud: Tempus.

Simonin, L. 1869. *Underground Life or Mines and Miners*, H.W. Bristow (tr.). London: Chapman & Hall.

Singh, R.D. 1997. 'Mining', in *History of Technology in India I*, A.K. Bag (ed.), 48–79. New Delhi: Indian National Science Academy.

Singh, R.N. 2004. 'Analyses of metal objects', in *Early Farming Communities of the Kaimur*, R.S. Bisht (ed.), 591–603. Jaipur: Publication Scheme.

Sisco, A.G. and Smith, C.S. (trs and eds) 1951. *Lazarus Ercker: Treatise on Ores and Assaying*. Chicago: University of Chicago Press.

Skibo, J.M., Schiffer, M.B. and Reid, K.C. 1989. 'Organic-tempered pottery: an experimental study', *American Antiquity* 54: 122–46.

Smith, C.S. and Gnudi, M.T. (trs and eds) 1942. *The Pirotechnia of Vannoccio Biringuccio*. Chicago: Basic Books,

Smith, E.A. 1918. *The Zinc Industry*. London: Longmans Green and Co.

Smith, E.A. 1923–4. 'Early methods of assaying', *Transactions of the Institute of Mining and Metallurgy* 33: 272–327.

Smith, E.A. 1947. *The Sampling and Assay of Precious Metals, Comprising Gold, Silver and Platinum, and the Platinum Group Metals in Ores Bullion and Products*, 2nd edn. London: Griffin.

Souza, G.B. 1986. *The Survival of Empire: Portuguese Trade and Society in China and the South China Sea 1630–1734*. Cambridge: Cambridge University Press.

Souza, G.B. 1991. 'Ballast goods: Chinese maritime trade in zinc and sugar in the seventeenth and eighteenth centuries', in *Emporia, Commodities and Entrepreneurs in Asian Maritime Trade c.1400–1750*, R. Ptak and D. Rothermund (eds), 291–315. Stuttgart: Steiner.

Srikantha Murthy, K.R. (ed.) 2004. *The* Bhāvaprakāśa Pūrvakha *of Bhāvamiśra*. Varanasi: Chowhamba Krishnadas Academy.

Srinivasan, N.R. 1998. 'Revival of rural ironmaking in Orissa', in *Archaeometallurgy in India*, V. Tripathi (ed.), 228–40. Delhi: Sharada Publishing House.

Staunton, Sir G. 1797. *An Authentic Account of an Embassy from the King of Great Britain to the Emperor of China*. London: G. Nicol.

Stephens, F.J. 1901–2. 'Geology and mineral resources of Kumaon and Garwhal', *Transactions of the Institution of Mining and Metallurgy* 10: 393–414.

Stillman, A.L. 1923. *Briquetting*. Easton, PA: Chemical Publishing Co.

Stós-Fertner, Z. 1980. 'Application of X-ray fluorescence analysis to the determination of pigments from illuminations and miniatures in medieval incunabula', in *Proceedings of the 16th International Symposium on Archaeometry and Archaeological Prospection, Edinburgh 1976*, E.A. Slater and J.O. Tate (eds), 126–38. Edinburgh: National Museum of Antiquities of Scotland.

Straczek, J.A. and Srikantan, B. 1966. *The Geology of the Zawar Zinc-Lead Area, Rajasthan, India*. New Delhi: Geological Survey of India.

Stuiver, M. and Pearson, G.L. 1986. 'High-precision calibration of the radiocarbon time scale AD 1950–500 BC', *Radiocarbon* 28(2B): 805–38.

Subbaiah, K.V. 1984. 'Lead rolls of the Satavahana period', *Journal of Archaeological Chemistry* 2: 31–2.

Subbarayappa, B.V. 1971. 'Chemical practices and alchemy', in *A Concise History of Science in India*, D.M. Bose, S.N. Sens and B.V. Subbarayappa (eds), 274–349. New Delhi: Indian National Science Academy.

Subbarayappa, B.V. 1999. 'Indian alchemy: its origin and ramification', in *History of Science, Philosophy and Culture in Indian Civilization IV (1): Chemistry and Chemical Techniques in India*, B.V. Subbarayappa (ed.), 262–92. New Delhi: Centre for Studies in Civilizations.

Sundaram, C.V., Raj, B. and Rajagopalan, C. 1999. 'History of metallurgy in India', in *History of Science, Philosophy and Culture in Indian Civilization IV (1): Chemistry and Chemical Techniques in India*, B.V. Subbarayappa (ed.), 23–53. New Delhi: Centre for Studies in Civilizations.

Tafel, V. and Wagenmann, K. 1953. *Lehrbuch der metallhüttenkunde*, Vol. II 2. Leipzig: Auflage.

Taggart, A.F. 1927. *Handbook of Ore Dressing*. New York: John Wiley & Sons.

Temin, P. 1964. *Iron and Steel in 19th Century America*. Cambridge, MA: MIT Press.

Tewari, D.N. 1992. *Tropical Forestry in India*. Dehra Dun: International Book Distributors.

Thapar, B.K. 1964–5. 'Prakash 1955', *Ancient India* 20–21: 5–167.

Thorpe, E. 1922. *A Dictionary of Applied Chemistry, III*. London: Longmans, Green & Co.

Tillotson, G. (ed.) 2007. *James Tod of Rajasthan*. Mumbai: Marg.

Timberlake, S. 2005. 'The experimental smelting of copper carbonate ores from Engine Vein, Alderley Edge', in *The Archaeology of Alderley Edge: Survey, Excavation and Experiment in an Ancient Mining Landscape*, S. Timberlake and A.J.N.W. Prag (eds), 198–216. Oxford: British Archaeological Reports 396.

Tite, M.S., Maniatis, Y., Meeks, N.D., Bimson, M., Hughes, M.J. and Leppard, S.C. 1982a. 'Technological studies of ancient ceramics from the Near East, Aegean and Southeast Europe', in *Early Pyrotechnology*, T.A. and S.F. Wertime (eds), 61–71. Washington, DC: Smithsonian Institution Press.

Tite, M.S., Freestone, I.C., Meeks, N.D. and Bimson, M. 1982b. 'The use of scanning electron microscopy in the technological examination of ancient ceramics', in *Archaeological Ceramics*, J.S. Olin and A.D. Franklin (eds), 109–20. Washington, DC: Smithsomian Institution Press.

Tite, M.S., Freestone, I.C., Meeks, N.D. and Craddock, P.T. 1985. 'The examination of refractory ceramics from metal production and metalworking sites', in *The Archaeologist and the Laboratory*, P. Phillips (ed.), 50–55. London: Council for British Archaeology.

Tite, M.S., Kilikoglou, V. and Vekinis, G. 2001. 'Strength, toughness and thermal shock resistance of ancient ceramics, and their influence on technological choice', *Archaeometry* 43(3): 301–24.

Tiwari, R.K. and Kavdia, N.K. 1984. 'Ancient mining activity around Agucha village, Bhilwara District, Rajasthan', *Man and Environment* 8: 81–7.

Tod, J. 1978. *The Annals and Antiquities of Rajputana*, 2 vols. London: Smith and Elder [originally published 1829 and 1832, reprinted 1978, New Delhi: M.N. Publishers].

Toll, C. (tr.) 1968. *Al Hamdānī:Kitābal-Gauharatain... Die beiden Edelmetalle Gold und Silber*. Uppsala: Studia Semetica Upsaliensia, Almqvist & Wiksell.

Tomber, R., Cartwright, C.R. and Gupta, S. 2011 'Rice temper: technological solutions and source identification in the Indian Ocean', *Journal of Archaeological Science* 38: 360–66.

Tripathi, I.D. 1978. *Rasarnavam Rastantram*. Varanasi: Chowkhambha Sanskrit Series.

Tripathi, V. (ed.) 1998. *Archaeometallurgy in India*. Delhi: Sharada Publishing House.

Tripathi, V. and Tripathi, A. 1994. 'Iron working in ancient India: an ethnoarchaeological study', in *From Sumer to Meluhha: Contributions to the Archaeology of South and West Asia in Memory of George F. Dales*, J.M. Kenoyer (ed.), 241–52. Madison: Department of Anthropology, University of Wisconsin.

Turner, P.J. 1989. *Roman Coins from India*. London: Royal Numismatic Society.

Tylecote, R.F. 1976. *A History of Metallurgy*. London: The Metals Society.

Tylecote, R.F. 1981. 'From pot bellows to tuyeres', *Levant* 13: 107–18.

Tylecote, R.F. 1987. *The Early History of Metallurgy in Europe*. London: Longman.

Tylecote, R.F., Ghaznavi, H.A. and Boydell, P.J. 1977. 'Partitioning of trace elements between the ores, fluxes, slags and metal during the smelting of copper', *Journal of Archaeological Science* 4(4): 305–33.

Ullah, S. and Marshall, J. 1951. 'Copper, bronze and lead objects', in *Taxila: An Illustrated Account of the Excavations Carried out at Taxila 1913–1934*, J. Marshall, II, 564–72. Cambridge: Cambridge University Press.

Untracht, O. 1997. *Traditional Jewelery of India*. London: Thames and Hudson.

Vasu, S.C. (ed. and tr.) 1899. *The Astādhyāyī of Pāṇini*. Benares: S.C. Bose [reprinted 1962 New Delhi: Motilal Banarasidass].

Viccajee, F.K. 1908. *Notes on the Hand Minting of Coins of India*. Hyderabad: A.V. Pillal & Son.

Vikram, K. 1984. *Stone Industry in India*. Jaipur: Stone Technology Centre.

Wagner, D.B. 1985. *Dabieshan*. London: Curzon Press.

Wagner, D.B. 1997. *The Traditional Chinese Iron Industry and its Modern Fate*. London: Curzon.

Wagner, D.B. 2008. *Science and Civilisation in China: Chemistry and Chemical Technology V 11: Ferrous Metallurgy*. Cambridge: Cambridge University Press.

Wahl, J. 1993. 'Trés Minas-Vorbericht über die archäologischen Ausgragungen in Bereich des Römischen Goldwerks 1986–7', in *Montanarchäologie in Europa*, H. Steuer and U. Zimmermann (eds), 123–52. Sigmaringen: Jan Thorbecke Verlag.

Walsh, R. 1929. 'Galen visits the Dead Sea and copper mines of Cyprus (166 AD)', *Bulletin of the Geographical Society of Philadelphia* 25: 92–110.

Wang, H., Cowell, M., Cribb, J. and Bowman, S. (eds) 2005. *Metallurgical Analysis of Chinese Coins at the British Museum*. London: British Museum Press.

Warmington, E.H. 1974. *The Commerce between the Roman Empire and India*, 2nd edn. Cambridge: Cambridge University Press.

Watson, A.M. 1967. 'Back to gold and silver', *Economic History Review* 20(1): 1–34.

Watson, R. 1796. 'Of *Lapis calaminaris* – blende-zinc-brass', *Chemical Essays IV*, 1–84. London: T. Evans.

Watson, T.C. 1834. 'Chirra Punji, and a detail of some of the favourable circumstances which render it an advantageous site for the erection of an iron and steel manufactory on an extensive scale', *Journal of the Asiatic Society of Bengal* 3(1): 25–33.

Weisgerber, G. and Roden, C. 1985. 'Römische Schmiedeszenen und ihre Gebläse', *Der Anschnitt* 37(1): 2–21.

Weisgerber, G. and Roden, C. 1986. 'Griechische Metallhandwerker und ihre Gebläse', *Der Anschnitt* 38(1): 2–26.

Weisgerber, G. and Willies, L. 2000. 'The use of fire in prehistoric and ancient mining: firesetting', *Paléorient* 26(2): 131–49.

Welter, J-M. 2003. 'The zinc content of brass: a chronological indicator?', *Techné* 18: 27–36.

Wertime, T.A. 1968. 'A metallurgical expedition through the Persian Desert 1968', *Science* 159(3818): 927–35.

Willies, L. 1983. 'Ancient mining in the Udaipur District of Rajasthan, India, a report of preliminary investigations', *Bulletin of the Peak District Mines Historical Society* 8(4): 262–8.

Willies, L. 1987. 'Ancient zinc-lead mining in Rajasthan, India', *Bulletin of the Peak District Mines Historical Society* 10(2): 81–123.

Willies, L. 1988. 'Early metal mining in India', in *Aspects of Ancient Mining and Metallurgy: Acta of a British School at Athens Centenary Conference, 1986*, J.E. Jones (ed.), 136–45. Bangor: Department of Classics, University College of North Wales.

Willies, L. 1989. 'Ancient lead-zinc-silver mines in Rajasthan', *Mining Magazine* January: 31–5.

Willies, L. 1991a. 'Ancient mining in Israel', *Bulletin of the Peak District Mines Historical Society* 11(4): 217–21.

Willies, L. 1991b. 'Lead ore preparation and smelting', in *The Industrial Revolution in Metals*, J. Day and R.F. Tylecote (eds), 84–130. London: Institute of Metals.

Willies, L. 1992a. 'Ancient and post-medieval mining sites in the Khetri Copper Belt and Kolar Gold Field, India', *Bulletin of the Peak District Mines Historical Society* 11(6): 285–95.

Willies, L. 1992b. 'A visit to the Kolar Gold Field, India', *Bulletin of the Peak District Mines Historical Society* 11(4): 217–21.

Willies, L. 1993. 'Development of lead smelting techniques', in *Archeologia delle attivitá estrattive e metallurgiche*, R. Francovich (ed.), 497–514. Florence: All'Insignia del Giglio.

Willies, L. 1994. 'Firesetting technology', in *Mining Before Powder*, T.D. Ford and L. Willies (eds), 1–8. Issued as *Bulletin of the Peak District Mines Historical Society* 12(3) and a Historical Metallurgy Society Special Publication.

Willies, L. 1997. 'Roman mining at Rio Tinto, Huelva, Spain, and appendix: the 1981 archaeological survey', *Mining History: Bulletin of the Peak District Mines Historical Society* 13(3): 1–30.

Willies, L. and Cranstone, D. (eds) 1992. *Boles and Smeltmills*. London: Historical Metallurgy Society [republished as *The Art of Converting Red, or Rosette Copper into Brass by Means of Calamine Stone*, A.P. Woolrich and A. den Ouden (eds and trs), n.d. Eindhoven: Der Archaeologische Pers.].

Willies, L., Craddock, P.T., Gurjar, L.K. and Hegde, K.T.M. 1984. 'Ancient lead and zinc mining in Rajasthan, India', *World Archaeology* 16(2): 222–33.

Winterhalder, B., Larsen, R. and Thomas, R.-B. 1974. 'Dung as an essential resource in highland Peruvian community', *Human Ecology* 2(2): 59–104.

Worthington, T. and Craddock, B. 1996. 'Modern stone tools', *Mining History: The Bulletin of the Peak District Mines Historical Society* 13(1): 58.

Wulff, H.E. 1966. *The Traditional Crafts of Persia*. Cambridge, MA: MIT Press.

Xu, L. 1998. 'Traditional zinc-smelting technology in the Guma District of Hezhang County', in *2000 Years of Zinc and Brass*, P.T. Craddock (ed.), 2nd rev. edn, 115–32. London: British Museum Press.

Yih, T.D. and de Kreek, J. 1993. 'Preliminary data on the metallic composition of some S.E. Asian coinage as revealed by X-ray fluorescence (XRF) analysis', *Seaby's Numismatic Circular* 101(1): 7–9.

Zeuner, F.E. 1959. 'On the origin of the cinder mounds of the Bellary District, India', *Bulletin of the Institute of Archaeology* 2: 37–44.

Zhou, W. 1993. 'A new study on the history of the use of zinc in China', *Bulletin of the Metals Museum of the Japan Institute of Metals* 19: 49–53.

Zhou, W. 1996. 'Chinese traditional zinc-smelting technology and the history of zinc production in China', *Bulletin of the Metals Museum of the Japan Institute of Metals* 25: 36–47.

Zhou, W. 2001. 'The emergence and development of brass smelting techniques in China', *Bulletin of the Metals Museum of the Japan Institute of Metals* 34: 87–98.

Zhou, W., Martinón-Torres, M., Chen, J. and Liu, H. 2012. 'Large scale zinc production in Fengdu, Chongqing, in the Ming Period', *Journal of Archaeological Science* 39(4): 908–21.

Zwicker, U. 1992. 'Nondestructive and other investigations of metal objects of the Archaeological Museum, Nicosia', in *Studies in Honour of Vassos Karageorghis*, G.C. Ioannides (ed.), 165–77. Nicosia: Society of Cypriot Studies.

Index